Principles and Special Purpose Applications of ELECTROMAGNETIC FIELD AND HIGH VOLTAGE

Package Three Special Purpose Applications-Part Two

Moayad Abdullah AlMayouf, Ph.D.

PARTRIDGE

Library of Congress Control Number: 2018946408
ISBN: Softcover 978-1-5437-4628-0
eBook 978-1-5437-4629-7

Print information available on the last page.

To order additional copies of this book, contact
Toll Free 800 101 2657 (Singapore)
Toll Free 1 800 81 7340 (Malaysia)
orders.singapore@partridgepublishing.com

www.partridgepublishing.com/singapore

Principles and Special Purpose Applications of Electromagnetic Field and High Voltage

Package Three
Special Purpose Applications-Part Two

Book 6: Particle Accelerator
Book 7: Nuclear Magnetic Resonance-NMR
Book 8: Electrostatic Precipitators
Book 9: Magnetic Bearing

Dr. Moayad Abdullah AlMayouf
July 2018

بسم الله الرحمن الرحيم

الثناء والرحمة الى الوالدين غفر الله لهما

الاعتراف بالجميل والفضل الى اختي شهزنان و اخي فاروق لرعايتهم خلال المسيرة التعليمية

الشكر الخاص والامتنان الى زوجتي فائزة لدعمها و صبرها خلال مرحلتي العمل و كتابة هذة المجموعة من الكتب

التحية الى كل من عملت معهم خلال الاربعون عاما الماضية

الى كل من ذكرت انفا والى ابنائي عبدالله و سارة و شهد و دعاء و الاحفاد والى الجيل الجديد من الاكاديميين والباحثين والطلبة اهدي و اقدم هذا العمل

Acknowledgment and Compliments

Mercy and AlRahma to my late Parents

Gratefulness to my sister Shahzanan and brother Farouq for their patronage during the educational career

Special thanks and gratitude to my wife Faiza for her support and patient during the course of the work and in writing the subscript of these books

Thanks and solute all that I had worked with during the last 40 years.

To all those whom I had mentioned including my son Abdullah and daughters Sarah, Shahad and Doaa as well as the Grandchildren and the new generation of Academics, Researchers and Students I present this work.

Preface

General Preface to the Packages of Books

"The Principles and Special Purpose Applications of Electromagnetic Field and High Voltage" is a series of books, which are been classified in three packages:

Package One: Principles of Electromagnetic Field and High Voltage
Book 1: Principles of Electromagnetic Field EMF
Book 2: Principles of High Voltage HV
Package Two: Special Purpose Applications-Part One
Book 3: Ion Implantation
Book 4: Electromagnetic Isotope Separation-EMIS
Book 5: Mass Spectrometer
Package Three: **Special Purpose Applications-Part Two**
Book 6: Particle Accelerator
Book 7: Nuclear Magnetic Resonance-NMR
Book 8: Electrostatic Precipitators
Book 9: Magnetic Bearing

In Package One, two books are presented. Book 1 covers the Principles of Electromagnetic Field, where the physical fundamentals of electromagnetic phenomena are been considered taking into consideration their technical applications. The book contains the general theory of the electromagnetic field necessary for the study of the principal applications in various domains. Book 2 covers Principles of High Voltage, where the principles of electric breakdown and Ionization process are described taking into consideration the various media of operations and technical applications. The book contains the general theory of the ionization and breakdown in solid and liquid dielectric as well as the gases with and without crossed magnetic field. In addition, this book covers the means of high voltage generation and measurements methods for Alternative Current AC, Direct Current DC and Impulse Voltage.

Packages Two and Three cover the fundamental, design principles, and operation of various machines and systems in which the principles mechanisms of both the electromagnetic field and high voltages are applied. The description for each application shall cover the theory & formulas, design and performance parameters, as well as the main fields of applications.

In Package Two, a set of special purpose applications covering the Ion Implantation, Electromagnetic Isotope Separation-EMIS and Mass Spectrometer are covered in three books 3, 4 and 5. Package Three includes another set of special purpose applications. In this package, Particle Accelerator, Nuclear Magnetic Resonance-NMR, Electrostatic Precipitators and Magnetic Bearing are been covered in four books 6, 7, 8 and 9 respectively.

This series of packages and books, designed for undergraduates and postgraduates in the fields of physics and electrical engineering, as well as field engineers in the related research and developing sectors.

The Author does not pretend neither claimed of editing all materials covered by this work. This work is simply an outgrowth of effort in editing, collecting, reforming and gathering the related topics, texts and images using the best available sources and references as listed at the end of each book. All this effort is been based on practical efforts and experiences on researches, studies, and lectures on undergraduate and postgraduate levels.

When making use of this book, all parties must rely at the end on their own skill and judgment. The author does not accept any liability whatsoever for any loss or damage caused by any error of quotation. Any and all such liability is disclaimed.

The author has spent around 8400 Man-hour during 2005-2018 for the course of editing this series of books. Hope this work will help users from various disciplines undergraduate and postgraduate for classroom and research purposes as well as the practices in the electrical engineering.

Preface to Package Three

In this Package, four books 6, 7, 8 and 9 are presented to cover the second part of special purpose applications of Electromagnetic Field and High Voltage.

Book 6 is covering the principle, design and operation of Particle Accelerator which includes Eight main chapters. The topics of each chapter are briefly covering the following:

Chapter One - Background: Introduction; History; Main Development; Operating Particle Accelerators Around the World

Chapter Two - Fundamentals of Particle Physics: Building Blocks of Matter; Fundamental Particles; Forces and Interaction; Special Relativity; Units of Mass, Energy, and Momentum

Chapter Three -Accelerator Physics and Principles: Fundamental and Physical Theories (Newton's laws of motion, Particle Acceleration, and Kinetic Energy); Equations of motion (Introduction, Kinematic equations for one particle, Dynamic equations of motion, and Analytical mechanics, Electrodynamics)

Chapter Four - Design of Accelerators: Design Principles and Goals; Particle Beam (Main components for particle-beam storage, Beam Dynamics and Focusing, and Machine Tolerances); Design Principles of Linear Accelerators (History of Linear Accelerator Design and Design Principles); Design Principles of Circular Accelerators (History of Circular Accelerator Design and Design Principles); Beam Diagnostics and Detectors Design

Chapter Five - Applications of Accelerators: Main Applications of accelerators; Applications on Energy levels (Low Energy Machine and High Energy Machines); Usages Technologies and limitations (Cyclotrons, Synchrotrons, and DESY)

Chapter Six-Types of Accelerators: Linear particle accelerators; Circular or cyclic accelerators (Cyclotrons, Synchrocyclotrons and Isochronous cyclotrons, Betatrons, Synchrotrons, Storage rings); Advantages versus disadvantages of Accelerators types

Chapter Seven - Generation of Higher Energies Accelerators

Chapter Eight- Annexes : Standard Model and Types of Interaction; General Theory of Relativity; Quantum Mechanics; Electron Gamma Shower EGS Application; List of accelerators in particle physics

Book 7 is covering the principle, design and operation of Nuclear Magnetic Resonance-NMR which includes Seven main chapters. The topics of each chapter are briefly covering the following:

Chapter One-Introduction and Background: Principles; History; NMR Spectroscopy;The Basics for NMR Physics and Engineering

Chapter Two-Theory of Nuclear Magnetic Resonance: The Physics and Engineering of NMR (Spin Physics, Nuclear spin and Magnets, Energy Level and Magnetic Field); Values of spin angular momentum (Spin behavior in a magnetic field, Magnetic resonance by nuclei. Shielding Effect)

Chapter Three - NMR Spectroscopy: Chemical Shift; Knight shift; Spin-Spin Coupling; The Time Domain NMR Signal; The +/- Frequency Convention

Chapter Four -NMR Spectroscopy (Performance & Technique Types): Continuous wave (CW) spectroscopy; Fourier transform spectroscopy; Multi-dimensional NMR Spectroscopy; Solid-state NMR spectroscopy; Sensitivity; Isotopes

Chapter Five-- NMR Hardware: Hardware Overview; Magnet; Field Lock; Shim Coils; Sample Probe; Gradient Coils; Quadrature Detector; Digital Filtering; Safety

Chapter Six - Applications: Introduction; Application Classification; Medicine; Chemistry; Non-destructive testing; Data acquisition in the petroleum industry; Flow probes for NMR spectroscopy; Process control; Earth's field NMR; Quantum computing; Magnetometers

Chapter Seven - Annexes: Fourier Transforms; Pulse Sequences; Carbon-13 NMR; Two Dimensional 2D technology; Software and Glossary of NMR

Book 8 is covering the principle, design and operation of Electrostatic Precipitators which includes Ten main chapters. The topics of each chapter are briefly covering the following:

Chapter One - Basics and Principles: Electrostatics; Precipitators
Chapter Two - Electrostatic Precipitator - Principles and Design: Introduction; Philosophy and Principles; Design Methods and Parameters; Design variables and Factors; Typical Ranges of Design Parameters; System Selection
Chapter Three - Electrostatic Precipitator - Principles and Design: Introduction; Philosophy and Principles; Design Methods and Parameters; Design variables and Factors; Typical Ranges of Design Parameters; System Selection
Chapter Four - Components of ESP: Discharge electrodes; Collection electrodes; High voltage equipment; Gas Distribution Systems; Rappers; Hoppers; Ductwork (Shell)
Chapter Five - Classification and Types of ESP: Tubular and Plate ESPs; Single-stage and Two-stage ESPs; Cold-side and Hot-side ESPs; Dry and Wet ESPs
Chapter Six- Electrostatic Precipitator- Operation and Performance: Introduction; ESP Operation; ESP Performance
Chapter Seven-Routine Maintenance and Safety Measures: Preventive maintenance checklist; ESP operating problems; Safety Measures
Chapter Eight- Examples of ESP Applications
Chapter Nine- Review Exercise and Cases of studies
Chapter Ten-Attachments

Book 9 is covering the principle, design and operation of Magnetic Bearing which includes Eight main chapters. The topics of each chapter are briefly covering the following:

Chapter One - Introduction and Development of Bearing: Development of Conventional Bearing; Development of Magnetic Bearing; Bearings Types

Chapter Two - Principles and Fundamentals of Electromagnetic Bearing: Features and Milestones of Magnetic bearing; Operating Principles; Main Performance Parameters (Load and forces, Static and dynamic stiffness and bandwidth, Speed, Losses, Other parameters and features); Categories of Magnetic Bearings (The Earnshaw stability criterion, Classical active magnetic bearings AMB, High temperature AMB, Permanent magnet bearings PMB, Superconducting bearings SCB)

Chapter Three - Magnetic Bearing Analysis: Introduction and Steps of Analysis; Magnetic Circuit (Magnetic circuit of an inner rotor bearing, Magnetic circuit of an outer rotor bearing, Magnetic circuit of an intermediate rotor bearing); Eddy Current Circuits (Induced eddy currents, Induced voltage, Circuit parameters, Eddy currents induced by lateral motion, Stray eddy currents)

Chapter Four - Performance Parameters: Bearing Forces; Bearing Losses; Stiffness and Damping(Stiffness, Damping) ;Rotor Dynamics and Stability(Equations of motion. Static equilibrium, Load range, Stability)

Chapter Five - Active Magnetic Bearing AMB: Working Principle and AMB components; Operation and Performance of AMB (Basic Operation, Performance and specification criteria); Characteristics of Magnetic Bearing Systems (Open-Loop behavior of rotor system, Bearing system load capacity, Limitations); Controllability and Compensation of the AMB(Controller requirements, Linear compensation

Chapter Six - Design of Magnetic Bearings: Background; Design Parameters and features; Magnetic Model; Current - Force Relationship; Types of Losses; High Temperature Aspects in AMB Design

Chapter Seven – Applications of Magnetic Bearings: Fields of Applications; Examples of AMB Applications; Special Design Applications

Chapter Eight-Annex: Earnshaw's Theorem

Principles and Special Purpose Applications of Electromagnetic Field and High Voltage

Package 3:Special Purpose Applications-Part Two

Book 6: Particle Accelerator

Dr. Moayad Abdullah AlMayouf

July 2018

Particle Accelerator

Contents

1.1. Introduction

A particle accelerator (or **atom smasher**, in the early 20th century) is a device that uses electromagnetic fields to propel ions or charged subatomic particles to high speeds and to contain them in well-defined beams. An ordinary CRT television set is a simple form of accelerator.

While it is possible to accelerate charged particles using electrostatic fields, like in a Cockcroft-Walton voltage multiplier, this method has limits given by electrical breakdown at high voltages. Furthermore, due to electrostatic fields being conservative, the maximum voltage limits the kinetic energy that is applicable to the particles.

To circumvent this problem, linear particle accelerators operate using time-varying fields. To control this fields using hollow macroscopic structures through which the particles are passing (*wavelength restrictions*), the frequency of such acceleration fields is located in the radio frequency region of the electromagnetic spectrum. The space around a particle beam is evacuated to prevent scattering with gas atoms, requiring it to be enclosed in a vacuum chamber (or *beam pipe*). Due to the strong electromagnetic fields that follow the beam, it is possible for it to interact with any electrical impedance in the walls of the beam pipe. This may be in the form of a resistive impedance (i.e., the finite resistivity of the beam pipe material) or an inductive/capacitive impedance (due to the geometric changes in the beam pipe's cross section).

These impedances will induce **wake fields** (a strong warping of the electromagnetic field of the beam) that can interact with later particles. Since this interaction may have negative effects, it is studied to determine its magnitude, and to determine any actions that may be taken to mitigate it.

Accelerators were invented in the 1930s to provide energetic particles to investigate the structure of the atomic nucleus. The accelerator main function is to speed up and increase the energy of a beam of particles by generating electric fields that accelerate the particles, and magnetic fields that steer and focus them. Since then, Accelerators have been used to investigate many aspects of particle physics.

Fig.1.1 A 1960s single stage 2 MeV linear Van de Graaff accelerator, here opened for maintenance
https://en.wikipedia.org/wiki/File:2mv_accelerator-MJC01.jpg

Description: 1960s vintage 2 MeV Van de Graaff linear accelerator. A single ended belt charging linear accelerator made by "High Voltage" used primarily to accelerate hydrogen and helium ions from a RF positive ion source. The machine was capable of terminal voltages above 2 million volts. This machine operated at the Australian National University from the early 1960s till 2000. It consists of an endless fabric belt, inside the tube, that carries charge to the inside of a large spherical capacitive electrode (not shown)

Particle accelerators come in two basic designs, linear (linac) and circular (synchrotron). The beam of particles of accelerators in the form of a ring (a circular accelerator), travels repeatedly round a loop, while in a straight line (a linear accelerator), travels from one end to the other.

A synchrotron achieves high energy by circulating particles many times before they hit their targets. Also the longer a linac is the higher the energy of the particles it can produce.

High-energy physics research has always been the driving force behind the development of particle accelerators. In general, accelerators give high energy to subatomic particles, which then collide with targets. Out of this interaction come many other subatomic particles that pass into detectors. From the information gathered in the detectors, physicists can determine properties of the particles and their interactions.

The higher the energy of the accelerated particles, the more closely we can probe the structure of matter. For that reason a major goal of researchers is to produce higher particle energies.

Particle accelerators are crucial to the study of quantum physics, as they allow for experiments that replicate high energies that existed in the early universe. Many types of particles can only be studied in particle accelerators, because they exist for such a short amount of time.

Accelerators are used in medicine as well as in physics research. There is also an increasing interest in radiation therapy in the medical world and industry has been a long-time user of ion implantation and many other applications. Consequently accelerators now constitute a field of activity in their own right with professional physicists and engineers dedicated to their study, construction and operation.

The principles and fundamental; design and engineering principles; uses of particle accelerators; types of accelerators as well as targets and detectors are considered in this document.

1.2. History

The earliest circular accelerators were cyclotrons, invented in 1929 by Ernest O. Lawrence at the University of California, Berkeley. Cyclotrons have a single pair of hollow 'D'-shaped plates to accelerate the particles and a single large dipole magnet to bend their path into a circular orbit. Lawrence's first cyclotron was a 4 inches (100 mm) in diameter. Later he built a machine with a 60 in dipole face, and planned one with a 184-inch dia, which was, however, taken over for World War II-related work connected with uranium isotope separation; after the war it continued in service for research and medicine over many years.

The first large proton synchrotron was the Cosmotron at Brookhaven National Laboratory, which accelerated protons to about 3 GeV. The Bevatron at Berkeley, completed in 1954, was specifically designed to accelerate protons to sufficient energy to create antiprotons, and verify the particle-antiparticle symmetry of nature, then only strongly suspected. The Alternating Gradient Synchrotron (AGS) at Brookhaven was the first large synchrotron with alternating gradient, "strong focusing" magnets, which greatly reduced the required aperture of the beam, and correspondingly the size and cost of the bending magnets. The Proton Synchrotron, built at CERN, was the first major European particle accelerator and generally similar to the AGS.

The Fermilab Tevatron has a ring with a beam path of 4 miles (6 km). The largest circular accelerator ever built was the LEP synchrotron at CERN with a circumference 26.6 kilometers, which was an electron/positron collider. It has been dismantled and the underground tunnel is being reused for a proton collider called the LHC, due to start operation in at the end of July 2008. However, after operating for a short time, 100 of the giant superconducting magnets failed and the LHC had to shut down in September 2008. According to a press release from CERN that was printed in Scientific American "the most likely cause of the problem was a faulty electrical connection between two magnets, which probably melted at a high current leading to mechanical failure."

An apparent later report by the BBC said "On Friday, a failure, known as a quench, caused around 100 of the LHC's super-cooled magnets to heat up by as much as 100 degrees." This was later described as a "massive magnet quench".

This caused a rupture of a high magnitude – leaking one ton of liquid helium into the LHC tunnels. The liquid helium is used for more efficient use of power and the super cooling of the liquid helium allows the electrical resistance of the superconducting magnets to be nonexistent – zero ohms. According to the BBC, before the accident its operating temperature was 1.9 kelvins (−271 °C; −456 °F), which is colder than deep space.

The aborted Superconducting Super Collider (SSC) in Texas would have had a circumference of 87 km. Construction was started in 1991, but abandoned in 1993. Very large circular accelerators are invariably built in underground tunnels a few meters wide to minimize the disruption and cost of building such a structure on the surface, and to provide shielding against intense secondary radiations that may occur. These are extremely penetrating at high energies.

Current accelerators such as the Spallation Neutron Source, incorporate superconducting Cryomodules. The Relativistic Heavy Ion Collider, and Large Hadron Collider also make use of superconducting magnets and RF cavity resonators to accelerate particles.

A majority of the world's particle accelerators are situated in the United States, either at major universities or national laboratories. In Europe the principal facility is at CERN near Geneva, Switzerland; in Russia important installations exist at Dubna and Serpukhov.

The early history of accelerators can be traced from three separate roots. Each root is based on an idea for a different acceleration mechanism and all three originated in the twenties. These history roots are abstracted in the following topics and categorization of accelerator development in both the research and engineering programmes:

A. The Main "History Line'

The first root is generally taken as the principal "history line", since it was the logical consequence of the vigorous physics research programme in progress at the turn of the last century. Indeed, particle physics research has always been the driving force behind accelerator development and it is therefore very natural to also consider high-energy physics as the birth place.

The line is started at the end of the last century to show the natural progression through atomic physics to nuclear physics and the inevitable need for higher energy and higher intensity "atomic projectiles" than those provided by natural radioactive sources. In this context, the particle accelerator was a planned development and it fulfilled its goal of performing the first man-controlled splitting of the atom. It was Ernest Rutherford, in the early twenties, who realised this need, but the electrostatic machines, then available, were far from reaching the necessary voltage and for a few years there was no advance.

Suddenly, the situation changed in 1928, when Gurney and Gamov independently predicted tunneling and it appeared that an energy of 500 keV might just suffice to split the atom. This seemed technologically feasible to Rutherford and he immediately encouraged Cockcroft and Walton to start designing a 500 kV particle accelerator. 1928 Cockcroft & Walton start designing an 800 kV generator encouraged by Rutherford. Four years later in 1932, they split the lithium atom with 400 keV protons. This was the first fully man-controlled splitting of the atom which earned them the Nobel prize in 1951.

1932 Generator reaches 700 kV and Cockcroft & Walton split lithium atom with only 400 keV protons. They received the Nobel Prize in 1951. Figure 1.2 (a) shows the original apparatus, which is now kept in the Science Museum, London. The top electrode contains the proton source and was held at 400 kV, the intermediate drift tube at 200 kV and final drift tube and target at earth potential. This structure can be seen inside the evacuated glass tube in Fig. 1.2 above the curtained booth in which the experimenter sat while watching the evidence of nuclear disintegrations on a scintillation screen.

The voltage generator, Fig. 1.2 (b), was at the limit of the in-house technology available to Cockcroft and Walton and the design voltage of 800 kV was never reached due to a persistent spark discharge which occurred at just over 700 kV. However, the famous atom-splitting experiment was carried out at 400 kV, well within the capabilities of the apparatus. The **Cockcroft Walton generator,** as it became known, was widely used for many years after as the input stage (up to 800 kV) for larger accelerators, since it could deliver a high current.

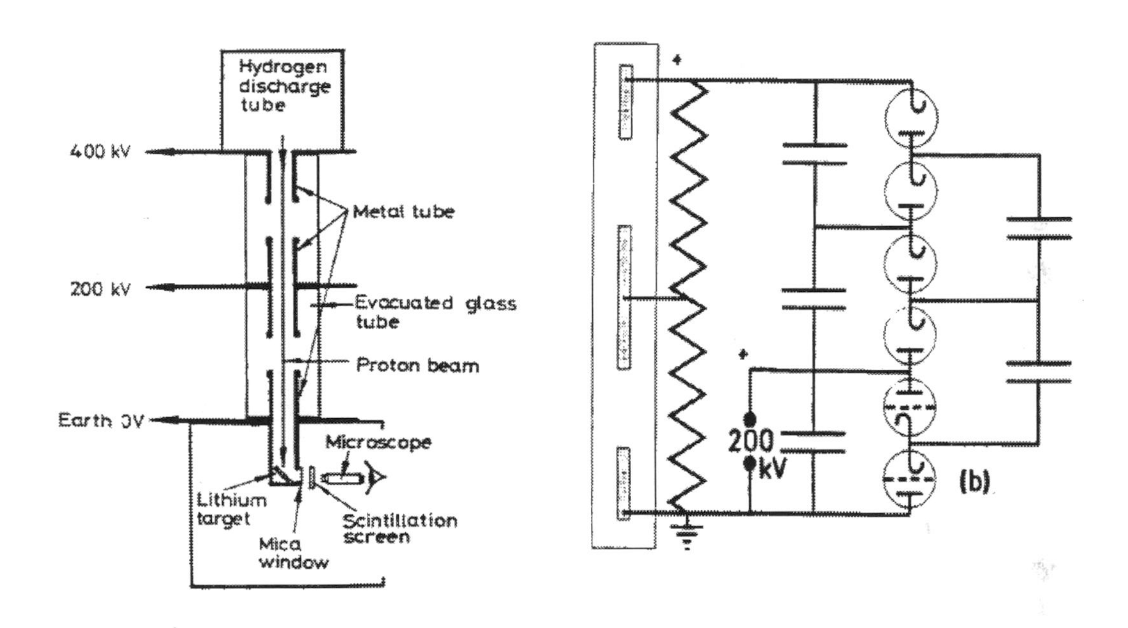

(a) Accelerating column (b) DC generator

Fig.1.2 Cockcroft and Walton's apparatus for splitting the lithium nucleus

At about the same time Van de Graaff, an American who was in Oxford as a Rhodes Scholar, invented an electrostatic generator for nuclear physics research and later in Princeton, he built his first machine, which reached a potential of [1.5 MV]. It took some time to develop the acceleration tube and this type of machine was not used for physics research until well after the atom had been split in 1932. The principle of this type of generator is shown in Fig. 1.3.

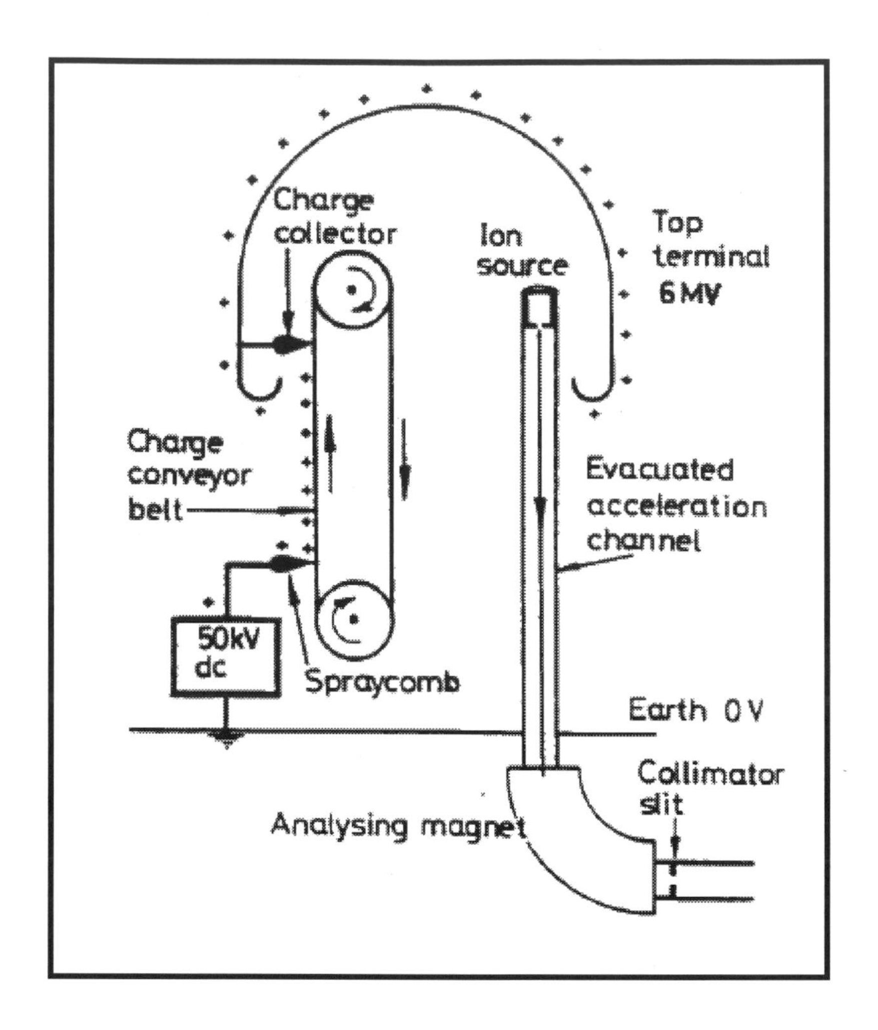

Fig. 1.3 Van de Graaff electrostatic generator

Two new features appeared in later versions of the **Van de Graaff generator**. Firstly, the sparking threshold was raised by putting the electrode system and accelerating tube in a high-pressure tank containing dry nitrogen, or Freon, at 9-10 atmospheres, which enables operation typically up to 10 MV. The second was a later development, which has the special name of the **Tandem** accelerator (see Fig. 1.4).

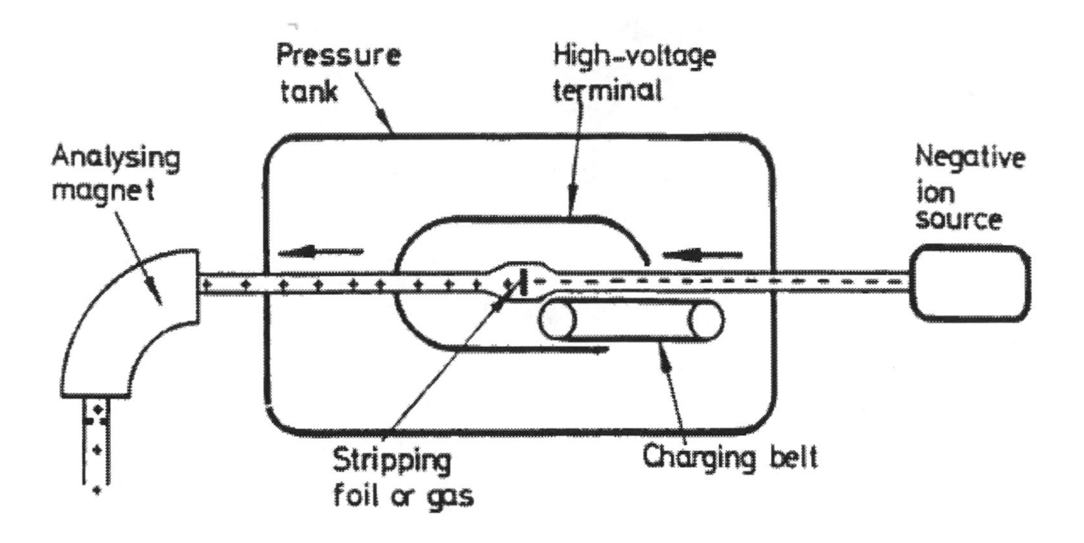

Fig. 1.4 Two-stage Tandem accelerator

The new feature in the Tandem accelerator was to use the accelerating voltage twice over. First an extra electron is attached to the neutral atoms to create negative ions. In recent years, a great deal of development has been done and it is now possible to obtain negative ion sources for almost all elements. The negative ion beam is injected at ground potential into the Tandem and accelerated up to the high-voltage terminal where it passes through a thin foil which strips at least two electrons from each negative ion converting them to positive ions. They are then accelerated a second time back to earth potential. The Van de Graaff generator and the Tandem provide beams of stable energy and small energy spread, but they are unable to provide as high currents as the Cockcroft-Walton generator.

The highest energy Tandem is at Oak Ridge National Laboratory with 24.5 MV on the central terminal. However, development is not at a standstill and there is a project (the Vivitron) underway at Strasbourg to build a Tandem operating at 35 MV.

B. The second "History Line"

The direct-voltage accelerators were the first to be exploited for nuclear physics research, but they were limited to the maximum voltage that could be generated in the system (except for the astute double use of the applied voltage in the Tandem). This limitation was too restrictive for the requirements of high-energy physics and an alternative was needed.

In fact, an alternative had already been proposed in 1924 in Sweden by Ising [Ising proposes time varying fields across drift tubes]. He planned to repeatedly apply the same voltage to the particle using alternating fields and his invention was to become the underlying principle of all of today's ultra-high-energy accelerators. This is known as resonant acceleration, which can achieve energies above that given by highest voltage in the system.

The difference between the acceleration mechanisms of Cockcroft and Walton and Ising depend upon whether the fields are static (i.e. conservative) or time-varying (i.e. non conservative).

The electric field can be expressed in a very general form as the sum of two terms, the first being derived from a scalar potential and the second from a vector potential,

$$\vec{E} = - \nabla\Phi - \frac{\partial}{\partial t}\vec{A} \qquad (1.1)$$

Where $$\vec{B} = \nabla \times \vec{A} \qquad (1.2)$$

The first term in Eq.1.1 describes the static electric field of the Cockcroft-Walton and Van de Graaff machines. When a particle travels from one point to another in an electrostatic field, it gains energy according to the potential difference, but if it returns to the original point, for example, by making a full turn in a circular accelerator, it must return to its original potential and will lose exactly the energy it has gained. Thus a gap with a DC voltage has no net accelerating effect in a circular machine.
The second term in Eq. 1.1 describes the time-varying field. This is the term that makes all the present-day high-energy accelerators function. The combination of (1.1) and (1.2) yields Faraday's law,

$$\nabla \times \vec{E} = -\frac{\partial}{\partial t}\vec{B} \qquad (1.3)$$

which relates the electric field to the rate of change of the magnetic field. There are two basic geometries used to exploit Faraday's Law for acceleration. The first of which is the basis of Ising's idea and the second "history line", and the second is the basis of the third "history line" to be described later.

Ising suggested accelerating particles with a linear series of conducting drift tubes and Wideröe built a 'proof-of-principle' linear accelerator in 1928. Alternate drift tubes are connected to the same terminal of an RF generator. The generator frequency is adjusted so that a particle traversing a gap sees an electric field in the direction of its motion and while the particle is inside the drift tube the field reverses so that it is again the direction of motion at the next gap. As the particle gains energy and speed the structure periods must be made longer to maintain synchronism (see Fig. 1.5).

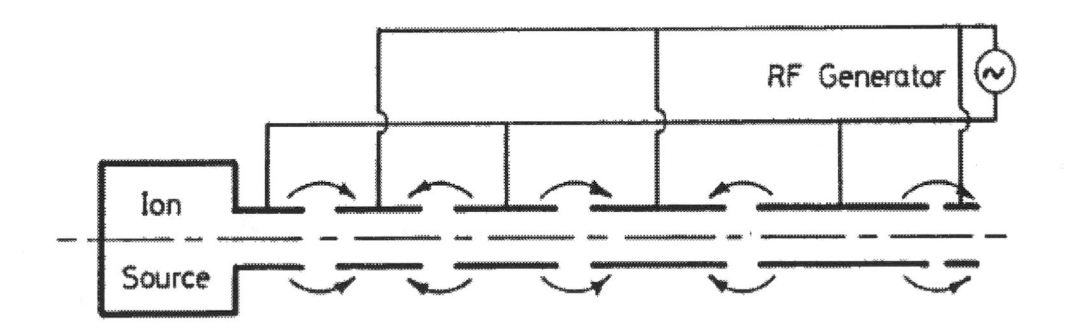

Fig. 1.5 RF linac

Clearly, as the velocity increases the drift tubes become inconveniently long, unless the frequency can be increased, but at high frequencies the open drift-tube structure is lossy. This problem is overcome by enclosing the structure to form a cavity (in a circular machine) or series of cavities (in a linear machine), working typically in the MHz range. The underlying principle remains unchanged, but there are several variants of the accelerating structure design.

Ising's original idea can be considered as the beginning of the 'true' accelerator.

Indeed, the next generation of linear colliders, which will be in the TeV range, will probably still be applying his principle of resonant acceleration, except that the frequency will probably be in the tens of GHz range.

Technologically the linear accelerator, or **linac** as it is known, was rather difficult to build and, during the 1930's, it was pushed into the background by a simpler idea conceived by Ernest Lawrence in 1929, the fixed-frequency cyclotron (see Fig. 1.6).

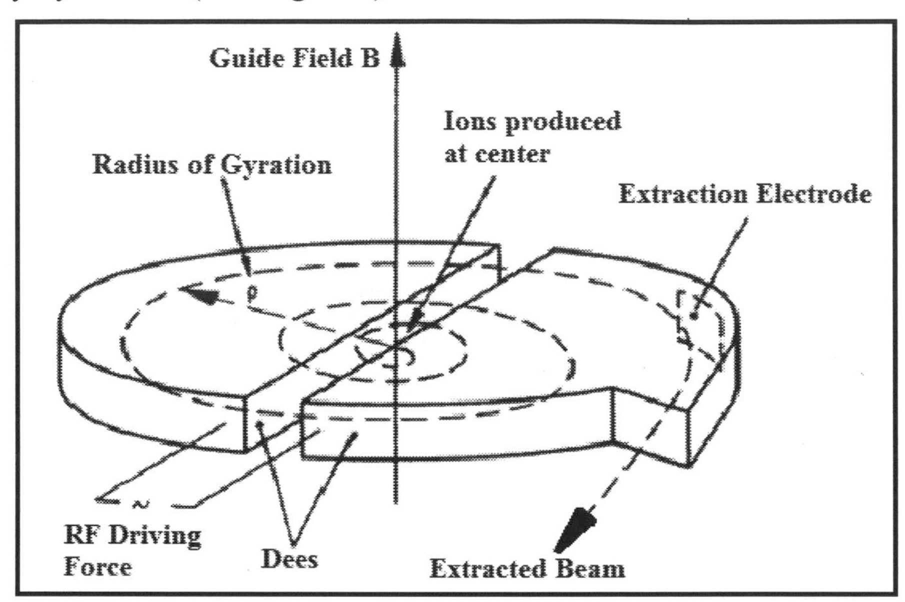

Fig. 1.6 Schematic cyclotron

Lawrence's idea was inspired by a written account of Wideröe's work and M. Livingston demonstrated the principle by accelerating hydrogen ions to 80 keV in 1931. Lawrence's first model worked in 1932.

It was less than a foot in diameter and could accelerate protons to 1.25 MeV. He split the atom only weeks after Cockcroft and Walton. Lawrence received the Nobel Prize in 1939, and by that year the University of California had a 5-foot diameter cyclotron (the 'Crocker' cyclotron) capable of delivering 20 MeV protons, twice the energy of the most energetic alpha particles emitted from radioactive sources. The cyclotron, however, was limited in energy by relativistic effects and despite the development of the synchrocyclotron, a new idea was still required to reach yet higher energies in order to satisfy the curiosity of the particle physicists. This new idea was to be the synchrotron, which will be described later.

C. The third and fainter 'History Line'

In the previous section, it was mentioned that there were two equipment configurations for exploiting Faraday's Law for acceleration. First, consider the application of Faraday's Law to the linac, which is made more evident by enclosing the gaps in cavities. For simplicity the fields in a single RF cavity are shown schematically in Fig. 1.7(a).

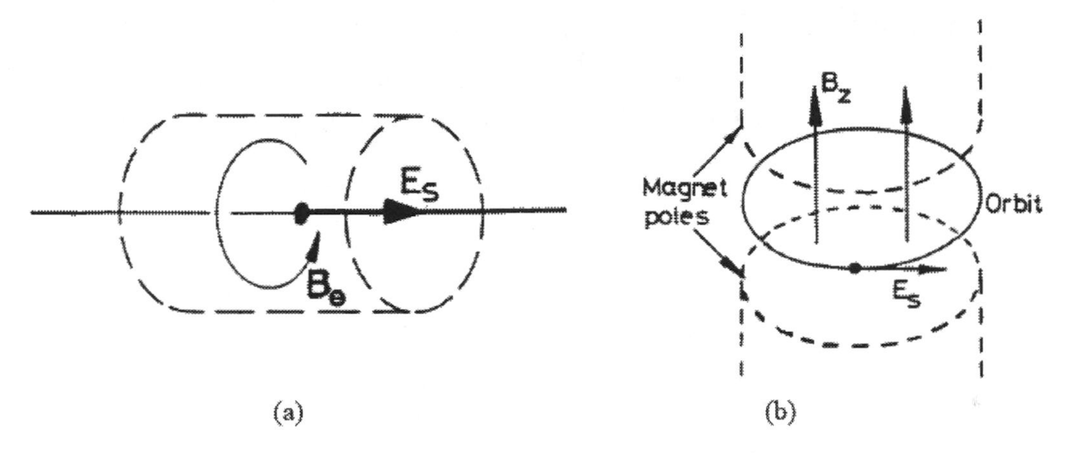

Fig. 1.7 Acceleration configurations

The azimuthal magnetic field is concentrated towards the outer wall and links the beam. Faraday's Law tells us the periodic rise and fall of this magnetic field induces an electric field on the cavity axis, which can be synchronised with the passage of the beam pulse.

Suppose now that the topology is transformed, so that the beam encircles the magnetic field as shown in Fig. 1.7(b). Wideröe suggested this configuration and the acceleration mechanism, now known as "Betatron acceleration". He called his idea a "strahlung transformator" or "ray transformer", because the beam effectively formed the secondary winding of a transformer (see Fig.1.8). As the flux through the magnet core is increased, it induces an azimuthal e.m.f. which drives the charged beam particles to higher and higher energies. The trick is to arrange for the increase in the magnetic field in the vicinity of the beam to correspond to the increase in particle energy, so that the beam stays on the same orbit (Known as the Wideröe condition, or 2-to-1 rule). This device, the Betatron, is insensitive to relativistic effects and was therefore ideal for accelerating electrons. The Betatron has also the great advantages of being robust and simple.

The one active element is the power converter that drives the large inductive load of the main magnet. The focusing and synchronisation of the beam energy with the field level are both determined by the geometry of the main magnet.

Wideröe put this idea in his laboratory notebook, while he was a student, but it remained unpublished only to re-surface many years later when Kerst built the first machine of this type. When in 1941 Kerst and Serber published a paper on the particle oscillation in their betatron, the term "betatron oscillation" became universally adopted for referring to such oscillations in all devices.

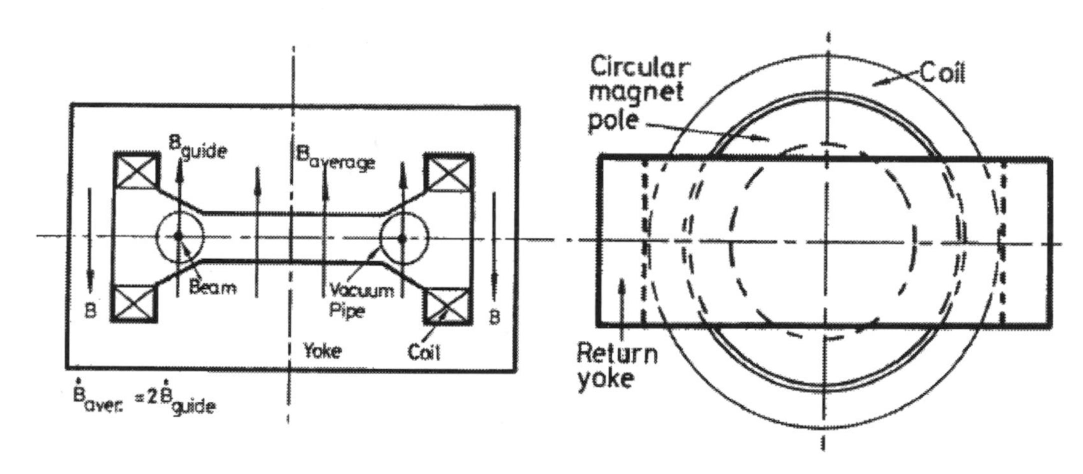

Fig. 1.8 Strahlung transformator or betatron

1950 Kerst builds the world's largest Betatron of 300 MeV. The development of Betatrons for high-energy physics was short, ending in 1950 when Kerst built the world's largest Betatron (300 MeV), but they continued to be built commercially for hospitals and small laboratories where they were considered as reliable and cheap. In fact the Betatron acceleration mechanism is still of prime importance. In the present-day synchrotron, there is a small contribution to the beam's acceleration which arises from the increasing field in the main dipoles. If an accurate description of the longitudinal motion is required, then the Betatron effect has to be included.

1.3. Main Developments

By the 1940's three acceleration mechanisms had been demonstrated, the DC acceleration, resonant acceleration and the Betatron mechanism. In fact, there were to be no new ideas for acceleration mechanisms until the mid-1960's, when collective acceleration was proposed in which heavy ions are accelerated in the potential well of an electron ring and the 1980's when there were several Workshops devoted entirely to finding new acceleration techniques. However, the acceleration mechanism is not sufficient by itself and other equally important developments are needed.

In order to accelerate particles to very high energies, it is also necessary to have focusing mechanisms in the transverse and longitudinal (energy) planes. This was not always appreciated. In the early cyclotrons, for example, the field was made as uniform as possible only to find that the beam was unstable. Livingston who was Lawrence's research student, told how they shimmed the magnet for each small step in energy to keep the beam stable, thus ending up with a field shape for transverse stability that decreased with radius. Theory has later shown that this decrease should be an inverse power law of the radius between zero and unity.

The cyclotron is limited by relativistic effects, which cause the particles to slow down and lose synchronism with the RF field. At first glance it would appear that one would only have to reduce the frequency in order to maintain synchronism, but this is a little too naïve since the spread in revolution frequency with energy would quickly exploit the natural energy spread in the beam and disperse the particles away from the peak of the RF voltage. In this case a longitudinal focusing mechanism is needed. This problem was overcome by E. McMillan and independently by V. Veksler who discovered the principle of phase stability in 1944 and invented the synchrotron. Phase stability is general to all RF accelerators except the fixed-frequency cyclotron. The effect is that a bunch of particles, with an energy spread, can be kept bunched throughout the acceleration cycle by simply injecting them at a suitable phase of the RF cycle. This focusing effect was strong enough that the frequency modulation in the synchro-cyclotron did not have to be specially tailored and was simply sinusoidal.

Synchro-cyclotrons can accelerate protons to about 1 GeV, a great improvement on the simple cyclotron, but the repetition rate reduces the particle yield.

In the synchrotron the guide field increases with particle energy, so as to keep the orbit stationary as in the Betatron, but acceleration is applied with an RF voltage via a gap or cavity. In 1946 F. Goward and D. Barnes were the first to make a synchrotron work, and in 1947 M. Oliphant, J. Gooden and G. Hyde proposed the first proton synchrotron for 1 GeV in Birmingham, UK. However, the Brookhaven National Laboratory, USA, built their 3 GeV Cosmotron by 1952, just one year ahead of the Birmingham group.

The mechanism known for focusing in the transverse plane was called weak, or constant-gradient focusing. In this case, the guide field decreases slightly with increasing radius and its gradient is constant all-round the circumference of the machine. The tolerance on the gradient is severe and sets a limit to the size of such an accelerator. The aperture needed to contain the beam also becomes very large and the magnet correspondingly bulky and costly. In the early fifties the limit was believed to be around 10 GeV. In the same year as the Cosmotron was finished (1952) E. Courant, M. Livingston and H. Snyder proposed strong focusing, also known as alternating-gradient (AG) focusing. The idea had been suggested earlier by Christofilos but it was not published.

This new principle revolutionized synchrotron design, allowing smaller magnets to be used and higher energies to be envisaged. It is directly analogous to a well-known result in geometrical optics that the combined focal length F of a pair of lenses of focal lengths f_1 and f_2 separated by a distance d is given by

$$\frac{1}{F} = \frac{1}{f_1} + \frac{1}{f_2} - \frac{d}{f_1 f_2} \qquad (1.4)$$

If the lenses have equal and opposite focal lengths, f1 = -f2 and the overall focal length F = f^2/d, which is always positive. In fact, F remains positive over quite a large range of values when $f_1$and f_2 have unequal values but are still of opposite sign. Thus within certain limits a series of alternating lenses will focus. Intuitively one sees that, although the beam may be defocused by one lens, it arrives at the following lens further from the axis and is therefore focused more strongly. Structures based on this principle are referred to as AG structures.

The synchrotron quickly overshadowed the synchrocyclotron and the betatron in the race for higher energies. The adoption of alternating gradient focusing for machines and transfer lines was even quicker. CERN for example immediately abandoned its already-approved project for a 10 GeV/c weak focusing synchrotron in favor of a 25 GeV/c AG machine, which it estimated could be built for the same price.

The next step was the storage ring collider. In physics experiments, the useful energy for new particle production is the energy that is liberated in the center-of-mass system. When an accelerator beam is used on a fixed target, only a fraction of the particle's energy appears in the center-of-mass system, whereas for two equal particles in a head-on collision, all of the particles' energy is available. This fundamental drawback of the fixed-target accelerator becomes more punitive as the energy increases. For example, it would have needed a fixed target accelerator of over 1TeV to match the center-of-mass energy available in the CERN ISR (2 x 26 GeV proton colliding rings).

The storage-ring collider now dominates the high-energy physics field. Single-ring colliders, using particles and antiparticles in the same magnetic channel, were the first type of collider to be exploited at Frascati in the AdA (Annelli di Accumulazione) project (1961). The first double-ring proton collider was the CERN ISR (Intersecting Storage Rings), 1972-1983.The highest-energy collisions obtained to date are 2 x 900 GeV in the Fermilab, single-ring, proton-antiproton collider.

Colliders have been very successful as physics research instruments. The J/y particle was discovered at SPEAR by B. Richter and at the same time by Ting at BNL – they shared the 1976 Nobel Prize. The CERN proton-antiproton storage ring was also the source of a Nobel Prize for C. Rubbia and S. van der Meer in 1984, following the discovery of the W and Z particles. The proton-antiproton colliders were only made possible by the invention of stochastic cooling by S. van der Meer for the accumulation of the antiprotons.

The use of superconductivity in proton machines has made the very highest energies possible. There has also been another change taking place, which has been called the Exogeographical transition (a phrase coined by Professor N. Cabibbo at a Workshop held at Frascati in 1984). This refers to the arrangements that have made it possible to bury the very large machines such as LEP and HERA deep under property which does not belong to the laboratory concerned. Without such agreements, Europe could not have maintained its leading position in the world accelerator league.

In order to fill in some of the bigger gaps in this brief history, it is now necessary to jump back in time to mention some of the other accelerators, which may not have featured as a high energy machine, but have found their place as injectors or as being suitable for some special application.

The Microtron, sometimes known as the electron cyclotron, was an ingenious idea due to Veksler (1945). The electrons follow circular orbits of increasing radius, but with a common tangent. An RF cavity positioned at the point of the common tangent supplies a constant energy increment on each passage. The relativistic mass increase slows the revolution frequency of the electrons, but by a constant increment on each passage. If this increment is a multiple of the RF oscillator frequency, the electrons stay in phase, but on a different orbit. Microtrons operate at microwave frequencies and are limited to tens of MeV. They are available commercially and are sometimes used as an injector to a larger machine.

The radio-frequency quadrupole (RFQ) suggested in 1970 by I. Kapchinski and V. Telyakov is useful at low energies and is increasingly replacing the Cockcroft-Walton as injector. The RFQ combines focusing and acceleration in the same RF field.

The electron storage rings have given birth to the synchrotron radiation sources, more usually referred to as light sources. These machines are now the fastest growing community in the accelerator world and the first commercially available compact synchrotron light source for lithography has just come onto the market.

The linear accelerator was eclipsed during the thirties by circular machines. However, the advances in ultra-high frequency technology during World War II (radar) opened up new possibilities and renewed interest in linac structures. Berkeley was first, with a proton linear accelerator of 32 MeV built by Alvarez in 1946. The Alvarez accelerator has become very popular as an injector for large proton and heavy-ion synchrotrons all over the world with energies in the range of 50–200 MeV, which is essentially non-relativistic particles. The largest proton linear accelerator to date is the 800 MeV 'pion factory' (LAMPF) at Los Alamos.

The first electron linear accelerators were studied at Stanford and at the Massachusetts Institute for Technology (MIT) in 1946. This type of accelerator has also had a spectacular development, up to the largest now in operation, the 50 GeV linear accelerator at the Stanford Linear Accelerator Centre (SLAC). Like betatrons they have become very popular in fields outside nuclear physics, particularly for medicine.

The "Livingston Chart" (see Fig.1.9) showing accelerator energy versus time (Years of Commissioning).

The chart shows, in a very striking way, how the succession of new ideas and new technologies has relentlessly pushed up accelerator beam energies over five decades at the rate of over one and a half orders of magnitude per decade. The filled circles indicate new or upgraded accelerators of each type.

From this chart one may derive the following:

- Exponential growth of Energy with time
- Increase of the energy by an order of magnitude every 6-10 years
- Each generation replaces previous one to get even higher energies
- In addition to Energy level, Further parameters such as intensity and size of the beam are considered as well in the development of the accelerators

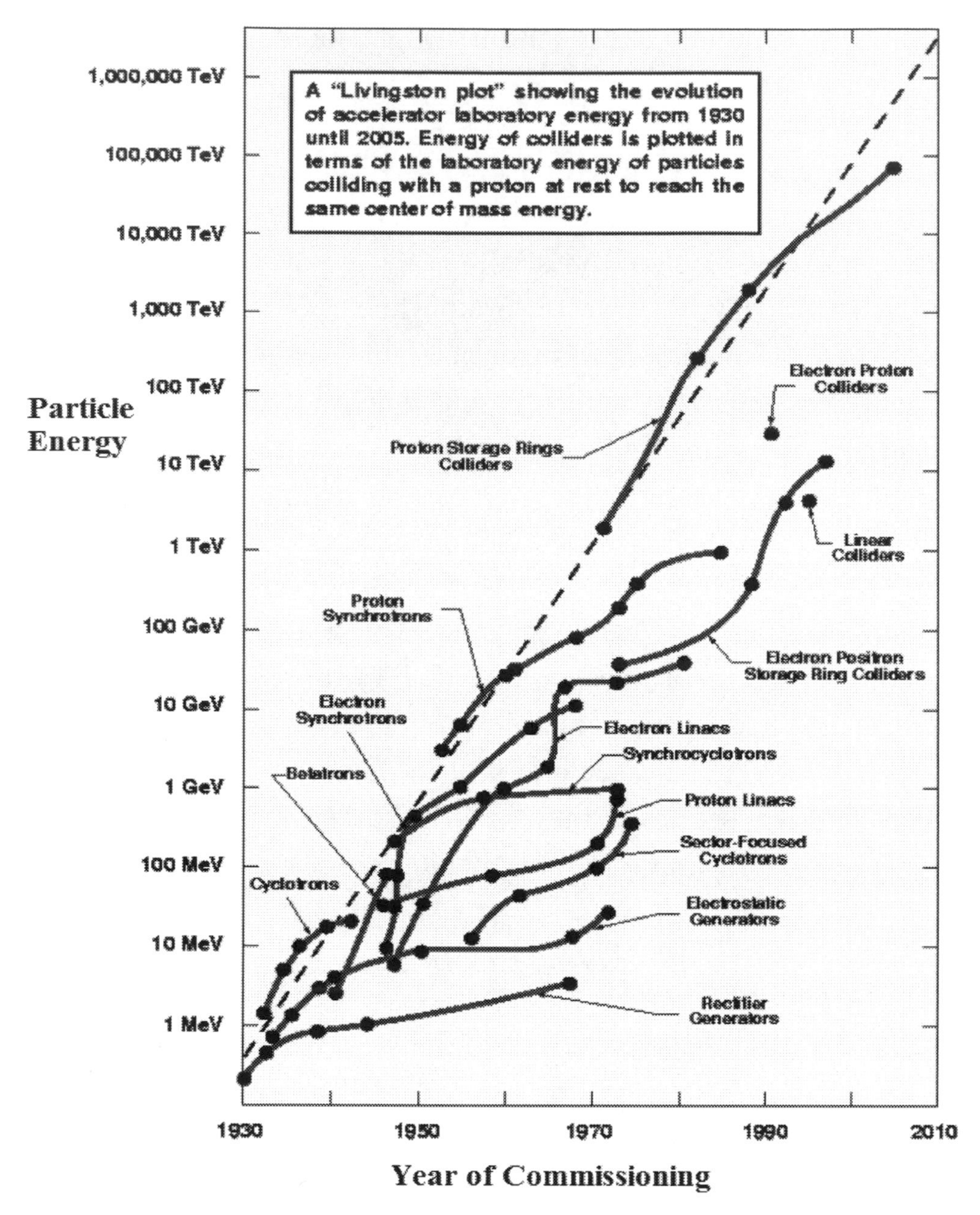

Fig.1.9 Livingston chart: Particle Energy verses Year of Commissioning

This relationship can be expressed quantitatively. To examine matter at the scale of an atom (about 10^{-8} centimeter), the energies required are in the range of a thousand electron volts. (An electron volt is the energy unit customarily used by particle physicists; it is the energy a particle acquires when it is accelerated across a potential difference of one volt.) At the scale of the nucleus, energies in the million electron volt-or MeV-range are needed. To examine the fine structure of the basic constituents of matter requires energies generally exceeding a billion electron volts, or 1 GeV. But there is another reason for using high energy. Most of the objects of interest to the elementary particle physicist today do not exist as free particles in Nature; they have to be created artificially in the laboratory.

The well-known Energy relationship ($E = mc^2$) governs the collision energy E required to produce a particle of mass m. Many of the most interesting particles are so heavy that collision energies of many GeV are needed to create them. In fact, the key to understanding the origins of many parameters, including the masses of the known particles, required to make today's theories consistent is believed to reside in the attainment of collision energies in the trillion electron volt, or TeV, range.

The chart originally produced by M. Stanley Livingston in 1954, shows how the energy laboratories of the particle beams produced by accelerators has increased. This plot has been updated by adding modern developments. One of the first things to notice is that the energy of man-made accelerators has been growing exponentially in time. Starting from the 1930s, the energy has increased-roughly speaking-by about a factor of 10 every six to eight years.

In 1969, physicists at CERN had studied the idea of building two intersecting storage rings that could be fed by the existing 28 GeV proton synchrotron (CERN-PS). Construction took place on the new Intersecting Storage Rings (ISR) between 1966 and 1971.

The **European Organization for Nuclear Research**, known as **CERN**; derived from the name *Conseil européen pour la recherche nucléaire*), is a European research organization that operates the largest particle physics laboratory in the world. CERN has come a long way since its foundation in 1954, the organization is based in a northwest suburb of Geneva on the Franco–Swiss border, (46°14′3″N 6°3′19″E) and has 22 member states. CERN is an official United Nations Observer. CERN's main function is to provide the particle accelerators and other infrastructure needed for high-energy physics research – as a result, numerous experiments have been constructed at CERN through international collaborations.

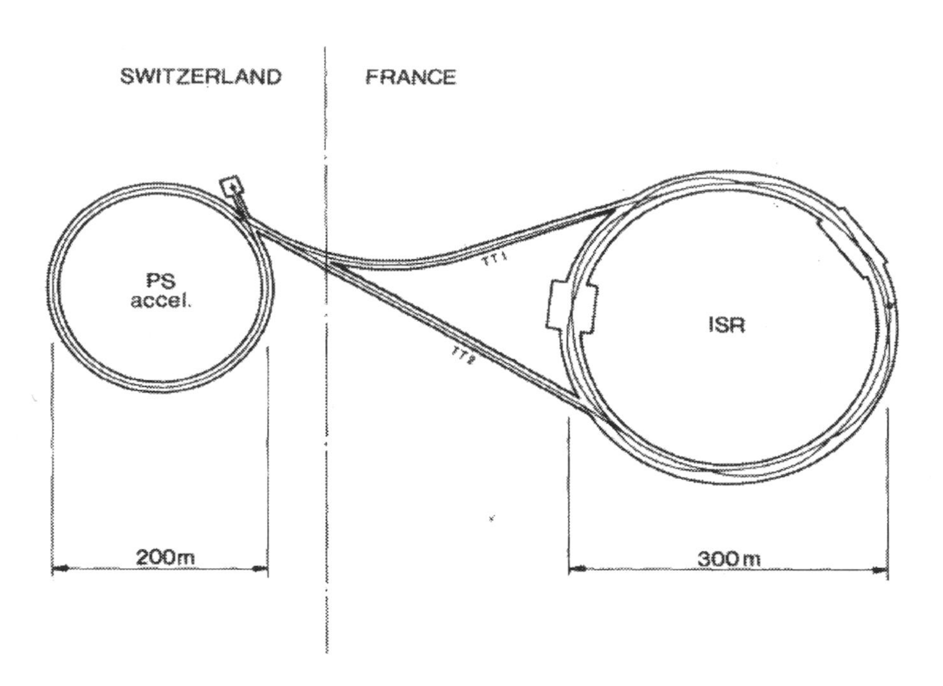

Fig.1.10 The two Intersecting Storage Rings ISR fed by the 28 GeV proton synchrotron PS - CERN

The two Intersecting Storage Rings ISR consisted of two concentric rings of magnets, 300 m in diameter. Located in an underground tunnel 200 m away from the 28 GeV proton synchrotron, the injected protons traveled in opposite directions. The two rings are not perfectly circular, and are interwoven with 8 intersection regions where the beams can brought to collide. Each ring contained a beam pipe surrounded by magnets to direct the circulating particles. Protons circulated in opposite directions and collided with a maximum centre-of-mass energy of 62 GeV. This is the equivalent of a 2000 GeV beam hitting a stationary target. On 27 January 1971, two beams of protons collided in the Intersecting Storage Rings (ISR) for the first time.

The ISR performed the first-ever proton-proton and proton-antiproton collisions. It was also where stochastic cooling was first developed. This technique reduces both the transverse dimension of the beam and the spread in the energy of the particles. The technique was adopted by the Super Proton Synchrotron for proton-antiproton collisions and is still used in the Antiproton Decelerator. The ISR paved the way for later accelerators as it hinted at the smaller elements of protons which are now known to be quarks and gluons.

1.4. Operating Particle Accelerators Around the World

The list of Operating Particle Accelerators Around the World is given by http://www-elsa.physik.uni-bonn.de/accelerator_list.html. (permission from Spokesperson of the ELSA board of directors Universität Bonn been taken on 19th June 2018). This list does not include accelerators which are used for medical or industrial purposes only.

For a detail on each Accelerator, Please visit also the WWW Virtual library of Beam Physics and Accelerator Technology, the Division of Physics of Beams of the American Physical Society, and the Los Alamos Accelerator Code Group.

This list is classified and sorted according to the Continents locations and Accelerators types as following:

A. Sorted by Continents Location

❖ Europe

AGOR	Accelerateur Groningen-ORsay, KVI Groningen, Netherlands
ALBA	Synchrotron Light Facility, Barcelona, Spain
ANKA	Ångströmquelle Karlsruhe, Karlsruhe, Germany
ARRONAX	Accelerator for Research in Radiochemistry and Oncology in Nantes Atlantique,Saint Herblain, France
BERLinPro	Berlin Energy Recovery Linac Project, Helmholtz-Zentrum Berlin für Materialien und Energie GmbH, Germany
BESSY II	Helmholtz-Zentrum Berlin für Materialien und Energie GmbH, Germany
CeBeTeRad	Institute of Nuclear Chemistry and Technology, Warszawa, Poland
CEMHTI	Conditions Extrêmes et Matériaux : Haute Température et Irradiation, Orléans, France
CERN	Centre Europeen de Recherche Nucleaire, Geneva, Suisse (LHC, PS-Division, SL-Division)
CMAM	Centro de Microanálisis de Materiales, Universidad Autonoma de Madrid, Spain
CNA	Centro Nacional de Aceleradores, Seville, Spain
COSY	Cooler Synchrotron, IKP, FZ Jülich, Germany (COSY Status)
CYCLONE	Cyclotron of Louvain la Neuve, Louvain-la-Neuve, Belgium
DELTA	Dortmunder ELekTronenspeicherring-Anlage, Zentrum für Synchrotronstrahlung der Technischen Universität Dortmund, Germany
DESY	Deutsches Elektronen Synchrotron, Hamburg, Germany (XFEL, PETRA III, FLASH, ILC, PITZ)

ELBE	ELectron source with high Brilliance and low Emittance, Helmholtz-Zentrum Dresden - Rossendorf e.V. (HZDR), Germany
ELETTRA	AREA Science Park, Trieste, Italy
ELSA	Electron Stretcher Accelerator, Bonn University, Germany (ELSA status)
ELU-6e	Institute of Applied Radiation Chemistry, Technical University of Lodz, Poland
ESRF	European Synchrotron Radiation Facility, Grenoble, France
ESSB	ESS-Bilbao, Zamudio, Spain
GANIL	Grand Accélérateur National d'Ions Lourds, Caen, France
GSI	Gesellschaft für Schwerionenforschung, Darmstadt, Germany
HISKP	Helmholtz-Institut für Strahlen- und Kernphysik, Bonn, Germany (Isochron Cyclotron)
IHEP	Institute for High Energy Physics, Protvino, Moscow region, Russian Federation
INFN	Istituto Nazionale di Fisica Nucleare, Italy, LNF - Laboratori Nazionali di Frascati (DAFNE, DAFNE beam test facility) LNL - Laboratori Nazionali di Legnaro (Tandem, CN Van de Graaff, AN 2000 Van de Graaff), LNS - Laboratori Nazionali del Sud, Catania, (Superconducting Cyclotron)
ISA	Institute for Storage Ring Facilities (ASTRID, ASTRID2, ELISA), Aarhus, Denmark
ISIS	Rutherford Appleton Laboratory, Oxford, U.K.
JINR	Joint Institute for Nuclear Research, Dubna, Russian Federation (NICA)
JYFL	Jyväskylän Yliopiston Fysiikan Laitos, Jyväskylä, Finland
MLL	Maier-Leibnitz-Laboratorium: Accelerator of LMU and TU Muenchen, Munich, Germany
MAMI	Mainzer Microtron, Universität Mainz, Germany
MAX IV	Lund University, Sweden
MPI-HD	Max Planck Institut für Kernphysik, Heidelberg, Germany
MIC	Microanalytical center at JSI, Ljubljana, Slovenia
MLS	Metrology Light Source, Physikalisch-Technische Bundesanstalt, Germany
PITZ	Photo Injector Test facility at DESY in Zeuthen, Germany
RUBION	Zentrale Einrichtung für Ionenstrahlen und Radionuklide, Universität Bochum, Germany
S-DALINAC	Superconducting Darmstadt linear accelerator, Technische Universität Darmstadt, Germany
SLS	Paul Scherrer Institut PSI, Villigen, Switzerland
SOLEIL	Synchrotron SOLEIL, GIF-SUR-YVETTE CEDEX, France
TSL	The Svedberg Laboratory, Uppsala University, Sweden

❖ North America

88" Cycl.	88-Inch Cyclotron, Lawrence Berkeley Laboratory (LBL), Berkeley, CA
ALS	Advanced Light Source, Lawrence Berkeley Laboratory (LBL), Berkeley, CA (ALS Status)
ANL	Argonne National Laboratory, Chicago, IL (Advanced Photon Source APS, Argonne Tandem Linac Accelerator System ATLAS)
BATES	Bates Linear Accelerator Center, Massachusetts Institute of Technology, USA
BNL	Brookhaven National Laboratory, Upton, NY (NSLS II, RHIC)
CAMD	Center for Advanced Microstructures and Devices, Louisiana State University
CENPA	Center for Experimental Nuclear Physics and Astrophysics, University of Washington, USA
CESR	Cornell Electron-positron Storage Ring, Cornell University, Ithaca, NY
CHESS	Cornell High Energy Synchrotron Source, Cornell University, Ithaca, NY
CLS	Canadian Light Source, U of Saskatchewan, Saskatoon, Canada
CNL	Crocker Nuclear Laboratory, University of California Davis, CA
FNAL	Fermi National Accelerator Laboratory , Batavia, IL
FSU	John D. Fox Superconducting Accelerator Laboratory, Florida State University, USA
IAC	Idaho accelerator center, Pocatello, Idaho
ISNAP	Institute for Structure and Nuclear Astrophysics, Notre Dame University, USA
IUCF	Indiana University Cyclotron Facility, Bloomington, Indiana
JLab	aka TJNAF, Thomas Jefferson National Accelerator Facility (formerly known as CEBAF), Newport News, VA
LAC	Louisiana Accelerator Center, U of Louisiana at Lafayette, Louisiana
LANL	Los Alamos National Laboratory
MIBL	Michigan Ion Beam Laboratory, University of Michigan
NSCL	National Superconducting Cyclotron Laboratory, Michigan State University
ORNL	Oak Ridge National Laboratory Oak Ridge, Tennessee
OUAL	John E. Edwards Accelerator Laboratory, Ohio University, USA
PBPL	Particle Beam Physics Lab (Neptune-Laboratory, PEGASUS - Photoelectron Generated Amplified Spontaneous Radition Source)
RPI	The Gaerttner LINAC Laboratory, MANE School of Enginering, USA
SLAC	Stanford Linear Accelerator Center, (SLC - SLAC Linear electron positron Collider, SSRL - Stanford Synchrotron Radiation Laboratory)
SNS	Spallation Neutron Source, Oak Ridge, Tennessee
SRC	Synchrotron Radiation Center, U of Wisconsin - Madison
SURF III	Synchrotron Ultraviolet Radiation Facility, National Institute of Standards and Technology (NIST), Gaithersburg, Maryland

TAMU	Cyclotron Institute, Texas A&M University, USA
TRIUMF	Canada's National Laboratory for Particle and Nuclear Physics, Vancouver, BC (Canada)
TUNL	Triangle Universities Nuclear Laboratory, USA
UMASS	University of Massachusetts Lowell Radiation Laboratory, USA
UNAM	Universidad Nacional Autónoma de México, Mexico
WMU	Van de Graaff Accelerator at the Physics Department of the Western Michigan University, Kalamazoo, Michigan
WNSL	Wright Nuclear Structure Laboratory, Yale University, USA

❖ South America

CAB	LINAC at Centro Atómico Bariloche, Argentina
LAFN	Laboratório Aberto de Física Nuclear, São Paulo, Brazil
LNLS	Laboratorio Nacional de Luz Sincrotron, Campinas SP, Brazil
RIBRAS	Radioactive Ion Beam in Brasil, São Paulo, Brazil
TANDAR	Tandem Accelerator, Buenos Aires, Argentina

❖ Asia

BEPC, BEPC II	Beijing Electron-Positron Collider, Beijing, China
HLS	Hefei Light Source, Univ. of Science & Technology of China, Hefei city, China
INDUS	Centre for Advanced Technology CAT, INDORE, India
KEK	National Laboratory for High Energy Physics ("Koh-Ene-Ken"), Tsukuba, Japan (KEK-B, 12 GeV proton synchrotron)
PAL	Pohang Accelerator Laboratory, Pohang, Korea
RIKEN	Institute of Physical and Chemical Research ("Rikagaku Kenkyusho"), Hirosawa, Wako, Japan
SESAME	Synchrotron-light for Experimental Science and Applications in the Middle East, Jordan (under construction)
SPring-8	Super Photon ring - 8 GeV, Japan
SSRF	Shanghai Synchrotron Radiation Facility, Shanghai, China
TPS	Taiwan Photon Source, Hsinchu, Taiwan
UAC	Inter-University Accelerator Centre, New Delhi, India
VECC	Variable Energy Cyclotron, Calcutta, India
VEPP	Budker Institute of Nuclear Physics, Novosibirsk, Russia (VEPP-3, VEPP-4M, VEPP-2000)

❖ Africa

iThemba	Laboratory for Accelerator Based Sciences, Cape Town, South Africa

❖ Australia

ANSTO	Australian Nuclear Science and Technology Organisation, Lucas Heights, Australia
ANU	Australian National University, Canberra, Australia
AS	Australian Synchotron, Melbourne, Victoria, Australia
MARC	Micro-Analytical Research Centre, University of Melbourne, Australia

B. Sorted by Accelerator Type

B1. Electrons

❖ Stretcher Ring/Continuous Beam facilities

ELSA (Bonn U), JLab, MAMI (Mainz U), MAX IV, SLAC

❖ Synchrotron Light Sources, Storage Rings

ALBA, ANKA, ALS (LBL), APS (ANL), AS, ASTRID (ISA), ASTRID2 (ISA), BESSY II, CAMD (LSU), CeBeTeRad, CHESS (Cornell Wilson Lab), CLS (U of Saskatchewan), DELTA (U of Dortmund), ELBE (HZDR), Elettra, ELISA (ISA), ELSA (Bonn U), ELU-6e, ESRF, HASYLAB (DESY), HLS, INDUS (CAT), MAX IV, MLS, LNLS, NSLS II, PAL, SESAME, SLS (PSI), SPEAR (SSRL, SLAC), SOLEIL, SPring-8, SRC (U of Wisconsin), SSRF, SURF III (NIST), TPS, TUNL

❖ Other

BATES, IAC, Neptune, PEGASUS, PITZ, S-DALINAC, UNAM, WMU

B2. Protons

ARRONAX, 88" Cyclotron (LBL), CERI, CNA, CNL (UC DAVIS), COSY (FZ Jülich), ISIS, IUCF, KEK, LHC (CERN), iThemba, PS (CERN), PSI, RHIC (BNL), SPS (CERN), TRIUMF, TSL

B3. Light and Heavy Ions

ARRONAX, 88" Cyclotron (LBL), AGOR, ANSTO, ANU, ASTRID (ISA), CENPA, CMAM, CNL (UC DAVIS), ATLAS (ANL), CERI, CYCLONE, ESSB, GANIL, GSI, HISKP, ININ, ISNAP, iThemba, IUCF, JYFL, LAFN, LAC, LHC (CERN), LNL (INFN), LNS (INFN), Maier-Leibnitz-Laboratorium, MIBL, MIC, MPI-HD, NSCL, ORNL, OUAL, PSI, RHIC (BNL), RIBRAS, RUBION, SNS, SPS (CERN), TAMU, TANDAR, TSL, TUNL, NICA(JINR), UAC, UMASS, UNAM, VECC

B4. Collider

BEPC II, CESR, DAFNE (LNF), KEK-B, LHC (CERN), RHIC (BNL), SLC (SLAC), TESLA (DESY), VEPP-3, VEPP-4M, VEPP-2000 (BINP)

2

FUNDAMENTALS OF PARTICLE PHYSICS

In general particle physics seeks to answer two main questions:

1. What are the fundamental (smallest) building blocks from which all matter is made?
2. What are the interactions between them that govern how they combine and decay?

The answers to the issues of fundamental , building blocks and interactions may be clarified in the following sections.

2.1. Building Blocks of Matter

1- Fundamental Building Blocks

The boxes below describe in general the building blocks of matter. More detail on the individual block is given in next section.

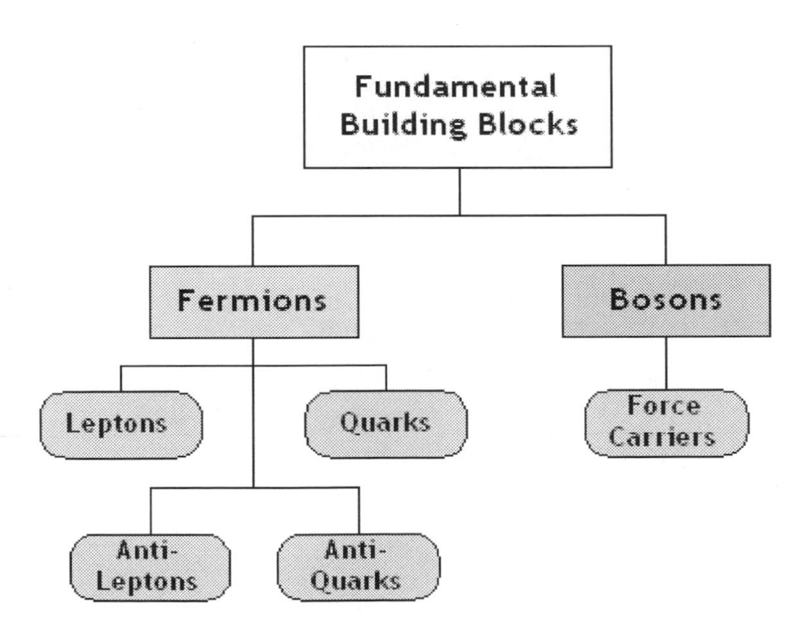

Fig.2.1 Building Blocks of Matter

In particle physics, every particle has a corresponding antiparticle. A particle and its antiparticle have identical Mass and Spin. Spin is the name for the angular momentum carried by a particle.

For composite particles, the spin is made up from the combination of the spins of the constituents plus the angular momentum of their motion around one-another.

For fundamental particles spin is an intrinsic and inherently quantum property, it cannot be understood in terms of motions internal to the object.

In quantum theories all angular momenta due to motion of one object around another are given as integer multiples of Planck's constant divided by (2pi). This quantity is called h-bar and is 6.58×10^{-25} GeV seconds.

Particles can be divided into two classes on the basis of their spins:

- **Bosons** are particles with integer spin in h-bar units.
- **Fermions** are particles with and odd number of half-integer h-bar units of spin. (1/2, 3/2, 5/2 ...)

Fermions obey a Pauli Exclusion Principle (No two of the same type can exist in the same state at the same place and time).

Bosons do not follow such a principle and, in a certain sense, favor states with many particles in the same state. For example, a laser beam is formed from a "coherent state" of many identical photons.

Matter particles (quarks and leptons) are fermions with 1/2 unit of spin.

Baryons are made from odd numbers of quarks and so are fermions.

Force carriers (gluons, W's, Z's, and photons) are bosons with 1 unit of spin. Mesons also are bosons.

More detail on the individual block is given in next sections.

2- Hadrons

Hadrons are particles made from quarks and/or gluons, bound together by their interactions, where the fundamental strong interactions occur between any two particles that have color charge, that is, quarks, antiquarks, and gluons.

Hadrons have no net strong charge (or color charge) but they do have residual strong interactions due to their color-charged substructure. Many types of hadrons were discovered in the 50's and 60's. The idea of the quarks was first proposed to explain the many observed hadrons. There are two classes of hadrons: Baryons and Mesons.

Color charge is the charge associated with strong interaction. "Color" is a whimsically named attribute of quarks and gluons that cannot be seen.

All hadrons are color-neutral objects. This means they have no net color charge.

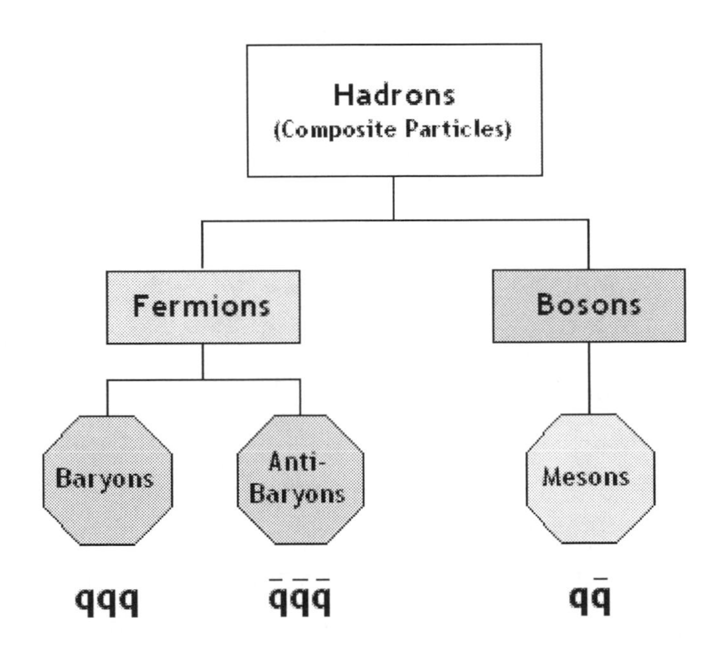

Fig.2.2 Classes of Hadrons

3- Baryons and Anti Baryons

Baryons are particles made from a basic structure of three quarks while anti-baryons are made from three antiquarks.

Baryons carry an odd half quantum unit of angular momentum (spin) and, hence, are **fermions,** which mean that they obey the Pauli Exclusion Principle.

Sample Fermionic Hadrons Baryons (qqq) and Anti-Baryon (qqq¯)					
Symbol	**Name**	**Quark Content**	**Electric Charge**	**Mass (GeV/c2)**	**Spin**
p	Proton	uud	1	0.938	½
$\bar{p}$	anti-proton	$\bar{u}\bar{u}\bar{d}$	-1	0.938	½
n	Neutron	udd	0	0.940	½
Λ	Lambda	uds	0	1.116	½
Ω^-	Omega	sss	-1	1.672	3/2

The proton is the only baryon that is stable in isolation. Its basic structure is two up quarks and one down quark. Neutrons are also baryons. Although neutrons are not stable in isolation, they can be stable inside certain nuclei. A neutron's basic structure is two down quarks and one up quarks.

More massive baryons may be made from any set of three quarks. Baryons containing more massive quarks are all unstable because these quarks decay via weak interactions. There are also more massive baryons that have the same quark content as a proton or a neutron but have additional angular momentum. These are typically very short-lived because they can decay to a proton or a neutron and a meson via residual strong interactions.

4- Mesons

Mesons are color-neutral particles with a basic structure of one quark and one antiquark. There are no stable mesons. Mesons have integer (or zero) units of spin, and hence are **bosons**, which means that they do not obey Pauli exclusion principle rules.

Sample Bosonic Hadrons Mesons ($q\bar{q}$)					
Symbol	**Name**	**Quark Content**	**Electric Charge**	**Mass (GeV/c2)**	**Spin**
π^+	Pion	$u\bar{d}$	+1	0.140	0
K^-	Kaon	$s\bar{u}$	-1	0.494	0
ρ^+	Rho	$u\bar{d}$	+1	0.770	1
D^+	D+	$c\bar{d}$	+1	1.869	0
η_c	eta-c	$c\bar{c}$	0	2.979	0

2.2. Fundamental Particles

The concept of a particle is a natural idealization of our everyday observation of matter. Dust particles or baseballs, under ordinary conditions, are stable objects that move according to simple laws of motion. Certainly we know that matter is indeed composed of the objects we call atoms. However we now understand that atoms are built up of smaller parts. These parts are electrons and a nucleus. The nucleus is much smaller than the atom and is itself composed of protons and neutrons.

In the 1930s, it seemed that protons, neutrons, and electrons were the smallest objects into which matter could be divided and they were termed "elementary particles". Again, later knowledge changed our understanding as physicists discovered yet another layer of structure within the protons and neutrons. It is now known that protons and neutrons are made up quarks. Over 100 other "elementary" particles were discovered between 1930 and the present time. These elementary particles are all made from quarks and/or antiquarks. These particles are called hadrons.

Once quarks were discovered, it was clear that all these hadrons were composite objects, so only in outdated text books are they still called "elementary". Leptons, on the other hand, still appear to be structure less. Today, quarks and leptons, and their antiparticles, are candidates for being the fundamental building blocks from which all else is made. Particle physicists call them the "fundamental" or "elementary" particles - both names denoting that, as far as current experiments can tell, they have no substructure.

In the modern theory, The Fundamental Particles are known as the Standard Model where there are 12 fundamental matter particle types and their corresponding antiparticles.

The matter particles divided into two classes: quarks and leptons. There are six particles of each class and six corresponding antiparticles. In addition, there are gluons, photons, and W and Z bosons, the force carrier particles that are responsible for strong, electromagnetic, and weak interactions respectively. These force carriers are also fundamental particles.

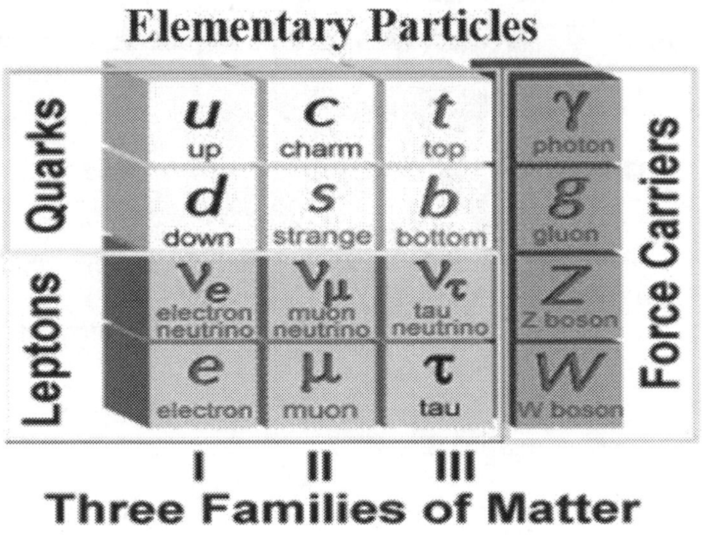

Fig. 2.3. "Fundamental" or "Elementary" particles

All we know is that quarks and leptons are smaller than 10^{-19} meters in radius. They have no internal structure or even any size. It is possible that future evidence will, once again, show this understanding to be an illusion and demonstrate that there is substructure within the particles that we now view as fundamental.

1- Electrons and Positrons:

The electron has a mass of 0.000511 GeV/c^2. The electron is the ***least*** massive charged particle of any type. It is absolutely stable because conservation of energy and electric charge together forbid any decay.

The antiparticle of the electron is called a **positron**. It has exactly the same mass as the electron, but the opposite sign (+1) for its electric charge. Positrons are also stable particles. However, positrons can annihilate when they meet an electron. Both the electron and the positron vanish, and their energy goes into photons and, possibly, more massive particles. Conversely, photons with sufficient energy ($E > 2x$ (mass of electron) x c^2) can produce an electron and a positron - this is called pair production.

2- Quarks

Quarks are fundamental matter particles that are constituents of neutrons and protons and other hadrons. There are six different types of quarks. Each quark type is called a flavor.

Flavor		Mass(GeV/c^2)	Electric Charge (e)
U	Up	0.004	+2/3
D	Down	0.008	-1/3
C	Charm	1.5	+2/3
S	Strange	0.15	-1/3
T	Top	176	+2/3
B	Bottom	4.7	-1/3

Quarks only exist inside hadrons because they are confined by the strong (or color charge) force fields. Therefore, we cannot measure their mass by isolating them. Furthermore, the mass of a hadron gets contributions from quark kinetic energy and from potential energy due to strong interactions. For hadrons made of the light quark types, the quark mass is a small contribution to the total hadron mass. For example, compare the mass of a proton (0.938 GeV/c^2) to the sum of the masses of two up quarks and one down quark (total of 0.02 GeV/c^2).

3- Leptons:

Leptons are fundamental particles that have no strong interactions. The six known types of leptons are shown in the table below. There are also six anti-lepton types, one for each lepton.

Flavor		Mass (GeV/c^2)	Electric Charge (e)
v_e	Electron Neutrino	<7 x 10-9	0
e^-	Electron	0.000511	-1
v_μ	Muon Neutrino	<0.0003	0
μ^-	Muon (mu-minus)	0.106	-1
v_τ	Tau Neutrino	<0.03	0
τ^-	Tau (tau-minus)	1.7771	-1

2.3. Forces and Interaction

All forces between objects are due to interactions. All particle decays are due to interactions. The **four** types fundamental interaction processes responsible for all observed processes are:

1. **Strong** interactions, responsible for forces between quarks and gluons and nuclear binding.
2. **Electromagnetic** interactions, responsible for electric and magnetic forces.
3. **Weak** interactions, responsible for the instability of all but the least massive fundamental particles in any class.
4. **Gravitational** interactions, responsible for forces between any two objects due to their energy (which, of course, includes their mass).

The Standard Model of particle physics is the relativistic quantum field theory of these the strong, electromagnetic, and weak interactions. In such theories, each type of interaction has a characteristic set of force carrier particles associated with quantum excitation of the force field related to that interaction.

The carrier particles either appear in intermediate stages or are produced during all particle processes involving that type of interaction. Forces between particles can be described in terms of static (unchanging) force fields and exchanges of force carrier particles between the affected particles.

- **Gluons** are the carrier particles of **strong** interactions.
- **Photons** are the carrier particles of **electromagnetic** interactions.
- **W and Z bosons** are the carrier particles of **weak** interactions.
- The name for the carrier particle of **gravitational** interactions is the **graviton**. Gravitons are not considered to be a part of the Standard Model.

In the Standard Model a fifth interaction type is needed to account for the masses of all particles. For further details on Standard model and the type of interaction processes see Annex A1.

2.4. Special Relativity

Newton's laws of motion give us a complete description of the behavior moving objects at low speeds. The laws are different at speeds reached by the particles at SLAC.

Einstein's Special Theory of Relativity describes the motion of particles moving at close to the speed of light. In fact, it gives the correct laws of motion for any particle. This doesn't mean Newton was wrong; his equations are contained within the relativistic equations. Newton's "laws" provide a very good approximate form, valid when v is much less than c. For particles moving at slow speeds (very much less than the speed of light), the differences between Einstein's laws of motion and those derived by Newton are tiny. That's why relativity doesn't play a large role in everyday life. Einstein's theory supersedes Newton's, but Newton's theory provides a very good approximation for objects moving at everyday speeds.

Einstein's theory is now very well established as the correct description of motion of relativistic objects, which are those traveling at a significant fraction of the speed of light.

Because most of us have little experience with objects moving at speeds near the speed of light, Einstein's predictions may seem strange. However, many years of high energy physics experiments have thoroughly tested Einstein's theory and shown that it fits all results to date.

Einstein's theory of special relativity results from two statements which are considered as the two basic postulates of special relativity as following:

1- **The speed of light is the same for all observers, no matter what their relative speeds.**
The first postulate (The Speed of Light is the same for all observers) is the crucial idea that led Einstein to formulate his theory. It means we can define a quantity c, the speed of light, which is a fundamental constant of nature.

This is quite different from the motion of ordinary, massive objects. For example, if I am driving down the freeway at 80 km. per hour relative to the road, a car traveling in the same direction at 90 kmph has a speed of only 10 kmph relative to me, while a car coming in the opposite direction at 90 kmph approaches me at a rate of 170 kmph. Their speed relative to me depends on my motion as well as on theirs.

2- **The laws of physics are the same in any inertial (i.e, non-accelerated) frame of reference.**

This means that the laws of physics observed by a hypothetical observer traveling with a relativistic particle must be the same as those observed by an observer who is stationary in the laboratory. This second postulate (Physics is the same for all inertial observers) is really a basic though unspoken assumption in all of science. This does not mean that things behave in the same way on the earth and in space, e.g. an observer at the surface of the earth is affected by the earth's gravity, but it does mean that the effect of a force on an object is the same independent of what causes the force and also of where the object is or what its speed is.

Einstein developed a theory of motion that could consistently contain both the same speed of light for any observer and the familiar addition of velocities described above for slow-moving objects. This is called the ***special theory of relativity***, since it deals with the **relative** motions of objects.

Given these two statements, Einstein showed how definitions of momentum and energy must be refined and how quantities such as length and time must change from one observer to another in order to get consistent results for physical quantities such as particle half-life. For further details see Attachment A2.

2.5. Units of Mass, Energy, and Momentum

Instead of using kilograms to measure mass, physicists use a unit of energy (electron volt). It is the energy gained by one electron when it moves through a potential difference of one volt.

By definition, one electron volt (**eV**) is equivalent to 1.6×10^{-19} joules.

Let's look at an example of how this energy unit works. The rest mass of an electron is 9.11×10^{-31} kg. Using $E = mc^2$ and a calculator we get:

$$E = 9.11 \times 10^{-31}\,\text{kg} \times (3 \times 10^{8}\,\text{m/s})^2 = 8.199 \times 10^{-14}\,\text{joules}$$

This gives us the energy equivalent of one electron. So, whether we say we have 9.11×10^{-31} kg or 8.199×10^{-14} joules, we really talking about the same thing (an electron). Physicists go one stage further and convert the joules to electron volts. This gives the mass of an electron as 0.511 MeV (about half a million **eV**).

Generally speaking, the high energy physicist definition for the mass of an electron is in terms of energy units. Therefore, you may notice that if you solve $E=mc^2$ for m, you get $m=E/c^2$, so the unit of energy should be eV/c^2.What happened to the c^2? It's very simple, particle physicists choose units of length so that the speed of light=1! How can we do that? All we have to do is use a conversion factor to get back the "real" (i.e. every day) units, if we want them.

Not only are mass and energy measured in **eV**, so is momentum. It makes life so much easier than dividing by c^2 or c all the time.

3

ACCELERATOR PHYSICS AND PRINCIPLES

3.1. Fundamental and Physical Theories

Accelerator physics is a branch of applied physics, concerned with designing, building and operating particle accelerators. As such, it can be circumscribed as the study of motion, manipulation and observation of relativistic charged particle beams and their interaction with accelerator structures by electromagnetic fields. The main principles and fundamentals and physical theories govern the physics and principles of accelerator are as briefly highlighted in the following subsections.

3.1.1. Newton's laws of motion

Newton's laws of motion are three physical laws that together laid the foundation for classical mechanics. These three laws of motion were first compiled by Isaac Newton in his (Mathematical Principles of Natural Philosophy), first published in 1687. Newton used them to explain and investigate the motion of many physical objects and systems. They describe the relationship between a body and the forces acting upon it, and its motion in response to said forces. They have been expressed in several different ways over nearly three centuries, and can be summarized as follows:

A. Newton's First law

The first law states that if the net force (the vector sum of all forces acting on an object) is zero, then the velocity of the object is constant. Velocity is a vector quantity which expresses both the object's speed and the direction of its motion; therefore, the statement that the object's velocity is constant is a statement that both its speed and the direction of its motion are constant.

The first law can be stated mathematically as $\sum F = 0 \Leftrightarrow \frac{dv}{dt} = 0$

Consequently,

- An object that is at rest will stay at rest unless an external force acts upon it.

- An object that is in motion will not change its velocity unless an external force acts upon it.

This is known as uniform motion. The first law of motion postulates the existence of at least one frame of reference called a Newtonian or inertial reference frame, relative to which the motion of a particle not subject to forces is a straight line at a constant speed. Newton's first law is often referred to as the law of inertia.

B. Newton's Second Law

The second law states that the net force on an object is equal to the rate of change (that is, the derivative) of its linear momentum p in an inertial reference frame:

$$F = \frac{dP}{dt} = \frac{d(mv)}{dt}$$

The second law can also be stated in terms of an object's acceleration. Since Newton's second law is only valid for constant-mass systems, mass can be taken outside the differentiation operator by the constant factor rule in differentiation. Thus,

$$F = m\,\frac{dv}{dt} = ma$$

where, F is the net force applied, *m* is the mass of the body, and **a** is the body's acceleration. Thus, the net force applied to a body produces a proportional acceleration. In other words, if a body is accelerating, then there is a force on it.

Consistent with the first law, the time derivative of the momentum is non-zero when the momentum changes direction, even if there is no change in its magnitude; such is the case with uniform circular motion. The relationship also implies the conservation of momentum: when the net force on the body is zero, the momentum of the body is constant. Any net force is equal to the rate of change of the momentum. Newton's second law requires modification if the effects of special relativity are to be taken into account, because at high speeds the approximation that momentum is the product of rest mass and velocity is not accurate.

C. Newton's Third Law

The third law states that all forces exist in pairs: if one object A exerts a force $\mathbf{F}_A$ on a second object B, then B simultaneously exerts a force $\mathbf{F}_B$ on A, and the two forces are equal and opposite:

$$\mathbf{F}_A = -\mathbf{F}_B$$

The third law means that all forces are interactions between different bodies, and thus that there is no such thing as a unidirectional force or a force that acts on only one body. This law is sometimes referred to as the action-reaction law, with $\mathbf{F}_A$ called the "action" and $\mathbf{F}_B$ the "reaction". The action and the reaction are simultaneous, and it does not matter which is called the action and which is called reaction; both forces are part of a single interaction, and neither force exists without the other.

Newton's laws of motion, together with his law of universal gravitation and the mathematical techniques of calculus, provided for the first time a unified quantitative explanation for a wide range of physical phenomena.

These three laws hold to a good approximation for macroscopic objects under everyday conditions. However, Newton's laws (combined with universal gravitation and classical electrodynamics) are inappropriate for use in certain circumstances, most notably at very small scales, very high speeds (in special relativity, the Lorentz factor must be included in the expression for momentum along with rest mass and velocity) or very strong gravitational fields. Therefore, the laws cannot be used to explain phenomena such as conduction of electricity in a semiconductor, optical properties of substances, and superconductivity. Explanation of these phenomena requires more sophisticated physical theories, including general relativity and quantum field theory.

In modern physics, the laws of conservation of momentum, energy, and angular momentum are of more general validity than Newton's laws, since they apply to both light and matter, and to both classical and non-classical physics. This can be stated simply, "Momentum, energy and angular momentum cannot be created or destroyed".

3.1.2. Particle Acceleration

The development of charged-particle accelerators has progressed along double paths which by the appearance of particle trajectories are distinguished as linear accelerators and circular accelerators. Particles travel in linear accelerators only once through the accelerator structure while in a circular accelerator they follow a closed orbit periodically for many revolutions accumulating energy at every traversal of the accelerating structure.

In general, acceleration of particles for linear or circular trajectories is governed by:

$$a = v^2/d \quad \text{for linear accelerator, or} \qquad (3.1)$$

$$a = \omega^2 r \quad \text{for circular accelerator} \qquad (3.2)$$

The idea of accelerators in particle physics is to give particles more momentum, hence reducing the de Broglie wavelength. The simplest way to accelerate a particle is to attract an electron that is boiled off by thermionic emission and attract it to an anode, as for example we have seen that done in the cathode ray tube in a TV set.

For a given speed, a proton which has 1800 times the mass of the electron can have 1800 times the momentum, and a de Broglie wavelength of 1800 times less.

Charged particles can be accelerated to high energies in high vacuum regions where beam-gas collisions are rare. Particles are extracted from their sources (ions from plasmas, electrons from cathodes), focused into narrow beams and accelerated by electric fields created by the high voltage generators. The conformation and strength of the fields are determined by the shapes and spacing of a series of electrodes with intermediate potentials. The use of multiple electrodes prevents spark discharges inside the acceleration tube, protects the insulating rings between the electrodes from the effects of scattered particles and permits the use of high accelerating potentials and strong electric fields.

3.1.3. Kinetic Energy

The speed of the particles can be easily measured by considering their kinetic energy given by:

$$E_k = ½mv^2 \qquad (3.3)$$

The energy acquired is:

$$\Delta E = eV \qquad (3.4)$$

For a typical accelerating voltage of 50 000 V, the energy gained is:

$$\Delta E = 1.6 \times 10^{-19} \text{ C} \times 50000 \text{ V} = 8.0 \times 10^{-15} \text{ J}$$

The speed can be worked out from:

$$v^2 = 2\ E_k\ /m \ = 2 \times 8.0 \times 10^{-15} \text{ J}/9.11 \times 10^{-31} \text{ kg}$$

$$v^2 = 1.71 \times 10^{16} \text{ m}^2 \text{ s}^{-2}$$

$$v = 1.31 \times 10^8 \text{ m s}^{-1}$$

If the voltage increased by 100 times to 5×10^6 V (5 MV), which is quite possible, we would increase the speed by10 times to 1.31×10^9 ms^{-1}. However Einstein's Special Relativity Theory tells us that nothing can travel faster than the speed of light. This tells us two things:

1. That laws and models that we use in normal physics do not apply to all situations.
2. There are strange things that happen when things travel towards the speed of light.

We find that as a particle approaches the speed of light, some of its kinetic energy is converted to mass. The particle gets heavier. This is called a relativistic effect.

This should not be a surprise since mass and energy are interchangeable according to Einstein's famous equation:

$$E = mc^2 \quad (3.5)$$

Joules and kilograms in this context are rather clumsy to work with. So we use **electron-volts** (eV).They are a unit of energy, not voltage. This should not be too difficult since we know that:

$$\text{Energy (J)} = \text{Charge (C)} \times \text{Voltage (V)}$$

The electron-volt is the energy gained by unit charge accelerated through a potential difference of 1 volt.

$$1 \text{ eV} = 1.6 \times 10^{-19} \text{ J}$$

The electron above accelerated through 50 000 V would have an energy of 50 000 eV or 50 keV. A lot easier than 8.0×10^{-15} J.

Many accelerators can accelerate particles to energies of Giga electron-volts (GeV) (10^9 eV) or even Tera-electron-volt (TeV) (10^{12} eV).

The units found in particle physics are often expressed in very odd looking units that are based on the rest mass of the particle. If we were able to convert all an electron's mass into energy according to eq. (3.2), we would get:

$$E_0 = 9.11 \times 10^{-31} \text{ kg} \times (3.0 \times 10^8 \text{ m s}^{-1})^2$$
$$= 8.20 \times 10^{-14} \text{ J} = 5.12 \times 10^5 \text{ eV}$$

This is 0.512 MeV. Since $E = mc^2$, it does not take a genius to rearrange this to give the mass:

$$m = E/c^2 \quad (3.6)$$

So we can say that the rest mass of the particle is 0.512 MeV/c^2. Heavier particles has rest masses of GeV/c^2. These strange looking units allow particle physicists to see if they have enough energy for a collision without having to do any calculations.

3.2. Equations of Motion

3.2.1. Introduction

In mathematical physics, **equations of motion** are equations that describe the behavior of a physical system in terms of its motion as a function of time. More specifically, the equations of motion describe the behavior of a physical system as a set of mathematical functions in terms of dynamic variables: normally spatial coordinates and time are used, but others are also possible, such as momentum components and time. The most general choice are generalized coordinates which can be any convenient variables characteristic of the physical system. The functions are defined in a Euclidean space in classical mechanics, but are replaced by curved spaces in relativity. If the dynamics of a system is known, the equations are the solutions to the differential equations describing the motion of the dynamics.

There are two main descriptions of motion: dynamics and kinematics. Dynamics is general, since momenta, forces and energy of the particles are taken into account. In this instance, sometimes the term refers to the differential equations that the system satisfies (e.g., Newton's second law or Euler–Lagrange equations), and sometimes to the solutions to those equations.

Historically, equations of motion initiated in classical mechanics and the extension to celestial mechanics, to describe the motion of massive objects. Later they appeared in electrodynamics, when describing the motion of charged particles in electric and magnetic fields. With the advent of general relativity, the classical equations of motion became modified. In all these cases the differential equations were in terms of a function describing the particle's trajectory in terms of space and time coordinates, as influenced by forces or energy transformations. However, the equations of quantum mechanics can also be considered equations of motion, since they are differential equations of the wave function, which describes how a quantum state behaves analogously using the space and time coordinates of the particles.

Equations of motion typically involve:

- a differential equation of motion, usually identified as some physical law and applying definitions of physical quantities, is used to set up an equation for the problem,
- setting the boundary and initial value conditions,
- a function of the position (or momentum) and time variables, describing the dynamics of the system,
- solving the resulting differential equation subject to the boundary and initial conditions.

The differential equation is a general description of the application and may be adjusted appropriately for a specific situation, the solution describes exactly how the system will behave for all times after the initial conditions, and according to the boundary conditions.

3.2.2. Kinematic Equations for one Particle

From the instantaneous position **r** = **r** (t), (instantaneous means at an instant value of time t), the instantaneous velocity **v** = **v** (t) and acceleration **a** = **a** (t) have the general, coordinate-independent definitions;

$$\mathrm{v} = \frac{dr}{dt} \quad \text{and} \quad \mathrm{a} = \frac{dv}{dt} = \frac{d^2 r}{dt^2} \qquad (3.7)$$

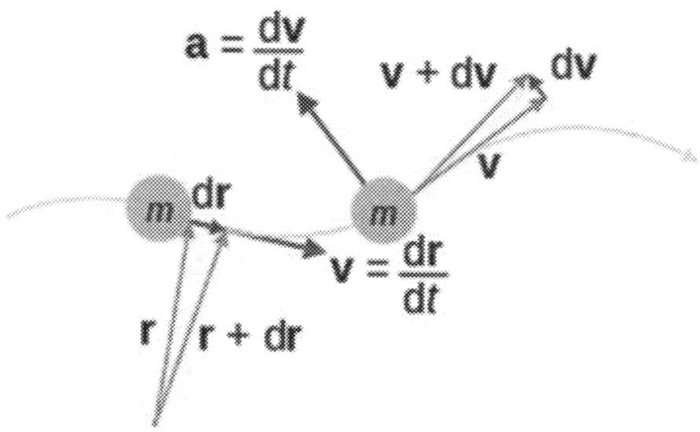

Fig. 3.1. Kinematic quantities of a classical particle of mass m: position r, velocity v, acceleration a.

Notice that velocity always points in the direction of motion and the acceleration is directed towards the center of curvature of the path. In other words for a curved path the velocity (tangent vector) is first order derivatives and the acceleration is a second order derivatives related to curvature.

The rotational analogues are the angular position (angle the particle rotates about some axis) $\theta = \theta(t)$, angular velocity $\omega = \omega(t)$, and angular acceleration $\alpha = \alpha(t)$:

$$\omega = \hat{n}\frac{d\theta}{dt} \quad \text{and} \quad \alpha = \frac{d\omega}{dt} = \hat{n}\,\frac{d^2\,\theta}{dt^2} \qquad (3.8)$$

Where $\hat{n} = \hat{e_r} \times \widehat{e_\theta}$ is a unit axial vector, pointing parallel to the axis of rotation, ($\hat{e_r}$) is the unit vector in direction of **r**, and ($\widehat{e_\theta}$)is the unit vector tangential to the angle. In these rotational definitions, the angle can be any angle about the specified axis of rotation. It is customary to use θ, but this does not have to be the polar angle used in polar coordinate systems.

The following relations hold for a point-like particle, orbiting about some axis with angular velocity **ω**:

$$v = \omega \times r \quad \text{and} \quad a = \alpha \times r + \omega \times v \qquad (3.9)$$

where r is a radial position, v the tangential velocity of the particle, and **a** the particle's acceleration. More generally, these relations hold for each point in a rotating continuum rigid body.

Elementary and frequent examples in kinematics involve projectiles, for example a ball thrown upwards into the air. Given initial speed *u*, one can calculate how high the ball will travel before it begins to fall. The acceleration is local acceleration of gravity *g*. At this point one must remember that while these quantities appear to be scalars, the direction of displacement, speed and acceleration is important. They could in fact be considered as uni-directional vectors. Choosing *s* to measure up from the ground, the acceleration *a* must be in fact (−*g*), since the force of gravity acts downwards, the acceleration in polar coordinate system on the ball is given by:

$$\frac{d^2\,\theta}{dt^2} = \sum_n \omega_n^2\theta \qquad (3.10)$$

3.2.3. Dynamic Equations of Motion

A. Newtonian Mechanics

It may be simple to write down the equations of motion in vector form using Newton's laws of motion, but the components may vary in complicated ways with spatial coordinates and time, and solving them is not easy. Often there is an excess of variables to solve for the problem completely, so Newton's laws are not the most efficient method for generally finding and solving for the motion of a particle. In simple cases of rectangular geometry, the use of Cartesian coordinate's works fine, but other coordinate systems can become dramatically complex.

The first developed and most famous is Newton's second law of motion, there are several ways to write and apply it, and the most general is:

$$F = \frac{dP}{dt}$$

where p = p(t) is the momentum of the particle and F = F(t) is the resultant external force acting on the particle (not any force the particle exerts) - in each case at time t. The law is also written more famously as:

$$F = ma$$

since m is a constant in Newtonian mechanics. However the momentum form is preferable since this is readily generalized to more complex systems, generalizes to special and general relativity, and since momentum is a conserved quantity; with deeper fundamental significance than the position vector or its time derivatives.

For a number of particles, the equation of motion for one particle i influenced by other particles is:

$$\frac{dP_i}{dt} = F_E + \sum_{i \neq j} F_{ij} \qquad (3.11)$$

where p_i = momentum of particle i, F_{ij} = force on particle i by particle j, and F_E = resultant external force (due to any agent not part of system). Particle i does not exert a force on itself.

For rigid bodies, Newton's second law for rotation takes the same form as for translation:

$$\tau = \frac{dL}{dt}$$

where L is the angular momentum. Analogous to force and acceleration:

$$\tau = I.\ \alpha$$

where I is the moment of inertia tensor. Likewise, for a number of particles, the equation of motion for one particle *i* is:

$$\frac{dL_i}{dt} = \tau_E + \sum_{i \neq j} \tau_{ij} \tag{3.12}$$

where L_i = angular momentum of particle *i*, τ_{ij} = torque on particle *i* by particle *j*, and τ_E = resultant external torque (due to any agent not part of system). Particle *i* does not exert a torque on itself.

Some examples of Newton's law include describing the motion of a pendulum:

$$-mg \sin\theta = m\frac{d^2(l\theta)}{dt^2} \quad \rightarrow \quad \frac{d^2\theta}{dt^2} = -\frac{g}{l}\sin\theta \tag{3.13}$$

a damped, driven harmonic oscillator:

$$F_0 \sin(\omega t) = m\left(\frac{d^2x}{dt^2} + 2\varsigma\,\omega_0\frac{dx}{dt} + \omega_0{}^2 x\right) \tag{3.14}$$

or a ball thrown in the air, in air currents (such as wind) described by a vector field of resistive forces R = R(r, *t*):

$$\frac{Gm\,M}{|\mathbf{r}|^2}\,\widehat{e_r} + \mathbf{R} = m\,\frac{d^2\mathbf{r}}{dt^2} + 0 \rightarrow \frac{d^2\mathbf{r}}{dt^2} = -\frac{G\,M}{|\mathbf{r}|^2}\,\widehat{e_r} + \mathbf{A} \tag{3.15}$$

where ***G*** = gravitational constant, *M* = mass of the Earth and **A** = **R**/*m* is the acceleration of the projectile due to the air currents at position **r** and time *t*. Newton's law of gravity has been used. The mass *m* of the ball cancels.

B. Eulerian Mechanics

Euler developed Euler's laws of motion, analogous to Newton's laws, for the motion of rigid bodies.

Euler's first law states that the linear momentum of a body, **p** (also denoted **G**) is equal to the product of the mass of the body *m* and the velocity of its center of mass $\mathbf{v}_{cm}$:

$$\mathbf{P} = m\,\mathbf{v}_{cm} \qquad (3.16)$$

Internal forces between the particles that make up a body do not contribute to changing the total momentum of the body. The law is also stated as:

$$\mathbf{F} = m\,a_{cm} \qquad (3.17)$$

where $\mathbf{a}_{cm} = d\mathbf{v}_{cm}/dt$ is the acceleration of the center of mass and $\mathbf{F} = d\mathbf{p}/dt$ is the total applied force on the body. This is just the time derivative of the previous equation (*m* is a constant).

Euler's second law states that the rate of change of angular momentum **L** (also denoted **H**) about a point that is fixed in an inertial reference frame or the mass center of the body, is equal to the sum of the external moments of force (torques) **M** (also denoted $\boldsymbol{\tau}$ or $\boldsymbol{\Gamma}$) about that point:

$$\mathrm{M} = \frac{d\mathbf{L}}{dt} \qquad (3.18)$$

Note that the above formula holds only if both **M, L** are computed with respect to a fixed inertial frame or a frame parallel to the inertial frame but fixed on the center of mass. For rigid bodies translating and rotating in only 2d, this can be expressed as:

$$\mathbf{M} = \mathbf{r}_{cm}\ \mathrm{X}\ \mathbf{a}_{cm}\, m + I \propto \qquad (3.19)$$

Where $\mathbf{r}_{cm}$ is the position vector of the center of mass with respect to the point about which moments are summed, $\boldsymbol{\alpha}$ is the angular acceleration of the body, and *I* is the moment of inertia.

C. Newton–Euler equations

The Newton–Euler equations combine Euler's equations into one.

Traditionally the Newton–Euler equations is the grouping together of Euler's two laws of motion for a rigid body into a single equation with 6 components, using column vectors and matrices. These laws relate the motion of the center of gravity of a rigid body with the sum of forces and torques (or synonymously moments) acting on the rigid body.

- **Center of mass frame**

With respect to a coordinate frame whose origin coincides with the body's center of mass, they can be expressed in matrix form as:

$$\begin{pmatrix} \mathbf{F} \\ \boldsymbol{\tau} \end{pmatrix} = \begin{pmatrix} m\mathbf{1} & 0 \\ 0 & \mathbf{I}_{\mathbf{cm}} \end{pmatrix} \begin{pmatrix} \mathbf{a}_{\mathbf{cm}} \\ \boldsymbol{\alpha} \end{pmatrix} + \begin{pmatrix} 0 \\ \boldsymbol{\omega} \times \mathbf{I}_{\mathbf{cm}}\, \boldsymbol{\omega} \end{pmatrix} \qquad (3.20)$$

where

$\mathbf{F}$ = total force acting on the center of mass
m = mass of the body
$\mathbf{1}$ = the 3×3 identity matrix
$\mathbf{a}_{cm}$ = acceleration of the center of mass
$\mathbf{v}_{cm}$ = velocity of the center of mass
$\boldsymbol{\tau}$ = total torque acting about the center of mass
$\mathbf{I}_{cm}$ = moment of inertia about the center of mass
$\boldsymbol{\omega}$ = angular velocity of the body
$\boldsymbol{\alpha}$ = angular acceleration of the body

The Newton–Euler equations are used as the basis for more complicated "multi-body" formulations (screw theory) that describe the dynamics of systems of rigid bodies connected by joints and other constraints. Multi-body problems can be solved by a variety of numerical algorithms.

- **Any reference frame**

With respect to a coordinate frame located at point **P** that is fixed in the body and not coincident with the center of mass, the equations assume the more complex form as given in the matrix form Eq.(3.21) below:

$$\begin{pmatrix} \mathbf{F} \\ \boldsymbol{\tau}_\mathrm{p} \end{pmatrix} = \begin{pmatrix} m\mathbf{1} & -m\lfloor \mathrm{C} \rfloor^{\mathrm{X}} \\ m\lfloor \mathrm{C} \rfloor^{\mathrm{X}} & \mathbf{I}_{\mathbf{cm}} - m\lfloor \mathrm{C} \rfloor^{\mathrm{X}} \lfloor \mathrm{C} \rfloor^{\mathrm{X}} \end{pmatrix} \begin{pmatrix} \mathbf{a}_\mathrm{p} \\ \boldsymbol{\alpha} \end{pmatrix} + \begin{pmatrix} m\lfloor \boldsymbol{\omega} \rfloor^{\mathrm{X}} \lfloor \boldsymbol{\omega} \rfloor^{\mathrm{X}}\, \mathrm{C} \\ \lfloor \boldsymbol{\omega} \rfloor^{\mathrm{X}} \left(\mathbf{I}_{\mathbf{cm}} - m\lfloor \mathrm{C} \rfloor^{\mathrm{X}} \lfloor \mathrm{C} \rfloor^{\mathrm{X}} \right) \boldsymbol{\omega} \end{pmatrix}$$

--------- (3.21)

where **C** is the location of the center of mass expressed in the body-fixed frame. $\lfloor \mathrm{C} \rfloor^{\mathrm{X}}$ and $\lfloor \boldsymbol{\omega} \rfloor^{\mathrm{X}}$ are denoted as skew-symmetric cross product matrices and approximated by:

$$\lfloor \mathrm{C} \rfloor^{\mathrm{X}} \equiv \begin{pmatrix} 0 & -C_z & C_y \\ C_z & 0 & -C_x \\ -C_y & C_x & 0 \end{pmatrix} \text{ and } \lfloor \boldsymbol{\omega} \rfloor^{\mathrm{X}} \equiv \begin{pmatrix} 0 & -\omega_z & \omega_y \\ \omega_z & 0 & -\omega_x \\ -\omega_y & \omega_x & 0 \end{pmatrix} \quad (3.22)$$

The inertial terms are contained in the spatial inertia matrix

$$\begin{pmatrix} m\mathbf{1} & -m\lfloor \mathrm{C} \rfloor^{\mathrm{X}} \\ m\lfloor \mathrm{C} \rfloor^{\mathrm{X}} & \mathbf{I}_{\mathbf{cm}} - m\lfloor \mathrm{C} \rfloor^{\mathrm{X}} \lfloor \mathrm{C} \rfloor^{\mathrm{X}} \end{pmatrix} \quad (3.23)$$

while the fictitious forces are contained in the term:

$$\begin{pmatrix} m\lfloor \boldsymbol{\omega} \rfloor^{\mathrm{X}} \lfloor \boldsymbol{\omega} \rfloor^{\mathrm{X}}\, \mathrm{C} \\ \lfloor \boldsymbol{\omega} \rfloor^{\mathrm{X}} \left(\mathbf{I}_{\mathbf{cm}} - m\lfloor \mathrm{C} \rfloor^{\mathrm{X}} \lfloor \mathrm{C} \rfloor^{\mathrm{X}} \right) \boldsymbol{\omega} \end{pmatrix} \quad (3.24)$$

When the center of mass is not coincident with the coordinate frame (that is, when **C** is nonzero), the translational and angular accelerations (**a** and $\boldsymbol{\alpha}$) are coupled, so that each is associated with force and torque components.

3.2.4. Analytical Mechanics

Analytical mechanics (or **theoretical mechanics**), developed in the 18th century and onward, are mathematical physics' refinements of classical mechanics, originally Newtonian mechanics, often termed vectorial mechanics. To model motion, analytical mechanics uses two scalar properties of motion—its kinetic energy and its potential energy—not Newton's vectorial forces. (A scalar is represented by a quantity, as denotes intensity, whereas a vector is represented by quantity plus direction.)

Principally Lagrangian mechanics (The introduction of generalized coordinates) and Hamiltonian mechanics(which is in terms of generalized coordinates and momenta):, both tightly intertwined, analytical mechanics efficiently extends the scope of classical mechanics to solve problems by employing the concept of a system's constraints and path integrals.

Using all three coordinates of 3d space is unnecessary if there are constraints on the system. Generalized coordinates $\mathbf{q}(t) = [q_1(t), q_2(t) \dots q_N(t)]$, where N is the total number of degrees of freedom the system has, are any set of coordinates used to define the configuration of the system, in the form of arc lengths or angles. They are a considerable simplification to describe motion since they take advantage of the intrinsic constraints that limit the system's motion - i.e. the number of coordinates is reduced to a minimum, rather than demanding rote algebra to describe the constraints and the motion using all three coordinates.

Corresponding to generalized coordinates are:

- their time derivatives, the generalized velocities: $\dot{q} = dq/dt$,
- conjugate "generalized" momenta: $P = \frac{\partial L}{\partial \dot{q}} = \frac{\partial S}{\partial q}$

where

- the ***Lagrangian*** is a function of the configuration **q**, the rate of change of configuration d**q**/dt, and time t:

$$L = L\,[\mathrm{q}(t), \dot{q}(t), t]\,, \qquad (3.25)$$

- the ***Hamiltonian*** is a function of the configuration **q**, motion **p**, & time t:

$$H = H\,[\mathrm{q}(t), \mathrm{p}(t), t]\,, \text{ and} \qquad (3.26)$$

- ***Hamilton's principal function***, also called the ***classical action*** is a functional of L:

$$S[\mathrm{q}, t] = \int_{t1}^{t2} L(\mathrm{q}, \dot{\mathrm{q}}, t)\; dt \qquad (3.27)$$

The Lagrangian or Hamiltonian function is set up for the system using the q and p variables, then these are inserted into the Euler–Lagrange or Hamilton's equations to obtain differential equations of the system. These are solved for the coordinates and momenta.

3.2.5. Electrodynamics

In electrodynamics, the force on a charged particle of charge q is the Lorentz force:

$$\mathbf{F} = q\,(\mathbf{E} + \mathbf{v} \times \mathbf{B}) \tag{3.28}$$

Where both the E field and B field vary in space and time.

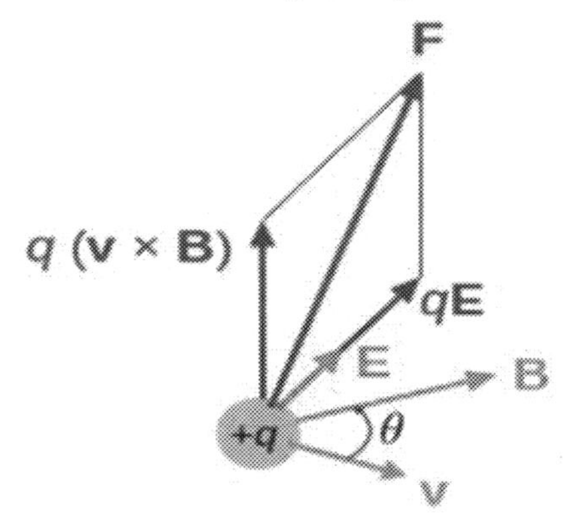

Fig.3.2. Lorentz force F on a charged particle (of charge q) in motion (instantaneous velocity v).

Combining with Newton's second law gives a first order differential equation of motion, in terms of position of the particle:

$$m\,\frac{d^2\mathrm{r}}{dt^2} = q\,(\mathrm{E} + \frac{d\mathrm{r}}{dx} \times \mathrm{B}) \tag{3.29}$$

or its momentum:

$$\frac{d\mathrm{P}}{dx} = q\,(\mathrm{E} + \frac{\mathrm{P} \times B}{m}) \tag{3.30}$$

The same equation can be obtained using the Lagrangian (and applying Lagrange's equations above) for a charged particle of mass m and charge q:

$$L = \frac{m}{2}\,\dot{\mathrm{r}}.\dot{\mathrm{r}} + q\,\mathrm{A}\,.\,\dot{\mathrm{r}} - q\phi \tag{3.31}$$

where **A** and ϕ are the electromagnetic scalar and vector potential fields.

The Lagrangian indicates an additional detail: the canonical momentum in Lagrangian mechanics is given by:

$$\mathrm{P} = \frac{\partial L}{\partial \dot{\mathrm{r}}} = m\,\dot{\mathrm{r}} + q\,\mathrm{A} \tag{3.32}$$

instead of just $m\mathbf{v}$, implying the motion of a charged particle is fundamentally determined by the mass and charge of the particle. The Lagrangian expression was first used to derive the force equation.

Alternatively the Hamiltonian (and substituting into the equations):

$$\boldsymbol{H} = \frac{(\mathbf{P} - q\,\mathbf{A})^2}{2m} - q\phi \tag{3.33}$$

can derive the Lorentz force equation.

4

ACCELERATOR DESIGN AND ENGINEERING

4.1. Design Principles and Goals

Particle accelerator, apparatus used in nuclear physics to produce beams of energetic charged particles and to direct them against various targets. The Particle accelerator, popularly called atom smashers, are designed and needed to observe objects as small as the atomic nucleus in studies of its structure and of the forces that hold it together. Accelerators are also needed to provide enough energy to create new particles. Besides pure research, accelerators have practical applications in medicine and industry, most notably in the production of radioisotopes.

There are many types of accelerator designs, although all have certain features in common. Only charged particles (most commonly protons and electrons, and their antiparticles; less often deuterons, alpha particles, and heavy ions) can be artificially accelerated; therefore, the first stage of any accelerator is an ion source to produce the charged particles from a neutral gas.

Accelerators solve two problems for physicists. First, since all particles behave like waves, physicists use accelerators to increase a particle's momentum, thus decreasing its wavelength enough that physicists can use it to poke inside atoms. Second, the energy of speedy particles is used to create the massive particles that physicists want to study.

Basically, an accelerator takes a particle, speeds it up using electromagnetic fields (steady, alternating, or induced) to contain and focus the beam, and bombards the particle into a target or other particles. Surrounding the collision point are detectors that record the many pieces of the event.

The output of a particle accelerator can generally be directed towards multiple lines of experiments, one at a given time, by means of a deviating electromagnet. This makes it possible to operate multiple experiments without needing to move things around or shutting down the entire accelerator beam.

In the linear particle accelerator (linac), the target is simply fitted to the end of the accelerator. The particle track in a cyclotron is a spiral outwards from the center of the circular machine, so the accelerated particles emerge from a fixed point as for a linear accelerator.

For the circular accelerator the situation is more complex. The purpose of an accelerator for synchrotron radiation sources, is to accelerate the particles to the desired energy for interaction with matter. Then, a fast acting dipole magnet is used to switch the particles out of the circular synchrotron tube and towards the target.

A variation commonly used for particle physics research is a collider, also called a storage ring collider. Two circular synchrotrons are built in close proximity – usually on top of each other and using the same magnets (which are then of more complicated design to accommodate both beam tubes). Bunches of particles travel in opposite directions around the two accelerators and collide at intersections between them. This can increase the energy enormously; whereas in a fixed-target experiment the energy available to produce new particles is proportional to the square root of the beam energy, in a collider the available energy is linear.

There are several different ways to design these accelerators, each with its benefits and drawbacks. Here's a quick list of the major accelerator design choices:

A. Accelerators can be arranged to provide collisions of two types:

A1. **Fixed target**: Shoot a particle at a fixed target.

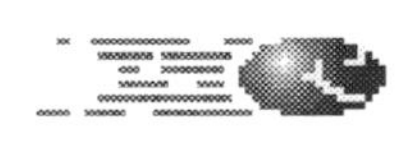 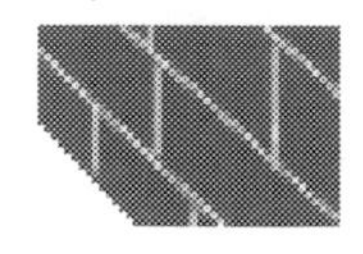

A2. **Colliding beams**: Two beams of particles are made to cross each other.

B. Accelerators are shaped in one of two ways; either linear or circular, the difference being whether the particle is shot like a bullet from a gun (the linear accelerator) or whether the particle is twirled in a very fast circle, receiving a bunch of little kicks each time around (the circular accelerator). Both types accelerate particles by pushing them with an electric-field wave.

B1. **Linear accelerators**: Accelerators in which the particle starts at one end and comes out the other, i.e. the particle path is a straight line. In **linear** accelerators, also called ***linacs***, the beam usually passes only once through the accelerating fields and magnetic focusing fields.

Linear accelerators (*linacs*) are used for fixed-target experiments, as injectors to circular accelerators, or as linear colliders.

Fixed target:

Injector to a circular accelerator:

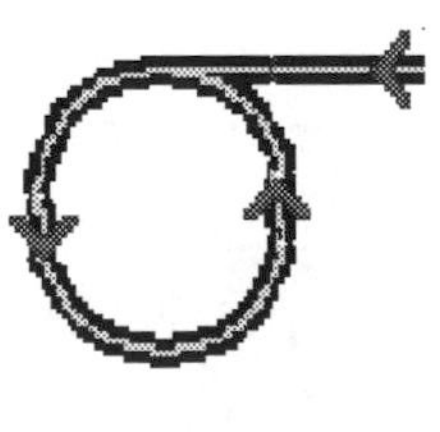

Linear collider:

B2. **Circular/Synchrotrons**: Accelerators built in a circle, in which the particle goes around many times, where the magnetic field is used to bend the particles in a circular or spiral path. In circular accelerators - cyclotrons and synchrotrons - magnetic fields bend the particles' path so that the particle beam passes repeatedly through the accelerating structures and focusing magnets.

Circular/Synchrotrons

In order to appreciate the differences in accelerator designs, the basic plans of the world's major accelerators are briefly as following:

SLAC: Stanford Linear Accelerator Center, in California, discovered the charm quark (also discovered at Brookhaven) and tau lepton; ran an accelerator producing huge numbers of B mesons.

Fermilab: Fermi National Laboratory Accelerator, in Illinois, where the bottom and top quarks and the tau neutrino were discovered.

CERN: European Laboratory for Particle Physics, crossing the Swiss-French border, where the W and Z particles were discovered.

BNL: Brookhaven National Lab, in New York, simultaneously with SLAC discovered the charm quark.

CESR: Cornell Electron-Positron Storage Ring, in New York. CESR performed detailed studies of the bottom quark.

DESY: Deutsches Elektronen-Synchrotron, in Germany; gluons were discovered here.

KEK: High Energy Accelerator Research Organization, in Japan, is now running an accelerator producing huge numbers of B mesons.

IHEP: Institute for High-Energy Physics, in the People's Republic of China, performs detailed studies of the tau lepton and charm quark.

4.2. Particles Beam

4.2.1. Main components for particle-beam storage

1- Magnets

A force must be applied to particles in such a way that they are constrained to move approximately in a specific path. This may be accomplished using either dipole electrostatic or dipole magnetic fields, but because most storage rings store relativistic charged particles it turns out that it is most practical to utilise magnetic fields produced by dipole magnets. However, electrostatic accelerators have been built to store very low energy particles, and quadrupole fields may be used to store (uncharged) neutrons; these are comparatively rare.

Dipole magnets alone only provide what is called weak focusing, and a storage ring composed of only these sorts of magnetic elements results in the particles having a relatively large beam size. Interleaving dipole magnets with an appropriate arrangement of quadrupole and sextupole magnets can give a suitable strong focusing system that can give a much smaller beam size.

Dipole and quadrupole magnets deflect different particle energies by differing amounts, a property called chromaticity by analogy with physical optics. The spread of energies that is inherently present in any practical stored particle beam will therefore give rise to a spread of transverse and longitudinal focusing, as well as contributing to various particle beam instabilities. Sextupole magnets (and higher order magnets) are used to correct for this phenomenon, but this in turn gives rise to nonlinear motion that is one of the main problems facing designers of storage rings.

2- Vacuum

As the bunches will travel many millions of kilometers (considering that they will be moving at near the speed of light for many hours), any residual gas in the beam pipe will result in many, many collisions. This will have the effect of increasing the size of the bunch, and increasing the energy spread. Therefore, a better vacuum yields better beam dynamics. Also, single large-angle scattering events from either the residual gas, or from other particles in the bunch (Touschek effect), can eject particles far enough that they are lost on the walls of the accelerator vacuum vessel. This gradual loss of particles is called beam lifetime, and means that storage rings must be periodically injected with a new complement of particles.

3- Particle injection and timing

Injection of particles into a storage ring may be accomplished in a number of ways, depending on the application of the storage ring. The simplest method uses one or more pulsed deflecting dipole magnets (injection kicker magnets) to steer an incoming train of particles onto the stored beam path; the kicker magnets are turned off before the stored train returns to the injection point, thus resulting in a stored beam. This method is sometimes called single-turn injection.

Multi-turn injection allows accumulation of many incoming trains of particles, for example if a large stored current is required. For particles such as protons where there is no significant beam damping, each injected pulse is placed onto a particular point in the stored beam transverse or longitudinal phase space, taking care not to eject previously-injected trains by using a careful arrangement of beam deflection and coherent oscillations in the stored beam. If there is significant beam damping, for example radiation damping of electrons due to synchrotron radiation, then an injected pulse may be placed on the edge of phase space and then left to damp in transverse phase space into the stored beam before injecting a further pulse. Typical damping times from synchrotron radiation are tens of milliseconds, allowing many pulses per second to be accumulated.

If extraction of particles is required (e.g. in a chain of accelerators), then single-turn extraction may be performed analogously to injection. Resonant extraction may also be employed.

4.2.2. Beam Dynamics and Focusing

The particles must be stored for very large numbers of turns potentially larger than 10 billion. This long term stability is challenging, and one must combine the magnet design with tracking codes and analytical tools in order to understand and optimize the long term stability.

In the case of electron storage rings, radiation damping eases the stability problem by providing a non-Hamiltonian motion returning the electrons to the design orbit on the order of the thousands of turns. Together with diffusion from the fluctuations in the radiated photon energies, an equilibrium beam distribution is reached.

Due to the high velocity of the particles, and the resulting Lorentz force for magnetic fields, adjustments to the beam direction are mainly controlled by magnetostatic fields that deflect particles. In most accelerator concepts (excluding compact structures like the cyclotron or betatron), these are applied by dedicated electromagnets with different properties and functions. An important step in the development of these types of accelerators was the understanding of strong focusing. Dipole magnets are used to guide the beam through the structure, while quadrupole magnets are used for beam focusing, and sextupole magnets are used for correction of dispersion effects.

A particle on the exact design trajectory (or design orbit) of the accelerator only experiences dipole field components, while particles with transverse position deviation $\boldsymbol{x(s)}$ are re-focused to the design orbit. For preliminary calculations, neglecting all fields components higher than quadrupolar, an inhomogenic Hill differential equation

$$\frac{d^2}{ds^2}x(s) + k(s)x(s) = \frac{1}{\boldsymbol{R}}\frac{\boldsymbol{\Delta p}}{\boldsymbol{p}} \qquad (4.1)$$

can be used as an approximation, with a non-constant focusing force $\boldsymbol{k(s)}$, including strong focusing and weak focusing effects the relative deviation from the design beam impulse $\Delta p/p$ the trajectory curvature radius R, and the design path length s, thus identifying the system as a parametric oscillator. Beam parameters for the accelerator can then be calculated using Ray transfer matrix analysis; e.g., a quadrupolar field is analogous to a lens in geometrical optics, having similar properties regarding beam focusing (but obeying Earnshaw's theorem).

Basically, an accelerator takes a particle, speeds it up using electromagnetic fields, and bashes the particle into a target or other particles. Surrounding the collision point are detectors that record the many pieces of the event.

It is fairly easy to obtain particles. Physicists get electrons by heating metals; heating a metal causes electrons to be ejected, where for example a television, like a cathode ray tube, uses this mechanism. Protons are obtained by ionizing hydrogen, i.e. by robbing hydrogen of its electron. To get antiparticles, first have energetic particles hit a target. Then pairs of particles and antiparticles will be created via virtual photons or gluons. Magnetic fields can be used to separate the particles and antiparticles, see fig.4.1.

Fig. 4.1. Particles and Antiparticles

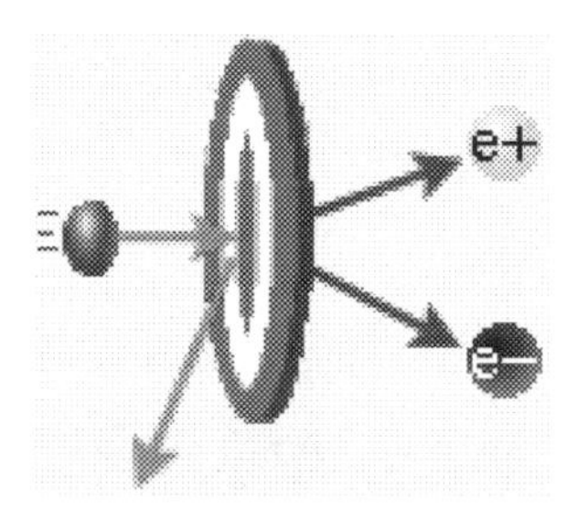

Accelerators speed up charged particles by creating large electric fields which attract or repel the particles. This field is then moved down the accelerator, "pushing" the particles along.

In a linear accelerator the field is due to traveling electromagnetic (E-M) waves. When an E-M wave hits a bunch of particles, those in the back get the biggest boost, while those in the front get less of a boost. In this fashion, the particles "ride" the front of the E-M wave like a bunch of surfers.

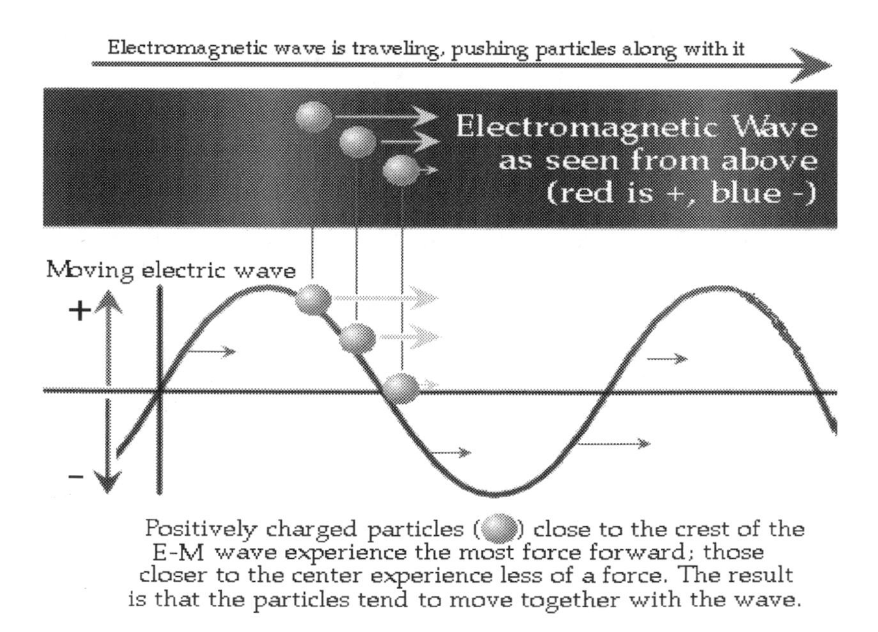

Fig. 4.2. Moving electric wave

4.2.3. Machine Tolerances

Errors in the alignment of components, field strength, etc., are inevitable in machines of this scale, so it is important to consider the tolerances under which a machine may operate.

Engineers will provide the physicists with expected tolerances for the alignment and manufacture of each component to allow full physics simulations of the expected behaviour of the machine under these conditions. In many cases it will be found that the performance is degraded to an unacceptable level, requiring either re-engineering of the components, or the invention of algorithms that allow the machine performance to be 'tuned' back to the design level.

This may require many simulations of different error conditions in order to determine the relative success of each tuning algorithm, and to allow recommendations for the collection of algorithms to be deployed on the real machine.

4.3. Design Principles of Linear Accelerators

4.3.1 History of Linear Accelerator Design

As stated earlier, the linear accelerator is a device for accelerating electrons or positively charged particles to a high energy, by the application of an alternating rather than a direct field, and in a straight line, rather than in an orbit curved by a magnetic field as in the cyclotron, betatron, or synchrotron.

A linear particle accelerator (often shortened to Linac) is a type of particle accelerator that greatly increases the kinetic energy of charged subatomic particles or ions by subjecting the charged particles to a series of oscillating electric potentials along a linear beamline; this method of particle acceleration was invented by Leó Szilárd. It was patented in 1928 by Rolf Widerøe, who also built the first operational device.

In a linear accelerator (Linac), particles are accelerated in a straight line with a target of interest at one end.

The early linear accelerators used high voltage to produce high-energy particles; a large static electric charge was built up, which produced an electric field along the length of an evacuated tube, and the particles acquired energy as they moved through the electric field. The Cockcroft-Walton accelerator produced high voltage by charging a bank of capacitors in parallel and then connecting them in series, thereby adding up their separate voltages. The Van de Graaff accelerator achieved high voltage by using a continuously recharged moving belt to deliver charge to a high-voltage terminal consisting of a hollow metal sphere. Today these two electrostatic machines are used in low-energy studies of nuclear structure and in the injection of particles into larger, more powerful machines. Linear accelerators can be used to produce higher energies, but this requires increasing their length.

Linear accelerators, in which there is very little radiation loss, are the most powerful and efficient electron accelerators; the largest of these, the Stanford linear accelerator (SLAC), completed in 1957, is 2 mile (3.2 km) long and produces 20-GeV (in particle physics energies are commonly measured in millions (MeV) or billions (GeV) of electron-volts (eV)).

Modern linear machines differ from earlier electrostatic machines in that they use electric fields alternating at radio frequencies to accelerate the particles, instead of using high voltage. The acceleration tube has segments that are charged alternately positive and negative. When a group of particles passes through the tube, it is repelled by the segment it has left and is attracted by the segment it is approaching. Thus the final energy is attained by a series of pushes and pulls. Recently, linear accelerators have been used to accelerate heavy ions such as carbon, neon, and nitrogen.

4.3.2. Design Principles

The essential principle of Linac is some form of loaded waveguide, in which an oscillating field of high amplitude can be set up, which can be analyzed into traveling waves, one of which travels with the velocity of the particle to be accelerated. Particles in the correct phase can then remain always in the phase of the traveling wave corresponding to acceleration, and can pick up energy continually, just as if they were in a constant field. If the properties of the field are arranged to vary, as the particle goes along the tube, so as to keep step with the acceleration of the particle, energies of any amount can in principle be acquired.

The design of a linac depends on the type of particle that is being accelerated: electrons, protons or ions. Linac range in size from a cathode ray tube (which is a type of linac) to the 3.2 km long linac at the Stanford Linear Accelerator SLAC, National Accelerator Laboratory in Menlo Park, California.

The Cathode Ray Tube (CRT) is an accelerator, because electrons are attracted from a cathode to an anode, and their speed increases. The electron microscope works on much the same principles, and it can resolve down to the atomic level, about 10^{-10} m.

To resolve further, we need to get down to de Broglie wavelengths of 10^{-15} m. Higher powered machines are needed, as shown for example in a linear accelerator (LINAC) below.

Ions have a heavier mass than electrons, so have a greater **momentum** (hence de Broglie wavelength). They are attracted to the first of the drift tubes by an opposite charge. The drift tube is connected to an **alternating** supply. As the ion passes through the **polarity** changes, so the ions is repelled by the first and attracted to the second. Then the polarity is changed again. The effect of this is that the little brutes are first pulled in then kicked up the backside out. By time they get to the end, they are shifting like greased lightning.

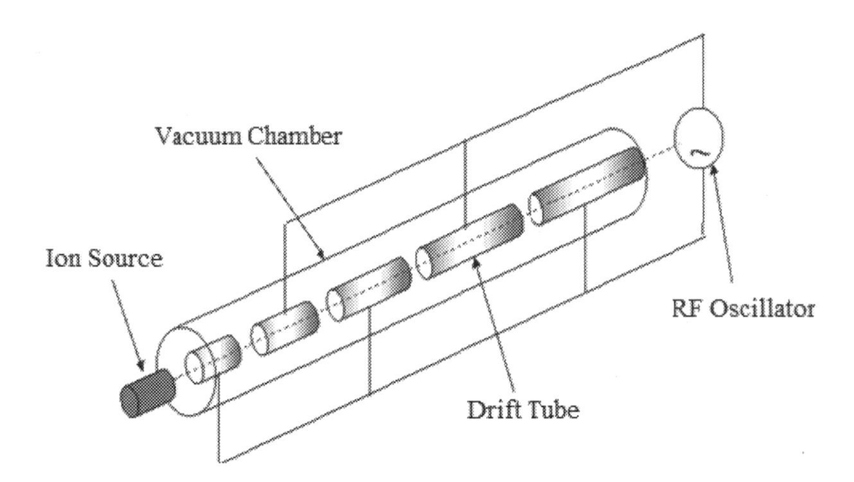

Fig. 4.3. The picture shows a linear accelerator (Linac)

The longest machine is 3 km long. For a machine that requires millimetric precision in setting up, building it across an active fault line did not show a great deal of foresight.

In accelerator physics all forces on particles originate from electromagnetic fields. The interaction of such fields with charged particles will be discussed in a very general way to point out the basic process of particle acceleration. For particle acceleration we consider only the electric-field term of the Lorentz force. The nature of the electric field can be static, pulsed, generated by a time varying magnetic field or a rf field. Both the electric and magnetic fields are connected by Maxwell's equation

$$\nabla \mathrm{x} \mathrm{E} = -\frac{1}{c}\frac{\mathrm{dB}}{\mathrm{dt}} \quad (4.2)$$

Using Stokes' integral theorem

$$\int \nabla \mathrm{xE}.\, \mathrm{d}a = \oint \mathrm{E}ds = -\frac{1}{c}\frac{\mathrm{d}}{\mathrm{dt}}\int \mathrm{B}.\, da \quad (4.3)$$

Where da is the differential surface vector directed normal to the surface.

From eq.(4.3) it becomes apparent that a changing magnetic flux generates an electric field surrounding this flux. Electromagnetic fields at all frequencies can be used for particle acceleration. On one end of the spectrum we have static electric fields which play a significant role for low-energy particle acceleration. Slowly varying magnetic fields are the basis for electric transformers and can be applied as well for the acceleration of particles in induction linear accelerators or betatrons. The most extensive use of electromagnetic waves utilizes rf fields in specially designed resonant cavities. At the other end of the frequency spectrum of electromagnetic waves are intense light beams, for example, in the form of laser beams which can provide very high electric fields. At this time, however, the technical problem of turning the purely transverse laser field into longitudinal accelerating fields has not been solved yet.

Linear high-energy accelerators use a linear array of plates (or drift tubes) to which an alternating high-energy field is applied. As the particles approach a plate they are accelerated towards it by an opposite polarity charge applied to the plate. As they pass through a hole in the plate, the polarity is switched so that the plate now repels them and they are now accelerated by it towards the next plate. Normally a stream of "bunches" of particles is accelerated, so a carefully controlled AC voltage is applied to each plate to continuously repeat this process for each bunch.

As the particles approach the speed of light the switching rate of the electric fields becomes so high that they operate at microwave frequencies, and so RF cavity resonators are used in higher energy machines instead of simple plates.

A linear particle accelerator consists of the following elements:

- **The particle source.** The design of the source depends on the particle that is being moved. Electrons are generated by a cold cathode, a hot cathode, a photocathode, or radio frequency (RF) ion sources. Protons are generated in an ion source, which can have many different designs. If heavier particles are to be accelerated, (e.g., uranium ions), a specialized ion source is needed.

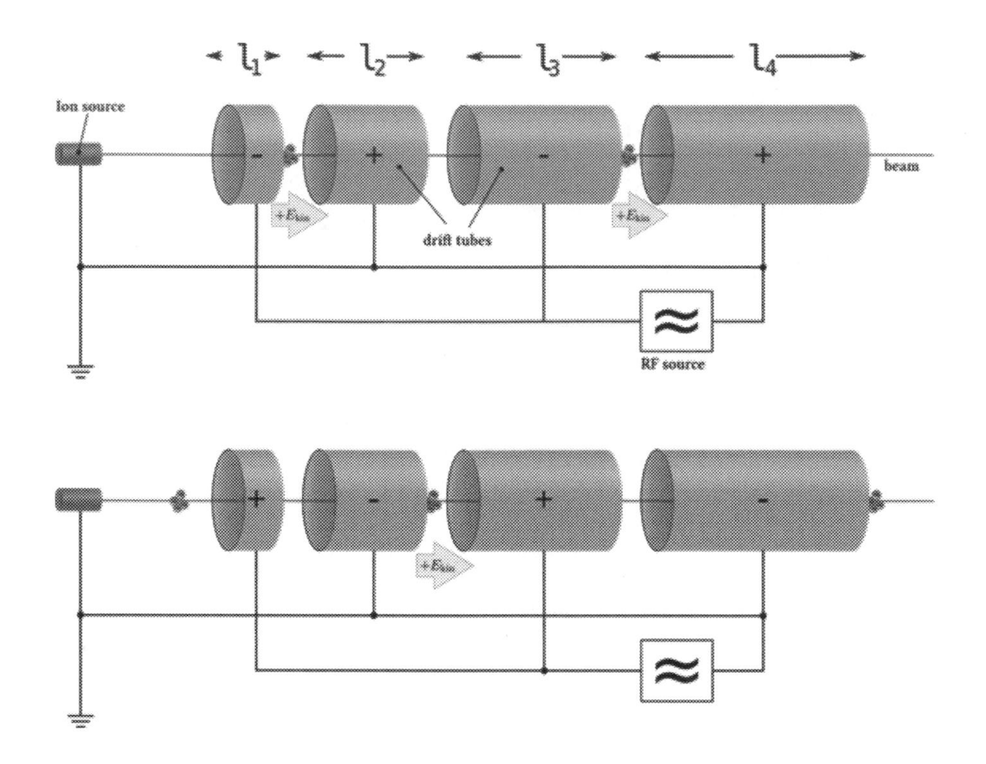

Fig. 4.4. Sketch of the Ising/Widerøe linear accelerator concept, employing oscillating fields (1928)

- **A high voltage source** for the initial injection of particles.

- **A hollow pipe vacuum chamber.** The length will vary with the application. If the device is used for the production of X-rays for inspection or therapy the pipe may be only 0.5 to 1.5 meters long. If the device is to be an injector for a synchrotron it may be about ten meters long. If the device is used as the primary accelerator for nuclear particle investigations, it may be several thousand meters long.

- Within the chamber, **electrically isolated cylindrical electrodes** are placed, whose length varies with the distance along the pipe. The length of each electrode is determined by the frequency and power of the driving power source and the nature of the particle to be accelerated, with shorter segments near the source and longer segments near the target. The mass of the particle has a large effect on the length of the cylindrical electrodes; for example an electron is considerably lighter than a proton and so will generally require a much smaller section of cylindrical electrodes as it accelerates very quickly. Likewise, because its mass is so small, electrons have much less kinetic energy than protons at the same speed. Because of the possibility of electron emissions from highly charged surfaces, the voltages used in the accelerator have an upper limit, so this can't be as simple as just increasing voltage to match increased mass.

- One or more sources of **radio frequency energy**, used to energize the cylindrical electrodes. A very high power accelerator will use one source for each electrode. The sources must operate at precise power, frequency and phase appropriate to the particle type to be accelerated to obtain maximum device power.

- **An appropriate target**. If electrons are accelerated to produce X-rays then a water cooled tungsten target is used. Various target materials are used when protons or other nuclei are accelerated, depending upon the specific investigation. For particle-to-particle collision investigations the beam may be directed to a pair of storage rings, with the particles kept within the ring by magnetic fields. The beams may then be extracted from the storage rings to create head on particle collisions.

4.4. The Design Principles of Circular Accelerators

4.4.1. History of Circular Accelerator Design

A selective types of Circular Accelerators are highlighted in the following sections:

A. Early Proposals

Parallel with the development of electrostatic and linear rf accelerators the potential of circular accelerators was recognized and a number of ideas for such accelerators have been developed over the years. Technical limitations for linear accelerators to produce high-power rf waves stimulated the search for alternative accelerating methods or ideas for accelerators that would use whatever little rf fields could be produced as efficiently as possible.

In order to reach high energy without the long paths required by linear accelerators, E. O. Lawrence proposed (1932) that particles could be accelerated to high energies in a small space by making them travel in a circular or nearly circular path.

Interest in circular accelerators quickly moved up to the forefront of accelerator design and during the thirties made it possible to accelerate charged particles to many million electron volts. Only the invention of the rf Klystron by the Varian brothers at Stanford in 1937 gave the development of linear accelerators the necessary boost to reach par with circular accelerators again. Since then both types of accelerators have been developed further and neither type has yet outperformed the other. In fact, both types have very specific advantages and disadvantages and it is mainly the application that dictates the use of one or the other.

Two approaches exist for exceeding the relativistic limit for cyclotrons. In the synchrocyclotron, the frequency of the accelerating electric field steadily decreases to match the decreasing angular velocity of the protons. In the isochronous cyclotron, the magnet is constructed so the magnetic field is stronger near the circumference than at the center, thus compensating for the mass increase and maintaining a constant frequency of revolution. The first synchrocyclotron, built at the Univ. of California at Berkeley in 1946, reached energies high enough to create pions, thus inaugurating the laboratory study of the meson family of elementary particles.

B. Proton Synchrotron

In April 1948, the Atomic Energy Commission approved a plan for a proton synchrotron to be built at Brookhaven. Early in Brookhaven Lab history, the consortium of universities responsible for founding the new research center, decided that Brookhaven should provide leading facilities for high energy physics research.

In this device, a ring of magnets surrounds a doughnut-shaped vacuum tank. The magnetic field rises in step with the proton velocities, thus keeping them moving in a circle of nearly constant radius, instead of the widening spiral of the cyclotron. The entire center section of the magnet is eliminated, making it possible to build rings with diameters measured in miles. Particles must be injected into a synchrotron from another accelerator. The synchrotron can be used to accelerate electrons but is inefficient. An electron moves much faster than a proton of the same energy and hence loses much more energy in synchrotron radiation. A circular machine used to accelerate electrons is the betatron, invented by Donald Kerst in 1939. Electrons are injected into a doughnut-shaped vacuum chamber that surrounds a magnetic field. The magnetic field is steadily increased, inducing a tangential electric field that accelerates the electrons.

The 6.2-GeV synchrotron (the bevatron) at the Lawrence Berkeley National Laboratory was used to discover the antiproton.

The 500-GeV synchrotron at the Fermi National Accelerator Laboratory at Batavia, was built to be the most powerful accelerator in the world in the early 1970s, with a ring circumference of approximately 4 mi (6 km). The machine was upgraded (1983) to accelerate protons and counter propagating antiprotons to such enormous speeds that the ensuing impacts delivered energies of up to 2 trillion electron-volts (TeV) - hence the ring was been dubbed the Tevatron. The Tevatron was an example of a so-called colliding-beams machine, which is really a double accelerator that causes two separate beams to collide, either head-on or at a grazing angle. Because of relativistic effects, producing the same reactions with a conventional accelerator would require a single beam hitting a stationary target with much more than twice the energy of either of the colliding beams.

C. Cosmotron (1952-1966)

The first proton synchrotron was the cosmotron at Brookhaven (N.Y.) National Laboratory, which began operation in 1952 and eventually attained an energy of 3 GeV.

The new machine would accelerate protons to previously unheard of energies—comparable to the cosmic rays showering the earth's outer atmosphere. It would be called the Cosmotron.

Fig.4.5. Cosmotron (1952-1966) -Dismantiled in 1969

By Pearson Scott Foresman - Archives of Pearson Scott Foresman, donated to the Wikimedia Foundation This file has been extracted from another file: PSF C-200002.png, Public Domain, https://commons.wikimedia.org/w/index.php?curid=5509550
https://en.wikipedia.org/wiki/Cosmotron#/media/File:Cosmotron_(PSF).png
https://www.bnl.gov/about/history/accelerators.php

Early in Brookhaven Lab history, the consortium of universities responsible for founding the new research center, decided that Brookhaven should provide leading facilities for high energy physics research. In April 1948, the Atomic Energy Commission approved a plan for a proton synchrotron to be built at Brookhaven. The new machine would accelerate protons to previously unheard of energies—comparable to the cosmic rays showering the earth's outer atmosphere. It would be called the Cosmotron.

The Cosmotron was the first accelerator in the world to send particles to energies in the billion electron volt, or GeV, region. The machine reached its full design energy of 3.3 GeV in 1953.

Not only was the Cosmotron the world's highest energy accelerator, it was also the first synchrotron to provide an external beam of particles for experimentation outside the accelerator itself.

The Cosmotron was the first machine to produce all the types of negative and positive mesons known to exist in cosmic rays, making possible the discoveries of the K0L meson and the first vector meson. It was also the first accelerator to produce heavy unstable particles, some of which were formerly called "V" particles, and this led directly to the experimental confirmation of the theory of associated production of strange particles.

After 14 years of service to the physics community, the Cosmotron ceased operation in 1966 and was dismantiled in 1969. Knowledge gained from the Cosmotron's experiments would lead to revolutionary design improvements and pave the way for construction of Brookhaven's next big accelerator: the Alternating Gradient Synchrotron.

Not only was the Cosmotron the world's highest energy accelerator, it was also the first synchrotron to provide an external beam of particles for experimentation outside the accelerator itself.

D. Alternating Gradient Synchrotron (1960-present)

The Alternating Gradient Synchrotron (AGS) was built on the innovative concept of the alternating gradient, or strong-focusing, principle, developed by Brookhaven physicists Ernst Courant, M. Stanley Livingston, and Hartland Snyder in the 1950s. This breakthrough concept in accelerator design allowed scientists to accelerate protons to energies that would have been otherwise unachievable. The AGS became the world's premiere accelerator when it reached its design energy of 33 billion electron volts (GeV) in July of 1960.

Fig. 4.6. Alternating Gradient Synchrotron AGS 33GeV (1960-Continue)

Until 1968, the AGS was the highest energy accelerator in the world, slightly higher than its 28 GeV sister machine, the Proton Synchroton at CERN, the European Organization for Nuclear Research. AGS dscoveries earned researchers three Nobel Prizes (the muon-neutrino, a symmetry-breaking called CP violation and the J/psi particle) and serves as the injector for Brookhaven's Relativistic Heavy Ion Collider. It also remains the world's highest intensity high-energy proton accelerator. The AGS and its accompanying Booster accelerator are the only U.S. heavy ion accelerators suitable for simulating the biological effects of space radiation.

One of the most famous early detectors used at the AGS to study sub-atomic phenomena was the 80-inch bubble chamber. When its first photograph of particle interactions was taken in June 1963, the 80-inch bubble chamber was the largest detector in the world. In 1964,a team of researchers led by future Laboratory director Nicholas Samios used the 80-inch to discover the omega-minus particle. This finding supported the first attempt by physicists to organize the increasingly long list of subatomic particles into an orderly pattern, similar to that used to arrange elements in the periodic table.

Later, another detector, the 7-foot bubble chamber, began routine operations in 1974. The following year, the 7-foot chamber was used to discover the charmed baryon, a particle composed of three quarks, one of which was the "charmed" quark. This result helped physicists confirm that new member of the quark family.

80-inch Bubble Chamber 7-foot Bubble Chamber

Fig.4.7. Two types of Buble Chambers

E. National Synchrotron Light Source (1982-2014)

One of the world's most widely used scientific research facilities, the National Synchrotron Light Source (NSLS) hosted about 2,400 researchers annually from more than 400 universities, laboratories, and companies. Beginning in the 1980s, research conducted at NSLS yielded advances in biology, physics, chemistry, geophysics, medicine, and materials science.

An accelerator takes stationary charged particles, such as electrons, and drives them to velocities near the speed of light. Forced by magnets to travel around a circular storage ring, the charged particles give off electromagnetic radiation and lose energy. This enrgy is emitted in the form of light—a phenomenon known as synchrotron radiation.

While it cannot be seen by the human eye, when used in certain ways and viewed by special detectors, this light reveals structures and features of individual atoms, molecules, crystals, cells and more—especially when the wavelength and corresponding energy of the light are matched to the size of the sample being viewed. Because synchrotron light is very intense and well focused, it is preferred to light produced by conventional laboratory sources.

When the U.S. Department of Energy recognized the need for "second generation" electron synchrotrons dedicated to the production of light, it budgeted construction funding for Brookhaven's National Synchrotron Light

Source (NSLS). Ground was broken for NSLS on September 28, 1978, and the vacuum ultraviolet (VUV) ring began operations in late 1982, while the x-ray ring was commissioned in 1984.

Research conducted at NSLS yielded advances in biology, physics, chemistry, geophysics, medicine, and materials science. Examples of research performed at NSLS include investigations into the chemical origins of nerve impulses, studies of the crystal structure of new materials such as high-temperature superconductors, studies of arthritis and osteoporosis, and techniques to make faster, smaller computer chips.

Two Brookhaven scientists, Renate Chasman and G. Kenneth Green were responsible for developing the technology that made the NSLS possible: the "double focusing achromat," more commonly known as the Chasman-Green lattice. The lattice is the periodic arrangement of magnets that bend, focus and correct the electron beam. When special magnets are inserted into the accelerator ring, the electron beam "wiggles" and emits even more intense synchrotron radiation. Chasman and Green's inclusion of these devices in their design of the storage rings enabled the NSLS to deliver world-class beams of light and their innovation is now a standard component of synchrotrons the world over.

Fig.4.8. NSLS under Inspection

F. Relativistic Heavy Ion Collider (2000-present)

Physicists at Brookhaven persevered, continuing to push for an advanced accelerator design. In 1984, the first proposal was submitted for the machine now known as Relativistic Heavy Ion Collider (RHIC). RHIC's main mission was to search for a state of matter called quark-gluon plasma, believed to have existed just moments after the Big Bang. It was considered very cost-effective to build RHIC at Brookhaven because of the existing accelerator infrastructure which could be used to inject protons and heavy ions into the machine as well as the fact that a tunnel (originally excavated for ISABELLE) was already completed. (In 1974 the U.S. High Energy Physics Advisory Panel (HEPAP) recommended that ISABELLE (the Intersecting **S**torage **A**ccelerator + "belle") be built at Brookhaven. It was to be a 200+200 GeV proton-proton system using superconducting magnets. Construction began in 1978. The following year a prototype magnet was successfully tested. In 1981, however, production models of magnets failed at less than the magnetic field intensity needed for operation. The technical problems were encountered in the fabrication of the superconducting magnets needed to power the machine, consequently this project was cancelled in July, 1983. HEPAP then shifted its focus to the Superconducting Supercollider, which would itself later be canceled).

Brookhaven received funding to proceed with the construction of RHIC in 1991 and physics achieved its first successful operation in the summer of 2000, capping ten years of development. However, the history of RHIC stretches back more than 30 years beginning with an idea for a machine called the Intersecting Storage Accelerator.

Fig. 4.9. The first ring of magnets being placed for RHIC in the 1990s

G. Large Hadron Collider (LHC)

In Nov., 2009, the Large Hadron Collider (LHC), a synchroton constructed by CERN, became operational, and in Mar., 2010, it accelerated protons to 3.5 TeV to produce collisions of 7 TeV, a new record. The LHC is located in central Europe, on the border between France and Switzerland, not far from Geneve.

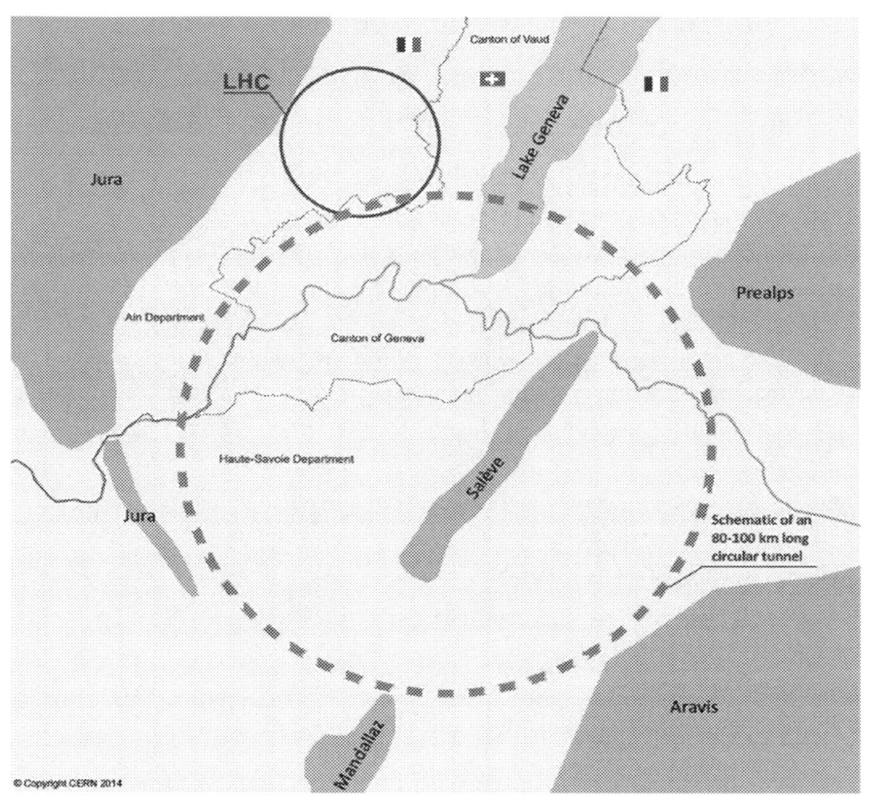

Fig. 4.10. location Map of LHC

Map of the Very Large Hadron Collider (VLHC) and its location compared to the Large Hadron Collider (LHC). Credit: CERN

The Large Hadron Collider (LHC) was built by the European Organisation for Nuclear Research (CERN) from 1998 until 2008. Described as "one of the great engineering milestones of mankind", it allowed physicists to test the predictions of different theories of particle physics and high-energy physics – and most importantly, to prove or disprove the existence of the long-theorised Higgs Boson, as well as the large family of new particles predicted by supersymmetric theories.

Fig.4.11. Tunnel Image inside the Large Hadron Collider
Cinema Mode vedios on LHC Collider (CERN) can be obtained at
WWW.OLOSCIENCE.COM

The Collider is a ring-shaped underground tunnel about 8486 meters in diameter. The LHC's main ring, which uses superconducting magnets, is housed in a circular tunnel some 17 mi (27 km); the tunnel was originally constructed for the Large Electron Positron Collider, which operated from 1989 to 2000.

The LHC is being used to investigate the Higgs particle as well as quarks, gluons, and other particles and aspects of physics' Standard Model. In 2012 CERN scientists announced the discovery of a new elementary particle consistent with a Higgs particle; they confirmed its discovery the following year. The Higgs was confirmed by data from the LHC in 2013, and in subsequent decades the LHC would continue to address many unsolved questions, improving knowledge of physical laws. An upgrade was completed in 2015, doubling its energy from 3.5 to 7 tera-electronvolts (7 TeV) per beam.

A further performance boost in the 2020s increased the luminosity of the machine by a factor of 10 – providing a better chance to see rare processes and improving statistically marginal measurements.

By smashing particles together in high-energy collisions, it is possible to recreate the conditions in the earliest moments of the universe. The higher the energy, the further back in time researchers can simulate, and the more likely it is that exotic interactions will be observed.

The Very Large Hadron Collider (VLHC) is the successor to the Large Hadron Collider (LHC). The detailed design and location choice to be finalised in the mid-2020s, with construction taking a decade after that. With a tunnel measuring 60 miles (100 km), the VLHC is by far the largest particle accelerator ever built, dwarfing the LHC. Reaching from the Jura mountains in the west, to the Alps in the east, its diameter is so huge that it requires excavation under Lake Geneva, see map in fig. 4.12. Its collision energy is over 50 tera-electronvolts (50 TeV) per beam, more than seven times that of its predecessor.

The VLHC leads to a revolution in particle physics – vastly improving our knowledge of dark matter, dark energy, string theory and supersymmetry (the latter is a theory that suggests a second, "superpartner" may be coupled to each and every Higgs boson). New information is gleaned on the structure and nature of extra dimensions and how these influence the universe, giving credence to theories beyond the Standard Model.

Longer term, the VLHC helps in the development of picotechnology enabling new applications at scales that are orders of magnitude smaller than nanotechnology. Particle accelerators continue to grow in size and power, eventually becoming too large for Earth to support them and requiring space-based locations. By the middle of the 4th millennium, the very earliest moment of the Big Bang can be simulated, demonstrating a state known as the Grand Unification Energy, in which fundamental forces are united into a single force.

H. Accelerator and Synchrotron Light Source UVX

The Brazilian Synchrotron Light Laboratory (LNLS) is responsible for the operation of the only synchrotron light source in Latin America. LNLS receives academic and industrial researchers from several countries and represents a mark in the institutional design of scientific research in Brazil. The Laboratory also developed locally the knowledge about accelerator and beamline construction, with the production of components and equipment made in Brazil as much as possible. LNLS Acts as an open, multi-user and multidisciplinary National Laboratory, and providing an infrastructure in the state of the art for research, development and innovation using the synchrotron light for the academic and business communities.

The Accelerators Division is responsible for establishing the design specifications of components for the accelerators of the new synchrotron light source, as well as the diagnostic, acceleration and injection systems of new light source. It is also responsible for the operation of the current synchrotron light source UVX.

Synchrotron Light, or Radiation, is a type of electromagnetic radiation that spans a wide range of the electromagnetic spectrum – from infrared light, to ultraviolet radiation and x-rays. Synchrotron light is produced when charged particles, accelerated to speeds approaching the speed of light, have their trajectory deflected by magnetic fields. The Synchrotron Light Source is a large machine, capable of controlling the movement of these charged particles, typically electrons, to produce Synchrotron Light.

To produce synchrotron light is necessary that an ultra-relativistic electron beam, that is, with velocity very close to the speed of light, be kept in stable conditions. The synchrotron light is emitted when these electrons are accelerated in curved trajectories. For this process to occur, a set of particles accelerators is used. Both for the current synchrotron light UVX source and for the new synchrotron light source SIRIUS, the electrons are produced and accelerated in an injector system, composed of a linear accelerator and a circular accelerator. Next, they are kept in stable orbits in the main storage ring for many hours emitting the synchrotron light that is directed to the experiment stations, called beamlines.

The UVX storage ring is a second-generation 1.37 GeV synchrotron light source. The injection system includes a 120 MeV linear accelerator and a 500 MeV booster synchrotron. The machine operates in decay mode and most of the beam lines are based on dipole magnets. Two RF stations are used to supply energy for the electron beam, each station including a single cell room temperature accelerating cavity and a 50 kW solid state amplifier operating at 476 MHz. The storage ring is housed in a temperature controlled tunnel and feedback systems are used to keep the beam orbit stable during the user shifts. Two diagnostics beam lines are used to monitor the beam.

Fig. 4.12. The synchrotron light source UVX

The UVX light source has a 6-fold symmetry DBA lattice.The magnetic lattice includes 12 1.67 T dipoles, 36 quadrupoles and 18 sextupoles. There are 6 straight sections of 2.95 m, 4 of them available for insertion devices, 3 of which are currently occupied with an EPU50 undulator, a 4T SCW60 superconducting wiggler and a 2T W80 hybrid wiggler. The machine is about 93 m long and the harmonic number is 148. A set of 18 horizontal and 24 vertical steering magnets is used by a 3 kHz fast orbit feedback system to keep the beam orbit within tolerances during the whole user beam shift.

The booster is a 2-fold symmetry racetrack synchrotron. The magnetic lattice includes 12 dipoles in two 180° FODO cell based achromatic arcs. The circumference of the booster is about one third that of the main ring and it operates at 0.2 Hz. The booster RF system includes a single cell cavity and a 2 kW solid state amplifier working at 476 MHz.

The second-generation synchrotron light source, named UVX, was designed and built by Brazilians with national technology, during the 80s and the 90s, and open in 1997. The UVX synchrotron has today 17 experimental stations, known as beamlines, which allow the execution of experiments in several techniques of microscopic analysis of materials with Infrared and UV radiation and X-rays. These beamlines are experimental stations where materials are analyzed and they function as complex microscopes that focus synchrotron radiation so that it illuminates the samples under study and allow the observation of their microscopic properties.

For example, the XRD1 beamline is an experimental facility dedicated to X-ray powder diffraction analysis in the hard x-rays energy range (5.5 to 14 keV). It focus on the determination of structural parameters of polycrystalline samples, with applications to physics, chemistry, materials science, materials engineering, geosciences, pharmacy, biology, etc. It is also possible to study the samples in a large range of temperatures (100 – 1070 K).

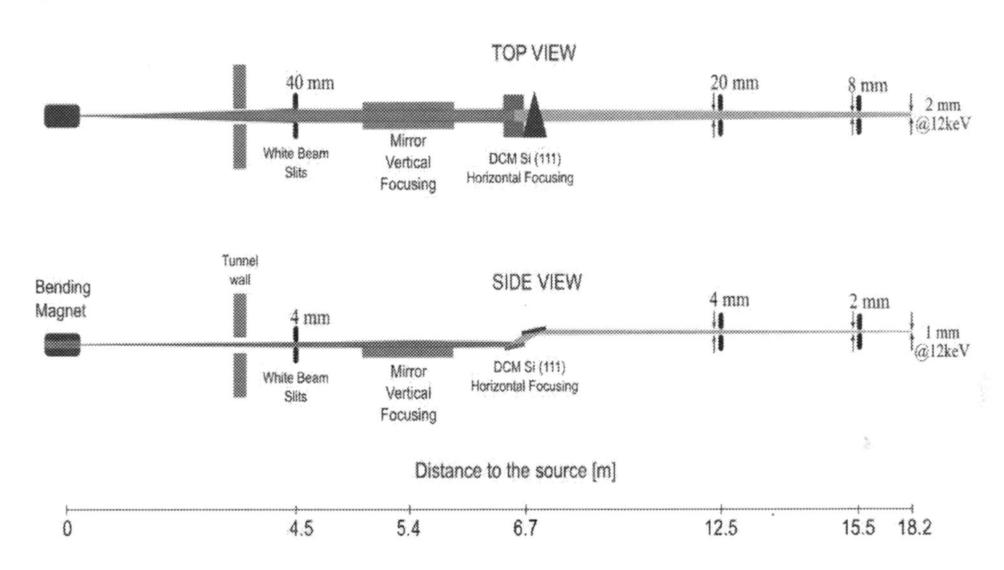

Fig.4.13. XRD1 beamline

Every year, the LNLS facilities benefit around 1200 Brazilian and foreign researchers, committed to over 400 studies that result in approximately 200 papers published in scientific journals. The Laboratory also foster partnerships with the Brazilian industry in research, development and innovation in energy, chemicals, pharmaceuticals and other products.

The LNLS is currently building SIRIUS, a fourth-generation synchrotron light source planned to be one of the most advanced in the world. Sirius will be the biggest and the most complex scientific infrastructure , which put Brazil at a worldwide leadership position in synchrotron light generation. The new synchrotron light source is designed to be the brightest among all the equipment in its energy class and to receive up to 40 beamlines. Sirius will open new research perspectives in many areas such as material science, nanotechnology, biotechnology, energy, food, environment, health, defense and many others.

Fig.4.14. SIRIUS, a fourth-generation synchrotron light source

The synchrotron light source is designed and focused on the determination of structural parameters of polycrystalline samples. The area of research and investication diagnosises cover (Ref.: https://www.lnls.cnpem.br/science/) the following:

1- **Materials for future electronics:** *Research investigates the formation of distinct phases of bismuth-based molecule in topological insulators.*Topological insulators are materials only a few atoms thick that behave as insulators in the inner atomic layers, but as conductors in the surface. The electrical conductivity of these superficial layers is remarkably resistant to the atomic disorder caused by the presence of impurities, which is not the case in other materials.

2- Nanotechnology for the Textile Industry: *Research develops new method for dyeing with natural dyes.*For thousands of years humanity has used substances extracted from plants, insects, soil and rocks to give color to fabrics, ceramics and other products. Nowadays, synthetic dyes, as an alternative to the natural dyes are used in a variety of industries, from textiles to cosmetics, and both production use of these substances can lead to environmental problems if they are not properly degraded or removed from industrial effluents.

3- **Nanodrug against vitiligo:** *Successful in-vitro test shows potential for new treatment.* Vitiligo is a disease characterized by loss of skin pigmentation due to the death of cells called melanocytes, responsible for melanin production. This disease afflicts up to 1% of the world population and current treatments are not effective and exhibit several side effects. According to the most accepted theory, the disease results from an autoimmune reaction. It is known that one of the proteins involved in melanin synthesis, called TyRP-1, also functions as a melanocyte differentiation antigen, marking the melanocytes to be attacked by the immune system. Therefore, a possible strategy to thwart the progression of the disease is to prevent the production of TyRP-1.

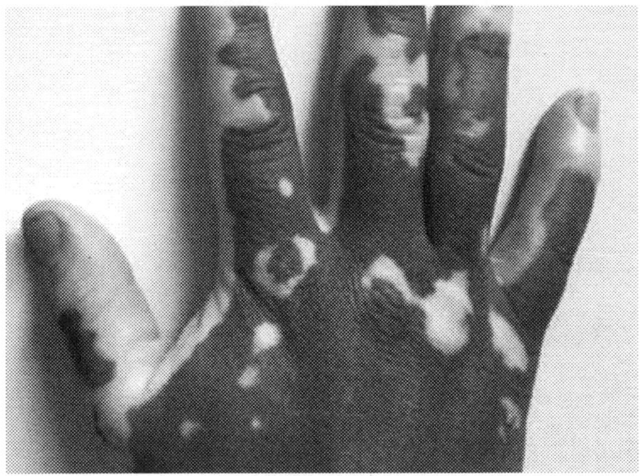

S Science 11/05/18

4- **The Amazonian biodiversity applied to the production of second generation bioethanol:** *Enzyme extracted from microorganisms from Amazonian lake shows potential for biofuel production.* The growing understanding that the rise in average temperature of the planet is caused by human action has, in recent years, intensified the search for clean, renewable and cheap energy sources. One alternative source are the second generation biofuels produced from agricultural waste, such as sugar cane straw and bagasse, which are mainly composed of cellulose. The production of second generation ethanol, for example, breaks down this cellulose into simpler sugars with the use of several enzymes, followed by fermentation into ethanol.

5- **Nanotechnology in plant fertilization:** *Research investigates mechanism of absorption and transport of zinc nanoparticle.* In agriculture, several of the nutrients needed for the growth and development of plants are supplied or supplemented by fertilizers. Some nutrients, such as phosphorus (P) and potassium (K), are needed in large quantities, but obtained from limited mineral sources. Others - such as manganese (Mn), copper (Cu) or zinc (Zn) - are only needed in small quantities and their excessive application can be toxic to plants or to important microorganisms present in the soil.

6- Recycling of pollutants for alternative fuel production:

Research analyzes the deactivation of nickel catalysts in dry reforming of methane. Hydrogen gas (H2) is one of the alternatives to fossil fuels since its combustion has as final product only water vapor. A promising way to produce hydrogen is from the so-called biogas: methane (CH4) and carbon dioxide (CO2) originated from the fermentation of organic matter in anaerobic environments, such as landfills. One possible process for this transformation is the dry reforming of methane (DRM), where CO2 and CH4 in the biogas react (in the presence of a catalyst) yielding a mixture of H2 and carbon monoxide (CO) known as the synthesis gas.

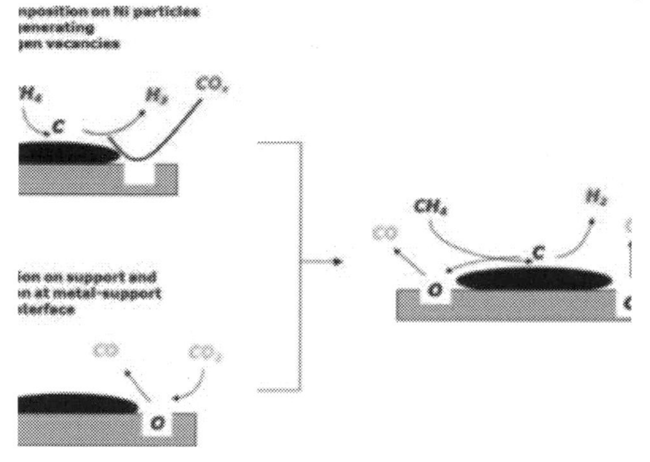

S Science 09/02/18

7- Structure and Catalytic Activity of Copper Nanoparticles:

Research investigates the addition of ceria on the activity of catalysts for the water-gas shift reaction. Catalysts are substances that promote and accelerate chemical reactions without being consumed during the process and are widely used in industrial processes to produce various chemicals. Catalysts based on copper nanoparticles dispersed in an oxide support benefit various reactions, such as the synthesis of methanol, the alcohol dehydrogenation, or the water gas shift (WGS) reaction which is one of the main processes for hydrogen production on an industrial scale. In this reaction, carbon monoxide reacts with water to produce carbon dioxide CO2 and hydrogen gas H2.

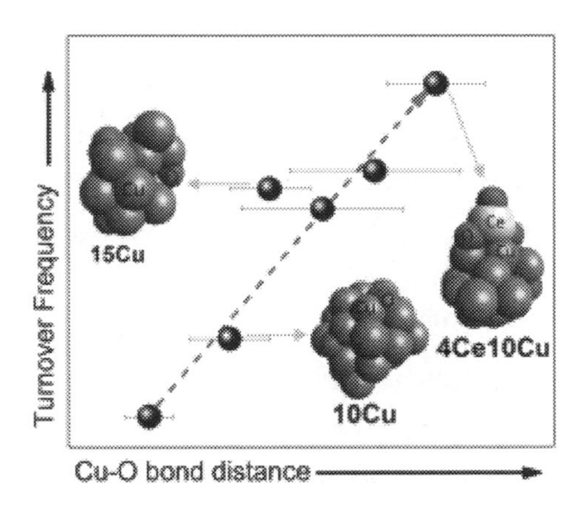

S Science 22/01/18

8- A new x-ray technique to unravel electronic properties of actinide compounds: *A new research demonstrates a direct and selective way to investigate 5f electrons in actinide compounds as well as their interaction with other valence electrons.* Actinides are a series of chemical elements that form the basis of nuclear fission technology, finding applications in strategic areas such as power generation, space exploration, diagnostics and medical treatments, and also in some special glass. Thorium (Th) and Uranium (U) are the most abundant actinides in the Earth's crust. A deeper understanding of the properties of uranium and other actinides is necessary not only for their more efficient use in existing applications but also for proposing new applications. Several open questions remain, progress in this area usually limited in part by the difficulty in handling these materials safely.

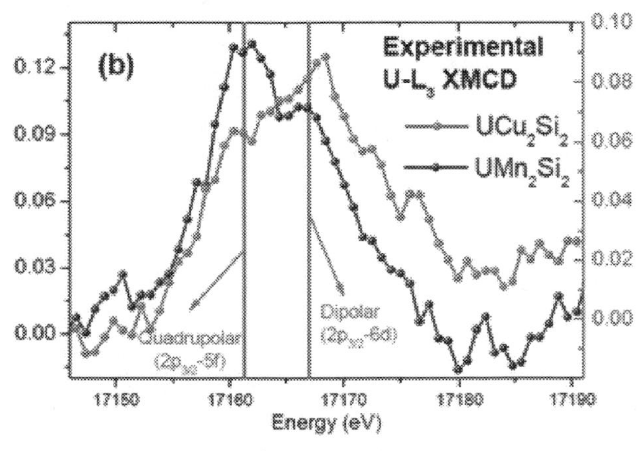

S Science 04/12/17

9- New Catalysts for the synthesis 0f 0rganic compounds:

Research proposes new mechanism for Suzuki-type C-C homocoupling reaction catalyzed by palladium nanocubes. The production of chemical compounds from simpler organic molecules is of great importance for various industrial processes. It is based on the bonding between carbons of the precursor organic compounds, aided by catalysts (typically transition metals). These reactions make it possible to obtain natural and synthetic substances for the development of new materials, such as polymers and pharmaceuticals. In particular, the so-called carbon-carbon (CC) cross-coupling reactions, in which two different precursor molecules are bound to form the final chemical compound, are of such importance that their development granted the 2010 Nobel Prize in chemistry to researchers Richard F. Heck, Ei-ichi Negishi and Akira Suzuki.

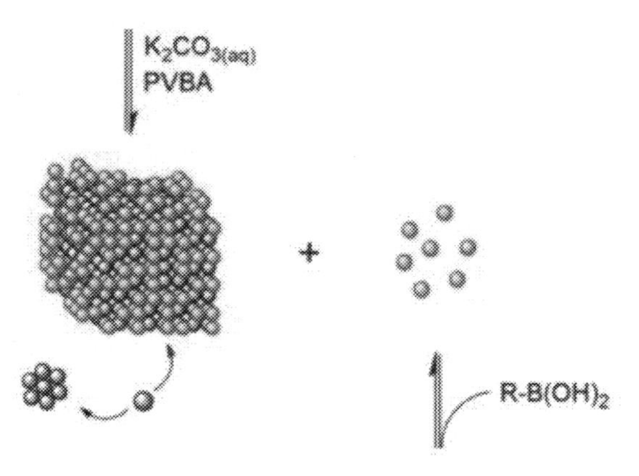

S Science 06/11/17

4.4.2 Design Principles

To keep any particle going in a circle, there needs to be a constant force on that particle towards the center of the circle. In a circular accelerator, an electric field makes the charged particle accelerate, while large magnets provide the necessary inward force to bend the particle's path in a circle. (In the image below, the particle's velocity is represented by the arrow directed to the right (East), while the inward force supplied by the magnet is the arrow directed to the south.

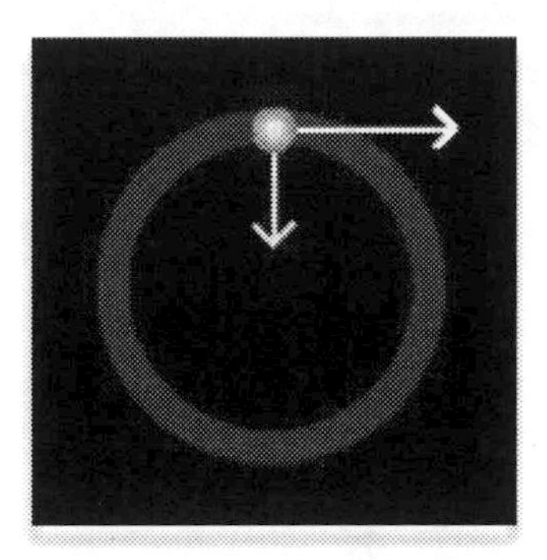

Fig. 4.15. Velocity and force directions on the particle

The presence of a magnetic field does not add or subtract energy from the particles. The magnetic field only bends the particles' paths along the arc of the accelerator. Magnets are also used to direct charged particle beams toward targets and to "focus" the beams, just as optical lenses focus light.

As stated earlier in the linear accelerators Eq. (4.2), the accelerating field is determined by integrating Maxwell's equation

$$\nabla \mathrm{x} \mathbf{E} = -\frac{1}{c}\frac{\mathrm{d}}{\mathrm{dt}}\mathrm{B}$$

and utilizing Stokes's theorem, we obtain the energy gain per turn

$$\oint \mathbf{E} ds = -\frac{1}{c}\frac{\mathrm{d}\phi}{\mathrm{dt}} \tag{4.4}$$

where ϕ is the magnetic flux enclosed by the integration path, which is identical to the design orbit of the beam.

Circular accelerators are based on the use of magnetic fields to guide the charged particles along a closed orbit. When we put charged particles into a magnetic field, we find that they are subject to a force at right angles to the direction of the travel. The force always remains at right angles, so the resulting path is circular. The radius depends on the square of the speed. Consider a mass m travelling at a velocity v. A force F is applied at 90°.

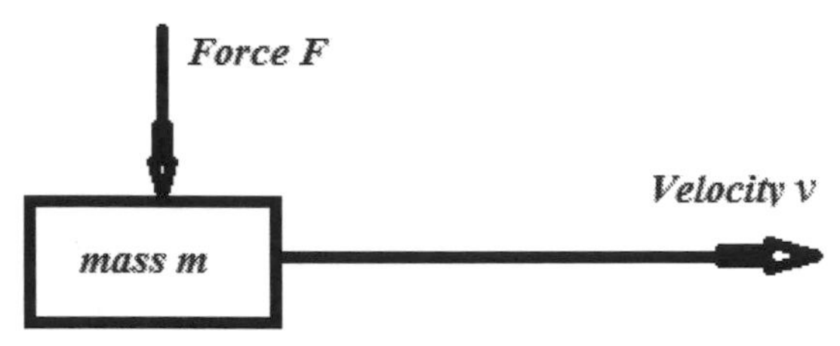

Fig. 4.16. Velocity and force directions on Mass *m*

If the force continues to be applied at 90°, the resulting path is circular. The linear speed remains constant, but the direction changes. There is a change in velocity, hence acceleration.

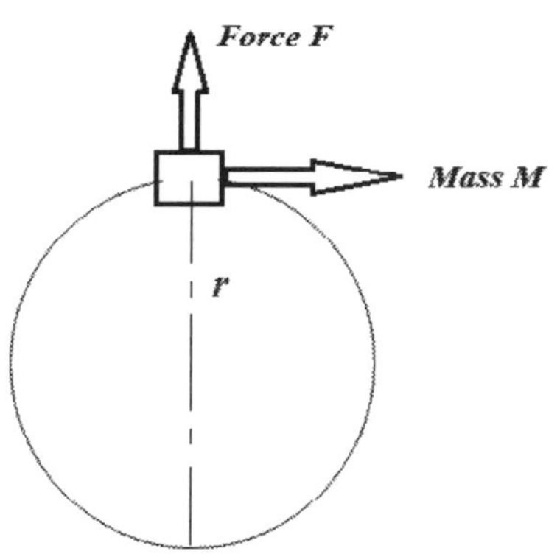

Fig. 4.17. Circular path of the particle due to magnetic field

In a cyclotron, the speed increases as the electron passes between the poles, so the path is a spiral with increasing radius. Then the charged particles come out tangentially in a straight line.

The angular velocity, physics code ω is how big an angle is turned in one second in radian.

- One radian is the angle that subtends an arc whose length is the same as the radius.
- The angle in radians for arc length s is: θ rad = s/r.

- 1 revolution is 2π radians.
- For small angles in radians, $\theta \approx \sin\theta \approx \tan\theta$.
- ω rad s^{-1} is the angular velocity. $\omega = 2\pi f \Rightarrow$ linear speed $v = \omega r = 2\pi f r$ m/s.
- The direction of the velocity is tangential.

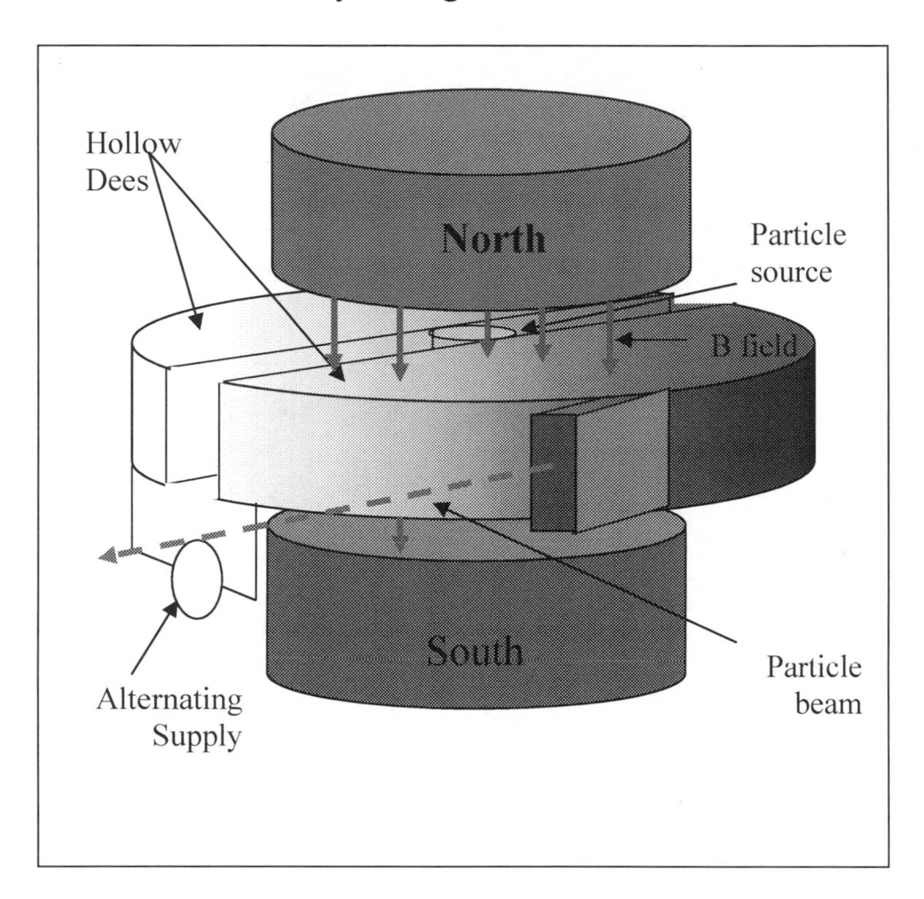

Fig. 4.18. Particle beam location with the magnet

A common bear-trap is to fail to convert revolutions per minute to radians per second. Divide the rpm by 60, then multiply the answer by 2π.

Acceleration is always towards the center of the circle and is given by:

$$a = \omega^2 r \text{, or } \quad a = v^2/r \tag{4.5}$$

Centripetal force is described by the formula:

$$F = mv^2/r \quad \text{, or} \quad F = m\omega^2 r \tag{4.6}$$

The force acts towards the center of the circle.

This force is balanced by the counter force applied on a charged particle in a B-field. The force on a charged particles is given by:

$$F = Bqv\ sin\theta \qquad (4.7)$$

Where

F : Force (N);
B : Magnetic field strength (T);
q : Charge (C);
v :Velocity (m/s);
θ : angle with the magnetic field (usually 90°, so sin θ = 1).

The **direction** of the force is given by Fleming's Left Hand Rule. Since the force is always at 90° to the direction of the velocity, the path is circular which is either clockwise or anticlockwise depending on:

- The direction of the field;
- The charge of the particles.

We can produce an expression to give us the radius of the track:

$$Bqv = mv^2/r, \quad or \quad Bq = mv/r \qquad (4.8)$$

So we can write:

$$r = mv/Bq \qquad (4.9)$$

Since the equation for momentum is:

$$p = mv \qquad (4.10)$$

We can write:

$$r = p/Bq \qquad (4.11)$$

The greater the momentum, the greater the radius of the path. Alternatively a bigger magnetic field is needed to keep it on a track of the same radius.

The simplicity of circular accelerators and the absence of significant synchrotron radiation for protons and heavier particles like ions has made circular accelerators the most successful and affordable principle to reach the highest possible proton energies for fundamental research in high energy physics.

At present protons are being accelerated up to 1000 GeV in the presently highest energy proton synchrotron at the Fermi National Accelerator Laboratory FNAL. Higher energy proton accelerators up to 20 TeV are under construction or in planning.

For electrons the principle of circular accelerators has reached a technical and economic limit at about 100 GeV due to synchrotron radiation losses, which make it increasingly harder to accelerate electrons to higher energies. Further progress in the attempt to reach higher electron energies is being pursued though the principle of linear colliders, where synchrotron radiation is avoided.

4.5. Beam Diagnostics and Detectors Design

A vital component of any accelerator is the diagnostic device that allow various properties of the particle bunches to be measured.

A particle can be fully identified and detected when its charge and its mass are recognized. In principle we can calculate the mass of a particle if we know its momentum and *either* its speed or its energy. However, for a particle moving close to the speed of light any small uncertainty in momentum or energy makes it difficult to determine its mass from these two, so we need to measure speed too.

To look for these various particles and decay products, physicists have designed multi-component detectors that test different aspects of an event. Each component of a modern detector is used for measuring particle energies and momenta, and/or distinguishing different particle types. When these components work together to detect an event, individual particles can be singled out from the multitudes for analysis.

Following each event, computers collect and interpret the vast quantity of data from the detectors and present the extrapolated results to the physicist.

Physicists are curious about the events that occur during and after a particle's collision. For this reason, they place detectors in the regions which will be showered with particles following an event. Detectors are built in different ways according to the type of collision they analyze. The beams from a circular accelerator (synchrotron) can be used for colliding-beam experiments or extracted from the ring for fixed-target experiments:

1. <u>Fixed-target experiment</u>

In a fixed-target experiment the particles are extracted and accelerated to hit a fixed target –detector, detectors are generally cone shaped and are placed "downstream."

In a fixed-target experiment (see sketches below) a charged particle such as an electron or a proton is accelerated by an electric field and collides with stationary targets. A detector determines the charge, momentum, mass, etc. of the resulting particles. Almost any type of target can be selected. Many different types of electrically charged particles can be accelerated, including ions from every element in the periodic table. In physics research, the target typically is a piece of metal, or a gas-filled or liquid filled tank. In other applications, it might be anything from a cancerous tumor needing treatment to a metal surface needing hardening.

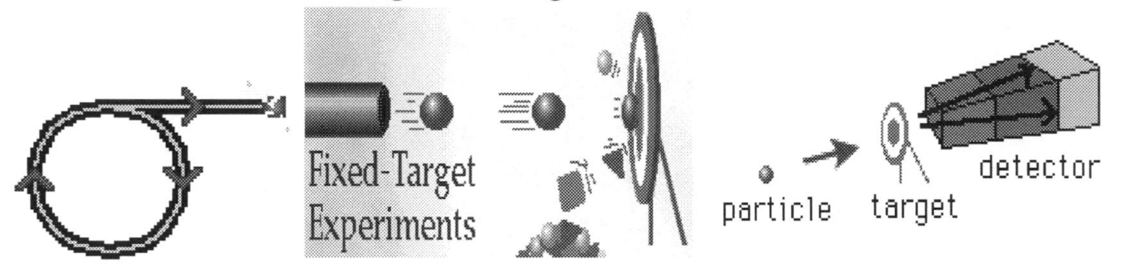

Fig. 4.19. Fixed-target experiment

An example of this process is Rutherford's gold foil experiment, in which the radioactive source provided high-energy alpha particles, which collided with the fixed target of the gold foil. The detector of zinc sulfide screen is used to test for the presence of invisible alpha particles in order to determine the path of alpha particles, modern physicists must look at particles' decay products, and from these deduce the particles' existence.

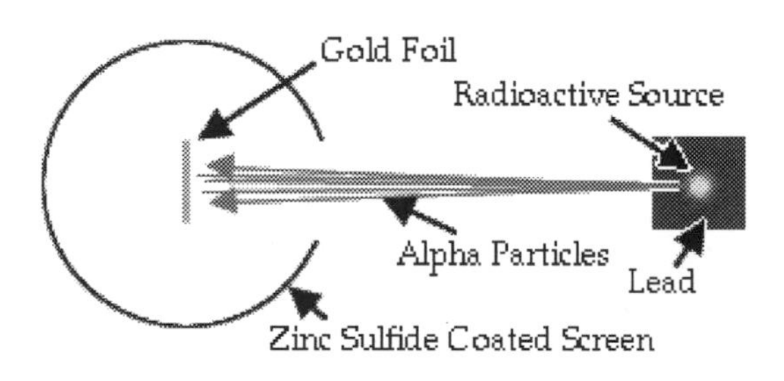

Fig. 4.20. Gold foil experiment

2. Colliding Beams Experiment:

In a colliding-beam experiment two beams of high-energy particles are made to cross each other. During a colliding-beam experiment, the particles radiate in all directions, so the detector is spherical or, more commonly, cylindrical.

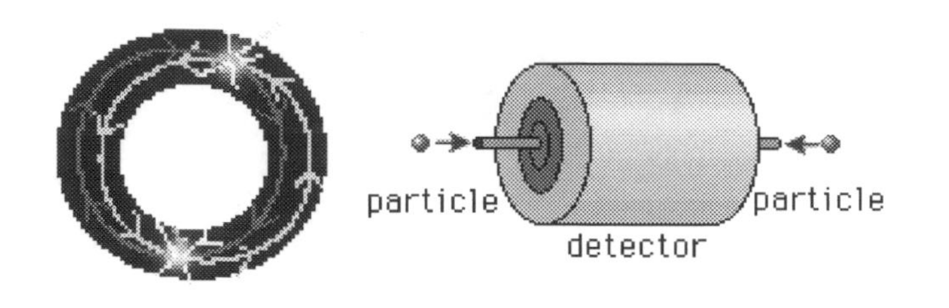

Fig. 4.21. Colliding beams experiment

The advantage of this arrangement is that both beams have significant kinetic energy, so a collision between them is more likely to produce a higher mass particle than would a fixed-target collision (with the one beam) at the same energy. Since we are dealing with particles with a lot of momentum, these particles have short wavelengths and make excellent probes.

Far higher energies can be generated in colliding-beam accelerators than in fixed-target scenarios, because the total energy in both beams is available. In a circular collider, beams orbiting in opposite directions collide at one or more locations in the ring. (However, the number of collisions per second is far less than with a fixed target.)

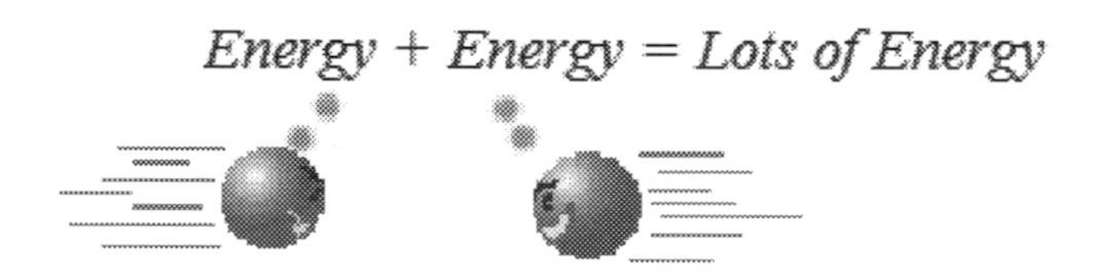

Fig. 4.22. Energy of colliding beams/particle

After an accelerator has pumped enough energy into its particles, they collide either with a target or each other. Each of these collisions is called an **event**. The physicist's goal is to isolate each event, collect data from it, and check whether the particle processes of that event agree with the theory they are testing.

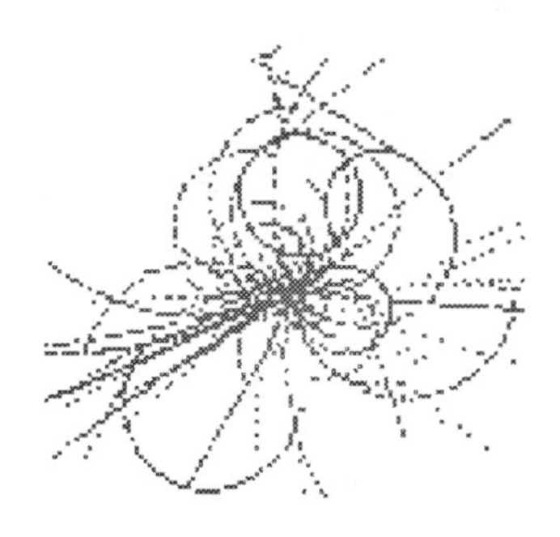

Fig. 4.23. Collide of particles, Events

Each event is very complicated since lots of particles are produced. Most of these particles have lifetimes so short that they go an extremely short distance before decaying into other particles, and therefore leave no detectable track.

A multi-layer detector is used to identify particles. Each layer gives different information about the "event." Computer calculations based on the information from all the layers reconstruct the positions of particle tracks and identify the momentum, energy, and speed of as many as possible of the particles produced in the event.

A typical machine may use many different types of measurement device in order to measure different properties. These include (but are not limited to) Beam Position Monitors (BPMs) to measure the position of the bunch, screens (fluorescent screens, Optical Transition Radiation (OTR) devices) to image the profile of the bunch, wire-scanners to measure its cross-section, and toroids or ICTs to measure the bunch charge (i.e., the number of particles per bunch).

While many of these devices rely on well understood technology, designing a device capable of measuring a beam for a particular machine is a complex task requiring much expertise. Not only is a full understanding of the physics of the operation of the device necessary, but it is also necessary to ensure that the device is capable of measuring the expected parameters of the machine under consideration.

Success of the full range of beam diagnostics often underpins the success of the particle accelerator as a whole.

The SLAC Large Detector (SLD) , see Fig. 4.25., made use of the unique capabilities of the Stanford Linear Collider (SLC) to perform studies of polarized Z particles (or Z^0 bosons) produced in collisions between polarized electrons and positrons. This detector stands six stories tall.

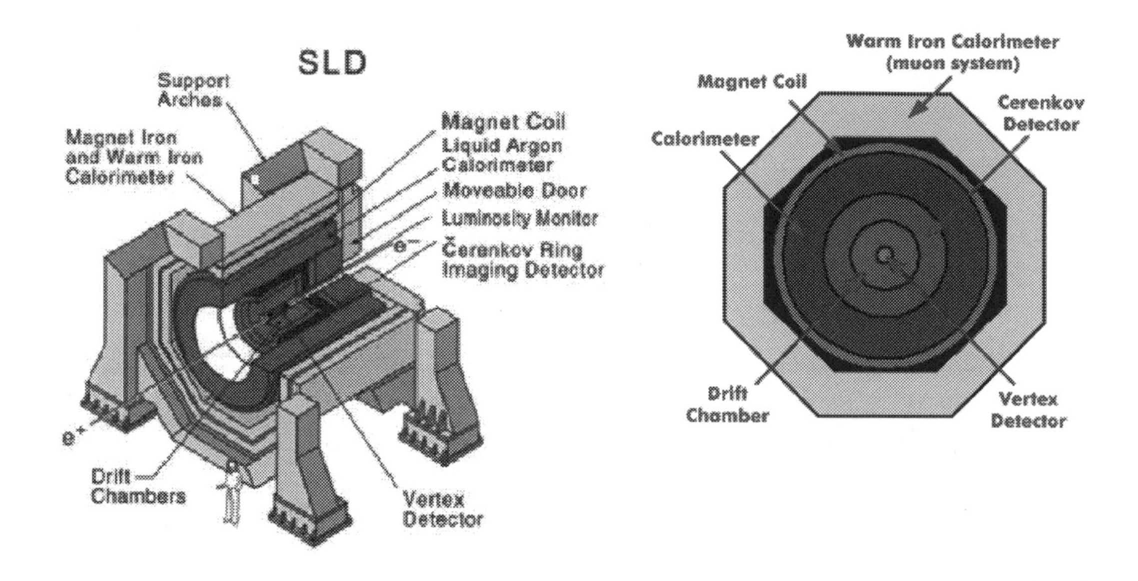

Fig. 4.24. Cross-Sectional View of the SLD Detector

This cutaway schematic shows all the SLAC Large Detector elements installed inside the massive steel barrel and end caps. The complete detector weights 4,000 tons and stands six stories tall.

The layers of the SLD Detector surrounding the Collision Point starting from the inner layer (centerline) to the outer layer are briefly described as following:

- The innermost layer, the vertex detector, about the size of a Coke can, gives the most accurate information on the position of the tracks. A vertex detector gives the most accurate location of any outgoing charged particles as they pass through it. The SLD vertex detector is an array of layers of postage-stamp-sized electronic chips surrounding the beam pipe on concentric cylinders. Each chip, called a Charge-Coupled Device (CCD), is like an electronic checkerboard with about 1000 squares on a side. (Similar CCDs form the recording elements in many video cameras.)
- The next layer, the drift chamber, detects the positions of charged particles at several points along the track. The curvature of the track in

the magnetic field reveals the particle's momentum. The drift chamber, a horizontal, thin-walled cylinder, six feet long and six feet in diameter, with a small tube through the center to accommodate the beam pipe and vertex detector. The drift chamber measures a particle's position to a few thousandths of an inch at eighty steps along its path from the inner to outer wall. Because these charged particles are in the strong magnetic field of the SLD, the path is a curve. Knowing this curvature, one can calculate the momentum of the particle.

- The middle layer, the Cerenkov detector, measures particle velocity. Outside the drift chamber is the Cerenkov detector, or more accurately, the Cerenkov Ring Imaging Detector (CRID). It is used to measure particle velocities through Cerenkov radiation. This form of radiation is somewhat like a sonic boom except it is light rather than sound. It occurs when a particle travels through a medium (here, freon gas) at a speed that is faster than the speed of light in the medium (but slower than the speed of light in a vacuum, of course), just as a sonic boom occurs when an object travels in a medium (air) faster than the speed of sound in the medium.
- The next layer, the liquid argon calorimeter, stops most of the particles and measures their energy. This is the first layer that records neutral particles. Calorimeters (calorie meter) in physics and chemistry experiments measure the total heat of a reaction or process, which of course is an energy measurement. Heat is just one form of energy; particle physicists use the term calorimeter for any device designed to measure energy.
- The large magnet coil separates the calorimeter and the next layer.
- The outermost layer (magnet iron and warm iron calorimeter) detects muons. The outermost layer of the detector is the warm iron calorimeter, consisting of hundreds of long, thin gas-filled boxes stuffed between the 14 iron plates that make up the outer body of the detector. These boxes are strung with high-voltage wires and act, much like the drift chamber, as a collection of Geiger-counter-like detectors that record passage of any charged particle. Metal strips along the plastic boxes pick up signals that show directly which wire a particle passes, and other strips show where along the wire it went by. The result is a point on the path of the particle.

5

TYPES OF ACCELERATORS

There are two basic classes of accelerators, known as electrostatic and oscillating field accelerators. Electrostatic accelerators use static electric field between two different fixed potentials to accelerate particles. A small-scale example of this class is the cathode ray tube in an ordinary old television set. Other examples are the Cockcroft–Walton generator and the Van de Graaf generator. The achievable kinetic energy for particles in these devices is limited by electrical breakdown. Oscillating field accelerators, on the other hand, use radio frequency electromagnetic fields and circumvent the breakdown problem. The electric field component of radio waves accelerates particles inside a partially closed conducting cavity acting as a RF cavity resonator. This class, which was first developed in the 1920s, is the basis for all modern accelerator concepts and large-scale facilities. Rolf Widerøe, Gustav Ising, Leó Szilárd, Donald Kerst and Ernest Lawrence are considered as pioneers of this field, conceiving and building the first operational linear particle accelerator, the betatron, and the cyclotron.

Alongside their best known use in particle physics as colliders (e.g. LHC, RHIC, Tevatron), particle accelerators are used in a large variety of applications, including particle therapy for oncological purposes, and as synchrotron light sources for fields such as condensed matter physics.

Because colliders can give evidence on the structure of the subatomic world, accelerators were commonly referred to as **atom smashers** in the 20th century. Despite the fact that most accelerators (but not ion facilities) actually propel subatomic particles, the term persists in popular usage when referring to particle accelerators in general.

Particle accelerators come in two basic designs, linear (linac) and circular (synchrotron).

5.1. Linear Accelerator

The linear accelerators (Linac) have many applications:

- Generates X-rays and high energy electrons for medicinal purposes in radiation therapy:
- Serves as particle injectors for higher-energy accelerators; and
- Directly to achieve the highest kinetic energy for light particles (electrons and positrons) for particle physics.

They are also used to provide an initial low-energy kick to particles before they are injected into circular accelerators.

Linear accelerators are also widely used in medicine, for radiotherapy and radiosurgery. Medical grade Linacs accelerate electrons using a klystron and a complex bending magnet arrangement which produces a beam of 6-30 million electron-volt (MeV) energy. The electrons can be used directly or they can be collided with a target to produce a beam of X-rays. The reliability, flexibility and accuracy of the radiation beam produced have largely supplanted the older use of Cobalt-60 therapy as a treatment tool.

Fig.5.1. The Linac within the Australian Synchrotron uses radio waves from a series of RF cavities at the start of the Linac to accelerate the electron beam in bunches to energies of 100 MeV.

Linear accelerators, in which there is very little radiation loss, are the most powerful and efficient electron accelerators; the largest of these, as stated earlier, the Stanford linear accelerator (**SLAC**), which is 3.2 km long and produces 20-GeV. **SLAC** is now used, however, not for particle physics but to produce a powerful X-ray laser.

Fig.5.2. Stanford Linear Accelerator, SLAC
Attribution: By Peter Kaminski (United States Geological Survey) [Public domain], via Wikimedia Commons

This long Accelerator particle**, SLAC** has produced three Nobel Prizes, including one for the discovery of the charm quark and one for the discovery of the tau lepton. It is the world's longest linear accelerator, and maybe even the world's straightest object.

The main advantages of using Linacs are:

- Linacs of appropriate design are capable of accelerating heavy ions to energies exceeding those available in ring-type accelerators, which are limited by the strength of the magnetic fields required to maintain the ions on a curved path.
- High power linacs are developed for production of electrons at relativistic speeds, required since fast electrons traveling in an arc will lose energy through synchrotron radiation; this limits the maximum power that can be imparted to electrons in a synchrotron of given size.
- Linacs are also capable of prodigious output, producing a nearly continuous stream of particles, whereas a synchrotron will only periodically raise the particles to sufficient energy to merit a "shot" at

the target. (The burst can be held or stored in the ring at energy to give the experimental electronics time to work, but the average output current is still limited.) The high density of the output makes the linac particularly attractive for use in loading storage ring facilities with particles in preparation for particle to particle collisions.

- The high mass output makes the device practical for the production of antimatter particles, which are generally difficult to obtain, being only a small fraction of a target's collision products. These may then be stored and further used to study matter-antimatter annihilation.
- As there are no primary bending magnets, the cost of an accelerator is reduced.

However the disadvantages may include the following:

- The device length limits the locations where one may be placed.
- A great number of driver devices and their associated power supplies are required, increasing the construction and maintenance expense of this portion.
- If the walls of the accelerating cavities are made of normally conducting material and the accelerating fields are large, the wall resistivity converts electric energy into heat quickly. On the other hand superconductors have various limits and are too expensive for very large accelerators. Therefore, high energy accelerators such as SLAC, still the longest in the world (in its various generations), are run in short pulses, limiting the average current output and forcing the experimental detectors to handle data coming in short bursts.

5.2. Circular or Cyclic Accelerators

In order to reach high energy without the prohibitively long paths required of linear accelerators, E. O. Lawrence proposed (1932) that particles could be accelerated to high energies in a small space by making them travel in a circular or nearly circular path.

In the circular accelerator, particles move in a circle until they reach sufficient energy. The particle track is typically bent into a circle using electromagnets. The advantage of circular accelerators over linear accelerators (linacs) is that the ring topology allows continuous acceleration, as the particle can transit indefinitely. Another advantage is that a circular accelerator is smaller than a linear accelerator of comparable power (i.e. a linac would have to be extremely long to have the equivalent power of a circular accelerator).

Depending on the energy and the particle being accelerated, circular accelerators suffer a disadvantage in that the particles emit synchrotron radiation. When any charged particle is accelerated, it emits electromagnetic radiation and secondary emissions. As a particle traveling in a circle is always accelerating towards the center of the circle, it continuously radiates towards the tangent of the circle. This radiation is called synchrotron light and depends highly on the mass of the accelerating particle. For this reason, many high energy electron accelerators are linacs. Certain accelerators (synchrotrons) are however built specially for producing synchrotron light (X-rays).

Since the special theory of relativity requires that matter always travels slower than the speed of light in a vacuum, in high-energy accelerators, as the energy increases the particle speed approaches the speed of light as a limit, but never attains it. Therefore particle physicists do not generally think in terms of speed, but rather in terms of a particle's energy or momentum, usually measured in electron volts (eV). An important principle for circular accelerators, and particle beams in general, is that the curvature of the particle trajectory is proportional to the particle charge and to the magnetic field, but inversely proportional to the (typically relativistic) momentum.

Table (5.1) Summary of Circular Machines

Machine	RF frequency f	Magnetic Field B	Orbit Radius ρ	Comment
Cyclotron	constant	constant	increases with energy	Particles out of synch with RF; low energy beam or heavy ions
Isochronous Cyclotron	constant	varies	increases with energy	Particles in synch, but difficult to create stable orbits
Synchro-cyclotron	varies	constant	increases with energy	Stable oscillations
Synchrotron	varies	varies	constant	Flexible machine, high energies possible
FFAG	varies	constant in time, varies with radius	increases with energy	Increasingly attraction option for 21st century designs

5.2.1. Betatrons

1- Background

As the first "circular electron accelerator" we may consider what has been invented and developed a hundred years ago in the form of the electrical current transformer. Here we find the electrons in the wire of a secondary coil accelerated by an electro motive force generated by a time varying magnetic flux penetrating the area enclosed by the secondary coil.

A betatron is a cyclic particle accelerator developed to accelerate electrons, which was the first betatron completed under the direction of the American physicist Donald W. Kerst at the University of Illinois in 1940. The concepts ultimately originate from Rolf Widerøe, whose development of an induction accelerator failed due to the lack of transverse focusing. Wideroe finally recognized the importance of a fixed orbit radius and formulated the Wideroe ½ - condition, which is a necessary although not sufficient condition for the successful operation of a beam transformer or betatron as it was later called, because it functions optimally only for the acceleration of beta rays or electrons. The betatron was the first important machine for producing high energy.

The name "betatron" (a reference to the beta particle, a fast electron) was chosen during a departmental contest. Other proposals were "rheotron", "induction accelerator", "induction electron accelerator", and perhaps "Extraordinarily high velocity electron generator.

Betatron, a type of particle accelerator that uses the electric field induced by a varying magnetic field to accelerate electrons (beta particles) to high speeds in a circular orbit. These machines, like synchrotrons, use a donut-shaped ring magnet with a cyclically increasing B field, but accelerate the particles by induction from the increasing magnetic field, as if they were the secondary winding in a transformer, due to the changing magnetic flux through the orbit.

The betatron consists of an evacuated tube formed into a circular loop and embedded in an electromagnet in which the windings are parallel to the loop. An alternating electric current in these windings produces a varying magnetic field that periodically reverses in direction. During one quarter of the alternating current cycle, the direction and strength of the magnetic field, as well as the rate of change of the field inside the orbit, have values appropriate for accelerating electrons in one direction.

Electron acceleration is controlled by two forces, one acting in the direction of the motion of the electrons and the other at right angles to that direction. The force in the direction of electron motion is exerted by the electric field produced via induction by the strengthening of the magnetic field within the circle; this force accelerates the electrons. The second - perpendicular - force arises as the electrons move through the magnetic field, and it maintains the electrons in a circular orbit within the closed loop.

At the beginning of the appropriate quarter-cycle, electrons are injected into the betatron, where they make hundreds of thousands of orbits, gaining energy all the while. At the end of the quarter-cycle, the electrons are deflected onto a target to produce X-rays or other high-energy phenomena. Large bettors have produced electron beams with energies greater than 340 megaelectron volts (MeV) for use in particle-physics research. Weight considerations place severe limitations on the construction of high-energy betatrons; the electromagnet of a 340-MeV unit weighs about 330 tons.

Lower-energy betatrons in the 7–20-MeV range, however, have been specially constructed to serve as sources of energetic "hard" X-rays for use in medical and industrial radiography. Portable betatrons, operating at energy levels of approximately 7 MeV, have been designed for specialized applications in industrial radiography - for example, to examine concrete, steel, and cast-metal construction for structural integrity.

Betatrons were historically employed in particle physics experiments to provide high energy beams of electrons-up to about 300 MeV. If the electron beam is directed at a metal plate, the betatron can be used as a source of energetic x-rays or gamma rays; these x-rays may be used in industrial and medical applications (historically in radiation oncology).

A small version of a Betatron was also used to provide electrons converted into neutrons by a target to provide prompt initiation of some nuclear weapons.

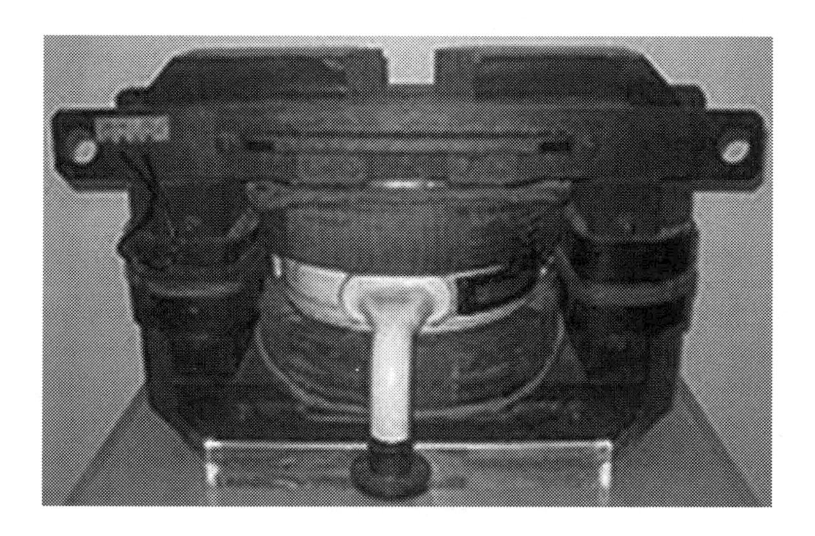

Fig.5.3. A 6 MeV Betatron (1942)

Fotografiert im Deutschen Museum Bonn

Betatron is essentially a transformer with a torus-shaped vacuum tube as its secondary coil. An alternating current in the primary coils accelerates electrons in the vacuum around a circular path. The betatron was the first important machine for producing high energy electrons.

Betatrons were historically employed in particle physics experiments to provide high energy beams of electrons—up to about 300 MeV. If the electron beam is directed at a metal plate, Betatron can be used as a source of energetic x-rays or gamma rays; these x-rays may be used in industrial and medical applications (historically in radiation oncology). A small version of a Betatron was also used to provide electrons converted into hard X-rays by a target to provide prompt initiation of some experimental nuclear weapons by means of photon-induced fission and photon->neutron reactions in the bomb core. The Radiation Center, the first private medical center to treat cancer patients with a betatron, was opened by Dr. O. Arthur Stiennon in a suburb of Madison, Wisconsin in the late 1950s.

The maximum energy that Betatron can impart is limited by the strength of the magnetic field due to the saturation of iron and by practical size of the magnet core. The next generation of accelerators, the synchrotrons, overcame these limitations.

2- Operation Principle

In Betatron, the changing magnetic field from the primary coil accelerates electrons injected into the vacuum torus, causing them to circle round the torus in the same manner as current is induced in the secondary coil of a transformer (Faraday's Law).

Betatron makes use of the transformer principle, where the secondary coil is replaced by an electron beam circulating in a closed doughnut shaped vacuum chamber. Alternating current in the primary coils accelerates electrons in the vacuum around a circular path. A time-varying magnetic field is enclosed by the electron orbit and the electrons gain an energy in each turn which is equal to the electro-motive force generated by the varying magnetic field. The principle arrangements of the basic components of a betatron are shown in Fig. 5.4.

In a betatron, the changing magnetic field from the primary coil accelerates electrons injected into the vacuum torus, causing them to circle round the torus in the same manner as current is induced in the secondary coil of a transformer (Faraday's Law).

The stable orbit for the electrons satisfies

$$\theta_0 = 2\,\pi\,r_0^2\,H_0 \tag{5.1}$$

where

θ_0 is the flux within the area enclosed by the electron orbit,
r_0 is the radius of the electron orbit, and
H_0 is the magnetic field at r_0.

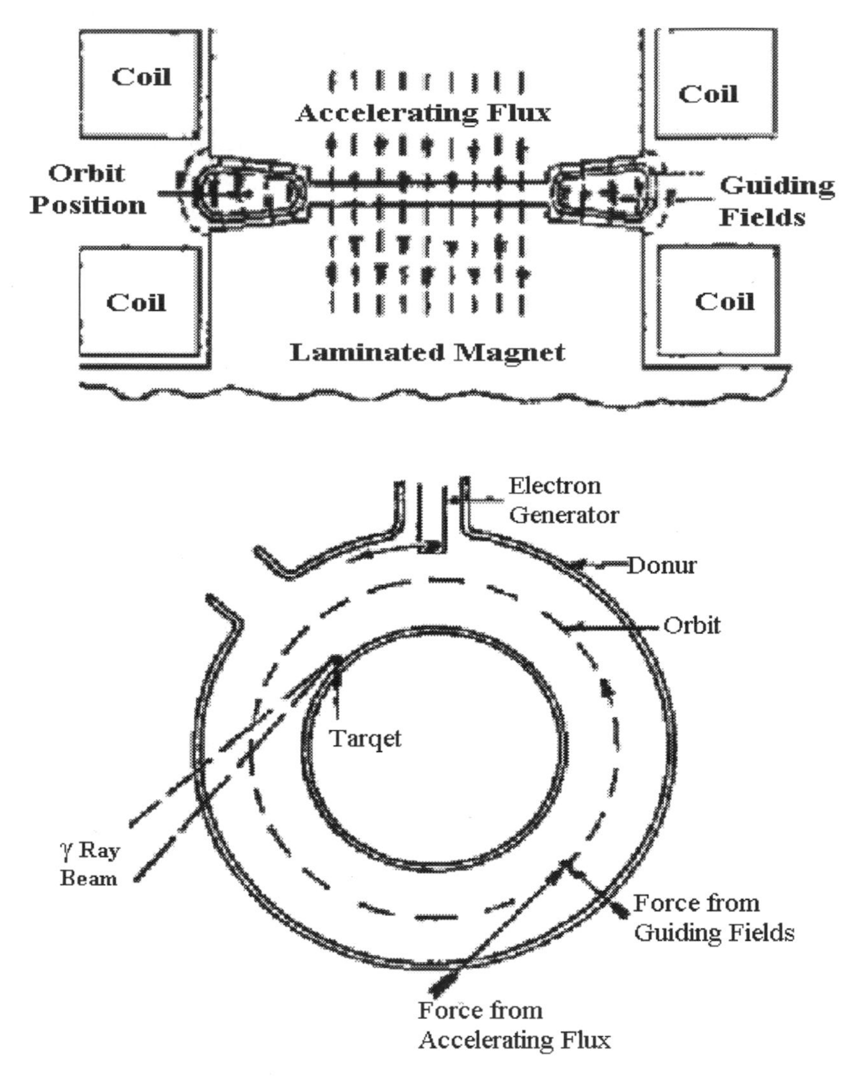

Fig.5.4. The principle arrangement of a Betatron

In other words, the magnetic field at the orbit must be half the average magnetic field over its circular cross section:

$$H_0 = \frac{\theta_0}{2\,\pi\,r_0^2} \tag{5.2}$$

This condition is often called Widerøe's condition.

The magnetic field used to make the electrons move in a circle is also the one used to accelerate them, although the magnet must be carefully designed so that the field strength at the orbit radius is equal to half the average field strength linking the orbit:

$$B_{orbit} = \frac{\bar{B}}{2} \quad (5.3)$$

If the magnetic field increases, there is a changing flux linking the loop of electrons and so an induced e.m.f. which accelerates the electrons. As the electrons get faster they need a larger magnetic field to keep moving at a constant radius, which is provided by the increasing field – the effects are proportional, so the field is always strong enough to keep the electrons in orbit!

The field is changed by passing an alternating current through the primary coils and particle acceleration occurs on the first quarter of the voltage sine wave's cycle. Although the last quarter of the cycle also has a changing field that would accelerate the electrons, it is in the wrong direction for them to move in the correct circle! The target is bombarded with pulses of particles at the frequency of the ac supply.

The particles have maximum energy when the magnetic field is at its strongest value but the formula used for the cyclotron will not work for betatrons because the electron will be relativistic. However, if the total energy is much greater than the rest energy then $\boldsymbol{E = pc}$ is a good approximation. As the centripetal force is again provided by the Lorentz force,

$$\frac{\boldsymbol{mv^2}}{\boldsymbol{r}} = q\boldsymbol{v}B_{orbit} \quad (5.4)$$

the maximum momentum will be $p = rq\,B_{orbit}$ and hence $E = rqc\,B_{orbit}$

An example of this type of machine is 315 MeV Betatron, built in 1949 at University of Chicago. With a magnetic field strength of 0.92 T and orbit radius of 1.22 m, the calculated energy is 340 MeV, in reasonable agreement, as synchrotron radiation losses have not been taken into account.

The formula for the electron's momentum can also be derived from Faraday's law of electromagnetic induction.

$$emf = N\frac{d\varphi}{dt} = NA\frac{d\bar{B}}{dt} \qquad (5.5)$$

where $N = 1$ and $A = \pi r^2$

Converting from voltage induced to electric field strength using $E = V/d$ gives

$$E\,2\,\pi r = \pi\, r^2 . \frac{d\bar{B}}{dt}$$

$$E.2\pi r = \pi r^2 . \frac{d\bar{B}}{dt}$$, and so

$$E = \frac{r}{2} . \frac{d\bar{B}}{dt} \quad \text{or} \quad E = \frac{R}{2} . \frac{d\bar{B}}{dt} \quad \text{for } r = R \qquad (5.6)$$

The force on the electron will be given by $F = qE$ so Newton's second law of motion gives

$$F = \frac{dp}{dt} = \frac{Rq}{2} . \frac{d\bar{B}}{dt}$$ and integrating gives

$$p = \frac{Rq\bar{B}}{2} = RqB_{orbit}$$ as given earlier. (5.7)

Betatrons are still used in industry and medicine as they are the very compact accelerators for electrons. Cyclotrons are similarly compact but cannot accelerate electrons to useful energies. The maximum energy that betatron can impart is limited by the strength of the magnetic field due to the saturation of iron and by practical size of the magnet core. The next generation of accelerators, the synchrotrons, overcame these limitations.

5.2.2. Cyclotron

1- Background

A **cyclotron** is a type of particle accelerator in which charged particles accelerate outwards from the center along a spiral path. The particles are held to a spiral trajectory by a static magnetic field and accelerated by a rapidly varying (radio frequency) electric field.

The cyclotron was invented and patented by Ernest Lawrence of the University of California, Berkeley, where it was first operated in 1932. A graduate student, M. Stanley Livingston, did much of the work of translating the idea into working hardware. Lawrence read an article about the concept of a drift tube linac by Rolf Widerøe, who had also been working along similar lines with the betatron concept. The first European cyclotron was constructed in Leningrad in the physics department of the Radium Institute, headed by Vitaly Khlopin. This instrument was first proposed in 1932 by George Gamow and Lev Mysovskii and was installed and became operative by 1937.

In the cyclotron, a cylindrical magnet bends the particle trajectories into a circular path whose radius depends on the mass of the particles, their velocity, and the strength of the magnetic field. The particles are accelerated within a hollow, circular, metal box that is split in half to form two sections, each in the shape of the capital letter ***D***. A radio-frequency electric field is impressed across the gap between the ***D*** 's so that every time a particle crosses the gap, the polarity of the ***D*** 's is reversed and the particle gets an accelerating "kick." The key to the simplicity of the cyclotron is that the period of revolution of a particle remains the same as the radius of the path increases because of the increase in velocity. Thus, the alternating electric field stays in step with the particles as they spiral outward from the center of the cyclotron to its circumference. However, according to the theory of relativity the mass of a particle increases as its velocity approaches the speed of light; hence, very energetic, high-velocity particles will have greater mass and thus less acceleration, with the result that they will not remain in step with the field. For protons, the maximum energy attainable with an ordinary cyclotron is about 10 million electron-volts.

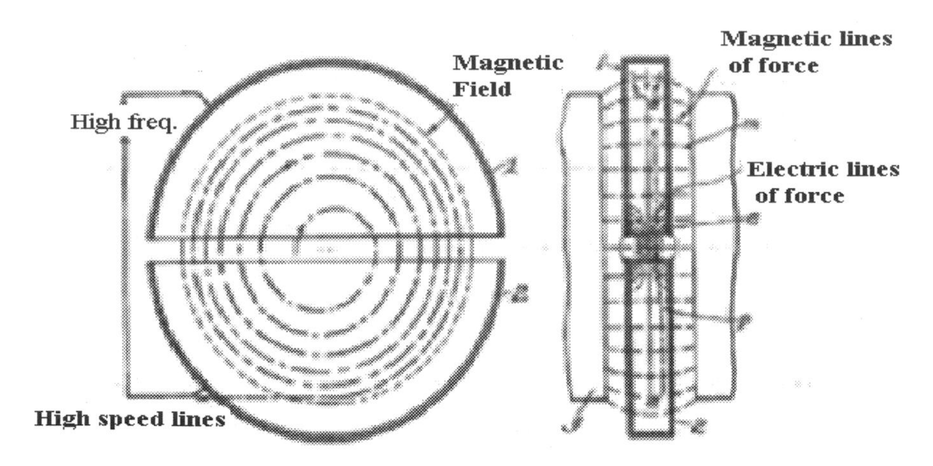

Fig.5.5. Diagram of cyclotron operation from Lawrence's 1934 patent.
The "D" shaped electrodes are enclosed in a flat vacuum chamber, which is installed in a narrow gap between the two poles of a large magnet.

2- Principle of Operation

Cyclotron consisted of two large dipole magnets designed to produce a semi-circular region of uniform magnetic field, pointing uniformly downward. These were called "dees" because of their D-shape. The two D's were placed back-to-back with their straight sides parallel but slightly separated.

An oscillating voltage was applied to produce an electric field across this gap. Particles injected into the magnetic field region of a D trace out a semicircular path until they reach the gap. The electric field in the gap then accelerates the particles as they pass across it.

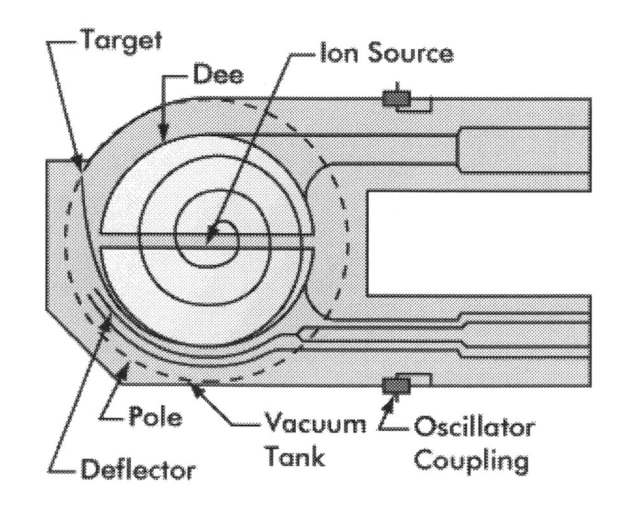

Fig.5.6. Particle accelerator- Cyclotron

The particles now have higher energy so they follow a semi-circular path in the next D with larger radius and so reach the gap again. The electric field frequency must be just right so that the direction of the field has reversed by their time of arrival at the gap. The field in the gap accelerates them and they enter the first D again. Thus the particles gain energy as they spiral around. The trick is that as they speed up, they trace a larger arc and so they always take the same time to reach the gap. This way a constant frequency electric field oscillation continues to always accelerate them across the gap. The limitation on the energy that can be reached in such a device depends on the size of the magnets that form the D's and the strength of their magnetic fields.

Cyclotrons reach an energy limit because of relativistic effects whereby the particles effectively become more massive, so that their cyclotron frequency drops out of synch with the accelerating RF. Therefore simple cyclotrons can accelerate protons only to an energy of around 15 million electron volts (15 MeV, corresponding to a speed of roughly 10% of c), because the protons get out of phase with the driving electric field. If accelerated further, the beam would continue to spiral outward to a larger radius but the particles would no longer gain enough speed to complete the larger circle in step with the accelerating RF. Cyclotrons are nevertheless still useful for lower energy applications.

Cyclotrons accelerate charged particle beams using a high frequency alternating voltage which is applied between two "D"-shaped electrodes. An additional static magnetic field $\boldsymbol{B}$ is applied in perpendicular direction to the electrode plane, enabling particles to re-encounter the accelerating voltage many times at the same phase. To achieve this, the voltage frequency must match the particle's cyclotron resonance frequency

$$f = \frac{qB}{2\,\pi\,m} \qquad (5.8)$$

with the relativistic mass m and its charge q. This frequency is given by equality of centripetal force and magnetic Lorentz force. The particles, injected near the center of the magnetic field, increase their kinetic energy only when recirculating through the gap between the electrodes; thus they travel outwards along a spiral path.

Their radius will increase until the particles hit a target at the perimeter of the vacuum chamber, or leave the cyclotron using a beam tube, enabling their use e.g. for particle therapy. Various materials may be used for a target, and the collisions will create secondary particles which may be guided outside of the cyclotron and into instruments for analysis.

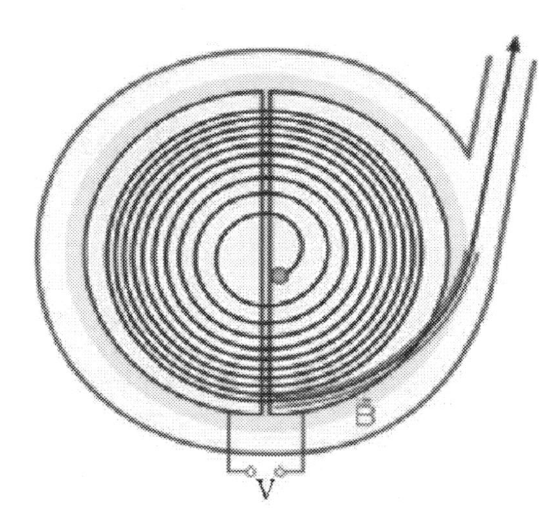

Fig.5.7. Sketch of a particle being accelerated in a cyclotron, and being ejected through a beam line.

5.2.3. Synchrotrons

1- Background

A **synchrotron** is a particular type of cyclic particle accelerator originating from the cyclotron in which the guiding magnetic field (bending the particles into a closed path) is time-dependent, being synchronized to a particle beam of increasing kinetic energy. The synchrotron is one of the first accelerator concepts that enable the construction of large-scale facilities, since bending, beam focusing and acceleration can be separated into different components.

Edwin McMillan constructed the first electron synchrotron in 1945, although Vladimir Veksler had already (unknown to McMillan) published the principle in a Soviet journal in 1944. The first proton synchrotron was designed by Sir Marcus Oliphant and built in 1952.

Circular electron accelerators fell somewhat out of favor for particle physics around the time that SLAC was constructed, because their synchrotron losses were considered economically prohibitive and because their beam intensity was lower than for the unpulsed linear machines.

Fig.5.8. Segment of an electron synchrotron at DESY

The Cornell Electron Synchrotron, built at low cost in the late 1960s, was the first in a series of high-energy circular electron accelerators built for fundamental particle physics, culminating in the LEP at CERN. A large number of electron synchrotrons have been built in the past two decades, specialized to be synchrotron light sources, of ultraviolet light and X rays.

The **Deutsches Elektronen-Synchrotron** (German Electron Synchrotron), commonly abbreviated **DESY**, is a national research center in Germany which operates particle accelerators used to investigate the structure of matter. It conducts a broad spectrum of inter-disciplinary scientific research in three main areas: particle and high energy physics; photon science; and the development, construction and operation of particle accelerators. DESY is publicly financed by the Federal Republic of Germany, the States of Germany, and the German Research Foundation (DFG). DESY is a member of the Helmholtz Association and operates at sites in Hamburg and Zeuthen.

DESY's function is to conduct fundamental research. It specializes in:

- **Particle accelerator** development, construction and operation.
- **Particle physics** research to explore the fundamental characteristics of matter and forces, including astroparticle physics

- **Photon science** research in surface physics, material science, chemistry, molecular biology, geophysics and medicine through the use of synchrotron radiation and free electron lasers

In addition to operating its own large accelerator facilities, DESY also provides consulting services to research initiatives, institutes and universities. It is closely involved in major international projects such as the European X-Ray Free-Electron Laser, the Large Hadron Collider in Geneva, the Ice Cube Neutrino Observatory at the South Pole and the International Linear Collider.

Once the synchrotron principle was developed, it was found to be a much cheaper way to achieve high energy particles than the cyclotron and so the original cyclotron method is no longer used.

In a synchrotron, brunches of charged particles (electrons) circulate at nearly the speed of light for several hours inside a long ring-shaped tube under vacuum. As magnets surrounding the tube bend their trajectories, the electrons emit (Synchrotron light), with wavelengths that range from infrared radiation to X-rays. The emitted light is collected by different beamlines-optical systems connected to the ring; thus, many experiments stations located on different beam lines can be run simultaneously.

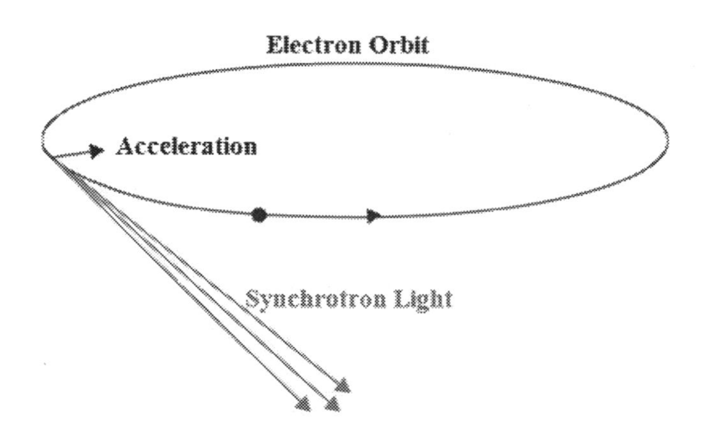

Fig.5.9. Emitted Synchrotron light

When electrons are accelerated, part of the energy in electromagnetic force field shall be emitted as electromagnetic radiation(e.g. radio waves). The electromagnetic field surrounding the electrons is unable to respond instantaneously when the electrons are deflected; some of the energy in the field keeps going, producing a tangential cone of synchrotron radiation. As the electrons' energy increases, the cone of radiation narrows, and the radiated power goes up dramatically.

A synchrotron light source is a combination of different electron accelerator types, including a storage ring in which the desired electromagnetic radiation is generated. In addition to the storage ring, a synchrotron light source usually contains a linear accelerator (linac) and another synchrotron which is sometimes called booster in this context. The linac and the booster are used to successively accelerate the electrons to their final energy before they are magnetically "kicked" into the storage ring. Synchrotron light sources in their entirety are sometimes called "synchrotrons", although this is technically incorrect.

A cyclic collider is also a combination of different accelerator types, including two intersecting storage rings and the respective pre-accelerators.

2- Principle of Operation

While a classical cyclotron uses both a constant guiding magnetic field and a constant-frequency electromagnetic field (and is working in classical approximation), its successor, the isochronous cyclotron, works by local variations of the guiding magnetic field, adapting the increasing relativistic mass of particles during acceleration.

In a synchrotron, this adaptation is done by variation of the magnetic field strength in time, rather than in space. For particles that are not close to the speed of light, the frequency of the applied electromagnetic field may also change to follow their non-constant circulation time. By increasing these parameters accordingly as the particles gain energy, their circulation path can be held constant as they are accelerated. This allows the vacuum chamber for the particles to be a large thin torus, rather than a disk as in previous, compact accelerator designs. Also, the thin profile of the vacuum chamber allowed for a more efficient use of magnetic fields than in a cyclotron, enabling the cost-effective construction of larger synchrotrons.

While the first synchrotrons and storage rings like the Cosmotron and ADA strictly used the toroid shape, the strong focusing principle independently discovered by Ernest Courant et al. and Nicholas Christofilos allowed the complete separation of the accelerator into components with specialized functions along the particle path, shaping the path into a round-cornered polygon. Some important components are given by radio frequency cavities for direct acceleration, dipole magnets (bending magnets) for deflection of particles (to close the path), and quadrupole / sextupole magnets for beam focusing.

Fig.5.10. A drawing of the Cosmotron

The combination of time-dependent guiding magnetic fields and the strong focusing principle enabled the design and operation of modern large-scale accelerator facilities like colliders and synchrotron light sources. The straight sections along the closed path in such facilities are not only required for radio frequency cavities, but also for particle detectors (in colliders) and photon generation devices such as Wigglers and Undulators (in third generation synchrotron light sources).

The maximum energy that a cyclic accelerator can impart is typically limited by the maximum strength of the magnetic fields and the minimum radius (maximum curvature) of the particle path. Thus one method for increasing the energy limit is to use superconducting magnets, these not being limited by magnetic saturation. electron/positron accelerators may also be limited by the emission of synchrotron radiation, resulting in a partial loss of the particle beam's kinetic energy. The limiting beam energy is reached when the energy lost to the lateral acceleration required to maintain the beam path in a circle equals the energy added each cycle.

More powerful accelerators are built by using large radius paths and by using more numerous and more powerful microwave cavities. Lighter particles (such as electrons) lose a larger fraction of their energy when deflected. Practically speaking, the energy of electron/positron accelerators is limited by this radiation loss, while this does not play a significant role in the dynamics of proton or ion accelerators. The energy of such accelerators is limited strictly by the strength of magnets and by the cost.

Fig.5.11. The interior of the Australian Synchrotron Facility 2006.

216 m circumference storage ring, with a beamline at front right. In the middle of the storage ring is the booster ring and linac. The storage ring's interior includes a synchrotron and a linac. The trolly carries different types of magnets as explained in section 4.2.1. (The larger one is a dipole magnet used to bend the electron beam and produce the synchrotron radiation. The front magnet is a sextupole magnet, the magnet (at back) a quadrupole magnet - these are used to focus and steer the electron beam).

Unlike in a cyclotron, synchrotrons are unable to accelerate particles from zero kinetic energy; one of the obvious reasons for this is that its closed particle path would be cut by a device that emits particles. Thus, schemes were developed to inject pre-accelerated particle beams into a synchrotron. The pre-acceleration can be realized by a chain of other accelerator structures like a linac, a microtron or another synchrotron; all of these in turn need to be fed by a particle source comprising a simple high voltage power supply, typically a Cockcroft-Walton .

Starting from an appropriate initial value determined by the injection energy, the field strength of the dipole magnets is then increased. If the high energy particle are emitted at the end of the acceleration procedure, e.g. to a target or to another accelerator, the field strength is again decreased to injection level, starting a new injection cycle. Depending on the method of magnet control used, the time interval for one cycle can vary substantially between different installations.

5.2.4 Synchrocyclotron and Isochronous Cyclotron

1- Background

There are ways of modifying the classic cyclotron to increase the relativistic limit for cyclotrons. This may be done in a continuous beam, constant frequency, and machine by shaping the magnet poles so to increase magnetic field with radius. Then higher energy particles travel a shorter distance in each orbit than they otherwise would, and can remain in phase with the accelerating field.

In the synchrocyclotron, the frequency of the accelerating electric field steadily decreases to match the decreasing angular velocity of the protons. The Synchrocyclotron, accelerates the particles in bunches, in a constant B field, but reduces the RF accelerating field's frequency so as to keep the particles in step as they spiral outward. This approach suffers from low average beam intensity due to the bunching, and again from the need for a huge magnet of large radius and constant field over the larger orbit demanded by high energy.

Another possibility, the Isochronous Cyclotron, the magnet is constructed so the magnetic field is stronger near the circumference than at the center, thus compensating for the mass increase and maintaining a constant frequency of revolution. The first synchrocyclotron, built at the Univ. of California at Berkeley in 1946, reached energies high enough to create pions, thus inaugurating the laboratory study of the meson family of elementary particles.

The advantage of Isochronous Cyclotrons is that they can deliver continuous beams of higher average intensity, which is useful for some applications. The main disadvantages are the size and cost of the large magnet needed, and the difficulty in achieving the higher field required at the outer edge.

2- Principle of Operation

A. Relativistic Considerations

In the nonrelativistic approximation, the frequency does not depend upon the radius of the particle's orbit, since the particle's mass is constant. As the beam spirals out, its frequency does not decrease, and it must continue to accelerate, as it is travelling a greater distance in the same time period.
In contrast to this approximation, as particles approach the speed of light, their relativistic mass increases, requiring either modifications to the frequency, leading to the ***synchrocyclotron***, or modifications to the magnetic field during the acceleration, which leads to the ***isochronous cyclotron***. The relativistic mass can be rewritten as

$$m = \frac{m_o}{\sqrt{1-(\frac{v}{c})^2}} = \frac{m_o}{\sqrt{1-\beta^2}} = \gamma\, m_o \qquad (5.9)$$

where

m_0 is the particle rest mass

$\beta = \frac{v}{c}$ is the relative velocity, and the Lorentz factor is given by

$$\gamma = \frac{1}{\sqrt{1-\beta^2}} = \frac{1}{\sqrt{1-(\frac{v}{c})^2}} \qquad (5.10)$$

The relativistic cyclotron frequency and angular frequency can be rewritten as

$$f = \frac{qB}{2\pi\gamma m_0} = \frac{f_0}{\gamma} = f_0\sqrt{1-\beta^2} = f_0\sqrt{1-(\frac{v}{c})^2} \quad (5.11)$$

$$\omega = 2\pi f = \frac{qB}{\gamma m_0} = \frac{\omega_0}{\gamma} = \omega_0\sqrt{1-\beta^2} = \omega_0\sqrt{1-(\frac{v}{c})^2} \quad (5.12)$$

Where in classical approximation

f_0 : the cyclotron frequency, and

ω_0: the cyclotron angular frequency

The gyro-radius for a particle moving in a static magnetic field is then given by

$$r = \frac{v}{\omega} = \frac{\beta C}{\beta} = \frac{\gamma\beta m_0 C}{qB} \quad (5.13)$$

where:

$\omega\tau = v = \beta C$

v : the (linear) velocity

B. Synchrocyclotron

A synchrocyclotron is a cyclotron in which the frequency of the driving RF electric field is varied to compensate for relativistic effects as the particles' velocity begins to approach the speed of light. This is in contrast to the classical cyclotron, where the frequency was held constant, thus leading to the synchrocyclotron operation frequency being

$$f = \frac{f_0}{\gamma} = f_0\sqrt{1-\beta^2} \quad (5.14)$$

where $\boldsymbol{f_o}$ is the classical cyclotron frequency and $\boldsymbol{\beta = \frac{v}{c}}$ again is the relative velocity of the particle beam.

The rest mass of an electron is 511 keV/c^2, so the frequency correction is 1% for a magnetic vacuum tube with a 5.11 keV/c^2 direct current accelerating voltage. The proton mass is nearly two thousand times the electron mass, so the 1% correction energy is about 9 MeV, which is sufficient to induce nuclear reactions.

C. Isochronous Cyclotron

An alternative to the synchrocyclotron is the ***isochronous cyclotron***, which has a magnetic field that increases with radius, rather than with time. Isochronous cyclotrons are capable of producing much greater beam current than synchrocyclotrons, but require azimuthal variations in the field strength to provide a strong focusing effect and keep the particles captured in their spiral trajectory. For this reason, an isochronous cyclotron is also called an "AVF (azimuthal varying field) cyclotron". This solution for focusing the particle beam was proposed by L. H. Thomas in 1938.

Recalling the relativistic gyro-radius

$$r = \frac{\gamma m_0 v}{qB} \tag{5.15}$$

and the relativistic cyclotron frequency $f = \frac{f_o}{\gamma}$, one can choose B to be proportional to the Lorentz factor , $B = \gamma B_o$. This results in the relation

$$r = \frac{m_o v}{qB_o} \tag{5.16}$$

which again only depends on the velocity v, like in the non-relativistic case. Also, the cyclotron frequency is constant in this case.

The transverse de-focusing effect of this radial field gradient is compensated by ridges on the magnet faces which vary the field azimuthally as well. This allows particles to be accelerated continuously, on every period of the radio frequency (RF), rather than in bursts as in most other accelerator types. This principle that alternating field gradients have a net focusing effect is called strong focusing. It was obscurely known theoretically long before it was put into practice.

Examples of isochronous cyclotrons abound; in fact almost all modern cyclotrons use azimuthally-varying fields. The TRIUMF cyclotron mentioned below is the largest with an outer orbit radius of 7.9 meters, extracting protons at up to 510 MeV, which is 3/4 of the speed of light. The PSI cyclotron reaches higher energy but is smaller because of using a higher magnetic field.

5.2.5. Storage Ring

1- Background

A storage ring is a special type of synchrotron in which the kinetic energy of the particles is kept constant, i.e. a storage ring is the same thing as a synchrotron, except that it is designed just to keep the particles circulating at a constant energy for as long as possible, not to increase their energy any further. However, the particles must still pass through at least one accelerating cavity each time they circle the ring, just to compensate for the energy they lose to synchrotron radiation.

Two storage rings have been built at SLAC; SPEAR, a 3 GeV ring completed in the early 70's and PEP a 9 GeV ring completed in the early 80's. SPEAR is now used solely by SSRL while PEP has been rebuilt as a two-ring facility known as the *B* Factory where results are being accumulated by the BaBar detector and studied by the BaBar collaboration.

For some applications, it is useful to store beams of high energy particles for some time (with modern high vacuum technology, up to many hours) without further acceleration. This is especially true for colliding beam accelerators, in which two beams moving in opposite directions are made to collide with each other, with a large gain in effective collision energy. Because relatively few collisions occur at each pass through the intersection point of the two beams, it is customary to first accelerate the beams to the desired energy, and then store them in storage rings, which are essentially synchrotron rings of magnets, with no significant RF power.

2- Principle of operation

A **storage ring** is a type of circular particle accelerator in which a continuous or pulsed particle beam may be kept circulating for a long period of time, up to many hours. Storage of a particular particle depends upon the mass, energy and usually charge of the particle being stored. Most commonly, storage rings are to store electrons, positrons, or protons.

The most common application of storage rings is to store electrons which then radiate synchrotron radiation. There are over 50 facilities based on electron storage rings in the world today, used for a variety of studies in chemistry and biology. Storage rings are used to produce polarized high-energy electron beams through the Sokolov-Ternov effect. Arguably the best known application of storage rings is their use in particle accelerators and in particle colliders, in which two counter-rotating beams of stored particles are brought into collision at discrete locations, the results of the subatomic interactions being studied in a surrounding particle detector. Examples of such facilities are LHC, LEP, PEP-II, KEKB, RHIC, Tevatron and HERA.

Technically speaking, a storage ring is a type of synchrotron. However, a conventional synchrotron serves to accelerate particles from a low to a high energy with the aid of radio-frequency accelerating cavities; a storage ring, as the name suggests, keeps particles stored at a constant energy, and radio-frequency cavities are only used to replace energy lost through synchrotron radiation and other processes.

Gerard K. O'Neill proposed the use of storage rings as building blocks for a collider in 1956. A key benefit of storage rings in this context is that the storage ring can accumulate a high beam flux from an injection accelerator that achieves a much lower flux.

5.3. Advantages and Disadvantages of Accelerators types

The advantage of a circular accelerator **design** over a linear accelerator is that the particles in a circular accelerator (synchrotron) go around many times, getting multiple kicks of energy each time around. Therefore, synchrotrons can provide very high-energy particles without having to be of tremendous length. Moreover, the fact that the particles go around many times means that there are many chances for collisions at those places where particle beams are made to cross.

On the other hand, linear accelerators are much easier to build than circular accelerators because they don't need the large magnets required to coerce particles into going in a circle. Circular accelerators also need an enormous radii in order to get particles to high enough energies, so they are expensive to build.

Another thing that physicists need to consider is that when a charged particle is accelerated, it radiates away energy. At high energies the radiation loss is larger for circular acceleration than for linear acceleration. In addition, the radiation loss is much worse for accelerating light electrons than for heavier protons. Electrons and anti-electrons (positrons) can be brought to high energies only in linear accelerators or in circular ones with large radii.

The main advantages and disadvantages of the linear accelerator are obvious from its form.

As advantages, we may list the saving of the expense of the large magnet, which is necessary in the circular machines; the fact that the size and expense of the machine is roughly proportional to the final energy of the particle, rather than to a higher power of the energy, as in circular machines, suggesting that at least for very high energies it may be a more economical device than the circular machines; and the fact that the particles will automatically emerge in a well-collimated bean, whereas beam ejection is one of the principal difficulties in the circular machines.

The principal disadvantage of the linear accelerator is the fact that an individual particle, instead of passing through the same alternating field again and again, and using the same power source and accelerating gaps many times, as in the cyclotron and other circular machines, must pass through a succession of alternating fields and a succession of power sources.

This multiplicity of sources and fields results in expense, in a large duplication of high'- frequency equipment; in complication, in the construction of a very long and elaborate tube with its adjacent power sources; and in a design difficulty, in the adjustment of the phase of the oscillating field over a great length of accelerator, so as to insure that the field will stay in step with the particles.

There is another design difficulty not shared by the circular machines, which does not appear until one makes a little mathematical analysis. In any accelerator, the particles must travel a very long distance, either in a straight line or a circle or a spiral, before they acquire the energies about which one talks in present discussion of hundreds of millions or billions of electron-volts. In this very long path, they are likely to spread from their ideal path, and become lost to the beam.

The spreading can be of two sorts. First, they can spread laterally, or be defocussed. Secondly, since the operation of all these devices except the betatron depends on having the particles bunched longitudinally, in bunches a wavelength apart, in the proper phase of a traveling wave to be accelerated, the particles can spread longitudinally, getting ahead or behind their bunches, and can be debunched. Now in the cyclotron and synchrotron, those particles which are in the correct phase to be bunched and accelerated are also in the right phase to be focussed, so that stability is automatic. On the other hand, in the linear accelerator, particles in the phase for bunching and acceleration are defocussed. This defocussing imposes a design problem of serious proportions for the linear accelerator.

If the circular machines in the billion-volt range were perfectly easy to build, there would be no question but that they would be the preferred devices, because the disadvantages and difficulties of the linear accelerator, as just enumerated, are formidable. But all the recent thinking about circular machines has shown that their difficulties are formidable as well. Among electron accelerators, present thinking indicates that the synchrotron is preferred to the betatron for energies more than a few hundred million electron-volts, but the synchrotron is believed to face a practical limit, on account of radiative losses by the electrons, at about a billion electron-volts.

No device except the linear accelerator has been suggested to go beyond this range. As for proton and other positive ion accelerators, the ordinary cyclotron meets relativistic difficulties at about a hundred million electron-volts. The frequency-modulated or synchro-cyclotron is ideally adapted for energies of several hundred million electron-volts. Attempts at design of higher energy machines of this type, however; show that the dee structure and the magnet design begin to face problems at about six hundred million electron-volts which become very serious indeed at a billion electron-volts, and probably are insurmountable much above that energy.

For the range of several billion electron-volt positive ions, present thinking suggests only one alternative to the linear accelerator: the proton synchrotron, a device in which both magnetic field and frequency are simultaneously modulated, during the cycle of acceleration of the particles, in such a way as to keep the radius of the particle's orbit constant, and yet to keep the particle in step with the radio-frequency field. This has the advantage that only an annular magnet is required, as in the synchrotron, so that the magnet cost does not go up so rapidly with energy as in the synchro-oyclotron, with its solid magnet. Furthermore, the accelerating electrodes, or dees, can be relatively small, and the frequencies encountered are so low that the frequency modulation is simple. Its difficulties are nevertheless very great. The magnet, even though annular, is exceedingly large and expensive.

The mutual adjustment, in time during the accelerating cycle, of frequency of oscillator and magnetic field, to keep the particles in step with the accelerating field, must be very accurate, and is probably in principle as difficult an adjustment as that of the field in the linear accelerator, to keep it in step with the particles. And-there is further difficulty, not so obvious at first glance, that the circumference of the orbit is very large, and in all present thinking there is only one accelerating unit around the circumference, which does not supply a very great increment of energy per revolution of the particle. Hence the particle receives much less energy, per unit length of its own path, than in the linear accelerator, where every effort is made to concentrate the accelerating units and make them as powerful as possible, so as to cut down the total length of the machine.

In building up its total energy, then, the particle in the proton synchrotron travels very much further than in the linear accelerator, and as a result the chances of spreading, straggling, and loss of particles from the beam are much greater, so that it is likely that the number of particles lost from the beam will be even greater in the proton synchrotron than in the linear accelerator, even in spite of the inherently defocussing nature of the latter device, which we have mentioned above.

6

APPLICATIONS OF ACCELERATORS

6.1. Introduction

In early particle accelerators a Cockcroft-Walton voltage multiplier was responsible for voltage multiplying. This piece of the accelerator helped in the development of the atomic bomb. Built in 1937 by Philips of Eindhoven it currently resides in the National Science Museum in London, England.

Beams of high-energy particles are useful for both fundamental and applied research in the sciences. For the most basic inquiries into the dynamics and structure of matter, space, and time, physicists seek the simplest kinds of interactions at the highest possible energies. These typically entail particle energies of many GeV, and the interactions of the simplest kinds of particles: leptons (e.g. electrons and positrons) and quarks for the matter, or photons and gluons for the field quanta. Since isolated quarks are experimentally unavailable due to color confinement, the simplest available experiments involve the interactions of, first, leptons with each other, and second, of leptons with nucleons, which are composed of quarks and gluons. To study the collisions of quarks with each other, scientists resort to collisions of nucleons, which at high energy may be usefully considered as essentially 2-body interactions of the quarks and gluons of which they are composed. Thus elementary particle physicists tend to use machines creating beams of electrons, positrons, protons, and anti-protons, interacting with each other or with the simplest nuclei (eg, hydrogen or deuterium) at the highest possible energies, generally hundreds of GeV or more.

Nuclear physicists and cosmologists may use beams of bare atomic nuclei, stripped of electrons, to investigate the structure, interactions, and properties of the nuclei themselves, and of condensed matter at extremely high temperatures and densities, such as might have occurred in the first moments of the Big Bang. These investigations often involve collisions of heavy nuclei – of atoms like iron or gold – at energies of several GeV per nucleon.

At lower energies, beams of accelerated nuclei are also used in medicine, as for the treatment of cancer.

Besides being of fundamental interest, high energy electrons may be coaxed into emitting extremely bright and coherent beams of high energy photons – ultraviolet and X ray – via synchrotron radiation, which photons have numerous uses in the study of atomic structure, chemistry, condensed matter physics, biology, and technology. Examples include the ESRF in Europe, which has recently been used to extract detailed 3-dimensional images of insects trapped in amber. Thus there is a great demand for electron accelerators of moderate (GeV) energy and high intensity.

The data which have been collected by W. Scarf and W. Wiesczycka (U. Amaldi Europhysics News, June 31, 2000) indicated that the World wide inventory of accelerators, in total 15,000. These are generally distributed on the following applications:

- Ion implanters and surface modifications: 7,000
- Accelerators in industry: 1,500
- Accelerators in non-nuclear research: 1,000
- Radiotherapy: 5,000
- Medical isotopes production: 200
- Hadron therapy: 20
- Synchrotron radiation sources: 70
- Nuclear and particle physics research: 110

6.2. Main Applications of Accelerators

In high-energy physics (also called particle physics), particles are accelerated to energies far higher than usually found on earth. Collisions then produce particles that did not exist in nature since the Big Bang, yielding data on the properties of these new particles - fundamental information about nature itself. Nuclear physics uses the energy of beams to study the internal properties of the atom's nucleus or the dynamics when nuclei interact. Sometimes this results in producing isotopes that do not normally exist in nature.

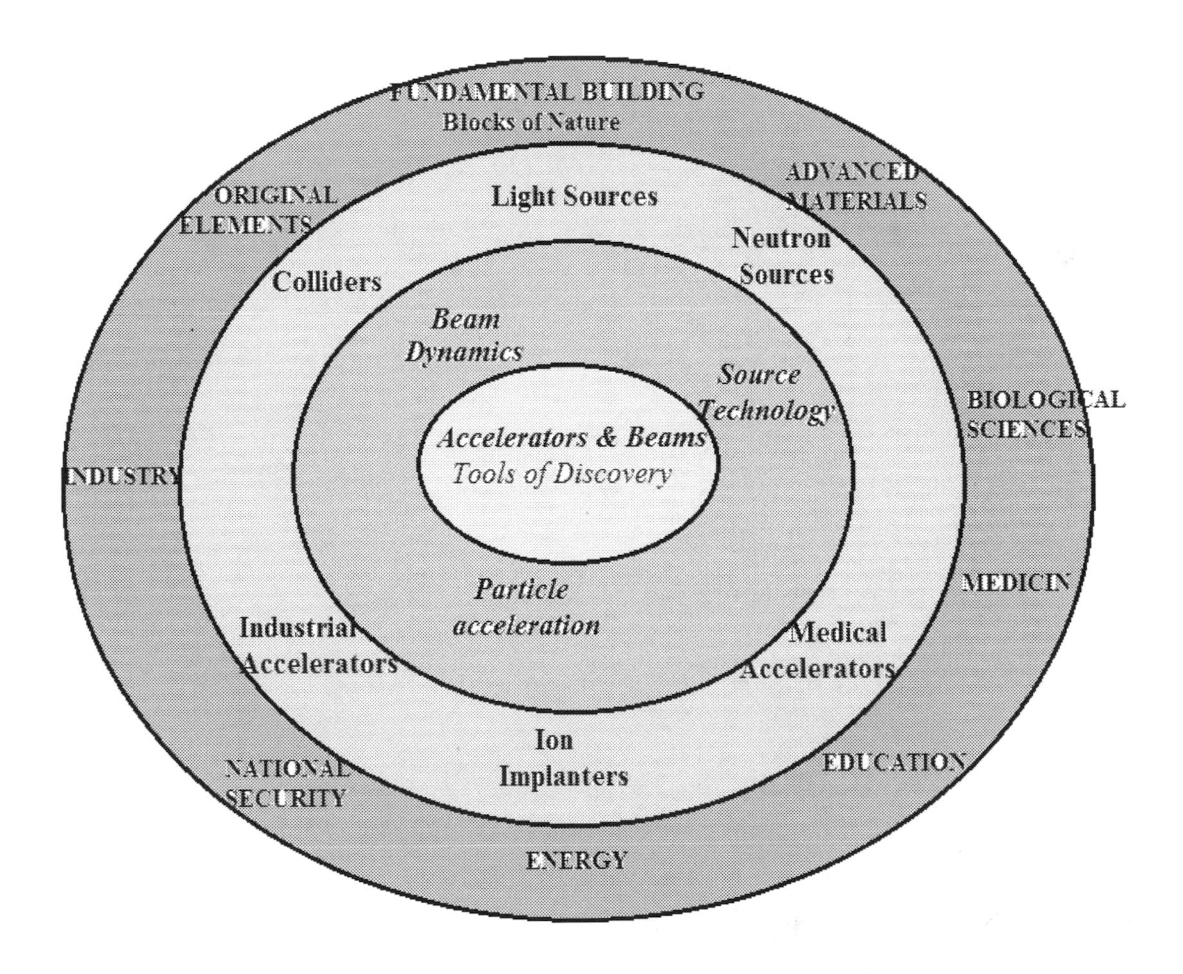

Fig. 6.1 Accelerators applications on wide range of services

Beams for both particle physics and nuclear physics artificially recreate conditions that existed when the universe was much hotter, generating the ambient temperatures from these earlier times. Such conditions enable pursuit of big questions, such as the meaning of mass, a concept humans have puzzled over for millennia. Why are particles like electrons nearly massless, and particles like protons or top quarks massive? Some 95% of our universe consists of things we don't understand: dark matter and dark energy. Accelerators offer the possibility of answers.

The world has only a few huge, expensive accelerators for research at the frontiers of knowledge, but there are many thousands of smaller accelerators.

The wide range of accelerator applications may cover the following:

- Accelerators for diagnosing illness and fighting cancer
- Accelerators to beat food-borne illness
- An accelerator makes our band-aids safe
- Accelerators and national security
- Accelerators validate nuclear weapons readiness
- Beams of light from beams of particles
- Accelerators energize a new kind of laser
- Free-electron laser applications
- Accelerators for improving materials' surfaces
- Accelerator-based neutron science yields payoffs
- Accelerators boost international cooperation
- Accelerators and astronomy and on museum

Examples of these applications are highlighted as following:

1- Accelerators for diagnosing illness and fighting cancer

Across the world, hospitals and doctors use thousands of medical accelerators. Cyclotrons, modern versions of the original Lawrence invention, produce many different kinds of radioactive substances, called isotopes, for diagnostic procedures and therapy.

Systems makes it possible to deliver radiation therapy in dramatically shorter treatment times. Many medical accelerators produce radiation for directly attacking cancer. Advances in proton and ion beam therapy are enabling doctors to avoid harming tissue near the cancer.

2- Accelerators to beat food-borne illness

Every day in the world, many people die from food-borne illness, even though electron accelerators can make food much safer, just as pasteurization makes milk much safer. Electron beams, or X-rays derived from them, can kill dangerous bacteria like E. coli, salmonella and listeria.

Food irradiation could join pasteurization, chlorination and immunization as pillars of public health technology. But even though food irradiation is completely safe and does not degrade wholesomeness, nutritional value, quality or taste, consumer acceptance has been slow. That word irradiation makes people wary.

Nevertheless food irradiation increasingly is gaining formal approval in various countries. Accelerator technology is already making food safer and increasing shelf life, which increases food supply to feed an ever-growing world population and reduce world hunger.

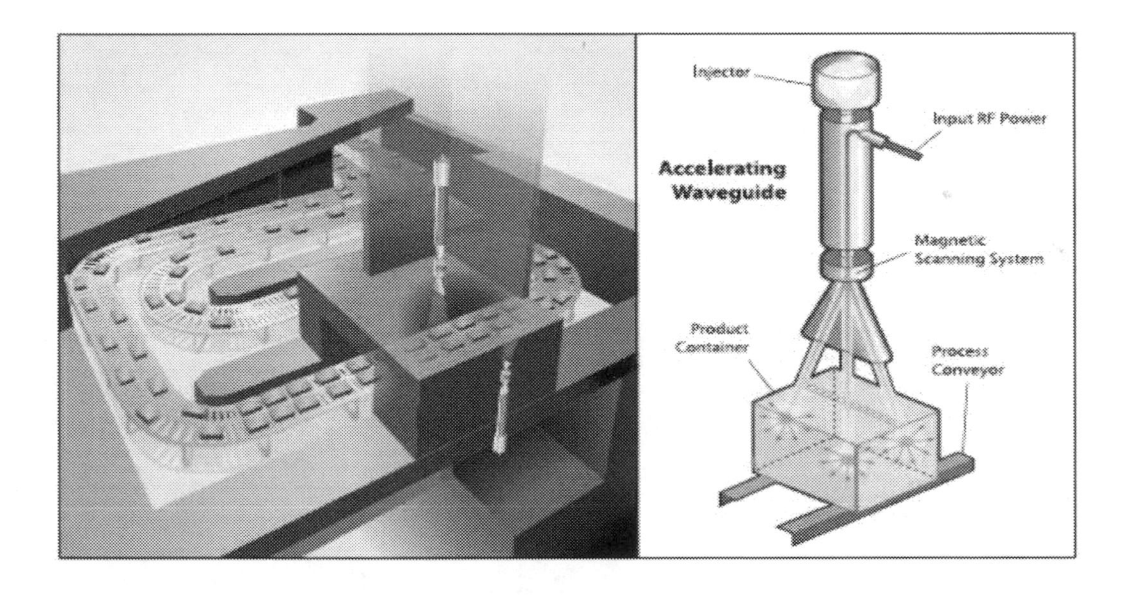

Fig.6.2 A production line conveys food products for irradiation by electron beams from two vertical accelerators, above and below the conveyor.

3- An accelerator makes our band-aids safe

Electron beams can sterilize products effectively, efficiently, and fast. They kill all bacteria. They penetrate packaging, and even the shipping cartons holding the packages, so that there is no danger of contamination during or after sterilization.

Because the beams generate no heat, they can be used on plastic medical supplies like catheters or cloth bandages.

They can also be used to kill bacteria in human blood platelets and skin grafts.

The “dose” from an electron beam with 10 to 50 kilowatts of power lets the conveyor run as fast as a few feet per minute, allowing sterilization of an entire truckload of boxes in only a few hours. A typical facility can process medical supplies so quickly that it needs several loading docks for offloading and reloading product.

Fig. 6.3 Electron beams can sterilize products such as plastic medical supplies like catheters or cloth bandages

4- <u>Accelerators and national security</u>

Millions of tractor-trailer-sized cargo containers are moving each year around the world. How can these large steel-walled boxes be inspected for several reasons and purposes and particularly for security and safety? Accelerators offer answers for scanning various kinds of cargo containers and vehicles effectively and efficiently.

Fig.6.4 High-density rail cargo imager scans fully loaded trains, tankers and double-stacked cars in a single pass using X-rays.

5- <u>Accelerators validate nuclear weapons readiness</u>

In a special facility at Los Alamos National Laboratory, two electron accelerators at right angles to each other let scientists monitor realistic but non-nuclear tests of replacement components for the nation's nuclear weapons. "Stockpile stewardship" is necessary because over time some components degrade, perhaps losing functionality, based on interactions from the natural radioactivity of the weapon itself.

Nuclear weapons testing moratoria preclude integrity tests. Therefore non-nuclear tests are necessary. To make the mockup non-nuclear, a heavy metal surrogate stands in for the nuclear fuel, but all other components can be exact replicas. In the test each electron beam is focused onto a metal target. This target converts the beam's kinetic energy into X-rays that generate images showing the dynamic events that trigger a nuclear detonation. Everything is real except the nuclear fuel. Because the surrogate fuel and the components being tested become hot enough to melt and flow like water, such a test is called hydrodynamic — leading to the name Dual-Axis Radiographic Hydrodynamic Test Facility, or DARHT Facility.

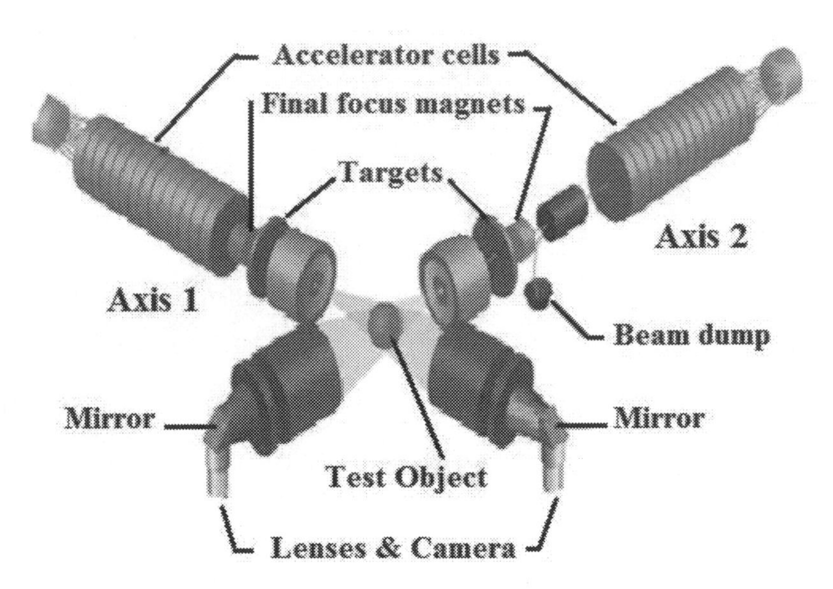

Fig. 6.5 In DARHT Facility, two electron accelerators work together to enable safe, realistic, non-nuclear monitoring of nuclear weapons readiness

6- Beams of light from beams of particles

In science and technology, the word light applies generally to electromagnetic radiation. Most wavelengths of light aren't visible. Light sources generate microwave, infrared, visible, ultraviolet, X-ray and gamma-ray light. An equivalent statement is that light sources generate beams of microwave, infrared, visible, ultraviolet, X-ray and gamma-ray photons. (See Fig.6.6A.)

Using the magnet in accelerator can generate intense, focused and steer the light beams. When a particle beam passes between the north and south poles of an accelerator magnet (N and S) in Fig.6.7B the beam not only changes direction, it also emits exceptionally intense, tightly focused light.

The magnets, not only can steer a particle beam around a circular accelerator or through a linear one, but they can tap the accelerator beam to get a specific light beams based on magnet beam intensity. This light can then be directed away from the accelerator, shining down beamlines to various scientific experiments or technological uses. See Fig.6.7.

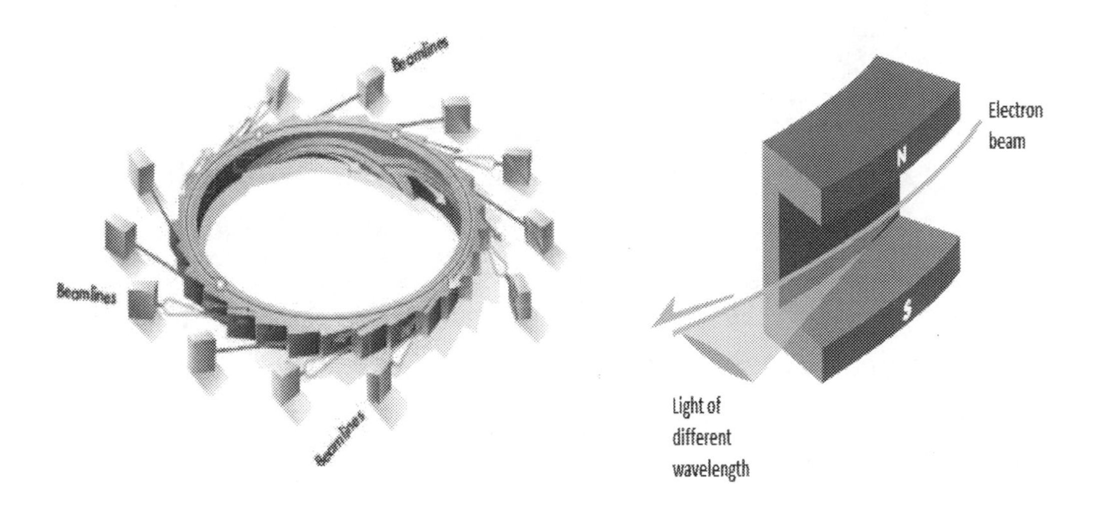

A. Accelerators beamlines gates B. Magnet Steering beamlines

Fig.6.6 Generation of Beams of light

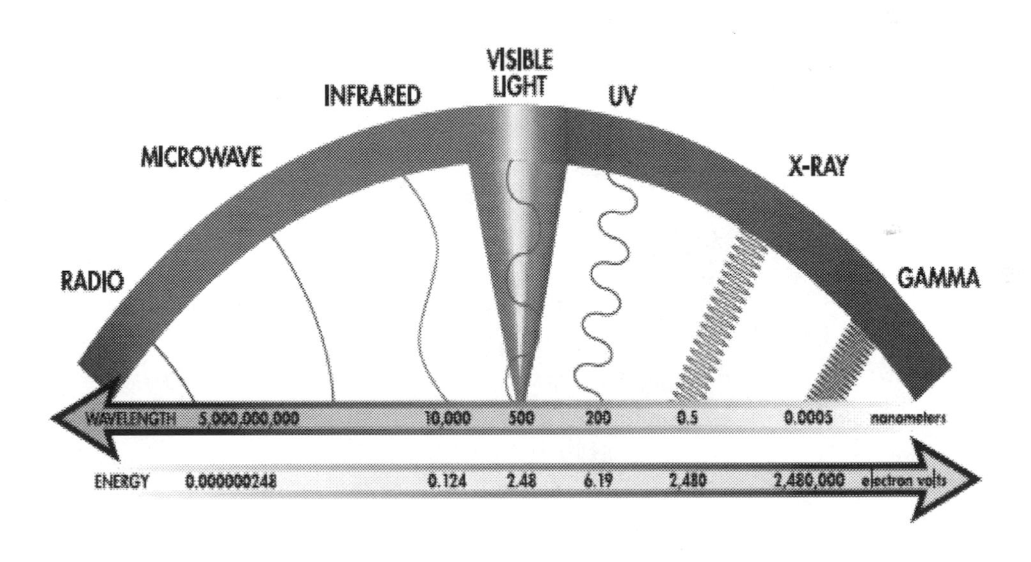

1centimeter =10,000,000 nanometer

Fig.6.7 light of different wavelength

7- <u>Accelerators energize a new kind of laser</u>

Conventional lasers make their extraordinarily useful kind of light by jiggling electrons that are bound in atoms. Accelerator-driven Free-Electron Lasers (FELs) make light by using magnets to jiggle electrons that are freed from atoms, as shown in Fig.6.8. With free electrons, the light's wavelength (color) can be selected. For many applications, operators can select the resulting light's wavelength by varying the beam's energy or the magnetic field strength.

An FEL's beam is delivered in pulses rather than a steady stream. These bursts of light can be timed, with pulse sequences shorter than a trillionth of a second. For many applications, this too is a very important feature.

At Stanford University's SLAC National Accelerator Laboratory, the Linac Coherent Light Source (LCLS), is the world's most powerful X-ray laser. The linac (linear accelerator) generates high-energy beams of free electrons for making this special kind of light.

The resulting laser light arrives in staccato bursts one-tenth of a trillionth of a second long. These intense, ultrafast pulses let researchers scrutinize complex, ultra-small structures by freeze framing atomic motions. The researchers get to see the fundamental processes of chemistry, drug development and life itself in a new light.

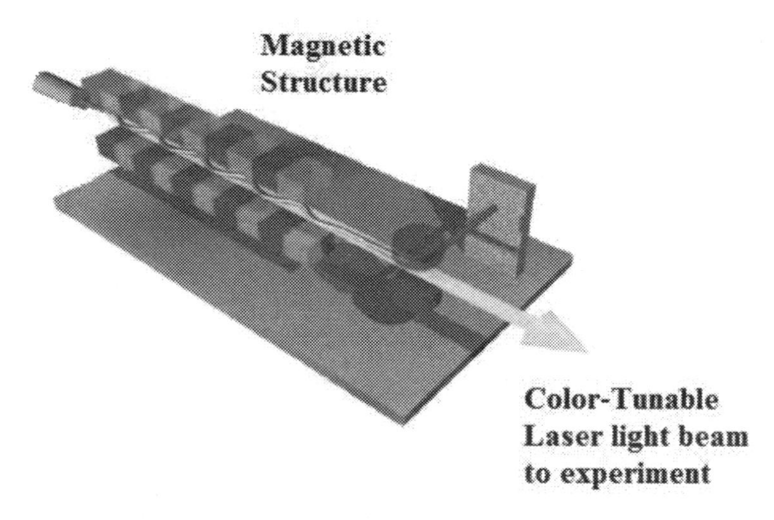

Fig.6.8 Powerful X-ray laser light from electron beams

8- <u>Accelerators for improving materials' surfaces</u>

An accelerator-based manufacturing technique called ion implantation modifies semiconductors' electrical properties precisely and cost-effectively, leading to better, cheaper electronics. Ions are atoms with positive or negative charge. Implanting the ions in metal surfaces means greater toughness and less corrosion which lead to longer life for the medical prostheses like artificial hips and in tools like drill bits, that means a longer working lifetime.

Chip manufacturers create integrated circuits by intentionally introducing impurities (boron or phosphorus ions) into silicon wafers. Using an accelerator they create a beam of high-energy boron or phosphorus ions, and then move the silicon wafers into the beam for implantation of the ions into the wafers.

9- <u>Accelerator-based neutron science yields payoffs</u>

Only negatively or positively charged particles can be accelerated. But accelerated particle beams can cause the release of neutrons. In turn, these electrically neutral particles can be formed into beams themselves, yielding practical payoffs.

The uncharged neutrons can go where charged particles can't, providing detailed snapshots of material structure and "movies" of molecules in motion. This neutron research improves a multitude of products, from medicine and food to electronics, cars, airplanes and bridges.

One particularly tantalizing neutron-research subject is superconductors, materials that conduct electricity with almost no energy loss. Neutron research shows how electromagnetic fields behave inside certain superconductors.

If superconductors can be developed for the electrical grid, a hydroelectric dam or a wind farm could provide cheaper power to distant cities.

10- <u>Portable accelerators</u>

The accelerator generates the neutrons by manipulating isotopes of hydrogen. It can discover oil deposits by detecting porosity, which shows the presence of liquid or gas. Electrical characteristics tell whether a liquid is water or oil.

Portable neutron generators (vacuum-tight accelerator) have other uses too. They can analyze metals and alloys, and they can detect explosives, drugs, or materials for nuclear weapons. See example of a mini-neutron tube in fig.

6.9, which is used for oil well analysis during drilling, as well as homeland security detection of explosives and fissile materials in luggage or cargo.

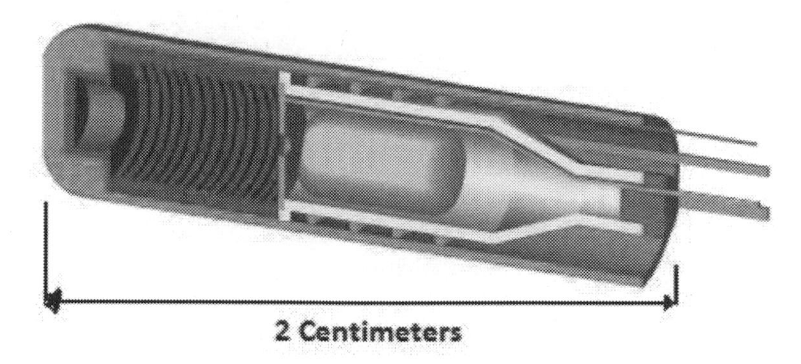

Fig. 6.9 A mini-neutron tube - a small deuteron-triton accelerator

11- Accelerators and Astronomy

- Accelerators are like microscopes. Microscopes reveal what's extremely small. Accelerators reveal information about what's millions of times smaller still.
- Accelerators are also like telescopes. Telescopes reveal the universe itself.
- Accelerators reveal information that helps astronomy.
- Accelerators are used to study nuclear processes that first occurred during the Big Bang and that continue in stars, novae and supernovae. This field is called nuclear astrophysics.
- Accelerator physicists work with nuclear astrophysicists to use high-intensity, low-energy beams to explore reactions at stellar energies.

The Facility for Rare Isotope Beams, under construction in USA, at Michigan State University, will provide intense beams of rare isotopes, short-lived atomic nuclei not normally found on earth. This facility will enable researchers to address questions such as: What is the origin of the elements we find in nature? Why do stars sometimes explode?

12- Accelerator on Museum

AGLAE, Accélérateur Grand Louvre d'Analyse Élémentaire in Paris, is the world's only accelerator facility fully dedicated to the study and investigation of works of art and archeological artifacts. It serves more than 1200 French museums. The 4-million-electronvolt proton beam delicately probes a large variety of materials: jewels, ceramics, glass, alloys, coins and statues, as well as paintings and drawings. These investigations provide information on the sources of the materials, the ancient formulas used to produce them, and the optimal ways to preserve these treasures.

The Story of Ishtar; In 1863, while excavating a tomb from the ancient civilization in Mesopotamia (200 BC – 200 AD), an amateur archeologist who was the French consul in Baghdad discovered this 5-inchtall alabaster figurine representing the goddess Ishtar. He donated it to the Louvre. Recently a Louvre curator asked the AGLAE team to analyze the figurine's red eyes and red navel. The inlays turned out to be exquisite rubies, a great mystery since rubies are only found in remote lands like India or Southeast Asia. Analysis of rubies with known provenance from Paris jewelers yielded trace-element fingerprints showing that Ishtar's rubies originated in Burma - testifying to an unreported trade network, perhaps by ship, between Babylon and Southeast Asia.

13- <u>Accelerators boost international cooperation</u>

CERN, the prestigious European Organization for Nuclear Research, and Fermilab, each with a long tradition of international cooperation, serve researchers who work cooperatively and represent scores of nations. These and many other accelerator laboratories link diverse societies and contribute to a culture of peace. Builders and users of accelerators and accelerator-driven light sources come together regularly at international conferences.

SESAME (Synchrotron-light for Experimental Science and Applications in the Middle East), a synchrotron light source is built in Al-Balqa, Jordan, set up under the auspices of UNESCO, is closely modeled on CERN with membership list includethe middle east countries: Bahrain, Cyprus, Egypt, Iran, Israel, Jordan, Pakistan, the Palestinian Authority and Turkey.

This advanced synchrotron will serve a wide spectrum of disciplines, ranging from biology and medical sciences through materials science, physics and chemistry to archeology. Sesame is a synchrotron – a large device that accelerates electrons around a circular tube of 130 metres in diameter, guided by magnets and other equipment, close to the speed of light. This generates radiation which is filtered and flows down beamlines –

essentially long pipes in which instruments are placed to collect the radiation and perform the various experiments.

Sesame's scientists plan to open the synchrotron with three main beamlines, though the project can house up to 20. The first is an X-ray beam which scientists say can be used to analyse soil samples and air particles, identifying contaminants in the environment, as well as, potentially, their sources, in a region suffering from high levels of pollution.

The second will be an infrared beamline, which will allow researchers to study living cells and tissue. Some preliminary tests at the centre have focused on studying the evolution of breast cancer cells, potentially opening avenues that would help with much earlier detection.

The last beamline, currently under construction, will be used in protein crystallography, a technique that would allow scientists, among other applications, to study in more depth the structure of viruses and develop drugs that are better able to target them.

14- Education

As accelerator science and technology have advanced, the use of accelerators which has been grown explosively. Accelerators are crucial to medicine and to the semiconductor industry and are likely to become even more central to many fields of research. We need to continue to educate accelerator scientists and support accelerator research centers. This field is one of the most effective for educating students who can then work productively in many fields and industries.

Because of the future promises on the developments and construction of accelerators facilities, the need for advancing the frontiers of knowledge on accelerator principles and functions are vitally required all over the world.

Worldwide accelerator-based science must have access to sufficient scientific and engineering talent with a broad array of technical skills. To meet this challenge, there should be more education and training for early-career scientists and engineers than ever.

6.2. Applications on Energy Levels

6.2.1. Low Energy Machines

Every day examples of particle accelerators are cathode ray tubes found in television sets and X-ray generators. These low-energy accelerators generators use a single pair of electrodes with a DC voltage of a few thousand volts between them. In an X-ray generator, the target itself is one of the electrodes. A low-energy particle accelerator called an ion implanter is used in the manufacture of integrated circuits.

The cathode ray tube (CRT) is a vacuum tube containing an electron gun (a source of electrons) and a fluorescent screen, with internal or external means to accelerate and deflect the electron beam, used to create images in the form of light emitted from the fluorescent screen. The image may represent electrical waveforms (oscilloscope), pictures (television, computer monitor), radar targets and others.

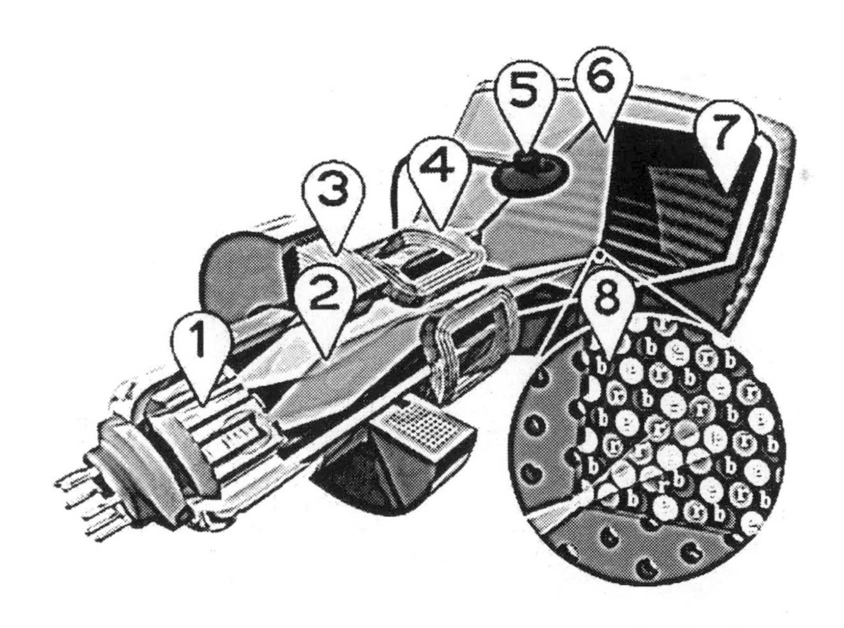

Fig.6.10 Cutaway rendering of a color CRT

1. Three Electron guns (for red, green, and blue phosphor dots), 2. Electron beams, 3. Focusing coils, 4. Deflection coils, 5. Anode connection, 6. Mask for separating beams for red, green, and blue part of displayed image, 7. Phosphor layer with red, green, and blue zones, 8. Close-up of the phosphor-coated inner side of the screen

Color CRTs have three separate electron guns (shadow mask-type) or electron guns that share some electrodes for all three beams (Sony Trinitron, and licensed versions)

The CRT uses an evacuated glass envelope which is large, deep, heavy, and relatively fragile. Display technologies without these disadvantages, such as flat plasma screens, liquid crystal displays, DLP, OLED displays have replaced CRTs in many applications and are becoming increasingly common as costs decline.

6.2.2. High Energy Machines

DC accelerator types capable of accelerating particles to speeds sufficient to cause nuclear reactions are Cockcroft-Walton generators or voltage multipliers, which convert AC to high voltage DC, or Van de Graaff generators that use static electricity carried by belts.

Fig.6.11 Beamlines leading from the Van de Graaf accelerator to various experiments, in the basement of the Jussieu Campus in Paris

In early particle accelerators a Cockcroft-Walton voltage multiplier was responsible for voltage multiplying. This piece of the accelerator was part of one of the early particle accelerators responsible for the development of the atomic bomb. Built in 1937 by Philips of Eindhoven it currently resides in the National Science Museum in London, England.

Fig.6.12 Cockcroft-Walton voltage multiplier

A Van de Graaff generator is an electrostatic generator which uses a moving belt to accumulate very high electrostatically stable voltages on a hollow metal globe on the top of the stand. The potential differences achieved in modern Van de Graaff generators can reach 5 megavolts. The Van de Graaff generator can be thought of as a constant-current source connected in parallel with a capacitor and a very large electrical resistance

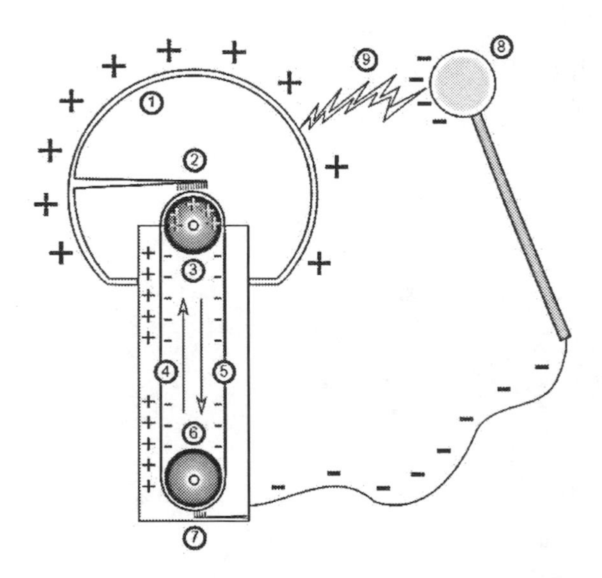

Fig.6.13 Schematic view of a classical Van de Graaff generator
1) hollow metal sphere, 2) upper electrode, 3) upper roller (metal), 4) side of the belt with positive charges, 5) opposite side of the belt with negative charges, 6) lower roller (for example in acrylic glass), 7) lower electrode (ground), 8) spherical device with negative charges, used to discharge the main sphere, 9) spark produced by the difference of potentials

The largest and most powerful particle accelerators, such as the Large Hadron Collider (LHC) the Relativistic Heavy Ion Collider (RHIC), and the Tevatron, are used for experimental particle physics.

A-The **Large Hadron Collider (LHC)**: LHC is the world's largest and highest-energy particle accelerator, intended to collide opposing particle beams, of either protons at an energy of 7 TeV per particle, or lead nuclei at an energy of 574 TeV per nucleus. The Large Hadron Collider was built by the European Organization for Nuclear Research (CERN) with the intention of testing various predictions of high-energy physics, including the existence of the hypothesized Higgs boson and of the large family of new particles predicted by supersymmetry. It lies in a tunnel 27 kilometers in circumference, as much as 175 meters beneath the Franco-Swiss border near Geneva, Switzerland. It is funded by and built in collaboration with over 10,000 scientists and engineers from over 100 countries as well as hundreds of universities and laboratories.

On 10 September 2008, the proton beams were successfully circulated in the main ring of the LHC for the first time. On 19 September 2008, the operations were halted due to a serious fault between two superconducting bending magnets. Due to the time required to repair the resulting damage and to add additional safety features, the LHC is scheduled to be operational in mid-November 2009.

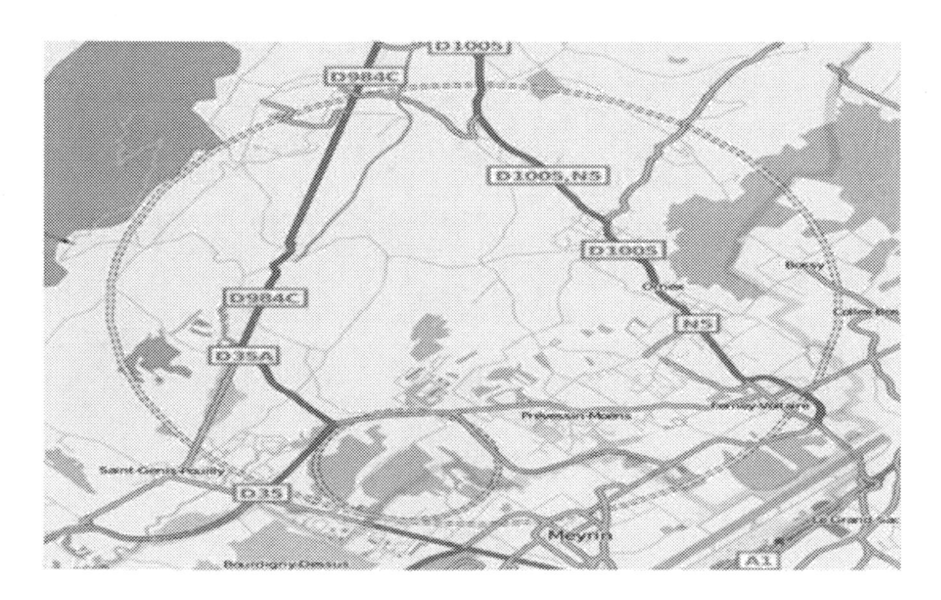

Fig. 6.14 Map of the Large Hadron Collider at CERN

B-<u>The Relativistic Heavy Ion Collider (RHIC)</u>: RHIC is a heavy-ion collider located at and operated by Brookhaven National Laboratory (BNL) in Upton, New York.

At present, RHIC is the most powerful heavy-ion collider in the world, although the LHC is expected to collide ions at higher energies in late 2009. It is also distinctive in its capability to collide spin-polarized protons.

Fig.6.15 The RHIC at Brookhaven National Laboratory

Some of the superconducting magnets were manufactured by Northrop Grumman Corp. at Bethpage, New York. Note especially the second, independent ring behind the blue striped one. Barely visible and between the white and red pipes on the left wall, is the orange Crash Cord, which should be used to stop the beam in the case a person is still left in the tunnel.

C-<u>Tevatron</u>: It is a circular particle accelerator at the Fermi National Accelerator Laboratory in Batavia, Illinois and remains the highest energy particle collider in the world until collisions begin at the Large Hadron Collider (LHC). The Tevatron is a synchrotron that accelerates protons and antiprotons in a 6.28 km ring to energies of up to 1 TeV, hence the name.

Fig. 6.16 The Tevatron (background) and *Main Injector* rings

The Tevatron was completed in 1983 at a cost of $120 million and has been regularly upgraded since then. (The 'Energy Doubler', as it was known then, produced its first accelerated beam - 512 GeV - on July 3, 1983.) The Main Injector was the most substantial addition, built over five years from 1994 at a cost of $290 million.

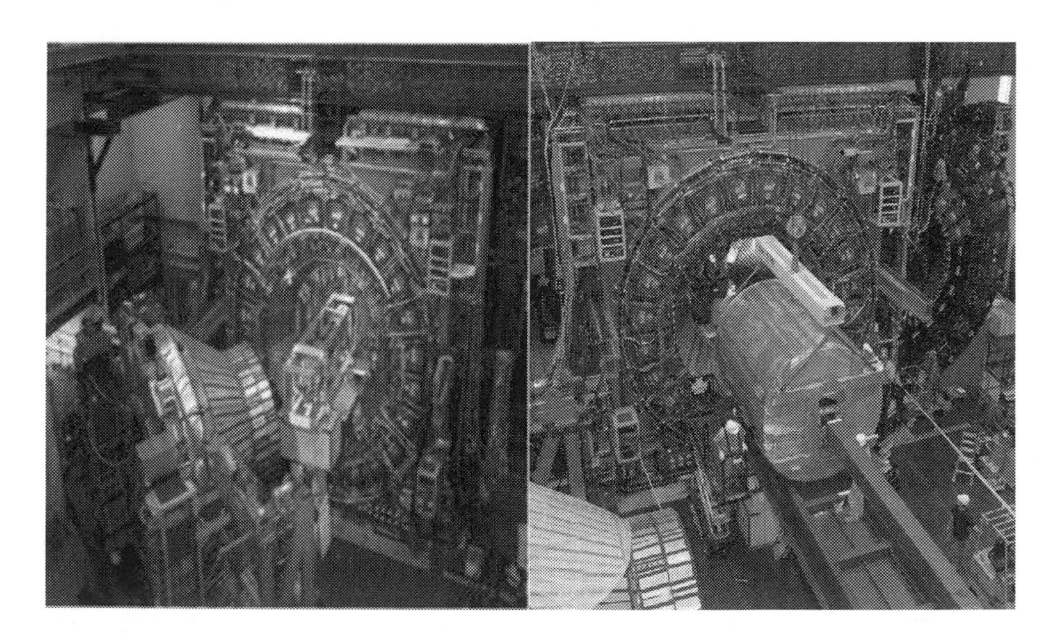

Fig. 6.17 Tevatron Accelerator Fermilab

When the Tevatron opened in 1983 at Fermi National Laboratory, outside Chicago, it was the world's most powerful particle accelerator, designed to smash protons and antiprotons together in order to see what makes up the universe. After its discovery in 1995 of the top quark, the most massive subatomic particle, the Tevatron was the toast of particle physics, and made Fermilab the center of the global high energy physics community. From there, it would begin the search for the Higgs. It was a search that might have continued in Texas, at the Superconducting Supercollider, a massive successor machine, if that project hadn't been cancelled in 1993. When the LHC came online in Switzerland in 2009, scientists at the Tevatron's two experiments, CDF and DZero, would join physicists at their European counterpart in the meticulous hunt. At times, it seemed that America's biggest science experiment might even beat the LHC to the punch.

Researchers at Fermilab just broke ground on a new research and industrialization facility called the Illinois Accelerator Research Center. When it's finished sometime in 2013, scientists and private businesses will use the new 42,000-square-foot building to develop new accelerator technology that can be used for industry, medicine and national security purposes. In addition to a new building, the project will use space once dedicated to the now-silent Tevatron.

Scientists from Fermilab and the Argonne National Laboratory will use the facility for research, but Fermilab says a major focus will be developing partnerships with private industry. Although we usually think of the Tevatron and the Large Hadron Collider, there are something like 30,000 other accelerators around the world, working on projects from paleontology and biology to food packaging.

A proposed new high-intensity proton accelerator, would spark the development of new accelerator tech, which could also feed into the IARC. A new neutrino project called LBNE would use the world's highest-intensity neutrino beam to determine why matter prevails over antimatter in the universe, and a new muon experiment called Mu2e will study whether muons can change into electrons, the way quarks and neutrinos can switch teams.

6.3. Usages Technologies and limitations

6.3.1 Cyclotrons

For several decades, cyclotrons were the best source of high-energy beams for nuclear physics experiments; several cyclotrons are still in use for this type of research. The results enable the calculation of various properties, such as the mean spacing between atoms and the creation of various collision products. Subsequent chemical and particle analysis of the target material may give insight into nuclear transmutation of the elements used in the target.

Cyclotrons can be used in particle therapy to treat cancer. Ion beams from cyclotrons can be used, as in proton therapy, to penetrate the body and kill tumors by radiation damage, while minimizing damage to healthy tissue along their path. Cyclotron beams can be used to bombard other atoms to produce short-lived positron-emitting isotopes suitable for PET imaging.

More recently cyclotrons currently installed at hospitals for particle therapy have been retrofitted to enable them to produce technetium-99.[10] Technetium-99 is a diagnostic isotope in short supply due to difficulties at Canada's Chalk River facility.

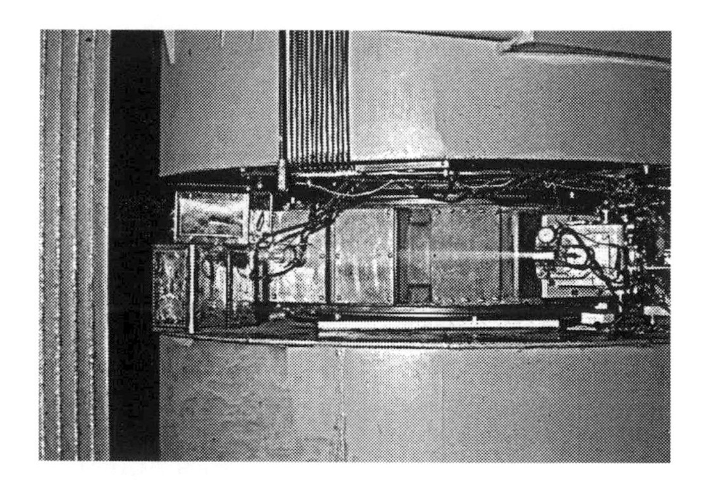

Fig. 6.18 A 60-inch cyclotron, circa 1939.
The beam of accelerated ions (likely protons or deuterons) escaping the accelerator and ionizing the surrounding air causing a blue glow

The cyclotron was an improvement over the linear accelerators (*linac*s) that were available when it was invented, being more cost- and space-effective due to the iterated interaction of the particles with the accelerating field. In the 1920s, it was not possible to generate the high power, high-frequency radio waves which are used in modern linacs (generated by klystrons). As such, impractically long linac structures were required for higher-energy particles. The compactness of the cyclotron reduces other costs as well, such as foundations, radiation shielding, and the enclosing building.

Cyclotrons have a single electrical driver, which saves both money and power. Furthermore, cyclotrons are able to produce a continuous stream of particles at the target, so the average power passed from a particle beam into a target is relatively high.

Fig. 6.19 The magnet portion of a 27" cyclotron

The grey object is the upper pole piece, routing the magnetic field in two loops through a similar part below. The white canisters held conductive coils to generate the magnetic field. The D electrodes are contained in a vacuum chamber that was inserted in the central field gap.

The spiral path of the cyclotron beam can only "sync up" with klystron-type (constant frequency) voltage sources if the accelerated particles are approximately obeying Newton's Laws of Motion. If the particles become fast enough that relativistic effects become important, the beam becomes out of phase with the oscillating electric field, and cannot receive any additional acceleration. The classical cyclotron is therefore only capable of accelerating particles up to a few percent of the speed of light. To accommodate increased mass the magnetic field may be modified by appropriately shaping the pole pieces as in the isochronous cyclotrons, operating in a pulsed mode and changing the frequency applied to the dees as in the synchrocyclotrons, either of which is limited by the diminishing cost effectiveness of making larger machines. Cost limitations have been overcome by employing the more complex synchrotron or modern, klystron-driven linear accelerators, both of which have the advantage of scalability, offering more power within an improved cost structure as the machines are made larger.

The world's largest cyclotron is at the RIKEN laboratory in Japan. Called the SRC, for Superconducting Ring Cyclotron, it has 6 separated superconducting sectors, and is 19 m in diameter and 8 m high. Built to accelerate heavy ions, its maximum magnetic field is 3.8 tesla, yielding a bending ability of 8 tesla-meters. The total weight of the cyclotron is 8,300 tones. It has accelerated uranium ions to 345 MeV per atomic mass unit.

TRIUMF, Canada's national laboratory for nuclear and particle physics, houses one of the world's largest cyclotrons. The 18 m diameter, 4,000 tonne main magnet produces a field of 0.46 T while a 23 MHz 94 kV electric field is used to accelerate the 300 μA beam. Its large size is partly a result of using negative hydrogen ions rather than protons. The advantage is that extraction is simpler; multi-energy, multi-beams can be extracted by inserting thin carbon stripping foils at appropriate radii. The disadvantage is that the magnetic field is limited: a magnetic field larger than about 0.5 tesla can prematurely strip the loosely-bound second electron. TRIUMF is run by a consortium of sixteen Canadian universities and is located at the University of British Columbia, Vancouver, Canada.

The spiraling of electrons in a cylindrical vacuum chamber within a transverse magnetic field is also employed in the magnetron, a device for producing high frequency radio waves (microwaves).

Fig. 6.20 A French cyclotron, produced in Zurich, Switzerland in 1937

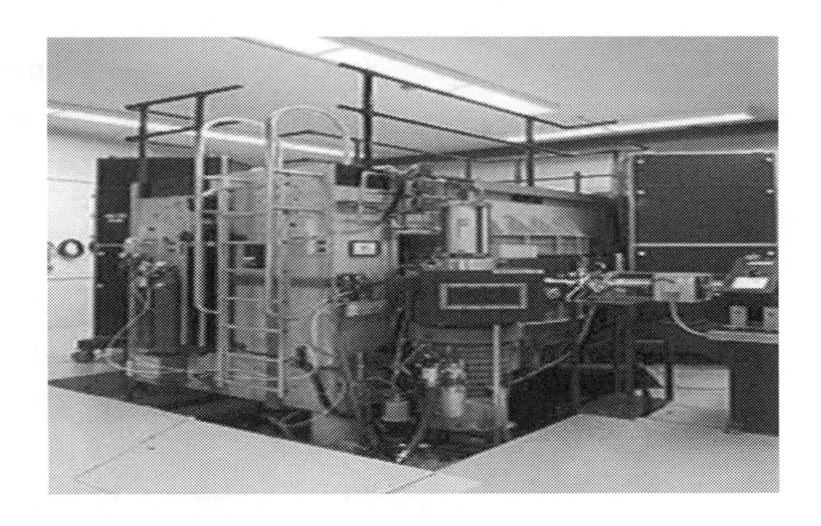

Fig. 6.21 A modern cyclotron for radiation therapy

6.3.2 Synchrotron

The synchrotron moves the particles through a path of constant radius, allowing it to be made as a pipe and so of much larger radius than is practical with the cyclotron and synchrocyclotron. The larger radius allows the use of numerous magnets, each of which imparts angular momentum and so allows particles of higher velocity (mass) to be kept within the bounds of the evacuated pipe. The magnetic field strength of each of the bending magnets is increased as the particles gain energy in order to keep the bending angle constant.

There are several circular accelerators at Fermi National Accelerator Laboratory. Particles pass through each cavity many times as they circulate around the ring, each time receiving a small acceleration, or increase in energy. When either the energy or the field strength changes so does the radius of the path of the particles.

Thus, as the particles increase in energy the strength of the magnetic field that is used to steer them must be changed with each turn to keep the particles moving in the same ring. The change in magnetic field must be carefully synchronized to the change in energy or the beam will be lost. Hence the name "synchrotron".

Fig. 6.22 Aerial photo of the Tevatron at Fermilab
(The main accelerator is the ring above; the one below (about one-third the diameter, despite appearances) is for preliminary acceleration, beam cooling and storage, etc.)

The range of energies over which particles can be accelerated in a single ring is determined by the range of field strength available with high precision from a particular set of magnets. To reach high energies, physicists sometimes use a sequence of different size synchrotrons, each one feeding the next bigger one. Particles are often pre-accelerated before entering the first ring, using a small linear accelerator or other device.

To reach still higher energies, with relativistic mass approaching or exceeding the rest mass of the particles (for protons, billions of electron volts GeV), it is necessary to use a synchrotron. This is an accelerator in which the particles are accelerated in a ring of constant radius. An immediate advantage over cyclotrons is that the magnetic field need only be present over the actual region of the particle orbits, which is very much narrower than the diameter of the ring. (The largest cyclotron built in the US had a 184 in dia magnet pole, whereas the diameter of the LEP and LHC is nearly 10 km. The aperture of the beam of the latter is of the order of centimeters.)

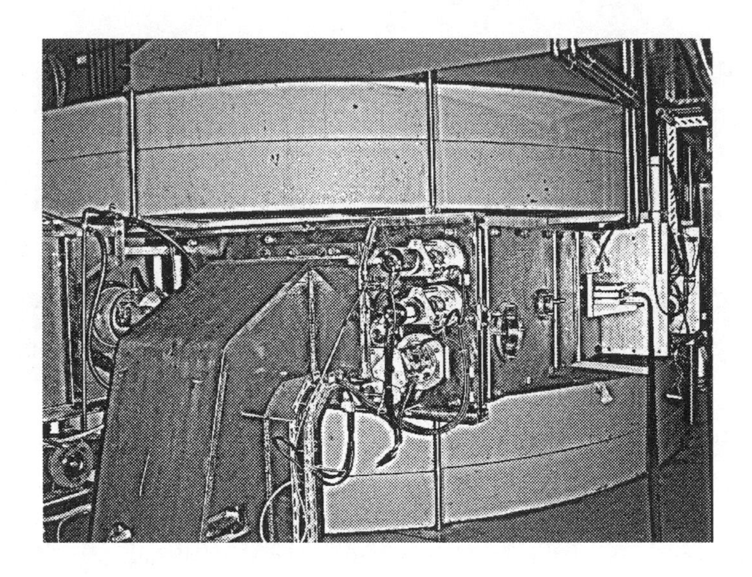

Fig. 6.23 A magnet in the Synchrocyclotron at the Orsay proton therapy Center

However, since the particle momentum increases during acceleration, it is necessary to turn up the magnetic field B in proportion to maintain constant curvature of the orbit. In consequence synchrotrons cannot accelerate particles continuously, as cyclotrons can, but must operate cyclically, supplying particles in bunches, which are delivered to a target or an external beam in beam "spills" typically every few seconds.

Since high energy synchrotrons do most of their work on particles that are already traveling at nearly the speed of light *c*, the time to complete one orbit of the ring is nearly constant, as is the frequency of the RF cavity resonators used to drive the acceleration.

Note also a further point about modern synchrotrons: because the beam aperture is small and the magnetic field does not cover the entire area of the particle orbit as it does for a cyclotron, several necessary functions can be separated. Instead of one huge magnet, one has a line of hundreds of bending magnets, enclosing (or enclosed by) vacuum connecting pipes. The design of synchrotrons was revolutionized in the early 1950s with the discovery of the strong focusing concept. The focusing of the beam is handled independently by specialized quadrupole magnets, while the acceleration itself is accomplished in separate RF sections, rather similar to short linear accelerators. Also, there is no necessity that cyclic machines are circular, but rather the beam pipe may have straight sections between magnets where beams may collide. This has developed into an entire separate subject, called "beam physics" or "beam optics".

More complex modern synchrotrons such as the Tevatron, LEP, and LHC may deliver the particle bunches into storage rings of magnets with constant B, where they can continue to orbit for long periods for experimentation or further acceleration. The highest-energy machines such as the Tevatron and LHC are actually accelerator complexes, with a cascade of specialized elements in series, including linear accelerators for initial beam creation, one or more low energy synchrotrons to reach intermediate energy, storage rings where beams can be accumulated or "cooled" (reducing the magnet aperture required and permitting tighter focusing; see beam cooling), and a last large ring for final acceleration and experimentation.

Fig. 6.24 Modern industrial-scale synchrotrons (Soleil near Paris)

One of the early large synchrotrons, now retired, is the Bevatron, constructed in 1950 at the Lawrence Berkeley Laboratory. The name of this proton accelerator comes from its power, in the range of 6.3 GeV (then called BeV for billion electron volts; the name predates the adoption of the SI prefix giga-). A number of transuranium elements, unseen in the natural world, were first created with this machine. This site is also the location of one of the first large bubble chambers used to examine the results of the atomic collisions produced here.

Another early large synchrotron is the Cosmotron built at Brookhaven National Laboratory which reached 3.3 GeV in 1953.

The main synchrotron applications include the following:

A. As part of colliders

Until August 2008, the highest energy collider in the world was the Tevatron, at the Fermi National Accelerator Laboratory, in the United States. It accelerated protons and antiprotons to slightly less than 1 TeV of kinetic energy and collided them together. The Large Hadron Collider (LHC), which has been built at the European Laboratory for High Energy Physics (CERN), has roughly seven times this energy (so proton-proton collisions occur at roughly 14 TeV). It is housed in the 27 km tunnel which formerly housed the Large Electron Positron (LEP) collider, so it will maintain the claim as the largest scientific device ever built. The LHC will also accelerate heavy ions (such as lead) up to an energy of 1.15 PeV.

The largest device of this type seriously proposed was the Superconducting Super Collider (SSC), which was to be built in the United States. This design, like others, used superconducting magnets which allow more intense magnetic fields to be created without the limitations of core saturation. While construction was begun, the project was cancelled in 1994, citing excessive budget overruns — this was due to naïve cost estimation and economic management issues rather than any basic engineering flaws. It can also be argued that the end of the Cold War resulted in a change of scientific funding priorities that contributed to its ultimate cancellation. However, the tunnel built for its placement still remains, although empty. While there is still potential for yet more powerful proton and heavy particle cyclic accelerators, it appears that the next step up in electron beam energy must avoid losses due to synchrotron radiation. This will require a return to the linear accelerator, but with devices significantly longer than those currently in use. There is at present a major effort to design and build the International Linear Collider (ILC), which will consist of two opposing linear accelerators, one for electrons and one for positrons. These will collide at a total center of mass energy of 0.5 TeV.

B. As part of synchrotron light sources

Synchrotron radiation also has a wide range of applications and many 2nd and 3rd generation synchrotrons have been built especially to harness it. The largest of those 3rd generation synchrotron light sources are the European Synchrotron Radiation Facility (ESRF) in Grenoble, France, the Advanced

Photon Source (APS) near Chicago, USA, and SPring-8 in Japan, accelerating electrons up to 6, 7 and 8 GeV, respectively.

Synchrotrons which are useful for cutting edge research are large machines, costing tens or hundreds of millions of dollars to construct, and each beamline (there may bje 20 to 50 at a large synchrotron) costs another two or three million dollars on average. These installations are mostly built by the science funding agencies of governments of developed countries, or by collaborations between several countries in a region, and operated as infrastructure facilities available to scientists from universities and research organisations throughout the country, region, or world. More compact models, however, have been developed, such as the Compact Light Source.

C. Further Applications of Synchrotrons

- Life sciences: protein and large-molecule crystallography
- LIGA based microfabrication
- Drug discovery and research
- "Burning" computer chip designs into metal wafers
- Analysing chemicals to determine their composition
- Observing the reaction of living cells to drugs
- Inorganic material crystallography and microanalysis
- Fluorescence studies
- Semiconductor material analysis and structural studies
- Geological material analysis
- Medical imaging
- Proton therapy to treat some forms of cancer

6.3.3 DESY

The German electron synchrotron DESY (Deutsches Elektronen-Synchrotron) in Hamburg is a center for research into the structure of matter. DESY offers a powerful high-energy accelerator, the hadron electron ring facility HERA (Hadron-Elektron-Ring-Anlage) The underground HERA tunnel has a circumference of 6,336 meters and accommodates two superimposed storage rings. At two points, the particles accelerated to high energies are directed towards each other. During these collisions, new particles are generated allowing statements on the structure of matter.

DESY uses the double ring storage DORIS and the accelerator PETRA (Positron-Elektron-Tandem-Ring-Anlage) as intense synchrotron radiation sources for a broad spectrum of tests with electromagnetic radiation. The experiments are coordinated, prepared and carried out by the synchrotron radiation laboratory in Hamburg HASYLAB (Hamburger Synchrotron strahlungs Lab.). With international cooperation, DESY is planning the future project TESLA, a 33 kilometer long super conducting electron-positron linear collider. 3,400 scientists from 35 nations are involved in the DESY experiments(30% of them carry out research in particle physics at HERA, other carry out research with synchrotron radiation at HASYLAB.

DESY operates in two locations. The primary location is in a suburb of Hamburg. In 1992, DESY expanded to a second site in Zeuthen near Berlin. The DESY Hamburg site is located in the suburb Bahrenfeld, west of the city. Most of DESY's research in high energy physics with elementary particles has been taking place here since 1960. The site is bounded by the ring of the PETRA particle accelerator and part of the larger HERA (Hadron Elektron Ring Anlage) ring. Besides these accelerators there is also a free electron laser called XFEL under development. This project is meant to secure DESY's future place among the top research centers of the world.

Following German reunification, the DESY expanded to a second site. The former Institute for High Energy Physics (German: Institut für Hochenergiephysik IfH) in Zeuthen, southeast of Berlin, was the high energy physics laboratory of the German Democratic Republic and belonged to the Academy of Sciences of the GDR. The institute was merged with the DESY on 1 January 1992.

Fig. 6.25 Segment of a particle accelerator at DESY

DESY's accelerators were not built all at once, but rather were added one by one to meet the growing demand of the scientists for higher and higher energies to gain more insight into particle structures. In the course of the construction of new accelerators the older ones were converted to pre-accelerators or to sources for synchrotron radiation for laboratories with new research tasks (for example for HASYLAB).

After the shutdown of the accelerator HERA in 2007, DESY's most important facilities are the high intensity source for synchrotron radiation, PETRA III, the synchrotron-research lab HASYLAB, the free-electron laser FLASH (previously called VUV-FEL), and the test facility for the planned European XFEL. The development of the different facilities will be described chronologically in the following sections:

1- <u>DESY</u>

The construction of the first particle accelerator **DESY** (Deutsches Elektronen Synchrotron, "German Electron Synchrotron") began in 1960. At that time it was the biggest facility of this kind and was able to accelerate electrons to 7.4 GeV. On 1 January 1964 the first electrons were accelerated in the synchrotron and the research on elementary particles began.

The international attention first focused on DESY in 1966 due to its contribution to the validation of quantum electrodynamics, which was achieved with results from the accelerator. In the following decade DESY established itself as a center of excellence for the development and operation of high-energy accelerators.

The synchrotron radiation, which comes up as a side effect, was first used in 1967 for absorption measurements. For the arising spectrum there had not been any conventional radiation sources beforehand. The European Molecular Biology Laboratory EMBL made use of the possibilities that arose with the new technology and in 1972 established a permanent branch at DESY with the aim of analyzing the structure of biological molecules by means of synchrotron radiation.

The electron-synchrotron DESY II and the proton-synchrotron DESY III were taken into operation in 1987 and 1988 respectively as pre-accelerators for HERA.

2- DORIS III

DORIS (**Do**ppel-**Ri**ng-**S**peicher, "double-ring storage"), built between 1969 and 1974, was DESY's second circular accelerator and its first storage ring with a circumference of nearly 300 m. Constructed as an electron-positron storage ring, one could conduct collision-experiments with electrons and their antiparticles at energies of 3.5 GeV per beam. In 1978 the energy of the beams was raised to 5 GeV each.

With evidence of the "excited charmonium states" DORIS made an important contribution to the process of proving the existence of heavy quarks. In the same year there were the first tests of X-ray lithography at DESY, a procedure which was later refined to X-ray depth lithography.

In 1987 the ARGUS detector of the DORIS storage ring was the first place where the conversion of a B-meson into its antiparticle, the anti-B-meson was observed. From this one could conclude that it was possible, for the second-heaviest quark - the bottom-quark - under certain circumstances to convert into a different quark. One could also conclude from this that the unknown sixth quark - the top quark - had to possess a huge mass. The top quark was found eventually in 1995 at the Fermilab in the USA.

After the commissioning of HASYLAB in 1980 the synchrotron radiation, which was generated at DORIS as a byproduct, was used for research there. While in the beginning DORIS was used only ⅓ of the time as a radiation source, from 1993 on the storage-ring solely served that purpose under the name DORIS III. In order to achieve more intense and controllable radiation, DORIS was upgraded in 1984 with Wigglers and Undulators. By means of a special array of magnets the accelerated electrons could now be brought onto a slalom course. By this the intensity of the emitted synchrotron radiation was increased a hundredfold in comparison to conventional storage ring systems.

DORIS III provides 33 photon beamlines, where 44 instruments are operated in circulation. The overall beam time per year amounts to 8 to 10 months.

3- OLYMPUS

The former site of ARGUS in DORIS is now the location of the OLYMPUS experiment, which began installation in 2010. OLYMPUS uses the toroidal magnet and pair of drift chambers from the MIT-Bates BLAST experiment along with refurbished time-of-flight detectors and multiple luminosity monitoring systems. OLYMPUS measures the positron-proton to electron-proton cross section ratio to precisely determine the size of two-photon exchange in elastic ep scattering. This may help resolve the proton form factor discrepancy between recent measurements made using polarization techniques and ones using the Rosenbluth separation method. OLYMPUS began taking data in early 2012.

4- PETRA II

PETRA (Positron-Elecctron-Tandem-Ring-Anlayzer) was built between 1975 and 1978. At the time of its construction it was the biggest storage ring of its kind and still is DESY's second largest synchrotron after HERA. PETRA originally served for research on elementary particles. The discovery of the gluon, the carrier particle of the strong nuclear force, in 1979 is counted as one of the biggest successes. PETRA can accelerate electrons and positrons to 19 GeV.

Research at PETRA lead to an intensified international use of the facilities at DESY. Scientists from China, England, France, Israel, the Netherlands,

Norway and the USA participated in the first experiments at PETRA alongside many German colleagues.

In 1990 the facility was taken into operation under the name PETRA II as a pre-accelerator for protons and electrons/positrons for the new particle accelerator HERA. In March 1995, PETRA II was equipped with undulators to create greater amounts of synchrotron radiation with higher energies, especially in the X-ray part of the spectrum. Since then PETRA serves HASYLAB as a source of high-energy synchrotron radiation and for this purpose possesses three test experimental areas. Positrons are accelerated to up to 12 GeV nowadays.

5- Petra III

Petra III is the third incarnation for the PETRA storage ring operating a regular user programme as the most brilliant storage ring based X-ray source worldwide since August 2010. The accelerator produces a particle energy of 6 GeV.

6- HASYLAB

The HASYLAB ("Hamburg Synchrotron radiation Laboratory") is used for research with synchrotron radiation at DESY. It was opened in 1980 with 15 experimental areas (today there are 42). The laboratory adjoins to the storage ring DORIS in order to be able to use the generated synchrotron radiation for its research. While in the beginning DORIS served only one third of the time as a radiation source for HASYLAB, since 1993 all its running time is available for experiments with synchrotron radiation. On top of the 42 experimental areas DORIS provides, there are also three test experimental areas available for experiments with high-energy radiation generated with the storage ring PETRA.

Fig. 6.26 The ARGUS detector at DESY

After the upgrade of DORIS with the first wigglers, which produced far more intense radiation, the first Moessbauer spectrum acquired by means of synchrotron radiation was recorded at HASYLAB in 1984.

In 1985 the development of more advanced X-ray technology made it possible to bring to light the structure of the influenza virus. In the following year researchers at HASYLAB were the first to successfully make the attempt of exciting singular grid oscillations in solid bodies. Thus it was possible to conduct analyses of elastic materials, which were possible prior to this only with nuclear reactors via neutron scattering.

In 1987 the workgroup for structural molecular biology of the Max Planck Society founded a permanent branch at HASYLAB. It uses synchrotron radiation to study the structure of ribosomes.

Nowadays many national and foreign groups of researchers conduct their experiments at HASYLAB: All in all 1900 scientists participate in the work. On the whole the spectrum of the research ranges from fundamental research to experiments in physics, material science, chemistry, molecular biology, geology and medicine to industrial cooperations.

One example is OSRAM, which since recently uses HASYLAB to study the filaments of their light bulbs. The gained insights helped to notably increase the life span of the lamps in certain fields of application.

In addition researchers at HASYLAB analysed among other things minuscule impurities in silicone for computer chips, the way catalysators work, the microscopic properties of materials and the structure of protein molecules.

7- <u>HERA</u>

HERA (**H**adron-**E**lektron-**R**ing-**A**nlage, "Hadron-Electron-Ring-Facility") was DESY's largest synchrotron and storage ring, with a circumference of 6336 meters. The construction of the subterranean facility began in 1984 and was an one of the first really internationally financed projects of this magnitude. In addition to Germany 11 further countries participated in the development of HERA. HERA project many international facilities consented to already help with the construction. All in all more than 45 institutes and 320 corporations participated with donations of money and/or materials in the construction of the facility, more than 20% of the costs were carried by foreign institutions. The accelerator began operation on November 8, 1990 and the first two experiments started taking data in 1992. HERA was mainly used to study the structure of protons and the properties of quarks. It was closed down June 30 in 2007.

HERA was the only accelerator in the world that was able to collide protons with either electrons or positrons. To make this possible HERA used mainly superconducting magnets, which was also a world first. At HERA it was possible to study the structure of protons up to 30 times more accurately than before. The resolution covered structures 1/1000 of the proton in size. In the years to come there were made a lot of discoveries concerning the composition of protons from quarks and gluons.

HERA's tunnels run 10 to 25 meters below ground level and have an inner diameter of 5.2 meters. For the construction the same technology was used as for the construction of subway tunnels. Two circular particle accelerators run inside the tube. One accelerated electrons to energies of 27.5 GeV the other one protons to energies of 920 GeV in the opposite direction. Both beams completed their circle nearly at the speed of light, making approximately 47 000 revolutions per second.

At two places of the ring the electron and the proton beam could be brought to collision. In the process electrons or positrons are scattered at the constituents of the protons, the quarks. The products of these particle collisions, the scattered lepton and the quarks, which are produced by the fragmentation of the proton, were registered in huge detectors. In addition to the two collision zones there are two more interaction zones. All four zones are placed in big subterraneous halls. A different international group of researchers were at work in each hall. These groups developed, constructed and run house-high, complex measurement devices in many years of cooperative work and evaluate enormous amounts of data.

The experiments in the four halls are as following:

1) H1 Experiment

H1 is a universal detector for the collision of electrons and protons and was located in DESY's HERA-Hall North. It had been active since 1992, measured 12 m × 10 m × 15 m and weighs 2 800 tons.

It was designed for the decryption of the inner structure of the proton, the exploration of the strong interaction as well as the search for new kinds of matter and unexpected phenomena in particle physics.

Following the example of HERA, many scientific projects of a large scale are financed jointly by several states. By now this model is established and international cooperation is pretty common with the construction of those facilities.

2) ZEUS Experiment

ZEUS is like H1 a detector for electron-proton collisions and was located in HERA-Hall South. Built in 1992 it measured 12 m × 11 m × 20 m and weighs 3600 tons. Its tasks resemble H1's.

3) HERA-B Experiment

HERA-B was an experiment in HERA-Hall West which collected data from 1999 to February 2003. By using HERA's proton beam, researchers at HERA-B conducted experiments on heavy quarks. It measured 8 m × 20 m × 9 m and weighed 1 000 tons.

4) HERMES Experiment

The HERMES experiment in HERA-Hall East was taken into operation in 1995. HERA's longitudinally polarised electron beam was used for the exploration of the spin structure of nucleons. For this purpose the electrons were scattered at energies of 27.5 GeV at an internal gas target. This target and the detector itself were designed especially with a view to spin polarised physics. It measured 3.5 m × 8 m × 5 m and weighs 400 tons.

8- FLASH

FLASH (**F**ree-electron -**LAS**er in **H**amburg) is a superconducting linear accelerator with a free electron laser for radiation in the vacuum-ultraviolet and soft X-ray range of the spectrum. It originated from the TTF (TESLA Test Facility), which was built in 1997 to test the technology that was to be used in the planned linear collider TESLA, a project which was replaced by the ILC (International Linear Collider). For this purpose the TTF was enlarged from 100 m to 260 m.

At the VUV-FEL technology for the future-project European XFEL is tested as well as for the ILC. Five test experimental areas have been in use since the commissioning of the facility in 2004.

9- European XFEL

The European x-ray free electron laser (European XFEL) is an X-ray laser currently under construction. It is a European project in collaboration with DESY and is as of 2012 slated to be operational by the end of 2015. The 3.4 km long tunnel will contain an 2.1 km long superconducting linear accelerator were electrons will be accelerated to an energy of up to 17.5 GeV. It will produce extremely short and powerful X-ray flashes which will have many applications. Construction of the tunnels was completed in summer 2012.

10- Further Accelerators

In addition to the larger ones, there are also several smaller particle accelerators which serve mostly as pre-accelerators for PETRA and HERA. Among these are the linear accelerators LINAC I (operated from 1964 to 1991 for electrons), LINAC II (operated since 1969 for positrons) and LINAC III (operated since 1988 as a pre-accelerator for protons for HERA).

7

GENERATION OF HIGHER ENERGIES

7.1. Background

At present the highest energy accelerators are all circular colliders, but it is likely that limits have been reached in respect of compensating for synchrotron radiation losses for electron accelerators, and the next generation will probably be linear accelerators 10 times the current length. An example of such a next generation electron accelerator is the 40 km long International Linear Collider ILC, due to be constructed between 2015-2020.

The **ILC** is a proposed linear particle accelerator. It is planned to have a collision energy of 500 GeV initially, and, if approved after the project has published its Technical Design Report, planned for 2012, could be completed in the late 2010s. A later upgrade to 1000 GeV is possible. The host country for the accelerator has not yet been chosen. Studies for a competing project called CLIC are also underway; it seems unlikely that both machines will be built.

The ILC would collide electrons with positrons. It will be between 30 km and 50 km (19-31 mi.) long, more than 10 times as long as the 50 GeV Stanford Linear Accelerator, the longest existing linear particle accelerator.

In the past century, physicists have explored smaller and smaller scales, cataloguing and understanding the fundamental components of the universe, trying to explain the origin of mass and probing the theory of extra dimensions. And in recent years, experiments and observations have pointed to evidence that we can only account for a surprising five percent of the universe.

Scientists believe that the remaining 95 percent is a mysterious dark matter and dark energy, revealing a universe far stranger and more wonderful than they ever suspected. The global particle physics community agrees that a precision machine—the proposed International Linear Collider—will answer these questions about what the universe is made of and provide exciting new insights into how it works. Using unprecedented technology, discoveries are within reach that could stretch our imagination with new forms of matter, new forces of nature, new dimensions of space and time and bring into focus Albert Einstein's vision of an ultimate unified theory.

As of 2005, it is believed that plasma wakefield acceleration in the form of electron-beam 'afterburners' and standalone laser pulsers will provide dramatic increases in efficiency within two to three decades. In plasma wakefield accelerators, the beam cavity is filled with a plasma (rather than vacuum). A short pulse of electrons or laser light either constitutes or immediately trails the particles that are being accelerated. The pulse disrupts the plasma, causing the charged particles in the plasma to integrate into and move toward the rear of the bunch of particles that are being accelerated. This process transfers energy to the particle bunch, accelerating it further, and continues as long as the pulse is coherent.

Energy gradients as steep as 200 GeV/m have been achieved over millimeter-scale distances using laser pulsers and gradients approaching 1 GeV/m are being produced on the multi-centimeter-scale with electron-beam systems, in contrast to a limit of about 0.1 GeV/m for radio-frequency acceleration alone. Existing electron accelerators such as SLAC could use electron-beam afterburners to greatly increase the energy of their particle beams, at the cost of beam intensity. Electron systems in general can provide tightly collimated, reliable beams; laser systems may offer more power and compactness. Thus, plasma Wakefield accelerators could be used-if technical issues can be resolved - to both increase the maximum energy of the largest accelerators and to bring high energies into university laboratories and medical centers.

7.2. Black hole production

In the future, the possibility of **Black Hole (BH)** production at the highest energy accelerators may arise if certain predictions of superstring theory are accurate. This and other exotic possibilities have led to public safety concerns that have been widely reported in connection with the LHC, which began operation in 2008. The various possible dangerous scenarios have been assessed as presenting "no conceivable danger" in the latest risk assessment produced by the LHC Safety Assessment Group. If they are produced, it is proposed that BHs would evaporate extremely quickly via the unconfirmed theory of Bekenstein-Hawking radiation. If colliders can produce BHs, cosmic rays (and particularly ultra-high-energy cosmic rays, UHECRs) must have been producing them for eons, but they have yet to harm us. It has been argued that to conserve energy and momentum, any BHs created in a collision between an UHECR and local matter would necessarily be produced moving at relativistic speed with respect to the Earth, and should escape into space, as their accretion and growth rate should be very slow, while BHs produced in colliders (with components of equal mass) would have some chance of having a velocity less than Earth escape velocity, 11.2 km per sec, and would be liable to capture and subsequent growth. Yet even on such scenarios the collisions of UHECRs with white dwarfs and neutron stars would lead to their rapid destruction, but these bodies are observed to be common astronomical objects. Thus if stable micro black holes should be produced, they must grow far too slowly to cause any noticeable macroscopic effects within the natural lifetime of the solar system.

Physicists working at the Large Hadron Collider report that after a series of tests, they have not seen any mini black holes, to the chagrin of string theorists and the relief of disaster theorists.

Researchers working on the Compact Muon Solenoid team have been crunching numbers to test a form of string theory that calls for the creation and instant evaporation of miniature black holes. They report that the telltale signs of these black holes are disappointingly absent, however.

Fig.7.1 The Compact Muon Solenoid CMS Black Hole seen under construction in late 2008

String theory is the most widely accepted attempt to unify the two major fields of physics, quantum mechanics and relativity. It holds that electrons and quarks are not objects, but one-dimensional strings whose oscillation gives them their observed qualities. It also says the universe has about a many dimensions, rather than the usual four (length, width, height and time).

In one version of string theory, if these dimensions exist, gravitons - hypothetical particles that transmit gravity - would leak into them, explaining why gravity is so much weaker than the other forces (As New Scientist explains it). It's not really weaker, it just seems weaker, because some of its particles are in another dimension we can't see. Happily, it takes a lot less energy to test this than it would to actually unify all the forces, and it just so happens it's is in the energy range that the LHC, the world's most powerful particle accelerator, is capable of testing.

If this is all true, particles that collided at energies beyond this graviton-leaking energy cutoff would get so close together that gravity would take over, and they would merge to form a tiny black hole. The black holes would instantly decay, so there would be no danger of Earth being swallowed whole, and the decay would be visible as jets of particles. But the researchers have so far seen no jets.

This doesn't disprove string theory - it just proves that mini black holes can't be produced at energies between 3.5 and 4.5 trillion electron volts. But they could still theoretically be produced at higher energies, so when the LHC fully fires up in 2013, string theorists will be holding their breath.

Meanwhile, the tests show the LHC is performing supremely well, so physicists aim to keep it running through 2012. This means they might be able to find the elusive Higgs boson sooner than expected.

In general relativity, a **black hole** is a region of space in which the gravitational field is so powerful that nothing, not even light, can escape. The black hole has a one-way surface, called an event horizon, into which objects can fall, but out of which nothing can come. It is called "black" because it absorbs all the light that hits it, reflecting nothing, just like a perfect black-body in thermodynamics. Quantum analysis of black holes shows them to possess a temperature and Hawking radiation.

In the picture below a simulated view of a black hole in front of the Large Magellanic Cloud. The ratio between the black hole Schwarzschild radius and the observer distance to it is 1:9. Of note is the gravitational lensing effect known as an Einstein ring, which produces a set of two fairly bright and large but highly distorted images of the Cloud as compared to its actual angular size.

Fig.7.2 Simulated view of a black hole in front of the large magellanic cloud

Despite its invisible interior, a black hole can reveal its presence through interaction with other matter. A black hole can be inferred by tracking the movement of a group of stars that orbit a region in space which looks empty. Alternatively, one can see gas falling into a relatively small black hole, from a companion star. This gas spirals inward, heating up to very high temperatures and emitting large amounts of radiation that can be detected from earthbound and earth-orbiting telescopes. Such observations have resulted in the scientific consensus that, barring a breakdown in our understanding of nature, black holes do exist in our galaxy.

7.3. High-energy physics

The largest particle accelerators with the highest particle energies, such as the RHIC, the Large Hadron Collider (LHC) at CERN (which came on-line in mid-November 2009) and the Tevatron, are used for experimental particle physics.

For the most basic inquiries into the dynamics and structure of matter, space, and time, physicists seek the simplest kinds of interactions at the highest possible energies. These typically entail particle energies of many GeV, and the interactions of the simplest kinds of particles: leptons (e.g. electrons and positrons) and quarks for the matter, or photons and gluons for the field quanta. Since isolated quarks are experimentally unavailable due to color confinement, the simplest available experiments involve the interactions of, first, leptons with each other, and second, of leptons with nucleons, which are composed of quarks and gluons. To study the collisions of quarks with each other, scientists resort to collisions of nucleons, which at high energy may be usefully considered as essentially 2-body interactions of the quarks and gluons of which they are composed. Thus elementary particle physicists tend to use machines creating beams of electrons, positrons, protons, and anti-protons, interacting with each other or with the simplest nuclei (e.g., hydrogen or deuterium) at the highest possible energies, generally hundreds of GeV or more. Nuclear physicists and cosmologists may use beams of bare atomic nuclei, stripped of electrons, to investigate the structure, interactions, and properties of the nuclei themselves, and of condensed matter at extremely high temperatures and densities, such as might have occurred in the first moments of the Big Bang. These investigations often involve collisions of heavy nuclei-of atoms like iron or gold-at energies of several GeV per nucleon.

Particle accelerators can also produce proton beams, which can produce proton-rich medical or research isotopes as opposed to the neutron-rich ones made in fission reactors; however, recent work has shown how to make ^{99}Mo, usually made in reactors, by accelerating isotopes of hydrogen, although this method still requires a reactor to produce tritium. An example of this type of machine is LANSCE at Los Alamos.

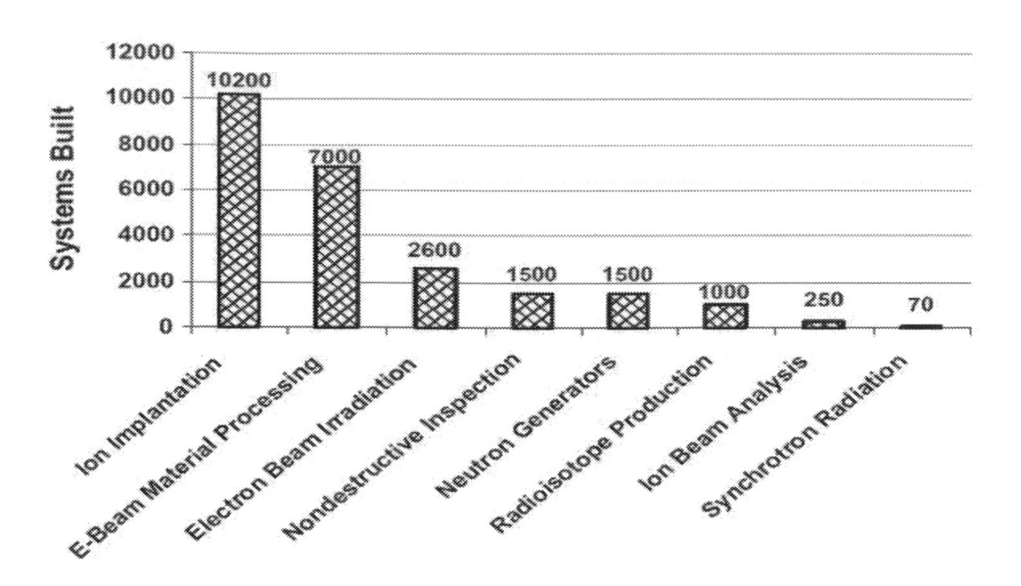

Fig.7.3 Breakdown of the cumulative number of industrial accelerators according to their applications

Annex A1 – Standard Model and Type of Interaction

A1.1.The Standard Model

The Standard Model is the name given to the current theory of fundamental particles and how they interact. This theory includes:

- Strong interactions due to the color charges of quarks and gluons.
- A combined theory of weak and electromagnetic interaction, known as electroweak theory, that introduces W and Z bosons as the carrier particles of weak processes, and photons as mediators to electromagnetic interactions.

The theory does not include the effects of gravitational interactions. These effects are tiny under high-energy Physics situations, and can be neglected in describing the experiments. Eventually, we seek a theory that also includes a correct quantum version of gravitational interactions, but this is not yet achieved.

The Standard Model was the triumph of particle physics of the 1970's . It incorporated all that was known at that time and has since then successfully predicted the outcome of a large variety of experiments. Today, the Standard Model is a well-established theory applicable over a wide range of conditions.

One part of the Standard Model is not yet well established. We do not know what causes the fundamental particles to have masses. The simplest idea is called the **Higgs mechanism.** This mechanism involves one additional particle, called the Higgs boson, and one additional force type, mediated by exchanges of this boson.

The Higgs particle has not yet observed. Today we can only say that if it exists, it must have a mass greater than about 80GeV/c^2. Searches for a more massive the Higgs boson are beyond the scope of the present facilities at SLAC or elsewhere. Future facilities, such as the Large Hadron Collider at CERN, or upgrades of present facilities to higher energies are intended to search for the Higgs particle and distinguish between competing concepts.

Thus, this one aspect of the Standard Model does not yet have the status of "theory" but still remains in the realm of hypothesis or model.

This theory describes the strong, weak, and electromagnetic interactions of quarks and gluons. It also describes the **weak** and **electromagnetic** interactions of leptons, the particles that do not participate in strong interactions. These interactions are
Described briefly as following:

A1.2. Types of Interactions

A1.2.1 Strong Interactions

Fundamental strong interactions occur between any two particles that have color charge, that is, quarks, antiquarks, and gluons.

Gluons

Gluons are the carrier particle for strong interactions. They are responsible for the binding force that confines all color-charged particles to form hadrons, such as protons, neutrons and pions. The resulting hadrons have no net color charge.

Nuclear Forces

Residual strong interactions between the color-charge neutral hadrons are responsible for the "strong nuclear force" ; the force that binds protons and neutrons together to form nuclei. All of modern particle physics was discovered in the effort to understand this force!

Residual strong interactions also are responsible for nuclear fission and fusion processes and for the most rapid decay processes of many hadrons. These residual strong interactions have short range. They occur via exchange of mesons, or because two hadrons come close enough together that they overlap and constituents of one hadron can directly feel forces from constituents of the other hadron.

A1.2.2 Weak Interactions

Fundamental weak interactions occur for all fundamental particles except gluons and photons. Weak interactions involve the exchange or production of W or Z bosons.

Weak forces are very short-ranged. In ordinary matter, their effects are negligible except in cases where they allow an effect that is otherwise forbidden. There are a number of conservation laws that are valid for strong and electromagnetic interactions, but broken by weak processes. So, despite their slow rate and short range, weak interactions play a crucial role in the make-up of the world we observe.

W Bosons

Any process where the number of particles minus the number of antiparticles of a given quark or lepton type changes is a weak decay process and involves a W-boson. Weak decays are thus responsible for the fact that ordinary stable matter contains only up and down type quarks and electrons. Matter containing any more massive quark or lepton types is unstable. If there were no weak interactions, then many more types of matter would be stable.

Z Bosons

Processes involving Z-bosons (called "neutral current processes") are even more elusive than W-boson effects, and were not recognized until after the electroweak theory had predicted they must exist in the early 1970s. Careful searches then found events that could not be explained without such processes.

Later, higher energy experiments produced W and Z bosons and studied their decays. Carlo Rubbia and Simon van der Meer were awarded the 1984 Nobel Prize for their development of the facility at CERN which first produced and detected W and Z bosons, and for their leadership of the discovery experiment.

Beta Decay: The First Known Weak Interaction

The weak interaction was first recognized in cataloging the types of nuclear radioactive decay chains, as alpha, beta, and gamma. decays. Alpha and gamma decays can be understood in terms of other known interactions (residual strong and electromagnetic, respectively). But, to explain beta decay required the introduction of an additional rare type of interaction -- called the weak interaction.

Beta decay is a process in which a neutron (two *down* quarks and one *up*) disappears and is replaced by a proton (two *up* quarks and one *down*), an electron, and an anti-electron neutrino. According to the Standard Model, a *down* quark disappears in this process and an *up* quark and a virtual W boson is produced. The W boson then decays to produce an electron and an anti-electron type neutrino. This can be represented by the Feynman diagram:

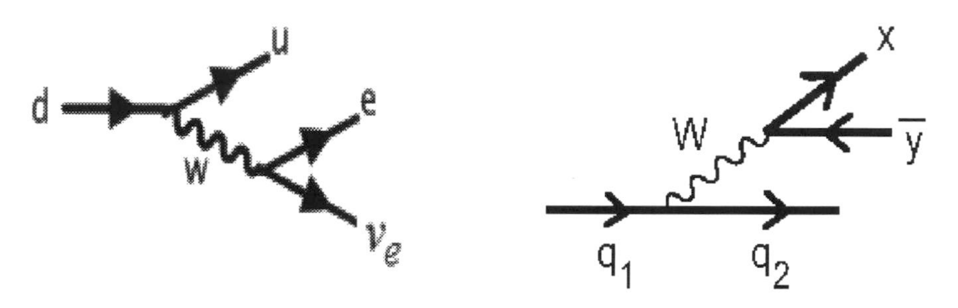

Fig. A1.1 Feynman diagram

Quark Flavor Changes

Quark flavor never changes *except* by weak interactions, like beta decay, that involve W bosons.

Any quark type can convert to any other quark type with a different electric charge by emitting or absorbing a W boson. The relative probabilities of the different possible transitions are shown schematically in the diagram shown in Fig. A1.2.

Decay processes always proceed from a more massive quark to a less massive quark, as indicated in the diagram, because the reverse process would violate conservation of energy. Scattering processes can involve the reverse transitions, provided sufficient energy is available.

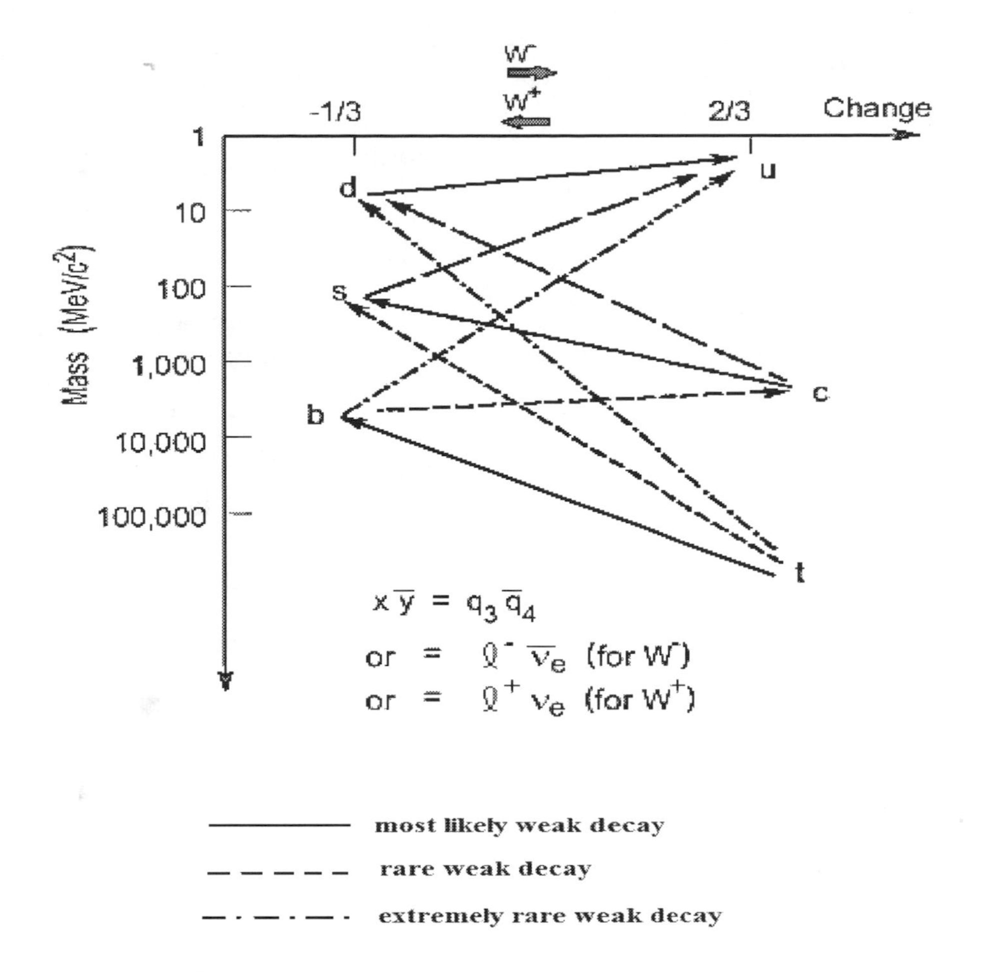

Fig. A1.2 Quarks and their weak decays

Lepton Number Conservation Rules

The set of observed lepton processes is much simpler. Each charged lepton is converted only to its own neutrino type by emitting or absorbing a W boson. This leads to the three lepton number conservation laws.

A1.2.3 Electromagnetic Interactions

Fundamental electromagnetic interactions occur between any two particles that have electric charge. These interactions involve the exchange or production of photons. Thus, photons are the carrier particles of electromagnetic interactions.

Electromagnetic decay processes can often be recognized by the fact that they produce one or more photons (also known as gamma particles). They proceed less rapidly than strong decay processes with comparable mass differences, but more rapidly than comparable weak decays.

Forces Within Atoms

Electromagnetic interactions are responsible for the binding force that causes negatively charged electrons to combine with positively charged nuclei to form atoms.

Forces Between Atoms

Residual electromagnetic interactions between electrically neutral atoms are responsible for the binding of atoms to form molecules and most of the forces (apart from gravity) that we experience in everyday life. Molecular binding effects result from atoms sharing and/or exchanging electrons.

The rigidity of the floor supporting you, the friction between your feet and the floor that allows you to walk, the pull of a rubber band on your finger, and the feel of the wind in your faces are all due to residual electromagnetic interactions. Forces such as these result from the changes in energy due to repositioning of electrons or atoms as material is deformed by contact with other material.

Electromagnetic Fields and Electromagnetic Waves

Electromagnetic interactions are also responsible for electric and magnetic field formation around electric charges and electric currents, and for traveling electromagnetic waves such as light, radio-waves, x-rays, and microwaves. All these phenomena are electromagnetic waves and differ only in wavelength.

In the quantum field theory, any changing electromagnetic fields or electromagnetic waves can be described in terms of photons. When there are many photons involved, the effects are equally correct (and more simply) given by the earlier non-quantum theory, namely Maxwell's equations.

Photons produced in radioactive decays are also called γ ("gamma") particles, originally called x-rays.

A1.2.4 Gravitational Interactions

Gravitational interactions have yet to be successfully incorporated into the quantum field theory and are a tiny effect in high energy particle collisions, so are ignored in the Standard Model.

Gravitational interactions occur between any two objects that have energy. Mass is just one possible form of this energy. (Photons are massless, but they experience gravitational forces.)

Gravitational interactions between fundamental particles are extremely weak, at least thirty orders of magnitude (that is $\approx 10^{-30}$) smaller than the weak interaction. Hence, gravitational effects can be ignored in particle physics processes involving small numbers of particles.

Why is Gravity so Obvious to Us?

The only reason we experience gravity as an important force is that there is no such thing as negative energy and, thus, the gravitational effects all objects without exception.

The earth exerts a much stronger gravitational pull on us than its electric pull. The electric charges in the earth are all balanced out (the positive charges of atomic nuclei screened by the negative charges of the electrons), but the masses of all the atoms in the earth add together to give a large gravitational effect on objects at the surface of the earth.

Quantum Gravity

The carrier particle for gravitational interactions has been named the graviton. However, no fully satisfactory quantum theory of gravitational interactions via graviton exchange has been identified. Thus the combination of gravity and particle physics remains a major outstanding problem. Much work in theoretical physics today is focused on this problem.

If we want to understand the big bang -- the very earliest moments in the history of the universe -- we will need to understand quantum gravity. The universe, at that time, was a very dense fluid of very-high energy particles. Gravitational interactions are comparable in strength to other particle interactions in that environment, so we need a consistent theory that can treat both interactions together to really understand that era.

String Theory

At this moment, the most promising approach to developing a quantum theory of gravity begins with the idea that elementary particles are not point like but rather are small lines or loops of energy, `strings'. It is very difficult to formulate a theory in which elementary particles have nonzero size which is consistent with relativity and with quantum mechanics.

String Theory is a mathematical theory of extended particles which has special properties that allow it to pass this consistency test. These properties predict the existence of force-carrying particles, such as photons and gluons --and gravitons--with the correct force laws. They also imply some more bizarre predictions, for example, the existence of seven extra space dimensions. Since we do not see any trace of them, even in the highest energy experiments to date, these extra dimensions of space must be very small closed loops.

Many theoretical physicists now believe that this radical modification of relativistic quantum mechanics could lead to a unified theory of all interactions.

A1.2.5 An Additional Force

A fifth force is needed to complete our understanding of particle masses.

In the late 1980s, experiments that looked at the effect of gravity as a function of distance from earth claimed to have found evidence for a new type of force.

All their results could be explained by ordinary gravitational forces -- because the expected behavior of gravity as you move away from the surface of the earth depends in detail on the profile of the rocks or soil nearby and their densities. Unexpected fissures or extra dense rocks could cause all the effects seen.

The Real Fifth Force, and More?

In the Standard Model, there is at least one additional type of interaction beyond the four known forces (weak, strong, electromagnetic, and gravitational). This force is needed to explain how all the fundamental particle masses are generated. This part of the theory is the least tested experimentally, so there are a number of different competing ideas on how it may work.

The simplest version introduced one more force--the Higgs force--and one more particle type--the Higgs particle--related to this force. Searches for this particle and efforts to learn more about how particle masses occur are one active area of particle studies. Other models introduce more complicated explanations for particle masses.

In addition, there are many speculations about physics beyond the Standard Model that introduce additional types of extremely-weak interactions. These interactions can only be observed if they mediate a process, such as proton decay, that is otherwise totally forbidden. So far, no experimental evidence for such processes has been found. However physicists like this idea, since such additional processes are predicted when we try to unify the strong, weak and electromagnetic interactions into a single "Grand Unified Theory." Such unification is suggested by the similarities of the underlying mathematical theories for these very different interactions.

Unification of all four force types, including gravity, is also a goal for particle physics. Gravity has a different mathematical structure, and so far no complete quantum theory of gravity has been developed. String Theory suggests possible answers, but much work remains to be done.

Annex A2- General Theory of Relativity

Particle physics theories are written in mathematical language called relativistic quantum field theory. This builds on the two major early twentieth century advances in physics, relativity and quantum mechanics. *Note that Einstein's General Theory of Relativity is a separate theory about a very different topic : the effects of gravity.*

A2.1 Relativistic Definitions

Physicists call particles with ***v/c*** comparable to 1 "relativistic" particles. Particles with ***v/c* << 1** (very much less than one) are "non-relativistic." At SLAC, we are almost always dealing with relativistic particles. Below we catalogue some essential differences between the relativistic quantities the more familiar non-relativistic or low-speed approximate definitions and behaviors.

Gamma (γ)

The measurable effects of relativity are based on gamma. Gamma depends only on the speed of a particle and is always larger than 1. By definition:

$$\gamma = \frac{1}{\sqrt{1-(v^2/c^2)}} \geq 1$$

Where **c** is the speed of light ***v*** is the speed of the object in question

For example, when an electron has traveled ten feet along the accelerator it has a speed of **0.99*c***, and the value of gamma at that speed is 7.09. When the electron reaches the end of the linac, its speed is 0.99999999995*c* where gamma equals 100,000.

What do these gamma values tell us about the relativistic effects detected at SLAC? Notice that when the speed of the object is very much less than the speed of light ($v << c$), gamma γ is approximately equal to 1. This is a non-relativistic situation (Newtonian).

Momentum

For non-relativistic objects Newton defined momentum, given the symbol p, as the product of mass and velocity

$p = m\,v.$

When speed becomes relativistic, we have to modify this definition

$p = \gamma\,(mv).$

Notice that this equation tells you that for any particle with a non-zero mass, the momentum gets larger and larger as the speed gets closer to the speed of light. Such a particle would have infinite momentum if it could reach the speed of light. Since it would take an infinite amount of force (or a finite force acting over an infinite amount of time) to accelerate a particle to infinite momentum, we are forced to conclude that a massive particle always travels at speeds less than the speed of light.

Some text books will introduce the definition m_0 for the mass of an object at rest, calling this the "rest mass" and define the quantity ($M = \gamma\, m_0$) as the mass of the moving object. This makes Newton's definition of momentum still true provided you choose the correct mass. In particle physics, when we talk about mass we always mean mass of an object at rest and we write it as ***m*** and keep the factor of gamma explicit in the equations.

Energy

Probably the most famous scientific equation of all time, first derived by Einstein is the relationship $E = mc^2$.

This tells us the energy corresponding to a mass m at rest. What this means is that when mass disappears, for example in a nuclear fission process, this amount of energy must appear in some other form. It also tells us the total energy of a particle of mass m sitting at rest.

Einstein also showed that the correct relativistic expression for the energy of a particle of mass m with momentum p is $E^2 = m^2c^4 + p^2c^2$. This is a key equation for any real particle, giving the relationship between its energy (E), momentum (p), and its rest mass (m).

If we substitute the equation for p into the equation for E above, with a little algebra, we get $E = \gamma\, mc^2$, so energy is gamma times rest energy. (*Notice again that if we call the quantity* $M = \gamma$ m *the mass of the particle then* $E = Mc^2$ *applies for any particle, but remember, particle physicists don't do that*).

Let's do a calculation. The rest energy of an electron is 0.511 MeV. As we saw earlier, when an electron has gone about 10 feet along the SLAC linac, it has a speed of 0.99c and a gamma of 7.09.

Therefore, using the equation E = *gamma* x *the rest energy*, we can see that the electron's energy after ten feet of travel is 7.09 x 0.511 MeV = 3.62 MeV. At the end of the linac, where gamma γ = 100,000, the energy of the electron is 100,000 x 0.511 MeV = 51.1 GeV.

The energy E is the total energy of a freely moving particle. We can define it to be the rest energy plus kinetic energy ($E = KE + mc^2$) which then defines a relativistic form for kinetic energy. Just as the equation for momentum has to be altered, so does the low-speed equation for kinetic energy ($KE = (1/2)mv^2$). Let's make a guess based on what we saw for momentum and energy and say that relativistically $KE = \gamma\ (1/2)mv^2$. A good guess, perhaps, but it's wrong.

Now here is an exercise for the interested reader. Calculate the quantity $KE = E - mc^2$ for the case of v very much smaller than c, and show that it is the usual expression for kinetic energy ($1/2\ mv^2$) plus corrections that are proportional to $(v/c)^2$ and higher powers of (v/c). The complicated result of this exercise points out why it is not useful to separate the energy of a relativistic particle into a sum of two terms, so when particle physicists say "the energy of a moving particle" they mean the total energy, not the kinetic energy.

Another interesting fact about the expression that relates E and p above ($E^2 = m^2c^4 + p^2c^2$), is that it is also true for the case where a particle has no mass (m=0). In this case, the particle always travels at a speed c, the speed of light. You can regard this equation as a definition of momentum for such a mass-less particle. Photons have kinetic energy and momentum, but no mass!

In fact Einstein's relationship tells us more, it says Energy and mass are interchangeable. Or, better said, rest mass is just one form of energy. For a compound object, the mass of the composite is not just the sum of the masses of the constituents but the sum of their energies, including kinetic, potential, and mass energy. The equation $\boldsymbol{E=mc^2}$ shows how to convert between energy units and mass units. Even a small mass corresponds to a significant amount of energy.

- In the case of an atomic explosion, mass energy is released as kinetic energy of the resulting material, which has slightly less mass than the original material.
- In any particle decay process, some of the initial mass energy becomes kinetic energy of the products.

Even in chemical processes there are tiny changes in mass which correspond to the energy released or absorbed in a process. When chemists talk about conservation of mass, they mean that the sum of the masses of the atoms involved does not change. However, the masses of molecules are slightly smaller than the sum of the masses of the atoms they contain (which is why molecules do not just fall apart into atoms). If we look at the actual molecular masses, we find tiny mass changes do occur in any chemical reaction.

At SLAC, and in any particle physics facility, we also see the reverse effect -- energy producing new matter. In the presence of charged particles a photon (which only has kinetic energy) can change into a massive particle and its matching massive antiparticle. The extra charged particle has to be there to absorb a little energy and more momentum, otherwise such a process could not conserve both energy and momentum. This process is one more confirmation of Einstein's special theory of relativity. It also is the process by which antimatter (for example the positrons accelerated at SLAC) is produced.

A2.2 Peculiar Relativistic Effects

Length Contraction and Time Dilation

One of the strangest parts of special relativity is the conclusion that two observers who are moving relative to one another, will get different measurements of the length of a particular object or the time that passes between two events.

Consider two observers, each in a space-ship laboratory containing clocks and meter sticks. The space ships are moving relative to each other at a speed close to the speed of light. Using Einstein's theory:

- Each observer will see the meter stick of the other as shorter than their own, by the same factor gamma (γ - defined above). This is called **length contraction**.

- Each observer will see the clocks in the other laboratory as ticking more slowly than the clocks in his/her own, by a factor gamma. This is called **time dilation**.

In particle accelerators, particles are moving very close to the speed of light where the length and time effects are large. This has allowed us to clearly verify that length contraction and time dilation do occur.

Time Dilation for Particles

Particle processes have an intrinsic clock that determines the half-life of a decay process. However, the rate at which the clock ticks in a moving frame, as observed by a static observer, is slower than the rate of a static clock. Therefore, the half-life of a moving particles appears, to the static observer, to be increased by the factor gamma.

For example, let's look at a particle sometimes created at SLAC known as a tau. In the frame of reference where the tau particle is at rest, its lifetime is known to be approximately 3.05×10^{-13} s. To calculate how far it travels before decaying, we could try to use the familiar equation distance equals speed times time. It travels so close to the speed of light that we can use ($c = 3 \times 10^8$ m/sec) for the speed of the particle. (As we will see below, the speed of light in a vacuum is the highest speed attainable.) If you do the calculation you find the distance traveled should **be** 9.15×10^{-5} meters.

$d = v\,t$; $d = (3 \times 10^8 \text{ m/sec})\,(3.05 \times 10^{-13} \text{ s}) = 9.15 \times 10^{-5}$ m

Here comes the tricky part - *we measure the tau particle to travel further than this*! Pause to think about that for a moment. This result is totally contradictory to everyday experience. If you are not puzzled by it, either you already know all about relativity or you have not been reading carefully.

What is the resolution of this apparent paradox? The answer lies in time dilation. In our laboratory, the tau particle is moving. The decay time of the tau can be seen as a moving clock. According to relativity, moving clocks tick more slowly than static clocks.

We use this fact to multiply the time of travel in the taus moving frame by gamma, this gives the time that we will measure. Then this time times c, the approximate speed of the tau, will give us the distance we expect a high energy tau to travel.

What is gamma γ in this case? It depends on the tau's energy. A typical SLAC tau particle has a gamma = 20. Therefore, we detect the tau to decay in an average distance of 20 x (9.15 x 10^{-5} m) = 1.8 x 10^{-3} m or approximately 1.8 millimeters. This is 20 times further than we expect it to go if we use classical rather than relativistic physics. (Of course, we actually observe a spread of decay times according to the exponential decay law and a corresponding spread of distances. In fact, we use the measured distribution of distances to find the tau half-life.)

Observations particles with a variety of velocities have shown that time dilation is a real effect. In fact the only reason cosmic ray muons ever reach the surface of the earth before decaying is the time dilation effect.

Length Contraction

Instead of analyzing the motion of the tau from our frame of reference, we could ask what the tau would see in its reference frame. Its half-life in its reference frame is 3.05 x 10^{-13} s. This does not change. The tau goes nowhere in this frame.

How far would an observer, sitting in the tau rest frame, see an observer in our laboratory frame move while the tau lives?

We just calculated that the tau would travel 1.8 mm in our frame of reference. Surely we would expect the observer in the tau frame to see us move the same distance relative to the tau particle. Not so says the tau-frame observer -- you only moved 1.8 mm/gamma = 0.09 mm relative to me. **This is length contraction.**

How long did the tau particle live according to the observer in the tau frame? We can rearrange $d = v \times t$ to read $t = d/v$. Here we use the same speed, Because the speed of the observer in the lab relative to the tau is just equal to (but in the opposite direction) of the speed of the tau relative to the observer in the lab, so we can use the same speed. So time = 0.09 x 10^{-3} m/(3 x 10^{8})m/sec = 3.0 x 10^{-13} sec. **This is the half-life of the tau as seen in its rest frame, just as it should be.**

Annex 3-Quantum Mechanics

Quantum mechanics is the description of physics at the scale of atoms, and the even smaller scales of fundamental particles.

Quantum theory is the language of all particle theories. It is formulated in a well-defined mathematical language. It makes predictions for the relative probabilities of the various possible outcomes, but not for which outcome will occur in any given case. Interpretation of the calculations, in words and images, often leads to statements that seem to defy common sense -- because our common sense is based on experience at scales insensitive to these types of quantum peculiarities.

Because we do not directly experience objects on this scale, many aspects of quantum behavior seem strange and even paradoxical to us. Physicists worked hard to find alternative theories that could remove these peculiarities, but to no avail.

The word quantum means a definite but small amount. The basic quantum constant *h*, known as Planck's constant, is 6.626069×10^{-34} Joule seconds.

Because the particles in high-energy experiments travel at close to the speed of light, particle theories are relativistic quantum field theories.

Let's look at just a few, of the many, quantum concepts that will be stated without explanation.

A3.1 Particle-Wave Duality

In quantum theories, energy and momentum have a definite relationship to wavelength. All particles have properties that are wave-like (such as interference) and other properties that are particle-like (such as localization). Whether the properties are primarily those of particles or those of waves, depends on how you observe them.

For example, photons are the quantum particles associated with electromagnetic waves. For any frequency, f, the photons each carry a definite amount of energy $(E = hf)$.

Only by assuming a particle nature for light with this relationship between frequency and particle energy could Einstein explain the photoelectric effect.

Conversely, electrons can behave like waves and develop interference patterns.

A3.2 Discrete Energy, Momenta, and Angular Momenta

In classical physics, quantities such as energy and angular momentum are assumed to be able to have any value. In quantum physics there is a certain discrete (particle-like) nature to these quantities.

For example, the angular momentum of any system can only come in integer multiples of $h/2$, where h is Planck's Constant. Values such as $(n+1/4)h$ are simply not allowed.

Likewise, if we ask how much energy a beam of light of a certain frequency, f, deposits on an absorbing surface during any time interval, we find the answer can only be nhf, where n is some integer. Values such as $(n+1/2)hf$ are not allowed.

To get some idea of how counter-intuitive this idea of discrete values is, imagine if someone told you that water could have only integer temperatures as you boiled it. For example, the water could have temperatures of 85°, 86° or 87°, but not 85.7° or 86.5°. It would be a pretty strange world you were living in if that were true.

The world of quantum mechanics is pretty strange when you try to use words to describe it.

A3.3 States and Quantum Numbers

In quantum mechanics, systems are described by the set of possible **states** in which they may be found. For example, the electron orbitals familiar in chemistry are the set of possible bound states for an electron in an atom.

Bound states are labeled by a set of **quantum numbers** that define the various conserved quantities associated with the state. These labels are pure numbers that count familiar discrete quantities, such as electric charge, as well as energy, and angular momentum, which can only have certain discrete values, bound quantum systems.

A3.4 Wave Function or Probability Amplitude

A state is described by a quantity that is called wave-function or probability amplitude. It is a complex-number-valued function of position, that is a quantity whose value is a definite complex number at any point in space. The probability of finding the particle described by the wave function (e.g. an electron in an atom) at that point is proportional to square of the absolute value of the probability amplitude.

The "pictures" of orbitals in an introductory chemistry textbook are a representation of the region within which the wave function gives a high probability of finding an electron for that state.

A3.5 Free States

We can also talk about quantum states for freely moving particles. These are states with definite momentum, *p*, and energy:

$$E = (p^2c^2 + m^2c^4)^{1/2}$$

The associated wave has frequency given by $f = pc/h$ where c is speed of light.

A3.6 Quantum Interference

Another peculiarly quantum property is that the wave-like nature of particles leads to interference effects that violate our usual notions of how probability works. Two processes, which when described in a particle language seem quite distinct, actually represent two different contributions to an overall probability amplitude.

The rule for probability in quantum mechanics is that probability is the square of the absolute value of the relevant probability amplitude.

Two processes that can be distinguished by measurement have separate probabilities, and these probabilities add in the usual way. The peculiarity comes about when the processes are not experimentally distinguishable, despite their different particle-language descriptions.

A3.7 An Example of Particle Interference

For example, consider the following Feynman diagrams. Both represent a process where an electron and a positron collide and an electron and a positron emerge traveling in different directions (what we call a scattering process).

We can read these diagrams as if they are a pictorial representation of a particle process that starts at the left and ends at the right of the picture.

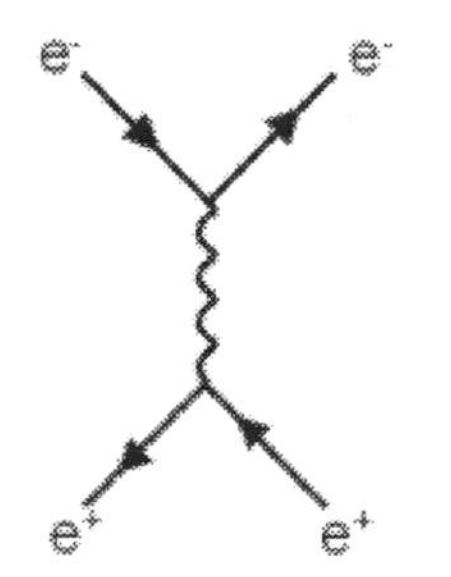

In the first diagram, the electron emits a photon and the positron absorbs it, thus changing the direction of both.

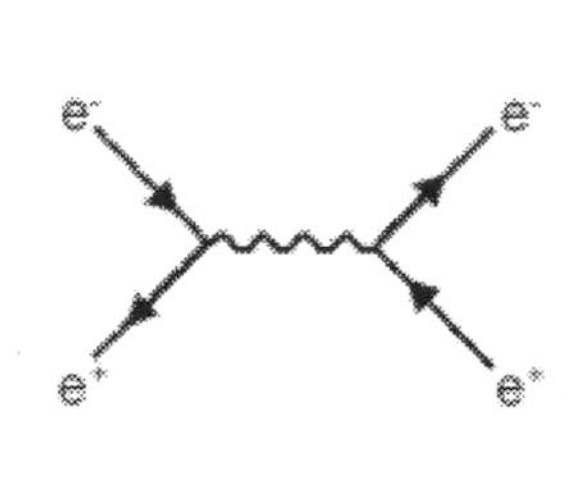

In the second diagram, the electron and positron meet and annihilate, or disappear. For a short time there is only a single virtual photon present, which then disappears to produce a new electron and positron pair, traveling apart in different directions from the initial pair.

These two pictures, and the words we use to "describe" what they represent, certainly appear to be two very different processes. But, we do not observe the intermediate stages and cannot do so without changing the outcome of the experiment.

A3.8. The Mathematics of Interference Calculations

Feynman's prescription assigns a complex number to each diagram, let us write these as **A** and **B**. (The values of **A** and **B** depend on the momenta and energy of the particles.) The probability of a given scattering occurring is given by $|\mathbf{A}+\mathbf{B}|^2$.

There is no way to say which of the two underlying processes represented by the two diagrams actually occurred. Furthermore, we cannot even say there is a probability of each process and then add the probabilities, since $|\mathbf{A}+\mathbf{B}|^2$ is not the same number as $|\mathbf{A}|^2 + |\mathbf{B}|^2$.

For example, let consider **A= 5** and B = -3. Then we might think the probability of the process represented by **A** was $|\mathbf{A}|^2$= **25**, while that represented by $|\mathbf{B}|^2$= **9**. Given this, we would be tempted to assign the probability 34 to having either the A or the B process occur -- but the quantum answer is $|\mathbf{A+B}|^2 = 2^2$= **4**. That is, the two processes interfere with one another and both contribute to make the net result smaller than it would be if either one alone were the only way to achieve the process!

This is the nature of quantum theories -- unobserved intermediate stages of a process cannot be treated by the ordinary rules of everyday experience.

In both diagrams, the photons that appear at intermediate stages are virtual particles that are not observable.

Annex A4 - Electron Gamma Shower EGS Application

A4.1 Radiation Effect

The "Electron-Gamma-Shower" (EGS) is now the most accurate method available to calculate radiation effects in body tissue and determine dosages.

One way EGS has been used in a "real world" application to improve radiation cancer treatments.

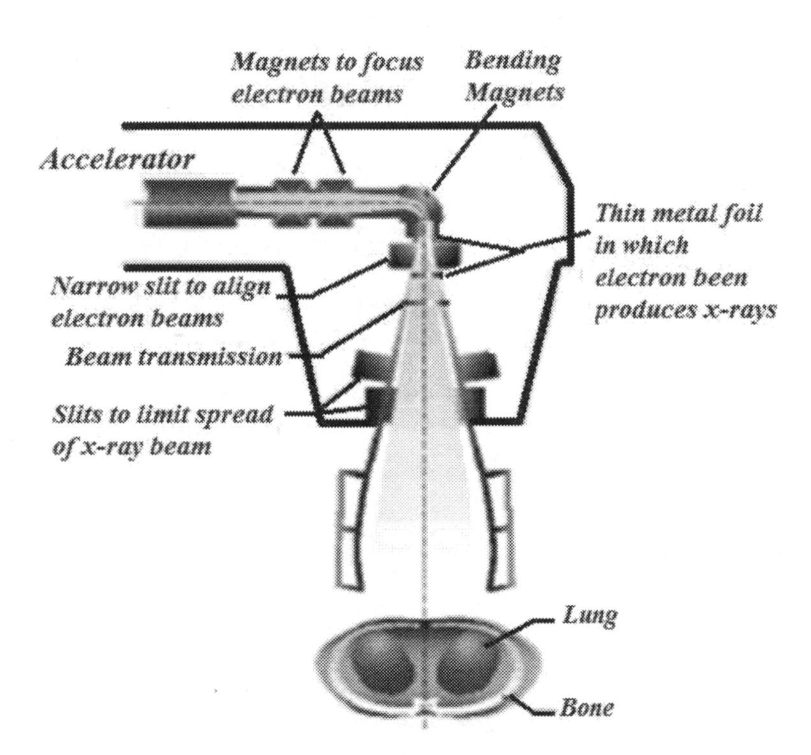

Fig. A4.1 Schematic diagram of a typical medical Accelerator used in Censer Radiotherapy

Approximately one in eight people receives radiation treatment for cancer. Electron accelerators produce the x-rays used in these treatments. The first accelerator for medical use was built at Stanford in 1956. Today every major hospital has one of them.

A4.2 The EGS Software

A medical physicist needs a computer program to calculate radiation effects in the body and plan treatments. The problem is how to give maximum radiation dose to the tumor with minimum damage to healthy tissue.

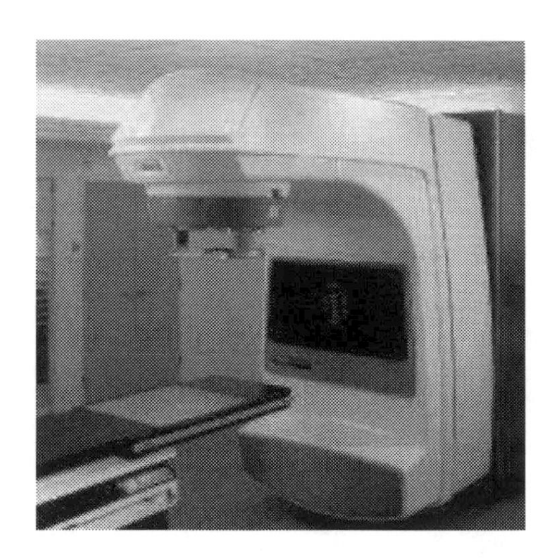

SLAC physicists originally developed the EGS program to model the passage of radiation through particle detectors. It has been "tuned" to match a wide variety of measured effects, adapted in collaboration with the National Research Council of Canada.

EGS uses the density and the locations of various tissue types in each patient's body in its calculations. Most body tissue has about the same density as water, but bones are twice as dense and lung tissue 1,000 times less dense. A model that treats all tissues alike would not do the job!

The EGS program is used as the "gold standard" to check whether simpler, faster, approximate calculation methods that are used by practitioners are doing the job accurately enough. As treatment techniques improve, more accurate calculations are needed.

A4.3 Acceleration of Electrons

Bunches of electrons are accelerated in the copper structure of the linac in much the same way as a surfer is pushed along by a wave.

In the linac, the wave is electromagnetic. That means it is made up of changing magnetic and electric fields.

Think of a magnetic field as a region of space where magnetic effects can be detected - one magnet pulling or pushing on another, for example. Similarly, an electric field is a region of space where electric effects can be detected. You can make an electric field by removing electrons from one substance and putting them on another. The region of space between the two substances then contains an electric field. An example is rubbing an inflated balloon on your hair. The effect is to make your hair stand on end.

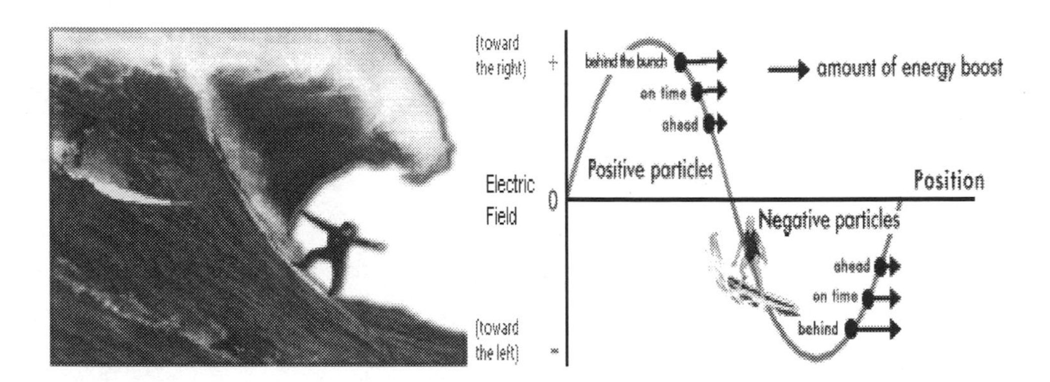

The electromagnetic waves that push the electrons in the linac are created by higher energy versions of the microwaves used in the microwave oven in your kitchen.

The microwaves from the klystrons in the Klystron Gallery are fed into the accelerator structure via the waveguides.

This creates a pattern of electric and magnetic fields, which form an electromagnetic wave traveling down the accelerator.

The 2-mile SLAC linear accelerator (linac) is made from over 80,000 copper discs and cylinders brazed together. Example of configuration and sizes of these rings are shown below

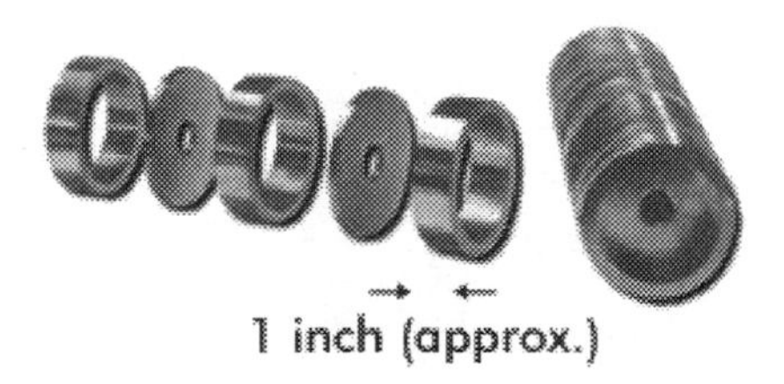

Inside the accelerator structure, the microwaves from the klystrons set up currents in the copper that cause oscillating electric fields pointing along the accelerator as well as oscillating magnetic fields in a circle around the interior of the accelerator pipe. The trick is to have the electrons or positrons arrive in each cell or cavity of the accelerator just at the right time to get maximum push from the electric field in the cavity. Of course, since positrons have opposite charge from electrons, they must arrive when the field is pointing the opposite way to be pushed in the same direction.

The size of the cavities in the accelerator is matched to the wavelength of the microwaves so that the electric and magnetic field patterns repeat every three cavities along the accelerator.

This means, in principle, there could be electron bunches following one another three cavities apart, and positron bunches half way in between. Usually the spacing between the bunches is kept somewhat larger (though always in multiples of three cavities for the same sign particles).

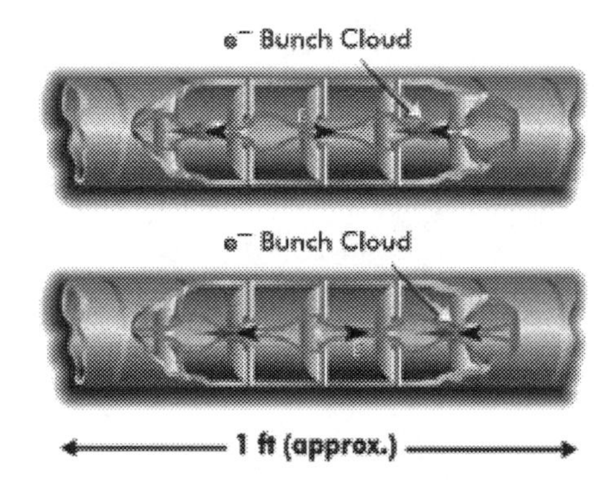

Notice how far the bunches have moved after just 1/20,000,000,000 of a second!

Annex A5 - List of accelerators in particle physics

An example of particle accelerators used for particle physics experiments are given in the tables below. Some early particle accelerators that more properly did nuclear physics, but existed prior to the separation of particle physics from that field, are also included. Although a modern accelerator complex usually has several stages of accelerators, only accelerators whose output has been used directly for experiments are listed. The list of Particle Accelerators is classified as following:

A5.1 Early accelerators

These all used single beams with fixed targets. They tended to have very briefly-run, inexpensive, and unnamed experiments.

A5.1.1 Cyclotrons

Accelerator	Location	Years of operation	Shape	Accel. Particle	Kinetic Energy	Notes and discoveries made
9-inch cyclotron	UC Berkeley	1931	Circular	H_2^+	1.0 MeV	Proof of concept
11-inch cyclotron	UC Berkeley	1932	Circular	Proton	1.2 MeV	
27-inch cyclotron	UC Berkeley	1932-1936	Circular	Deuteron	4.8 MeV	Investigated deuteron-nucleus interactions
37-inch cyclotron	UC Berkeley	1937-1938	Circular	Deuteron	8 MeV	Discovered many isotopes
60-inch cyclotron	UC Berkeley	1939-1941	Circular	Deuteron	16 MeV	Discovered many isotopes
184-inch	Berkeley	1942-	Circular	Various	>100	Research

cyclotron	Rad Lab[1]				MeV	on uranium isotope separation
Calutrons	Oak Ridge National Laboratory	1943-	"Horses hoe"	Uranium nuclei		Used to separate isotopes for the Manhattan project

[1] First accelerator built at the current Lawrence Berkeley National Laboratory site, then known as the Berkeley Radiation Laboratory ("Rad Lab" for short)

A5.1.2 Other early accelerator types

Accelerator	Location	Years of operation	Shape and size	Accel. particle	Kinetic Energy	Notes and discoveries made
Cockcroft and Walton's electrostatic accelerator	Cavendish Laboratory	1932	See Cockroft-Walton generator	Proton	0.7 MeV	First to artificially split the nucleus (Lithium)

A5.1.3 Synchrotrons

Accelerator	Location	Years of operation	Shape and size	Accel. Particle	Kinetic Energy	Notes and discoveries made
Cosmotron	Brookhaven National Laboratory	1953-1968	Circular ring (72 meters around)	Proton	3.3 GeV	Discovery of V particles, first artificial production of some mesons.
Birmingham Synchrotron	University of Birmingham	1953-1967		Proton	1 GeV	
Bevatron	Berkeley Rad Lab ie LBNL	1954-~1970	"Race track"	Proton	6.2 GeV	strange particle experiments, Antiproton and antineutron discovered, resonances

						discovered
Bevalac, combination of SuperHILAC linear accelerator, a diverting tube, then the Bevatron	Berkeley Rad Lab ie LBNL	~1970-1993	linear accelerator followed by "Race track"	any and all sufficiently-stable nuclei could be accelerated		observation of compressed nuclear matter. Depositing ions in tumors in cancer research.
Saturne	Saclay, France				3 GeV	
Synchrophasotron	Dubna, Russia	December 1949-present			10 GeV	
Zero Gradient Synchrotron	Argonne National Laboratory				12.5 GeV	
Proton Synchrotron	CERN	1959-present	Circular ring (600 meters around)	Proton	28 GeV	Used to feed ISR, SPS, LHC
Alternating Gradient Synchrotron	Brookhaven National Laboratory	1960-		Proton	33 GeV	J/Ψ, muon neutrino, CP violation in kaons, injects protons into RHIC

A5.2 Fixed-target accelerators

More modern accelerators that were also run in fixed target mode; often, they will also have been run as colliders, or accelerated particles for use in subsequently-built colliders.

Accelerator	Location	Years of operation	Shape and size	Accel. particle	Kinetic Energy	Experiments	Notes
SLAC Linac	Stanford Linear Accelerator center	1966-present	3 km linear accelerator	Electron / Positron	50 GeV		Repeatedly upgraded, used to feed PEP, SPEAR, SLC,

							and PEP-II
Fermilab Booster	Fermilab	1970-present	Circular Synchrotron	Protons	8 GeV	MiniBooNE	
Fermilab Main Injector	Fermilab	1995-present	Circular Synchrotron	Protons and antiprotons	150 GeV	MINOS	
Fermilab Main Ring	Fermilab	1970-1995	Circular Synchrotron	Protons and antiprotons	400 GeV (until 1979), 150 GeV thereafter		
Super Proton Synchrotron	CERN	1980-present	Circular Synchrotron	Protons and ions	480 GeV	OPERA and ICARUS at Laboratori Nazionali del Gran Sasso	
Bates Linear Accelerator	Middleton, MA	1967-2005	500 MeV recirculating linac and storage ring	polarized electrons	1 GeV		
CEBAF	Thomas Jefferson National Accelerator Facility, Newport News, VA	1984-present	5.75 GeV recirculating linac (upgrading to 12 GeV)	polarized electrons			
ELSA	Physikalisches Institut der Universität Bonn, Germany	1987-present	synchrotron and stretcher	(polarized) electrons	3.5 GeV	Crystal Barrel	
ISIS neutron source	Rutherford Appleton	1984-present	H- Linac followed by proton	Protons	800 MeV		Highest power operatio

	Laboratory, Didcot, Oxon		RCS				nal pulsed proton beam in the world
MAMI	Mainz, Germany		855 MeV accelerator	polarized electrons			
Tevatron	Fermilab	1978-present	Superconducting Circular Synchrotron	Protons	980 GeV		
Spallation Neutron Surce	Oak Ridge National Laboratory	2006 - Present	Linear (335 m) and Circular (248 m)	Protons	800 MeV - 1 GeV		

A5.3 Colliders

A5.3.1 Electron-positron colliders

Accelerator	**Location**	**Years of operation**	**Shape and circumference**	**Electron energy**	**Positron energy**	**Experiments**	**Notable Discoveries**
AdA	Frascati, Italy; Orsay, France	1961-1964	Circular, 3 meters	250 MeV	250 MeV		Touschek effect (1963); first e+e- interactions recorded (1964)
Princeton-Stanford (e-e-)	Stanford, California	1962-1967	Two-ring, 12 m	300 MeV	300 MeV		e+e- pair production
VEP-1 (e-e-)	Novosibirsk, Soviet Union	1964-1968	Two-ring, 2.70 m	130 MeV	130 MeV		e-e- scattering; QED radiative effects confirmed
VEPP-2, VEPP-2M	Novosibirsk, Soviet Union	1965-1999	Circular, 17.88 m	700 MeV	700 MeV	OLYA, ND, CMD; SND, CMD-2	multihadron production (1966), e+e- -> phi

							(1966), e+e- -> gamma gamma (1971), round beams (1990s)
SPEAR	SLAC					Mark I, Mark II, Mark III	Discovery of Charmonium states
PEP	SLAC					Mark II	
SLC	SLAC		Addition to SLAC Linac	45 GeV	45 GeV	SLD, Mark II	
LEP	CERN	1989-2000	Circular, 27km	104 GeV	104 GeV	Aleph, Delphi, Opal, L3	Only 3 light () weakly-interacting neutrinos exist, implying only three generations of quarks and leptons
DORIS	DESY	1974-1993	Circular, 300m	5 GeV	5 GeV	ARGUS, Crystal Ball, DASP, PLUTO	Oscillation in neutral B mesons
PETRA	DESY	1978-1986	Circular, 2km	20 GeV	20 GeV	JADE, MARK-J, PLUTO, TASSO	Discovery of the gluon in three jet events
CESR	Cornell University	1979-2002	Circular, 768m	6 GeV	6 GeV	CUSB, CHESS, CLEO, CLEO-2, CLEO-2.5, CLEO-3	First observation of B decay, charmless and "radiative penguin" B decays
CESR-c	Cornell University	2002-2008	Circular, 768m	6 GeV	6 GeV	CHESS, CLEO-c	
PEP-II	SLAC	1998-	Circular,	9	3.1 GeV	Babar	Discovery

		2008	2.2 km	GeV			of CP violation in B meson system
KEKB	KEK	1999-2008?	Circular, 3km	8.0 GeV	3.5 GeV	Belle	Discovery of CP violation in B meson system
VEPP-2000	Novosibirsk	2006-	Circular, 24m	1.0 GeV	1.0 GeV		
VEPP-4M	Novosibirsk	1994-	Circular, 366m	6.0 GeV	6.0 GeV		
BEPC	China	1989-2004	Circular, 240m	2.2 GeV	2.2 GeV	Beijing Spectrometer (I and II)	
DAΦNE	Frascati, Italy	1999-	Circular, 98m	0.7 GeV	0.7 GeV	KLOE	
BEPC II	China	2008-	Circular, 240m	3.7 GeV	3.7 GeV	Beijing Spectrometer III	

A5.3.2 Hadron colliders

Accelerator	Location	Years of operation	Shape and size	Particles collided	Beam energy	Experiments
Intersecting Storage Rings	CERN	1971-1984	Circular rings (948 m around)	Proton/ Proton	31.5 GeV	
(Super Proton Synchrotron)	CERN	1981-1984	Circular ring (6.9 km around)	Proton/ Antiproton		UA1, UA2
Tevatron Run I	Fermilab	1992-1995	Circular ring (6.3 km around)	Proton/ Antiproton	900 GeV	CDF, D0
Tevatron Run II	Fermilab	2001-present	Circular ring (6.3 km around)	Proton/ Antiproton	980 GeV	CDF, D0
RHIC proton+proton mode	BNL	2000-present	Circular ring (3.8 km)	Polarized Proton/ Proton	100 Gev to 250 GeV	PHENIX, STAR
Large Hadron	CERN	2008-present	Circular rings	Proton/ Proton	7 TeV	ALICE, ATLAS,

Collider			(27 km around)			CMS, LHCb, TOTEM

A5.3.3 Electron-proton colliders

Accelerator	**Location**	**Years of operation**	**Shape and size**	**Electron energy**	**Proton energy**	**Experiments**
HERA	DESY	1992-2007	Circular ring (6336 meters around)	27.5 GeV	920 GeV	H1, ZEUS, HERMES, HERA-B

A5.3.4 Ion colliders

Accelerator	**Location**	**Years of operation**	**Shape and size**	**Ions collided**	**Ion energy**	**Experiments**
Relativistic Heavy Ion Collider	Brookhaven National Laboratory, New York	2000-present	3.8 km	Au-Au; Cu-Cu; d-Au; polarized pp	0.1 TeV per nucleon	STAR, PHENIX, Brahms, Phobos
Large Hadron Collider, ion mode	CERN	2008-present	Circular rings (27 km around)	Pb-pb	2.76 TeV per nucleon	ALICE

9

GLOSSARY AND DEFINITIONS

- A -

Absorption
The transfer of energy to a medium, such as body tissues, as a radiation beam passes through the medium.

Accelerating cavity
Accelerating cavities produce the electric field that accelerates the particles inside particle accelerators. Because the electric field oscillates at radio frequency, these cavities are also referred to as radio-frequency cavities.

Accelerator
A machine in which beams of charged particles are accelerated to high energies. Electric fields are used to accelerate the particles while magnets steer and focus them. Beams can be made to collide with a static target or with each other. A collider is a special type of circular accelerator where beams travelling in opposite directions are accelerated and made to interact at designated collision points. A linear accelerator (or linac) is often used as the first stage in an accelerator chain. A synchrotron is an accelerator in which the magnetic field bending the orbits of the particles increases with the energy of the particles. This makes the particles move in a circular path.

AD
The Antiproton Decelerator, the CERN research facility that produces the low-energy antiprotons.

ALICE (A Large Ion Collider Experiment)
One of the four large experiments that will study the collisions at the LHC.

Angstrom
A unit of length equal to 10^{-10} meter.

Annihilation
A process in which a particle meets its corresponding antiparticle and both disappear. Their energy and momentum appears in some other form, producing other particles together with their antiparticles and providing their motion.

Antimatter
The particles that are common in our universe are defined as matter and their antiparticles as antimatter. Every kind of matter particle has a corresponding antiparticle. Charged antiparticles have the opposite electric charge to their matter counterparts. Although antiparticles are extremely rare in the Universe today, matter and antimatter are believed to have been created in equal amounts at the Big Bang. In the particle theory there is almost no *a priori* distinction between matter and antimatter. Their interactions are almost identical. The asymmetry of the universe between these two classes of particles is a deep puzzle which is yet to be fully understood.

Antiparticles
In particle physics every particle with any type of charge or fermion label has a corresponding antiparticle type. Any particle and its antiparticle have identical mass and *spin* but opposite charges. For example the antiparticle of an electron is a positron. It has exactly the same mass as an electron but positive charge.

Some particles are their own antiparticles, the antiparticle of a photon is a photon for instance. Conserved quantities such as baryon number and lepton number are further types of "charges" that are reversed for particle and antiparticle. Thus an electron and an electron neutrino both have electron number +1 while their antiparticles the positron and the anti-electron-neutrino have electron number -1.

Antiproton

The antiparticle of the proton.

ATLAS

One of the four large experiments that will study the collisions at the LHC.

Atom

All ordinary matter is made up of atoms, which are themselves composed of a nucleus and electrons. The protons and neutrons in the nucleus are made of quarks, the smallest known matter particles

Attenuation

The process by which a compound is reduced in concentration over time, through adsorption, degradation, dilution, and/or transformation. Radiologically, it is the reduction of the intensity of radiation upon passage through a medium. The attenuation is caused by absorption and scattering.

- B –

Backscattering

Primary radiation deflected or secondary radiation emitted in the general direction of the incident radiation beam.

Barrier

Radiation-absorbing material, such as lead or concrete, used to reduce radiation exposure. A primary barrier attenuates useful beam to the required degree. A secondary barrier attenuates stray radiation to the required degree.

Baryon

A hadron made from a basic structure of three quarks. The proton and the neutron are both baryon. The antiproton and the antineutron are antibaryons.

Beam

The particles in an accelerator are grouped together in a beam. Beams can contain billions of particles and can be divided into discrete portions called bunches. Each bunch is typically several centimeters long and just a few microns wide.

Bhabha Scattering

Scattering of positrons by electrons.

Big Bang

The name given to the explosive origin of the Universe.

Boson

The collective name given to the particles that carry forces between particles of matter. The general name for any particle with a spin of an integer number (0,1 or 2...) of quantum units of angular momentum. (named for Indian physicist S.N. Bose). The carrier particles of all interactions are bosons. *Mesons* are also bosons.

Bottom Quark or B Quark

The fifth flavor of quark (in order of increasing mass), with electric charge -1/3.

Bound State

This is a state in which a particle is confined within a composite system, for example an atom or a nucleus, because it does not have enough energy to escape. An electron in a atom is bound because of its electrical attraction to the nucleus, which makes the mass of the atom slightly less than the sum of the masses of the electron plus the rest of the atom without that electron.

Brachytherapy

This involves placing the source of radiation directly within the tumor and employs radioactive plaques, needles, tubes, wires, or small "seeds" made of radionuclides. These radioactive materials are placed over the surface of the tumor or implanted within the tumor, or placed within a body cavity surrounded by the tumor.

Bremsstrahlung

X-rays emitted when a charged particle (such as an electron) is decelerated by passing through matter. The word bremsstrahlung is German for "braking radiation".

Bubble Chamber

A chamber filled with liquid at low pressure chosen so that small bubbles form along the path of any charged particle. After each beam pulse a photographic record is made of the chamber and then it is depressurized to clear the bubbles.

- C -

Cathode Ray Tube (CRT)

An evacuated tube containing an anode and a cathode that generates cathode rays (electrons) when operated at a high voltage. The cathode rays produce an image on a screen when they strike phosphors on the screen, causing them to glow.

Calorimeter

An instrument for measuring the amount of energy carried by a particle. In particular, the electromagnetic calorimeter measures the energy of electrons and photons, whereas the hadronic calorimeter determines the energy of hadrons, that is, particles such as protons, neutrons, pions and kaons.

Carrier Particle

A fundamental boson associated with quantum excitations of the force field corresponding to some interaction. Gluons are carrier particles for strong interactions (color force fields), photons are carrier particles of electromagnetic interactions, and the W and Z bosons are carrier particles for weak interactions.

CARE (Co-ordinated Accelerator Research in Europe)

An EU-supported activity to generate a structured and integrated area in accelerator research and development in Europe.

CERN

The major European International Accelerator Laboratory located near Geneva, Switzerland. (European Organisation for Nuclear Research, originally called l'Organisation Européenne pour la Recherche Nucléair). The WWW was created at CERN.

Cerenkov Radiation

A charged particle emits Cerenkov radiation (light) in a cone around its direction of travel when it travels through any medium faster than the speed of light through that medium. (Cerenkov - is the name of the scientist who first recognized the nature of this effect and its possible use for distinguishing particle types.) Although the speed of light in a vacuum is the fastest speed that any particle or light can have, in a medium of any type light travels more slowly because of its interactions with the electric fields of the atoms in the medium and so it is possible for a high energy particle to be faster than light in some material . The blue light in the pools of water you may have seen in pictures of nuclear power plants is Cerenkov radiation from particles produced in the reactor.

Charge

A quantity carried by a particle that determines its participation in an interactions process. A particle with electric charge has electrical interactions; one with strong charge (or color charge) has strong interactions, etc.

Charm

One "flavor" of quarks. Also known as the C quark.

CLIC (Compact Linear Collider)

A site-independent feasibility study aiming at the development of a realistic technology at an affordable cost for an electron–positron linear collider for physics at multi-TeV energies.

Clinac®

Varian trade name for a range of linear accelerator models used in cancer treatment and stereotactic radiosurgery.

CMS (Compact Muon Solenoid)

One of the four large experiments that will study the collisions at the LHC.

CNGS (CERN Neutrinos to Gran Sasso)

A project that aims at the first observation of the tau neutrino by sending a beam of muon neutrinos from CERN to the Laboratori Nazionali del Gran Sasso in Italy.

Coherence

A term that's applied to electromagnetic waves. When they "wiggle" up and down together (in phase) they are said to be coherent. A laser is a good example of coherent light. An ordinary light bulb produces incoherent light much like the random waves produced when many raindrops hit a puddle. Electromagnetic radiation is coherent when the photons are produced in such a way that they are in phase with one another and incoherent when the phases of the photons are random. Partial coherence is an intermediate situation where there a significant fraction of the photons have related phase, but not all of them.

Collider

Special type of accelerator where counter-rotating beams are accelerated and interact at designated collision points. The collision energy is twice that of an individual beam, which allows higher energies to be reached than in fixed target accelerators.

Collimation

The alignment of the direction of the photons, so the beam of radiation can be directed at a well-defined part of a target material.

Collimator

A mechanical device, sometimes called a "slit", installed along the trajectory of a beam to reduce the size of the beam. Collimators are also useful for removing stray radiation.

Color Charge

The charge associated with strong interactions. Quarks and gluons have color charge and consequently participate in strong interactions. Leptons, photons and W and Z bosons do not have color charge and therefore do not participate in strong interactions.

Compton Scattering

The scattering of photons from charged particles is called Compton scattering after Arthur Compton who was the first to measure photon-electron scattering in 1922.

Cosmos

The universe regarded as an orderly, harmonious whole.

Cosmic ray

A high-energy particle that strikes the Earth's atmosphere from space, producing many secondary particles, also called cosmic rays.

CP violation
A subtle effect observed in the decays of certain particles that betrays Nature's preference for matter over antimatter.
Cryogenic distribution line (QRL)
The system used to transport liquid helium around the LHC at very low temperatures. This is necessary to maintain the superconducting state of the magnets that guide the particle beam.
Cryostat
A refrigerator used to maintain extremely low temperatures.
Crystalline
A regularly repeated crystal-like substructure
CT or CAT Scan (also called Computed Tomography or CT)
Detailed pictures of areas of the body created by a computer linked to an x-ray machine. Also called computed tomography (CT) scan or computed axial tomography (CAT) scan.

- D -
Dark matter
Only 4% of the matter in the Universe is visible. The rest is known as dark matter — 26%, and dark energy —70%. Finding out what it consists of is a major challenge for modern science.
Decay
Any process in which a particle disappears and in its place two or more different particles appear.
Detector
A device used to measure properties of particles. Some detectors measure the tracks left behind by particles, others measure energy. The term 'detector' is also used to describe the huge composite devices made up of many smaller detector elements. In the large detectors at the LHC each layer has a very specific task.
Dipole, Dipole Magnet
A magnet with two poles, like the north and south poles of a horseshoe magnet. In an accelerator the dipole magnets are used to steer a particle beam to the left or right by placing one pole above and the other below the beam pipe.
Dipoles are used in particle accelerators to keep particles moving in a circular orbit. In the LHC there are 1232 dipoles, each 15 m long.
Dose (Absorbed Dose)
More specifically referred to as "absorbed dose", this is a measure of the energy deposited within a given mass of a patient. Absorbed dose is quantified by the unit called the "rad".
Dose Calibration
Determining the response of a dosimeter to a known radiation exposure or known absorbed dose. For a beam of radiation, this means determining the absorbed dose rate at a calibrated point in the beam under a specified set of conditions. Normally, such a determination is carried out with a number of beams under different specified conditions.
Dose Equivalent (DE)
Parameter used to express the risk of the deleterious effects of ionization radiation upon living organisms. For radiation protection purposes, the quantity of the effective irradiation incurred by exposed persons, measured on a common scale in sievert (SI) or rem (non-SI).
Dose Rate

A measure of the dose delivered per unit time.

Dosimeter

A radiation sensitive device, e.g., film, monitor ion chamber, TLD, etc., with a known sensitivity that is placed in the beam path of radiation to dose.

Dosimetry

The calculations, measurements and other activities required for determining the radiation dose to be delivered.

Down Quark or D Quark

The second flavor of quark (in order of increasing mass), with electric charge -1/3.

- E -

Electric Field

A force field which defines what acceleration an electric charge placed at rest at any point in space will feel. Electric charges cause electric fields around them, which then apply a force to any other electric charge placed in the field. The electric field E has both a magnitude and a direction at each point in space, and the magnitude and direction of the resulting force on a charge q at that point is given by F= qE. When you get a shock from a door handle after scuffing your feet on a carpet you feel the effect of an electric field accelerating electrons.

Electromagnetic force

The electromagnetic force binds negative electrons to the positive nuclei in atoms, and underlies the interactions between atoms that give rise to molecules and to solids and liquids. Unlike gravity, it can produce both attractive and repulsive effects. Opposite electric charges (positive and negative) and opposite magnetic poles (north and south) attract, but charges or poles of the same type repel each other.

Electromagnetic Interaction

The interaction due to electric charge; this includes magnetic effects that have to do with moving electric charges.

Electromagnetic (em) wave

Electromagnetic waves make up the electromagnetic spectrum.Visible light, ultraviolet, infrared, radio and TV signals are all examples of "everyday" em waves. X-rays, microwaves and high energy photons or gamma rays are also electromagnetic waves.

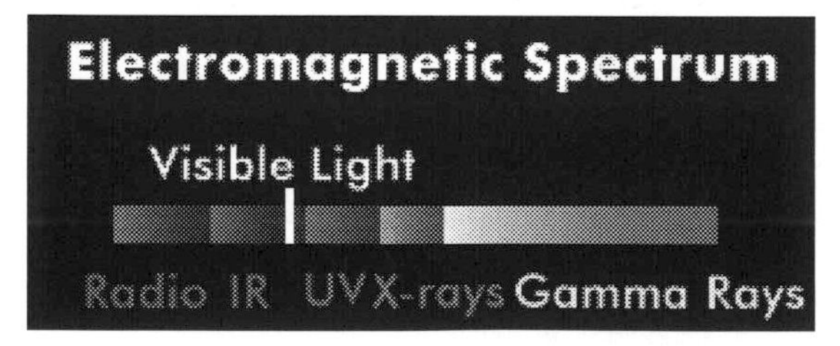

Simplified EM spectrum, representing energy per photon, not total beam energy

Electron

The least massive electrically charged particle, therefore absolutely stable. It is the most common lepton with charge -1. An electron is one of the fundamental particles in nature. Fundamental means that, as far as we know, an electron cannot be broken down into smaller particles. (This concept is one of the things SLAC physicists always challenge by looking for other particles.) Electrons are responsible for many of the phenomena that we observe in everyday life. Mutual repulsion between electrons in the atoms of the floor and those within your shoes keeps you from sinking and disappearing into the floor!!! Electrons carry electrical current and successful manipulation of electrons allows electronic devices, such as the one you are using, to function.

Electron Accelerator
Electrons carry electrical charge and successful manipulation of allows electronic devices to function. The picture and text on the video terminal in front of you is caused by electrons being accelerated and focused onto the inside of the screen, where a phosphor absorbs the electrons and light is produced. A television screen is a simple, low-energy example of an electron accelerator. A typical medical electron accelerator used in medical radiation therapy is about 1000 times more powerful than a color television set, while the electron accelerator at SLAC is about 2,000,000 times more powerful than a color TV. One example of an electron accelerator used in radiotherapy is the Clinac, manufactured by Varian Associates in Palo Alto, CA.
Electron Beam
The stream of electrons generated by the electron gun and accelerated by the accelerator guide.
Electron Beam Therapy
Treatment by electrons accelerated to high energies in a linear accelerator. Primarily used for lesions situated at or near the surface.
Electronic Structure
The distribution of electrons in the material and the energies related to changes in this distribution.
Electronvolt (eV)
A unit of energy or mass used in particle physics. One eV is extremely small, and units of a million electronvolts, MeV, or thousand million electronvolts, GeV, are more common. The latest generation of particle accelerators reaches up to several million million electronvolts, TeV. One TeV is about the energy of motion of a flying mosquito.
Elementary Particles
The name given to protons, neutrons and electrons before it was discovered that protons and neutrons had substructure (quarks). Today we use the term "fundamental" for the six types of quarks and the six leptons and their antiparticles, which have no known substructure. Gluons, photons and W and Z bosons are also fundamental particles. All other particles are composite, that is made from combinations of fundamental particles.
Enabling Grids for E-SciencE (EGEE) project
An EU-funded project led by CERN, now involving more than 90 institutions in over 30 countries worldwide, to provide a seamless Grid infrastructure that is available to scientists 24 hours a day.
End-cap
Detector placed at each end of a barrel-shaped detector to provide the most complete coverage in detecting particles.
eV -Electronvolt
The basic *unit* of energy used in high energy physics. It is the energy gained by one electron when it moves through a potential difference of one volt. By definition an eV is equivalent to 1.6 x 10^{-19} joules. This is a very small amount of energy and the more commonly used multiples are MeV (million eV), GeV (billion eV or giga-electronvolt) and TeV (trillion eV).
Evacuated
Sealed and pumped down to a pressure very much below atmospheric pressure. Typical pressures inside accelerators or waveguides are about 10^{-12} times atmospheric pressure.
Event

An event occurs when two particles collide or a single particle decay. Particle theories predict the probabilities of various events occurring when many similar collisions or decays are studied. They cannot predict the outcome for a single collision or decay.

Excitation

The addition of energy to a system, transferring it from its ground state to an excited state. Excitation of a nucleus, an atom, or a molecule can result from absorption of photons or from inelastic collisions with other particles.

- F -

Fermilab

Fermi National Accelerator Laboratory in Batavia, Illinois (named for particle physics pioneer Enrico Fermi).

Fermion

General name for a particle that is a matter constituent, characterized by spin in odd half integer quantum units (1/2,3/2,5/2...). Named for Italian physicist Enrico Fermi. Quarks, leptons and baryons are all fermions.

Flavor

The name used for the different quark types and the different lepton types. The six flavors of quarks are *up, down, strange, charm, bottom, top*, in increasing order of mass. The flavors of charged leptons are *,* *muon* and *tau*, again in increasing order of mass. For each charged lepton flavor there is a corresponding *neutrino* flavor.

Forces

There are four fundamental forces in nature. Gravity is the most familiar to us, but it is the weakest. Electromagnetism is the force responsible for thunderstorms and carrying electricity into our homes. The two other forces, weak and strong, are confined to the atomic nucleus. The strong force binds the nucleus together, whereas the weak force causes some nuclei to break up. The weak force is important in the energy-generating processes of stars, including the Sun. Physicists would like to find a theory that can explain all these forces. A big step forward was made in the 1960s when the electroweak theory uniting the electromagnetic and weak forces was proposed. This was later confirmed in a Nobel-prize-winning experiment at CERN.

Fundamental Interaction

The known fundamental interactions are the *strong, electromagnetic, weak* and gravitational interaction. These interactions explain all observed physical processes but do not explain particle masses. Any force between two objects is due to one or another of these interactions. All known particle decays can be understood in terms of these strong, electromagnetic or weak interactions.

Fundamental Particle

A particle with no internal substructure. In the *Standard Model*, the quarks, leptons, photons, Gluons ,W-boson and Z-bosons are fundamental. All other objects are made from these particles.

- G -

Gamma Rays

Gamma rays are electromagnetic waves or photons emitted from the nucleus (center) of an atom. See also: photon.

Geiger Counter

A device that detects the passage of charged particles via the ionization of gas that they cause as they pass through a region. Used to detect the particles produced in certain forms of radioactivity.

GeV (Giga Electron Volt)

Unit of energy equal to that acquired by a particle with one electronic charge in passing through a potential difference of one billion volts.

Gluon

The carrier particle of the strong interaction. Gluon is a special particle, called boson, that carries the strong force, one of the four fundamental forces, or interactions, between particles.

Gravitational Interaction

An attractive force between any two objects or particles. The "charge" that determines the strength of the gravitational interaction is energy. For a static object it is mass-energy but in fact all forms of energy both cause and feel gravitational effects.

Gray (Gy)

The SI unit for absorbed dose equal to an energy absorbed of one joule/kilogram in the stated medium.

- H -

Hadron

Any particle made of quarks ,antiquarks, and Gluons, i.e. a meson or a baryon. All such particles have no strong charge (i.e. are strong charge neutral objects) but participate in residual strong interactions due to the strong charges of their constituents.

Half-life

The time required for half the nuclei in a sample of a specific isotopic species to undergo radioactive decay.

Heterogeneous

Materials made from more than one type of substructure. Opposite of homogeneous, a material with a uniform composition. (Note that a material can be chemically complex yet homogeneous.)

Higgs boson

A particle predicted by theory. It is linked with the mechanism by which physicists think particles acquire mass.

High-Energy Physics

A branch of science that tries to understand the interactions of the fundamental particles, such as , photons, neutrons and protons (and many others than can be created). These particles are the basic building blocks of everyday matter, making up the human body as well as the entire universe. This type of physics is called *high-energy* because very powerful machines, such as the Two-Mile Accelerator at SLAC, are created to make these particles go very fast so that they can probe deeply into other particles and try to understand what they are made of.

- I -

Induced Radioactivity

Radioactivity produced in materials, especially metals, exposed to high-energy photons or neutrons.

Injector

System that supplies particles to an accelerator. The injector complex for the LHC consists of several accelerators acting in succession.

Intensity
The amount of radiation, for example, the number of photons arriving in a given time
Interaction
A process in which a particle decays or it responds to a force due to the presence of another particle (as in a collision).
Ion
An ion is an atom with one or more electrons removed (positive ion) or added (negative ion).
Ionization
The process by which a neutral atom or molecule acquires a positive or negative charge.
Ionizing Radiation
Radiation that has enough energy to eject from electrically neutral atoms, leaving behind charge atoms or ions. There are four basic types of ionizing radiation: Alpha particles (helium nuclei), beta particles (electrons), neutrons, and gamma rays (high frequency electromagnetic waves, x-rays, are generally identical to gamma rays except for their place of origin.) Neutrons are not themselves ionizing but their collisions with nuclei lead to the ejection of other charged particles that do cause ionization.
Isotope
Slightly different versions of the same element, differing only in the number of neutrons in the atomic nucleus—the number of protons is the same.

- J -
Jet
The name physicists give to a cluster of particles emerging from a collision or decay event all traveling in roughly the same direction and carrying a significant fraction of the energy in the event. The particles in the jet are chiefly hadrons.

K -
Kaon
A meson containing a strange quark (or antiquark). Neutral kaons come in two kinds, long-lived and short-lived. The long-lived ones occasionally decay into two pions, a CP-violating process. (See also Particles.)
Kelvin
A unit of temperature. One kelvin is equal to one degree Celsius. The Kelvin scale begins at absolute zero, –273.15°C, the coldest temperature possible.
Klystron
An evacuated tube used as an oscillator or amplifier at microwave frequencies. In the klystron, an electron beam is velocity modulated (periodically bunched) to produce large amounts of power.

- L -
LCG (LHC Computing Grid)
The mission of the LCG is to build and maintain a data-storage and analysis infrastructure for the entire high-energy physics community that will use the LHC.
LEP
The Large Electron–Positron Collider, which ran at CERN until 2000.
Lepton
A class of elementary particle that includes the electron. Lepton is

a fundamental matter particle that does not participate in strong interactions. The charge leptons are the (e), the muon (μ), the tau (τ) and their antiparticles. Neutral leptons are called neutrinos (η).

Linear Accelerator

A type of particle accelerator in which charged particles are accelerated in a straight line, either by a steady electrical field or by means of radiofrequency electric fields. In the latter variety, the passage of the particle is synchronized with the phase of the accelerating field. The SLAC Linear Accelerator (linac) is a two-mile long accelerator, consisting of a cylindrical, disc-loaded, copper waveguide placed on concrete girders in a tunnel about 25 feet underground.

LHC

The Large Hadron Collider, CERN's biggest accelerator.

LHCb (Large Hadron Collider beauty)

One of the four large experiments that will study the collisions at the LHC.

Linac

An abbreviation for linear accelerator.

- M -

Mammography

Mammography is a low-dose x-ray procedure that creates an image of the breast. The x-ray image is called a mammogram.

Matter

We call the commonly observed particles such as protons, neutrons and matter particles, and their *antiparticles* are then antimatter.

Mean-life

The *average* time during which a system, such as an atom, nucleus, or elementary particle, exists in a specified form. Also known as the average life.

Medical Physics

This is a very diverse field that applies the knowledge gained in other areas of physics (such as high-energy Physics) to heal people. Radiation therapy is one example. CAT scans, mammography, and other x-ray imaging techniques are diagnostic techniques that have also been developed by physicists working in medicine. Another important example is the technology that went into building the particle accelerator at SLAC, which has been adapted for use in hospitals to treat cancer patients with beams of and x-rays.

Meson

A hadron with the basic structure of one quark and one antiquark.

MeV (Mega Electron Volt)

Energy equal to that acquired by a particle with one electronic charge in passing through a potential difference of one million volts.

Micron

One millionth of a meter; also known as a micrometer.

Microwaves

Waves of electromagnetic radiation that oscillate from approximately 10^9 to 3 x 10^{11}Hz. (cycles per second). SLAC's klystrons produce microwaves of 2,856 MHz.

Model / Scientific model

The scientific model is a widely used tool in many fields of modern science. Scientists construct and develop 'models' to describe a scientific theory in the context of related phenomena. In general, a model is based on a theory (a set of hypothesis), acting on a set of parameters obtained from actual experimental data and/or from observations.

Computer simulations may sometimes be used to test the reliability of a model. If it was found to be reasonably reliable, the simulation can even be used to predict what would happen if the initial parameters were different.

Møller Scattering

Scattering of electrons by electrons.

Momentum

Momentum is a property of any moving object. For a slow moving object it is given by the mass times the velocity of the object. For an object moving at close to the speed of light this definition gets modified. The total momentum is a conserved quantity in any process. Physicists use the letter p to represent momentum, presumably because m was already used for mass, n for number, and o is too much like zero.

Monte Carlo Calculations

There is a gaming aspect to Monte Carlo calculations. Every simulation is based up events that happen randomly, and so the outcome of a calculation is not always absolutely predictable. This element of chance reminds one of gambling and so the originators of the Monte Carlo technique, Ulam and von Neumann, both respectable scientists, called the technique *Monte Carlo* to emphasize its gaming aspect.

Muon

A particle similar to the electron, but some 200 times more massive. (See also Particles.)

Muon chamber

A device that identifies muons, and together with a magnetic system creates a muon spectrometer to measure momenta. The detector chamber is designed so that the only charged particles that can get out to this layer are muons.

- N -

Neutrino

A neutral particle that hardly interacts at all. A lepton with no electric charge. Neutrinos participate only in weak (and gravitational) interactions and therefore are very difficult to detect. There are three known types of neutrino, all of which have very low or possibly even zero mass.

Neutron

A baryon with electric charge zero; it is a hadron with a basic structure of two down quarks and one up quark (held together by gluons).

Nucleus ,Nucleon

A collection of protons and neutrons that form the core of an atom (plural: *nuclei).*

- P -

Particles

In "particle physics", a subatomic object with definite mass and charge. There are two groups of elementary particles, quarks and leptons. The quarks are up and down, charm and strange, top and bottom. The leptons are electron and electron neutrino, muon and muon neutrino, tau and tau neutrino. There are four fundamental forces, or interactions, between particles, which are carried by special particles called bosons. Electromagnetism is carried by the photon, the weak force by the charged W and neutral Z bosons, the strong force by the gluon; gravity is probably carried by the graviton, which has not yet been discovered. Hadrons are particles that feel the strong force. They include mesons, which are composite particles made up of a quark–antiquark pair and baryons, which are particles containing three quarks. Pions and kaons are types of meson. Neutrons and protons (the constituents of ordinary matter) are baryons; neutrons contain one up and two down quarks; protons two up and one down quark.

Pair Production and Annihilation

Whenever sufficient energy is available to provide the mass-energy, a particle and its matching antiparticle can be produced (pair production). When a particle collides with its matching antiparticle they may annihilate -- which means they both disappear and their energy appears as some other particles -- with balanced number of particles and antiparticles for each type. All conservation laws are obeyed in these processes.

Pauli Exclusion Principle

No two fermions of the same type can exist in the same state at the same place and time.

PET Scan

Positron Emission Tomography scanning uses an array of stationary detectors around the patient and using the spatial 180 degree opposing properties of the 0.511-MeV annihilation radiation from positron-emitting radiopharmaceuticals deposited in the organ or region of interest. The name tomography refers to the fact that the scanner computes a "slice" of the scanned object, not just a flat image. Each slice really is a volumetric (tomo-) image (-graphy).

Phosphor

A substance that emits light when excited by radiation.

Photon

The carrier particle of the electromagnetic interaction. Depending on its frequency (and therefore its energy) photons can have different names such as visible light, X rays and gamma rays. We describe light in several ways. When we talk about "photons" we generally think of uncharged particles with out mass that carry energy (but be careful, there are other particles like this!). Photons of light are known by other names too, such as *gamma rays* and x-rays. Low-energy forms are called *ultraviolet rays*, *infrared rays*, even *radio waves*! A photon is one of the fundamental particle in nature and it plays an important role involving electron interactions. Photons are the most familiar particles in everyday existence. The light we see, the radiant heat we feel, microwaves we cook with, are make use of photons of different energies. An x-ray is simply a name given to the most energetic of these particles.

Pion

The lightest type of mesons. They are copiously produced in high energy particle collisions.

Planck's Constant

A fundamental physical constant, the elementary quantum of action, It is the ratio of the energy of a photon to its frequency and is equal to 6.626069 x 10^{-34} Joule seconds. Symbolized by h.

Polarization

A polarized particle beam is a beam of particles whose spins are aligned in a particular direction. The polarization of the beam is the fraction of the particles with the desired alignment.

Positron

Antiparticle of the electron.

Proton

The most common hadron, a baryon with electric charge +1 and opposite to that of the electron. Protons contain a basic structure of two up quarks and one down quark . The nucleus of a hydrogen atom is a proton. A nucleus with atomic number Z contains Z protons; therefore the number of protons is what distinguishes the different chemical elements.

PS

The Proton Synchrotron, backbone of CERN's accelerator complex.

- Q -

Quadrupole

A magnet with four poles, used to focus particle beams rather as glass lenses focus light. There are 392 main quadrupoles in the LHC.

Quantum

When used as a noun (plural quanta): a discrete quantity of energy, momentum or angular momentum, given in units involving Planck's constant h. For example electromagnetic radiation of a given frequency f is composed of quanta (also called photons) with energy hf.

When used as an adjective (as in quantum theory, quantum mechanics, quantum field theory): defines the theory as involving quantities which depend on Planck's constant h. In such theories radiation comes in discrete quanta as described above; angular momenta must be integer units of h, except that the intrinsic angular momenta of fundamental particles are integer multiples of $1/2h$; and solutions for the possible states of a particle in a potential (such as the states of an electron the electrical potential due to an atomic nucleus) occur only for certain discrete energies.

Quantum Mechanics

The laws of physics that apply on very small scales. The essential feature is that energy, momentum and angular momentum as well as charge come in discrete amounts called quanta.

Quantum Number

A number that labels a state, it denotes the number of quanta of a particular type that the state contains. Electric charge given as an integer multiple of the electron's charge is an example of a quantum number.

Quantum electrodynamics (QED)

The theory of the electromagnetic interaction.

Quantum chromodynamics (QCD)

The theory for the strong interaction, analogous to QED.

Quark

A fundamental matter particle that has strong interactions. Quarks have an electric charge of either +2/3 (up, charm and top) or -1/3 (down, strange and bottom) in units where the proton charge is 1.

Quark–gluon plasma (QGP)

A new kind of plasma in which protons and neutrons are believed to break up into their constituent parts. QGP is believed to have existed just after the Big Bang.

Quench

A quench occurs in a superconducting magnet when the superconductor warms up and ceases to superconduct.

- R -

Rad

One rad is equal to an energy absorption of 100 ergs in a gram of any material. An "erg" is a unit for quantifying energy (just like a "mile" is used for measuring distance).

Radiation

Radiation is energy in transit in the form of high speed particles and electromagnetic waves. Radiation is further defined into ionizing and non-ionizing radiation.

- *Ionizing radiation* is radiation with enough energy so that during an interaction with an atom, it can remove tightly bound electrons from their orbits, causing the atom to become charged or ionized. Examples are X-rays and electrons.
- *Non-ionizing radiation* is radiation without enough energy to remove tightly bound electrons from their orbits around atoms. Examples are microwaves and visible light.

Radiation Oncology
Treatment of tumors with ionization radiation.

Radiation Sickness
The syndrome associated with intense acute exposure to ionizing radiation.

Radiation Transport
The field of nuclear science dealing with the prediction and measurement of the movement of electromagnetic radiation or particles through matter.

Radioactivity
The property of spontaneously emitting alpha, beta, and/or gamma radiation as a result of nuclear disintegration.

Radiotherapy (also called Radiation Therapy)
Medical therapy consisting of one or more treatments with ionization radiation.

Radio Frequency (RF)
Any ac frequency at which coherent electromagnetic radiation of energy is possible. Usually considered to denote frequencies above 150 kilohertz and extending up to the infrared range.

Radiography
The making of shadow images on a photographic emulsion by the action of ionization radiation. The image is the result of the differential attenuation of the radiation in its passage through the object being radiographed.

Radioimmunotherapy
This type of radiotherapy is the application of monoclonal antibodies that have been tagged with high activities of suitable radionuclides. These tumor-specific antibodies are derived from the patient's own cancer and, hence, they selectively target this tumor when injected into the patient. Also known as monoclonoal antibody therapy.

Radiology
The branch of medicine that deals with the diagnostic and therapeutic applications of radiation.

Radionuclide
Radionuclides are materials that produce ionization radiation, such as X-rays, gamma rays, alpha particles, and beta particles.

Relativistic
Traveling at a significant fraction of the speed of light.

Residual Interaction
Interaction between objects that do not carry a charge but that contain constituents that do have a charge. Although some chemical substances involve electrically-charged ions, much of chemistry is due to residual electromagnetic interactions between electrically neutral atoms. The residual strong interaction between protons and neutrons, due to the strong charges of their quark constituents, is responsible for the binding of the nucleus.

Ring Imaging C Herenkov (RICH) counter
A kind of particle detector that uses the light emitted by fast-moving particles as a means of identifying them.

- S -

Scalar
Any quantity that has only magnitude as opposed to both magnitude and direction. For example mass is scalar quantity. By convention in physics the word speed is a scalar quantity, having only magnitude, while the word velocity is used to denote both the speed and the direction of the motion and is thus a *vector* quantity.
Scintillation
The flash of light emitted by an electron in an excited atom falling back to its ground state.
Sextupole
A magnet with six poles, used to apply corrections to particle beams. At the LHC, eight- and ten-pole magnets will also be used for this purpose.
Shield
A mass of attenuating material used to prevent or reduce the passage of radiation or particles.
Shower (also called Electromagnetic Cascade Shower)
It can create photons by interacting with a medium. In a similar way, photons can create electrons and their antiparticles, positrons, by interacting with a medium. So, imagine a very high-energy electron, of the sort used at SLAC, impinging on some material. The electron can set photons into motion and these photons can, in turn, set electrons and positrons into motion, and this process can continue to repeat. One high-energy electron can set thousands of particles into motion. Albert Einstein's famous relation governing the equivalence of matter and energy ($E = mc^2$) governs this process -- namely, matter (electrons and positrons) can be creased from pure energy and vice versa. The particle creation process only stops when the energy runs out.
Silicon Vertex Detector
This is a silicon based detector similar to that in a digital camera. It provides precision particle tracking by connecting the dots due to a particle passing through its multiple layers. This allows one to reconstruct any vertex from which two or more tracks emerge. Such a vertex, if outside the beam collision region, indicates the position of a particle decay.
SLAC
SLAC National Accelerator Laboratory - where this virtual visitor center is located, along with many real facilities.
SPEAR
Stanford Positron Electron Accelerating Ring.
Spectrometer
In particle physics, a detector system containing a magnetic field to measure momenta of particles.
SPECT
Single-Photon Emission Computerized Tomography involves scanning involving the rotation of detectors around a patient and acquires information on the concentration of radionuclides introduced to the patient's body. This is analogous to CT imaging with x-rays. See also PET.
Spectral Range

The range of wavelengths of the electromagnetic radiation that can be produced.

Spin

The name given to the angular momentum carried by a particle. For composite particles the spin is made up from the combination of the spins of the constituents plus the angular momentum of their motion around one-another. For fundamental particles spin is an intrinsic and inherently quantum property, it cannot be understood in terms of motions internal to the object.

SPS

The Super Proton Synchrotron. An accelerator that provides beams for experiments at CERN, as well as preparing beams for the LHC.

Standard Model

A collection of theories that embodies all of our current understanding about the behaviour of fundamental particles.

Strong force

The strong force holds quarks together within protons, neutrons and other particles. It also prevents the protons in the nucleus from flying apart under the influence of the repulsive electrical force between them (because they all have positive charge). Unlike the more familiar effects of gravity and electromagnetism where the forces become weaker with distance, the strong force becomes stronger with distance.

Superconductivity

A property of some materials, usually at very low temperatures, that allows them to carry electricity without resistance. If you start a current flowing in a superconductor, it will keep flowing forever—as long as you keep it cold enough.

Superfluidity

A phase of matter characterized by the complete absence of resistance to flow.

Supersymmetry

A theory that predicts the existence of heavy 'superpartners' to all known particles. It will be tested at the LHC.

Synchrotron

A particle accererator in which a magnetic field bends the orbits of the particles, which increases their energy. The particles travel in a circular path.

Stable

Does not decay.

Standard Model

Physicists' name for the current theory of fundamental particles and their interactions.

Steradian

The unit of "stereo angle" or solid angle, the angle in three dimensions. It is related to the radian, which is the unit of angle in two dimensions (abbreviated "sterad").

Sterotactic Radiosurgery

This involves the use of multiple small pencil of radiation fired from many different directions and all aimed at the tumor. Machines used include the "gamma-knife," with several hundred small, high-activity Cobalt-60 sources and conventional medical linear accelerators equipped with specially designed sterotactic hardware.

Storage Ring

A circular (or near circular) structure in which either high energy and/or positrons, or protons and/or antiprotons can be circulated many times and thus "stored". Used to achieve high energy collisions. Because of the very different masses of protons and electrons a storage ring must be designed for one or the other type and cannot work for both.

Strange Quark
The third flavor of quark (in order of increasing mass), with electric charge -1/3.
String Theory
A theory of elementary particles incorporating relativity and quantum mechanics in which the particles are viewed not as points but as extended objects. String theory is a possible framework for constructing unified theories which include both the microscopic forces and gravity.
Strong Force
The fundamental strong force is the force between quarks and Gluons that makes them combine to form the observed hadrons, such as protons and neutrons. It also causes forces between hadrons, such as the strong nuclear force that makes protons and neutrons bind together to form nuclei.
Strong Interaction
The interaction responsible for binding quarks and Gluons to make hadrons. Residual strong interactions provide the nuclear binding force. In nuclear physics the term strong interaction is also used for this residual effect. (As a parallel, the force between electrically charged particles is an electromagnetic interaction, the force between neutral atoms that leads to the formation of molecules is a residual electromagnetic effect.)
Subatomic Particle
Any particle that is small compared to the size of the atom.
Sun's Corona
The luminous irregular envelope of highly ionized gas outside the chromosphere of the sun.
Supernova
A star that explodes and becomes extremely luminous in the process.
Synchrotron Radiation
Whenever a charged particle undergoes accelerated motion it radiates electromagnetic energy. A common example is the emission of radio waves when electrons move back and forth in a radio antenna. A charged particle traveling in the arc of a circle is also undergoing acceleration, due to its change in direction. The radiation emitted by such particles is called synchrotron radiation and it is particularly intense and very directional when electrons traveling at close to the speed of light are bent in magnetic fields.

- T -
Target
A metallic object placed in the beam of to produce x-rays.
Tau
The third charged lepton (in order of increasing mass), with electric charge -1.
TeV (Tera Electron Volt)
Unit of energy equal to that acquired by a particle with one electronic charge in passing through a potential difference of one trillion volts.
Top Quark
The sixth flavor of quark (in order of increasing mass) with electric charge +2/3.
Track
The record of the path of a particle traversing a detector.
Tracking Chamber
A section of a particle detector capable of detecting the passage of electrically charged particles.

Transfer line
Carries a beam of particles, e.g., protons, from one accelerator to another using magnets to guide the beam.
Trigger
An electronic system for spotting potentially interesting collisions in a particle detector and triggering the detector's read-out system to record the data resulting from the collision.

- U -
Unified Field Theory
A unified field theory is one that attempts to combine any two or more of the known interaction types (strong, electromagnetic, weak and gravitational) in a single theory so that the two distinct types of interaction are seen as two different aspects of a single mathematical structure. A 'grand unified' theory (or GUT) unifies three of the four types (strong, weak and electromagnetic interactions) in this way. The benefit is that the unification gives a simpler overall theory and predicts relationships between parameters that are otherwise independent.
Units
The units one uses should be of a size that makes sense for the particular subject at hand. It is easiest to define units in each area of science and then relate them to one another than to go around measuring particle masses in grams or cheese in proton mass units.
In particle physics the standard unit is the unit of energy GeV. One eV (electron Volt) is the amount of energy that an gains when it moves through a potential difference of 1 Volt (in a vacuum). G stands for Giga, or 10^9. Thus a GeV is a billion (in US counting) electron Volts. The mass-energy of a proton or neutron is approximately 1 GeV.
Unstable
Matter that is capable of undergoing spontaneous change, as in a radioactive nuclide or an excited nuclear system. An unstable particle is any elementary particle that spontaneously decays into other particles.
Up Quark
The first flavor of quark (in order of increasing mass), with electric charge +2/3.

- V -
Vacuum
A volume of space that is substantively empty of matter, so that gaseous pressure is much less than standard atmospheric pressure.
Vector
Any quantity that has both magnitude and direction. Velocity is a vector. An example might be 55 mph south. 55 is the magnitude and south is the direction.

- W -
Waveguid
An evacuated rectangular copper tube that provides a path for microwaves to travel along. They are very carefully designed for a particular wavelength microwave, so as to transmit as much energy as possible.
W Boson
A carrier particle of the weak interaction.

Weak force

The weak force acts on all matter particles and leads to, among other phenomena, the decay of neutrons (which underlies many natural occurrences of radioactivity) and allows the conversion of a proton into a neutron (responsible for hydrogen burning in the centre of stars). It can be either an attractive or a repulsive force.

Weak Interaction

The interactions responsible for all processes in which flavor changes; hence for the instability of heavy leptons and quarks, and particles that contain them. Weak interactions that do not change flavor have also been observed.

- X -

X-rays

X-rays are electromagnetic waves (photons of light) emitted by energy changes of . These energy changes are either in the electron orbital shells that surround an atom or are due to the slowing down (i.e., interaction) of electrons in matter, such as a "target" in an x-ray machine.

- Z -

Z Boson

Also known as a Z Particle. A carrier particle of weak interactions. It is involved in weak processes that do not change flavor.

10

REFERENCES

1. US patent 1948384, Ernest O. Lawrence, "Method and apparatus for the acceleration of ions", issued 1934-02-20
2. Alonso, M.; Finn, E. (1996). Physics. Addison Wesley.
3. Widerøe, R. (17 December 1928). "Ueber Ein Neues Prinzip Zur Herstellung Hoher Spannungen". Archiv fuer Elektronik und Uebertragungstechnik (in German) 21 (4): 387.
4. "Breaking Through: A Century of Physics at Berkeley. 2. The Cyclotron.". Bancroft Library, UC Berkeley. 2012-02-25. Archived from the original on 2012-02-25.
5. Emelyanov, V. S. (November 1971). "Nuclear Energy in the Soviet Union". Science and Public Affairs: Bulletin of the Atomic Scientists. XXVII (9): 39. "State Institute of Radium, founded in 1922, now known as V. G. Khlopin Radium Institute"
6. V.G. Khlopin Radium Institute. History / Memorial and History / Chronology. Retrieved 25 February 2012.
7. A Timeline Of Major Particle Accelerators. Andrew Robert Steere,2005, A Thesis for the degree of M.SC in Physics,Department of Physics and Astronomy , Michigan State University.
8. Thomas, L. H. (1938). "The Paths of Ions in the Cyclotron I. Orbits in the Magnetic Field". Physical Review 54 (8): 580–588. doi:10.1103/PhysRev.54.580. edit
9. "In a breakthrough, Canadian researchers develop a new way to produce medical isotopes". The Globe And Mail (Vancouver). Tuesday, Feb. 21, 2012.
10. http://accelconf.web.cern.ch/accelconf/Cyclotrons2010/papers/tum2cio01.pdf
11. Courant, E. D.; Snyder, H. S. (Jan 1958). "Theory of the alternating-gradient synchrotron". Annals of Physics 3 (1): 1–48. doi:10.1006/aphy.2000.6012. edit
12. Wille, Klaus (2001). Particle Accelerator Physics: An Introduction. Oxford University Press. ISBN 978-0-19-850549-5. (slightly different notation)
13. Schopper, Herwig F. (1993). Advances of accelerator physics and technologies. World Scientific. ISBN 981-02-0957-6. Retrieved March 9, 2012.

14. Wiedemann, Helmut (1995). Particle accelerator physics 2. Nonlinear and higher-order beam dynamics. Springer. ISBN 0-387-57564-2. Retrieved March 9, 2012.
15. Lee, Shyh-Yuan (2004). Accelerator physics (2nd ed.). World Scientific. ISBN 978-981-256-200-5.
16. Chao, Alex W.; Tigner, Maury, eds. (1999). Handbook of accelerator physics and engineering (2nd ed.). World Scientific. ISBN 9810238584.
17. Chao, Alex W. (2012). Reviews of Accelerator Science and Technology Volume 4. World Scientific. ISBN 978-981-438-398-1.
18. Gerard K. O'Neill, Storage-Ring Synchrotron: Device for High-Energy Physics Research, Physical Review, Vol. 102 (1956); pages 1418-1419.
19. Widerøe, R. (17 December 1928). "Ueber Ein Neues Prinzip Zur Herstellung Hoher Spannungen". Archiv fuer Elektronik und Uebertragungstechnik 21 (4): 387.
20. Ising, Gustav (1928). "Prinzip Einer Methode Zur Herstellung Von Kanalstrahlen Hoher Voltzahl". Arkiv Fuer Matematik, Astronomi Och Fysik 18 (4).
21. "CERN reports on progress towards LHC restart". CERN . 19 June 2009. http://press.web.cern.ch/press/PressReleases/Releases2009/PR09.09E.html. 2009-07-21.
22. Matthew Early Wright (April 2005). "Riding the Plasma Wave of the Future". Symmetry: Dimensions of Particle Physics (Fermilab/SLAC), p. 12.
23. Briezman, et al. "Self-Focused Particle Beam Drivers for Plasma Wakefield Accelerators." 13 May 2005.
24. Phys. Rev. D 66, 091901 (2002) http://prola.aps.org/abstract/PRD/v66/i9/e091901
25. "Review of the Safety of LHC Collisions: LHC Safety Assessment Group" http://lsag.web.cern.ch/lsag/LSAG-Report.pdf.
26. "R. Jaffe et al., Rev. Mod. Phys. **72**, 1125–1140 (2000).". http://arxiv.org/abs/hep-ph/9910333.
27. Ellis J, Giudice G, Mangano ML, Tkachev I, Wiedemann U (LHC Safety Assessment Group) (5 September 2008). "Review of the Safety of LHC Collisions". "Journal of Physics G: Nuclear and Particle Physics. 35, 115004 (18pp). doi:10.1088/0954-3899/35/11/115004. CERN record. arXiv:0806.3414.
28. Stanley Humphries (1999) Principles of Charged Particle Acceleration Particle Accelerators around the world

29. Wolfgang K. H. Panofsky: The Evolution of Particle Accelerators & Colliders, (PDF), Stanford, 1997
30. P.J. Bryant, A Brief History and Review of Accelerators (PDF), CERN, 1994.
31. Heilbron, J.L.; Robert W. Seidel (1989). Lawrence and His Laboratory: A History of the Lawrence Berkeley Laboratory. Berkeley: University of California Press. ISBN 0-520-06426-7. http://ark.cdlib.org/ark:/13030/ft5s200764/.
32. David Kestenbaum, Massive Particle Accelerator Revving Up NPR's Morning Edition article on 9 April 2007
33. Ragnar Hellborg (ed.), ed (2005). Electrostatic Accelerators: Fundamentals and Applications. Springer. http://books.google.com/books?.
34. "Cathode Ray Tube". Medical Discoveries. Advameg, Inc.. 2007. http://www.discoveriesinmedicine.com/Bar-Cod/Cathode-Ray-Tube-CRT.html. Retrieved 2008-04-27.
35. Wong, May (October 22, 2006). "Flat Panels Drive Old TVs From Market". AP via USA Today. http://www.usatoday.com/tech/products/gear/2006-10-22-crt-demise_x.htm. Retrieved 2008-10-08.
36. "End of an era". The San Diego Union-Tribune. 2006-01-20. http://www.signonsandiego.com/uniontrib/20060120/news_1n20sony.html. Retrieved 2008-06-12.
37. "The basics of TV power". CNET. 2007. http://reviews.cnet.com/4520-6475_7-6400401-2.html. Retrieved 2007-01-13.
38. Wilson, R.R. (1978), The Tevatron, Batavia, Illinois: Fermilab, FERMILAB-TM-0763, http://lss.fnal.gov/archive/test-tm/0000/fermilab-tm-0763.shtml
39. "Fermilab physicists discover "doubly strange" particle". Fermilab. September 3, 2008. http://www.fnal.gov/pub/presspass/press_releases/Dzero_Omega-sub-b.html. Retrieved 2008-09-04.
40. "Observation of the doubly strange b baryon Omega_b-, Fermilab-Pub-08/335-E". Fermilab. http://arxiv.org/abs/0808.4142. Retrieved 2008-09-05.
41. R.W. Gurney and E.U. Condon, Nature 122, 439 (1928) and G. Gamov, Zeit f. Phys. 51, 204 (1928).
42. J.D. Crockcroft and E.T.S. Walton, "Experiments with high velocity ions", Proc. Royal ,Soc., Series A 136 (1932), 619–30.
43. R.J. Van de Graaff, "A 1,500,000 volt electrostatic generator", Phys. Rev., 387, (Nov.1931), 1919–20.

44. G. Ising, Arkiv för Matematik, Astronomi och Fysik, 18 (1924), 1–4.
45. E.O. Lawrence and N.E. Edlefsen, Science, 72 (1930), 376–7.
46. E.O. Lawrence and M.S. Livingston, "The production of high speed light ions without the use of high voltages", Phys. Rev., 40 (April 1932), 19–35.
47. W. Paul, "Early days in the development of accelerators", Proc. Int. Symposium in Honour of Robert R. Wilson, Fermilab, 1979 (Sleepeck Printing Co. Bellwood, Ill.,1979), 25–688.
48. Kerst, D. W. (1940). "Acceleration of Electrons by Magnetic Induction". Physical Review 58 (9): 841. Bibcode:1940PhRv...58..841K. doi:10.1103/PhysRev.58.841.
49. Kerst, D. W. (1941). "The Acceleration of Electrons by Magnetic Induction". Physical Review 60: 47–53. Bibcode:1941PhRv...60...47K. doi:10.1103/PhysRev.60.47.
50. Kerst, D. W.; Serber, R. (Jul 1941). "Electronic Orbits in the Induction Accelerator". Physical Review 60 (1): 53–58. Bibcode:1941PhRv...60...53K. doi:10.1103/PhysRev.60.53.
51. Wideröe, R. (17 Dec 1928). "Über ein neues Prinzip zur Herstellung hoher Spannungen". Archiv für Elektrotechnik (in German) 21 (4): 387–406. doi:10.1007/BF01656341.
52. Dahl, F. (2002). From nuclear transmutation to nuclear fission, 1932-1939. CRC Press. ISBN 978-0-7503-0865-6.
53. Hinterberger, Frank (2008). Physik der Teilchenbeschleuniger und Ionenoptik. Springer. doi:10.1007/978-3-540-75282-0. ISBN 978-3-540-75281-3.
54. "Physics and national socialism: an anthology of primary sources", Klaus Hentschel. Birkhäuser, 1996. ISBN 3-7643-5312-0, ISBN 978-3-7643-5312-4. p. 350.
55. Wille, Klaus (2001). Particle Accelerator Physics: An Introduction. Oxford University Press. ISBN 978-0-19-850549-5.
56. Science Service (1942). "Shall New Machine Be Named Betatron or Rheotron". The Chemistry Leaflet 15 (7-12).
57. Celia Elliot. "Physics in the 1940s: The Betatron". Physics Illinois: Time Capsules. Urbana-Champaign, IL: University of Illinois. Retrieved 13 April 2012.
58. R.A. Kingery; R.D. Berg; E.H. Schillinger (1967). "Electrons in Orbit". Men and Ideas in Engineering: Twelve Histories from Illinois. Urbana, IL: University of Illinois Press. p. 68. ASIN B002V8WB8I.
59. "The Biggest Betatron in the World". Life: 131. March 20, 1950.
60. R. Wideröe, "Some memories and dreams from the childhood of particle accelerators", Europhysics News, Vol. 15, No. 2 (Feb. 1984), 9–11.

61. D.W. Kerst, "The acceleration of electrons by magnetic induction", Phys. Rev., 60 (Jul 1942), 47–53.
62. D.W. Kerst and R. Serber, "Electronic orbits in the induction accelerator", Phys. Rev.,60 (July 1941), 53–58.
63. G.S. James, R.H. Levy, H.A. Bethe and B.T. Fields, "On a new type of accelerator for heavy ions", Phys. Rev., 145, 925 (1966).
64. M.S. Livingston, "Particle accelerators: A brief history", Harvard University Press,Cambridge, Massachusetts (1969), 32.
65. E.M. McMillan, "The synchrotron - a proposed high-energy particle accelerator", Phys.Rev., Letter to the editor, 68 (Sept. 1945), 1434.
66. V. Veksler, J. of Phys, USSR, 9 (1945), 153–8.
67. F.K. Goward and D.E. Barnes, Nature, 158 (1946), 413.
68. M.L. Oliphant, J.S. Gooden and G.S. Hyde, Proc. Phys. Soc., 59 (1947), 666.
69. E.D. Courant, M.S. Livingston and H.S. Snyder, "The strong-focusing synchrotron – a new high-energy accelerator", Phys. Rev., 88 (Dec. 1952), 1190-6, and E.D. Courant and H.S. Snyder, "Theory of the alternating-gradient synchrotron", Annals of Physics, No. 3 (1958), 1–48.
70. N.C. Christofilos, Unpublished report (1950), and U.S. Patent no. 2.736,799, filed March 10, 1950, issued February 28, 1956.
71. S. van der Meer, "Stochastic damping of betatron oscillations in the ISR", CERN/ISRPO/72–31 (August, 1972).
72. Livingston chart. First published in Livingston's book: "High-energy accelerators", Interscience Publishers Inc., New York (1954).
73. H. Hahn, "The relativistic heavy ion collider at Brookhaven", Proc. 1st European Part. Accel. Conf. (EPAC), Rome 1988 (World Scientific, Singapore, 1989), 109–11.
74. E. Picasso, "The LEP project", Proc. 1st European Part. Accel. Conf. (EPAC), Rome 1988 (World Scientific, Singapore, 1989) 3–6.
75. M. Tigner, "A possible apparatus for electron clashing-beam experiments", Letter to the editor, Nuovo Cim., 37 (1965), 1228–31.
76. DESY About retrieved 24-May-2012.
77. DESY Portrait retrieved 24-May-2012.
78. "OLYMPUS - Deutsches Elektronen-Synchrotron DESY". Retrieved 16 August 2012.
79. "OLYMPUS Collaboration". Retrieved 16 August 2012.
80. "PETRA III". Hasylab, Desy. Retrieved 2012-09-04.
81. Last run of HERA. Read September 30th 2007.
82. "European X-Ray Free-Electron Laser tunnel construction completed".
83. "European XFEL facts & figures". Retrieved 2009-11-27.

84. R.W. Gurney and E.U. Condon, Nature 122, 439 (1928) and G. Gamov, Zeit f. Phys. 51, 204 (1928).
85. J.D. Crockcroft and E.T.S. Walton, "Experiments with high velocity ions", Proc. Royal Soc., Series A 136 (1932), 619–30.
86. R.J. Van de Graaff, "A 1,500,000 volt electrostatic generator", Phys. Rev., 387, (Nov.1931), 1919–20.
87. G. Ising, Arkiv för Matematik, Astronomi och Fysik, 18 (1924), 1–4.
88. E.O. Lawrence and M.S. Livingston, "The production of high speed light ions without the use of high voltages", Phys. Rev., 40 (April 1932), 19–35.
89. R. Stiening, "The status of the Stanford Linear Collider", Proc. 1987 IEEE Part. Accel. Conf., Washington, March 1987, (IEEE, 1987), 1–7.
90. G. Brianti, "The large hadron collider (LHC) in the LEP tunnel", Ibid. 218–222.
91. K. Johnsen, "Long term future", ECFA Study Week on Instr. Tech. for High Luminosity Hadron Colliders, Barcelona, 1989, CERN 89–10, ECFA 89–124, Vol. 1 (Nov. 1989), 25-34.
92. J.P. Blewett, "200 TeV Intersecting Storage Accelerators", 8th Int. Conf. on High-Energy Accel., CERN, 1971 (CERN, Geneva, 1971), 501–4.

93. Handbook of Accelerator Physics and Engineering(books.google.jo/books?isbn=9810235003 Alex Chao, M. Tigner-1999 - Edited by internationally recognized authorities in the field, this handbook focuses on Linacs, Synchrotrons and Storage Rings and is intended as a vade mecum for professional engineers and physicists engaged in these subjects.)
94. Accelerator Physics books.google.jo/books?isbn=9812562001 Shyh-Yuan Lee - 2004.
95. Particle Accelerator Physics books.google.jo/books?isbn=3540490434 Helmut Wiedemann - 2007.
96. Contemporary Accelerator Physics books.google.jo/books?isbn=9812389008 Stephan I. Tzenov - 2004.
97. An Introduction to the Physics of Particle Accelerators (books.google.jo/books?isbn=9812779604 Mario Conte, William W. MacKay – 2008.
98. Advances of Accelerator Physics and Technologies (books.google.jo/books?isbn=9810209584 Herwig F. Schopper – 1993

99. The Physics of Particle Accelerators: An Introduction (books.google.jo/books?isbn=0198505493 Klaus Wille (prof.) - 2000 .
100. Nonlinear Problems in Accelerator Physics, Proceedings (books.google.jo/books?isbn=0750302380 Berz - 1993 .
101. General accelerator physics: proceedings - Volumes 1-2 (books.google.jo/books?id=BIkfAQAAMAAJ Paul James Bryant, CERN Accelerator School, Stuart Turner - 1985 - Snippet view)
102. Second advanced accelerator physics course: proceedings (books.google.jo/books?id=_4kfAQAAMAAJ Stuart Turner, CERN Accelerator School, Stuart Turner - 1989 - Snippet view)
103. Encyclopaedia of Physics (second Edition), R.G. Lerner, G.L. Trigg, VHC Publishers, 1991, ISBN (Verlagsgesellschaft) 3-527-26954-1 (VHC Inc.) 0-89573-752-3
104. Analytical Mechanics, L.N. Hand, J.D. Finch, Cambridge University Press, 2008, ISBN 978-0-521-57572-0
105. Halliday, David; Resnick, Robert; Walker, Jearl (2004-06-16). Fundamentals of Physics (7 Sub ed.). Wiley. ISBN 0-471-23231-9.
106. Classical Mechanics, T.W.B. Kibble, European Physics Series, 1973, ISBN 0-07-084018-0
107. Dynamics and Relativity, J.R. Forshaw, A.G. Smith, Wiley, 2009, ISBN 978-0-470-01460-8
108. M.R. Spiegel, S. Lipcshutz, D. Spellman (2009). Vector Analysis. Schaum's Outlines (2nd ed.). McGraw Hill. p. 33. ISBN 978-0-07-161545-7.
109. Essential Principles of Physics, P.M. Whelan, M.J. Hodgeson, second Edition, 1978, John Murray, ISBN 0-7195-3382-1
110. Hanrahan, Val; Porkess, R (2003). Additional Mathematics for OCR. London: Hodder & Stoughton. p. 219. ISBN 0-340-86960-7.
111. Keith Johnson (2001). Physics for you: revised national curriculum edition for GCSE (4th ed.). Nelson Thornes. p. 135. ISBN 978-0-7487-6236-1. The 5 symbols are remembered by "suvat". Given any three, the other two can be found.
112. 3000 Solved Problems in Physics, Schaum Series, A. Halpern, Mc Graw Hill, 1988, ISBN 978-0-07-025734-4
113. Jump up to: [a] [b] The Physics of Vibrations and Waves (3rd edition), H.J. Pain, John Wiley & Sons, 1983, ISBN 0-471-90182-2
114. Jump up to: [a] [b] An Introduction to Mechanics, D. Kleppner, R.J. Kolenkow, Cambridge University Press, 2010, p. 112, ISBN 978-0-521-19821-9
115. Encyclopaedia of Physics (second Edition), R.G. Lerner, G.L. Trigg, VHC publishers, 1991, ISBN (VHC Inc.) 0-89573-752-3

116. "Mechanics, D. Kleppner 2010"
117. "Relativity, J.R. Forshaw 2009"
118. R. Penrose (2007). The Road to Reality. Vintage books. p. 474. ISBN 0-679-77631-1.
119. Classical Mechanics (second edition), T.W.B. Kibble, European Physics Series, 1973, ISBN 0-07-084018-0
120. Electromagnetism (second edition), I.S. Grant, W.R. Phillips, Manchester Physics Series, 2008 ISBN 0-471-92712-0
121. Classical Mechanics (second Edition), T.W.B. Kibble, European Physics Series, Mc Graw Hill (UK), 1973, ISBN 0-07-084018-0.
122. Misner, Thorne, Wheeler, Gravitation
123. C.B. Parker (1994). McGraw Hill Encyclopaedia of Physics (second ed.). p. 1199. ISBN 0-07-051400-3.
124. C.B. Parker (1994). McGraw Hill Encyclopaedia of Physics (second ed.). p. 1200. ISBN 0-07-051400-3.
125. J.A. Wheeler, C. Misner, K.S. Thorne (1973). Gravitation. W.H. Freeman & Co. pp. 34–35. ISBN 0-7167-0344-0.
126. H.D. Young, R.A. Freedman (2008). University Physics (12th Edition ed.). Addison-Wesley (Pearson International). ISBN 0-321-50130-6.
127. Armstrong, A.G.A.M. , Fan, M.W., Simkin, J., Trowbridge, C.W.: Automated optimization of magnet design using the boundary integral method, IEEE Transactions on Magnetics, 1982
128. Bathe, K.-J: Finite Element Procedures, Prentice Hall, 1996
129. Bossavit, A.: Computational Electromagnetism, Academic Press, 1998
130. Binns, K.J., Lawrenson, P.J., Trowbridge, C.W.: The analytical and numerical solution of electric and magnetic fields, John Wiley & Sons, 1992
131. C.J. Collie: Magnetic fields and potentials of linearly varying current or magnetization in a plane bounded region, Proc. of the 1st Compumag Conference on the Computation of Electromagnetic Fields, Oxford, U.K., 1976
132. Biro, O., Preis, K., Paul. C.: The use of a reduced vector potential Ar Formulation for the calculation of iron induced field errors, Proceedings of the First international ROXIE user's meeting and workshop, CERN, March 1998.
133. Brebbia, C.A.: The boundary element method for engineers, Pentech Press, 1978.
134. Fetzer, J., Kurz, S., Lehner, G.: Comparison between different formulations for the solution of 3D nonlinear magnetostatic problems using BEM-FEM coupling, IEEE Transactions on Magnetics,Vol. 32, 1996

135. Halbach, K.: A Program for Inversion of System Analysis and its Application to the Design of Magnets, Proceedings of the International Conference on Magnet Technology (MT2), The Rutherford Laboratory, 1967
136. Halbach, K., Holsinger R.: Poisson user manual, Techn. Report, Lawrence Berkeley Laboratory, Berkeley, 1972
137. Hornsby, J.S.: A computer program for the solution of elliptic partial differential equations, Technical Report 63-7, CERN, 1967
138. Kurz, S., Fetzer, J., Rucker, W.M.: Coupled BEM-FEM methods for 3D field calculations with iron saturation, Proceedings of the First International ROXIE users meeting and workshop, CERN, March 16-18, 1998
139. Kurz, S., Russenschuck, S.: The application of the BEM-FEM coupling method for the accurate calculation of fields in superconducting magnets, Electrical Engineering, 1999
140. S. Kurz and S. Russenschuck, The application of the BEM-FEM coupling method for the accurate calculation of fields in superconducting magnets, Electrical Engineering, 1999.
141. Mess, K.H., Schm¨user, P., Wolff S.: Superconducting Accelerator Magnets, World Scientific, 1996
142. G.D. Nowottny, Netzerzeugung durch Gebietszerlegung und duale Graphenmethode, Shaker Verlag, Aachen, 1999.
143. Russenschuck, S.: ROXIE: Routine for the Optimization of magnet X-sections, Inverse field calculation and coil End design, Proceedings of the First International ROXIE users meeting and workshop, CERN 99-01, ISBN 92-9083-140-5.
144. Design of accelerator magnets, S. Russenschuck, CERN, 1211 Geneva 23, Switzerland
145. https://en.wikipedia.org/wiki/File:2mv_accelerator-MJC01.jpg
146. https://commons.wikimedia.org/w/index.php?curid=5509550
147. https://en.wikipedia.org/wiki/Cosmotron#/media/File:Cosmotron_(PSF).png
148. https://www.bnl.gov/about/history/accelerators.php
149. https://www.futuretimeline.net/21stcentury/2035.
150. https://en.wikipedia.org/wiki/Australian_Synchrotron
151. WWW.OLOSCIENCE.COM
152. https://www.lnls.cnpem.br/uvx-en/machine/
153. https://www.lnls.cnpem.br/science/
154. https://en.wikipedia.org/wiki/SLAC_National_Accelerator_Laboratory

155. https://commons.wikimedia.org/wiki/File:Stanford-linear-accelerator-usgs-ortho-kaminski-5900.jpg
156. https://upload.wikimedia.org/wikipedia/commons/8/83/Betatron_6MeV_(1942).jpg
157. https://en.wikipedia.org/wiki/File:DESY1.jpg
158. https://en.wikipedia.org/wiki/Synchrotron#/media/File:Aust.-Synchrotron-Interior- Panorama,-14.06.2007.jpg
159. https://upload.wikimedia.org/wikipedia/commons/7/7c/Particle_accelerator_DSC09089.jpg
160. https://upload.wikimedia.org/wikipedia/commons/c/c2/Van_de_graaf_generator.svg
161. http://upload.wikimedia.org/wikipedia/commons/2/25/Large_Hadron_Collider_at_CERN
162. http://upload.wikimedia.org/wikipedia/commons/c/c9/Rhictunnel.jpg
163. https://upload.wikimedia.org/wikipedia/commons/3/3f/Fermilab.jpg
164. https://en.wikipedia.org/wiki/File:Cyclotron_with_glowing_beam.jpg
165. https://en.wikipedia.org/wiki/File:LawrenceCyclotronMagnet.jpg
166. https://en.wikipedia.org/wiki/File:1937-French-cyclotron.jpg
167. https://en.wikipedia.org/wiki/File:Cyclotron.jpg
168. http://www.sciencephoto.com/contributor/fla
169. https://en.wikipedia.org/wiki/File:Orsay_proton_therapy_dsc04444.jpg
170. https://en.wikipedia.org/wiki/File:SOLEIL_le_01_juin_2005.jpg
171. https://en.wikipedia.org/wiki/File:DESY1.jpg
172. https://en.wikipedia.org/wiki/File:BH_LMC.png

ELECTRICAL ENGINEERING

Principles and Special Purpose Applications of Electromagnetic Field and High Voltage

Package 3:Special Purpose Applications-Part Two

Book 7: Nuclear Magnetic Resonance-NMR

Dr. Moayad Abdullah AlMayouf

July 2018

Nuclear Magnetic Resonance-NMR

Content

1

INTRODUCTION AND BACKGROUND

1.1. Principles

Nuclear **M**agnetic **R**esonance **NMR** is a branch of spectroscopy that deals with the phenomenon found in assemblies of large number of nuclei of atoms that possess both "magnetic moments" and "angular momentum" and is subjected to external magnetic field. Therefore **NMR** is the name given to a technique which exploits the magnetic properties of certain nuclei. NMR is a phenomenon which occurs when the nuclei of certain atoms are immersed in a static magnetic field and exposed to a second oscillating magnetic field. The Resonance implies that we are in tune with a natural frequency of the nuclear magnetic system in the magnetic field.

Some nuclei experience this phenomenon, and others do not, dependent upon whether they possess a property called spin. In principle, NMR is applicable to any nucleus possessing spin.

Most of the matters you can examine with NMR are composed of molecules. Molecules are composed of atoms. For water molecules, each water molecule has one oxygen and two hydrogen atoms. If we zoom into one of the hydrogen's past the electron cloud we see a nucleus composed of a single proton.

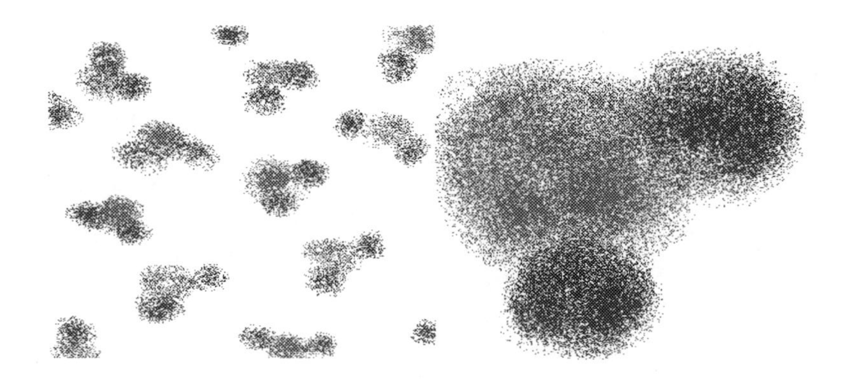

Fig.1.1. Water molecules, H2O

The proton possesses a property called spin which can be thought of as a small magnetic field, and will cause the nucleus to produce an NMR signal.

Fig.1.2. Poles of Magnets

Not all nuclei possess the property called spin. A list of these nuclei will be presented in Chapter 2 on spin physics.

NMR is a property that magnetic nuclei have in a magnetic field and applied electromagnetic (EM) pulse, which cause the nuclei to absorb energy from the EM pulse and radiate this energy back out. The energy radiated back out is at a specific resonance frequency which depends on the strength of the magnetic field and other factors. This allows the observation of specific quantum mechanical magnetic properties of an atomic nucleus.

A key feature of NMR is that the resonance frequency of a particular substance is directly proportional to the strength of the applied magnetic field. It is this feature that is exploited in imaging techniques; if a sample is placed in a non-uniform magnetic field then the resonance frequencies of the sample's nuclei depend on where in the field they are located. Since the resolution of the imaging techniques depends on how big the gradient of the field is, many efforts are made to develop more powerful magnets, often using superconductors. The effectiveness of NMR can also be improved using hyperpolarization, and/or using two-dimensional, three-dimensional and higher dimension multi-frequency techniques.

The principle of NMR usually involves two sequential steps:

- The alignment (polarization) of the magnetic nuclear spins in an applied, constant magnetic field $\mathbf{H}_0$.
- The perturbation of this alignment of the nuclear spins by employing an electro-magnetic field, usually radio frequency (RF) pulse. The required perturbing frequency is dependent upon the static magnetic

field (H_0) and the nuclei of observation.

To maximizes the NMR signal strength the two fields are usually chosen to be perpendicular to each other. The resulting response by the total magnetization (**M**) of the nuclear spins is the phenomenon that is exploited in NMR spectroscopy and magnetic resonance imaging. Both use intense applied magnetic fields (H_0) in order to achieve dispersion and very high stability to deliver spectral resolution, the details of which are described by the following:

1- **Chemical shifts**; describes the dependence of nuclear magnetic energy levels on the electronic environment in a molecule. Chemical shifts are relevant in NMR spectroscopy techniques such as proton NMR and carbon-13 NMR.

2- **Zeeman effect**; is the splitting of a spectral line into several components in the presence of a static magnetic field. It is analogous to the Stark effect, the splitting of a spectral line into several components in the presence of an electric field. The Zeeman effect is very important in applications such as nuclear magnetic resonance spectroscopy, electron spin resonance spectroscopy, magnetic resonance imaging (MRI) and Mössbauer spectroscopy. It may also be utilized to improve accuracy in Atomic absorption spectroscopy. When the spectral lines are absorption lines, the effect is called Inverse Zeeman effect. The Zeeman effect is named after the Dutch physicist Pieter Zeeman.

3- **Knight shifts (in metals)**; is a shift in the nuclear magnetic resonance frequency of a paramagnetic (substance first published in 1949 by the American physicist Walter David Knight).The Knight shift is due to the conduction electrons in metals.

Many scientific techniques exploit NMR phenomena to study molecular physics, crystals and non-crystalline materials through NMR spectroscopy. The impact of NMR spectroscopy on the natural sciences has been substantial. It can be applied to a wide variety of samples, both in the solution and the solid state.

Many types of information can be obtained from an NMR spectrum such as:

1) Identifying functional groups
2) Analyzing of NMR spectrum provides information on the number and type of chemical entities in a molecule.
3) Studying the mixtures of analytes, to understand dynamic effects such as change in temperature and reaction mechanisms, and
4) Understanding protein and nucleic acid structure and function.
5) NMR is also routinely used in advanced medical imaging techniques, such as in Magnetic Resonance Imaging (MRI).
6) NMR phenomena are also utilized in low-field NMR, NMR spectroscopy and MRI in the Earth's magnetic field (referred to as Earth's field NMR), and in several types of magnetometers.

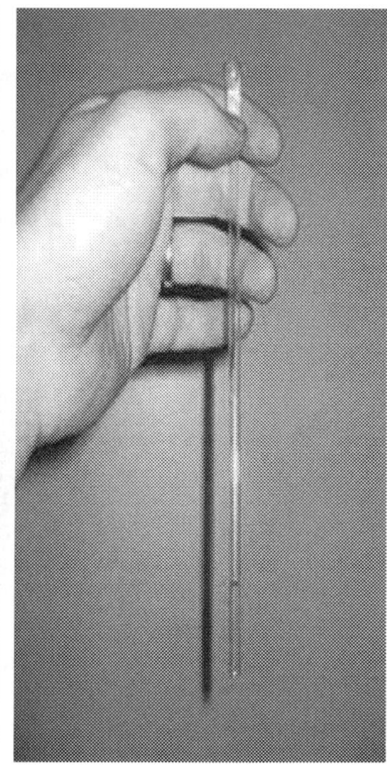

HWB-NMRv900 NMR Sample

Fig.1.3. NMR Set

1.2. History

Nuclear Magnetic Resonance NMR was first described and measured in molecular beams by Isidor Rabi in 1938. In 1945, Purcell, Torrey, and Pound (Harvard, Cambridge, Massachusetts) detected weak radio-frequency signals generated by the nuclei of atoms in about 1 kg of paraffin wax placed in a magnetic field. The milstones on the history line of developing the functions and uses of NMR may be traced throughout the list below:

1937 **Rabi** predicts and observes nuclear magnetic resonance
1946 Bloch, Purcell first nuclear magnetic resonance of bulk sample
1953 Overhauser NOE (nuclear Overhauser effect)
1966 Ernst, Anderson Fourier transform NMR
1975 Jeener, Ernst 2D NMR
1984 Nicholson NMR metabolomics
1985 Wüthrich first solution structure of a small protein (BPTI) from NOE derived distance restraints
1987 3D NMR + ^{13}C, ^{15}N isotope labeling of recombinant proteins (resolution)
1990 Pulsed field gradients (artifact suppression)
1996/7 Residual Dipolar Couplings (RDC) from partial alignment in liquid crystalline media TROSY (molecular weight > 100 kDa)
2000s Dynamic Nuclear Polarization (DNP) to enhance NMR sensitivity
2009 World's First (1000 MHz, 23.5 T) NMR Spectrometer-tested in Germany and installed in Lyon, France

Nobel prizes
1944 *Physics;* Rabi (Columbia)
1952 *Physics;* Bloch (Stanford), Purcell (Harvard)
1991 *Chemistry;* Ernst (ETH)
2002 *Chemistry;* Wüthrich (ETH)
2003 *Medicine;* Lauterbur (University of Illinois in Urbana), Mansfield (University of Nottingham)

Felix Bloch and Edward Mills Purcell refined, in 1946, the technique for use on liquids and solids, for which they shared the Nobel Prize in physics in 1952.

Purcell had worked on the development and radar applications during World War II at Massachusetts Institute of Technology's Radiation Laboratory. His work during that project on the production and detection of RF energy, and on the absorption of such RF energy by matter, preceded his discovery of NMR.

They noticed that magnetic nuclei, like ^{1}H and ^{31}P, could absorb RF energy when placed in a magnetic field of a strength specific to the identity of the nuclei. When this absorption occurs, the nucleus is described as being in resonance. Different atomic nuclei within a molecule resonate at different (radio) frequencies for the same magnetic field strength. The observation of such magnetic resonance frequencies of the nuclei present in a molecule allows any trained user to discover essential, chemical and structural information about the molecule.

The development of nuclear magnetic resonance as a technique of analytical chemistry and biochemistry parallels the development of electromagnetic technology and its introduction into civilian use.

Fig.1.4. World's First 1 GHz NMR Spectrometer -tested in Karlsruhe, Germany (1000 MHz, 23.5 T) - installed at the new 'Centre de RMN à Très Hauts Champs' in Lyon, France in 2009

1.3. NMR Spectroscopy

Nuclear Magnetic Resonance spectroscopy is a powerful and theoretically complex analytical tool. It is important to bear in mind that, with NMR, we are performing experiments on the **nuclei** of atoms, not the electrons. The chemical environment of specific nuclei is deduced from information obtained about the nuclei.

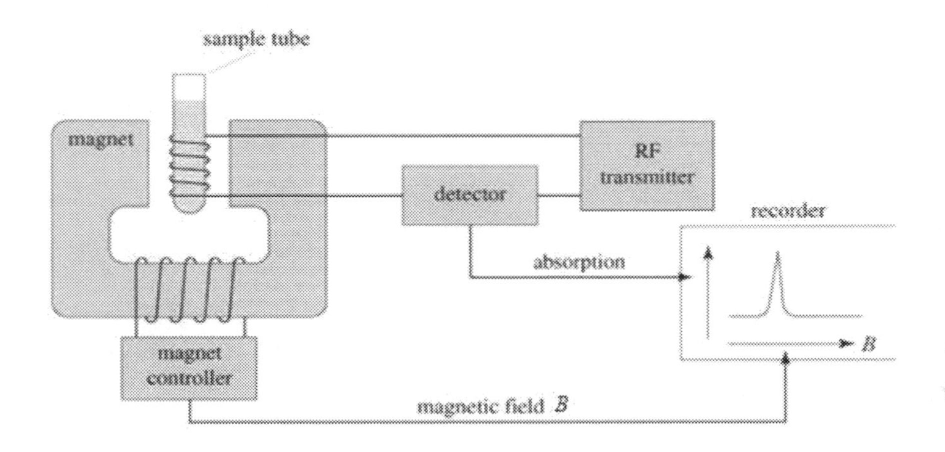

Fig.1.5. The NMR Spectrometer

Spectroscopy is the study of the interaction of electromagnetic radiation with matter. Nuclear magnetic resonance spectroscopy is the use of the NMR phenomenon to study physical, chemical, and biological properties of matter. The definition, purpose and task of Nuclear Magnetic Resonance (NMR) may be summarized by the following topics:

- Nuclear magnetic resonance (NMR) spectroscopy is a Radio Frequency Absorption Spectra of atomic nuclei in substances which are subjected to magnetic fields.
- It is based on the measurement of absorption of electromagnetic radiation in the radio-frequency region of roughly 1 to 1000 MHz.
- Nuclei of atoms rather than outer electrons are involved in the absorption process.
- Spectral Dispersion is Sensitive to the chemical environment via "coupling" to the electrons surrounding the nuclei.
- In order to cause nuclei to develop the energy states required for absorption to occur, it is necessary to place the analyte in an intense magnetic field.
- Nuclear magnetic resonance spectroscopy is one of the most powerful

tools for elucidating the structure of chemical species.

- Interactions can be interpreted in terms of structure, bonding, reactivity

As a consequence, NMR spectroscopy finds applications in several areas of science such as:

1- NMR spectroscopy is routinely used by chemists to study chemical structure using simple one-dimensional techniques.

2- Two-dimensional techniques are used to determine the structure of more complicated molecules. These techniques are replacing x-ray crystallography for the determination of protein structure.

3- Time domain NMR spectroscopic techniques are used to probe molecular dynamics in solutions.

4- Solid state NMR spectroscopy is used to determine the molecular structure of solids.

5- Other scientists have developed NMR methods of measuring diffusion coefficients.

1.4. The Basics Physics and Engineering for NMR

The mathematics and functions needed to describe and identify the Physics and Engineering of NMR cover the following:

- Exponential Functions
- Trigonometric Functions
- Differentials and Integrals
- Vectors
- Matrices
- Coordinate Transformations
- Convolutions
- Imaginary Numbers
- The Fourier Transform

These functions are briefly described as following:

A. <u>Exponential Functions</u>

Logarithms based on powers of (e) are called natural logarithms.

If $x = e^{y}$, then $ln(x) = y$, where $e^{x} = \exp(x) = (2.71828183)^{x}$

Many of the dynamic NMR processes are exponential in nature. For example, signals decay exponentially as a function of time. It is therefore essential to understand the nature of exponential curves. Three common exponential functions are given in fig.1.6. where t is a constant.

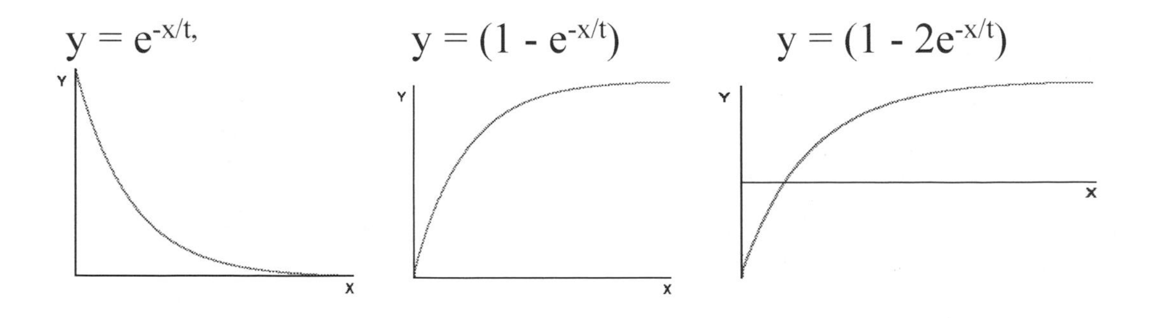

Fig. 1.6. The common exponential functions

B. <u>Trigonometric Functions</u>

The basic trigonometric functions sine and cosine describe sinusoidal functions which are 90° out of phase.

The trigonometric of the perpendicular triangle are :
Sin(θ) = Opposite / Hypotenuse
Cos(θ) = Adjacent / Hypotenuse
Tan(θ) = Opposite / Adjacent

The function sin(x) / x occurs often and is called sinc(x).

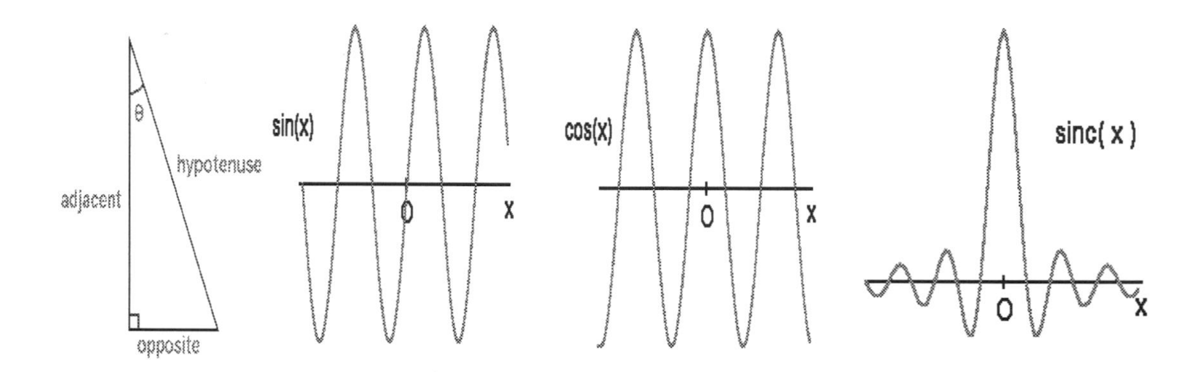

Fig. 1.7. The basic trigonometric functions

C. Differentials and Integrals

A differential can be thought of as the slope of a function at any point. For the function

$$y = x^2 + \frac{1}{3}x + 1$$

the differential of y with respect to x is

$$\frac{\partial y}{\partial x} = 2x + \frac{1}{3}$$

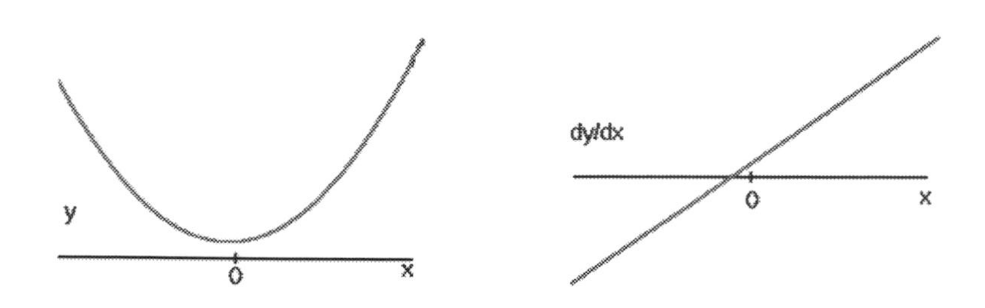

Fig.1.8. Function and its differential

An integral is the area under a function between the limits of the integral.

$$y = \int_0^2 (x+3)dx = \left(\frac{1}{2}x^2 + 3x\right)\Bigg|_0^2 = 8$$

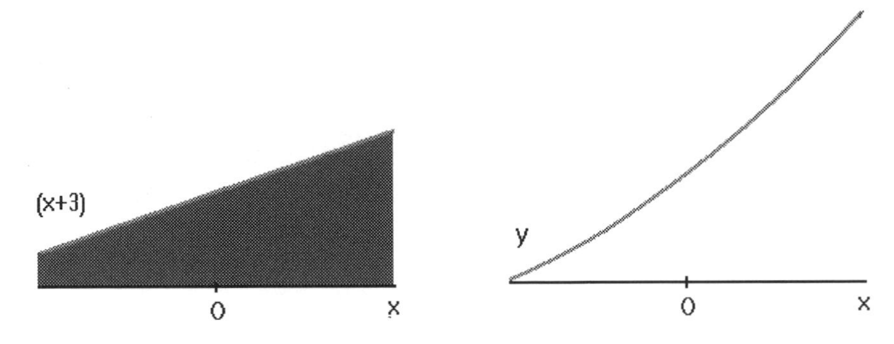

Fig.1.9. Function and its integral

An integral can also be considered a summation; in fact most integration is performed by computers by adding up values of the function between the integral limits.

D. Vectors

A vector is a quantity having both a magnitude and a direction. In fig.1.10. the vector is in the XY plane between the +X and +Y axes. The vector has X and Y components and a magnitude of $(X^2 + Y^2)^{1/2}$.

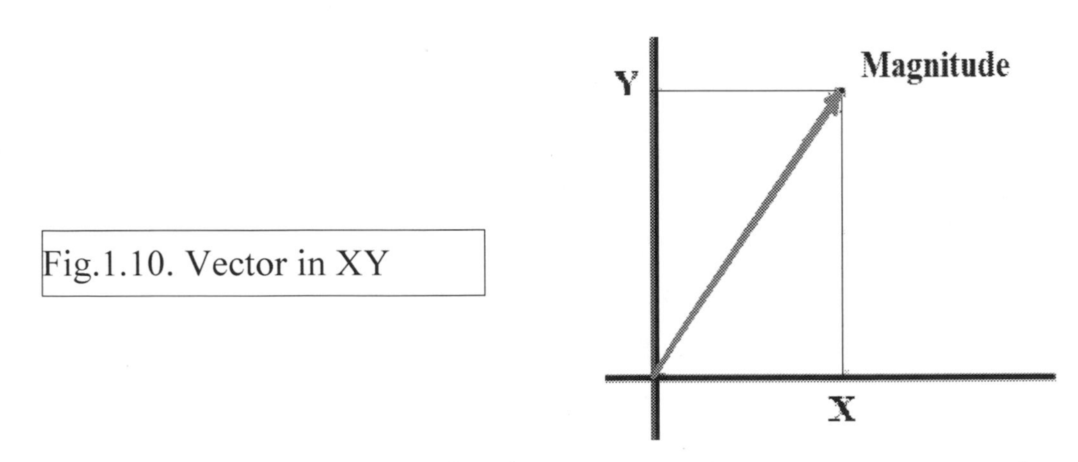

Fig.1.10. Vector in XY

The magnetization from nuclear spins is represented as a vector emanating from the origin of the coordinate system. Here it is in fig. 1.11. along the +Z axis.

Fig. 1.11. Magnetization Vector from

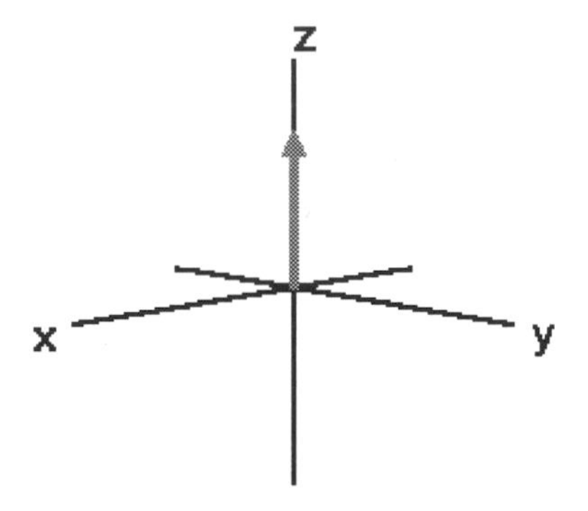

E. Matrices

A matrix is a set of numbers arranged in a rectangular array. E.g. a matrix has 3 rows and 4 columns and is said to be a 3 by 4 matrix, as below:

$$\begin{bmatrix} 7 & 1 & 6 & 0 \\ 7 & 7 & 5 & 0 \\ 2 & 9 & 0 & 4 \end{bmatrix}$$

You can multiply two matrices if, and only if, the number of columns in the first matrix equals the number of rows in the second matrix. Otherwise, the product of two matrices is undefined. See the example given below:

$$\begin{vmatrix} a_{11} & a_{12} & a_{13} \\ a_{21} & a_{22} & a_{23} \end{vmatrix} \begin{vmatrix} b_{11} & b_{12} \\ b_{21} & b_{22} \\ b_{31} & b_{32} \end{vmatrix} = \begin{vmatrix} c_{11} & c_{12} \\ c_{21} & c_{22} \end{vmatrix}$$

The elements of c Matrix are obtained by summing the following multiplication:

$$c_{11} = a_{11} \times b_{11} + a_{12} \times b_{21} + a_{13} \times b_{31}$$
$$c_{12} = a_{11} \times b_{12} + a_{12} \times b_{22} + a_{13} \times b_{32}$$
$$c_{21} = a_{21} \times b_{11} + a_{22} \times b_{21} + a_{23} \times b_{31}$$
$$c_{22} = a_{21} \times b_{12} + a_{22} \times b_{22} + a_{23} \times b_{32}$$

Where:

$$c_{ik} = \sum a_{ij}\, b_{jk}$$

If we multiply a 2×3 matrix with a 3×4 matrix, the product matrix is 2×4

F. Coordinate Transformations

A coordinate transformation is used to convert the coordinates of a vector in one coordinate system (XY) to that in another coordinate system (X"Y").

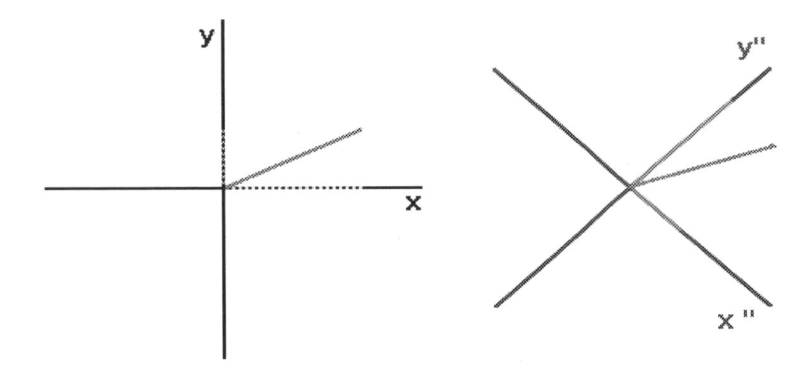

Fig.1.12. Transformation of coordinates

G. Convolution

The convolution of two functions is the overlap of the two functions as one function is passing over the second. The convolution symbol is ⊗. The convolution of h(t) and g(t) is defined mathematically as

$$f(t) = h(t) \otimes g(t) = \int h(\tau)\ g(t-\tau)d\tau$$

The above equation is depicted for rectangular shaped h(t) and g(t) functions in this animation.

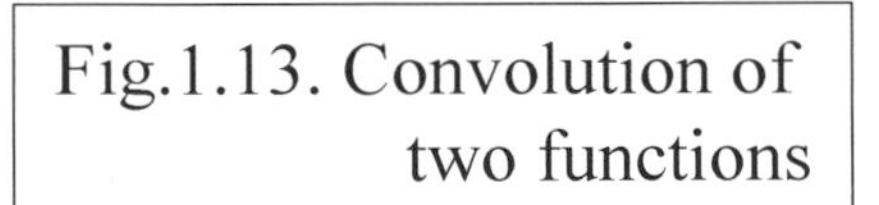

Fig.1.13. Convolution of two functions

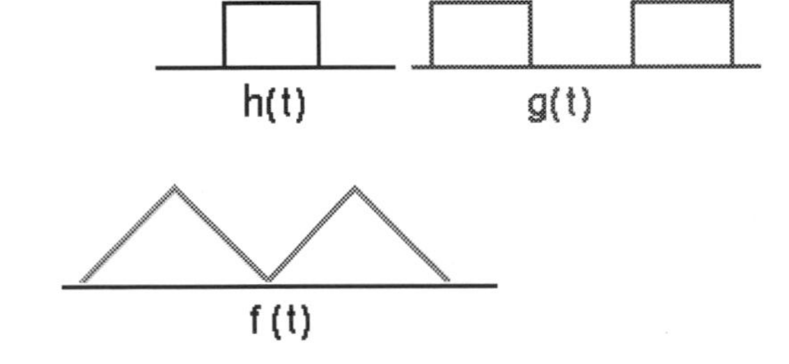

H. Imaginary Numbers

Imaginary numbers are those which result from calculations involving the square root of -1. Imaginary numbers are symbolized by *j*. A complex number is one which has a real (RE) and an imaginary (IM) part. The real and imaginary parts of a complex number are orthogonal.

Two useful relations between complex numbers and exponentials are

$e^{+jx} = \cos(x) + j\sin(x)$ and $e^{-jx} = \cos(x) - j\sin(x)$.

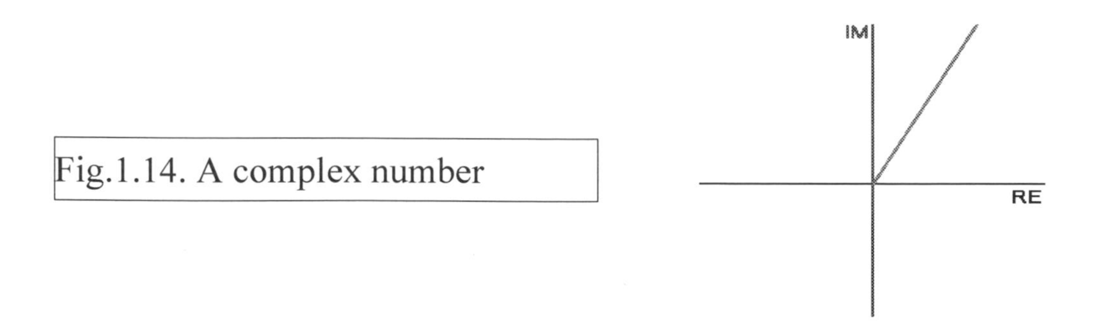

Fig.1.14. A complex number

i- Fourier Transforms

The Fourier transform (FT) is a mathematical technique for converting time domain data to frequency domain data, and vice versa. More detail on the Fourier Transform is given in Annex 1.

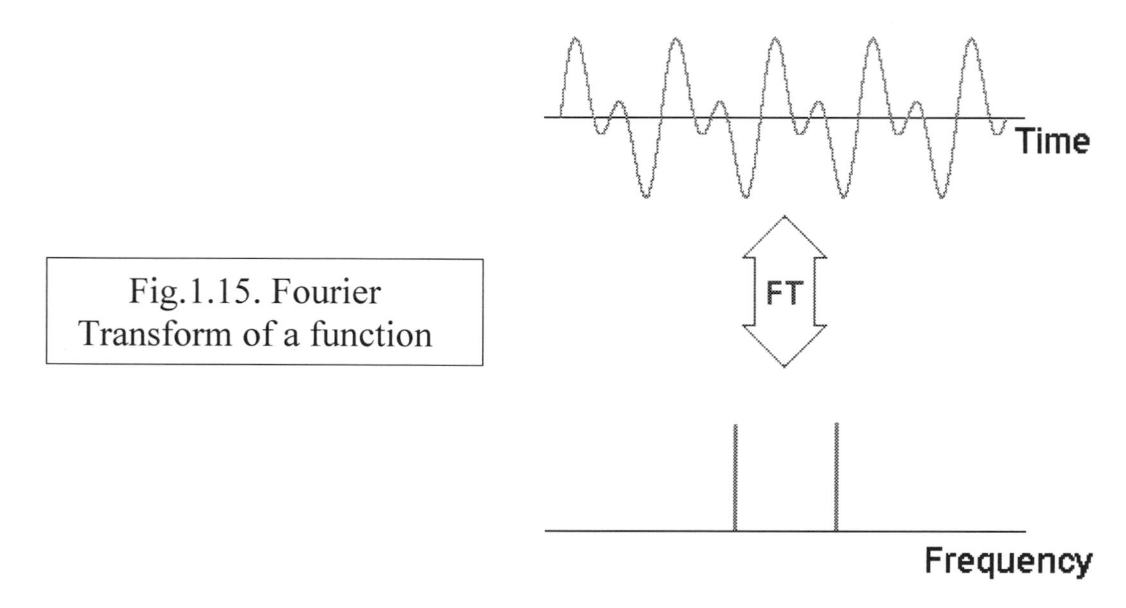

Fig.1.15. Fourier Transform of a function

THEORY OF NUCLEAR MAGNETIC RESONANCE

2.1. The Physics and Engineering of NMR

The topics and parameters describing the Physics and Engineering of NMR are lying under the following classification and understanding:

- Spin Physics
- Nuclear spin and Magnets
- Energy Level and Magnetic Field

These are briefly described in the following sections.

2.1.1 Spin physics

A rotating object possesses a quantity called angular momentum given by the right hand thumb rule. **Spin** is a type of angular momentum that does not vanish even at absolute zero! So a physical motion to represent this type of angular momentum is not without error. **Spin** is a fundamental property of electron and nucleus like mass, electric charge, and magnetism.

The shell model for the nucleus tells us that nucleons, just like electrons, fill orbitals. When the number of protons or neutrons equals 2, 8, 20, 28, 50, 82, and 126, orbitals are filled. Because nucleons have spin, just like electrons do, their spin can pair up when the orbitals are being filled and cancel out. Almost every element in the periodic table (see fig.2.1) has an isotope with a non-zero nuclear spin.

NMR can only be performed on isotopes whose natural abundance is high enough to be detected. Some of the nuclei routinely used in NMR are listed in the table 2.1.

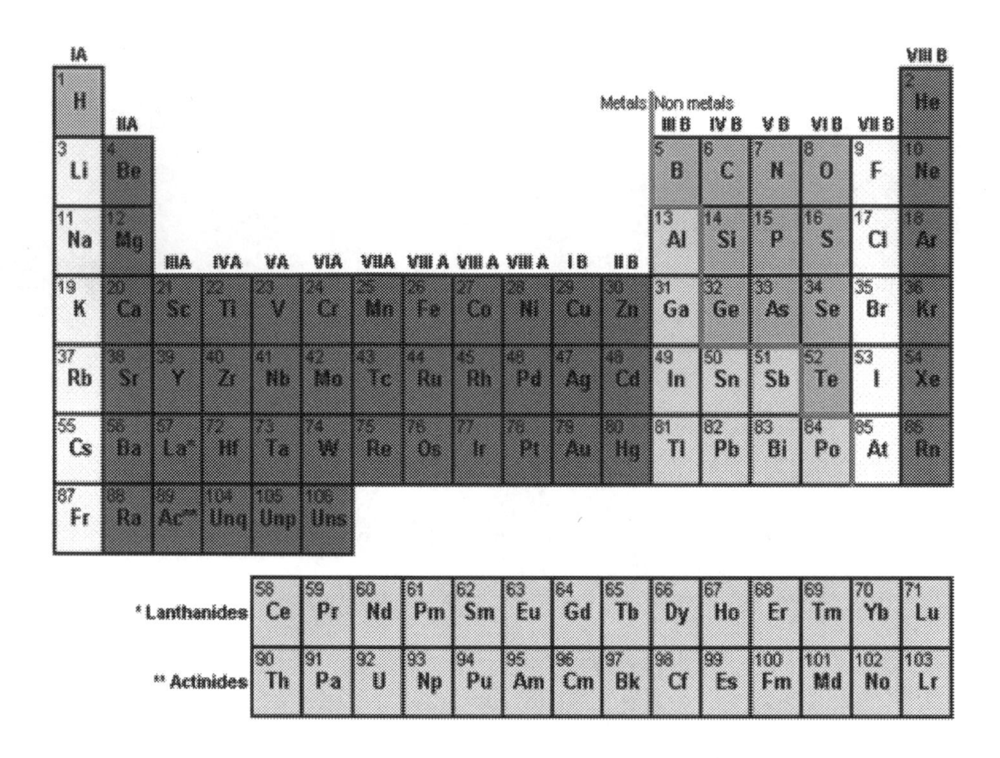

Fig.2.1. Periodic Table of the Elements (The given numbers are related to the element No. in the periodic table)

Table 2.1. Elements used in NMR

Nuclei	Unpaired Protons	Unpaired Neutrons	Net Spin	γ(MHz/T)
^{1}H	1	0	1/2	42.58
^{2}H	1	1	1	6.54
^{31}P	1	0	1/2	17.25
^{23}Na	1	2	3/2	11.27
^{14}N	1	1	1	3.08
^{13}C	0	1	1/2	10.71
^{19}F	1	0	1/2	40.08

The Spin is a fundamental property of nature like electrical charge or mass. Spin comes in multiples of 1/2 and can be + or -. Protons, electrons, and neutrons possess spin. Individual unpaired electrons, protons, and neutrons each possess a spin of 1/2.

In the deuterium atom (^{2}H), with one unpaired electron, one unpaired proton, and one unpaired neutron, the total electronic spin = 1/2 and the total nuclear spin = 1.

Two or more particles with spins having opposite signs can pair up to eliminate the observable manifestations of spin. An example is helium. In nuclear magnetic resonance, it is unpaired nuclear spins that are of importance.

Subatomic particles (electrons, protons and neutrons) can be imagined as spinning on their axes. In many atoms (such as ^{12}C) these spins are paired against each other, such that the nucleus of the atom has no overall spin. However, in some atoms (such as ^{1}H and ^{13}C) the nucleus does possess an overall spin. The rules for determining the net spin of a nucleus are as follows:

1. If the number of neutrons **and** the number of protons are both even, then the nucleus has **NO** spin.
2. If the number of neutrons **plus** the number of protons is odd, then the nucleus has a half-integer spin (i.e. 1/2, 3/2, 5/2)
3. If the number of neutrons **and** the number of protons are both odd, then the nucleus has an integer spin (i.e. 1, 2, 3)

When a particle placed in a magnetic field of strength **B**, a particle with a net spin can absorb a photon, of frequency υ. The frequency υ depends on the gyromagnetic ratio, γ of the particle as given by:

$\nu = \gamma B$, for hydrogen, $\gamma = 42.58$ MHz / T.

The gyromagnetic (or magnetogyric) ratio γ, is a fundamental nuclear constant which has a different value for every nucleus (can be positive or negative). For hydrogen, $\gamma = 42.58$ MHz / T.

2.1.2 Nuclear spin and magnets

All nucleons, that are neutrons and protons, composing any atomic nucleus, have the intrinsic quantum property of spin. The overall spin of the nucleus is determined by the spin quantum number S. If the number of both the protons and neutrons in a given nuclide are even then $S = 0$, i.e. there is no overall spin; just as electrons pair up in atomic orbital's, so do even numbers of protons or even numbers of neutrons (which are also spin - ½ particles and hence fermions) pair up giving zero overall spin.

However, a proton and neutron will have lower energy when their spins are parallel, **not anti-parallel**, as this parallel spin alignment does not infringe upon the Pauli Exclusion Principle, but instead has to do with the quark fine structure of these two nucleons. Therefore, the spin ground state for the deuteron (the deuterium nucleus, or the ^2H isotope of hydrogen) --that has only a proton and a neutron--corresponds to a spin value of **1**, not of zero; the single, isolated deuteron is therefore exhibiting an NMR absorption spectrum characteristic of a quadrupolar nucleus of spin **1**, which in the `rigid' state at very low temperatures is a characteristic (Pake) doublet, (not a singlet as for a single, isolated ^1H, or any other isolated fermion or dipolar nucleus of spin 1/2).

The **Pauli exclusion principle** is the quantum mechanical principle that no two identical fermions (particles with half-integer spin) may occupy the same quantum state simultaneously. A more rigorous statement is that the total wave function for two identical fermions is anti-symmetric with respect to exchange of the particles. The principle was formulated by Austrian physicist Wolfgang Pauli in 1925.

For example, no two electrons in a single atom can have the same four quantum numbers; if n, l, and m_l are the same, m_s must be different such that the electrons have opposite spins, and so on.

On the other hand, because of the Pauli principle, the (radioactive) tritium isotope has to have a pair of anti-parallel spin neutrons (of total spin zero for the neutron spin couple), plus a proton of spin 1/2; therefore, the character of the tritium nucleus (`triton') is again magnetic dipolar, not quadrupolar-- like its non-radioactive deuteron neighbor, and the tritium nucleus total spin value is again ½.

Just like for the simpler, abundant hydrogen isotope, ^{1}H nucleus (the *proton*). The NMR absorption (radio) frequency for tritium is however slightly higher for tritium than that of ^{1}H because the tritium nucleus has a slightly higher gyromagnetic ratio than ^{1}H. In many other cases of *non-radioactive* nuclei, the overall spin is also non-zero. For example, the ^{27}Al nucleus has an overall spin value $S = 5/2$.

A non-zero spin is thus always associated with a non-zero magnetic moment (μ) via the relation $\mu = \gamma S$, where γ is the gyromagnetic ratio. It is this magnetic moment that allows the observation of NMR absorption spectra caused by transitions between nuclear spin levels. Most nuclides (with some rare exceptions) that have both even numbers of protons and even numbers of neutrons, also have zero nuclear magnetic moments-and also have zero magnetic dipole and quadrupole moments; therefore, such nuclides do not exhibit any NMR absorption spectra. Thus, ^{18}O is an example of a nuclide that has no NMR absorption, whereas ^{13}C, ^{31}P, ^{35}Cl and ^{37}Cl are nuclides that do exhibit NMR absorption spectra; the last two nuclei are quadrupolar nuclei whereas the preceding two nuclei (^{13}C and ^{31}P) are dipolar ones.

Electron spin resonance (ESR) is a related technique which detects transitions between electron spin levels instead of nuclear ones. The basic principles are similar; however, the instrumentation, data analysis and detailed theory are significantly different. Moreover, there is a much smaller number of molecules and materials with unpaired electron spins that exhibit ESR (or electron paramagnetic resonance (EPR)) absorption than those that have NMR absorption spectra. Significantly also, is the much greater sensitivity of ESR and EPR in comparison with NMR.

Electron paramagnetic resonance (EPR) or **electron spin resonance** (ESR) spectroscopy is a spectroscopic technique for detecting and studying materials with unpaired electrons. A surprisingly large number of materials have unpaired electrons. These include free radicals, many transition metal ions, and defects in materials. Free electrons are often short-lived, but still play crucial roles in many processes such as photosynthesis, oxidation, catalysis, and polymerization reactions. The basic concepts of EPR are analogous to those of nuclear magnetic resonance (NMR), but it is electron spins that are excited instead of the spins of atomic nuclei. Because most stable molecules have all their electrons paired, the EPR technique is less widely used than NMR. However, this limitation also means that EPR offers great specificity, since ordinary chemical solvents and matrices do not give rise to EPR spectra. As a result EPR crosses several disciplines including: chemistry, physics, biology, materials science, medical science and many more.

EPR is a magnetic resonance technique very similar to NMR (Nuclear Magnetic Resonance). However, instead of measuring the nuclear transitions in our sample, we are detecting the transitions of unpaired electrons in an applied magnetic field. Like a proton, the electron has "spin", which gives it a magnetic property known as a magnetic moment. The magnetic moment makes the electron behave like a tiny bar magnet similar to one you might put on your refrigerator. When we supply an external magnetic field, the paramagnetic electrons can either orient in a direction parallel or antiparallel to the direction of the magnetic field. This creates two distinct energy levels for the unpaired electrons and allows us to measure them as they are driven between the two levels.

Initially, there will be more electrons in the lower energy level (i.e., parallel to the field) than in the upper level (antiparallel). We use a fixed frequency of microwave irradiation to excite some of the electrons in the lower energy level to the upper energy level. In order for the transition to occur we must also have the external magnetic field at a specific strength, such that the energy level separation between the lower and upper states is exactly matched by our microwave frequency. In order to achieve this condition, we sweep the external magnet's field while exposing the sample to a fixed frequency of microwave irradiation. The condition where the magnetic field and the microwave frequency are "just right" to produce an EPR resonance (or absorption) is known as the resonance condition. Photos of a typical magnet of EPR spectrometer are shown in fig.2.2.

EPR was first observed in Kazan State University by Soviet physicist Yevgeny Zavoisky in 1944, and was developed independently at the same time by Brebis Bleaney at the University of Oxford.

Fig.2.2. EPR spectrometer-Selected Magnet Systems

Furthermore, ferromagnetic materials and thin films may exhibit `very unusual', highly resolved ferromagnetic resonance (FMR) spectra, or ferromagnetic spin wave resonance (FSWR) excitations in non-crystalline solids such as ferromagnetic metallic glasses, well beyond the common single-transitions of most routine NMR, FMR and EPR studies.

The physics of spin can be realized by understanding the following parameters and definitions:

1- Boltzmann Statistics

When a group of spins is placed in a magnetic field, each spin aligns in one of the two possible orientations.

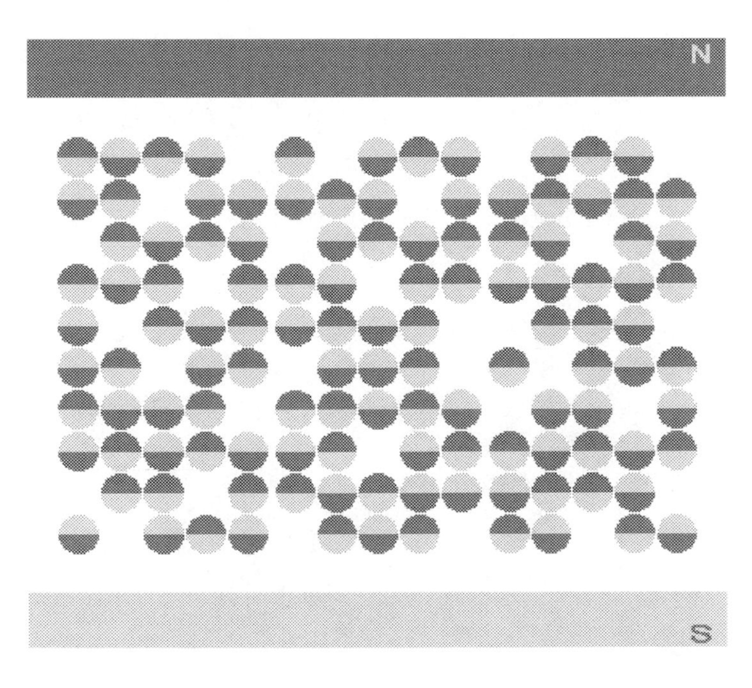

Fig.2.3. Spin Orientations

At room temperature, the number of spins in the lower energy level, N^+, slightly outnumbers the number in the upper level, N^-. Boltzmann statistics tells us that

$$N^-/N^+ = e^{-\frac{\Delta E}{k_B TP}} \qquad (2.1)$$

where ΔE is the energy difference between the spin states; k_B is Boltzmann's constant, 1.3805×10^{-23} J/Kelvin; and TP is the temperature in Kelvin.

As the temperature decreases, so does the ratio N^- / N^+. As the temperature increases, the ratio approaches one. The signal in NMR spectroscopy results from the difference between the energy absorbed by the spins which make a transition from the lower energy state to the higher energy state, and the energy emitted by the spins which simultaneously make a transition from the higher energy state to the lower energy state. The signal is thus proportional to the population difference between the states. NMR is a rather sensitive spectroscopy since it is capable of detecting these very small population differences. It is the resonance, or exchange of energy at a specific frequency between the spins and the spectrometer, which gives NMR its sensitivity.

2- Spin Packets

It is cumbersome to describe NMR on a microscopic scale. A macroscopic picture is more convenient. The first step in developing the macroscopic picture is to define the spin packet. A spin packet is a group of spins experiencing the same magnetic field strength. In this example, the spins within each grid section represent a spin packet.

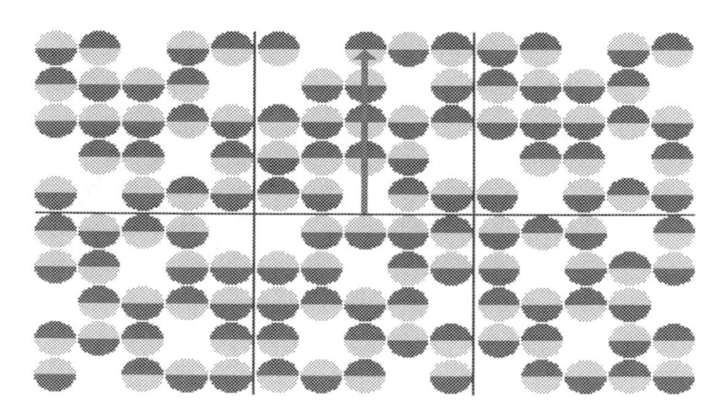

Fig.2.4. Spin Packets

At any instant in time, the magnetic field due to the spins in each spin packet can be represented by a magnetization vector. The size of each vector is proportional to $(N^+ - N^-)$. The vector sum of the magnetization vectors from all of the spin packets is the net magnetization. In order to describe pulsed NMR is necessary from here on to talk in terms of the net magnetization.

Adapting the conventional NMR coordinate system, the external magnetic field and the net magnetization vector at equilibrium are both along the Z axis.

3- T_1 Processes

At equilibrium, the net magnetization vector lies along the direction of the applied magnetic field B_o and is called the equilibrium magnetization M_o. In this configuration, the Z component of magnetization M_Z equals M_o. M_Z is referred to as the longitudinal magnetization. There is no transverse (M_X or M_Y) magnetization here.

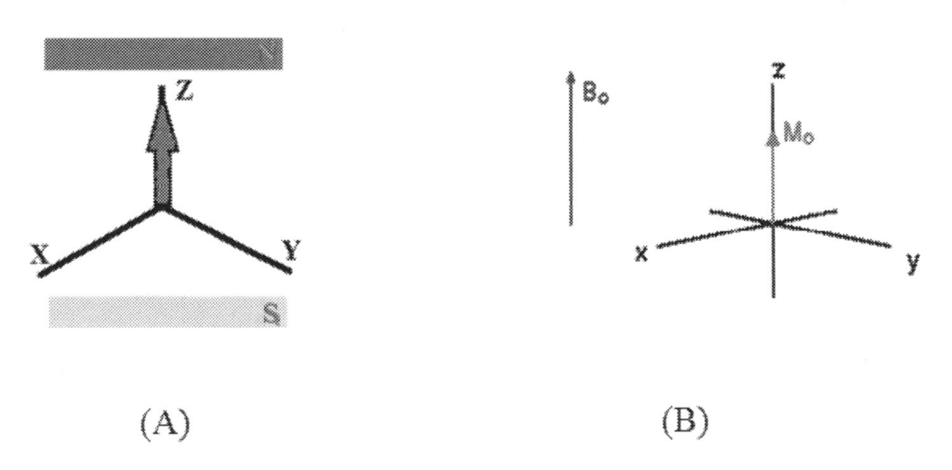

Fig.2.5. The external magnetic field and the net magnetization vector at equilibrium are both along the Z axis

It is possible to change the net magnetization by exposing the nuclear spin system to energy of a frequency equal to the energy difference between the spin states. If enough energy is put into the system, it is possible to saturate the spin system and make $M_Z=0$.

The time constant which describes how M_Z returns to its equilibrium value is called the spin lattice relaxation time (T_1). The equation governing this behavior as a function of the time t after its displacement is:

$$M_z = M_o (1 - e^{-t/T1}) \quad (2.2)$$

The Spin Lattice Relaxation Time (T_1) is therefore defined as the time required to change the Z component of magnetization by a factor of e, see figs.2.6. and 2.7.

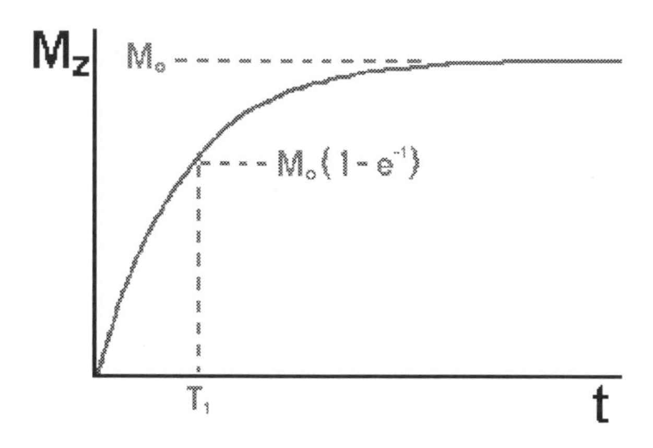

Fig.2.6. Spin Lattice Relaxation Time (T_1)

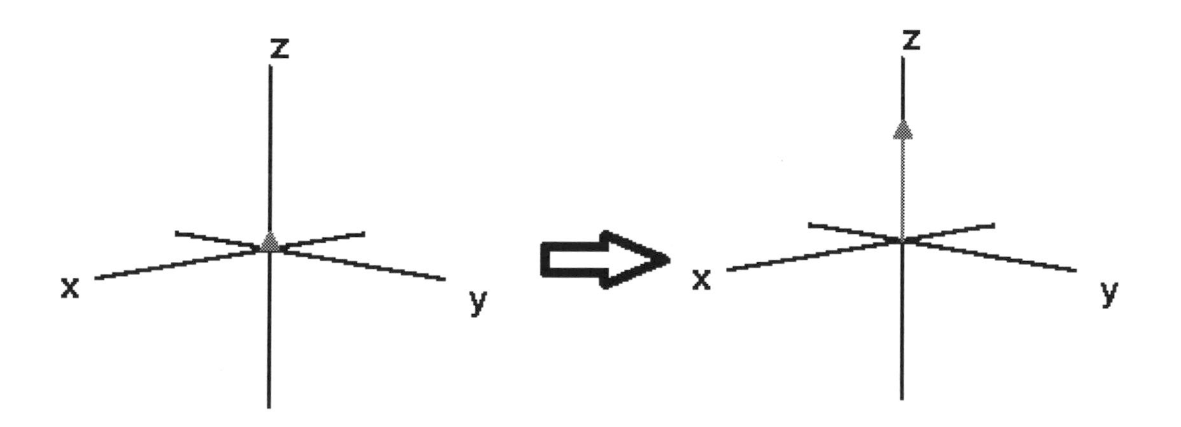

Fig.2.7. The Change in the net magnetization along the z-axis

If the net magnetization is placed along the – Zaxis, it will gradually return to its equilibrium position along the +Z axis at a rate governed by T_1. The equation governing this behavior is:

$$M_z = M_o (1 - 2e^{-t/T1}) \quad (2.3)$$

The spin-lattice relaxation time (T_1) is the time to reduce the difference between the longitudinal magnetization (M_Z) and its equilibrium value by a factor of e.

4- Precession

Nuclei precess around the magnetic field for essentially the same reasons that tops or gyroscopes precess around a gravitational field. Both gyroscopes and nuclei possess *angular momentum.* For the gyroscope, angular momentum results from a flywheel rotating about its axis. For the nucleus, angular momentum results from an intrinsic quantum property (spin).

The resultant circular motion is called *precession.* Precession occurs at a specific frequency denoted either by ω_o (called the *angular frequency*, measured in radians/sec) or f_o (called the *cyclic frequency*, measured in cycles/sec or Hertz [Hz]). Since 2π radians = 360° = 1 cycle (revolution), angular and cyclic frequencies can easily be converted by the equation:

$$\omega_o = 2\pi f_o \qquad (2.4)$$

The precession frequency of a gyroscope is a function of the mass and shape of the wheel, the speed of wheel rotation, and the strength of the gravitational field. The precession frequency of a nucleus is proportional to the strength of the magnetic field (B_o) and the *gyromagnetic ratio* (γ), a particle-specific constant incorporating size, mass, and spin. In NMR and MRI, the resonance frequency, frequency of absorption or the Larmor frequency υ is defined by:

$$\upsilon = f_o = \gamma B_o / 2\pi \qquad (2.5)$$

In other words, if the net magnetization is placed in the XY plane it will rotate about the Z axis at a frequency (Larmor frequency) equal to the frequency of the photon which would cause a transition between the two energy levels of the spin.

Precession is experienced by all particles with non-zero spin when placed in a static external magnetic field. As the earth's magnetic field is "always on", nearly every magnetically receptive nucleus on earth, including hydrogen in every drop of water in the ocean, is at this moment quietly precessing away.
Resonance. Although hydrogen nuclei are always precessing, they are not constantly undergoing nuclear magnetic *resonance* (NMR). Resonance is a natural phenomenon characterized by an oscillating response over a narrow

range of frequencies to the external input of energy. All resonant systems(mechanical or electromagnetic) are able to transfer energy between two or more different storage modes. However, slight energy losses always occur from cycle to cycle, so eventually the resonance dies out.

Magnetic resonance occurs when external energy is injected into a nuclear spin system near the Larmor (resonance) frequency. In both NMR and MRI, the primary source of energy input is from a rotating magnetic field generated by passing alternating current through a nearby radiofrequency (RF) coil.

The precessing nuclei can fall out of alignment with each other (returning the net magnetization vector to a non-precessing field) and stop producing a signal. This is called T_2 or transverse relaxation. Because of the difference in the actual relaxation mechanisms involved (for example, intermolecular versus intramolecular magnetic dipole-dipole interactions), T_1 is usually (except in rare cases) longer than T_2 (that is, slower spin-lattice relaxation, for example because of smaller dipole-dipole interaction effects).

5- T_2 Processes

In addition to the rotation, the net magnetization starts to dephase because each of the spin packets making it up is experiencing a slightly different magnetic field and rotates at its own Larmor frequency. The longer the elapsed time, the greater the phase difference. Here the net magnetization vector is initially along +Y. For this and all dephasing examples think of this vector as the overlap of several thinner vectors from the individual spin packets.

The time constant which describes the return to equilibrium of the transverse magnetization, M_{XY}, is called the spin-spin relaxation time, T_2.

$$M_{XY} = M_{XYo}\, e^{-t/T2} \qquad (2.6)$$

T_2 is always less than or equal to T_1. The net magnetization in the XY plane goes to zero and then the longitudinal magnetization grows in until we have M_o along Z.

Any transverse magnetization behaves the same way. The transverse component rotates about the direction of applied magnetization and dephases. T_1 governs the rate of recovery of the longitudinal magnetization.

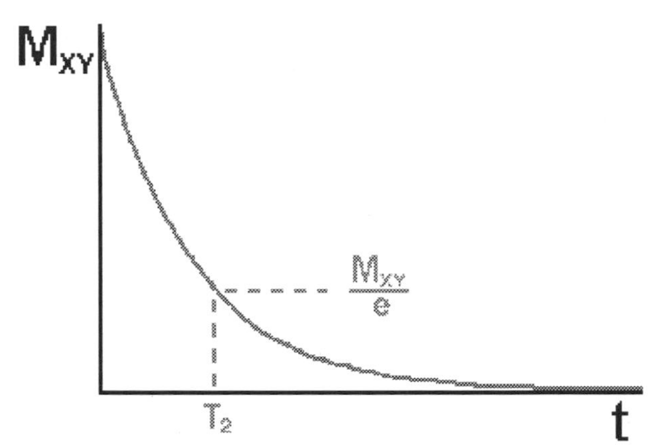

Fig.2.8. Spin-Spin Relaxation Time (T_2)

In summary, the spin-spin relaxation time, T_2, is the time to reduce the transverse magnetization by a factor of e. In the previous sequence, T_2 and T_1 processes are shown separately for clarity. That is, the magnetization vectors are shown filling the XY plane completely before growing back up along the Z axis. Actually, both processes occur simultaneously with the only restriction being that T_2 is less than or equal to T_1.

Two factors contribute to the decay of transverse magnetization; these are:

1) Molecular interactions (said to lead to a pure T_2 molecular effect)
2) Variations in B_o (said to lead to an inhomogeneous T_2 effect)

The combination of these two factors is what actually results in the decay of transverse magnetization. The combined time constant is called T_2 star and is given the symbol T_2*. The relationship between the T_2 from molecular processes and that from inhomogeneities in the magnetic field is as follows.

$$1/T_2^* = 1/T_2 + 1/T_{2\,inhomo} \qquad (2.7)$$

In practice, the value of T_2* which is the actually observed decay time of the observed NMR signal, or free induction decay (to 1/e of the initial amplitude immediately after the resonant RF pulse), also depends on the static magnetic field inhomogeneity, which is quite significant. (There is also a smaller but significant contribution to the observed FID shortening from the RF inhomogeneity of the resonant pulse). In the corresponding FT-NMR spectrum - meaning the Fourier transform of the free induction decay – the T_2* time is inversely related to the width of the NMR signal in frequency units. Thus, a nucleus with a long T_2 relaxation time gives rise to a very sharp NMR peak in the FT-NMR spectrum for a very homogeneous ("well-shimmed") static magnetic field, whereas nuclei with shorter T_2 values give rise to broad FT-NMR peaks even when the magnet is shimmed well. Both T_1 and T_2 depend on the rate of molecular motions as well as the gyromagnetic ratios of both the resonating and their strongly interacting, next-neighbor nuclei that are not at resonance.

A Hahn echo decay experiment can be used to measure the dephasing time, as shown in the animation below. The size of the echo is recorded for different spacing of the two pulses. This reveals the decoherence which is not refocused by the 180° pulse. In simple cases, an exponential decay is measured which is described by the T_2 time.

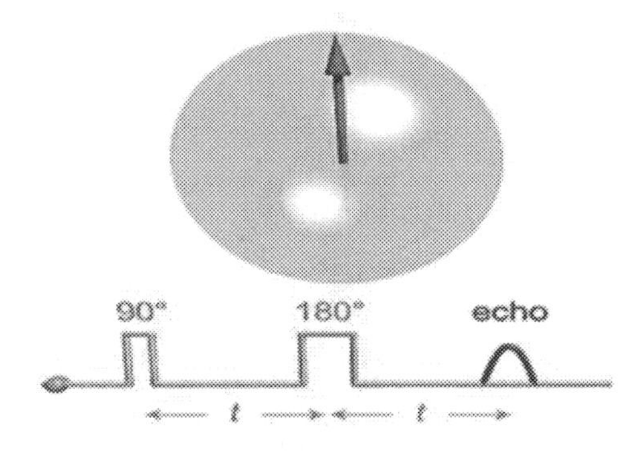

Fig.2.9 Echo recording for different spacing of two pulses

6- Rotating Frame of Reference

We have just looked at the behavior of spins in the laboratory frame of reference. It is convenient to define a rotating frame of reference which rotates about the Z axis at the Larmor frequency.

We distinguish this rotating coordinate system from the laboratory system by primes on the X and Y axes, X'Y'.

A magnetization vector rotating at the Larmor frequency in the laboratory frame appears stationary in a frame of reference rotating about the Z axis. In the rotating frame, relaxation of MZ magnetization to its equilibrium value looks the same as it did in the laboratory frame.

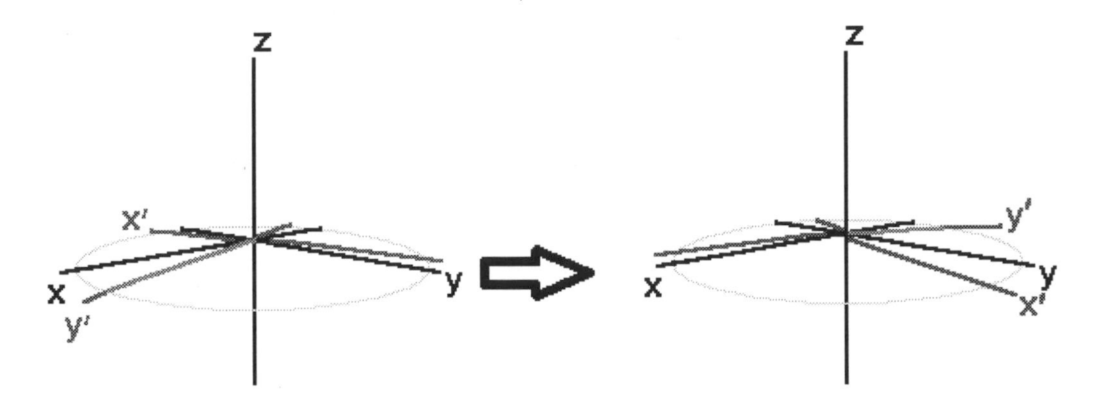

Fig.2.10. Rotating coordinate system.

A transverse magnetization vector rotating about the Z axis at the same velocity as the rotating frame will appear stationary in the rotating frame. A magnetization vector traveling faster than the rotating frame rotates clockwise about the Z axis. A magnetization vector traveling slower than the rotating frame rotates counter-clockwise about the Z axis.

7- Pulsed Magnetic Fields

A coil of wire placed around the X axis will provide a magnetic field along the X axis when a direct current is passed through the coil. An alternating current will produce a magnetic field which alternates in direction.

In a frame of reference rotating about the Z axis at a frequency equal to that of the alternating current, the magnetic field along the X' axis will be constant, just as in the direct current case in the laboratory frame.

This is the same as moving the coil about the rotating frame coordinate system at the Larmor Frequency. In magnetic resonance, the magnetic field created by the coil passing an alternating current at the Larmor frequency is called the B_1 magnetic field. When the alternating current through the coil is turned on and off, it creates a pulsed B_1 magnetic field along the X' axis.

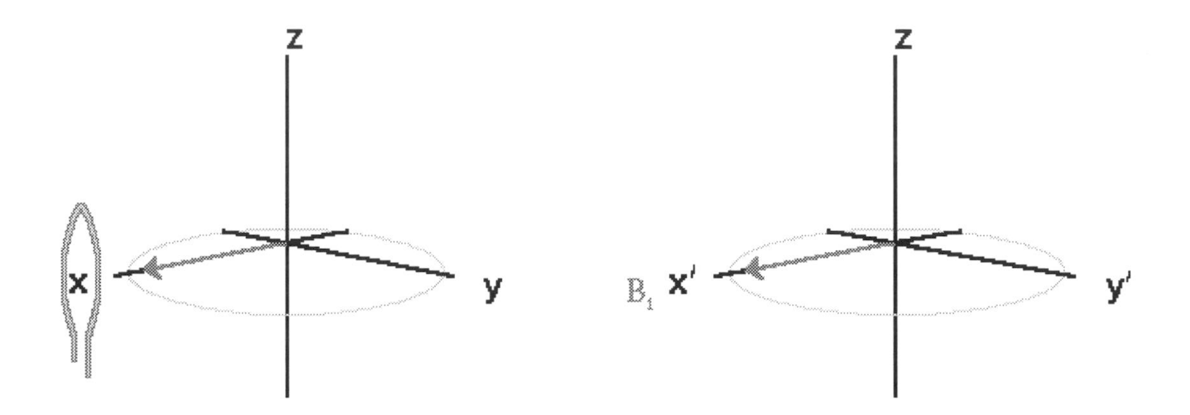

Fig.2.11. The magnetic field at stationary and rotating frame of reference

The spins respond to this pulse in such a way as to cause the net magnetization vector to rotate about the direction of the applied B_1 field. The rotation angle depends on the length of time the field is on, τ, and its magnitude B_1.

$$\theta = 2 \pi \gamma \tau B_1 \tag{2.8}$$

In our examples, τ will be assumed to be much smaller than T_1 and T_2.

A 90° pulse is one which rotates the magnetization vector clockwise by 90 degrees about the X' axis. A 90° pulse rotates the equilibrium magnetization down to the Y' axis. In the laboratory frame the equilibrium magnetization spirals down around the Z axis to the XY plane. You can see why the rotating frame of reference is helpful in describing the behavior of magnetization in response to a pulsed magnetic field.

A 180° pulse will rotate the magnetization vector by 180 degrees. A 180° pulse rotates the equilibrium magnetization down to along the -Z axis.

The net magnetization at any orientation will behave according to the rotation equation. For example, a net magnetization vector along the Y' axis will end up along the -Y' axis when acted upon by a 180° pulse of B_1 along the X' axis.

A net magnetization vector between X' and Y' will end up between X' and Y' after the application of a 180° pulse of B_1 applied along the X' axis.

Here θ is the rotation angle about the X' axis, [X', Y', Z] is the initial location of the vector, and [X'', Y'', Z''] the location of the vector after the rotation.

$$\begin{bmatrix} X'' \\ Y'' \\ Z'' \end{bmatrix} = \begin{bmatrix} 1 & 0 & 0 \\ 0 & Cos\theta & Sin\theta \\ 0 & -Sin\theta & Cos\theta \end{bmatrix} \begin{bmatrix} X' \\ Y' \\ Z \end{bmatrix} \qquad (2.9)$$

8- Spin Relaxation

How do nuclei in the higher energy state return to the lower state? Emission of radiation is insignificant because the probability of re-emission of photons varies with the cube of the frequency. At radio frequencies, re-emission is negligible. We must focus on non-radiative relaxation processes (thermodynamics).

Ideally, the NMR spectroscopic would like relaxation rates to be fast - but not too fast. If the relaxation rate is fast, then saturation is reduced. If the relaxation rate is too fast, line-broadening in the resultant NMR spectrum is observed.

There are two major relaxation processes;

- Spin - lattice (longitudinal) relaxation
- Spin - spin (transverse) relaxation

The process called population relaxation refers to nuclei that return to the thermodynamic state in the magnet. This process is also called T_1, "spin-lattice" or "longitudinal magnetic" relaxation, where T_1 refers to the mean time for an individual nucleus to return to its thermal equilibrium state of the spins. Once the nuclear spin population is relaxed, it can be probed again, since it is in the initial, equilibrium (mixed) state.

Nuclei in an NMR experiment are in a sample. The sample in which the nuclei are held is called the *lattice*. Nuclei in the lattice are in vibrational and rotational motion, which creates a complex magnetic field. The magnetic field caused by motion of nuclei within the lattice is called the *lattice field*. This lattice field has many components. Some of these components will be equal in frequency and phase to the Larmor frequency of the nuclei of interest. These components of the lattice field can interact with nuclei in the higher energy state, and cause them to lose energy (returning to the lower state). The energy that a nucleus loses increases the amount of vibration and rotation within the lattice (resulting in a tiny rise in the temperature of the sample).

The relaxation time, T_1 (the average life time of nuclei in the higher energy state) is dependent on the magnetogyric ratio of the nucleus and the mobility of the lattice. As mobility increases, the vibrational and rotational frequencies increase, making it more likely for a component of the lattice field to be able to interact with excited nuclei. However, at extremely high mobilities, the probability of a component of the lattice field being able to interact with excited nuclei decreases.

Spin - spin relaxation describes the interaction between neighboring nuclei with identical frequencies but differing magnetic quantum states. In this situation, the nuclei can exchange quantum states; a nucleus in the lower energy level will be excited, while the excited nucleus relaxes to the lower energy state. There is no net change in the populations of the energy states, but the average lifetime of a nucleus in the excited state will decrease. This can result in line-broadening.

Motions in solution which result in time varying magnetic fields cause spin relaxation. Time varying fields at the Larmor frequency cause transitions between the spin states and hence a change in M_Z. This screen depicts the field at the green hydrogen on the water molecule as it rotates about the external field B_o and a magnetic field from the blue hydrogen. Note that the field experienced at the green hydrogen is sinusoidal, see Fig.2.12.

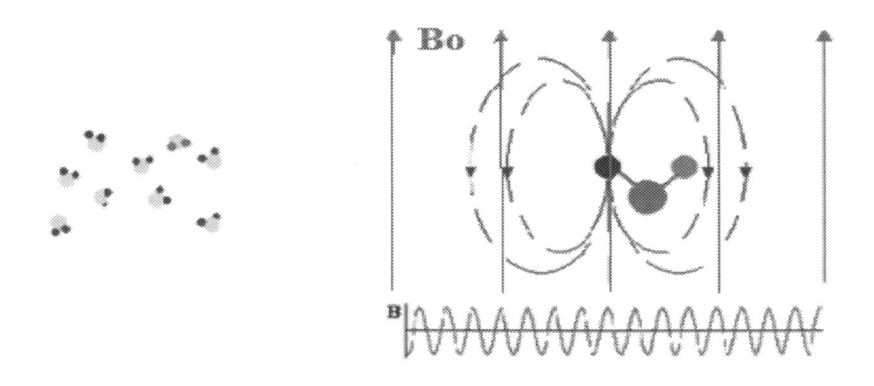

Fig.2.12. Field distribution at the hydrogen on water molecule

There is a distribution of rotation frequencies in a sample of molecules. Only frequencies at the Larmor frequency affect the relaxation time, T_1. Since the Larmor frequency is proportional to B_o, T_1 will therefore vary as a function of magnetic field strength. In general, T_1 is inversely proportional to the density of molecular motions at the Larmor frequency, see Fig.2.13.

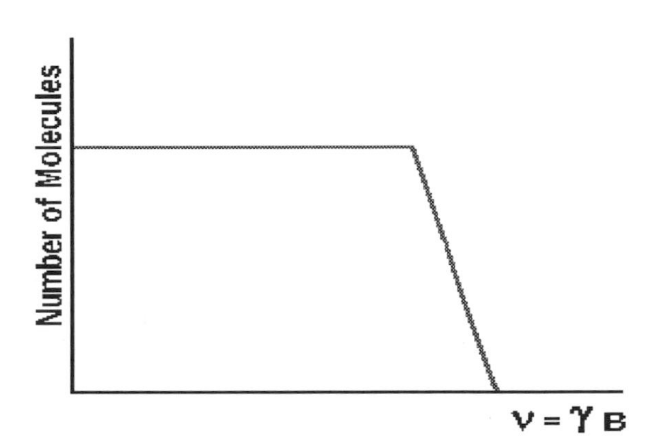

Fig.2.13 Pattern of Lamour frequency verses density of Molecular

The rotation frequency distribution depends on the temperature and viscosity of the solution. Therefore T_1 will vary as a function of temperature.

At the Larmor frequency indicated by ν_o, T_1 (280 K) < T_1 (340 K). The temperature of the human body does not vary by enough to cause a significant influence on T_1. The viscosity does however vary significantly from tissue to tissue and influences T_1 as is seen in the following molecular motion plot, fig.2.14 B.

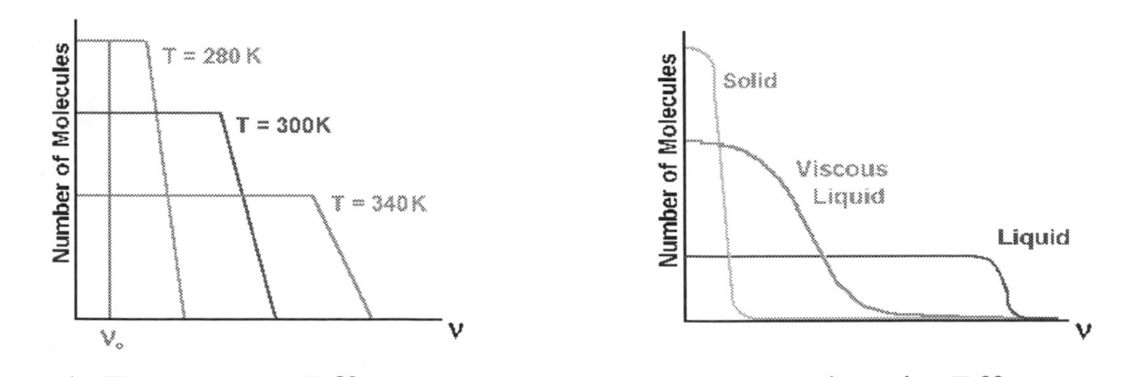

A. Temperature Effect B. Viscosity Effect

Fig.2.14. Lamour frequency with temperature and viscosity of the solution

Fluctuating fields which perturb the energy levels of the spin states cause the transverse magnetization to dephase. This can be seen by examining the plot of B_o experienced by the red hydrogens on the following water molecule. The number of molecular motions less than and equal to the Larmor frequency is inversely proportional to T_2.

In general, relaxation times get longer as B_o increases because there are fewer relaxation-causing frequency components present in the random motions of the molecules.

9- Spin Exchange

Spin exchange is the exchange of spin state between two spins.

For example, if we have two spins, A and B, and A is spin up and B is spin down, spin exchange between A and B can be represented with the following equation.

A(●) + B(●) ⇆ A(●) + B(●)

The bidirectional arrow indicates that the exchange reaction is reversible.

The energy difference between the upper and lower energy states of A and of B must be the same for spin exchange to occur. On a microscopic scale, the spin in the upper energy state (B) is emitting a photon which is being absorbed by the spin in the lower energy state (A). Therefore, B ends up in the lower energy state and A in the upper state.

Spin exchange will not affect T_1 but will affect T_2. T_1 is not affected because the distribution of spins between the upper and lower states is not changed. T_2 will be affected because phase coherence of the transverse magnetization is lost during exchange.

Another form of exchange is called chemical exchange. In chemical exchange, the A and B nuclei are from different molecules. Consider the chemical exchange between water and ethanol.

$CH_3CH_2OH_A + HOH_B \leftrightarrows CH_3CH_2OH_B + HOH_A$

Here the B hydrogen of water ends up on ethanol, and the A hydrogen on ethanol ends up on water in the forward reaction. There are four senarios for the nuclear spin, represented by the four equations.

$CH_3CH_2OH_A$(◒) + HOH_B(◓) ⇆ $CH_3CH_2OH_B$(◓) + HOH_A(◒)

$CH_3CH_2OH_A$(◓) + HOH_B(◒) ⇆ $CH_3CH_2OH_B$(◒) + HOH_A(◓)

$CH_3CH_2OH_A$(◒) + HOH_B(◒) ⇆ $CH_3CH_2OH_B$(◒) + HOH_A(◒)

$CH_3CH_2OH_A$(◓) + HOH_B(◓) ⇆ $CH_3CH_2OH_B$(◓) + HOH_A(◓)

Chemical exchange will affect both T_1 and T_2. T_1 is now affected because energy is transferred from one nucleus to another. For example, if there are more nuclei in the upper state of A, and a normal Boltzmann distribution in B, exchange will force the excess energy from A into B. The effect will make T_1 appear smaller. T_2 is effected because phase coherence of the transverse magnetization is not preserved during chemical exchange.

10- Bloch Equations

As a background for the definition of Bloch equations the behavior of bulk magnetization "M" in a magnetic field that arises from all of the magnetic moments in a sample is briefly described herewith.

M experiences a torque when placed in a magnetic field according to

$$\frac{d\mathbf{U}(t)}{dt} = \mathrm{M(t)} \times \mathrm{B(t)} \quad (2.10)$$

where **U**(t) is the bulk spin angular momentum. Note that all three vector quantities in Equation (2.9) are time dependent. The time-dependence of the magnetic field comes about when the radio frequency (*rf*) pulses applied along the x- or y-axis.

Equation (2.10) is essentially identical to an equation that describes the motion of a gyroscope:

$$\frac{d\mathbf{L}(t)}{dt} = \mathrm{r} \times \mathrm{mg} \quad (2.11)$$

where **L**(t) is the gyroscope's angular momentum, **r** the radius from the fixed point of rotation, *m* the mass, and **g** gravity. Thus, a nuclear spin in a magnetic field will behave much like a gyroscope in a gravitational field.

To make Equation (2.10) more useful, we use the relationship of the magnetic moment μ, given in the next section,

$$\mu = \gamma I = \frac{\gamma m h}{2\pi} = \gamma m \hbar$$

and multiply each side the gyromagnetic ratio γ to yield

$$\frac{dM(t)}{dt} = M(t) \text{ x } \gamma \text{ B(t)} \tag{2.12}$$

Equation (2.12) is the basis of the Bloch equations.

In 1946 Felix Bloch formulated a set of equations that describe the behavior of a nuclear spin in a magnetic field under the influence of *rf* pulses. He modified Equation (2.12) to account for the observation that nuclear spins "relax" to equilibrium values following the application of *rf* pulses. Bloch assumed they relax along the z-axis and in the x-y plane at different rates but following first-order kinetics. These rates are designated 1/T1 and 1/T2 for the z-axis and x-y plane, respectively. T1 is called spin-latice relaxation and T2 is called spin-spin(transvers) relaxation, as given in the previous sections. By considering the relaxation, Equation (2.12) becomes

$$\frac{dM(t)}{dt} = M(t) \text{ x } \gamma \text{ B}(t)\text{-R}(M(t)\text{-}M_0) \tag{2.13}$$

where **R** is the "relaxation matrix". Equation 2.13 is best understood by considering each of its components:

$$\frac{dM_z(t)}{dt} = \gamma\left[M_x(t)B_y(t) - M_y(t)B_x(t)\right] - \frac{(M_z(t)-M_0)}{T_1}$$

$$\frac{dM_x(t)}{dt} = \gamma\left[M_y(t)B_z(t) - M_z(t)B_y(t)\right] - \frac{M_x(t)}{T_2} \tag{2.14}$$

$$\frac{dM_y(t)}{dt} = \gamma[M_z(t)B_x(t) - M_x(t)B_z(t)] - \frac{M_y(t)}{T_2}$$

The terms in Equation (2.14) that do not involve either T1 or T2 are the result of the cross product in Equation (2.13). Equation (2.14) describes the motion of magnetization in the "laboratory frame", an ordinary coordinate system that is stationary. Mathematically (and conceptually) the laboratory frame is not the simplest coordinate system, because the magnetization is moving at a frequency ($\omega_0 = \gamma B_0$) in the x-y (transverse) plane. A simpler coordinate system is the "rotating frame", in which the x-y plane rotates around the z-axis at a frequency $\Omega = -\gamma B_0$.

In the rotating frame, magnetization "on resonance" does not precess in the transverse plane. The transformation of Equation (2.14) to the rotating frame is achieved by replacing each B_z (defined as B_0) by Ω/γ:

$$\frac{dM_z(t)}{dt} = \gamma\left[M_x(t)B_y^r(t) - M_y(t)B_x^r(t)\right] - \frac{(M_z(t)-M_0)}{T_1}$$

$$\frac{dM_x(t)}{dt} = \Omega M_y(t) - \gamma M_z(t)B_y^r(t) - \frac{M_x(t)}{T_2} \qquad (2.15)$$

$$\frac{dM_y(t)}{dt} = \gamma\, M_z(t)B_x^r(t) - \Omega M_x(t) - \frac{M_y(t)}{T_2}$$

In Equation (2.15), the components of **B** have been written with r superscripts to denote that it is a rotating frame. From this point onward, the rotating frame will be assumed without the superscript.

The Physical interpretation of Bloch Equations can be illustrated by a Single pulse experiment, where the behavior of Equation 2.15 is examined under two different limiting conditions, the effect of a short *rf* pulse and free precession. The *rf* pulse will be assumed to be very short compared to either relaxation times T1 and T2 as well as the angular frequency Ω. This assumption is valid for many typical pulsed NMR experiments in which the pulse lengths can be as short as 5 μs.

We will apply the *rf* pulse along the x-axis. These conditions allow us to neglect terms in Equation (2.15) that contain T1, T2, Ω, and B_y.

$$\frac{dM_z(t)}{dt} = -\gamma M_y(t)B_x^r(t)$$

$$\frac{dM_x(t)}{dt} = 0 \quad (2.16)$$

$$\frac{dM_y(t)}{dt} = \gamma\, M_z(t) B_x^r(t)$$

2.1.3 Energy Level and Magnetic Field

When the nucleus is in a magnetic field, the initial populations of the energy levels are determined by thermodynamics, as described by the Boltzmann distribution, section 2.1.2 Nuclear spin and Magnets. This is very important, and it means that the lower energy level will contain slightly more nuclei than the higher level. It is possible to excite these nuclei into the higher level with electromagnetic radiation. The frequency of radiation needed is determined by the difference in energy (transition energy) between the energy levels.

Some atomic nuclei have an intrinsic property called "spin". This was first demonstrated in 1922 by the Stern-Gerlach experiment, in which a beam of silver atoms were passed through a magnet field and split into two beams. These two beams represent the two states, α and β, of the silver (spin 1/2) nuclei.

The nuclear spin has an intrinsic angular momentum, a vector that is represented by the symbol **I** (vectors will be in bold). Vectors have 3 orientations (x, y, and z) and a length. However, the Heisenberg Uncertainty Principle tells that we can only know one orientation and the length simultaneously. By convention in NMR, we say that we know the z-orientation of the angular momentum of a nucleus in a magnetic field. The square of the magnitude of **I** is given by

$$|\mathbf{I}^2| = \hbar\,[I(I+1)] \quad (2.17)$$

where I is the spin quantum number (e.g. 1/2) and $\hbar$ is **the** reduced (Planck's constant h) ($\hbar = h/2\pi$) and called h bar or h cross.

The z-component of I is given by

$$I_z = \hbar\, m \quad (2.18)$$

where m is the quantum number with values $m = (-I, -I+1, ..., I-1, I)$. In the case of spin 1/2 nuclei, $I = 1/2$ and m can be -1/2 and 1/2.

A diagram illustrating these principles is given in fig.2.15.:

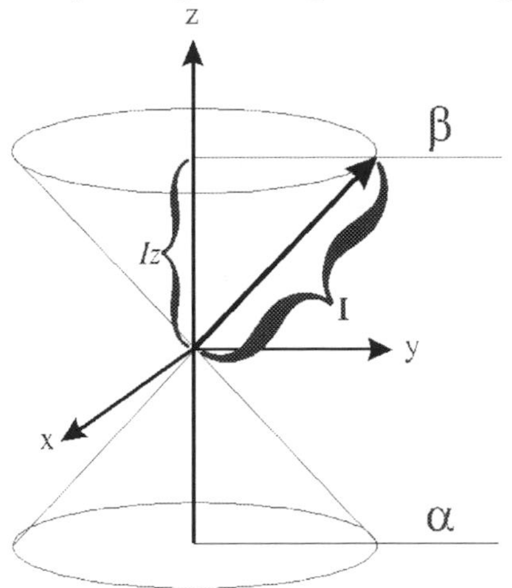

Fig. 2.15. Illustration diagram of Nuclear Spin

In this diagram, the coordinate system has been placed in the center of two cones. The angular momentum vector I is shown on the edge of the upper cone. The projection of I onto the z-axis is I_z, but I is completely undetermined in the (x,y) plane.

The overall spin, I, is important. Quantum mechanics tells us that a nucleus of spin I will have $2I + 1$ possible orientations. A nucleus with spin 1/2 will have 2 possible orientations. In the absence of an external magnetic field, these orientations are of equal energy. If a magnetic field is applied, then the energy levels split. Each level is given a magnetic quantum number, m.

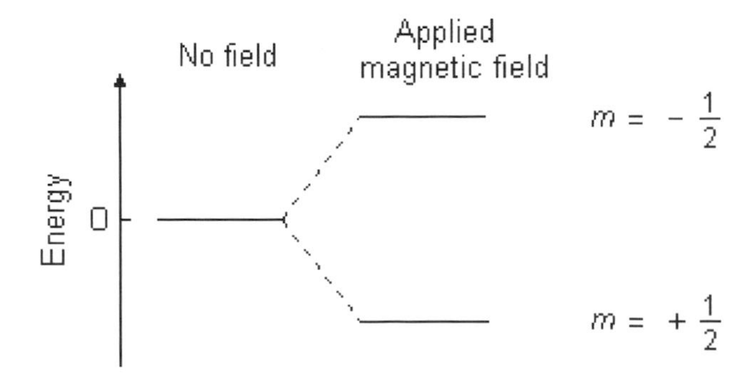

Fig.2.16. Energy levels for a nucleus with spin quantum number 1/2

The physics characteristics of spin can be realized by understanding the following parameters and definitions which are related to Energy Level and magnetic field:

1- Transition Energy

The nucleus has a positive charge and is spinning. This generates a small magnetic field. The nucleus therefore possesses a magnetic moment, μ, which is proportional to its spin, i.e. magnetic quantum number, m.

$$\mu = \frac{\gamma m h}{2\pi} = \gamma m \hbar, \quad \text{(See previous section)} \tag{2.19}$$

The constant, γ, is the gyromagnetic.

To conduct an NMR experiment, a sample is first placed into a static magnetic field. In this case, the energy of the magnetic moment of a nuclear spin in a magnetic field of NMR magnet is given by;

$$E = \mu B \tag{2.20}$$

and the energy associated with a particular quantum number m is

$$E = E_m = \frac{\gamma m h}{2\pi} B = \gamma m \hbar B \tag{2.21}$$

Where $\boldsymbol{B}$ is the strength of the magnetic field at the nucleus (in general **B** is a vector quantity). By convention, the direction of the static magnetic field is along the z-axis, and the magnitude of the magnetic field is given by B (no longer a vector since it points only along the z-axis).

The difference in energy between levels (the transition energy) can be found from

$$\Delta E = \frac{\gamma h}{2\pi} B = \gamma \hbar B \tag{2.22}$$

This means that if the magnetic field, B, is increased, so is ΔE. It also means that if a nucleus has a relatively large gyromagnetic ratio, then ΔE is correspondingly large.

This nucleus can undergo a transition between the two energy states by the absorption of a photon. A nucleus in the lower energy state absorbs a photon and ends up in the upper energy state. The energy of this photon must exactly match the energy difference between the two states.

NMR transition energies are very small. These small energies translate into low sensitivity. In the absence of a magnetic field, the α and β states are equally populated, leading to no net magnetization (that is why we need large static magnets to do NMR). When samples are placed into a magnetic field, α small excess of nuclei fall into the α state. This excess of spins in the α over the β states accounts for the entire net magnetization which is used in the NMR experiment. The ratio of the number of spins in the α state to those in the β state is given by a Boltzman distribution stated in the last section (2.1.2) and rewritten in terms of α and β numbers:

$$\frac{N_\alpha}{N_\beta} = e^{\frac{\Delta E}{K_B TP}} \qquad (2.23)$$

Typical values of magnetic field strengths produce as little as 1 in 10,000 excess α spins over β. Higher magnetic fields produce correspondingly larger differences in spin states, leading to greater sensitivity.

The energy, E, of a photon is also related to its frequency, υ and Planck's constant ($h = 6.626\text{x}10^{-34}$ J s) and is given by Planck's law:

$$E = h\upsilon \qquad (2.24)$$

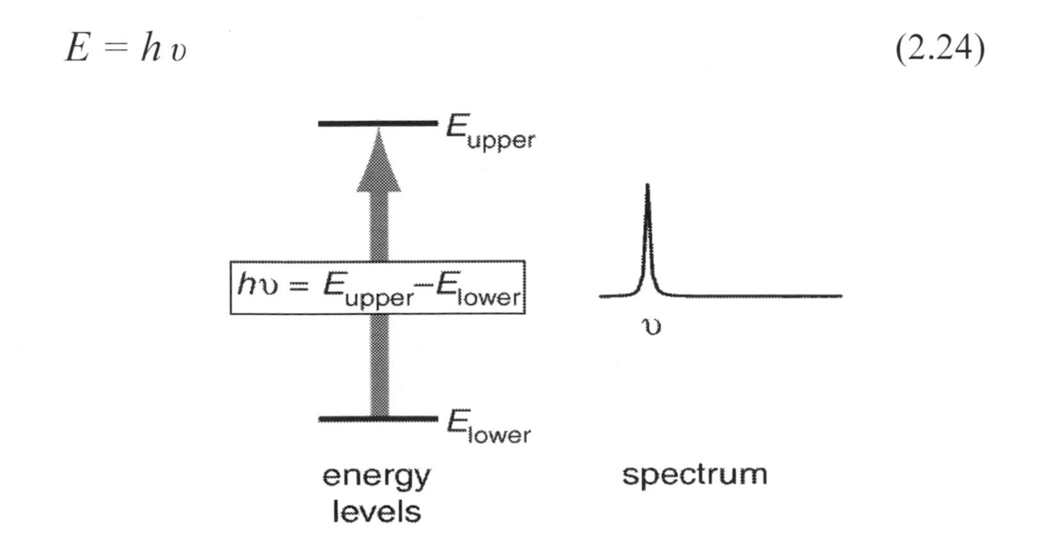

Fig.2.17. Energy levels

From Planck's law the resonance frequency (frequency of absorption or the Larmor frequency) υ, of an NMR transition in a magnet field is given by:

$$\upsilon = \gamma B / 2\pi \quad (2.25)$$

The units of υ are radians/second. More commonly, NMR frequencies are expressed in Hertz (Hz), from $\upsilon = \omega/2\pi$, yielding the expression

$$\omega = \gamma B \quad (2.26)$$

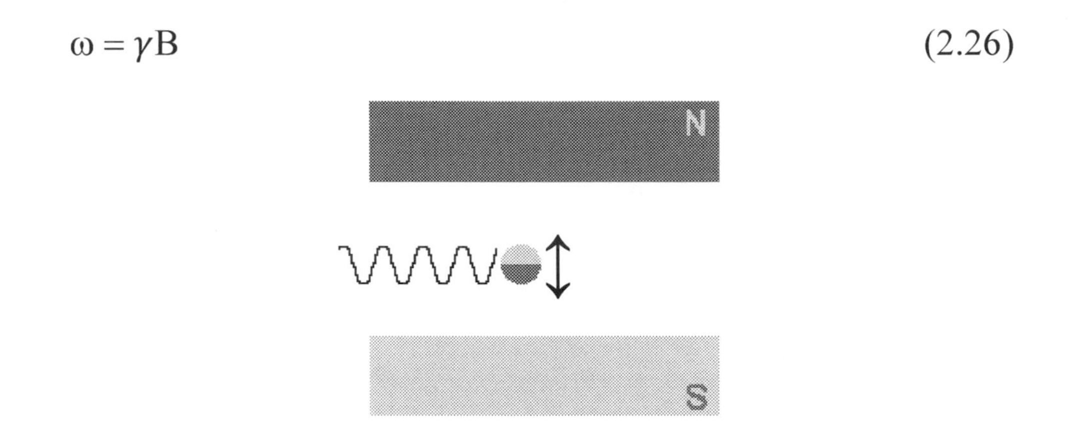

Fig.2.18 The Larmor frequency υ, of an NMR transition in a magnet field

2- Energy Level Diagrams

To understand how particles with spin behave in a magnetic field, consider a proton. This proton has the property called spin. Think of the spin of this proton as a magnetic moment vector, causing the proton to behave like a tiny magnet with a north and south pole.

Fig.2.19. Proton like tiny magnet

When the proton is placed in an external magnetic field, the spin vector of the particle aligns itself with the external field, just like a magnet would. There is a low energy configuration or state where the poles are aligned N-S-N-S and a high energy state N-N-S-S.

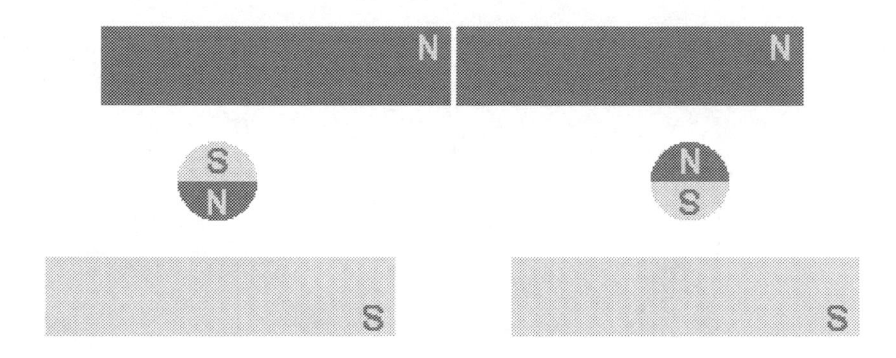

Fig.2.20. Configuration of low and high energy state

The energy of the two spin states can be represented by an energy level diagram.

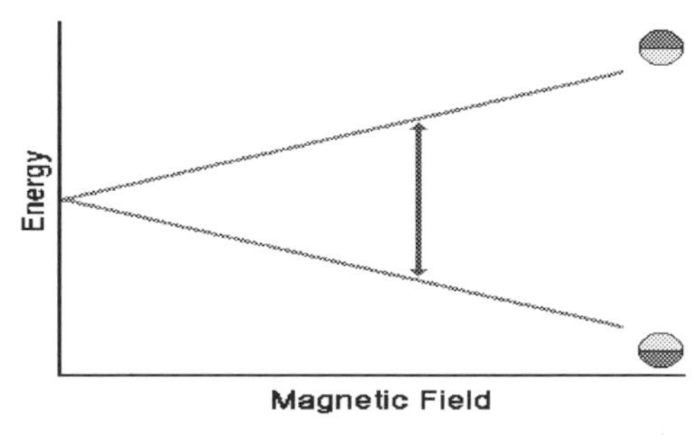

Fig.2.21. Energy Level Diagram.

We have seen that $\upsilon = \gamma B$ and $E = h \upsilon$, therefore the energy of the photon needed to cause a transition between the two spin states is

$$E = h \gamma B \qquad (2.27)$$

When the energy of the photon matches the energy difference between the two spin states an absorption of energy occurs.

In the NMR experiment, the frequency of the photon is in the radio frequency (RF) range. In NMR spectroscopy, υ is between 60 and 800 MHz for hydrogen nuclei. In clinical MRI, υ is typically between 15 and 80 MHz for hydrogen imaging.

3- CW NMR Experiment

The simplest NMR experiment is the Continuous Wave (CW) experiment. There are two ways of performing this experiment. In the first, a constant frequency, which is continuously on, probes the energy levels while the magnetic field is varied. The energy of this frequency is represented by the blue line in the energy level diagram.

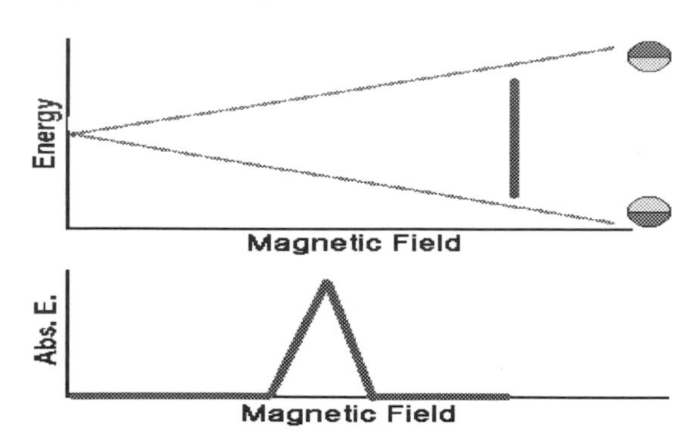

Fig.2.22. CW Experiment of constant frequency with variable magnetic field

The CW experiment can also be performed with a constant magnetic field and a frequency which is varied. The magnitude of the constant magnetic field is represented by the position of the vertical blue line in the energy level diagram.

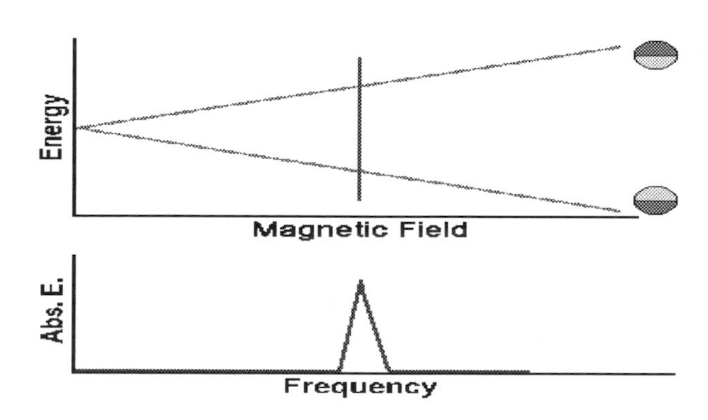

Fig.2.23. CW Experiment of constant magnetic field with variable frequency

2.2. Values of Spin Angular Momentum

The angular momentum associated with nuclear spin is quantized. This means both that the magnitude of angular momentum is quantized (i.e. S can only take on a restricted range of values), and also that the orientation of the associated angular momentum is quantized. The associated quantum number is known as the magnetic quantum number, m, and can take values from $+S$ to $-S$, in integer steps. Hence for any given nucleus, there is a total of $2S + 1$ angular momentum states.

The z-component of the angular momentum vector (**S**) is therefore $S_z = m\hbar$, The z-component of the magnetic moment is simply:

$$\mu_z = \gamma S_z = \gamma m\hbar \qquad (2.28)$$

Where γ stand for gyromagnetic ratio (can be positive or negative).

2.2.1 Spin Behavior in a Magnetic Field

Consider nuclei which have a spin of one-half, like ^1H, ^{13}C or ^{19}F. The nucleus has two possible spin states: $m = ½$ or $m = -½$ (also referred to as spin-up and spin-down, or sometimes α and β spin states, respectively). These states are degenerate, i.e. they have the same energy. Hence the number of atoms in these two states will be approximately equal at thermal equilibrium.

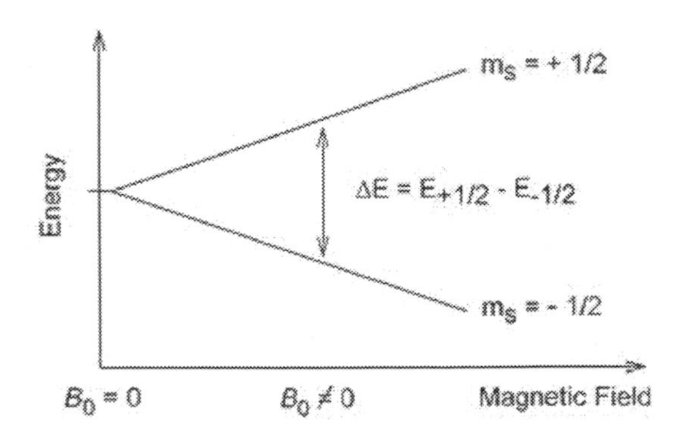

Fig.2.24.Splitting of nuclei spin states in an external magnetic field

If a nucleus is placed in a magnetic field, however, the interaction between the nuclear magnetic moment and the external magnetic field mean the two states no longer have the same energy. As stated earlier, the energy of a magnetic moment μ is given by:

$$E = -\mu B = -\mu_z B\cos(\theta) \qquad (2.29)$$

Usually B is chosen to be aligned along the z axis, therefore $\cos(\theta)=1$:

$$E = -\mu_z B, \quad \text{or alternatively: } E = -\gamma m\hbar B \qquad (2.30)$$

As a result the different nuclear spin states have different energies in a non-zero magnetic field. In hand-waving terms, we can talk about the two spin states of a spin ½ as being aligned either with or against the magnetic field. If γ is positive (true for most isotopes) then $m = ½$ is the lower energy state. The energy difference between the two states is:

$$\Delta E = \gamma\hbar B, \qquad (2.31)$$

and this difference results in a small population bias toward the lower energy state.

2.2.2 Magnetic Resonance by Nuclei

Electromagnetic radiation (**EM radiation** or **EMR**) is a form of energy emitted and absorbed by charged particles, which exhibits wave-like behavior as it travels through space. EMR has both electric and magnetic field components, which stand in a fixed ratio of intensity to each other, and which oscillate in phase perpendicular to each other and perpendicular to the direction of energy and wave propagation. In a vacuum, electromagnetic radiation propagates at a characteristic speed, the speed of light.

Electromagnetic radiation is a particular form of the more general electromagnetic field (EM field), which is produced by moving charges. Electromagnetic radiation is associated with EM fields that are far enough away from the moving charges that produced them, that absorption of the EM radiation no longer affects the behavior of these moving charges. These two types or behaviors of EM field are sometimes referred to as the near and far field.

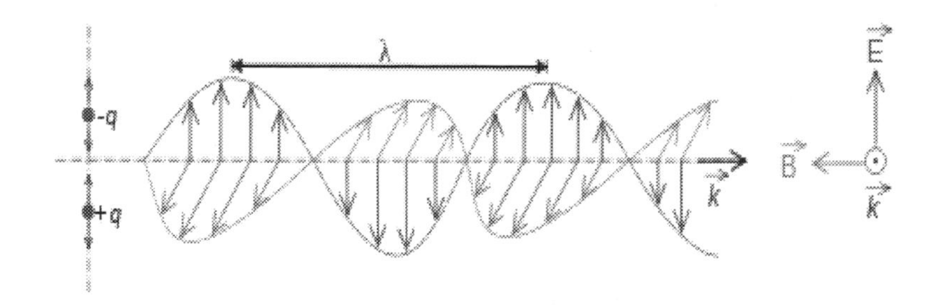

Fig.2.25. Electromagnetic radiation waves

The electromagnetic waves that compose electromagnetic radiation can be imagined as a self-propagating transverse oscillating wave of electric and magnetic fields. This diagram shows a plane linearly polarized EMR wave propagating from left to right. The electric field is in a vertical plane and the magnetic field in a horizontal plane. The two types of fields in EMR waves are always in phase with each other, and no matter how powerful, have a ratio of electric to magnetic intensity which is fixed and never varies.

In this section, the "classical" view of the behavior of the nucleus - that is, the behavior of a charged particle in a magnetic field is considered.

Imagine a nucleus (of spin 1/2) in a magnetic field. This nucleus is in the lower energy level (i.e. its magnetic moment does not oppose the applied field). The nucleus is spinning on its axis. In the presence of a magnetic field, this axis of rotation will precess around the magnetic field. If energy is absorbed by the nucleus, then the angle of precession, θ, will change. For a nucleus of spin 1/2, absorption of radiation "flips" the magnetic moment so that it opposes the applied field (the higher energy state).

The potential energy of the precessing nucleus as given by Eq.(2.27);

$$E = - \mu B \cos \theta$$

where θ is the angle between the direction of the applied field and the axis of nuclear rotation.

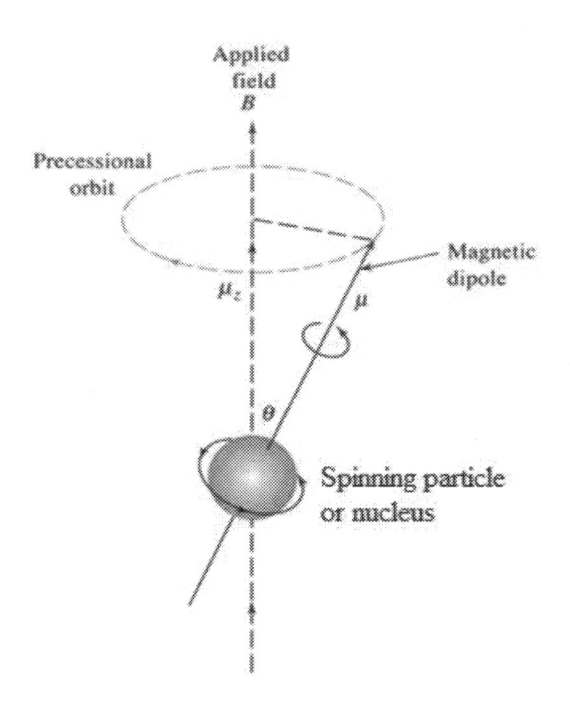

Fig.2.26. spinning particles or nucleus

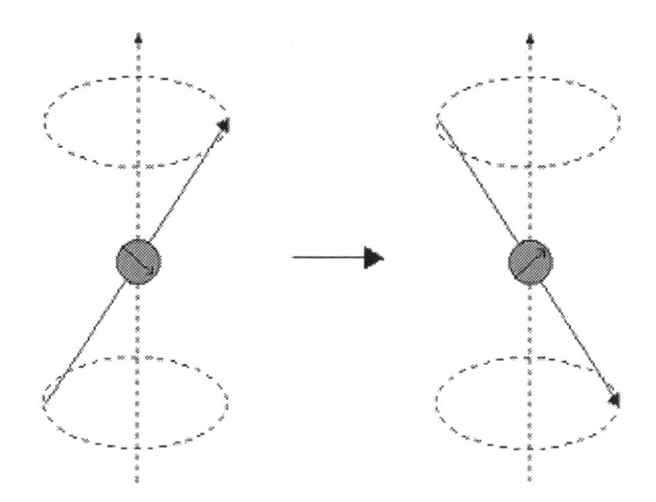

Fig.2.27. Flips the magnetic moment of spinning particles or nucleus

where the frequency of precession or Larmor frequency, which is, as the transition frequency given earlier (Eq.2.25) :

$$\upsilon = \gamma B / 2\pi$$

Resonant absorption by nuclear spins will occur only when electromagnetic radiation of the correct frequency (e.g., equaling the Larmor precession rate) is being applied to match the energy difference between the nuclear spin levels in a constant magnetic field of the appropriate strength. Such magnetic resonance frequencies typically correspond to the radio frequency (or RF) range of the electromagnetic spectrum for magnetic fields up to ~20 T. It is this magnetic resonant absorption which is detected in NMR.

It is important to realize that only a small proportion of "target" nuclei are in the lower energy state (and can absorb radiation). There is the possibility that by exciting these nuclei, the populations of the higher and lower energy levels will become equal. If this occurs, then there will be no further absorption of radiation. The spin system is saturated. The possibility of saturation means that we must be aware of the relaxation processes which return nuclei to the lower energy state.

2.2.3 The Shielding Effect

The magnetic field at the nucleus is **not** equal to the applied magnetic field; electrons around the nucleus shield it from the applied field. The difference between the applied magnetic field and the field at the nucleus is termed the nuclear shielding. Electrons in an atom can shield each other from the pull (attraction) of the nucleus. The shielding effect, describes the decrease in attraction between an electron and the nucleus in any atom with more than one electron shell. The more electron shells there are, the greater the shielding effect experienced by the outermost electrons.

Shielding effect can be defined also, as a reduction in the nuclear charge on the electron cloud, due to a difference in the attraction forces of the electrons on the nucleus. It is also referred to as the screening effect, atomic or nuclear shielding.

The shielding effect explains why valence shell electrons are more easily removed from the atom. The nucleus can pull the valence shell in tighter when the attraction is strong and less tight when the attraction is weakened. The more shielding that occurs, the further the valence shell can spread out. As a result, atoms will be larger.

It might appear from the above that all nuclei of the same nuclide would resonate at the same frequency. This is not the case. The most important perturbation of the NMR frequency for applications of NMR is the 'shielding' effect of the surrounding electrons. In general, this electronic shielding reduces the magnetic field at the nucleus (which is what determines the NMR frequency).

When an external magnetic field is applied it affects the motion of the electrons in a molecule, inducing a magnetic field within the molecule. The direction of the induced magnetic field is opposite to that of the applied field.

The induced field shields the nuclei (e.g. ^{13}C and ^{1}H) from the applied field. A stronger external field is needed in order for energy difference between spin states to match energy of rf radiation.

The size of the shielding effect is difficult to calculate precisely due to effects from quantum mechanics. As an approximation, we can estimate the effective nuclear charge on each electron by the following:

$$Z_{eff} = Z - S \qquad (2.32)$$

Where Z is the number of protons in the nucleus (Atomic Number) and S is the average number of electrons between the nucleus and the electron in question (number of shielding electrons).

S can be found by using quantum chemistry and the Schrödinger equation, or by using Slater's empirical formulas.

In Rutherford backscattering spectroscopy the correction due to electron screening modifies the Coulomb repulsion between the incident ion and the target nucleus at large distances.

In conclusion the shielding effect can be defined as the Shielding of outer electrons by inner electrons and by preventing outer electrons from “feeling” effective nuclear charge, see fig.2.28.sketch below.

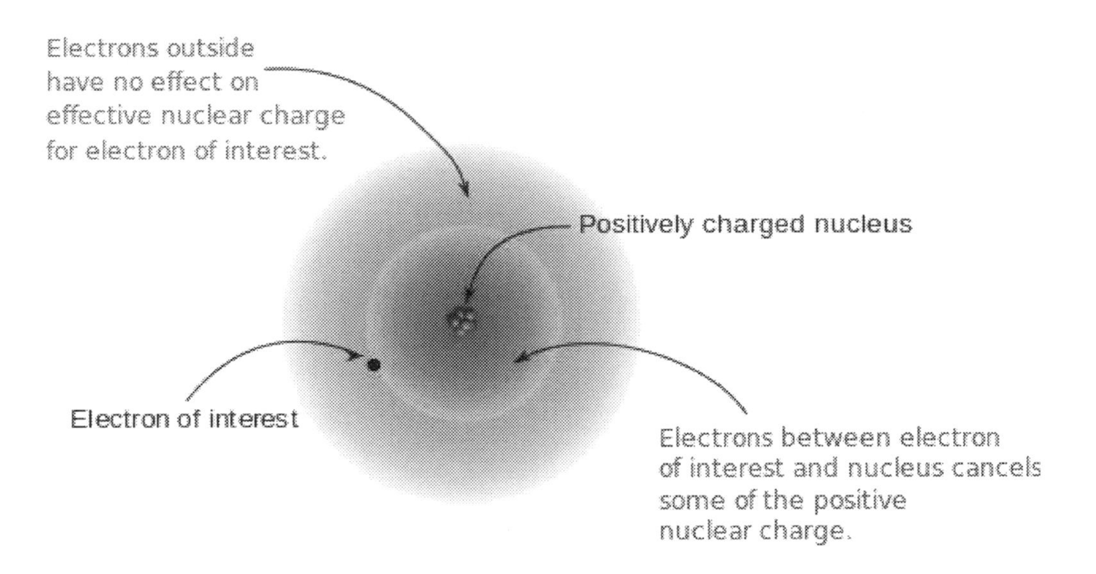

Fig.2.28. Sketch of electrons shield

Protons in different environments experience different degrees of shielding and have different chemical shifts. Chemical shift (see next section for definition) is a measure of the degree to which a nucleus in a molecule is shielded.

In hydrogen, or any other atom in group 1A of the periodic table (those with only one valence electron), the net force on the electron is just as large as the electromagnetic attraction from the nucleus. However, when more electrons are involved, each electron (in the n-shell) experiences not only the electromagnetic attraction from the positive nucleus, but also repulsion forces from other electrons in shells from 1 to n. This causes the net force on electrons in outer shells to be significantly smaller in magnitude; therefore, these electrons are not as strongly bonded to the nucleus as electrons closer to the nucleus. This phenomenon is often referred to as the orbital penetration effect.

The shielding theory also contributes to the explanation of why valence-shell electrons are more easily removed from the atom.

3

NMR SPECTROSCOPY

3.1. Chemical Shift

A. Conceptual Definition

In nuclear magnetic resonance (NMR) spectroscopy, the **chemical shift** is the resonant frequency of a nucleus relative to a standard. Often the position and number of chemical shifts are diagnostic of the structure of a molecule. Chemical shifts are also used to describe signals in other forms of spectroscopy such as photoemission spectroscopy.

The Chemical shift describes the dependence of nuclear magnetic energy levels on the electronic environment in a molecule. Chemical shifts are relevant in NMR spectroscopy techniques such as proton NMR and carbon-13 NMR.

An atomic nucleus can have a magnetic moment (nuclear spin), which gives rise to different energy levels and resonance frequencies in a magnetic field. The total magnetic field experienced by a nucleus includes local magnetic fields induced by currents of electrons in the molecular orbitals (note that electrons have a magnetic moment themselves). The electron distribution of the same type of nucleus (e.g. ^{1}H, ^{13}C, ^{15}N) usually varies according to the local geometry (binding partners, bond lengths, angles between bonds, ...), and with it the local magnetic field at each nucleus. This is reflected in the spin energy levels (and resonance frequencies). The variation of nuclear magnetic resonance frequencies of the same kind of nucleus, due to variations in the electron distribution, is called the chemical shift. The size of the chemical shift is given with respect to a reference frequency or reference sample, usually a molecule with a barely distorted electron distribution.

The magnetic field at the nucleus is **not** equal to the applied magnetic field; electrons around the nucleus shield it from the applied field. The difference between the applied magnetic field and the field at the nucleus is termed the nuclear shielding, as explained in the last section.

The signal frequency that is detected in NMR spectroscopy is proportional to the magnetic field applied to the nucleus. This would be a precisely determined frequency if the only magnetic field acting on the nucleus was the externally applied field. But the response of the atomic electrons to that externally applied magnetic field is such that their motions produce a small magnetic field at the nucleus which usually acts in opposition to the externally applied field. This change in the effective field on the nuclear spin causes the NMR signal frequency to shift. The magnitude of the shift depends upon the type of nucleus and the details of the electron motion in the nearby atoms and molecules. This shift in the NMR frequency due to the electrons' molecular orbital coupling to the external magnetic field is called chemical shift, and it explains why NMR is able to probe the chemical structure of molecules which depends on the electron density distribution in the corresponding molecular orbitals.

The precision of NMR spectroscopy allows this chemical shift to be measured, and the study of chemical shifts has produced a large store of information about the chemical bonds and the structure of molecules.

Consider the **e**-electrons in a molecule, see sketch in fig.3.1. They have spherical symmetry and circulate in the applied field, producing a magnetic field which opposes the applied field. This means that the applied field strength must be increased for the nucleus to absorb at its transition frequency. If a nucleus in a specific chemical group is shielded to a higher degree by a higher electron density of its surrounding molecular orbital, then its NMR frequency will be shifted "up field" (that is, a lower chemical shift). This up field shift is also termed diamagnetic shift.

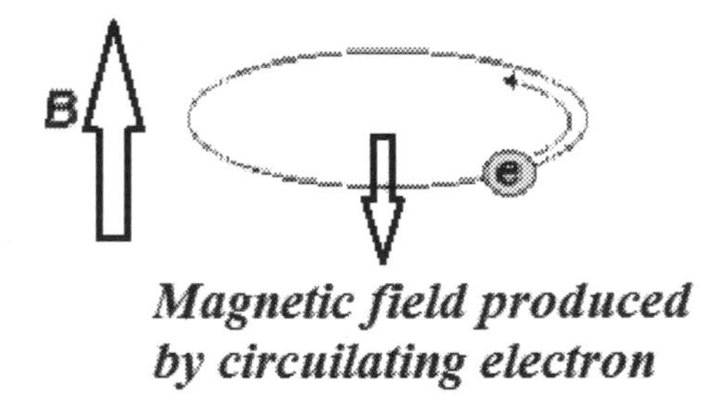

Fig.3.1. External and the produced magnetic fields of circulating electron

Electrons in p-orbitals have **no** spherical symmetry. They produce comparatively large magnetic fields at the nucleus, which give a low field shift. i.e. if it is less shielded by such surrounding electron density, then its NMR frequency will be shifted "downfield" (that is, a higher chemical shift). This "deshielding" is termed paramagnetic shift.

In proton (^1H) NMR, p-orbitals play no part (there aren't any!), which is why only a small range of chemical shift (10 ppm) is observed. We can easily see the effect of S-electrons on the chemical shift by looking at substituted methanes, CH_3X. As X becomes increasingly electronegative, so the electron density around the protons decreases, and they resonate at lower field strengths (increasing δ_H values).

Chemical shift is defined as nuclear shielding / applied magnetic field. Chemical shift is a function of the nucleus and its environment. It is measured relative to a reference compound. For ^1H NMR, the reference is usually tetramethylsilane, $Si(CH_3)_4$.

B. Mathematical Representation

Fortunately for the chemist, all proton resonances do not occur at the same position.

As defined in section 2.1.3, Eq.2.23 the **Larmor precession frequency** is given by:

$\upsilon_0 = \gamma B_0 / 2\pi$, where γ: gyromagnetic ratio

Larmor precession frequency varies because the actual magnetic field B at the nucleus is always **less** than the external field, say B_o. The origin of this effect is the "superconducting" circulation of electrons in the molecule, which occurs in such a way that a local magnetic field B_e is created, which opposes B_o (B_e is proportional to B_o).

Thus, $B = B_o - B_e$ (Magnetic field at nucleus) (3.1)

When an atom is placed in a magnetic field, its electrons circulate about the direction of the applied magnetic field. This circulation causes a small magnetic field at the nucleus which opposes the externally applied field.

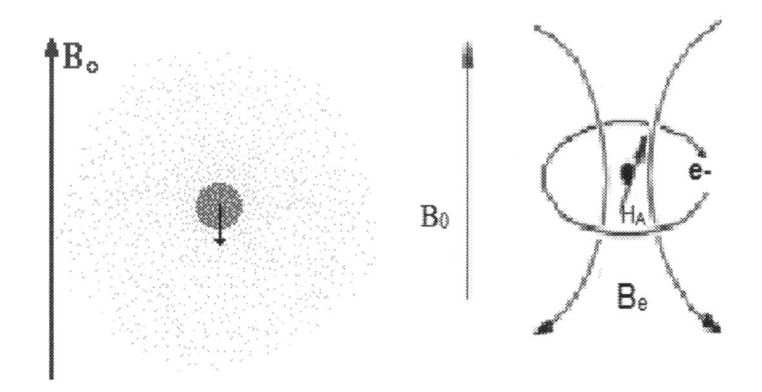

Fig.3.2. Lamour frequency of H_A -varies as B changes

The magnetic field at the nucleus (the effective field) is generally less than the applied field by a fraction σ, consequently can be expressed in terms of the externally applied field B_0 by the expression:

$$B = B_o (1\text{-}\sigma) \quad (3.2)$$

where σ is called the shielding factor or screening factor. The factor σ is small - typically 10^{-5} for protons and $<10^{-3}$ for other nuclei (Becker).

We therefore say that the nucleus is shielded from the external magnetic field. The extent of shielding is influenced by many structural features within the molecule and it is proportional to the external magnetic field B_o.

In practice the chemical shift is indicated by a symbol δ and defined in terms of a standard reference, independent units quoted as ppm:

$$\delta = \frac{(\upsilon_S - \upsilon_{REF})\text{x}10^6}{\upsilon_{REF}} \quad (3.3)$$

The signal shift is very small, parts per million, but the great precision with which frequencies can be measured permits the determination of chemical shift to three or more significant figures. Since the signal frequency is related to the shielding by:

$$\upsilon_0 = \frac{\gamma \; Bo\,(1-\sigma)}{2\pi} \quad (3.4)$$

Therefore, the chemical shift can also be expressed in terms of shielding factor or screening factor as following:

$$\delta = \frac{(\sigma_{REF} - \sigma_S)}{1 - \sigma_{REF}} \mathrm{x}10^6 \approx (\sigma_{REF} - \sigma_S)\mathrm{x}10^6 \qquad (3.5)$$

In some cases, such as the benzene molecule, the circulation of the electrons in the aromatic π orbitals creates a magnetic field at the hydrogen nuclei which enhances the B_o field. This phenomenon is called deshielding.

In this example, see fig.3.3., the B_o field is applied perpendicular to the plane of the molecule. The ring current is traveling clockwise if you look down at the plane.

The electron density around each nucleus in a molecule varies according to the types of nuclei and bonds in the molecule. The opposing field and therefore the effective field at each nucleus will vary. This is called the chemical shift phenomenon.

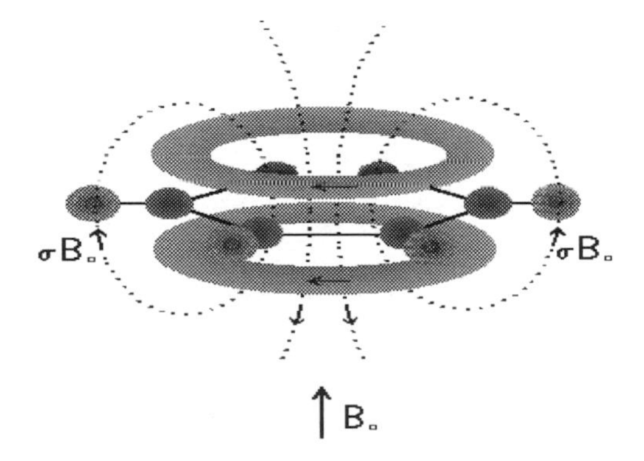

Fig.3.3. Chemical shift (Deshielding phenomenon) on benzene molecule

Consider the methanol molecule. The resonance frequency of two types of nuclei in this example differ.

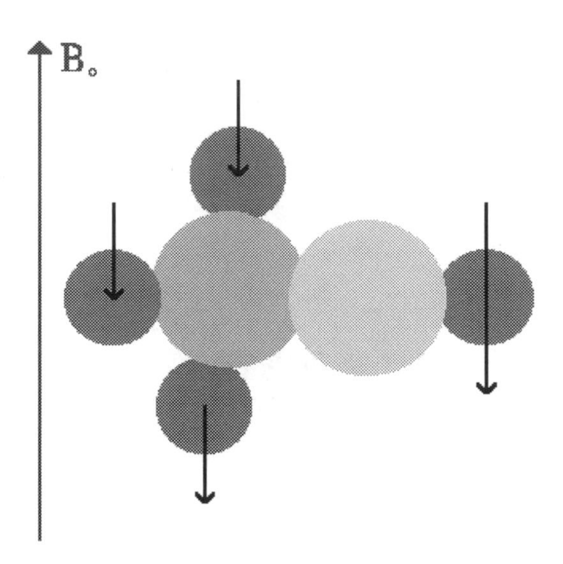

Fig.3.4. Chemical shift on methanol molecule

This difference will depend on the strength of the magnetic field, B_o, used to perform the NMR spectroscopy. The greater the value of B_o, the greater the frequency difference. This relationship could make it difficult to compare NMR spectra taken on spectrometers operating at different field strengths. The term chemical shift was developed to avoid this problem.

C. Experimental Verification

In NMR spectroscopy, experimentally measured proton chemical shifts are referenced to the ^1H signal of TetraMethylSilane (Me_4Si), this standard is often called TetraMethylSilane, $Si(CH_3)_4$, abbreviated TMS. For NMR studies in aqueous solution, where Me_4Si is not sufficiently soluble. For aqueous solution of cationic substrates (e.g., amino acids) where there may be interactions between the anionic reference compound and the substrates. Proton chemical shifts cover a range of over 30 ppm, but the vast majority appear in the region δ 0-10 ppm, where the origin is the chemical shift of TetraMethylSilane, see fig.3.5.

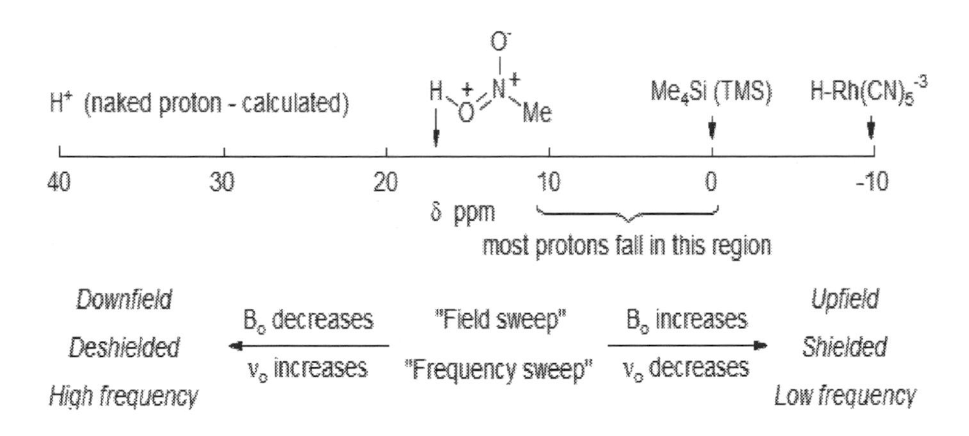

Fig.3.5. Chemical shift of TetraMethylSilane

The chemical shift is a very precise metric of the chemical environment around a nucleus. For example, the hydrogen chemical shift of a CH_2 hydrogen next to a Cl will be different than that of a CH_3 next to the same Cl. It is therefore difficult to give a detailed list of chemical shifts in a limited space. The animation window displays a chart of selected hydrogen chemical shifts of pure liquids and some gasses.

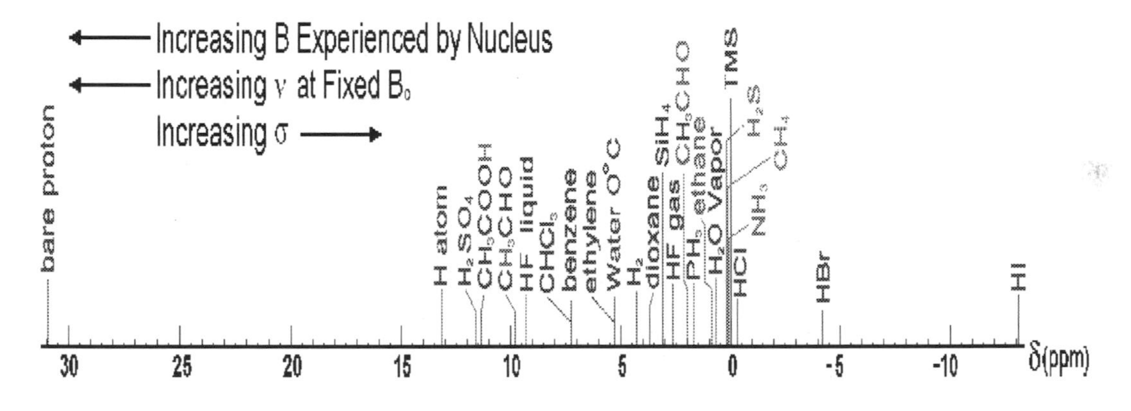

Fig.3.6. Chemical shift of hydrogen

The magnitude of the screening depends on the atom. For example, carbon-13 chemical shifts are much greater than hydrogen-1 chemical shifts.

The following tables present a few selected chemical shifts of fluorine-19 containing compounds, carbon-13 containing compounds, nitrogen-14 containing compounds, and phosphorous-31 containing compounds. These shifts are all relative to the bare nucleus. The reference signal usually used is DSS (Me_3Si-CH_2CH_2-$SO_3^-Na^+$, Tiers, *J. Org. Chem.* **1961**, *26*, 2097), an alternative reference standard, DSA ($Me_3Si-CH_2CH_2-NH_3^+$ $CF_3CO_2^-$, Nowick *Org. Lett.* **2003**, *5*, 3511).

1. Fluorine-19 Chemical Shifts

Fluorine-19 Environment	Chemical-Shift Range (ppm)
UF_6	-540
FNO	-269
F_2	-210
bare nucleus	0
$C(CF_3)_4$	284
$CF_3(COOH)$	297
fluorobenzene	333
F-	338
BF_3	345
HF	415

2. Carbon-13Chemical Shifts

Carbon-13* Environment	Chemical Shift Range (ppm)
$(CH_3)_2C^*O$	-12
CS_2	0
CH_3C^*OOH	16
C_6H_6	65
CHCl=CHCl (cis)	71
CH_3C^*N	73
CCl_4	97
dioxane	126
C^*H_3CN	196
CHI_3	332

3. Nitrogen-14 Chemical Shifts

Nitrogen-14* Environment	Chemical Shift Range (ppm)
NO_2Na	-355
NO_3- (aqueous)	-115
N_2 (liquid)	-101
pyridine	-93
bare nucleus	0
CH_3CN	25
CH_3CONH_2 (aqueous)	152
NH_4+ (aqueous)	245
NH_3 (liquid)	266

4. Phosphorous-31 Chemical Shifts

Phosphorous-31 Environment	Chemical Shif Range (ppm)
PBr_3	-228
$(C_2H_5O)_3$ P	-137
PF_3	-97
85% phosphoric acid	0
PCl_5	80
PH_3	238
P_4	450

3.2. Knight Shift

The **Knight Shift** is a shift in the nuclear magnetic resonance frequency of a paramagnetic substance first published in 1949 by the American physicist Walter David Knight. The Knight Shift is due to the conduction electrons in metals. They introduce an "extra" effective field at the nuclear site, due to the spin orientations of the conduction electrons in the presence of an external field. This is responsible for the shift observed in the nuclear magnetic resonance. The shift comes from two sources, one is the Pauli paramagnetic spin susceptibility, and the other is the s-component wave functions at the nucleus.

For an N spins in a magnetic induction field $\vec{B}$, the nuclear Hamiltonian for the Knight Shift is expressed in Cartesian form by:

$$\widehat{\mathcal{H}}_{KS} = -\sum_i^N \gamma_i \, , \hat{\vec{I}}_i \, , \widehat{\mathrm{K}}_i \, , \vec{B} \qquad (3.6)$$

Where, for the ith spin γ_i is the gyromagnetic ratio, $\hat{\vec{I}}_i$ is a vector of the Cartesian nuclear angular momentum operators, and the matrix $\widehat{\mathrm{K}}_i$ is a second-rank tensor similar to the chemical shift shielding tensor,

$$\widehat{\mathrm{K}}_i = \begin{bmatrix} k_{xx} & k_{xy} & k_{xz} \\ k_{yx} & k_{yy} & k_{yz} \\ k_{zx} & k_{zy} & k_{zz} \end{bmatrix} \qquad (3.7)$$

The Knight shift refers to the relative shift K in NMR frequency for atoms in a metal (e.g. sodium) compared with the same atoms in a nonmetallic environment (e.g. sodium chloride). The observed shift reflects the local magnetic field produced at the sodium nucleus by the magnetization of the conduction electrons. The average local field in sodium augments the applied resonance field by approximately one part per 1000. In nonmetallic sodium chloride the local field is negligible in comparison.

The Knight Shift is due to the conduction electrons in metals. They introduce an "extra" effective field at the nuclear site, due to the spin orientations of the conduction electrons in the presence of an external field. This is responsible for the shift observed in the nuclear magnetic resonance. The shift comes from two sources, one is the Pauli paramagnetic spin susceptibility, and the other is the s-component wave functions at the nucleus.

Depending on the electronic structure, the Knight Shift may be temperature dependent. However, in metals which normally have a broad featureless electronic density of states, Knight Shifts are temperature independent.

The Knight Shift for atoms in a metal (e.g. sodium) compared with the same atoms in a nonmetallic environment (e.g. sodium chloride). The observed shift reflects the local magnetic field produced at the sodium nucleus by the magnetization of the conduction electrons. The average local field in sodium augments the applied resonance field by approximately one part per 1000. In nonmetallic sodium chloride the local field is negligible in comparison.

3.3. Spin-Spin Coupling

Nuclei experiencing the same chemical environment or chemical shift are called equivalent. Those nuclei experiencing different environment or having different chemical shifts are nonequivalent. Nuclei which are close to one another exert an influence on each other's effective magnetic field. This effect shows up in the NMR spectrum when the nuclei are nonequivalent. If the distance between non-equivalent nuclei is less than or equal to three bond lengths, this effect is observable. This effect is called spin-spin coupling or J coupling. In molecular geometry, bond length or bond distance is the average distance between nuclei of two bonded atoms in a molecule. It is a transferable property of a bond between atoms of fixed types, relatively independent of the rest of the molecule. The length of the bond is determined by the number of bonded electrons (the bond order). The higher the bond order, the stronger the pull between the two atoms and the shorter the bond length. Generally, the length of the bond between two atoms is approximately the sum of the covalent radii of the two atoms.

Consider the following example. There are two nuclei, A and B, three bonds away from one another in a molecule.

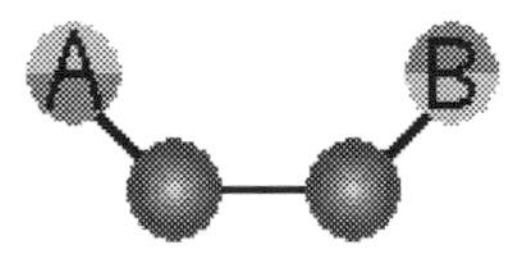

Fig.3.8. Two atoms/nuclei with three bonds away

The spin of each nucleus can be either aligned with the external field such that the fields are N-S-N-S, called spin up ●, or opposed to the external field such that the fields are N-N-S-S, called spin down ●. The magnetic field at nucleus A will be either greater than B_o or less than B_o by a constant amount due to the influence of nucleus B.

There are a total of four possible configurations for the two nuclei in a magnetic field. Arranging these configurations in order of increasing energy gives the arrangement shown in fig.3.9.

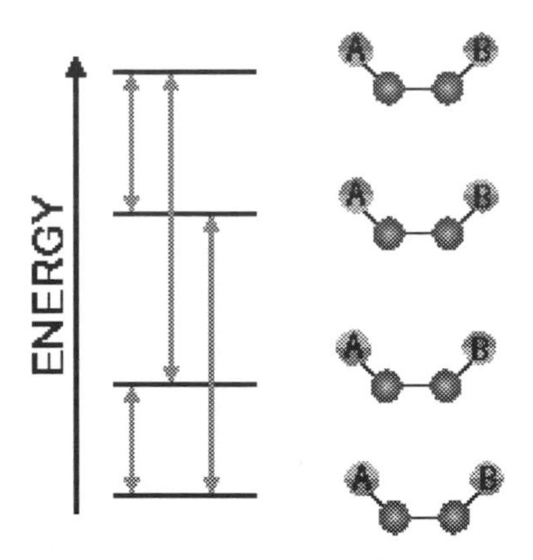

Fig.3.9. Allowed transitions between energy levels
(Four possible configurations for the two nuclei in a magnetic field)

The vertical lines in this diagram represent the allowed transitions between energy levels. In NMR, an allowed transition is one where the spin of one nucleus changes from spin up to spin down, or spin down to spin up. Absorptions of energy where two or more nuclei change spin at the same time are not allowed. There are two absorption frequencies for the A nucleus and two for the B nucleus represented by the vertical lines between the energy levels in this diagram.

The NMR spectrum for nuclei A and B reflects the splittings observed in the energy level diagram. The A absorption line is split into 2 absorption lines centered on δ_A, and the B absorption line is split into 2 lines centered on four possible configurations for the two nuclei in a magnetic field δ_B. The distance between two split absorption lines is called the J coupling constant or the spin-spin splitting constant and is a measure of the magnetic interaction between two nuclei.

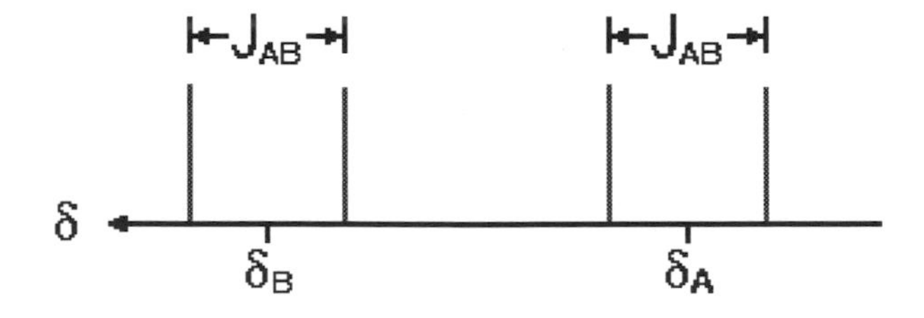

Fig.3.10. Splitting in the energy level diagram

For the next example, consider a molecule with three spin 1/2 nuclei, one type A and two type B.

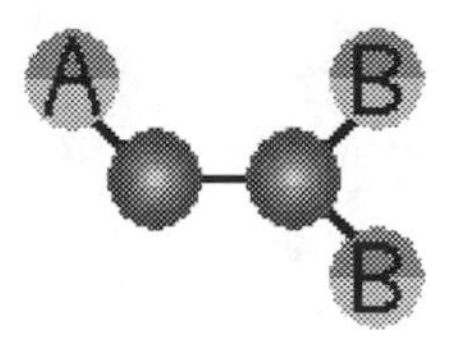

Fig.3.11. Molecule with three spin 1/2 nuclei

The type **B** nuclei are both three bonds away from the type **A** nucleus. The magnetic field at the **A** nucleus has three possible values due to four possible spin configurations of the two B nuclei.

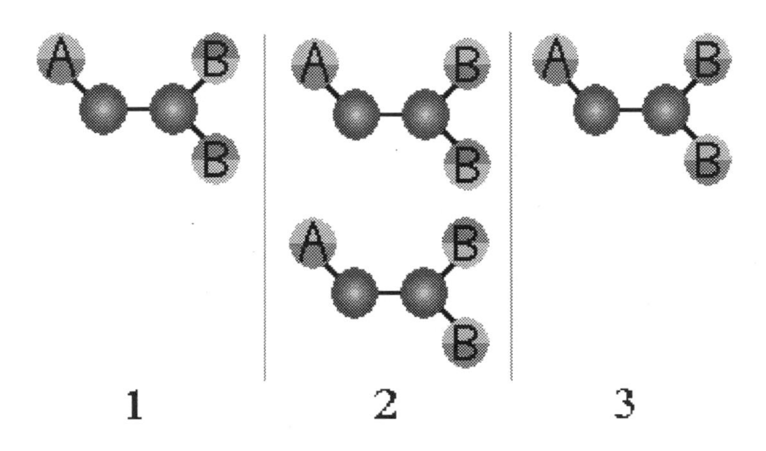

Fig.3.12. The magnetic field at the **A** nucleus has three possible values

The magnetic field at a **B** nucleus has two possible values.

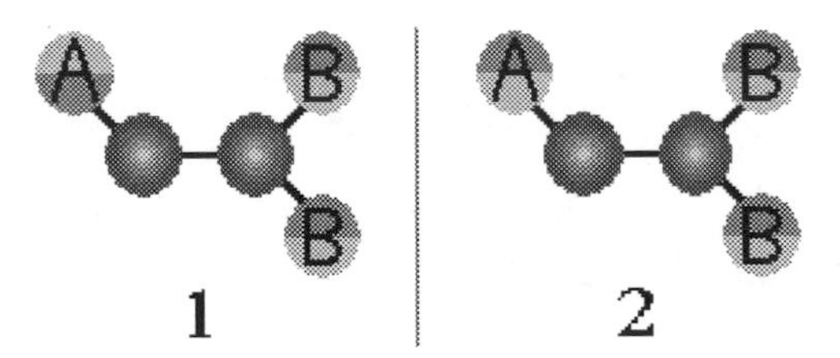

Fig.3.13. The magnetic field at the **B** nucleus has two possible values

The energy level diagram for this molecule has six states or levels because there are two sets of levels with the same energy.

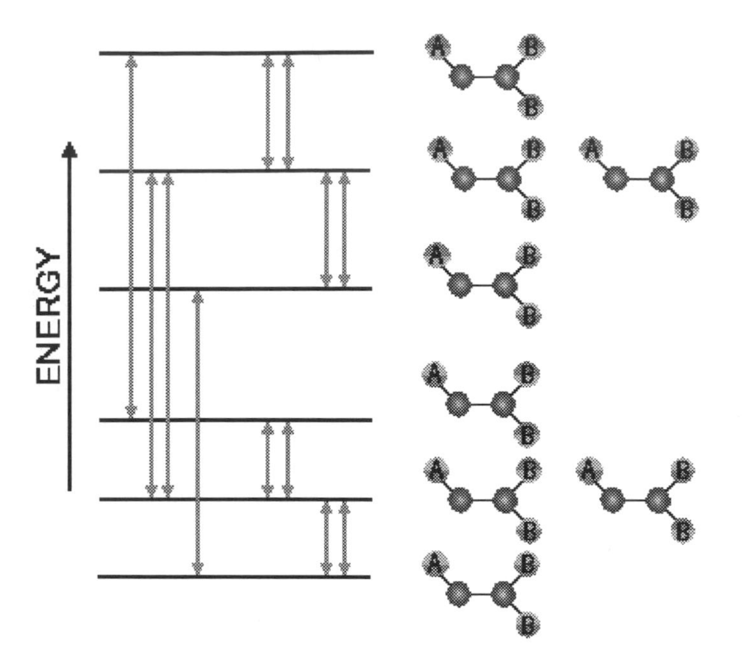

Fig.3.14. Allowed transitions between energy levels
(Six possible configurations for the Molecule with three spin 1/2 nuclei)

Energy levels with the same energy are said to be degenerate. The vertical lines represent the allowed transitions or absorptions of energy. Note that there are two lines drawn between some levels because of the degeneracy of those levels.

The resultant NMR spectrum is depicted in the animation window.

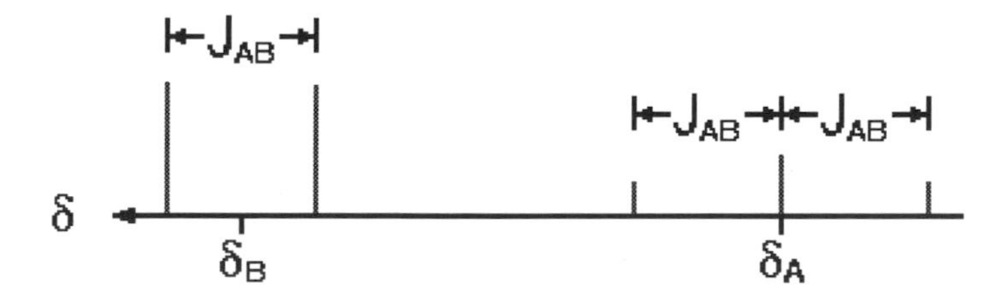

Fig.3.15. Splitting in the energy level diagram for the Molecule with three spin 1/2 nuclei

Note that the center absorption line of those centered at δ_A is twice as high as the either of the outer two. This is because there were twice as many transitions in the energy level diagram for this transition. The peaks at δ_B are taller because there are twice as many B type spins than A type spins.

The complexity of the splitting pattern in a spectrum increases as the number of B nuclei increases. The following table contains a few examples.

Configuration	Peak Ratios
A	1
AB	1:1
AB_2	1:2:1
AB_3	1:3:3:1
AB_4	1:4:6:4:1
AB_5	1:5:10:10:5:1
AB_6	1:6:15:20:15:6:1

This series is called Pascal's triangle and can be calculated from the coefficients of the expansion series; $(x+1)^n$, where n is the number of B nuclei in the above table.

When there are two different types of nuclei three bonds away there will be two values of J, one for each pair of nuclei.

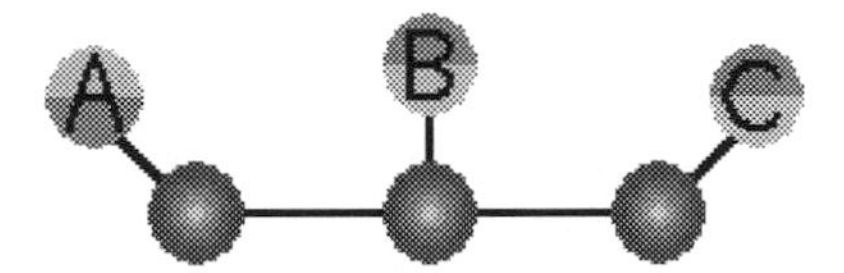

Fig.3.16. Two different types of atoms/nuclei(three nuclei) with three Bonds away

By now you get the idea of the number of possible configurations and the energy level diagram for these configurations, so we can skip to the spectrum. In the following example J_{AB} is greater J_{BC}.

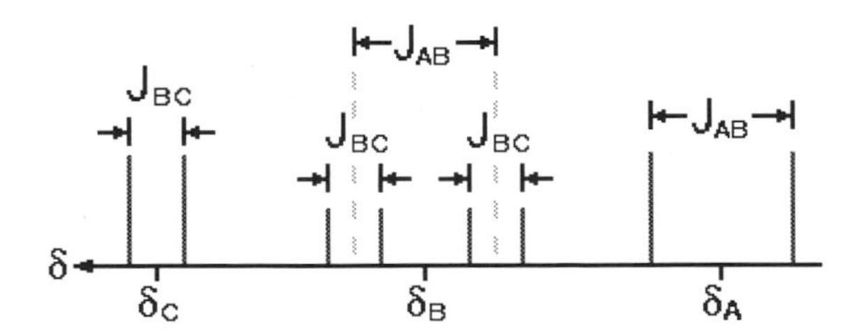

Fig.3.15. Splitting in the energy level diagram for the three nuclei

Consider the structure of ethanol;

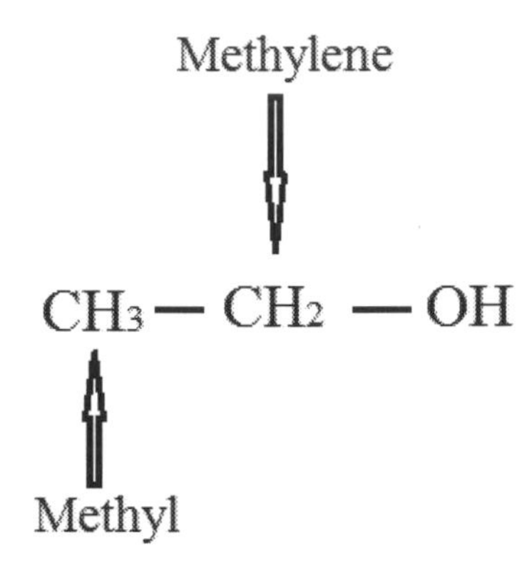

Fig.3.16. The structure of Ethanol

The ^{1}H NMR spectrum of ethanol (figure 3.17.) shows the methyl peak has been split into three peaks (a triplet) and the methylene peak has been split into four peaks (a quartet). This occurs because there is a small interaction (coupling) between the two groups of protons. The spacings between the peaks of the methyl triplet are equal to the spacings between the peaks of the methylene quartet. This spacing is measured in Hertz and is called the coupling constant, J.

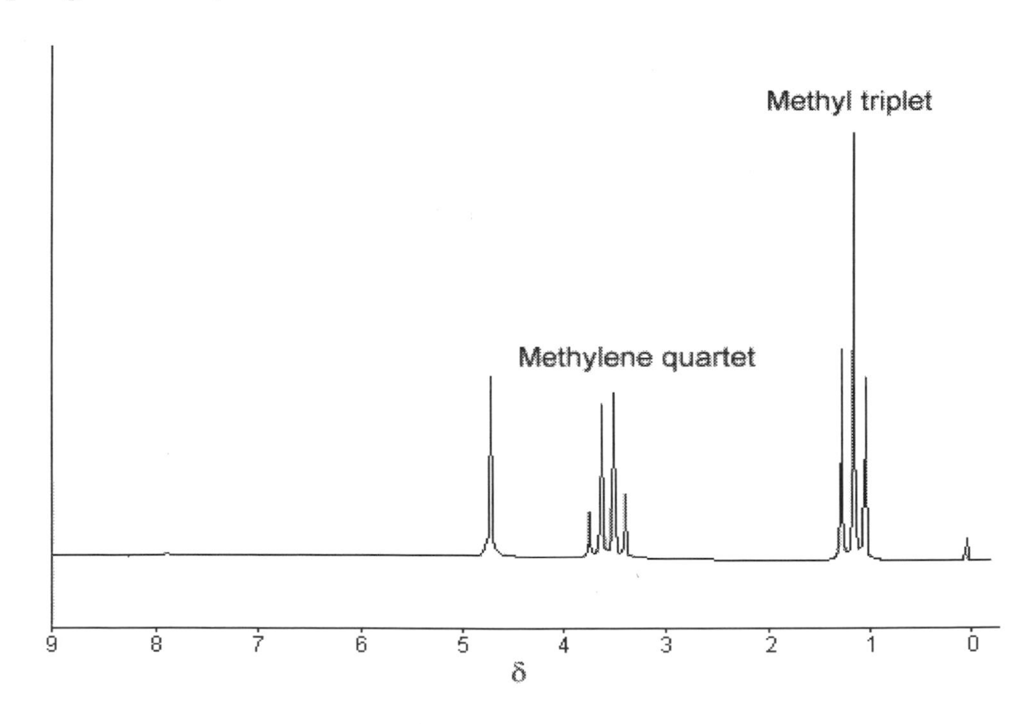

Fig.3.17. The ^{1}H NMR spectrum of ethanol

To see why the methyl peak is split into a triplet, let's look at the **methylene** protons. There are two of them, and each can have one of two possible orientations (aligned with or opposed against the applied field). This gives a total of four possible states; see fig.3.18.

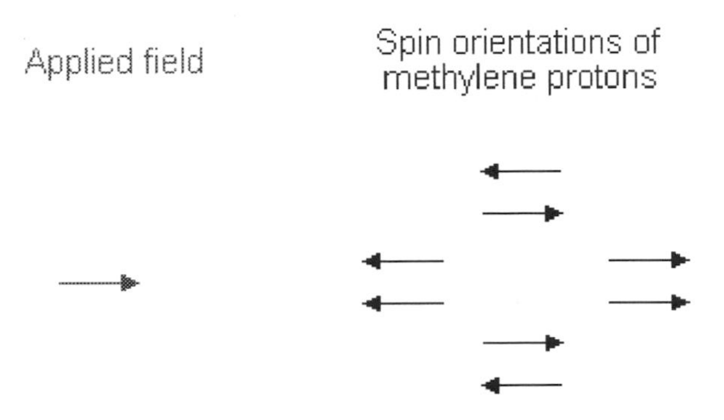

Fig.3.18. Four possible states of spin orientation for Methylene protons

In the first possible combination, spins are paired and opposed to the field. This has the effect of reducing the field experienced by the **methyl** protons; therefore a slightly higher field is needed to bring them to resonance, resulting in an upfield shift. Neither combination of spins opposed to each other has an effect on the methyl peak. The spins paired in the direction of the field produce a downfield shift. Hence, the methyl peak is split into three, with the ratio of areas 1:2:1.

Similarly, the effect of the methyl protons on the methylene protons is such that there are eight possible spin combinations for the three methyl protons;

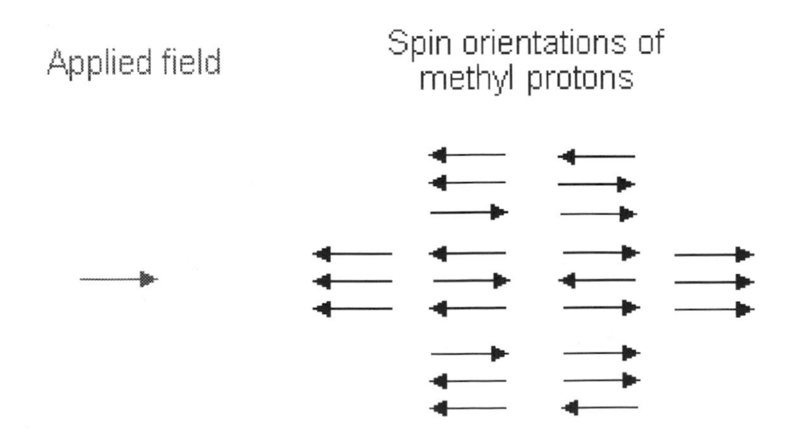

Fig.3.19. Eight possible states of spin orientation for Methyl protons

Out of these eight groups, there are two groups of three magnetically equivalent combinations. The methylene peak is split into a quartet. The areas of the peaks in the quartet have the ration 1:3:3:1.

In a first-order spectrum (where the chemical shift between interacting groups is much larger than their coupling constant), interpretation of splitting patterns is quite straightforward;

- The multiplicity of a multiplet is given by the number of equivalent protons in neighboring atoms plus one, i.e. *the n + 1 rule*
- Equivalent nuclei do not interact with each other. The three methyl protons in ethanol cause splitting of the neighboring methylene protons; they do not cause splitting among themselves
- The coupling constant is not dependent on the applied field. Multiplets can be easily distinguished from closely spaced chemical shift peaks.

3.4. The Time Domain NMR Signal

An NMR sample may contain many different magnetization components, each with its own Larmor frequency. These magnetization components are associated with the nuclear spin configurations joined by an allowed transition line in the energy level diagram. Based on the number of allowed absorptions due to chemical shifts and spin-spin couplings of the different nuclei in a molecule, an NMR spectrum may contain many different frequency lines.

In pulsed NMR spectroscopy, signal is detected after these magnetization vectors are rotated into the XY plane. Once a magnetization vector is in the XY plane it rotates about the direction of the B_o field, the +Z axis. As transverse magnetization rotates about the Z axis, it will induce a current in a coil of wire located around the X axis.

Plotting current as a function of time gives a sine wave.

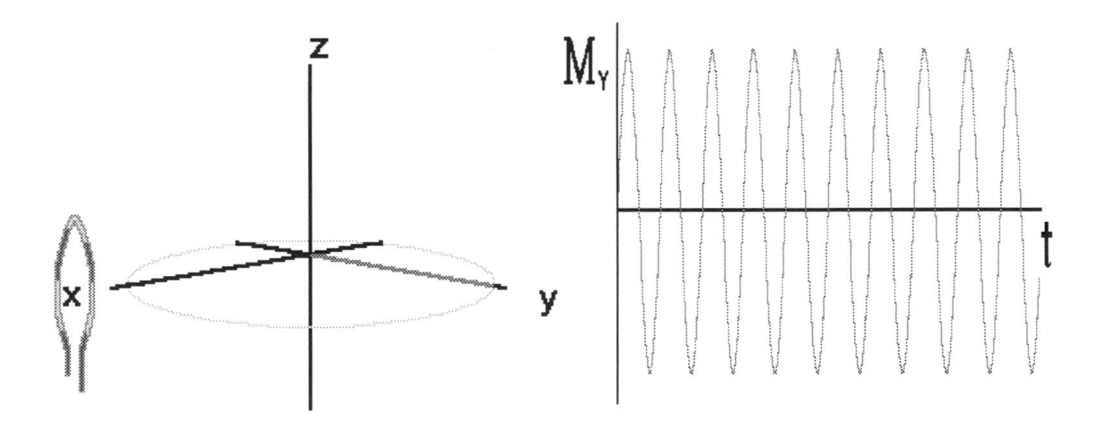

Fig.3.20. Free Induction Decay Signal

This wave will, of course, decay with time constant T_2* due to dephasing of the spin packets. This signal is called a Free Induction Decay (FID).

3.5. The +/- Frequency Convention

Transverse magnetization vectors rotating faster than the rotating frame of reference are said to be rotating at a positive frequency relative to the rotating frame (+υ). Vectors rotating slower than the rotating frame are said to be rotating at a negative frequency relative to the rotating frame (-υ).

It is worthwhile noting here that in most NMR spectra, the resonance frequency of a nucleus, as well as the magnetic field experienced by the nucleus and the chemical shift of a nucleus, increase from right to left.

4

NMR PERFORMANCE AND TECHNIQUE TYPES

NMR spectroscopy is one of the principal techniques used to obtain physical, chemical, electronic and structural information about molecules due to either the chemical shift Zeeman effect, or the Knight shift effect, or a combination of both, on the resonant frequencies of the nuclei present in the sample. It is a powerful technique that can provide detailed information on the topology, dynamics and three-dimensional structure of molecules in solution and the solid state. Thus, structural and dynamic information is obtainable (with or without "magic angle" spinning (MAS)) from NMR studies of quadrupolar nuclei (that is, those nuclei with spin $S > \frac{1}{2}$) even in the presence of magnetic dipole-dipole interaction broadening (or simply, dipolar broadening) which is always much smaller than the quadrupolar interaction strength because it is a magnetic vs. an electric interaction effect.

Additional structural and chemical information may be obtained by performing double-quantum NMR experiments for quadrupolar nuclei such as ^2H. Also, nuclear magnetic resonance is one of the techniques that has been used to design quantum automata, and also build elementary quantum computers.

An overview and description of the components of the NMR hardware are described in next chapter.

4.1. Continuous Wave (CW) Spectroscopy

In its first few decades, nuclear magnetic resonance spectrometers used a technique known as continuous-wave spectroscopy (CW spectroscopy). Although NMR spectra could be, and have been, obtained using a fixed magnetic field and sweeping the frequency of the electromagnetic radiation, this more typically involved using a fixed frequency source and varying the current (and hence magnetic field) in an electromagnet to observe the resonant absorption signals. This is the origin of the anachronistic, but still common, "high" and "low" field terminology for low frequency and high frequency regions respectively of the NMR spectrum.

CW spectroscopy is inefficient in comparison to Fourier techniques (see below) as it probes the NMR response at individual frequencies in succession. As the NMR signal is intrinsically weak, the observed spectra suffer from a poor signal-to-noise ratio.

Signal-to-noise ratio (often abbreviated **SNR** or **S/N**) is a measure used in science and engineering that compares the level of a desired signal to the level of background noise. It is defined as the ratio of signal power to the noise power. A ratio higher than 1:1 indicates more signal than noise. While SNR is commonly quoted for electrical signals, it can be applied to any form of signal (such as isotope levels in an ice core or biochemical signaling between cells).

The signal-to-noise ratio, the bandwidth, and the channel capacity of a communication channel are connected by the Shannon–Hartley theorem.

Signal-to-noise ratio is sometimes used informally to refer to the ratio of useful information to false or irrelevant data in a conversation or exchange. For example, in online discussion forums and other online communities, off-topic posts and spam are regarded as "noise" that interferes with the "signal" of appropriate discussion.

This can be mitigated by signal averaging i.e. adding the spectra from repeated measurements. While the NMR signal is constant between scans and so adds linearly, the random noise adds more slowly - as the square-root of the number of spectra (see Random walk). Hence the overall ratio of the signal to the noise increases as the square-root of the number of spectra measured.

A **random walk** is a mathematical formalization of a path that consists of a succession of random steps. For example, the path traced by a molecule as it travels in a liquid or a gas, the search path of a foraging animal, the price of a fluctuating stock and the financial status of a gambler can all be modeled as random walks, although they may not be truly random in reality. The term random walk was first introduced by Karl Pearson in 1905. Random walks have been used in many fields: ecology, economics, psychology, computer science, physics, chemistry, and biology. Random walks explain the observed behaviors of processes in these fields, and thus serve as a fundamental model for the recorded stochastic activity.

Example of eight random walks in one dimension starting at 0. The plot shows the current position on the line (vertical axis) versus the time steps (horizontal axis).

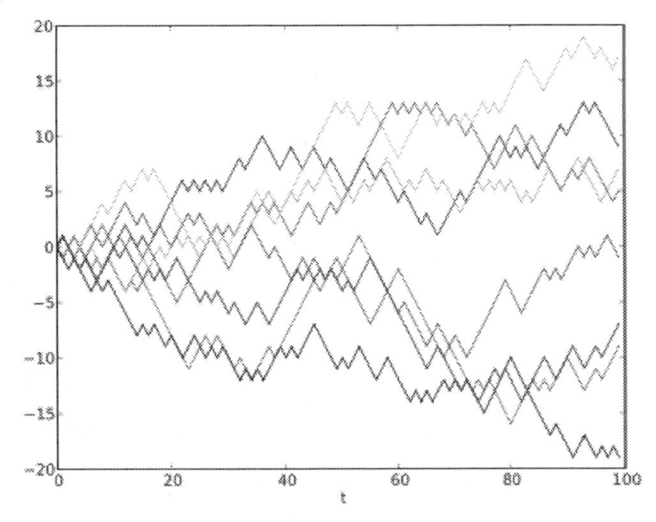

Fig.4.1.Example of eight random walks in one dimension starting at 0

4.2. Fourier Transform Spectroscopy

Most applications of NMR involve full NMR spectra, that is, the intensity of the NMR signal as a function of frequency. Early attempts to acquire the NMR spectrum more efficiently than simple CW methods involved irradiating simultaneously with more than one frequency. A revolution in NMR occurred when short pulses of radio-frequency were used (centered at the middle of the NMR spectrum). In simple terms, a short square pulse of a given "carrier" frequency "contains" a range of frequencies centered about the carrier frequency, with the range of excitation (bandwidth) being inversely proportional to the pulse duration (the Fourier transform (FT) of an approximate square wave contains contributions from all the frequencies in the neighborhood of the principal frequency). The restricted range of the NMR frequencies made it relatively easy to use short (millisecond to microsecond) radiofrequency (RF) pulses to excite the entire NMR spectrum.

Applying such a pulse to a set of nuclear spins simultaneously excites all the single-quantum NMR transitions. In terms of the net magnetization vector, this corresponds to tilting the magnetization vector away from its equilibrium position (aligned along the external magnetic field). The out-of-equilibrium magnetization vector precesses about the external magnetic field vector at the NMR frequency of the spins. This oscillating magnetization vector induces a current in a nearby pickup coil, creating an electrical signal oscillating at the NMR frequency. This signal is known as the free induction decay (FID) and contains the vector-sum of the NMR responses from all the excited spins. In order to obtain the frequency-domain NMR spectrum (NMR absorption intensity vs. NMR frequency) this time-domain signal (intensity vs. time) must be FTed. Fortunately the development of FT NMR coincided with the development of digital computers and Fast Fourier Transform algorithms.

Richard R. Ernst was one of the pioneers of pulse (FT) NMR and won a Nobel Prize in chemistry in 1991 for his work on FT NMR and his development of multi-dimensional NMR (see below).

See Annex 1 for Details of Fourier Transform Spectroscopy

4.3. Multi-Dimensional NMR Spectroscopy

The use of pulses of different shapes, frequencies and durations in specifically-designed patterns or pulse sequences allows the spectroscopist to extract many different types of information about the molecule. Multi-dimensional nuclear magnetic resonance spectroscopy is a kind of FT NMR in which there are at least two pulses and, as the experiment is repeated, the pulse sequence is varied. In multidimensional nuclear magnetic resonance there will be a sequence of pulses and, at least, one variable time period. In three dimensions, two time sequences will be varied. In four dimensions, three will be varied.

There are many such experiments. In one, these time intervals allow (amongst other things) magnetization transfer between nuclei and, therefore, the detection of the kinds of nuclear-nuclear interactions that allowed for the magnetization transfer. Interactions that can be detected are usually classified into two kinds. There are through-bond interactions and through-space interactions, the latter usually being a consequence of the nuclear Overhauser effect. Experiments of the nuclear Overhauser variety may be employed to establish distances between atoms, as for example by 2D-FT NMR of molecules in solution. See Annex 4

Although the fundamental concept of 2D-FT NMR was proposed by Professor Jean Jeener from the Free University of Brussels at an International Conference, this idea was largely developed by Richard Ernst who won the 1991 Nobel prize in Chemistry for his work in FT NMR, including multi-dimensional FT NMR, and especially 2D-FT NMR of small molecules. Multi-dimensional FT NMR experiments were then further developed into powerful methodologies for studying biomolecules in solution, in particular for the determination of the structure of biopolymers such as proteins or even small nucleic acids.

Kurt Wüthrich shared (with John B. Fenn) in 2002 the Nobel Prize in Chemistry for his work in protein FT NMR in solution.

See Annex 2 for Details of pulse Spectroscopy

4.4. Solid-State NMR Spectroscopy

This technique complements biopolymer X-ray crystallography in that it is frequently applicable to biomolecules in a liquid or liquid crystal phase, whereas crystallography, as the name implies, is performed on molecules in a solid phase.

X-ray crystallography is a method of determining the arrangement of atoms within a crystal, in which a beam of X-rays strikes a crystal and causes the beam of light to spread into many specific directions. From the angles and intensities of these diffracted beams, a crystallographer can produce a three-dimensional picture of the density of electrons within the crystal, see fig.4.2. From this electron density, the mean positions of the atoms in the crystal can be determined, as well as their chemical bonds, their disorder and various other information.

Fig.4.2. Three-dimensional picture of the density of electrons within the crystal

X-ray crystallography can locate every atom in a zeolite, an aluminosilicate with many important applications, such as water purification.

Though nuclear magnetic resonance is used to study solids, extensive atomic-level biomolecular structural detail is especially challenging to obtain in the solid state. There is little signal averaging by thermal motion in the solid state, where most molecules can only undergo restricted vibrations and rotations at room temperature, each in a slightly different electronic environment, therefore exhibiting a different NMR absorption peak. Such a variation in the electronic environment of the resonating nuclei results in a blurring of the observed spectra—which is often only a broad Gaussian band for non-quadrupolar spins in a solid- thus making the interpretation of such "dipolar" and "chemical shift anisotropy" (CSA) broadened spectra either very difficult or impossible.

Professor Raymond Andrew at Nottingham University in UK pioneered the development of high-resolution solid-state nuclear magnetic resonance. He was the first to report the introduction of the MAS (Magic Angle Sample spinning; MASS) technique that allowed him to achieve spectral resolution in solids sufficient to distinguish between chemical groups with either different Chemical Shifts (see section 3.1) or distinct Knight Shifts (see section 3.2).

In nuclear magnetic resonance, **magic angle spinning** (MAS) is a technique often used to perform experiments in solid-state NMR spectroscopy.

By spinning the sample (usually at a frequency of 1 to 70 kHz) at the magic angle θ_m (ca. 54.74°, where $\cos^2\theta_m=1/3$) with respect to the direction of the magnetic field, the normally broad lines become narrower, increasing the resolution for better identification and analysis of the spectrum.

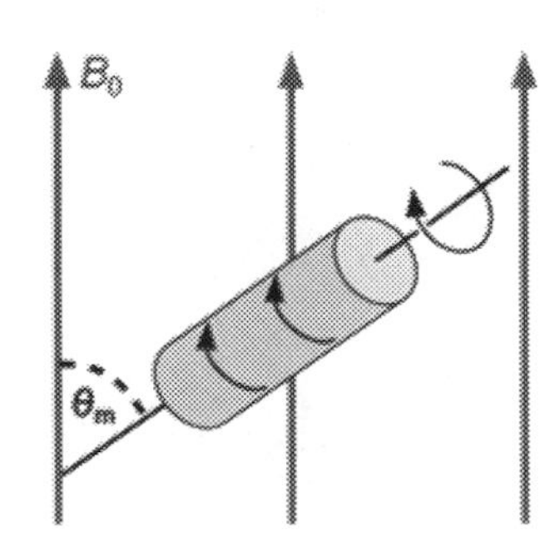

Fig.4.3. Magic-angle spinning

The sample (blue) is rotating with high frequency inside the main magnetic field (B_0). The axis of rotation is tilted by the magic angle θ_m with respect to the direction of B_0.

In MAS, the sample is spun at several kilohertz around an axis that makes the so-called magic angle θ_m (which is ~ 54.74°, where $\cos^2\theta_m = 1/3$, or $3\cos^2\theta_m - 1 = 0$) with respect to the direction of the static magnetic field $\mathbf{B_0}$; as a result of such magic angle sample spinning, the chemical shift anisotropy bands are averaged to their corresponding average (isotropic) chemical shift values. The above expression involving $\cos^2\theta_m$ has its origin in a calculation that predicts the magnetic dipolar interaction effects to cancel out for the specific value of θ_m called the magic angle. One notes that correct alignment of the sample rotation axis as close as possible to θ_m is essential for cancelling out the dipolar interactions whose strength for angles sufficiently far from θ_m is usually greater than ~10 kHz for C-H bonds in solids, for example, and it is thus greater than their CSA values.

A concept developed by Sven Hartmann and Erwin Hahn was utilized in transferring magnetization from protons to less sensitive nuclei (popularly known as cross-polarization) by M.G. Gibby, Alex Pines and John S. Waugh. Then, Jake Schaefer and Ed Stejskal demonstrated also the powerful use of cross-polarization under MAS conditions which is now routinely employed to detect low-abundance and low-sensitivity nuclei.

Solid-state NMR (SSNMR) spectroscopy is a kind of nuclear magnetic resonance (NMR) spectroscopy, characterized by the presence of anisotropic (directionally dependent) interactions.

Fig.4.4. Solid-state 900 MHz (21.1 T) NMR spectrometer at the Canadian National Ultrahigh-field NMR Facility for Solids.

4.5. Sensitivity

Because the intensity of nuclear magnetic resonance signals and, hence, the sensitivity of the technique depends on the strength of the magnetic field the technique has also advanced over the decades with the development of more powerful magnets. Advances made in audio-visual technology have also improved the signal-generation and processing capabilities of newer machines.

As noted above, the sensitivity of nuclear magnetic resonance signals is also dependent on the presence of a magnetically-susceptible nuclide and, therefore, either on the natural abundance of such nuclides or on the ability of the experimentalist to artificially enrich the molecules, under study, with such nuclides. The most abundant naturally-occurring isotopes of hydrogen and phosphorus (for example) are both magnetically susceptible and readily useful for nuclear magnetic resonance spectroscopy. In contrast, carbon and nitrogen have useful isotopes but which occur only in very low natural abundance.

Other limitations on sensitivity arise from the quantum-mechanical nature of the phenomenon. For quantum states separated by energy equivalent to radio frequencies, thermal energy from the environment causes the populations of the states to be close to equal. Since incoming radiation is equally likely to cause stimulated emission (a transition from the upper to the lower state) as absorption, the NMR effect depends on an excess of nuclei in the lower states. Several factors can reduce sensitivity, including

- Increasing temperature, which evens out the population of states. Conversely, low temperature NMR can sometimes yield better results than room-temperature NMR, providing the sample remains liquid.
- Saturation of the sample with energy applied at the resonant radiofrequency. This manifests in both CW and pulsed NMR; in the first, CW, case this happens by using too much continuous power that keeps the upper spin levels completely populated; in the second case, saturation occurs by pulsing too frequently-without allowing time for the nuclei to return to thermal equilibrium through spin-lattice relaxation. For nuclei such as ^{29}Si this is a serious practical problem as the relaxation time is measured in seconds; for protons in "pure" ice,

or ^{19}F in high-purity (undoped) LiF crystals the spin-lattice relaxation time can be on the order of an hour or longer. The use of shorter RF pulses that tip the magnetization by less than 90° can partially solve the problem by allowing spectral acquisition without the complete loss of NMR signal.

- Non-magnetic effects, such as electric-quadrupole coupling of spin-1 and spin-$^3/_2$ nuclei with their local environment, which broaden and weaken absorption peaks. ^{14}N, an abundant spin-1 nucleus, is difficult to study for this reason. High resolution NMR instead probes molecules using the rarer ^{15}N isotope, which has spin-$^1/_2$.

4.6. Isotopes

Many chemical elements can be used for NMR analysis.

Commonly used nuclei:

1. ^{1}H, the most commonly used fermionic nucleus in NMR investigation, has been studied using many forms of NMR. Hydrogen is highly abundant, especially in biological systems. It is the most nucleus most sensitive to NMR signal (apart from tritium) (^{3}H) (which is not commonly used as it is unstable and radioactive.) Proton NMR produces narrow chemical shift with sharp signals. The ^{1}H signal has been the sole diagnostic nucleus used for clinical magnetic resonance imaging.
2. ^{2}H, a **boson** nucleus commonly utilized as signal-free medium in the form of deuterated solvents during proton NMR, to avoid signal interference from hydrogen-containing solvents in measurement of ^{1}H solutes. Also used in determining the behavior of lipids in lipid membranes and other solids or liquid crystals as it is a relatively non-perturbing label which can selectively replace ^{1}H. Alternatively, ^{2}H can be detected in media specially labeled with ^{2}H.
3. ^{3}He, is very sensitive to NMR. There is a very low percentage in natural helium, and subsequently has to be purified from ^{4}He. It is used mainly in studies of endohedral fullerenes, where its chemical inertness is beneficial to ascertaining the structure of the entrapping fullerene.
4. ^{10}B, lower sensitivity than ^{11}B. Quartz tubes must be used as borosilicate glass interferes with measurement.
5. ^{11}B, more sensitive than ^{10}B, yields sharper signals. Quartz tubes must be used as borosilicate glass interferes with measurement.
6. ^{13}C is widely used, despite its relative paucity in naturally-occurring carbon (approximately 1%). It is stable to nuclear decay. Since there is a low percentage in natural carbon, spectrum acquisition on samples which have not been experimentally enriched in ^{13}C takes a long time. Frequently used for labeling of compounds in synthetic and metabolic studies. Has low sensitivity and wide chemical shift, yields sharp signals. Low percentage makes it useful by preventing spin-spin couplings and makes the spectrum appear less crowded. Slow relaxation means that spectra are not integrable unless long acquisition times are used. See Annex 3.

7. ^{14}N, medium sensitivity nucleus with wide chemical shift. Its large quadrupole moment interferes in acquisition of high resolution spectra, limiting usefulness to smaller molecules and functional groups with a high degree of symmetry such as the headgroups of lipids.
8. ^{15}N, relatively commonly used. Can be used for labeling compounds. Nucleus very insensitive but yields sharp signals. Low percentage in natural nitrogen together with low sensitivity requires high concentrations or expensive isotope enrichment.
9. ^{17}O, low sensitivity and very low natural abundance. Used in metabolic and biochemical studies.
10. ^{19}F, relatively commonly measured. Sensitive, yields sharp signals, has wide chemical shift.
11. ^{31}P, 100% of natural phosphorus. Medium sensitivity, wide chemical shift range, yields sharp lines. Used in biochemical studies.
12. ^{35}Cl and ^{37}Cl, broad signal. ^{35}Cl significantly more sensitive, preferred over ^{37}Cl despite its slightly broader signal. Organic chlorides yield very broad signals, its use is limited to inorganic and ionic chlorides and very small organic molecules.
13. ^{43}Ca, used in biochemistry to study calcium binding to DNA, proteins, etc. Moderately sensitive, very low natural abundance.
14. ^{195}Pt, used in studies of catalysts and complexes.

Other nuclei (usually used in the studies of their complexes and chemical binding, or to detect presence of the element):

- ^{6}Li, ^{7}Li ,^{9}Be ;^{19}F ;^{21}Ne ;^{23}Na ;^{25}Mg ;^{27}Al ;^{29}Si ;^{31}P ;^{33}S ;^{39}K, ^{40}K, ^{41}K ;^{45}Sc ;^{47}Ti, ^{49}Ti ;^{50}V, ^{51}V ;^{53}Cr ;^{53}Mn ;^{57}Fe ;^{59}Co ;^{61}Ni ;^{63}Cu, ^{65}Cu ;^{67}Zn ;^{69}Ga, ^{71}Ga ;^{73}Ge ;^{77}Se ;^{81}Br ;^{87}Rb ;^{87}Sr ;^{95}Mo ;^{109}Ag ;^{113}Cd ;^{125}Te ;^{127}I ;^{133}Cs ;^{135}Ba, ^{137}Ba ;^{139}La ;^{183}W ;^{199}Hg

5

NMR HARDWARE

5.1. Overview

The diagram below shows a schematic representation of the major systems and components of a Nuclear Magnetic Resonance Spectrometer NMRS and this chapter briefly states the function of each component.

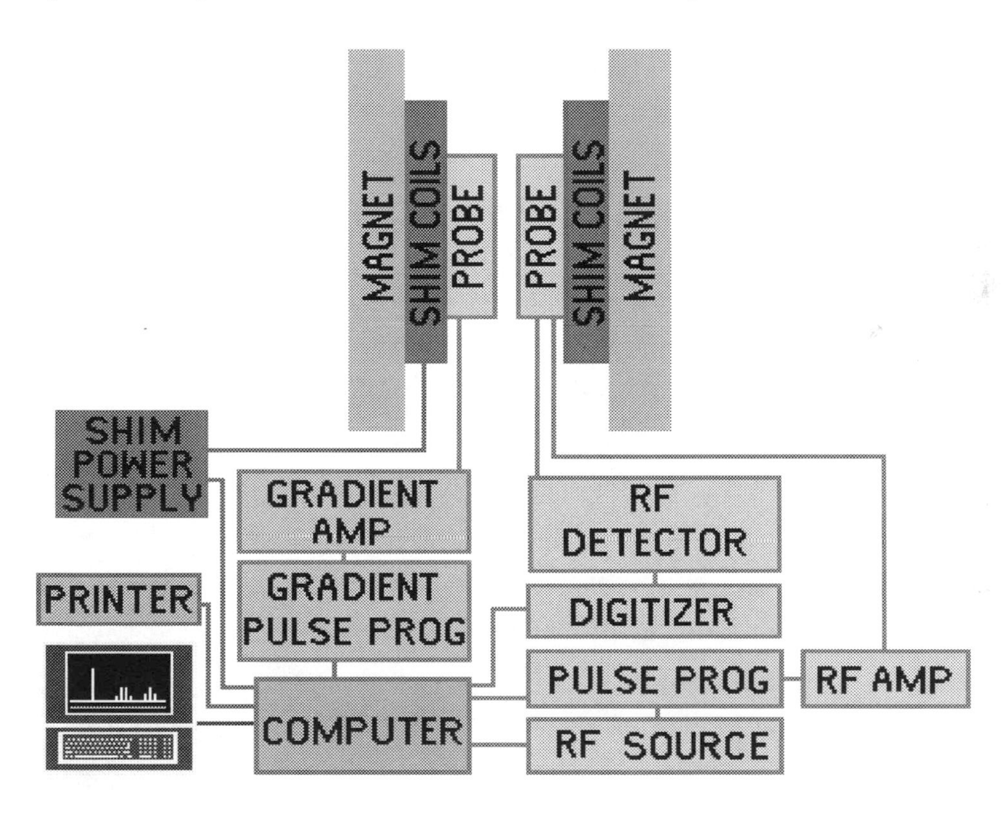

Fig.5.1. Schematic representation of the major systems and components of a Nuclear Magnetic Resonance Spectrometer NMRS

The superconducting magnet of the NMRS produces the B_o field necessary for the NMR experiments. Immediately within the bore of the magnet (airgap between the two faces of magnets) are the shim coils for homogenizing the B_o field. Within the shim coils is the probe. The probe contains the RF coils for producing the B_1 magnetic field necessary to rotate the spins by 90° or 180°. The RF coil also detects the signal from the spins within the sample. The sample is positioned within the RF coil of the probe. Some probes also contain a set of gradient coils. These coils produce a gradient in B_o along the X, Y, or Z axis. Gradient coils are used for gradient enhanced spectroscopy

The computer controls all of the components of the spectrometer. The RF components under control of the computer are the RF frequency source and pulse programmer. The source produces a sine wave of the desired frequency. The pulse programmer sets the width, and in some cases the shape, of the RF pulses. The RF amplifier increases the pulses power from milli Watts to tens or hundreds of Watts. The computer also controls the gradient pulse programmer which sets the shape and amplitude of gradient fields. The gradient amplifier increases the power of the gradient pulses to a level sufficient to drive the gradient coils.

The operator of the NMRS gives input to the computer through a console terminal with a mouse and keyboard. Some spectrometers also have a separate small interface for carrying out some of the more routine procedures on the spectrometer. A pulse sequence is selected and customized from the console terminal. The operator can see spectra on a video display located on the console and can make hard copies of spectra using a printer.

The next sections of this chapter go into more detail concerning the magnet, lock, shim coils, gradient coils, RF coils, and RF detector of nuclear magnetic resonance spectrometer.

5.2. NMR Magnet

1- Magnet Core

The NMR magnet is one of the most expensive components of the nuclear magnetic resonance spectrometer system. Most magnets are of the superconducting type. NMR magnet technology has evolved considerably since the development of NMR. Early NMR magnets were iron core permanent or electromagnets producing magnetic fields of less than 1.5 T. Today, most NMR magnets are of the superconducting type. Superconducting NMR magnets range in field strength from approximately 6 to 23.5 T.

The diagram chart below shows the history evolution of NMR magnets. In this chart the history of the growth of magnetic field strength in Tesla for using iron core and superconducting magnets where, the threshold of 7.0 Tesla is indicated on using superconducting magnet from an NMR spectrometer.

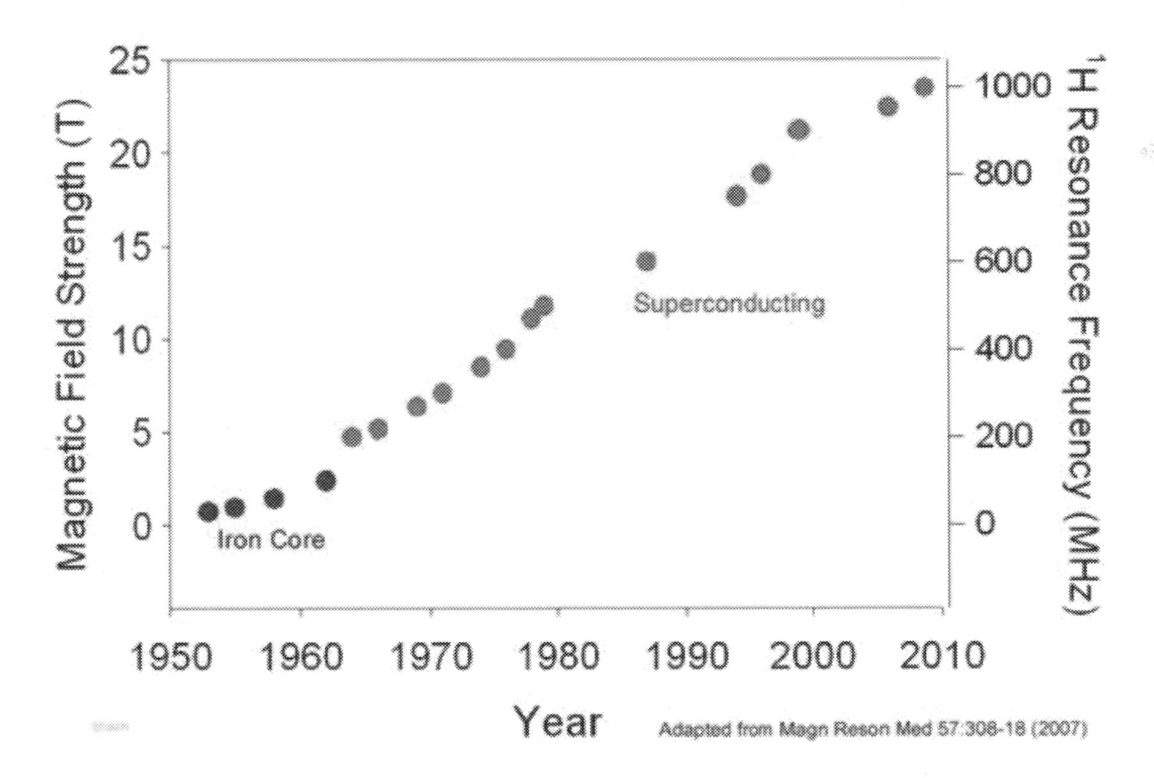

Fig.5.2. History of development for NMR magnets

A superconducting magnet has an electromagnet made of superconducting wire. Superconducting wire has a resistance approximately equal to zero when it is cooled to a temperature close to absolute zero (-273.15° C or 0 K) by immersing it in liquid helium. Once current is caused to flow in the coil it will continue to flow for as long as the coil is kept at liquid helium temperatures. (Some losses do occur over time due to the infinitesimally small resistance of the coil. These losses are on the order of a ppm of the main magnetic field per year.)

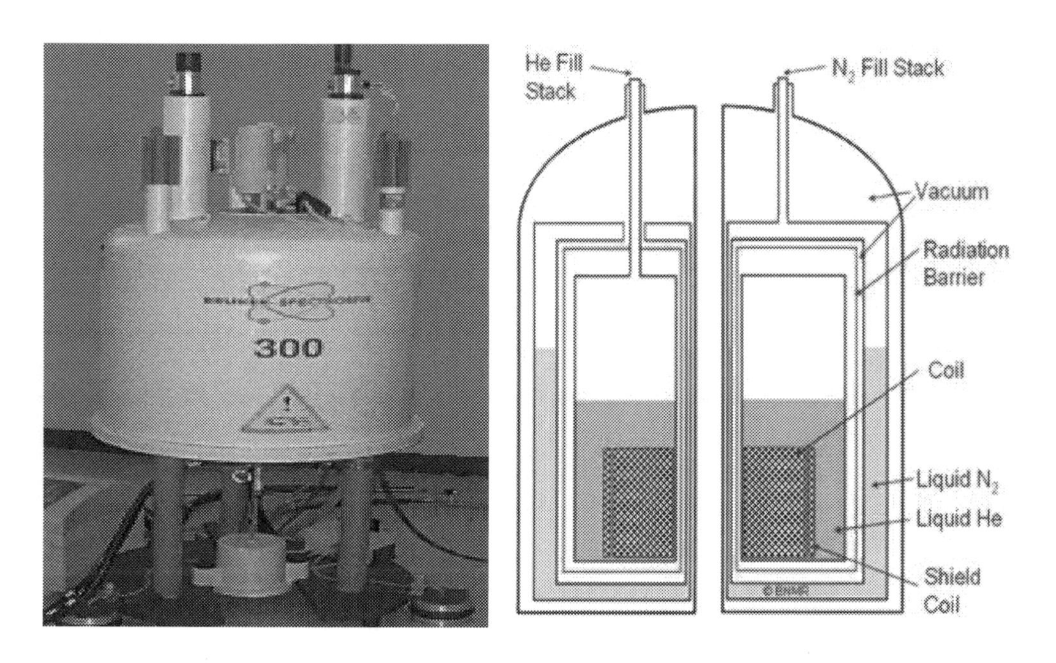

Fig.5.3. Example of superconducting magnet

The length of superconducting wire in the magnet is typically several miles. The superconducting elements of the wire are made of $(NbTaTi)_3Sn$. This material is brittle and therefore is embedded in copper for strength. The Cu has a high resistance compared to the superconductor which is carrying the current. This wire is wound into a multi-turn solenoid or coil. The coil of wire is kept at a temperature of 4.2K or less by immersing it in liquid helium. The coil and liquid helium are kept in a large dewar. This dewar is typically surrounded by a liquid nitrogen (77.4K) dewar, which acts as a thermal buffer between the room temperature air (293K) and the liquid helium. An example of cross sectional view of the superconducting magnet, depicting the concentric dewars, is shown in fig.5.4.

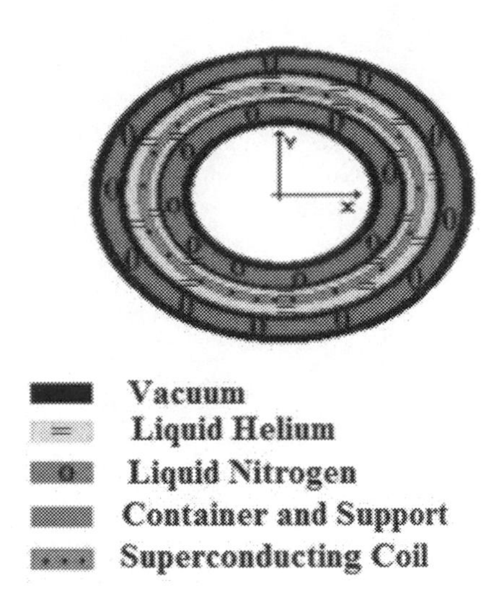

Fig.5.4. Example of cross sectional view of the superconducting magnet

Most NMR magnets today are shielded magnets. These magnets have an additional superconducting coil outside of the main coil which cancel out much of the fringe field from the main coil. As a consequence, the stray field outside the magnet is very small. Having a small fringe field becomes important in higher field magnets where safety concerns become more important.

The following image (see fig.5.5) is an actual cut-away view of a superconducting magnet. The magnet is supported by three legs, and the concentric nitrogen and helium dewars are supported by stacks coming out of the top of the magnet. A room temperature bore hole extends through the center of the assembly. The sample probe and shim coils are located within this bore hole. Also depicted in this picture is the liquid nitrogen level sensor, an electronic assembly for monitoring the liquid nitrogen level.

The vacuum region is filled with several layers of a reflective mylar film. The function of the mylar is to reflect thermal photons, and thus diminish heat from entering the magnet. Within the inside wall of the liquid nitrogen reservoir, we see another vacuum filled with some reflective mylar. The liquid helium reservoir comes next. This reservoir houses the superconducting solenoid or coil of wire.

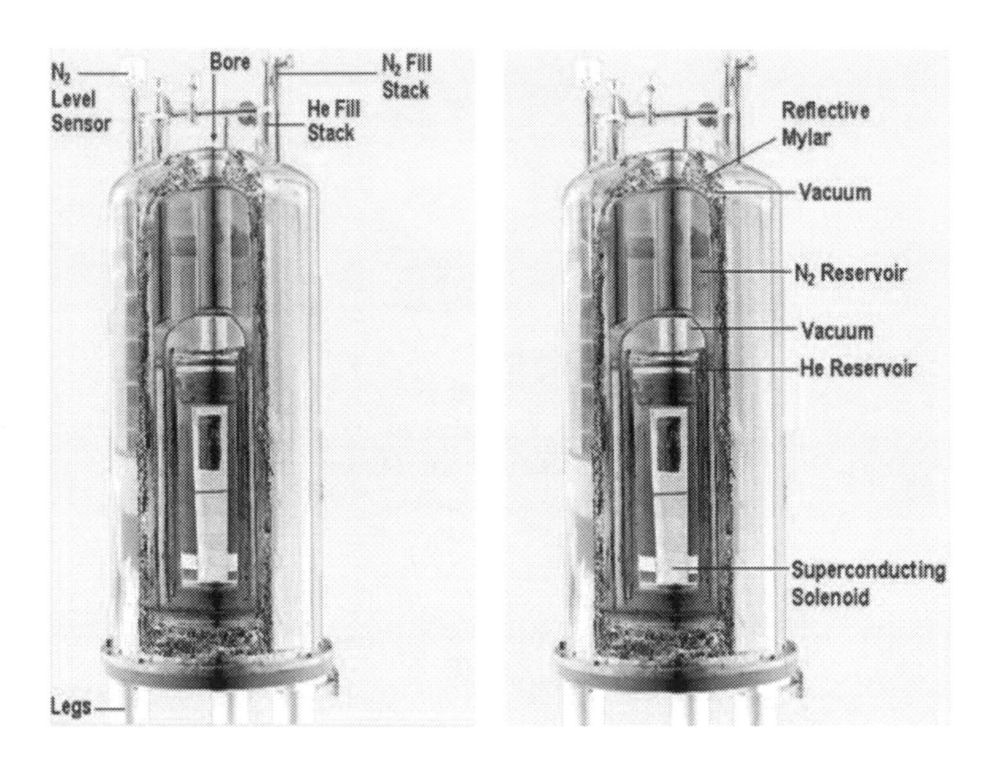

Fig.5.5. Actual cut-away view of a superconducting magnet.
Going from the outside of the magnet to the inside, we see a vacuum region followed by a liquid nitrogen reservoir.

2- Magnetic Field

A constant and spatially homogeneous magnetic field is one of the main parameter to maintain for the NMR operation. In order to compensate for the variation of the main magnetic fields the following means of adjustment are usually considered as part of NMR hardware set of components:

A. Shim Coils

The purpose of shimming a magnet is to make the magnetic field more homogeneous and to obtain better spectral resolution. Shimming can be performed manually or by computer control.

The purpose of shim coils on a spectrometer is to correct minor spatial inhomogeneities in the B_o magnetic field. These inhomogeneities could be caused by the magnet design, materials in the probe, variations in the thickness of the sample tube, sample permeability, and ferromagnetic materials around the magnet. A shim coil is designed to create a small magnetic field which will oppose and cancel out an inhomogeneity in the B_o

magnetic field.

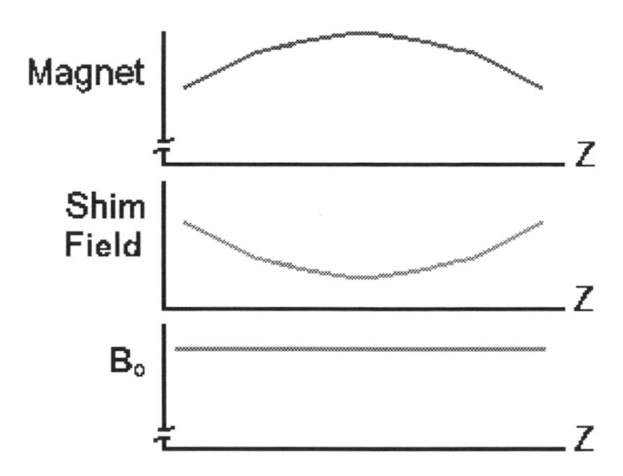

Fig.5.6. Shim field to maintain a constant magnetic field B_o

By passing the appropriate amount of current through each coil a homogeneous B_o magnetic field can be achieved. The optimum shim current settings are found by either minimizing the linewidth, maximizing the size of the Free Induction Decay (FID), or maximizing the signal from the field lock. On most spectrometers, the shim coils are controllable by the computer. A computer algorithm has the task of finding the best shim value by maximizing the lock signal.

Because these variations may exist in a variety of functional forms (linear, parabolic, etc.), shim coils are needed which can create a variety of opposing fields. Examples Some the Shim Coil Functional Forms are listed in the table(5.1) .

Table 5.1.
Shim Coil Functional Forms

Shim type		Form of Function	Shim type		Form of Function
01	Z^0	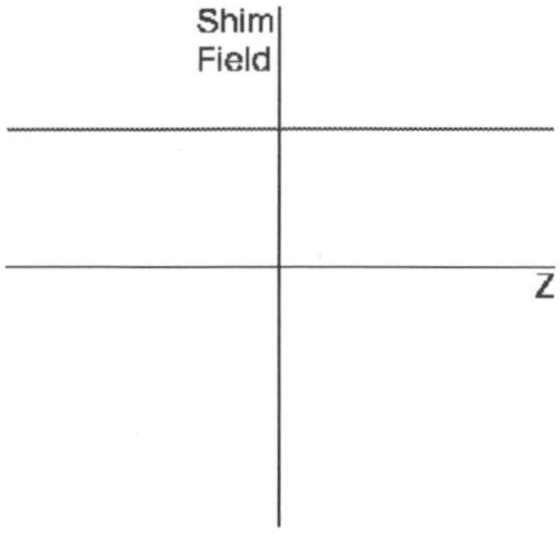	11	XY	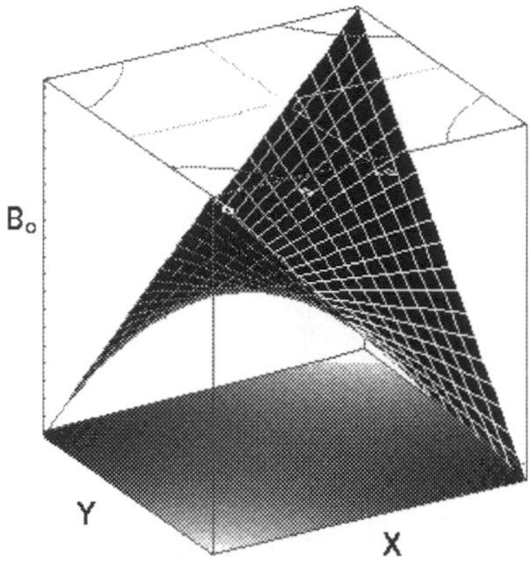
02	Z	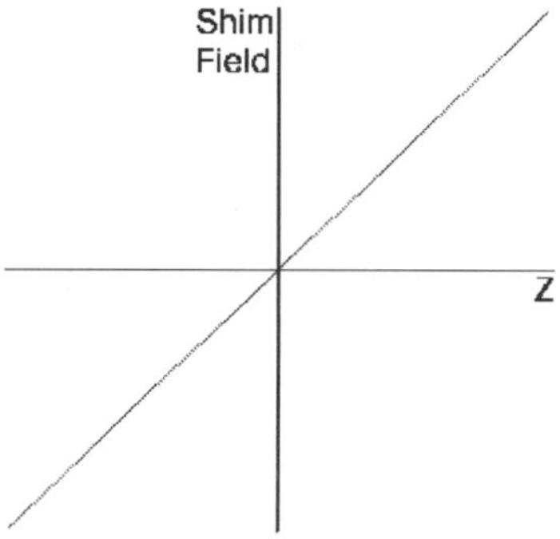	12	Y	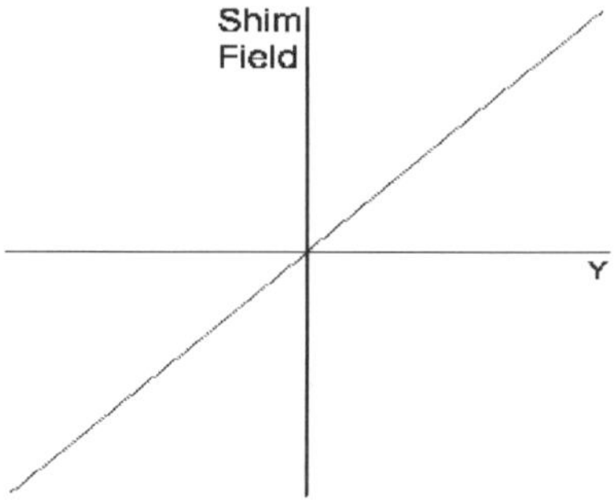
03	Z^2	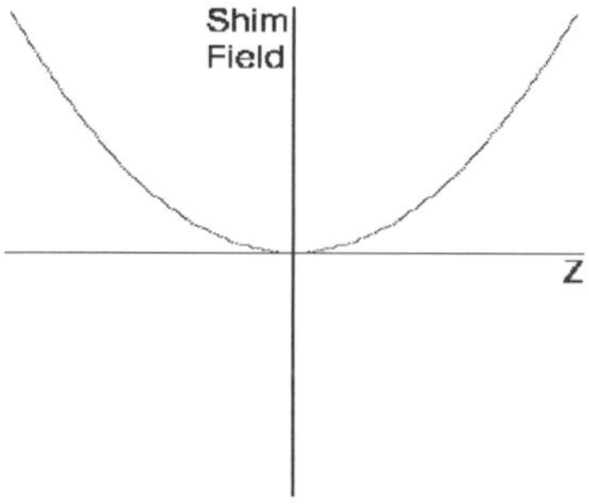	13	YZ	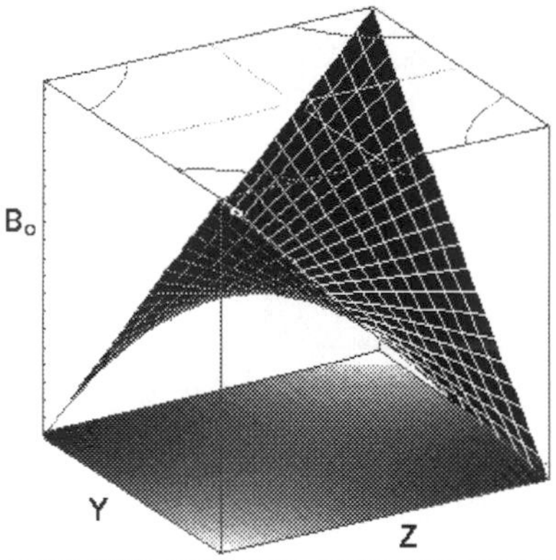

04	Z^3	Shim Field / Z	14	YZ^2	B_o / Z / Y
05	Z^4	Shim Field / Z	15	XZ^3	B_o / Z / X
06	Z^5	Shim Field / Z	16	X^2Y^2Z	X^2Y^2Z Cannot be reproduced on a 2-dimensional surface.
07	X	Shim Field / X	17	YZ^3	B_o / Z / Y

08	XZ	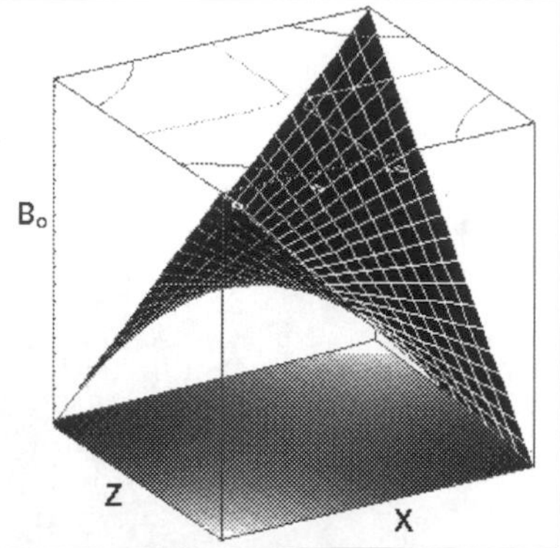	18	XYZ	XYZ Can not be reproduced on a 2-dimensional surface.
09	XZ^2	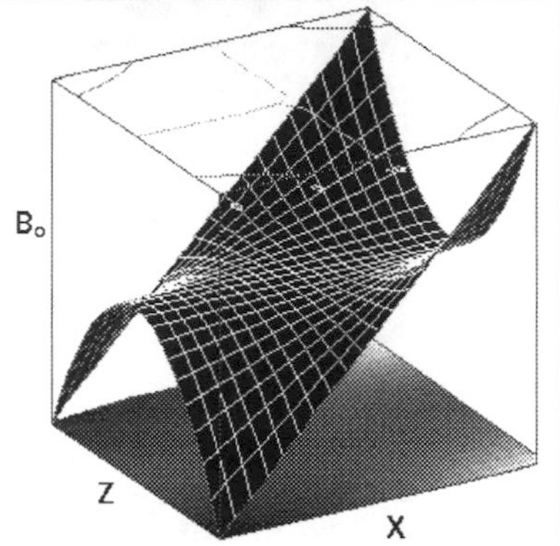	19	X^3	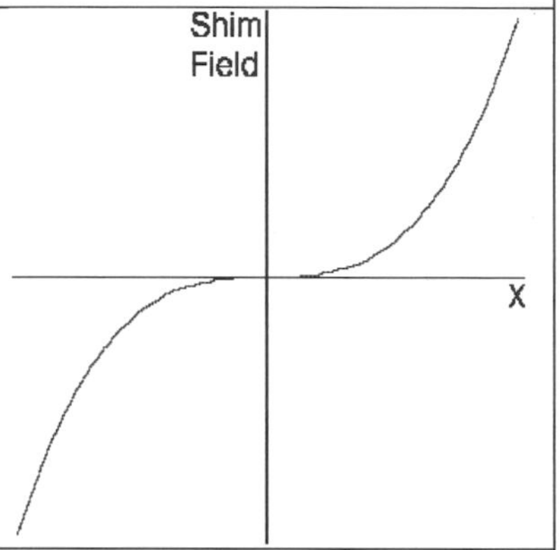
10	X^2Y^2	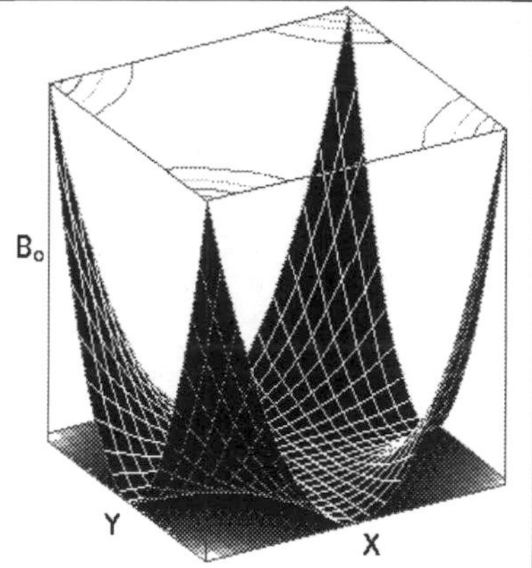	20	Y^3	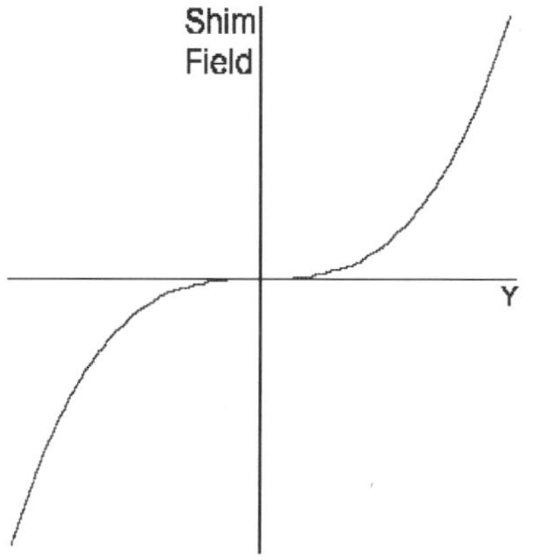

B. Field Lock

In order to produce a high resolution NMR spectrum of a sample, especially one which requires signal averaging or phase cycling, you need to have a temporally constant and spatially homogeneous magnetic field. The field strength might vary over time due to aging of the magnet, movement of metal objects near the magnet, and temperature fluctuations. Here is an example of a one line NMR spectrum of cyclohexane recorded while the B_o magnetic field was drifting a very significant amount. The field lock can compensate for these variations.

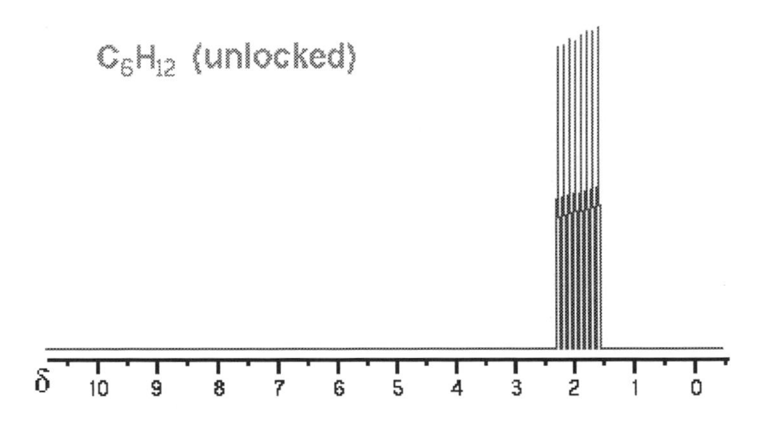

Fig.5.7. Drift on the magnetic field without compensation

The field lock is a separate NMR spectrometer within the spectrometer. This spectrometer is typically tuned to the deuterium NMR resonance frequency. It constantly monitors the resonance frequency of the deuterium signal and makes minor changes in the B_o magnetic field to keep the resonance frequency constant. The deuterium signal comes from the deuterium solvent used to prepare the sample. The Fig. below contains plots of the deuterium resonance lock frequency, the small additional magnetic field used to correct the lock frequency, and the resultant B_o field as a function of time while the magnetic field is drifting.

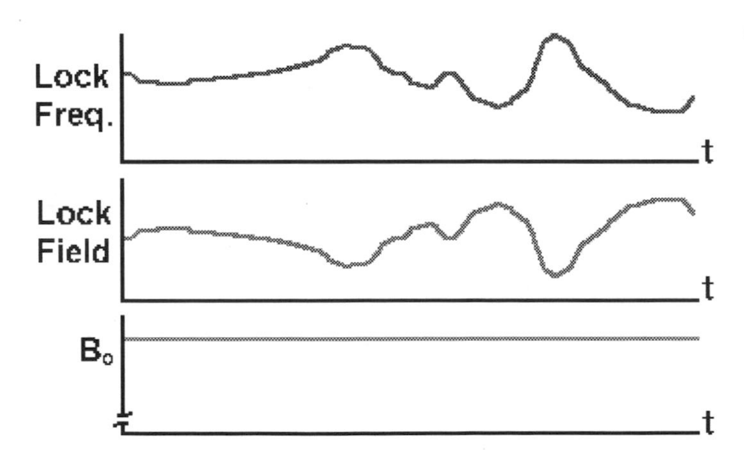

Fig.5.8. Magnetic field distribution with locking field

The lock frequency plot displays the frequency without correction. In reality, this frequency would be kept constant by the application of the lock field which offsets the drift.

5.3. Sample Probe

The sample probe is the name given to that part of the spectrometer which accepts the sample, sends RF energy into the sample, and detects the signal emanating from the sample. It contains sample spinner, the RF coil, and gradient coils.

A. Sample Spinner

The purpose of the sample spinner is to rotate the NMR sample tube about its axis. In doing so, each spin in the sample located at a given position along the Z axis and radius from the Z axis, will experience the average magnetic field in the circle defined by this Z and radius. The net effect is a narrower spectral linewidth. To appreciate this phenomenon, consider the following examples.

Picture an axial cross section of a cylindrical tube containing sample. In a very homogeneous B_o magnetic field this sample will yield a narrow spectrum.

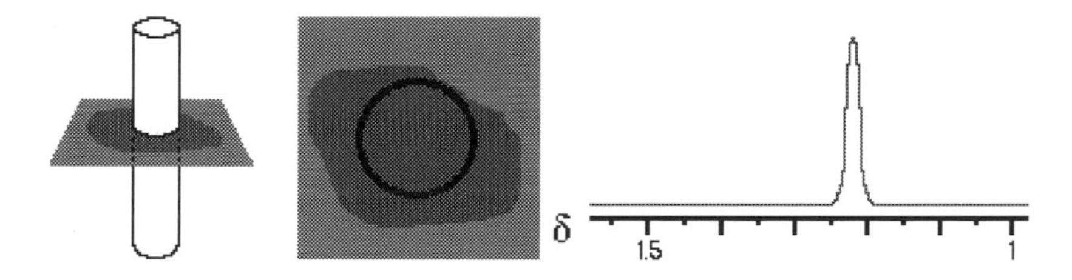

Fig.5.9. Sample yielding a narrow spectrum with homogeneous B_o magnetic field

In a more inhomogeneous field the sample will yield a broader spectrum due to the presence of lines from the parts of the sample experiencing different B_o magnetic fields.

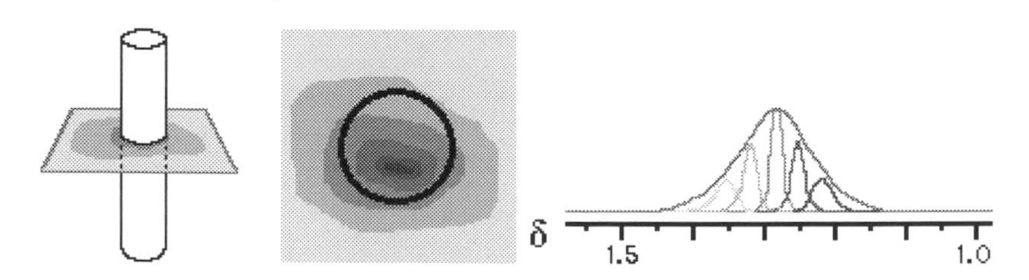

Fig.5.10. Sample yielding a broader spectrum with inhomogeneous field

When the sample is spun about its z-axis, inhomogeneities in the X and Y directions are averaged out and the NMR line width becomes narrower.

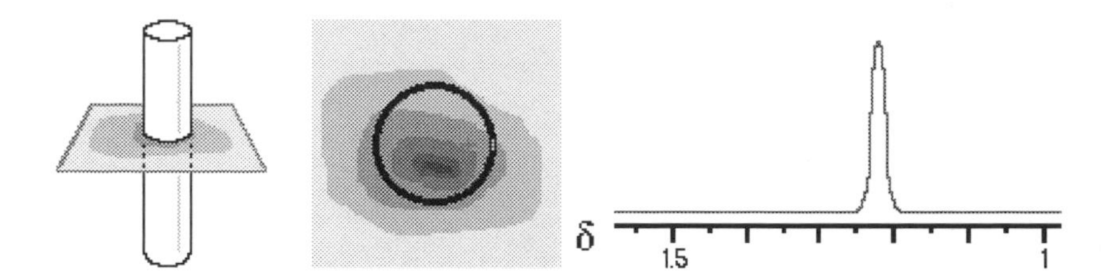

Fig.5.11. Sample spun about its z-axis yielding a narrow spectrum to compensate the inhomogeneities in the X and Y directions

In order to examine properties of the samples as a function of temperature many other instrumentation have to be used and facilitated to maintain the temperature of the sample above and below room temperature, such as the following:

- Feeding of Air or Nitrogen over the sample to heat or cool the sample.
- Monitoring the temperature at the sample with the aid of a thermocouple and electronic circuitry
- Maintaining the temperature by increasing or decreasing the temperature of the gas passing over the sample.

B. RF Coils

RF coils create the B_1 field which rotates the net magnetization in a pulse sequence. They also detect the transverse magnetization as it precesses in the XY plane. Most RF coils on NMR spectrometers are of the saddle coil design and act as the transmitter of the B_1 field and receiver of RF energy from the sample. You may find one or more RF coils in a probe. Each of these RF coils must resonate, that is they must efficiently store energy, at the Larmor frequency of the nucleus being examined with the NMR spectrometer.

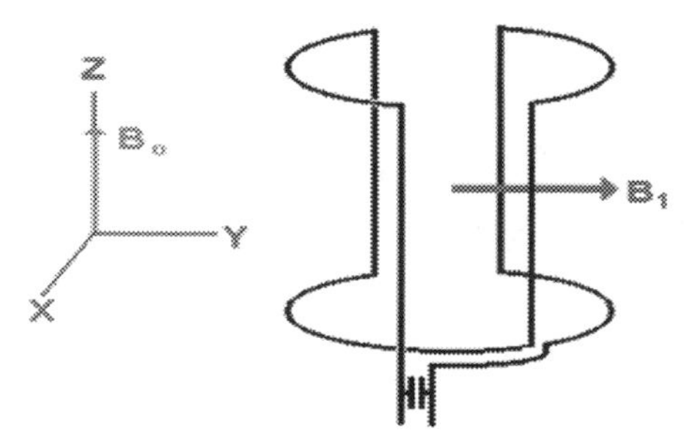

Fig.5.12. RF Coil - Saddle Coil

All NMR coils are composed of an inductor, or inductive elements, and a set of capacitive elements. The resonant frequency, ν, of an RF coil is determined by the inductance (L) and capacitance (C) of the inductor capacitor circuit.

$$\nu = \frac{1}{2\pi\sqrt{LC}} \qquad (5.1)$$

RF coils used in NMR spectrometers need to be tuned for the specific sample being studied. An RF coil has a bandwidth or specific range of frequencies at which it resonates. When you place a sample in an RF coil, the conductivity and dielectric constant of the sample affect the resonance frequency. If this frequency is different from the resonance frequency of the nucleus you are studying, the coil will not efficiently set up the B_1 field nor efficiently detect the signal from the sample. You will be rotating the net magnetization by an angle less than 90 degrees when you think you are rotating by 90 degrees. This will produce less transverse magnetization and less signal. Furthermore, because the coil will not be efficiently detecting the signal, your signal-to-noise ratio will be poor.

The B_1 field of an RF coil must be perpendicular to the B_o magnetic field. Another requirement of an RF coil in an NMR spectrometer is that the B_1 field needs to be homogeneous over the volume of your sample. If it is not, you will be rotating spins by a distribution of rotation angles and you will obtain strange spectra.

C. Gradient Coils

The gradient coils produce the gradients in the B_o magnetic field needed for performing gradient enhanced spectroscopy, diffusion measurements, and NMR microscopy. The gradient coils are located inside the RF probe. Not all probes have gradient coils, and not all NMR spectrometers have the hardware necessary to drive these coils.

The gradient coils are room temperature coils (i.e. do not require cooling with cryogens to operate) which, because of their configuration, create the desired gradient. Since the vertical bore superconducting magnet is most common, the gradient coil system will be described for this magnet.

Assuming the standard magnetic resonance coordinate system, a gradient in B_o in the Z direction is achieved with an antihelmholtz type of coil. Current in the two coils flow in opposite directions creating a magnetic field gradient between the two coils. The B field at the center of one coil adds to the B_o field, while the B field at the center of the other coil subtracts from the B_o field.

The X and Y gradients in the B_o field are created by a pair of (Look like figure-8) coils. The X axis of (Look like figure-8) coils create a gradient in B_o in the X direction due to the direction of the current through the coils. The Y axis (Look like figure-8) coils provides a similar gradient in B_o along the Y axis.

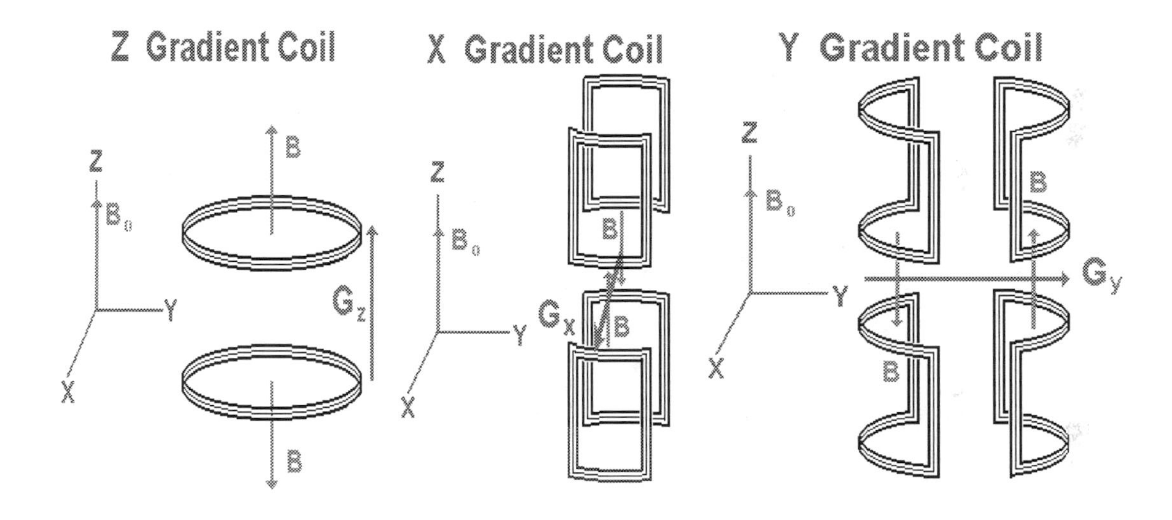

Fig.5.13. Z,X and Y- Gradient Coils

5.4. Filtration and Detection

A. Digital Filtering

Many newer spectrometers employ a combination of oversampling, digital filtering, and decimation to eliminate the wrap around artifact. Oversampling creates a larger spectral or sweep width, but generates too much data to be conveniently stored. Digital filtering eliminates the high frequency components from the data, and decimation reduces the size of the data set. The following flowchart summarizes the effects of the three steps by showing the result of performing an Fourier Transform (FT) after each step.

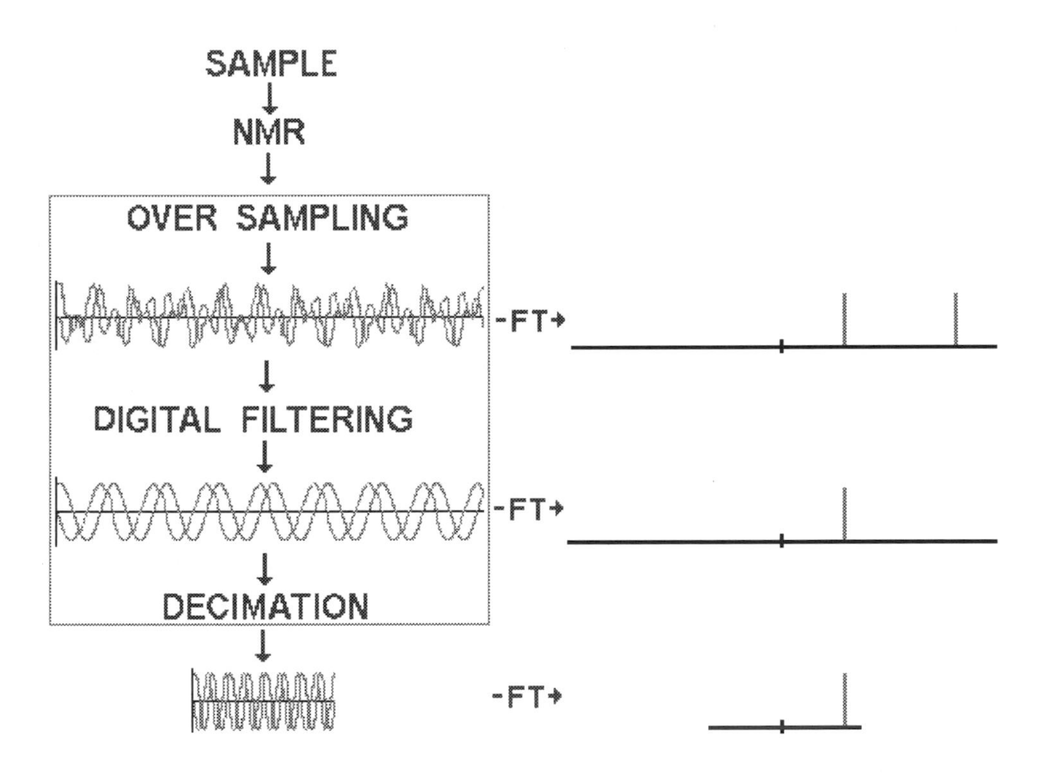

Fig.5.14. NMR Steps to eliminate the wrap around artifact

Let's examine oversampling, digital filtering, and decimation in more detail to see how this combination of steps can be used to reduce the wrap around problem.

Oversampling is the digitization of a time domain signal at a frequency much greater than necessary to record the desired spectral width. For example, if the sampling frequency, f_s, is increased by a factor of 10, the sweep width will be 10 times greater, thus eliminating wraparound. Unfortunately digitizing at 10 times the speed also increases the amount of raw data by a factor of 10, thus increasing storage requirements and processing time.

Filtering is the removal of a select band of frequencies from a signal. For an example of filtering, consider the frequency domain signal shown in fig.5.15. Frequencies above fo could be removed from this frequency domain signal by multipling the signal by this rectangular function. In NMR, this step would be equivalent to taking a large sweep width spectrum and setting to zero intensity those spectral frequencies which are farther than some distance from the center of the spectrum.

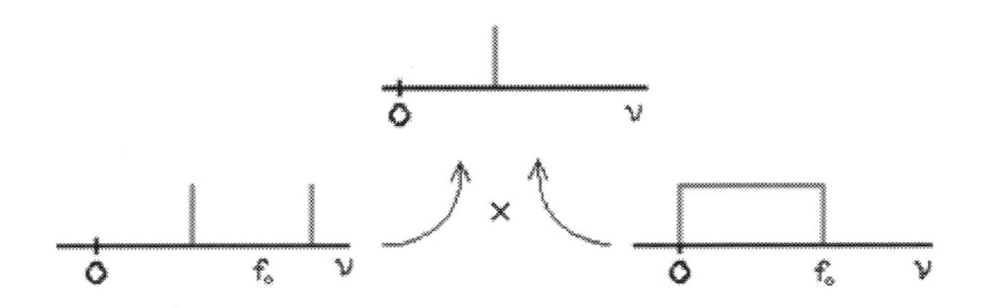

Fig.5.15. NMR Filtering

Digital filtering is the removal of these frequencies using the time domain signal. If two functions are multiplied in one domain (*i.e.* frequency), we must convolve the Fourier Transform FT of the two functions together in the other domain (*i.e.* time). To filter out frequencies above fo from the time domain signal, the signal must be convolved with the Fourier transform of the rectangular function, a sinc function.

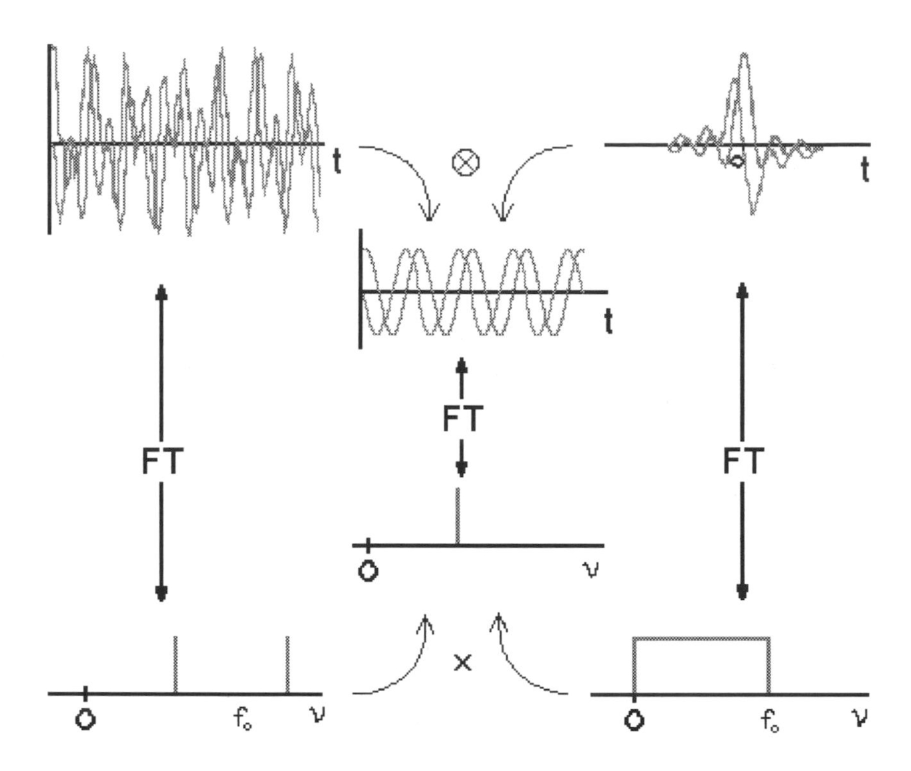

Fig.5.16. The signal convolved with the Fourier transform

This process eliminates frequencies greater than fo from the time domain signal. Fourier transforming the resultant time domain signal yields a frequency domain signal without the higher frequencies. In NMR, this step will remove spectral components with frequencies greater than $+f_o$ and less than $-f_o$.

Decimation is the elimination of data points from a data set. A decimation ratio of 4/5 means that 4 out of every 5 data points are deleted, or every fifth data point is saved. Decimating the digitally filtered data above, followed by a Fourier transform, will reduce the data set by a factor of five.

High speed digitizers, capable of digitizing at 2 MHz, and dedicated high speed integrated circuits, capable of performing the convolution on the time domain data as it is being recorded, are used to realize this procedure.

B. Quadrature Detector

The quadrature detector is a device which separates out the $M_{x'}$ and $M_{y'}$ signals from the signal from the RF coil. For this reason it can be thought of as a laboratory to rotating frame of reference converter. The heart of a quadrature detector is a device called a doubly balanced mixer. The doubly balanced mixer has two inputs and one output. If the input signals are Cos(A) and Cos(B), the output will be 1/2 Cos(A+B) and 1/2 Cos(A-B). For this reason the device is often called a product detector since the product of Cos(A) and Cos(B) is the output.

The quadrature detector typically contains two doubly balanced mixers, two filters, two amplifiers, and a 90° phase shifter. There are two inputs and two outputs on the device. Frequency $\nu$and ν_o are put in and the $M_{X'}$ and $M_{Y'}$ components of the transverse magnetization come out. There are some potential problems which can occur with this device which will cause artifacts in the spectrum. One is called a DC offset artifact and the other is called a quadrature artifact.

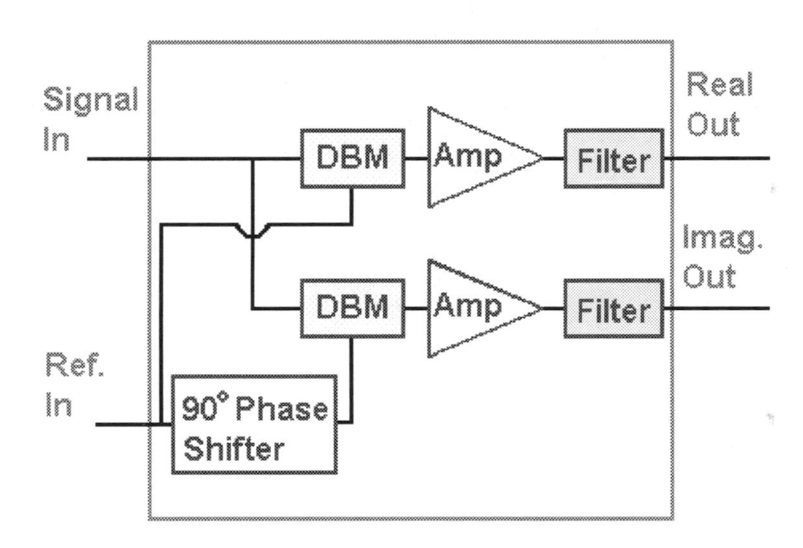

Fig.5.17 Components of The quadrature detector block

5.5. Safety

There are some important safety considerations which one should be familiar with before using an NMR spectrometer. These concern the use of strong magnetic fields and cryogenic liquids.

Magnetic fields from high field magnets can literally pick up and pull ferromagnetic items into the bore of the magnet. Caution must be taken to keep all ferromagnetic items away from the magnet because they can seriously damage the magnet, shim coils, and probe. The force exerted on the concentric cryogenic dewars within a magnet by a large metal object stuck to the magnet can break dewars and magnet supports. The kinetic energy of an object being sucked into a magnet can smash a dewar or an electrical connector on a probe. Small ferromagnetic objects are just as much a concern as larger ones. A small metal sliver can get sucked into the bore of the magnet and destroy the homogeneity of the magnet achieved with a set of shim settings.

There are additional concerns regarding the effect of magnetic fields on electronic circuitry, specifically pacemakers. An individual with a pacemaker walking through a strong magnetic field can induce currents in the pacemaker circuitry which will cause it to fail and possibly cause death. A person with a pacemaker must not be able to inadvertently stray into a magnetic field of five or more Gauss. Although not as important as a pacemaker, mechanical watches and some digital watches will also be affected by magnetic fields. Magnetic fields of approximately 50 Gauss will erase credit cards and magnetic storage media.

The liquid nitrogen and liquid helium used in NMR spectrometers are at a temperature of 77.4 K and 4.2 K respectively. These liquids can cause frostbite, which is not a concern unless you are filling the magnet. If you are filling the magnet or if you are operating the spectrometer, suffocation is another concern you need to be aware of. If the magnet quenches, or suddenly stops being a superconductor, it will rapidly boil off all its cryogens, and the nitrogen and helium gasses in a confined space can cause suffocation.

6

APPLICATIONS OF NMR SPECTROSCOPY

6.1. Introduction

Today, Nuclear Magnetic Resonance NMR has become a powerful analytical technology that has found a variety of applications in many disciplines of scientific research, medicine, and various industries.

NMR Spectroscopy is a nondestructive analytical technique that is used to probe the nature and characteristics of molecular structure. A simple NMR experiment produces information in the form of a spectrum, which is able to provide details about:

- The types of atoms present in the sample
- The relative amounts of atoms present in a sample
- The specific environments of atoms within a molecule
- The purity and composition of a sample
- Structural information about a molecule, including constitutional and conformational isomerization

There are a number of important characteristics of NMR spectroscopy which makes it favorable to a wide array of industrial, commercial and research applications:

- There are many nuclei which are NMR active. Some of the more common NMR active nuclei include ^{1}H, ^{2}H, ^{13}C, ^{11}B, ^{15}N, ^{19}F, ^{31}P and ^{195}Pt.
- NMR spectroscopy is generally a non-destructive technique, meaning that samples can be recovered.
- Only a small quantity of material is required for analysis; sample sizes of (5-20)mg are generally sufficient for most NMR experiments.
- Sample preparation is simple and minimal. Typically, samples are simply dissolved in an appropriate solvent.

NMR Spectroscopy is a technique used by most modern chemical laboratories. It has applications in a wide range of disciplines, and development of new applied methods for NMR is an active area of research. Methods in NMR spectroscopy have particular relevance to the following disciplines and industries:

- Chemical research and development: organic, inorganic and physical chemistry
- Chemical manufacturing industry
- Biological and biochemical research
- Food industry
- Pharmaceutical development and production
- Agrochemical development and production
- Polymer industry

6.2. Application Classification

Modern NMR spectroscopy has been emphasizing the application in biomolecular systems and plays an important role in structural biology. With developments in both methodology and instrumentation in the past two decades, NMR has become one of the most powerful and versatile spectroscopic techniques for the analysis of biomacromolecules, allowing characterization of biomacromolecules and their complexes up to 100 kDa. Together with X-ray crystallography, NMR spectroscopy is one of the two leading technologies for the structure determination of biomacromolecules at atomic resolution. In addition, NMR provides unique and important molecular motional and interaction profiles containing pivotal information on protein function. The information is also critical in drug development.

The common applications of NMR Spectroscopy include:

- Structure elucidation
- Chemical composition determination
- Formulations investigation
- Raw materials fingerprinting
- Mixture analysis
- Sample purity determination
- Quality assurance and control
- Quantitative analysis
- Compound identification and confirmation
- Analysis of inter- and intramolecular exchange processes
- Molecular characterisation
- Reaction kinetics examination
- Reaction mechanism investigation

Some of the applications of NMR spectroscopy are defined and listed below:

- **Solution structure:** The only method for atomic-resolution structure determination of biomacromolecules in aqueous solutions under near physiological conditions or membrane mimeric environments.
- **Molecular dynamics:** The most powerful technique for quantifying motional properties of biomacromolecules.
- **Protein folding:** The most powerful tool for determining the residual

structures of unfolded proteins and the structures of folding intermediates.

- **Ionization state:** The most powerful tool for determining the chemical properties of functional groups in bio-macro molecules, such as the ionization states of ionizing groups at the active sites of enzymes.
- **Weak intermolecular interactions:** Allowing weak functional interactions between macro-bio molecules (e.g., those with dissociation constants in the micro-molar to milli-molar range) to be studied, which is not possible with other technologies.
- **Protein hydration:** A power tool for the detection of interior water and its interaction with bio-macro molecules.
- **Hydrogen bonding:** A unique technique for the DIRECT detection of hydrogen bonding interactions.
- **Drug screening and design:** Particularly useful for identifying drug leads and determining the conformations of the compounds bound to enzymes, receptors, and other proteins.
- **Native membrane protein:** Solid state NMR has the potential for determining atomic-resolution structures of domains of membrane proteins in their native membrane environments, including those with bound ligands.
- **Metabolite analysis:** A very powerful technology for metabolite analysis.
- **Chemical analysis:** A matured technique for chemical identification and conformational analysis of chemicals whether synthetic or natural.
- **Material science:** A powerful tool in the research of polymer chemistry and physics.

However, the typical classification of NMR applications includes the following:

1) Structural (chemical) elucidation
    - Natural product chemistry
    - Synthetic organic chemistry (analytical tool of choice of synthetic chemists used in conjunction with MS and IR)
2) Study of dynamic processes
    - Reaction kinetics

- Study of equilibrium (chemical or structural)

3) Structural (three-dimensional) studies
    - Proteins, Protein-ligand complexes
    - DNA, RNA, Protein/DNA complexes
    - Polysaccharides
4) Metabolomics
5) Drug Design
    - Structure Activity Relationships by NMR
6) Medicine
    - MRI

The NMR Facility provides a range of flexible services which can be adapted to best suit the needs of the applications. In the following sections a selective applications are briefly described.

6.3. Medical Diagnosis

The use of nuclear magnetic resonance best known to the general public is magnetic resonance imaging for medical diagnosis and MR Microscopy in research settings, however, it is also widely used in chemical studies, notably in NMR spectroscopy such as proton NMR, carbon-13 NMR, deuterium NMR and phosphorus-31 NMR. Biochemical information can also be obtained from living tissue (e.g. human brain tumours) with the technique known as in vivo magnetic resonance spectroscopy or chemical shift NMR Microscopy.

These studies are possible because nuclei are surrounded by orbiting electrons, which are also spinning charged particles such as magnets and, so, will partially shield the nuclei. The amount of shielding depends on the exact local environment. For example, a hydrogen bonded to an oxygen will be shielded differently than a hydrogen bonded to a carbon atom. In addition, two hydrogen nuclei can interact via a process known as spin-spin coupling, if they are on the same molecule, which will split the lines of the spectra in a recognizable way.

Magnetic resonance imaging (MRI) is a medical imaging technique used in radiology to form pictures of the anatomy and the physiological processes of the body in both health and disease. MRI scanners use strong magnetic fields, electric field gradients, and radio waves to generate images of the organs in the body. MRI does not involve X-rays or the use of ionizing radiation, which distinguishes it from CT or CAT scans. Magnetic resonance imaging is a medical application of nuclear magnetic resonance (NMR). NMR can also be used for *imaging* in other NMR applications such as NMR spectroscopy.

While the hazards of X-rays are now well-controlled in most medical contexts, MRI may still be seen as a better choice than a CT scan. MRI is widely used in hospitals and clinics for medical diagnosis, staging of disease and follow-up without exposing the body to radiation. However, MRI may often yield different diagnostic information compared with CT.

MRI was originally called 'NMRI' (nuclear magnetic resonance imaging) and is a form of NMR, though the use of 'nuclear' in the acronym was dropped to avoid negative associations with the word.[1] Certain atomic nuclei are able to absorb and emit radio frequency energy when placed in an external magnetic field. In clinical and research MRI, hydrogen atoms are most often used to generate a detectable radio-frequency signal that is received by antennas in close proximity to the anatomy being examined. Hydrogen atoms are naturally abundant in people and other biological organisms, particularly in water and fat. For this reason, most MRI scans essentially map the location of water and fat in the body. Pulses of radio waves excite the nuclear spin energy transition, and magnetic field gradients localize the signal in space. By varying the parameters of the pulse sequence, different contrasts may be generated between tissues based on the relaxation properties of the hydrogen atoms therein.

To perform a study, the person is positioned within an MRI scanner that forms a strong magnetic field around the area to be imaged. In most medical applications, protons (hydrogen atoms) in tissues containing water molecules create a signal that is processed to form an image of the body. First, energy from an oscillating magnetic field temporarily is applied to the patient at the appropriate resonance frequency. The excited hydrogen atoms emit a radio frequency signal, which is measured by a receiving coil. The radio signal may be made to encode position information by varying the main magnetic field using gradient coils. As these coils are rapidly switched on and off they create the characteristic repetitive noise of an MRI scan. The contrast between different tissues is determined by the rate at which excited atoms return to the equilibrium state. Exogenous contrast agents may be given to the person to make the image clearer.

The major components of an MRI scanner are the main magnet, which polarizes the sample, the shim coils for correcting shifts in the homogeneity of the main magnetic field, the gradient system which is used to localize the MR signal and the RF system, which excites the sample and detects the resulting NMR signal. The whole system is controlled by one or more computers.

MRI requires a magnetic field that is both strong and uniform. The field strength of the magnet is measured in teslas – and while the majority of systems operate at 1.5 T, commercial systems are available between 0.2 to 7 T. Most clinical magnets are superconducting magnets, which require liquid helium. Lower field strengths can be achieved with permanent magnets, which are often used in "open" MRI scanners for claustrophobic patients.[5] Recently, MRI has been demonstrated also at ultra-low fields, i.e., in the microtesla-to-millitesla range, where sufficient signal quality is made possible by prepolarization (on the order of 10-100 mT) and by measuring the Larmor precession fields at about 100 microtesla with highly sensitive superconducting quantum interference devices (SQUIDs).

Magnetic resonance spectroscopy (MRS) is used to measure the levels of different metabolites in body tissues. The MR signal produces a spectrum of resonances that corresponds to different molecular arrangements of the isotope being "excited". This signature is used to diagnose certain metabolic disorders, especially those affecting the brain, and to provide information on tumor metabolism.

MRI has the advantages of having very high spatial resolution and is very adept at morphological imaging and functional imaging. MRI does have several disadvantages though. First, MRI has a sensitivity of around 10^{-3} mol/L to 10^{-5} mol/L, which, compared to other types of imaging, can be very limiting. This problem stems from the fact that the population difference between the nuclear spin states is very small at room temperature. For example, at 1.5 teslas, a typical field strength for clinical MRI, the difference between high and low energy states is approximately 9 molecules per 2 million. Improvements to increase MR sensitivity include increasing magnetic field strength, and hyperpolarization via optical pumping or dynamic nuclear polarization. There are also a variety of signal amplification schemes based on chemical exchange that increase sensitivity.

In the UK, the price of a clinical 1.5-tesla MRI scanner is around £920,000/US$1.4 million, with the lifetime maintenance cost broadly similar to the purchase cost. In the Netherlands, the average MRI scanner costs around €1 million, with a 7-T MRI having been taken in use by the UMC Utrecht in December 2007, costing €7 million. Pre-polarizing MRI (PMRI) systems using resistive electromagnets have shown promise as a low-cost alternative and have specific advantages for joint imaging near metal implants, however they are likely unsuitable for routine whole-body or neuroimaging applications.

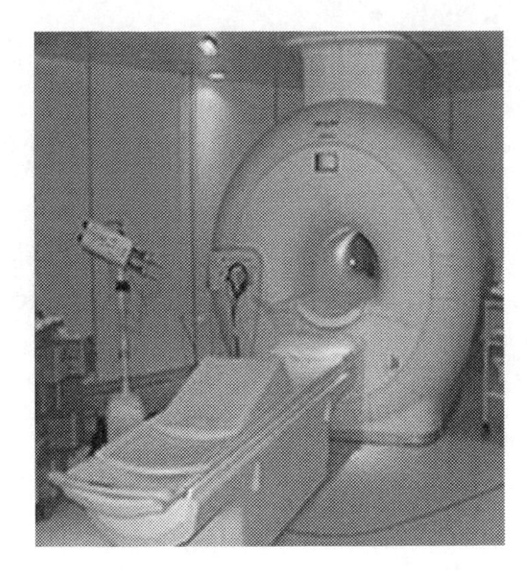

A 3 tesla clinical MRI scanner.

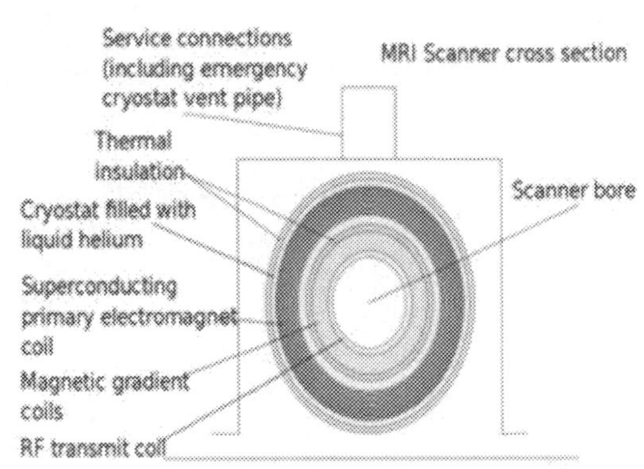

Cross section- Schematic of a cylindrical superconducting MRI scanner.

Fig.6. Magnetic Resonance Imaging MRI Scanner

6.4. Chemistry

By studying the peaks of nuclear magnetic resonance spectra, chemists can determine the structure of many compounds. It can be a very selective technique, distinguishing among many atoms within a molecule or collection of molecules of the same type but which differ only in terms of their local chemical environment.

A chemist can determine the identity of a compound by comparing the observed nuclear precession frequencies to known frequencies. Further structural data can be elucidated by observing spin-spin coupling, a process by which the precession frequency of a nucleus can be influenced by the magnetization transfer from nearby nuclei. Spin-spin coupling is most commonly observed in NMR involving common isotopes, such as Hydrogen-1 (^{1}H NMR).

Because the nuclear magnetic resonance timescale is rather slow, compared to other spectroscopic methods, changing the temperature of the experiment can also give information about fast reactions, such as the Cope rearrangement or about structural dynamics, such as ring-flipping in cyclohexane. At low enough temperatures, a distinction can be made between the axial and equatorial hydrogens in cyclohexane.

An example of nuclear magnetic resonance being used in the determination of a structure is that of buckminsterfullerene (often called "buckyballs", composition C_{60}). This now famous form of carbon has 60 carbon atoms forming a sphere.

The carbon atoms are all in identical environments and so should see the same internal H field. Unfortunately, buckminsterfullerene contains no hydrogen and so ^{13}C nuclear magnetic resonance has to be used. ^{13}C spectra require longer acquisition times since carbon-13 is not the common isotope of carbon (unlike hydrogen, where ^{1}H is the common isotope). However, in 1990 the spectrum was obtained by R. Taylor and co-workers at the University of Sussex and was found to contain a single peak, confirming the unusual structure of buckminsterfullerene.

Many of the molecules studied by NMR contain carbon. Unfortunately, the carbon-12 (^{12}C) nucleus does not have a nuclear spin, but the carbon-13 (^{13}C) nucleus does due to the presence of an unpaired neutron. Carbon-13 nuclei make up approximately one percent of the carbon nuclei on earth. Therefore, carbon-13 NMR spectroscopy will be less sensitive (have a poorer SNR) than hydrogen NMR spectroscopy. With the appropriate concentration, field strength, and pulse sequences, however, carbon-13 NMR spectroscopy can be used to supplement the previously described hydrogen NMR information.

For further details Annex 3 describes Carbon -13NMR.

6.5. Non-Destructive Testing

Nuclear magnetic resonance is extremely useful for analyzing samples non-destructively. Radio waves and static magnetic fields easily penetrate many types of matter and anything that is not inherently ferromagnetic. For example, various expensive biological samples, such as nucleic acids, including **RiboNucleic Acid** RNA and **DeoxyriboNucleic acid** DNA, or proteins, can be studied using nuclear magnetic resonance for weeks or months before using destructive biochemical experiments. This also makes nuclear magnetic resonance a good choice for analyzing dangerous samples.

RNA is part of a group of molecules known as the nucleic acids, which are one of the four major macromolecules (along with lipids, carbohydrates and proteins) essential for all known forms of life. Like DNA, RNA is made up of a long chain of components called nucleotides.

DNA is a nucleic acid containing the genetic instructions used in the development and functioning of all known living organisms (with the exception of RNA viruses). The DNA segments carrying this genetic information are called genes. Likewise, other DNA sequences have structural purposes, or are involved in regulating the use of this genetic information. Along with RNA and proteins, DNA is one of the three major macromolecules that are essential for all known forms of life.

The chemical structure of RNA is very similar to that of DNA, with two differences: (a) RNA contains the sugar ribose, while DNA contains the slightly different sugar deoxyribose (a type of ribose that lacks one oxygen atom), and (b) RNA has the nucleobase uracil while DNA contains thymine. Unlike DNA, most RNA molecules are single-stranded and can adopt very complex three-dimensional structures.

6.6. Data Acquisition in the Petroleum Industry

Another use for nuclear magnetic resonance is data acquisition in the petroleum industry for petroleum and natural gas exploration and recovery.

Data acquisition is the process of sampling signals that measure real world physical conditions and converting the resulting samples into digital numeric values that can be manipulated by a computer. Data acquisition systems (abbreviated with the acronym **DAS** or **DAQ**) typically convert analog waveforms into digital values for processing. The components of data acquisition systems include:

- Sensors that convert physical parameters to electrical signals.
- Signal conditioning circuitry to convert sensor signals into a form that can be converted to digital values.
- Analog-to-digital converters, which convert conditioned sensor signals to digital values.

Data acquisition applications are controlled by software programs developed using various general purpose programming languages such as BASIC, C, Fortran, Java, Lisp, Pascal.

A borehole is drilled into rock and sedimentary strata into which nuclear magnetic resonance logging equipment is lowered. Nuclear magnetic resonance analysis of these boreholes is used to measure rock porosity, estimate permeability from pore size distribution and identify pore fluids (water, oil and gas). These instruments are typically low field NMR spectrometers.

Low field NMR is a branch of nuclear magnetic resonance (NMR) that is either related to Earth's field NMR, or to NMR at a man-made very low magnetic field and shielding from (or compensating of) the Earth's magnetic field. With magnetic fields on the order of μT or nT, SQUIDs are typically used as detectors.

6.7. Flow Probes for NMR Spectroscopy

Probe is a device or agent used to detect or explore a substance (for example, a molecule used to detect the presence of a specific fragment of DNA or RNA or of a specific bacterial colony). Recently, real-time applications of NMR in liquid media have been developed using specifically designed flow probes (flow cell assemblies) which can replace standard tube probes. This has enabled techniques that can incorporate the use of high performance liquid chromatography (HPLC) or other continuous flow sample introduction devices.

HPLC is excellent for separating mixtures but generally poor at identification of compounds. But, since the Mass spectrometry (MS) provides us with uniquely valuable information: molecular mass (via mass to charge (m/z) ratio); molecular structural information and quantitative data, all at high sensitivity. However, it is best to apply separation techniques to complex mixtures before mass spectrometry is undertaken. The combination of these two techniques (LC/MS) thus provides an extraordinarily powerful analytical tool.

Liquid chromatography–mass spectrometry (LC-MS) is an analytical chemistry technique that combines the physical separation capabilities of liquid chromatography (or HPLC) with the mass analysis capabilities of mass spectrometry (MS). Coupled chromatography - MS systems are popular in chemical analysis because the individual capabilities of each technique are enhanced synergistically. While liquid chromatography separates mixtures with multiple components, mass spectrometry provides structural identity of the individual components with high molecular specificity and detection sensitivity. This tandem technique can be used to analyze biochemical, organic, and inorganic compounds commonly found in complex samples of environmental and biological origin. Therefore, LC-MS may be applied in a wide range of sectors including biotechnology, environment monitoring, food processing, and pharmaceutical, agrochemical, and cosmetic industries.

6.8. Process Control

NMR has now entered the arena of real-time process control and process optimization in oil refineries and petrochemical plants. Two different types of NMR analysis are utilized to provide real time analysis of feeds and products in order to control and optimize unit operations. These are:

- Time-domain NMR (TD-NMR) spectrometers operating at low field (2–20 MHz for ^1H) yield free induction decay data that can be used to determine absolute hydrogen content values, rheological information, and component composition. These spectrometers are used in mining, polymer production, cosmetics and food manufacturing as well as coal analysis.
- High resolution Fourier Transform FT-NMR spectrometers operating in the 60 MHz range with shielded permanent magnet systems yield high resolution ^1H NMR spectra of refinery and petrochemical streams. In FT NMR (also called pulse-FT NMR) the signal is generated by a (90°) rf pulse and then picked up by the receiver coil as a decaying oscillation with the spins resonance frequency ω.

The variation observed in these spectra with changing physical and chemical properties is modelled utilizing chemometrics to yield predictions on unknown samples. The prediction results are provided to control systems via analogue or digital outputs from the spectrometer.

6.9. Earth's Field NMR

Conventional high field NMR instruments require an extremely uniform magnetic field for polarisation of nuclear spins and the detection of the resultant magnetisation. It is difficult and expensive to make a magnetic field of the required homogeneity (much less than 1 ppm in some cases) and so sample sizes are usually quite small. However, because the polarisation and detection fields are large, the resulting signal to noise ratio (S/N) can also be large. Earth's field NMR (EFNMR) uses the globally available, homogeneous Earth's magnetic field for detection. The homogeneity of this field means that large samples can be used – partially compensating for the very low detection field. To improve the initial magnetization, a relatively crude electromagnet with a field about 350 times larger than the Earth's field is used to polarise the sample. This pre-polarization method greatly enhances the S/N of the EFNMR signal.

Earth's magnetic field (also known as the *geomagnetic field*) is the magnetic field that extends from the Earth's inner core to where it meets the solar wind, a stream of energetic particles emanating from the Sun. Its magnitude at the Earth's surface ranges from 25 to 65 microteslas (0.25 to 0.65 gauss). It is approximately the field of a magnetic dipole tilted at an angle of 11 degrees with respect to the rotational axis—as if there were a bar magnet placed at that angle at the center of the Earth. However, unlike the field of a bar magnet, Earth's field changes over time because it is generated by the motion of molten iron alloys in the Earth's outer core (the geodynamo).

The Magnetic North Pole wanders, but slowly enough that a simple compass remains useful for navigation. At random intervals (averaging several hundred thousand years) the Earth's field reverses (the north and south geomagnetic poles change places with each other). These reversals leave a record in rocks that allow paleomagnetists to calculate past motions of continents and ocean floors as a result of plate tectonics.

The region above the ionosphere, and extending several tens of thousands of kilometers into space, is called the magnetosphere. This region protects the Earth from cosmic rays that would strip away the upper atmosphere, including the ozone layer that protects the earth from harmful ultraviolet radiation.

In the Earth's magnetic field, NMR frequencies are in the audio frequency range. Earth's field NMR (EFNMR) is typically stimulated by applying a relatively strong dc magnetic field pulse to the sample and, following the pulse, analysing the resulting low frequency alternating magnetic field that occurs in the Earth's magnetic field due to free induction decay (FID). These effects are exploited in some types of magnetometers, EFNMR spectrometers, and MRI imagers. Their inexpensive portable nature makes these instruments valuable for field use and for teaching the principles of NMR and MRI.

Two of the most significant advantages of EFNMR over conventional NMR systems for teaching and some research applications are low purchase and operational costs and portability. This is because a large homogeneous magnet is not required for an EFNMR system. Despite the low detection field in EFNMR, the pre-polarisation methodology sufficiently increases sensitivity such that it is possible to perform a large range of modern and sophisticated NMR techniques.

Computer simulation of the Earth's field in a normal period between reversals is shown below. The tubes represent magnetic field lines, blue when the field points towards the center and yellow when away. The rotation axis of the Earth is centered and vertical. The dense clusters of lines are within the Earth's core.

Fig.6.2. Computer simulation of the Earth's field

6.10. Quantum Computing

In general the physics phenomenon and problems can be approached from two sides, from the world of macroscopic phenomena using some general principles or from the microscopic world of molecules, atoms, elementary particles. A macroscopic approach is used for examples in thermodynamics, while the microscopic approach constitutes the framework of statistical mechanics.

The main aim of statistical mechanics is to describe macroscopic properties of a system on the basis of the behavior of its constituents (atoms, molecules, . . .). The number of particles in a macroscopic volume is usually huge. For instance, 1 mm^3 of air at ambient pressure and temperature contains about 3×10^{16} molecules. It is pointless to even try to follow the evolution of such a big number of particles or to try to describe the state at a certain time, because it is not only impossible to achieve that computationally, but quite generally macroscopic measurements are coarse-grained and thus depend only on the average behavior of the constituents. Statistical mechanics uses probabilistic methods for linking the microscopic and macroscopic worlds. This approach is justified by the law of large numbers, which states that probabilistic results are essentially exact if the number of variables involved is very large.

Quantum statistical mechanics is based on the quantum-mechanical description of many-particle systems; i.e. Quantum statistical mechanics is statistical mechanics applied to quantum mechanical systems.

NMR differs from other implementations of quantum computers in that it uses an ensemble of systems, in this case molecules. The ensemble is initialized to be the thermal equilibrium state (quantum statistical mechanics). In other words, a statistical ensemble is a probability distribution for the state of the system.

In mathematical parlance, this state is given by the density matrix. The density matrix is a representation of a linear operator called the **density operator**. The density matrix is obtained from the density operator by choice of basis in the underlying space. In practice, the terms *density matrix* and *density operator* are often used interchangeably. Both matrix and operator are self-adjoint (or Hermitian), positive semi-definite, of trace one, and may be infinite-dimensional

In quantum mechanics a statistical ensemble (probability distribution over possible quantum states) is described by a **density operator *S*,** which is a non-negative, self-adjoint, and must always have a trace operator of 1 (Tr *S*=1) on the **Hilbert space *H*** describing the quantum system. This can be shown under various mathematical formalisms for quantum mechanics:

$$\boldsymbol{S} = \frac{e^{-\beta H}}{\mathrm{Tr}\,(e^{-\beta H})}$$

where ***H*** is the hamiltonian matrix (the energy operator ***H*** (Hamiltonian)) of an individual molecule and

$$\beta = \frac{1}{KT}$$

where *K* is the Boltzmann constant and *T* the temperature.

NMR quantum computing uses the spin states of molecules as qubits. Initially the approach was to use the spin properties of atoms of particular molecules in a liquid sample as qubits - this is known as liquid state NMR (LSNMR). This approach has since been superseded by solid state NMR (SSNMR) as a means of quantum computation. The ideal picture of liquid state NMR (LSNMR) quantum information processing (QIP) is based on a molecule of which some of its atom's nuclei behave as spin-½ systems. Solid state NMR (SSNMR) differs from LSNMR in that we have a solid state sample, for example a nitrogen vacancy diamond lattice rather than a liquid sample. This has many advantages such as lack of molecular diffusion decoherence, lower temperatures can be achieved to the point of suppressing phonon decoherence and a greater variety of control operations that allow us to overcome one of the major problems of LSNMR that is initialisation.

6.11. Magnetometers

Various magnetometers use NMR effects to measure magnetic fields, including proton precession magnetometers (PPM) (also known as proton magnetometers), and Overhauser magnetometers. See also Earth's field NMR.

Proton precession magnetometers, also known as proton magnetometers, PPMs or simply mags, measure the resonance frequency of protons (hydrogen nuclei) in the magnetic field to be measured, due to nuclear magnetic resonance (NMR). Because the precession frequency depends only on atomic constants and the strength of the ambient magnetic field, the accuracy of this type of magnetometer can reach 1 ppm.

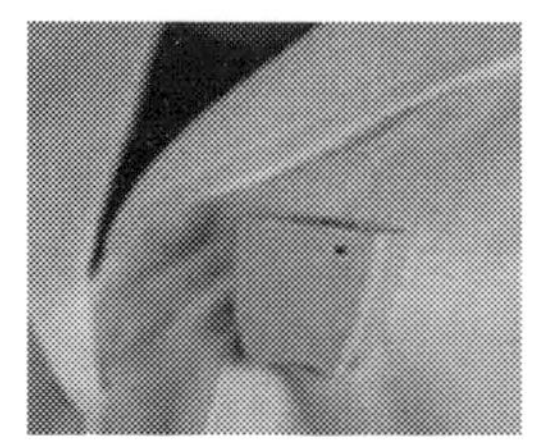

Fig.6.3. NMR pocket monitoring device containing three Hall-effect sensors.

A **magnetometer** is a measuring instrument used to measure the strength or direction of magnetic fields. Some countries, such as the USA, Canada and Australia classify the more sensitive magnetometers as military technology, and control their distribution.

The SI unit of measure for magnetic field strength is the tesla. As this is a very large unit for most practical uses, scientists commonly use the nanotesla (nT) as their working unit of measure. Engineers often measure magnetic fields in Gauss (1 Gauss = 100,000nT, 1 Gauss = 100,000 gamma).

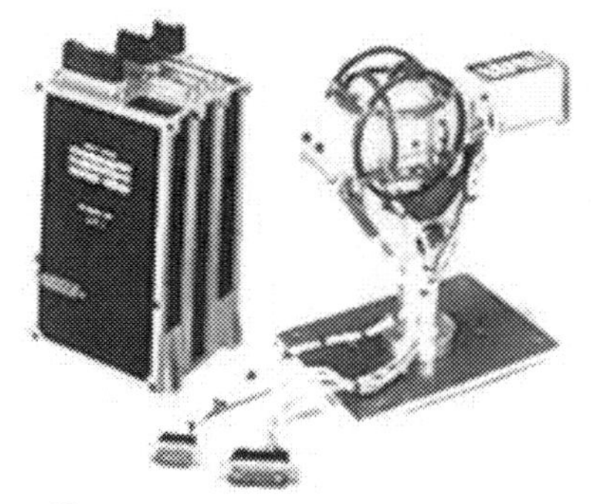

Fig.6.4. Helium Vector Magnetometer (HVM)

7

ANNEXES

Annex 1- Fourier Transforms

Annex 2- Pulse Sequences

Annex 3- Carbon-13 NMR

Annex 4- Two Dimensional 2D technology

Annex 5- Software and Glossary of NMR

Annex 1- Fourier Transforms

A1.1 Introduction

The Fourier Transform FT is a mathematical technique for converting time domain data to frequency domain data, and vice versa. A Fourier transform converts functions from time to frequency domains. An Inverse Fourier Transform IFT converts from the frequency domain to the time domain.

A1.2 The + and - Frequency Problem

A magnetization vector, starting at +x, is rotating about the Z axis in a clockwise direction. The plot of M_x as a function of time is a cosine wave. Fourier transforming this gives peaks at both $+\nu$ and $-\nu$ because the FT cannot distinguish between a $+\nu$ and a $-\nu$ rotation of the vector from the data supplied.

A plot of M_y as a function of time is a -sine function. Fourier transforming this gives peaks at $+\nu$ and $-\nu$ because the FT cannot distinguish between a positive vector rotating at $+\nu$ and a negative vector rotating at $-\nu$ from the data supplied.

The solution is to input both the M_x and M_y into the FT. The FT is designed to handle two orthogonal input functions called the real and imaginary components or Linear and Quadrature Detection

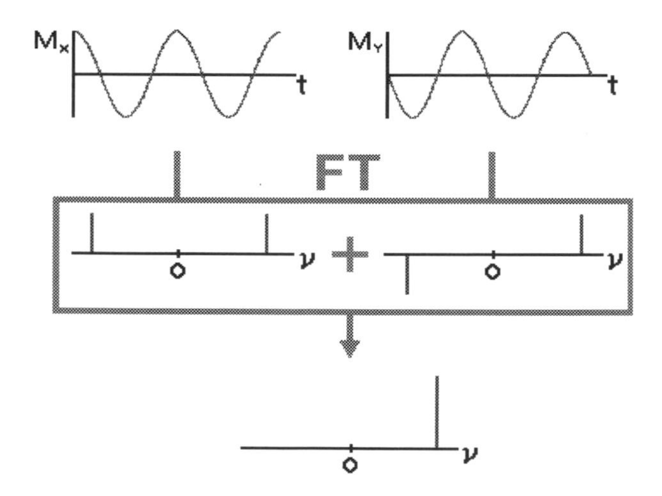

Detecting just the M_x or M_y component for input into the FT is called linear detection. This was the detection scheme on many older NMR spectrometers

and some magnetic resonance imagers. It required the computer to discard half of the frequency domain data.

Detection of both M_x and M_y is called quadrature detection and is the method of detection on modern spectrometers and imagers. It is the method of choice since now the FT can distinguish between $+\nu$ and $-\nu$, and all of the frequency domain data be used.

A1.3 The Fourier Transform

An FT is defined by the integral

$$f(\omega) = \int_{-\infty}^{+\infty} f(t)e^{-i\omega t}dt = \int_{-\infty}^{+\infty} f(t)[\cos(\omega t) - i\sin(\omega t)]dt$$

Think of $f(\omega)$ as the overlap of $f(t)$ with a wave of frequency ω.

$$f(\omega) = \sum_{-\infty}^{+\infty} f(t)[\cos(\omega t) - i\sin(\omega t)]$$

This is easy to picture by looking at the real part of $f(\omega)$ only.

$$f(\omega) = \sum_{-\infty}^{+\infty} f(t)\cos(\omega t)$$

Consider the function of time, $f(t) = \cos(4t) + \cos(9t)$.

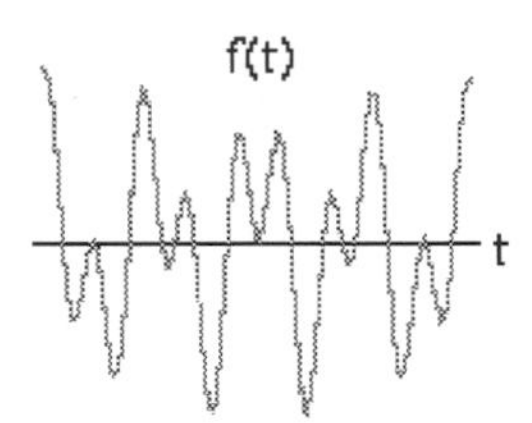

To understand the FT, examine the product of $f(t)$ with $\cos(\omega t)$ for ω values between 1 and 10, and then the summation of the values of this product between 1 and 10 seconds. The summation will only be examined for time values between 0 and 10 seconds (ω=1-10)

See the list of drawing for cases of (ω=1 to10)as given below:

list of drawing for cases of (ω=1 to10)

1) ω =1

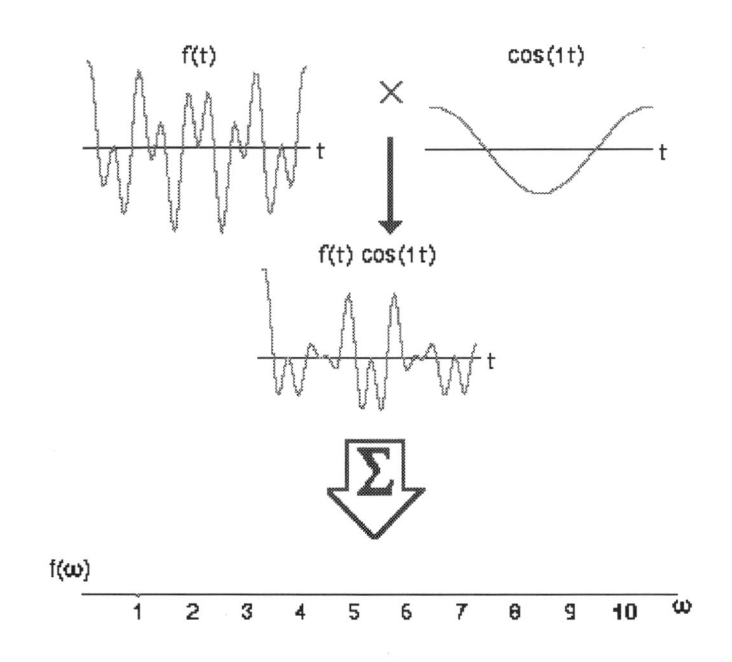

2) ω =2

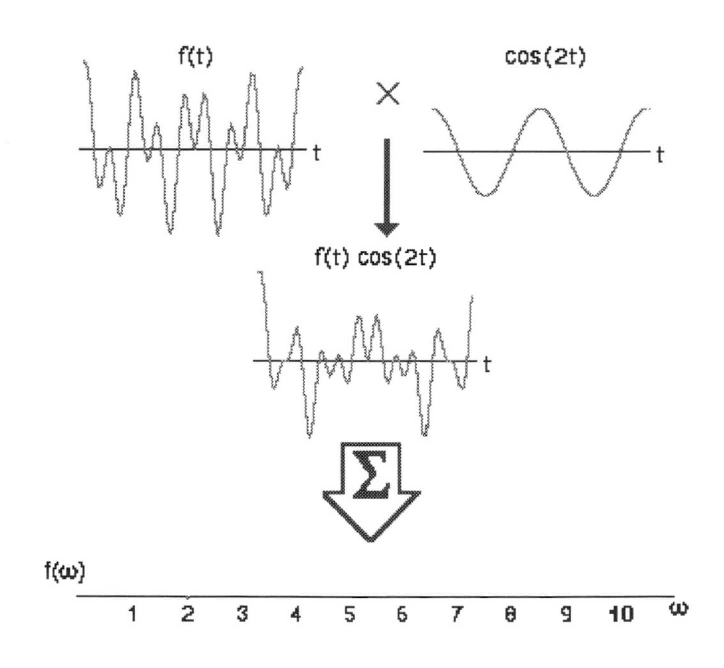

3) ω =3

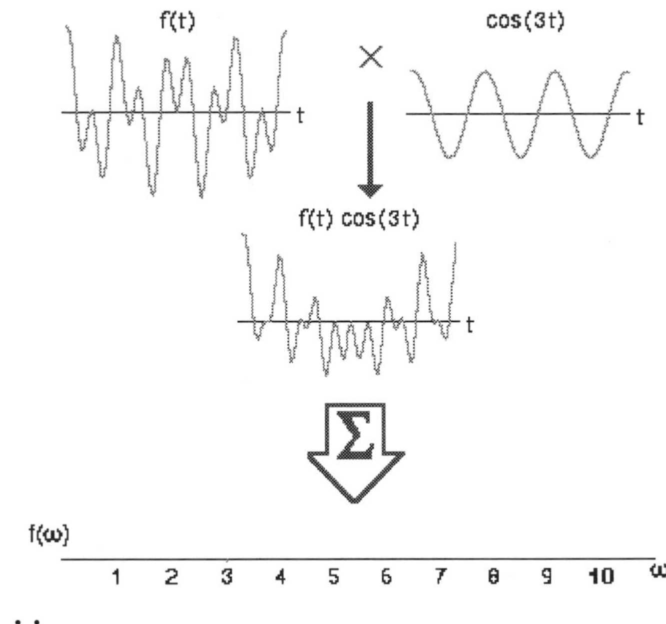

f(t)
cos(3t)
t
t
f(t) cos(3t)
t
Σ
f(ω)
1 2 3 4 5 6 7 8 9 10
ω

: : : : :
: : : : :
: : : : :

10) ω =1

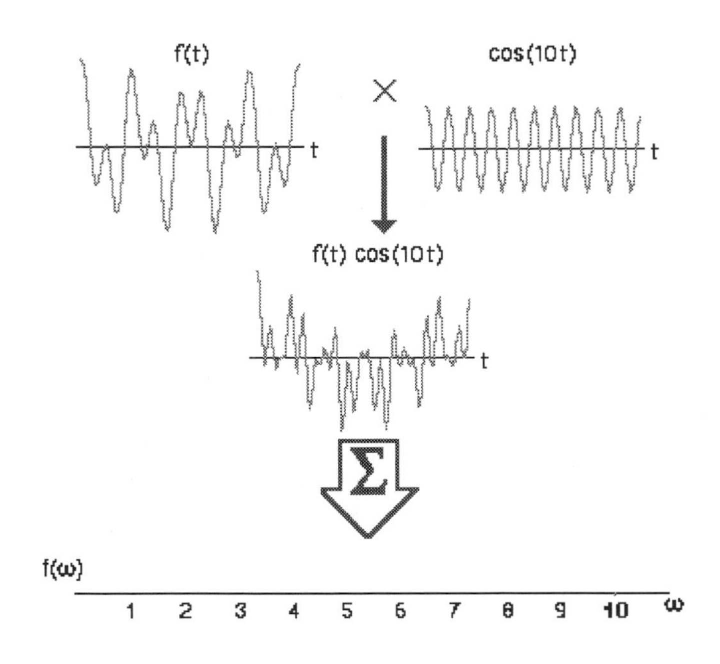

f(t)
cos(10t)
t
t
f(t) cos(10t)
t
Σ
f(ω)
1 2 3 4 5 6 7 8 9 10
ω

A1.4 Phase Correction

The actual FT will make use of an input consisting of a Real and an Imaginary part. You can think of M_x as the Real input, and M_y as the Imaginary input. The resultant output of the FT will therefore have a Real & an Imaginary component, too. Consider the following function:

$$f(t) = e^{-at}\, e^{-i2pnt}$$

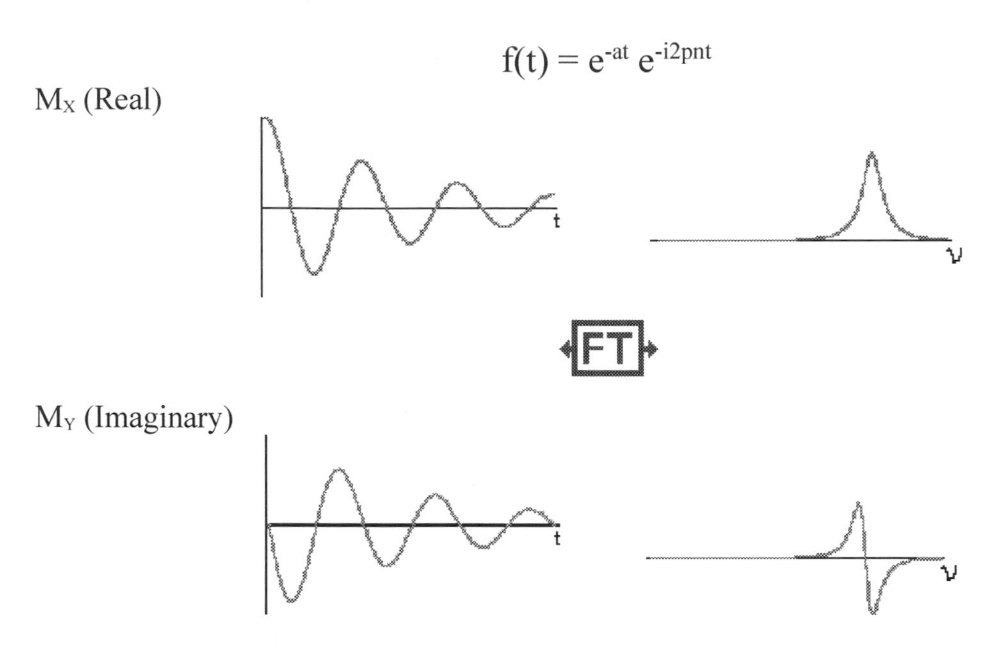

In FT NMR spectroscopy, the real output of the FT is taken as the frequency domain spectrum. To see an esthetically pleasing (absorption) frequency domain spectrum, we want to input a cosine function into the real part and a sine function into the imaginary parts of the FT. This is what happens if the cosine part is input as the imaginary and the sine as the real.

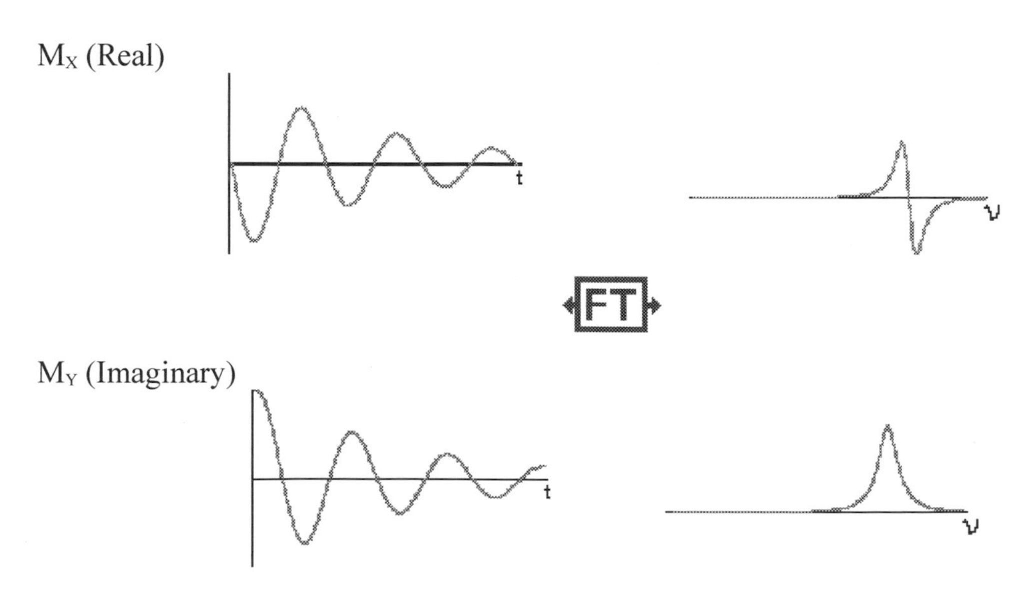

In an ideal NMR experiment all frequency components contained in the recorded FID have no phase shift. In practice, during a real NMR experiment a phase correction must be applied to either the time or frequency domain spectra to obtain an absorption spectrum as the real output of the FT.

$$\begin{bmatrix} RE'' \\ IM'' \end{bmatrix} = \begin{bmatrix} Cos(\phi) & Sin(\phi) \\ -Sin(\phi) & Cos(\phi) \end{bmatrix} \begin{bmatrix} RE \\ IM \end{bmatrix}$$

If the above mentioned FID is recorded such that there is a 45° phase shift in the real and imaginary FIDs, the coordinate transformation matrix can be used with ϕ = - 45°. The corrected FIDs look like a cosine function in the real and a sine in the imaginary.

Fourier transforming the phase corrected FIDs gives an absorption spectrum for the real output of the FT.

The phase shift also varies with frequency, so the NMR spectra require both constant and linear corrections to the phasing of the Fourier transformed signal.

$$\Phi = m\,n + b$$

Constant phase corrections, b, arise from the inability of the spectrometer to detect the exact M_x and M_y. Linear phase corrections, m, arise from the inability of the spectrometer to detect transverse magnetization starting immediately after the RF pulse. The following drawing depicts the greater loss of phase in a high frequency FID when the initial yellow section is lost.

From the practical point of view, the phase correction is applied in the frequency domain rather then in the time domain because we know that a real frequency domain spectrum should be composed of all positive peaks. We can therefore adjust b and m until all positive peaks are seen in the real output of the Fourier transform.

In magnetic resonance, the M_x or M_y signals are displayed. A magnitude signal might occasionally be used in some applications. The magnitude signal is equal to the square root of the sum of the squares of M_x and M_y.

A1.5 Fourier Pairs

To better understand FT NMR functions, you need to know some common Fourier pairs. Here are a few Fourier pairs which are useful in NMR. The amplitude of the Fourier pairs has been neglected since it is not relevant in NMR.

A. Constant value at all time

A DC offset or constant value.

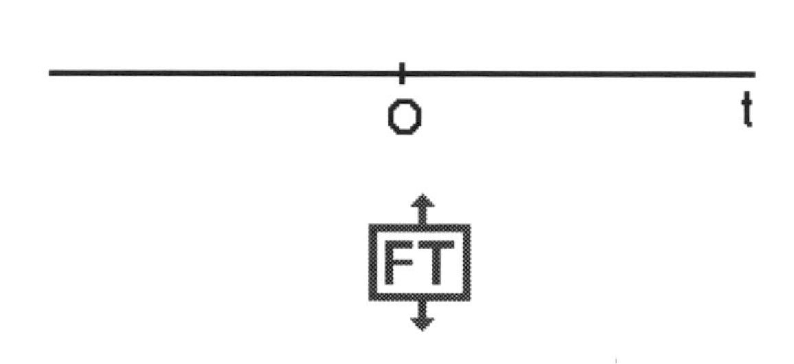

A delta function at zero.

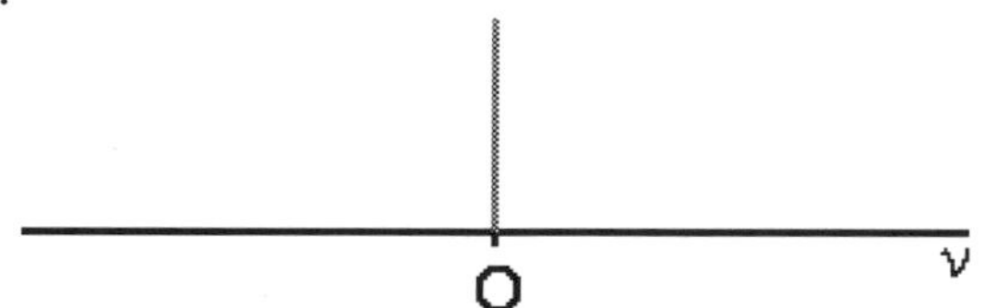

B. Real: cos(2π ν t), Imaginary: -sin(2π ν t)

Real: cos(2π ν t), Imaginary: -sin(2π ν t)

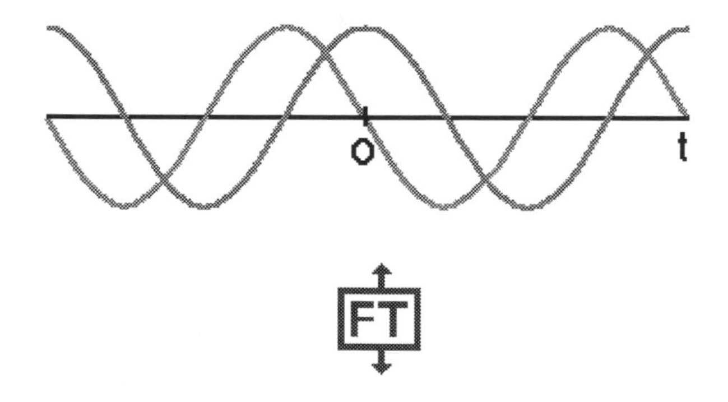

A delta function at ν.

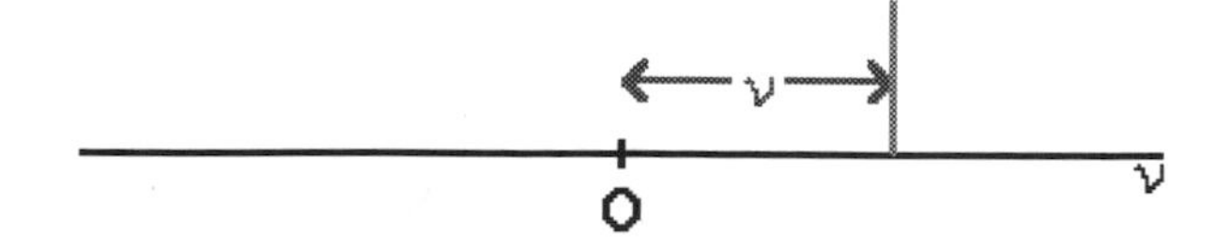

C. Comb Function (A series of delta functions separated by T)

Comb Function (A series of delta functions separated by T)

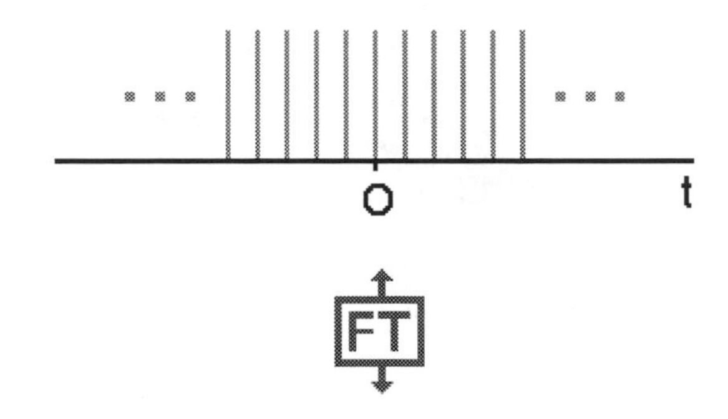

A comb function with separation 1/T.

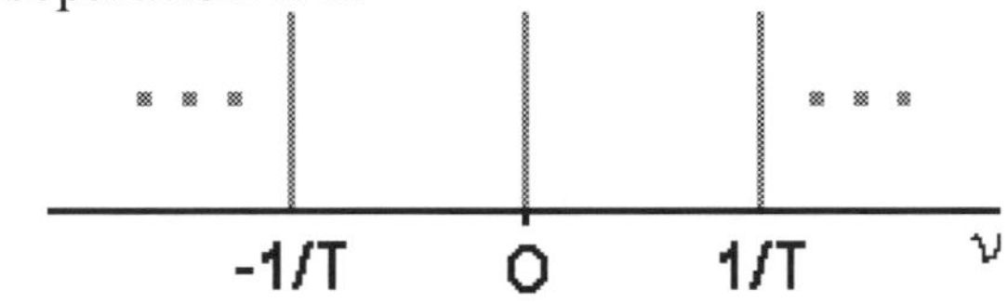

D. Exponential Decay: e^{-at} for $t > 0$.

e^{-at} for $t > 0$

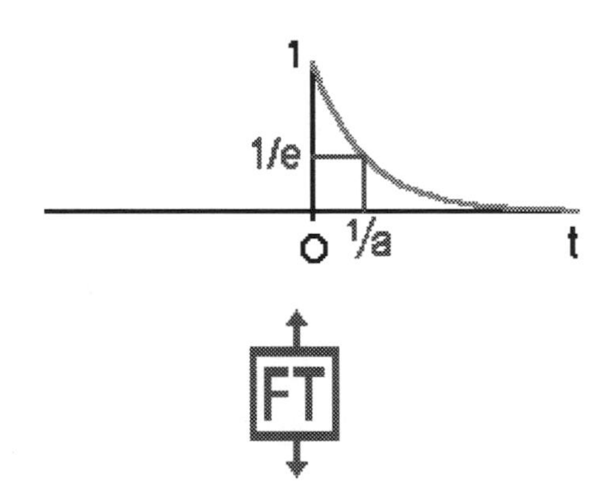

Lorentzian
RE: $a^2/(a^2 + 4\pi^2\nu^2)$
IM: $2a^2\pi\nu/(a^2 + 4\pi^2\nu^2)^2$

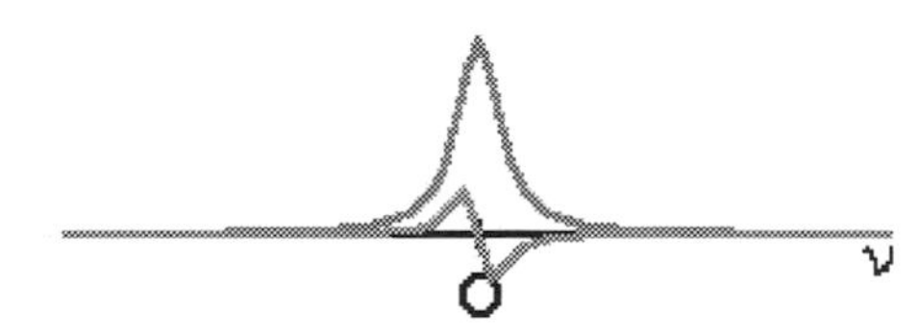

E. A square pulse starting at 0 that is T seconds long.

Rectangular pulse of width T starting at 0.

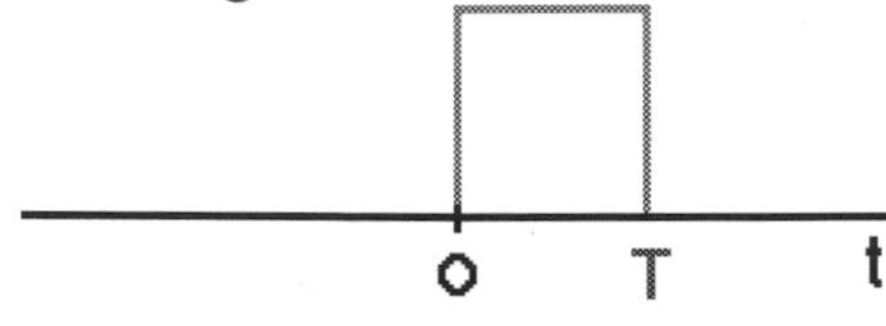

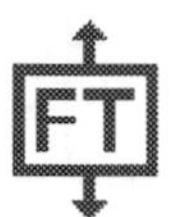

Sinc RE: $(\sin(2\pi\nu t))/(2\pi\nu t)$ IM: $-(\sin^2(2\pi\nu t))/(\pi\nu t)$

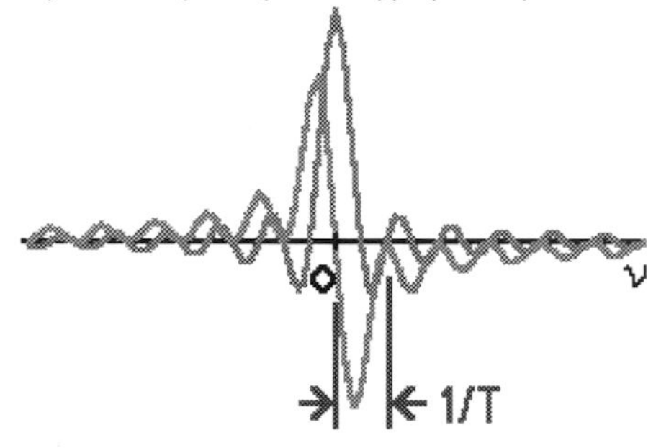

F. Gaussian: exp(-at²)

Gaussian: $\exp(-at^2)$

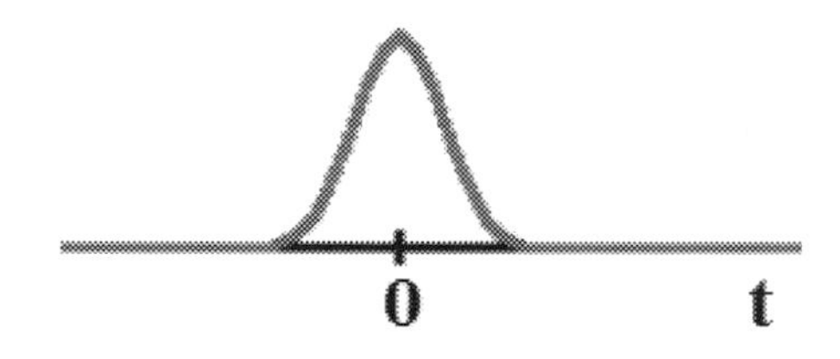

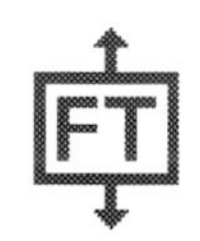

Gaussian: $\exp(-\pi^2\nu^2/a)$

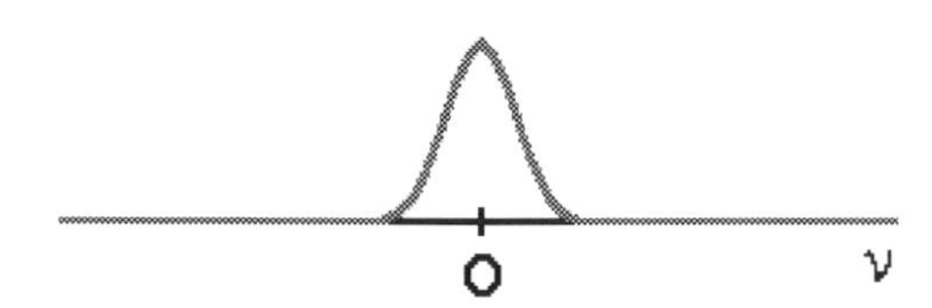

A1.6 Convolution Theorem

To the magnetic resonance scientist, the most important theorem concerning Fourier transforms is the convolution theorem. The convolution theorem says that the FT of a convolution of two functions is proportional to the products of the individual Fourier transforms, and vice versa.

If $f(\omega) = FT(\ f(t)\ )$ and $g(\omega) = FT(\ g(t)\ )$

then $f(\omega)\ g(\omega) = FT(\ g(t) \otimes f(t)\ )$ and $f(\omega) \otimes g(\omega) = FT(\ g(t)\ f(t)\ )$

It will be easier to see this with pictures. In the animation window we are trying to find the FT of a sine wave which is turned on and off. The convolution theorem tells us that this is a sinc function at the frequency of the sine wave.

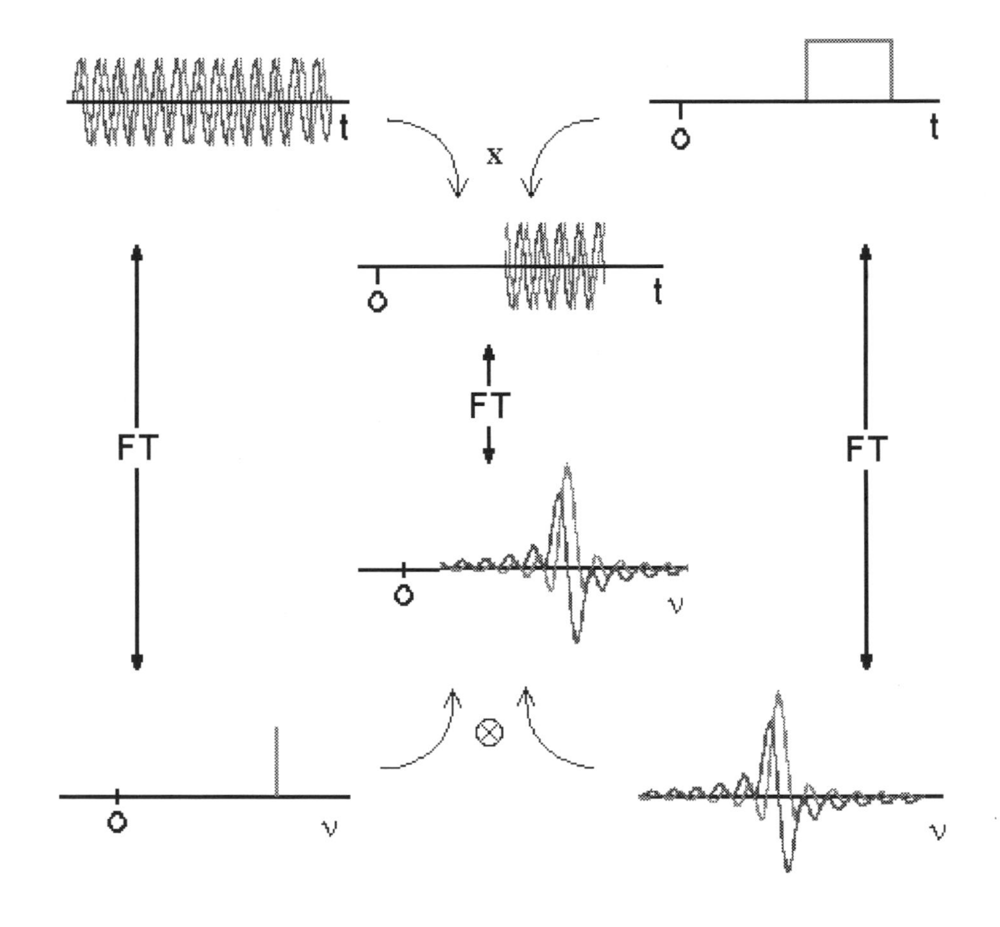

Another application of the convolution theorem is in noise reduction. With the convolution theorem it can be seen that the convolution of an NMR spectrum with a Lorentzian function is the same as the Fourier Transform of multiplying the time domain signal by an exponentially decaying function.

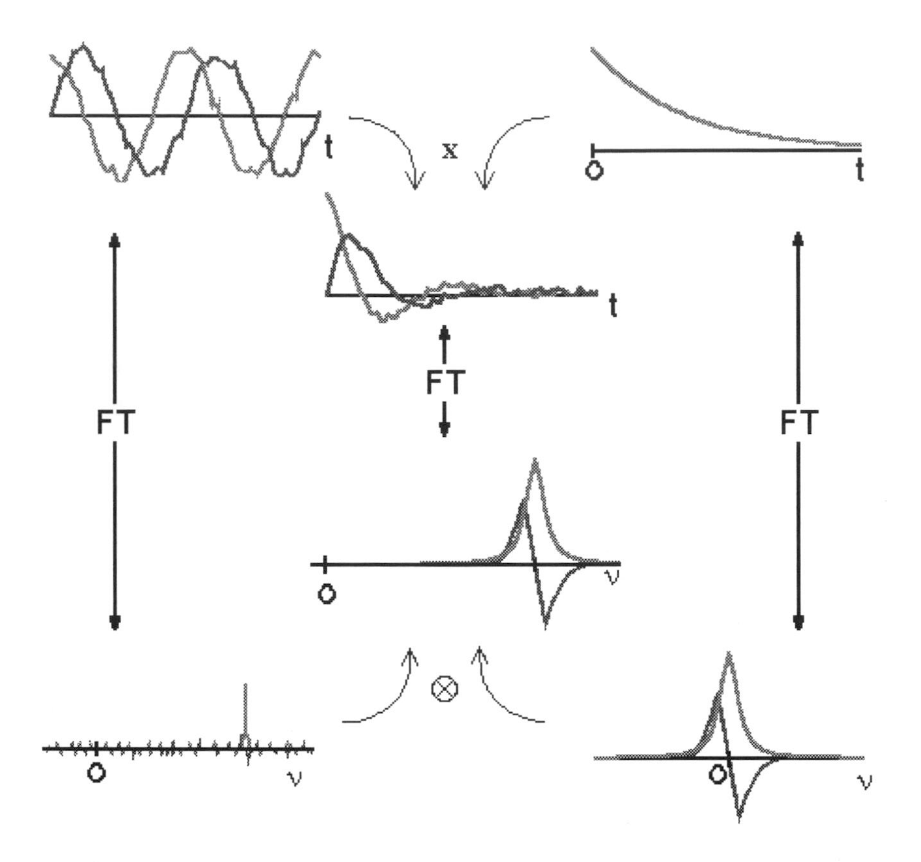

A1.7 The Digital FT

In a nuclear magnetic resonance spectrometer, the computer does not see a continuous FID, but rather an FID which is sampled at a constant interval. Each data point making up the FID will have discrete amplitude and time values. Therefore, the computer needs to take the FT of a series of delta functions which vary in intensity.

Original continuous FID.

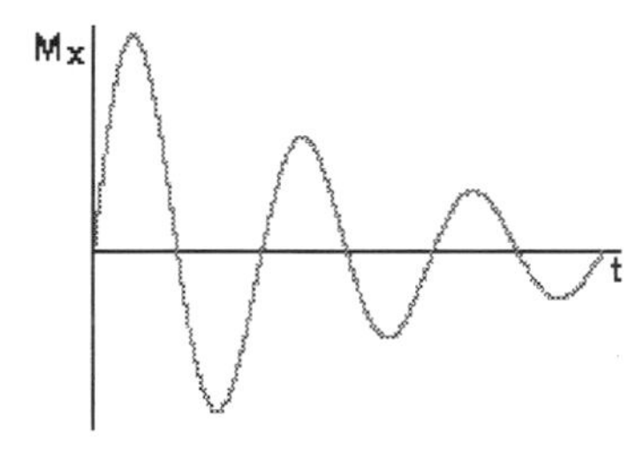

Sampled FID seen by FT algorithm in computer.

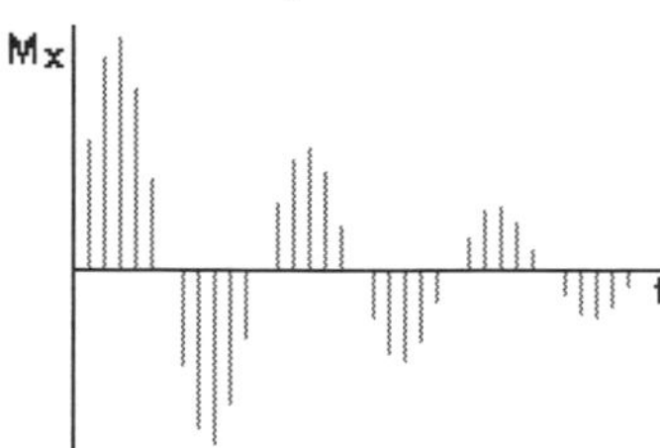

What is the FT of a signal represented by this series of delta functions? The answer will be addressed in the next heading, but first some information on relationships between the sampled time domain data and the resultant frequency domain spectrum. An n point time domain spectrum is sampled at dt and takes a time t to record. The corresponding complex frequency domain spectrum that the discrete FT produces has n points, a width f, and resolution df. The relationships between the quantities are as follows.

$$f = (1/dt) \text{ and } df = (1/t)$$

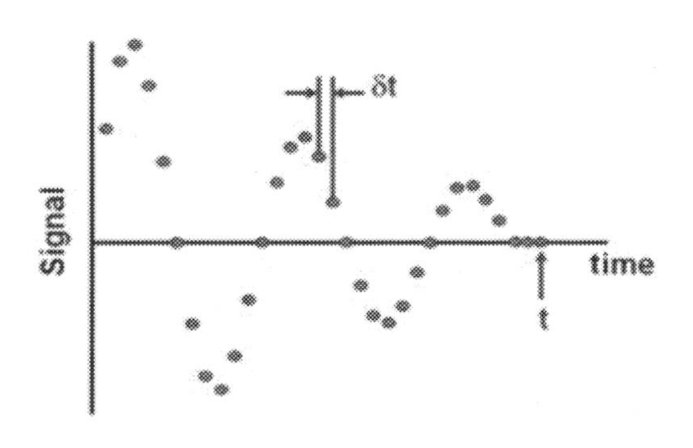

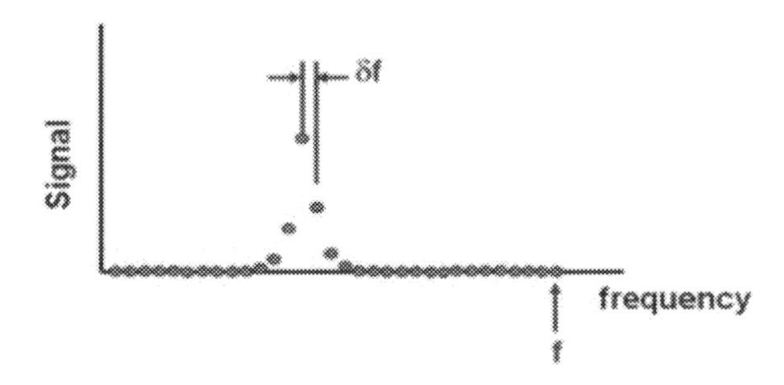

A1.8 Sampling Error

The wrap around problem or artifact in a nuclear magnetic resonance spectrum is the appearance of one side of the spectrum on the opposite side. In terms of a one dimensional frequency domain spectrum, wrap around is the occurrence of a low frequency peak which occurs on the high frequency side of the spectrum.

The convolution theorem can explain why this problem results from sampling the transverse magnetization at too slow a rate. First, observe what the FT of a correctly sampled FID looks like. With quadrature detection, the spectral width is equal to the inverse of the sampling frequency, or the width of the green box in the animation window.

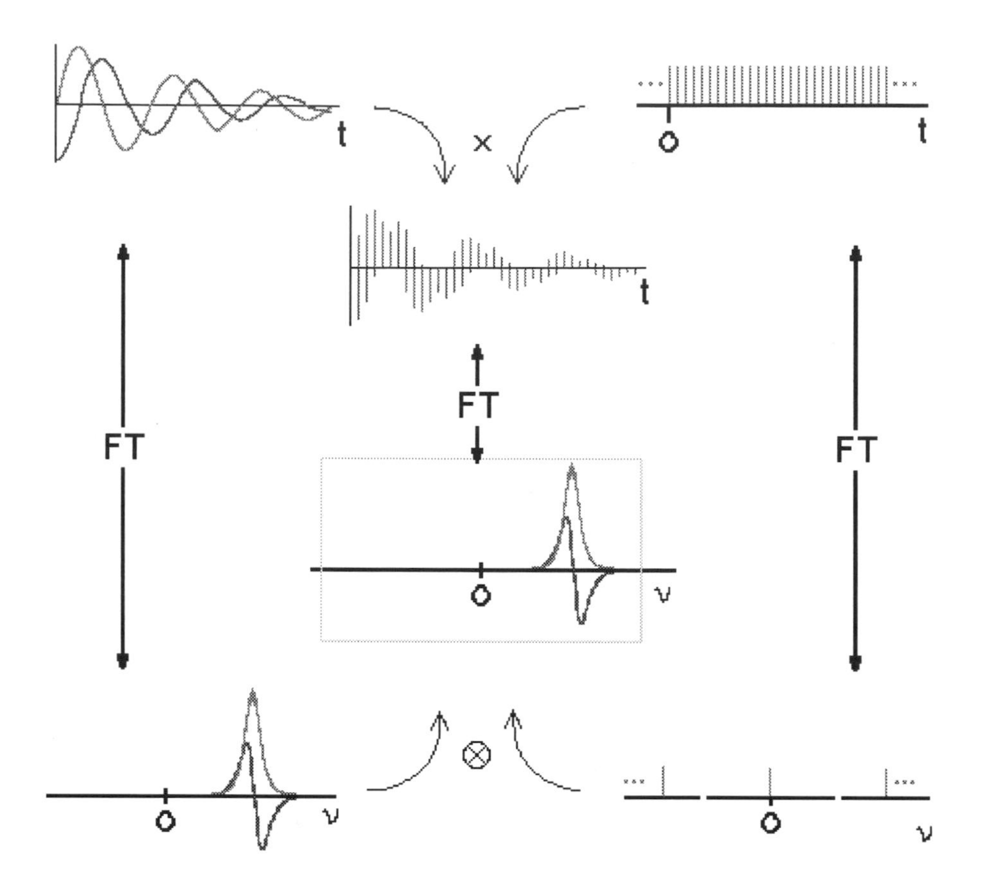

When the sampling frequency is less than the spectral width, wrap around occurs.

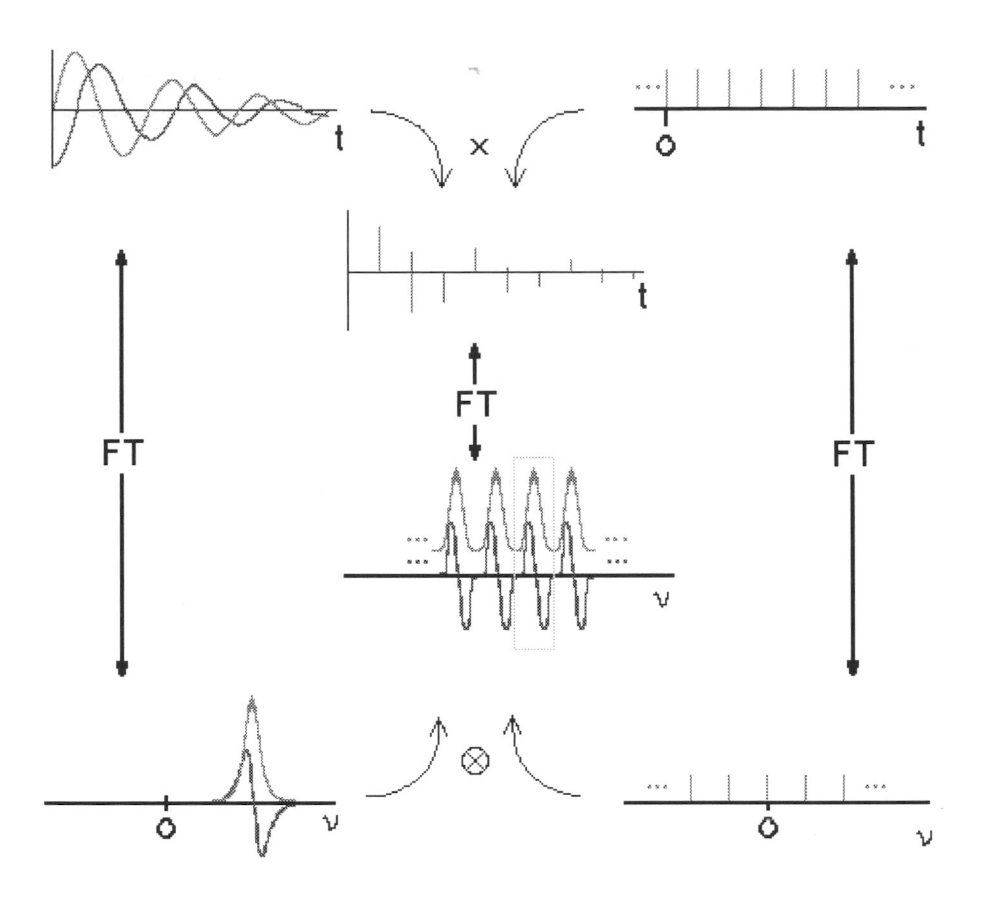

A1.9 The Two-Dimensional FT

The two-dimensional Fourier transform (2-DFT) is an FT performed on a two dimensional array of data.

Consider the two-dimensional array of data depicted in the animation window.

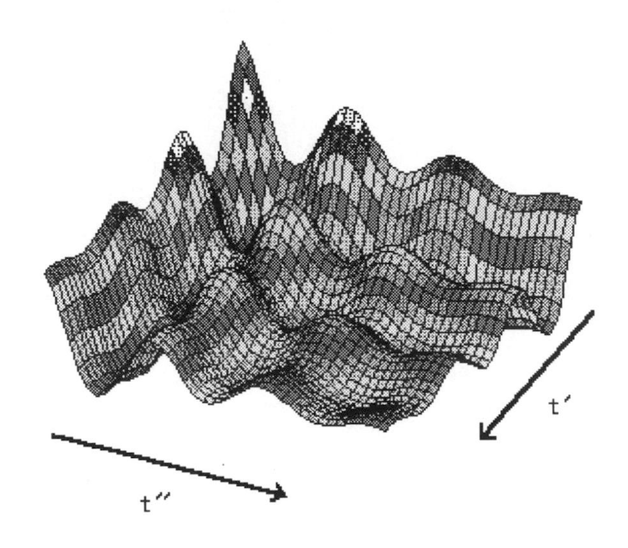

This data has a t' and a t" dimension. A FT is first performed on the data in one dimension and then in the second. The first set of Fourier transforms are performed in the t' dimension to yield an f' by t" set of data.

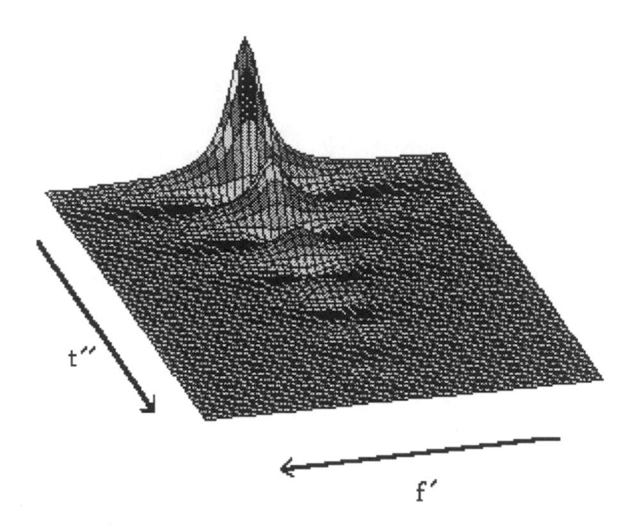

The second set of Fourier transforms is performed in the t" dimension to yield an f' by f" set of data.

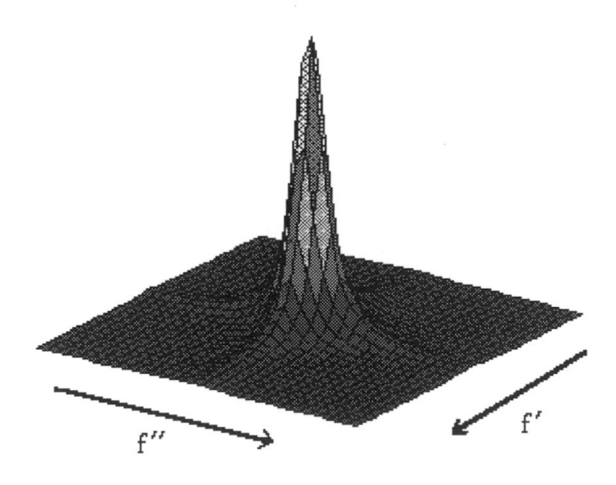

The 2-DFT is required to perform state-of-the-art MRI. In MRI, data is collected in the equivalent of the t' and t" dimensions, called k-space. This raw data is Fourier transformed to yield the image which is the equivalent of the f' by f" data described above.

Annex 2- Pulse Sequences

A2.1 Introduction

You have seen in Annex 1 how a time domain signal can be converted into a frequency domain signal. In this chapter you will learn a few of the ways that a time domain signal can be created. Three methods are presented here, but there are an infinite number of possibilities. These methods are called pulse sequences. A pulse sequence is a set of RF pulses applied to a sample to produce a specific form of NMR signal.

A2.2 The 90-FID Sequence

In the 90-FID pulse sequence, net magnetization is rotated down into the X'Y' plane with a 90° pulse.

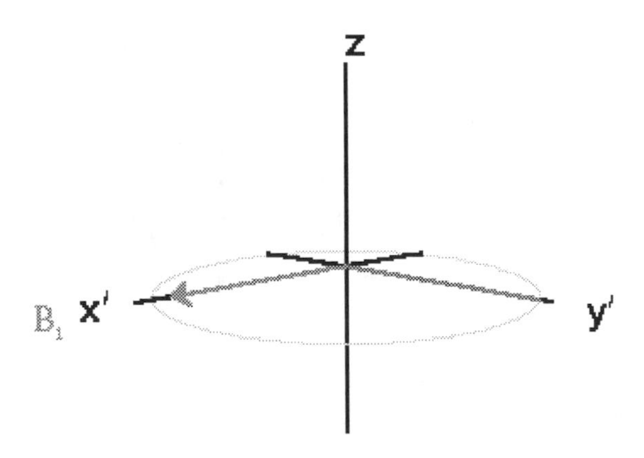

The net magnetization vector begins to process about the +Z axis. The magnitude of the vector also decays with time.

A timing diagram is a multiple axis plot of some aspect of a pulse sequence versus time. A timing diagram for a 90-FID pulse sequence has a plot of RF energy versus time and another for signal versus time.

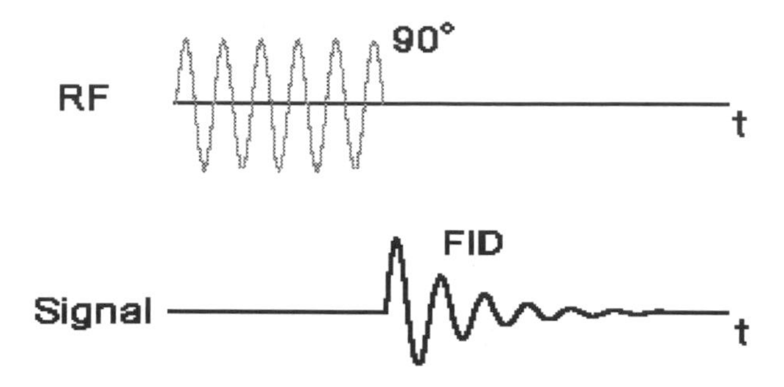

When this sequence is repeated, for example when signal-to-noise improvement is needed, the amplitude of the signal after being Fourier transformed (S) will depend on T_1 and the time between repetitions, called the repetition time (TR), of the sequence. In the signal equation below, *k* is a proportionality constant and ρ is the density of spins in the sample.

$$S = k\, \rho(1 - e^{-TR/T1})$$

A2.3 The Spin-Echo Sequence

Another commonly used pulse sequence is the spin-echo pulse sequence. Here a 90° pulse is first applied to the spin system. The 90° degree pulse rotates the magnetization down into the X'Y' plane. The transverse magnetization begins to dephase. At some point in time after the 90° pulse, a 180° pulse is applied. This pulse rotates the magnetization by 180° about the X' axis. The 180° pulse causes the magnetization to at least partially rephase and to produce a signal called an echo. A timing diagram shows the relative positions of the two radio frequency pulses and the signal.

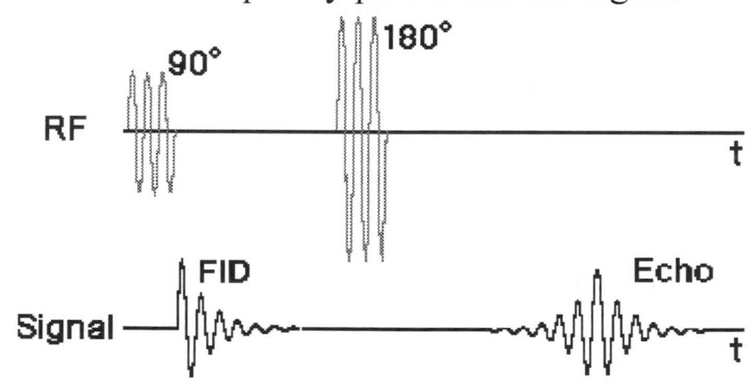

The signal equation for a repeated spin echo sequence as a function of the repetition time, TR, and the echo time (TE) defined as the time between the 90° pulse and the maximum amplitude in the echo is

$$S = k\, \rho(1 - e^{-TR/T1})\, e^{-TE/T2}$$

A2.4 The Inversion Recovery Sequence

An inversion recovery pulse sequence can also be used to record an NMR spectrum. In this sequence, a 180° pulse is first applied. This rotates the net magnetization down to the -Z axis. The magnetization undergoes spin-lattice relaxation and returns toward its equilibrium position along the +Z axis. Before it reaches equilibrium, a 90° pulse is applied which rotates the longitudinal magnetization into the XY plane. In this example, the 90° pulse is applied shortly after the 180° pulse. Once magnetization is present in the XY plane it rotates about the Z axis and dephases giving a FID.

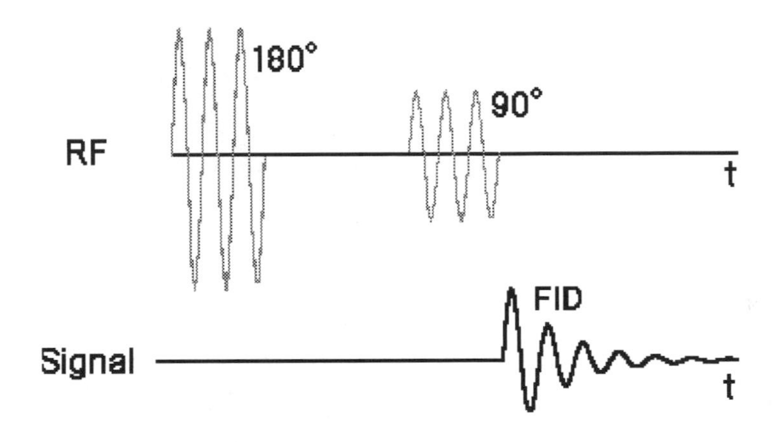

Once again, the timing diagram shows the relative positions of the two radio frequency pulses and the signal. The signal as a function of TI when the sequence is not repeated is

$$S = k\,\rho(\,1 - 2e^{-TI/T1}\,)$$

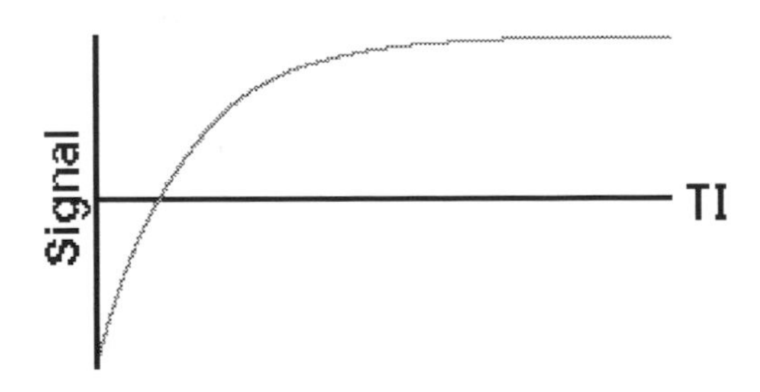

It should be noted at this time that the zero crossing of this function occurs for TI = T_1 *ln*2.

Annex 3- Carbon-13 NMR

A3.1 Introduction

Many of the molecules studied by NMR contain carbon. Unfortunately, the carbon-12 nucleus does not have a nuclear spin, but the carbon-13 (C-13) nucleus does due to the presence of an unpaired neutron. Carbon-13 nuclei make up approximately one percent of the carbon nuclei on earth. Therefore, carbon-13 NMR spectroscopy will be less sensitive (have a poorer SNR) than hydrogen NMR spectroscopy. With the appropriate concentration, field strength, and pulse sequences, however, carbon-13 NMR spectroscopy can be used to supplement the previously described hydrogen NMR information. Advances in superconducting magnet design and RF sample coil efficiency have helped make carbon-13 spectroscopy routine on most NMR spectrometers.

The *sensitivity* of an NMR spectrometer is a measure of the minimum number of spins detectable by the spectrometer. Since the NMR signal increases as the population difference between the energy levels increases, the sensitivity improves as the field strength increases. The sensitivity of carbon-13 spectroscopy can be increased by any technique which increases the population difference between the lower and upper energy levels, or increases the density of spins in the sample. The population difference can be increased by decreasing the sample temperature or by increasing the field strength. Several techniques for increasing the carbon-13 signal have been reported in the NMR literature.

Unfortunately, or fortunately, depending on your perspective, the presence of spin-spin coupling between a carbon-13 nucleus and the nuclei of the hydrogen atoms bonded to the carbon-13, splits the carbon-13 peaks and causes an even poorer signal-to-noise ratio. This problem can be addressed by the use of a technique known as decoupling, addressed in the next section.

A3.2 Decoupling

The signal-to-noise ratio in an NMR spectrometer is related to the population difference between the lower and upper spin state. The larger this difference the larger the signal. We know from chapter 3 that this difference is proportional to the strength of the B_o magnetic field.

To understand decoupling, consider the familiar hydrogen NMR spectrum of $HC\text{-}(CH_2CH_3)_3$.

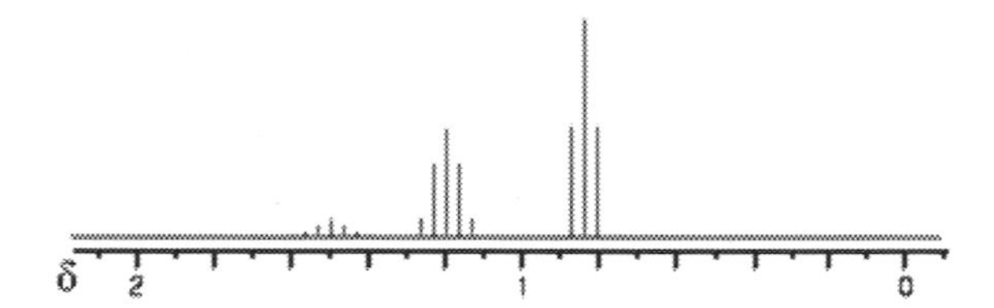

The HC hydrogen peaks are difficult to see in the spectrum due to the splitting from the 6 $-CH_2-$ hydrogens. If the effect of the 6 $-CH_2-$ hydrogens could be removed, we would lose the 1:6:15:20:15:6:1 splitting for the HC hydrogen and get one peak.

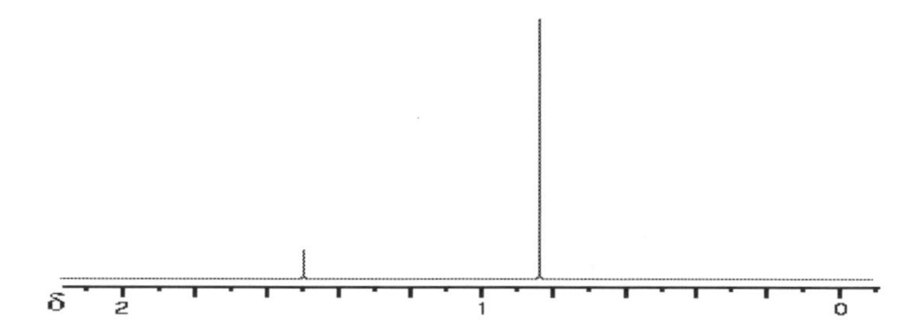

We would also lose the 1:3:1 splitting for the CH_3 hydrogens and get one peak. The process of removing the spin-spin splitting between spins is called decoupling. Decoupling is achieved with the aid of a saturation pulse. If the affect of the HC hydrogen is removed, we see the following spectrum.

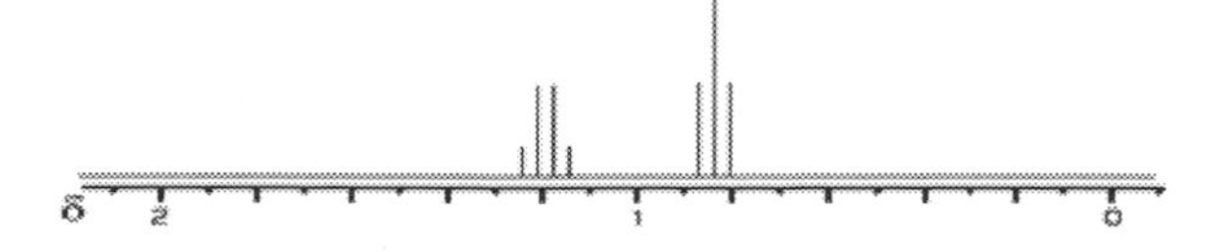

Similarly, if the affect of the $-CH_3$ hydrogens is removed, we see this spectrum.

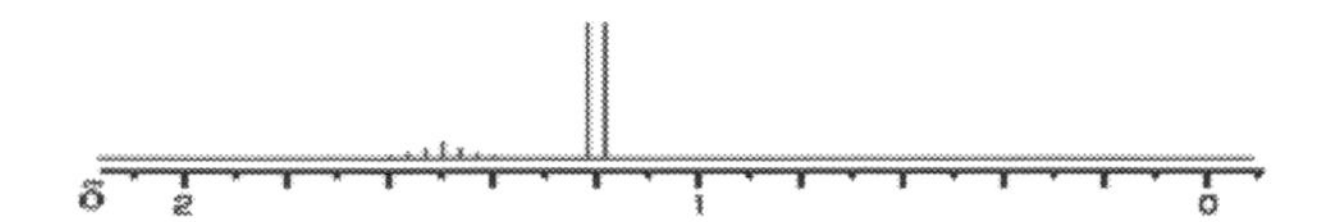

A saturation pulse is a relatively low power B_1 field left on long enough for all magnetization to disappear. A saturation pulse applied along X' rotates magnetization clockwise about X' several times. As the magnetization is rotating, T_2 processes cause the magnetization to dephase. At the end of the pulse there is no net Z, X, or Y magnetization. It is easier to see this behavior with the use of plots of M_Z, $M_{X'}$, and $M_{Y'}$ as a function of time.

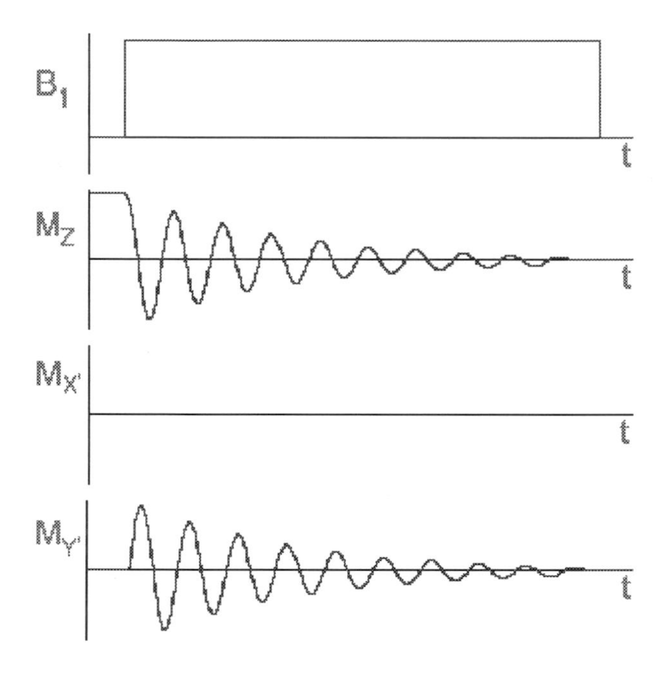

Since the B_1 pulse is long, its frequency content is small. It therefore can be set to coincide with the location of the $-CH_2-$ quartet and saturate the $-CH_2-$ spin system. By saturating the $-CH_2-$ spins, the $-CH_2-$ peaks and the splittings disappear, causing the height of the now unsplit HC- and $-CH_3$ peaks to be enhanced.

Now that the concept of decoupling has been introduced, consider the carbon-13 spectrum from CH_3I. The NMR spectrum from the carbon-13 nucleus will yield one absorption peak in the spectrum.

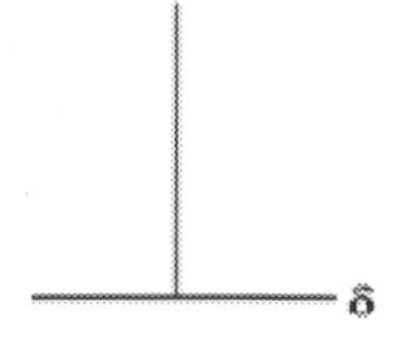

Adding the nuclear spin from one hydrogen will split the carbon-13 peak into two peaks.

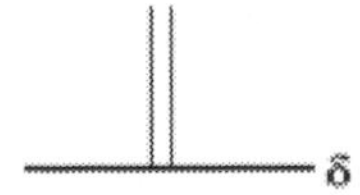

Adding one more hydrogen will split each of the two carbon-13 peaks into two, giving a 1:2:1 ratio.

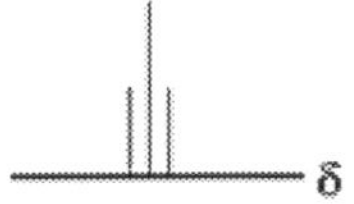

The final hydrogen will split each of the previous peaks, giving a 1:3:3:1 ratio.

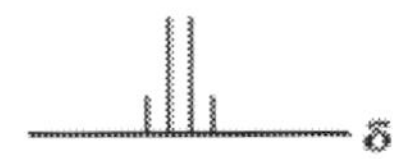

If the hydrogen spin system is saturated, the four lines collapse into a single line having an intensity which is eight times greater than the outer peak in the 1:3:3:1 quartet since 1+3+3+1=8 .

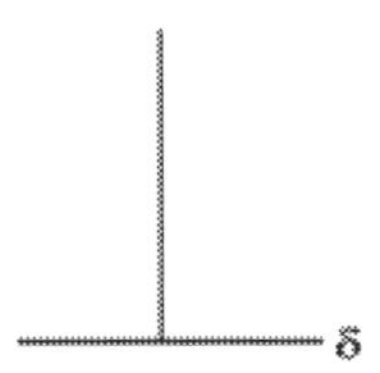

In reality, we see a single line with a relative intensity of 24. Where did the extra factor of three come from?

A3.3 NOE

The answer to the question raised in the previous paragraph is the nuclear Overhauser effect (NOE). To understand the NOE, consider a set of coupled hydrogen and carbon-13 nuclei.

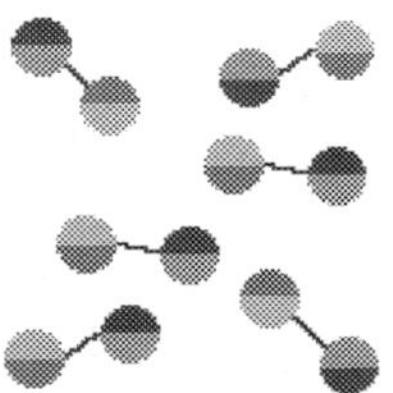

Assume that the red-green nuclei are carbon-13 and the blue-pink nuclei are hydrogen.

T_{1CC} is T_1 relaxation due to interactions between carbon-13 nuclei. T_{1HH} is T_1 relaxation due to interactions between hydrogen nuclei. T_{1CH} is T_1 relaxation due to interactions between carbon-13 and hydrogen nuclei.

$M_Z(C)$ is the magnetization due to carbon-13 nuclei. $M_o(C)$ is the equilibrium magnetization of carbon-13. $M_Z(H)$ is the magnetization due to hydrogen nuclei. $M_o(H)$ is the equilibrium magnetization of hydrogen.

The equations governing the change in the Z magnetization with time are:

$$\frac{dM_z(H)}{dt} = -\frac{M_z(H) - M_0(H)}{T_{1HH}} - \frac{M_z(C) - M_0(C)}{T_{1CH}}$$

$$\frac{dM_z(C)}{dt} = -\frac{M_z(C) - M_0(C)}{T_{1CC}} - \frac{M_z(H) - M_0(H)}{T_{1CH}}$$

If we saturate the hydrogen spins, $M_Z(H) = 0$.

$$\frac{dM_z(H)}{dt} = -\frac{0 - M_0(H)}{T_{1HH}} - \frac{M_z(C) - M_0(C)}{T_{1CH}}$$

$$\frac{dM_z(C)}{dt} = -\frac{M_z(C) - M_0(C)}{T_{1CC}} - \frac{0 - M_0(H)}{T_{1CH}}$$

Letting the system equilibrate, $d\,M_Z(C)/dt = 0$.

$$\frac{dM_z(H)}{dt} = +\frac{M_0(H)}{T_{1HH}} - \frac{M_z(C) - M_0(C)}{T_{1CH}}$$

$$0 = -\frac{M_z(C) - M_0(C)}{T_{1CC}} + \frac{M_0(H)}{T_{1CH}}$$

Rearranging the previous equation, we obtain an equation for $M_Z(C)$.

$$M_z(C) = M_0(C) + \frac{M_0(H)\,T_{1CC}}{T_{1CH}}$$

Note that $M_Z(C)$ has increased by $M_o(H)\ T_{1CC} / T_{1CH}$ which is approximately 2 $M_o(C)$, giving a total increase of a factor of 3 relative to the total area of the undecoupled peaks. This explains the extra factor of three (for a total intensity increase of 24) for the carbon-13 peak when hydrogen decoupling is used in the carbon-13 spectrum of CH_3I.

The following spin-echo sequence has been modified to decouple the hydrogen spins from the carbon-13 spins. The signal is recorded as the second half of the echo.

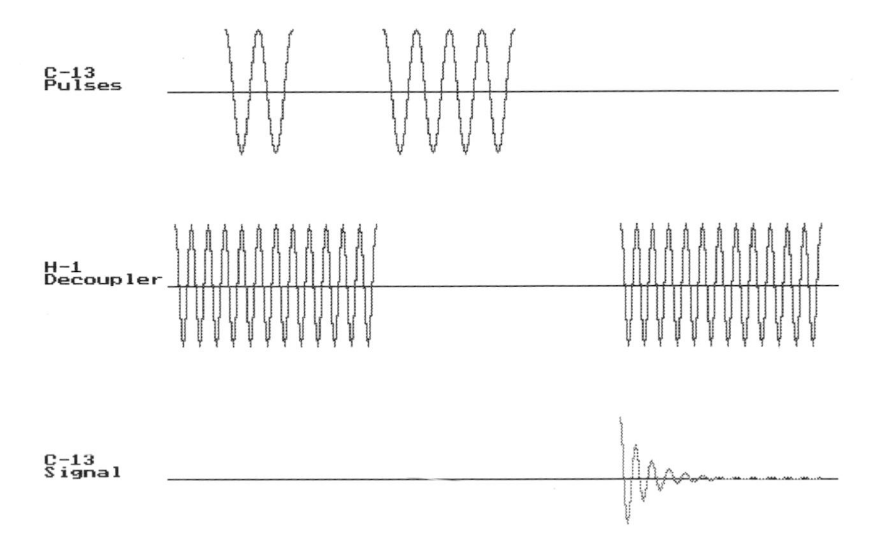

A3.4 Population Inversion

Another method of improving the NMR signal in systems with spin-spin coupling is population inversion. To understand the concept of a population inversion, as described in Chapter 2 that Boltzmann statistics tell us that there are more spins in the lower spin state than the upper one of a two spin state system. Population inversion is the interchange of the populations of these two spin states so that there are more spins in the upper state then the lower one.

To understand how a population inversion improves the signal-to-noise ratio in a spectrum, consider the CHI_3 molecule. CHI_3 will have four energy levels (L_1, L_2, L_3, and L_4) due to C-H spin-spin coupling.

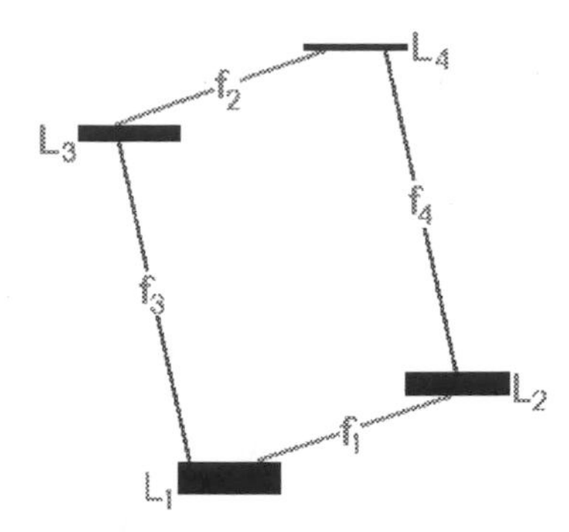

There are two carbon-13 absorption frequencies f_1 and f_2 and two hydrogen absorption frequencies f_3 and f_4. The population distribution between the four levels is such that the lowest state has the greatest population and the highest the lowest population. The two intermediate states will have populations between the outer two as indicated by the thickness of the levels in the accompanying diagram. The four lines in the spectrum will have intensities related to the population difference between the two levels spanned by the frequency. The two carbon-13 absorption lines (f_1 and f_2) will have a lower intensity than the hydrogen lines (f_3 and f_4) due to the smaller population difference between the two states joined by f_1 and f_2.

If the populations of L_3 and L_1 are inverted or interchanged with a frequency selective 180 degree pulse at f_3, the signal at f_2 will be enhanced because of the greater population difference between the states joined by f_2.

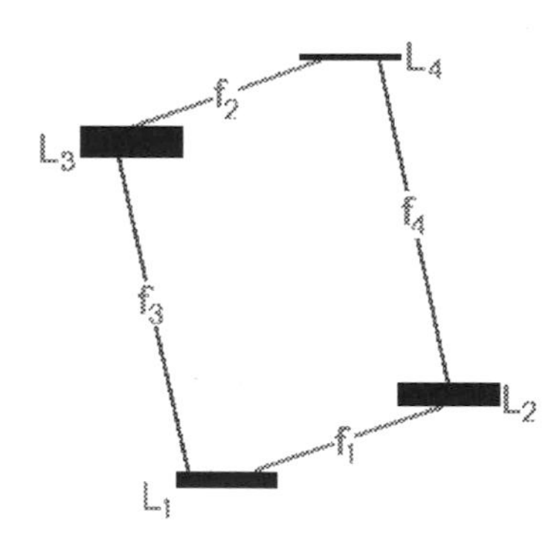

It should be noted that the signal at f_1 will be inverted because the upper state of the two joined by f_1 has a greater population than the lower one. An example of a population inverting pulse sequence designed to enhance the carbon-13 spectral lines is depicted in the animation window.

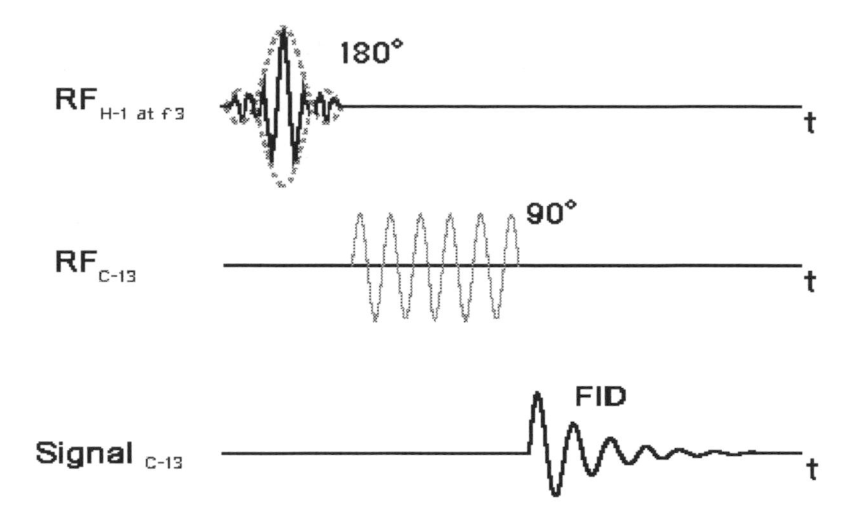

The 180 degree pulse at f_3 has a narrow band of frequencies centered on f_3 that selectively rotates only the magnetization at f_3 by 180 degrees.

A3.5 One Dimensional 1D C-13 Spectra

The following table of compounds contains links to their corresponding one-dimensional carbon-13 NMR spectra. The spectra were recorded on a 300 MHz NMR spectrometer with a delay time between successive scans of two seconds. This relatively short delay time may cause differences in the peak heights due to variations in T_1 values. Other differences may be caused by variations in the nuclear Overhauser effect. In spectra recorded with deuterated chloroform ($CDCl_3$) as the lock solvent, the three peaks at $\delta = 75$ are due to splitting of the $CDCl_3$ carbon-13 peak by the nuclear spin = 1 deuterium nucleus.

Molecule	Formula	Solvent	Spectrum
Cyclohexane	C_6H_{12}	$CDCl_3$	δ 200 100 0

Benzene	C_6H_6	$CDCl_3$	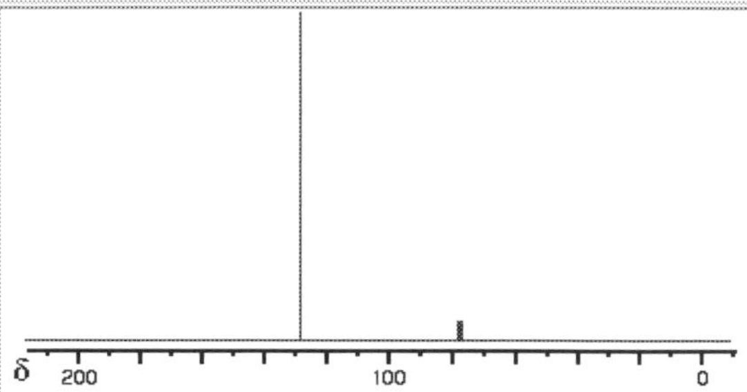 δ 200 100 0
Toluene	$C_6H_5CH_3$	$CDCl_3$	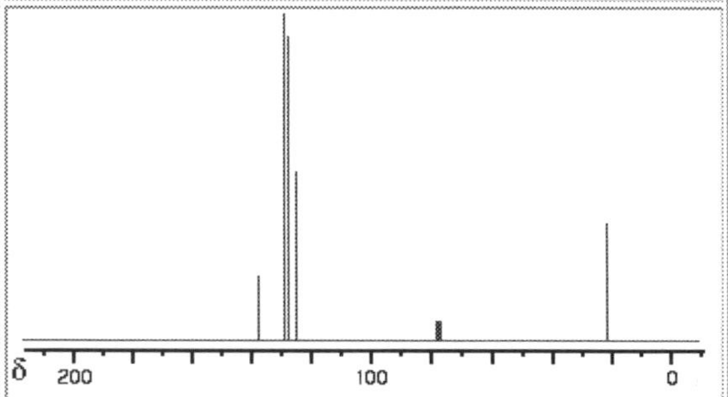 δ 200 100 0
ethyl benzene	$C_6H_5CH_2CH_3$	$CDCl_3$	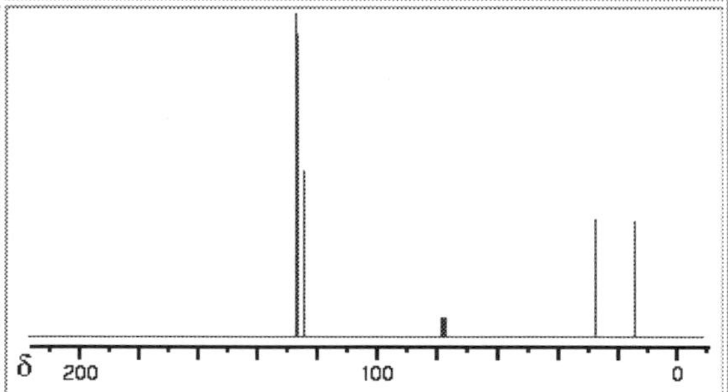 δ 200 100 0
Acetone	$CH_3(C{=}O)CH_3$	$CDCl_3$	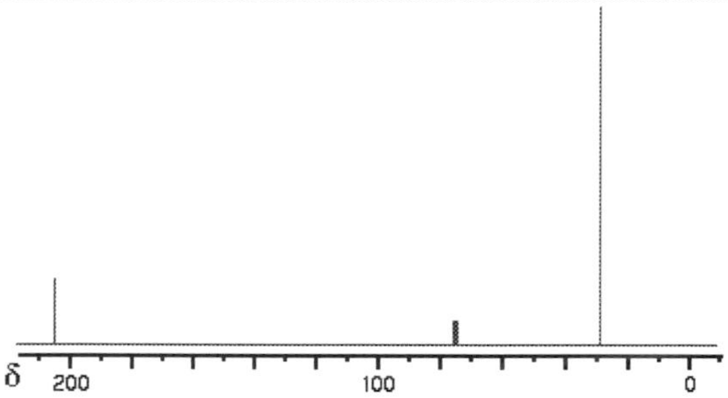 δ 200 100 0
methyl ethyl ketone	$CH_3(C{=}O)CH_2CH_3$	$CDCl_3$	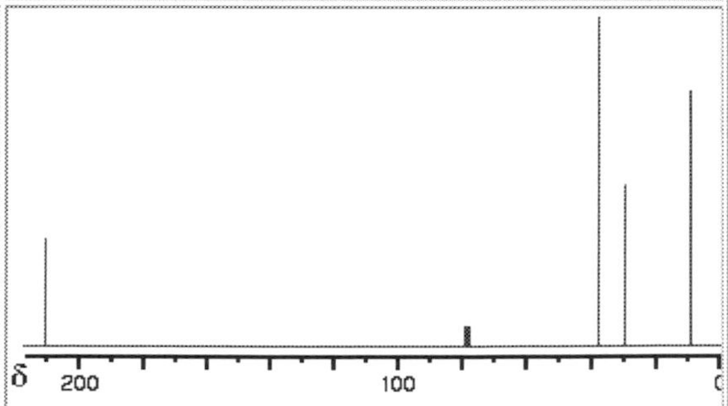 δ 200 100

Ethanol	CH_3CH_2OH	$CDCl_3$	δ 200 100 0
Ethanol	CH_3CH_2OH	D_2O	δ 200 100 0
1-propanol	$CH_3CH_2CH_2OH$	$CDCl_3$	δ 200 100 0
2-propanol	$(CH_3)_2CHOH$	$CDCl_3$	δ 200 100 0
t-butanol	$(CH_3)_3COH$	$CDCl_3$	δ 200 100 0

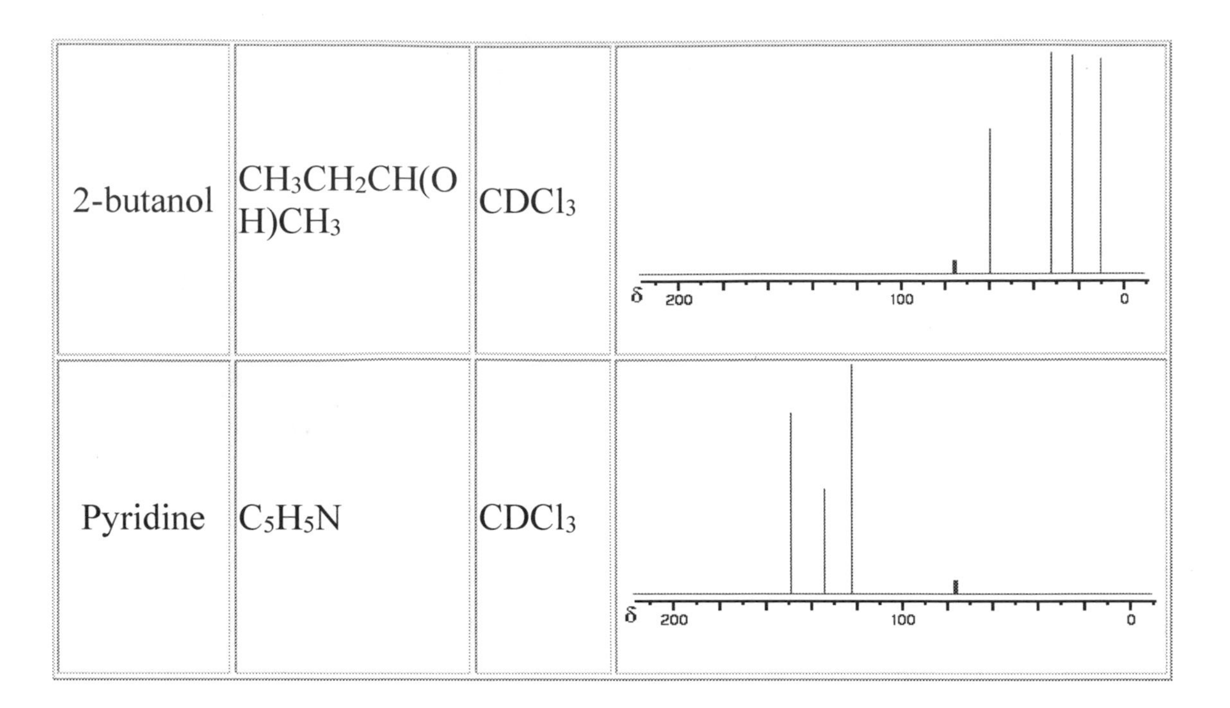

2-butanol	$CH_3CH_2CH(OH)CH_3$	$CDCl_3$	δ 200 100 0
Pyridine	C_5H_5N	$CDCl_3$	δ 200 100 0

Annex 4- Two Dimensional Techniques

A4.1 Introduction

In Annex 2 we saw the mechanics of the spin echo sequence. Recall that a 90 degree pulse rotates magnetization from a single type of spin into the XY plane. The magnetization dephases, and then a 180 degree pulse is applied which refocusses the magnetization.

When a molecule with J coupling (spin-spin coupling) is subjected to a spin-echo sequence, something unique but predictable occurs. Look at what happens to the molecule A_2-C-C-B where A and B are spin-1/2 nuclei experiencing resonance.

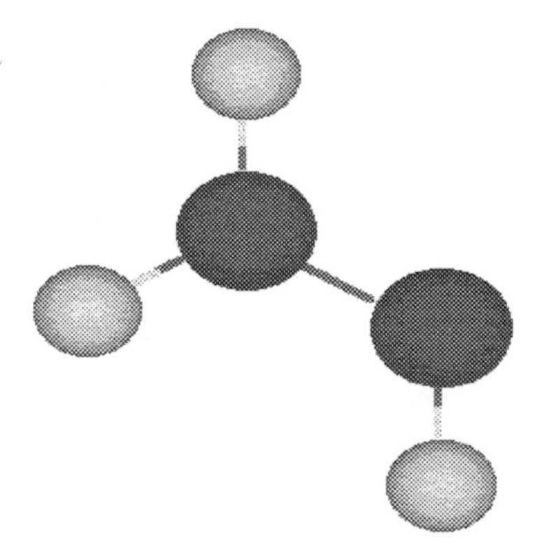

The NMR spectrum from a 90-FID sequence looks like this.

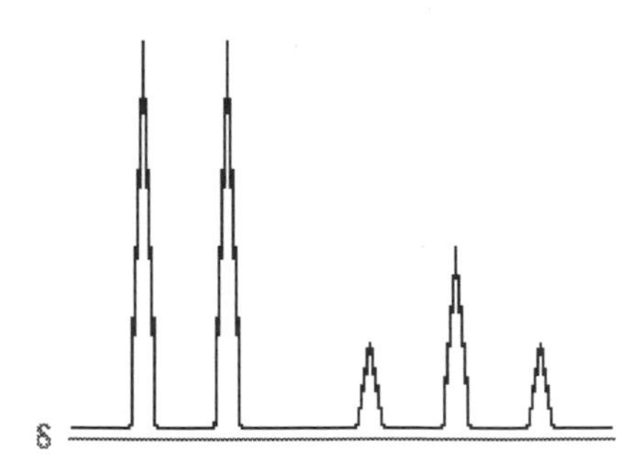

With a spin-echo sequence this same molecule gives a rather peculiar spectrum once the echo is Fourier transformed. Here is a series of spectra recorded at different TE times. The amplitude of the peaks have been standardized to be all positive when TE=0 ms.

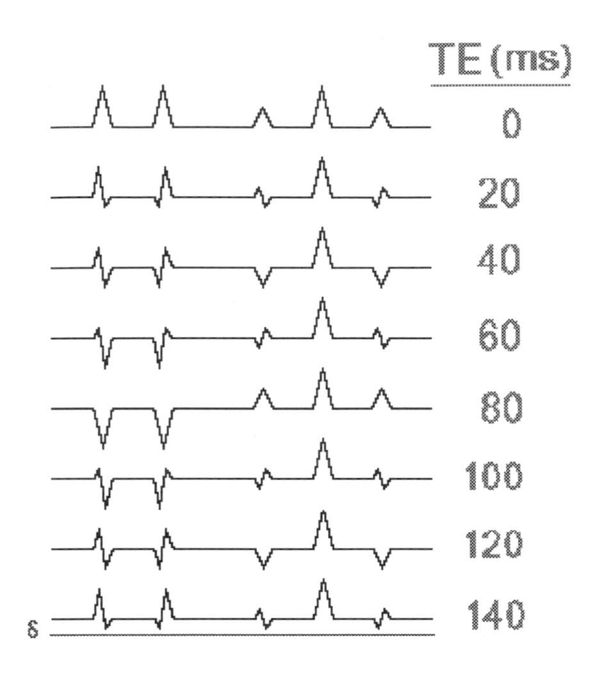

To understand what is happening, consider the magnetization vectors from the *A* nuclei. There are two absorptions lines in the spectrum from the *A* nuclei, one at ν+J/2 and one at ν-J/2. At equilibrium, the magnetization vectors from the ν+J/2 and ν-J/2 lines in the spectrum are both along +Z.

A 90 degree pulse rotates both magnetization vectors into the XY plane. Assuming a rotating frame of reference at $\nu_o = \nu$, the vectors precess according to their Larmor frequency and dephase due to T_2*. When the 180 degree pulse is applied, it rotates the magnetization vectors by 180 degrees about the X' axis. In addition the ν+J/2 and ν-J/2 magnetization vectors change places because the 180 degree pulse also flips the spin state of the *B* nucleus which is causing the splitting of the *A* spectral lines.

The two groups of vectors will refocus as they evolve at their own Larmor frequency. In this example the precession in the XY plane has been stopped when the vectors have refocussed. You will notice that the two groups of vecotrs do not refocus on the -Y axis. The phase of the two vectors on refocussing varies as a function of TE. This phase varies as a function of TE at a rate equal to the size of the spin-spin coupling frequency. Therefore, measuring this rate of change of phase will give us the size of the spin-spin coupling constant. This is the basis of one type of two-dimensional (2-D) NMR spectroscopy.

A4.2 J-resolved

In a 2-D J-resolved NMR experiment, time domain data is recorded as a function of TE and time. These two time dimensions will referred to as t_1 and t_2. For the *A_2-C-C-B* molecule, the complete time domain signals look like this.

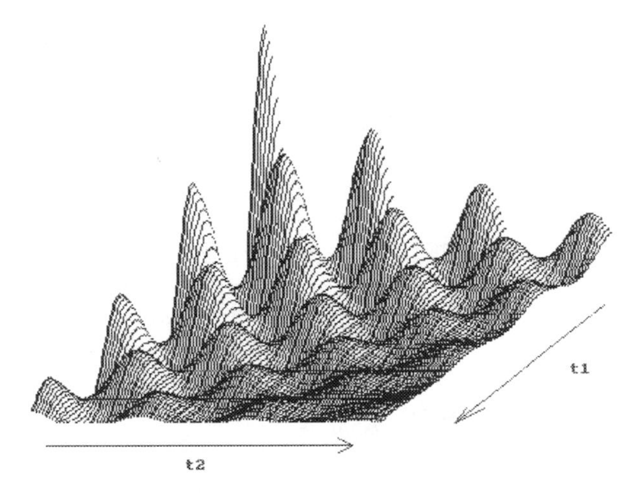

This data is Fourier transformed first in the t_2 direction to give an f_2 dimension,

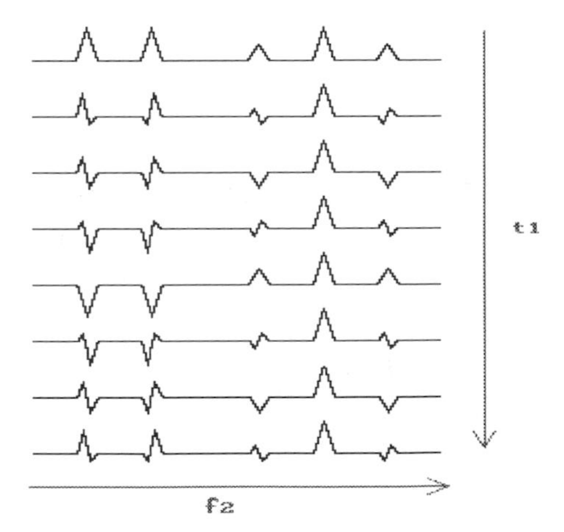

and then in the t_1 direction to give an f_1 dimension.

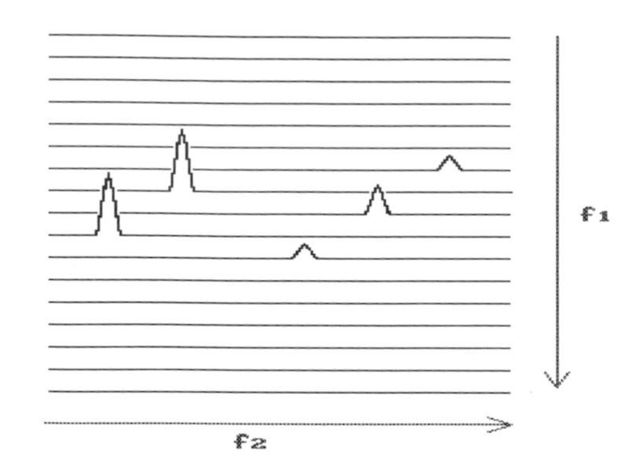

Displaying the data as shaded contours, we have the following two-dimensional data set.

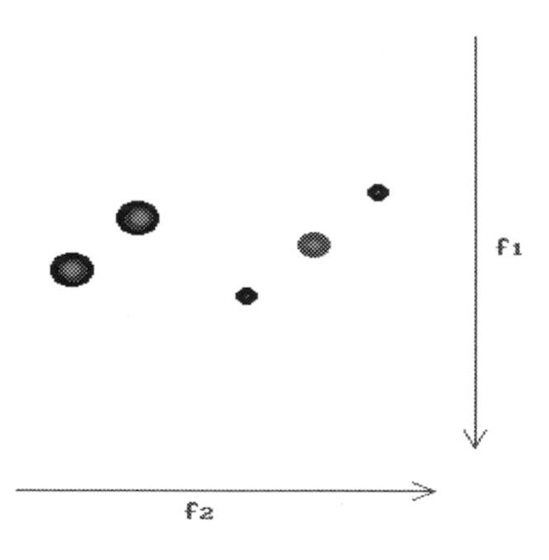

Rotating the data by 45 degrees makes the presentation clearer.

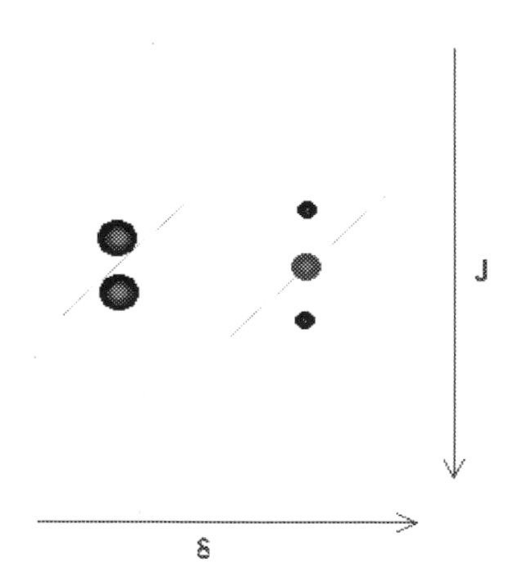

The f_1 dimension gives us J coupling information while the f_2 dimension gives chemical shift information. This type of experiment is called homonuclear J-Resolved 2-D NMR.

A4.3 COSY

The application of two 90 degree pulses to a spin system will give a signal which varies with time t_1 where t_1 is the time between the two pulses.

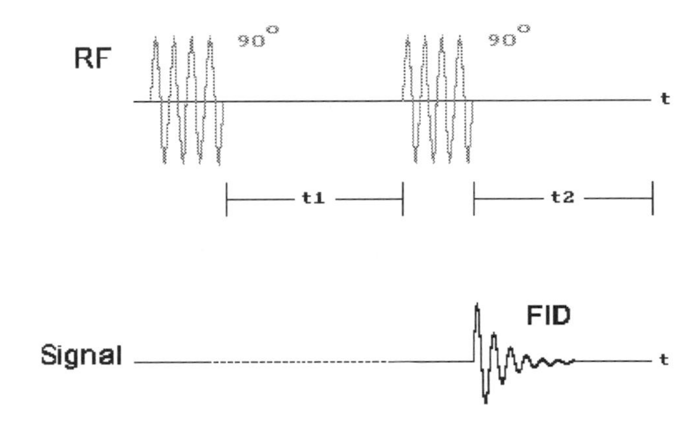

The Fourier transform of both the t_1 and t_2 dimensions gives us chemical shift information. The 2-D hydrogen correlated chemical shift spectrum of ethanol will look like this.

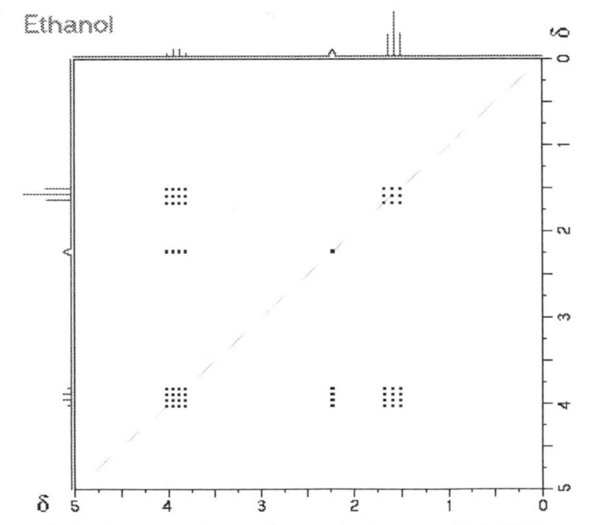

There is a wealth of information found in a COSY spectrum. A normal (chemical shift) 1-D NMR spectrum can be found along the top and left sides of the 2-D spectrum. Cross peaks exist in the 2-D COSY spectrum where there is spin-spin coupling between hydrogens. There are cross peaks between OH and CH_2 hydrogens.

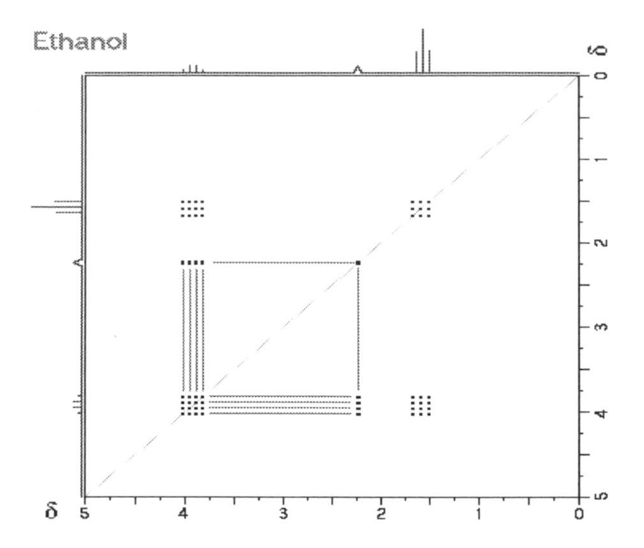

and also between CH_3 and CH_2 hydrogens

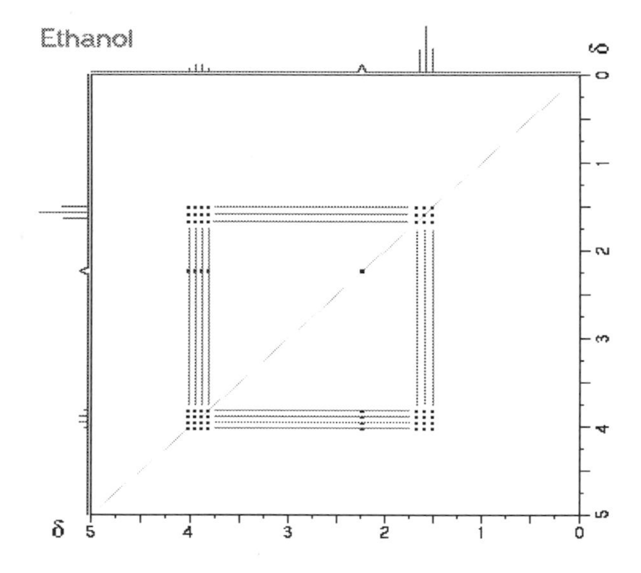

There are no cross peaks between the CH_3 and OH hydrogens because there is no coupling between the CH_3 and OH hydrogens.

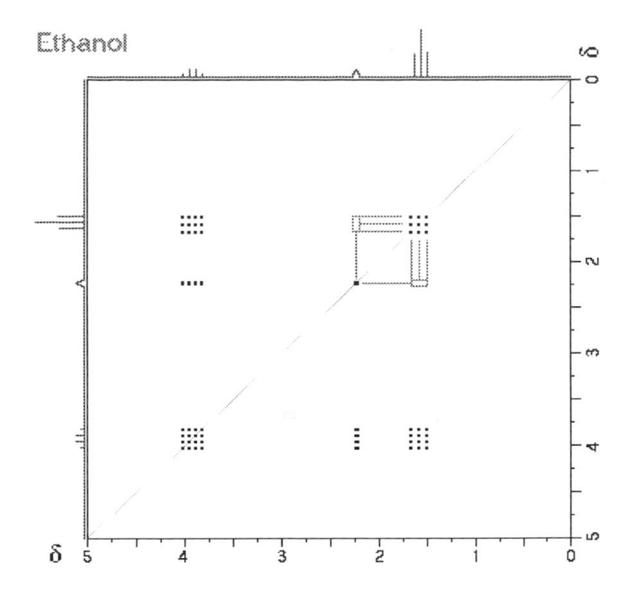

Heteronuclear correlated 2-D NMR is also possible and useful.

A4.4 Examples

The following table presents some of the hundreds of possible 2-D NMR experiments and the data represented by the two dimensions. The interested reader is directed to the NMR literture for more information.

2-D Experiment (Acronym)	Information	
	f_1	f_2
Homonuclear J resolved	J	δ
Heteronuclear J resolved	J_{AX}	δ_X
Homoculclear correlated spectroscopy (COSY)	δ_A	δ_A
Heteronuclear correlated spectroscopy (HETCOR)	δ_A	δ_X
Nuclear Overhauser Effect (2D-NOE)	δ_H, J_{HH}	δ_H, J_{HH}
2D-INADEQUATE	$\delta_A + \delta_X$	δ_X

The following table of molecules contains links to their corresponding two-dimensional NMR spectra. The spectra were recorded on a 300 MHz NMR spectrometer with $CDCl_3$ as the lock solvent.

Molecule	Formula	Type	Spectrum

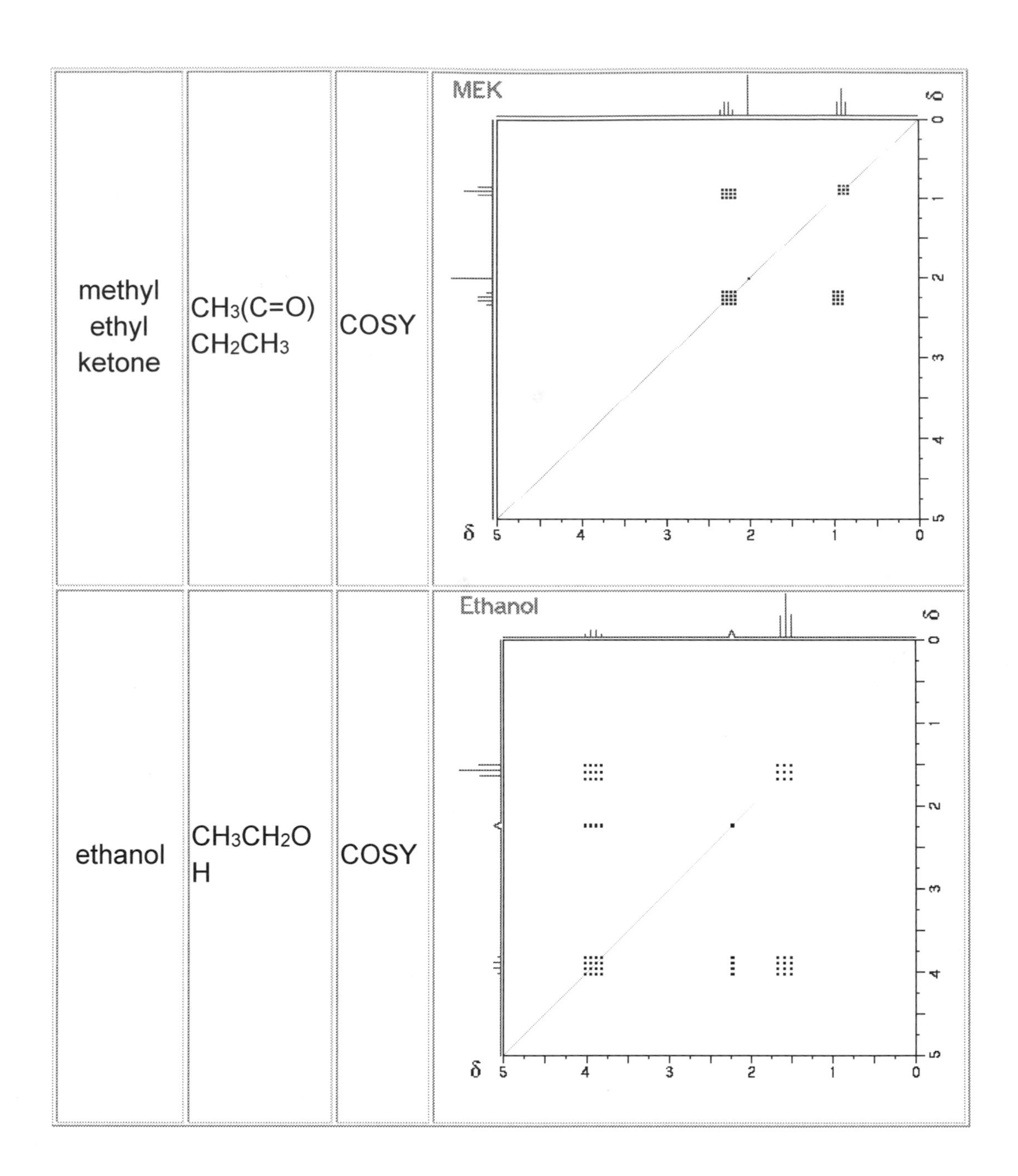

methyl ethyl ketone	$CH_3(C=O)CH_2CH_3$	COSY	MEK COSY spectrum
ethanol	CH_3CH_2OH	COSY	Ethanol COSY spectrum

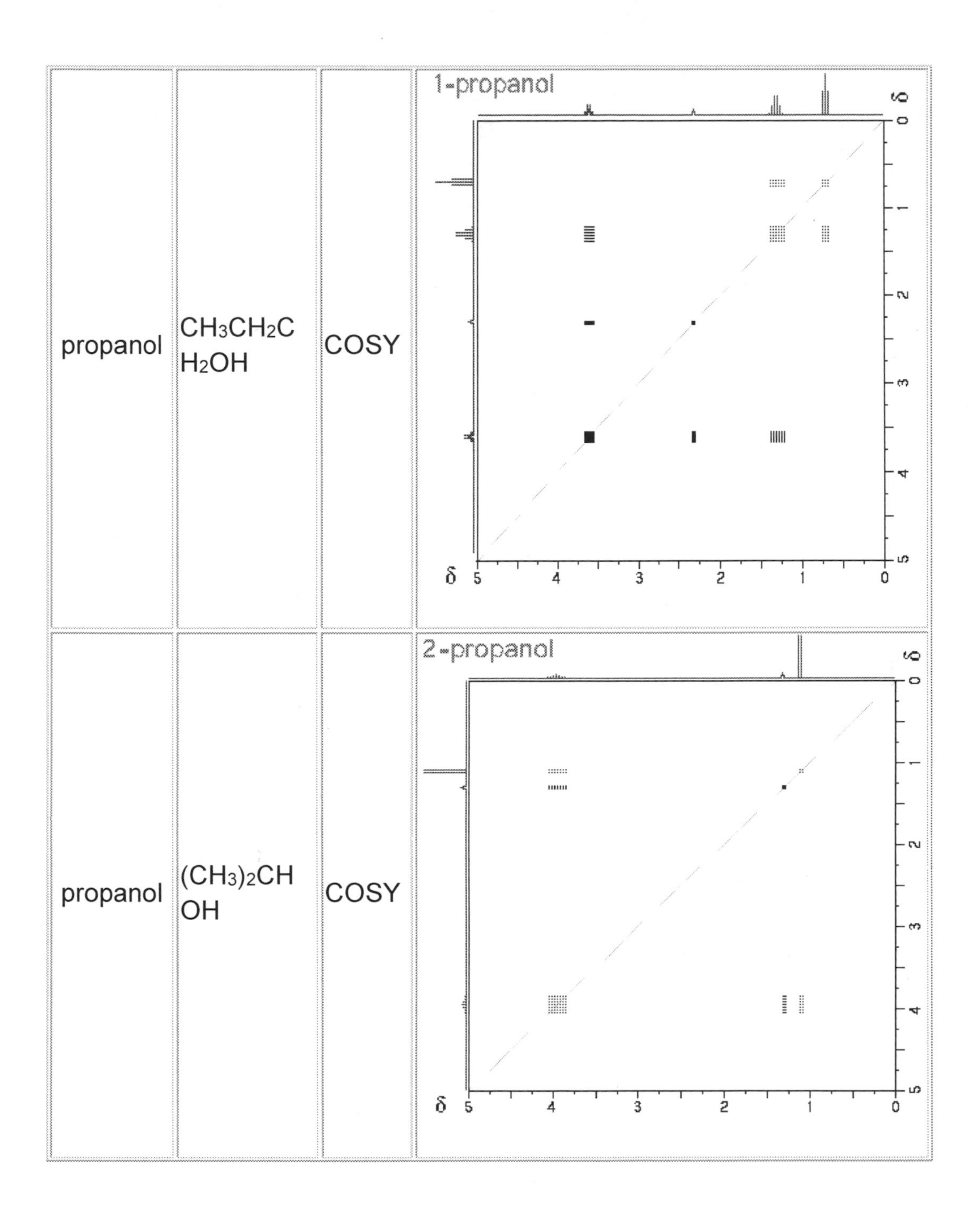

propanol	$CH_3CH_2CH_2OH$	COSY	1-propanol
propanol	$(CH_3)_2CHOH$	COSY	2-propanol

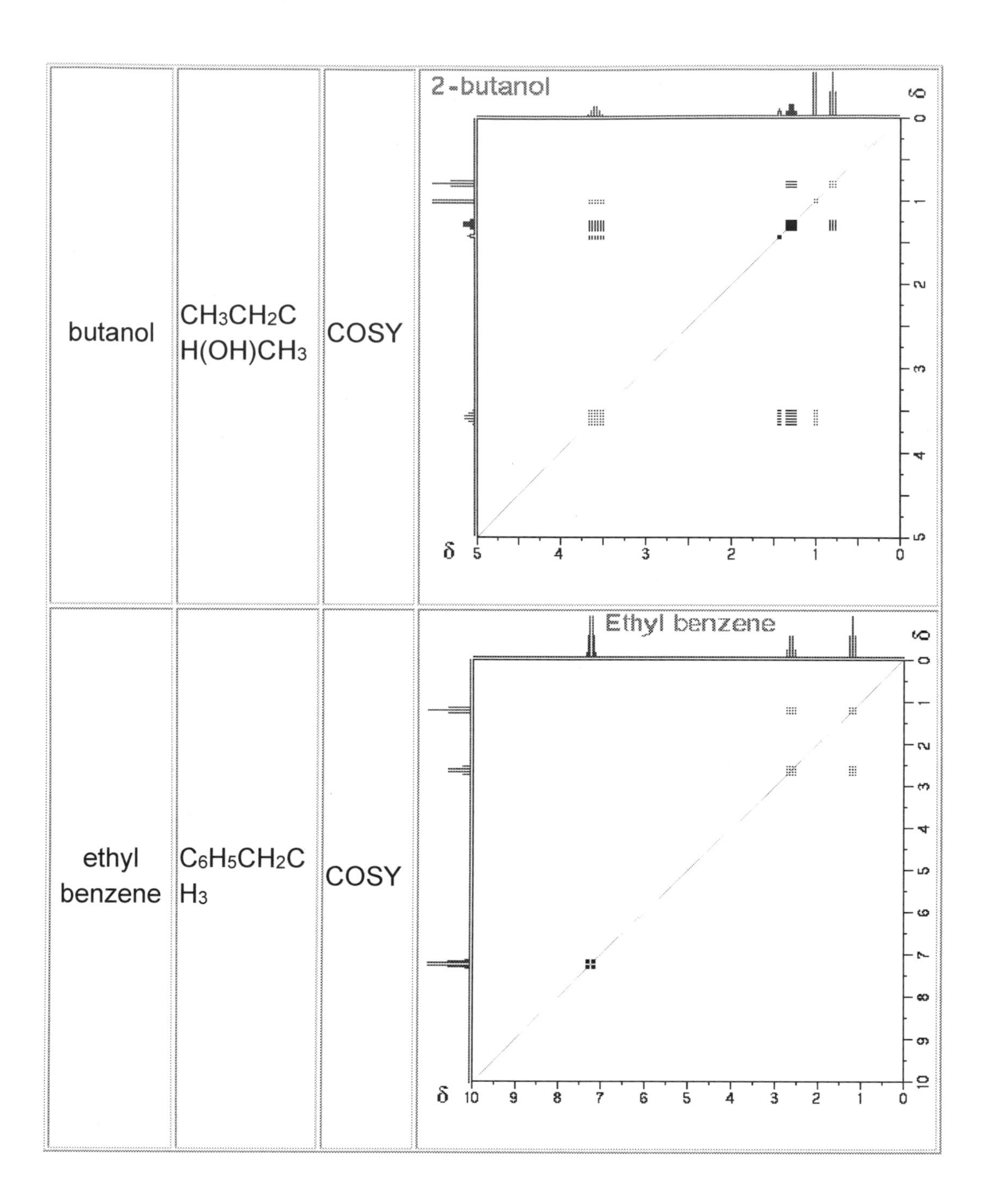

butanol	$CH_3CH_2CH(OH)CH_3$	COSY	2-butanol
ethyl benzene	$C_6H_5CH_2CH_3$	COSY	Ethyl benzene

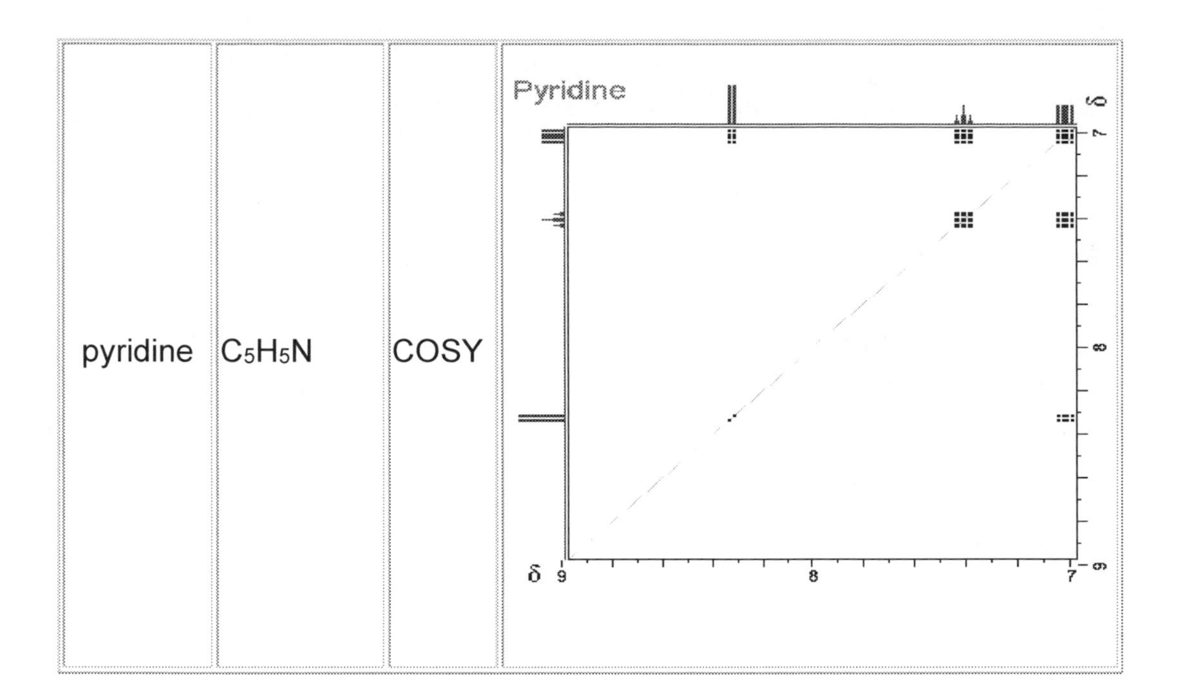

pyridine	C_5H_5N	COSY	

Annex 5 - Software and Glossary of NMR

A 5.1 Software

Below a list of NMR processing software which are used by NMR users:

- **CARA** - Computer Aided Resonance Assignment, freeware, developed at the group of Prof. Kurt Wüthrich
- **CCPN** NMR software suite from community led Collaborative Computing Project for NMR.
- **Janocchio** Conformation-dependent coupling and NOE prediction for small molecules.
- NMR processing software from **ACD/Labs** for 1D and 2D NMR spectra. DB interface available.
- NMR Prediction software **ACD/NMR** Predictors
- NMR simulation software **QSim**
- Free software for simulation of spin coupled multiplets and DNMR spectra **WINDNMR-Pro**
- NMR processing software **NMRPipe**
- **RMN** - An NMR data processing program for the Macintosh.

A 5.2 Glossary

Artifact:

A feature which appears in an NMR spectrum of a molecule which should not be present based on the chemical structure and pulse sequence used.

Chemical Screening:

The screening of an applied magnetic field experienced by a nucleus due to the electron cloud around an atom or molecule.

Chemical Shift:

A variation in the resonance frequency of a nuclear spin due to the chemical environment around the nucleus. Chemical shift is reported in ppm.

Coil:

One or more loops of a conductor used to create a magnetic field. In NMR, the term generally refers to the radiofrequency coil.

Convolution:

A mathematical operation between two functions.

Complex Data:

Numerical data with a real and an imaginary component.

Continuous Wave (CW):

A form of spectroscopy in which a constant amplitude electromagnetic wave is applied.

Coordinate Transformation:

A change in the axes used to represent some spatial quantity.

Cryopumping:

The condensation of air onto a surface cooled by a cryogenic liquid .

Dephasing Gradient:

A magnetic field gradient used to dephase transverse magnetization.

Digital Filtering:

A feature found on may newer spectrometers which eliminates wraparound artifacts by filtering out the higher frequency components in the time domain spectrum.

Doubly balanced mixer:

An electrical device, often referred to as a product detector, which is used in NMR to convert signals from the laboratory frame of reference to the rotating frame of reference.

Echo:

A form of magnetic resonance signal from the refocusing of transverse magnetization.

Echo Time (TE):

The time between the 90 degree pulse and the maximum in the echo in a spin-echo sequence.

Exchange, Chemical:

The interchange of chemically equivalent components on a molecule.

Exchange, Spin:

The interchange of spin state between two nuclei.

Figure-8 Coil:

A magnetic field gradient coil shaped like the number eight.

Free induction decay (FID):

A form of magnetic resonance signal from the decay of transverse magnetization.

Fourier transform (FT):

A mathematical technique capable of converting a time domain signal to a frequency domain signal and vice versa.

Gradient (G):

A variation in some quantity with respect to another. In the context of NMR, a magnetic field gradient is a variation in the magnetic field with respect to distance.

Gyromagnetic Ratio:

The ratio of the resonance frequency to the magnetic field strength for a given nucleus.

Imaginary Component:

The component of a signal perpendicular to the real signal.

Imaging Sequence:

A specific set of RF pulses and magnetic field gradients used to produce an image.

Inversion Recovery Sequence:

A pulse sequence producing signals which represent the longitudinal magnetization present after the application of a 180° inversion RF pulse.

Inversion Time (TI):

The time between the inversion pulse and the sampling pulse(s) in an inversion recovery sequence.

K-Space:

That image space represented by the time and phase raw data. The Fourier transform of k-space is the magnetic resonance image.

Larmor frequency:

The resonance frequency of a spin in a magnetic field. The rate of precession of a spin packet in a magnetic field. The frequency which will cause a transition between the two spin energy levels of a nucleus.

Longitudinal Magnetization:

The Z component of magnetization.

Lorentzian Lineshape:

A function obtained from the Fourier transform of an exponential function.

Magnitude:

The length of a magnetization vector. In NMR, the square root of the sum of the squares of the Mx and My components, *i.e.* the magnitude of the transverse magnetization.

Magnetic Resonance Imaging (MRI):

An imaging technique based on the principles of NMR.

Negative Frequency Artifact:

The appearance of smaller in amplitude peaks in one half of the spectrum which are the mirror image of ones in the opposite half.

Net Magnetization Vector:

A vector representing the sum of the magnetization from a spin system.

Nuclear Magnetic Resonance (NMR):

A spectroscopic technique used by scientists to elucidate chemical structure and molecular dynamics

Pixel:

Picture element.

Precess:

A rotational motion of a vector about the axis of a coordinate system where the polar angle is fixed and the azmuthal angle changes steadily.

Proportionality Constant:

A constant used to convert one set of units to another.

Pulse Sequence:

A series of RF pulses and/or magnetic field gradients applied to a spin system to produce a signal whose behavior gives information about some property of the spin system.

Quadrature Detection:

Detection of Mx and My simultaneously as a function of time.

Radio Frequency:

A frequency band in the electromagnetic spectrum with frequencies in the millons of cycles per second.

Raw data:

The Mx and My data as a function of time and/or other parameters in an NMR pulse sequence. This is also called k-space data.

Real component:

The component of a signal perpendicular to the imaginary signal.

Repetition Time:

The time between repetitions of the basic sequence in a pulse sequence.]

Resonance:

An exchange of energy between two systems at a specific frequency.

RF Coil:

An inductor-capacitor resonant circuit used to set up B_1 magnetic fields in the sample and to detect the signal from the sample.

RF Pulse:

A short burst of RF energy which has a specific shape.

Rotation Matrix:

A matrix used to describe the rotation of a vector.

Sample Probe:

That portion of the NMR spectrometer containing the RF coils and into which the sample is placed.

Saddle Coil:

A coil geometry which has two loops of a conductor wrapped around opposite sides of a cylinder.

Spin:

A fundamental property of matter responsible for NMR and MRI.

Spin Density:

The concentration of spins.

Spin-Echo:

An NMR sequence whose signal is an echo resulting from the refocusing of magnetization after the application of 90° and 180° RF pulses.

Spin-Lattice Relaxation:

The return of the longitudinal magnitization to its equilibrium value along the +Z axis.

Spin-Lattice Relaxation Time (T_1):

The time to reduce the difference between the longitudinal magnitization and its equilibrium value by a factor of e.

Spin Packet:

A group of spins experiencing the same magnetic field.

Spin-Spin Relaxation:

The return of the transverse magnitization to its equilibrium value (zero).

Spin-Spin Relaxation Time:

The time to reduce the transverse magnetization by a factor of e.

Sinc Pulse:

An RF pulse shaped like Sin(x)/x.

Superconduct:

To have no resistance. A perfect superconductor can carry an electrical current without losses.

T_2* :

Pronounced T-2-star. The spin-spin relaxation time composed of contributions from molecular interactions and inhomogeneities in the magnetic field.

Timing Diagram:

A multiaxis plot of some aspects of a pulse sequence as a function of time.

Transverse magnetization:

The XY component of the net magnetization.

8

REFERENCES

1. Joseph P. Hornak, Ph.D.,The Basics of NMR, 1997-2008 J.P. Hornak.
2. E. Fukushima and S.B.W. Roeder, Experimental Pulse NMR, Addison-Wesley, Reading, MA 1981.
3. T.C. Farrar, An Introduction To Pulse NMR Spectroscopy, Farragut Press, Chicago, 1987.
4. R.C. Jennison, Fourier Transforms and Convolutions, Pergamon Press, NY 1961.
5. E.O.Brigham, The Fast Fourier Transform, Prentice-Hall, Englewood Cliffs, NJ 1974.
6. A. Carrington and A.D. McLachlan, Introduction To Magnetic Resonance, Chapman and Hall, London 1967.
7. B. Noble and J.W. Daniel, Applied Linear Algebra, Prentice-Hall Englewood Cliffs, NJ.
8. G.B. Thomas, Calculus, Addison-Wesley, Reading, MA 1969.
9. H. Gunther, "Modern pulse methods in high-resolution NMR spectroscopy." Angew. Chem.. Int. Ed. Engl.22:350-380 (1983).
10. R.K. Harris, Nuclear Magnetic Resonance Spectroscopy, Pitman, London 1983.
11. T.C. Farrar, An Introduction to Pulse NMR Spectroscopy, Farragut Press, Chicago 1987.
12. T.C. Farrar and E.D. Becker, Pulse And Fourier Transform NMR, Academic Press, 1971.
13. 13. D. Shaw, Fourier Transform NMR Spectroscopy, Elsevier, NY 1976.
14. J.W.Akitt, NMR and Chemistry, An Introduction to the Fourier Transform - Multinuclear Era. Chapman and Hall, London, 1983.
15. R. Freeman, A Handbook of Nuclear Magnetic Resonance, Longman Scientific & Technical, Essex, England, 1988.
16. Woodrow W. Conover, "Magnet Shimming," Technical Report, Nicolet Magnetics Corporation
17. C.E. Hayes, W.A. Edelstein, J.F. Schenck, "Radio Frequency Resonators." in Magnetic Resonance Imaging, ed. by C.L. Partain, R.R. Price, J.A. Patton, M.V. Kulkarni, A.E. James, Saunders, Philadelphia, 1988.
18. G.M. Bydder, "Clinical Applications of Gadolinium-DTPA." in in Magnetic Resonance Imaging, ed. by D.D. Stark and W.G. Bradley, C.V. Mosby Co., St. Louis, MO 1988.
19. J. Granot, J. Magn. Reson. **70:**488-492 (1986).
20. S.R. Thomas, L.J. Busse, J.F. Schenck, "Gradient Coil Technology." in

Magnetic Resonance Imaging, ed. by C.L. Partain, R.R. Price, J.A. Patton, M.V. Kulkarni, A.E. James, Saunders, Philadelphia, 1988.

21. J. Gong and J.P. Hornak, "A Fast T_1 Algorithm." Magn. Reson. Imaging **10:**623-626 (1992).
22. X. Li and J.P. Hornak, "T_2 Calculations in MRI: Linear versus Nonlinear Methods." J. Imag. Sci. & Technol. **38:**154-157 (1994).
23. Peter Stilbs, "Fourier transform pulsed-gradient spin-echo studies of molecular diffusion." Progress in NMR Spectroscopy **19**:1-45 (1987).
24. J.R. Dyer, Applications of Absorption Spectroscopy by Organic Compounds, Prentice-Hall, Englewood Cliffs, NJ 1965.
25. D.A. Skoog, F.J. Holler, T.A. Nieman, Principles of Instrumental Analysis, 5th ed., Saunders College Publishing, New York, 1992.
26. D.E. Leyden, R.H. Cox, Analytical Applications of NMR, Wiley, NY 1977.
27. A.E. Derome, NMR Techniques for Chemistry Research, Pergamon Press, NY 1987.
28. Atta-ur-Rahman, Nuclear Magnetic Resonance, Basic Principles, Springer-Verlag, NY 1986.
29. F. Noack, "NMR field-cycling spectroscopy: principles and applications." Progress in NMR Spectroscopy, **18**171-276 (1986).
30. Field Cycling NMR Relaxometry: Review of Technical Issues and Applications, STELAR s.r.l., via E. Fermi 4, 27035 Mede (PV), Italy. © 2001, (http://www.stelar.it/)
31. E. Anoardo, G. Galli, G. Ferrante, "Fast Field Cycling NMR: Applications and Instrumentation." Appl. Magn. Reson. **20:**365-404 (2001)
32. F.A. Bovey, L. Jelinski, P.A. Mirau, Nuclear Magnetic Resonance Spectroscopy, Academic Press, NY, 1988.
33. Gary E. Martin, A. S. Zektzer (1988). Two-Dimensional NMR Methods for Establishing Molecular Connectivity. New York: Wiley-VCH. p. 59. ISBN 0-471-18707-0. http://books.google.com/books?id=9ysYrpe_NoEC&printsec=frontcover .
34. J.W. Akitt, B.E. Mann (2000). NMR and Chemistry. Cheltenham, UK: Stanley Thornes. pp. 273, 287. ISBN 0-7487-4344-8.
35. J.P. Hornak. "The Basics of NMR". http://www.cis.rit.edu/htbooks/nmr/. Retrieved 2009-02-23.
36. J. Keeler (2005). Understanding NMR Spectroscopy. John Wiley & Sons. ISBN 0-470-01786-4.
37. Kurt Wüthrich (1986). NMR of Proteins and Nucleic Acids. New York (NY), USA: Wiley-Interscience. ISBN 0-471-11917-2.
38. J.M Tyszka, S.E Fraser, R.E Jacobs (2005). "Magnetic resonance

microscopy: recent advances and applications". Current Opinion in Biotechnology **16** (1): 93–99. doi:10.1016/j.copbio.2004.11.004. PMID 15722021.

39. J.C. Edwards. "Principles of NMR". Process NMR Associates. http://www.process-nmr.com/pdfs/NMR%20Overview.pdf. Retrieved 2009-02-23.
40. R.L Haner, P.A. Keifer (2009). Encyclopedia of Magnetic Resonance. John Wiley. doi:10.1002/9780470034590.emrstm1085.
41. G.E Martin, A.S. Zekter (1988). Two-Dimensional NMR Methods for Establishing Molecular Connectivity. New York: VCH Publishers. p. 59.
42. J.W. Akitt, B.E. Mann (2000). NMR and Chemistry. Cheltenham, UK: Stanley Thornes. pp. 273, 287.
43. J.P. Hornak. "The Basics of NMR". http://www.cis.rit.edu/htbooks/nmr/. Retrieved 2009-02-23.
44. J. Keeler (2005). Understanding NMR Spectroscopy. John Wiley & Sons. ISBN 0470017864.
45. K. Wuthrich (1986). NMR of Proteins and Nucleic Acids. New York (NY), USA: Wiley-Interscience.
46. J.M Tyszka, S.E Fraser, R.E Jacobs (2005). "Magnetic resonance microscopy: recent advances and applications". "Current Opinion in Biotechnology **16** (1): 93–99. doi:10.1016/j.copbio.2004.11.004.
47. J.C. Edwards. "Principles of NMR". Process NMR Associates. http://www.process-nmr.com/pdfs/NMR%20Overview.pdf. Retrieved 2009-02-23.
48. I.I. Rabi, J.R. Zacharias, S. Millman, P. Kusch (1938). "A New Method of Measuring Nuclear Magnetic Moment". Physical Review **53**: 318. doi:10.1103/PhysRev.53.318.
49. Filler, AG: The history, development, and impact of computed imaging in neurological diagnosis and neurosurgery: CT, MRI, DTI: Nature Precedings DOI: 10.1038/npre.2009.3267.4.
50. I.C. Baianu, K.A. Rubinson, J. Patterson (1979). "Ferromagnetic Resonance and Spin Wave Excitations in Metallic Glasses". J. Phys. Chem. Solids **40** (12): 941–950. doi:10.1016/0022-3697(79)90122-7.
51. I.C. Baianu (1979). "Ferromagnetic resonance and spin wave excitations in metallic glasses: The effects of thermal ageing and long-range magnetic ordering". Solid State Communications **29**: i–xvi. doi:10.1016/0038-1098(79)91190-6.
52. David P. DiVincenzo (2000). "The Physical Implementation of Quantum Computation": Experimental Proposals for Quantum Computation. **arXiv:quant-ph/0002077**. p. 1 through 10. doi:10.1103/PhysRev.53.318.
53. "Nuclear Magnetic Resonance Fourier Transform Spectroscopy"

54. I.C. Baianu. "Two-dimensional Fourier transforms". 2D-FT NMR and MRI. PlanetMath. http://planetmath.org/encyclopedia/TwoDimensionalFourierTransforms.html. Retrieved 2009-02-22.
55. R. Taylor, J.P. Hare, A.K. Abdul-Sada, H.W. Kroto (1990). "Isolation, separation and characterization of the fullerenes C60 and C70: the third form of carbon". Journal of the Chemical Society, Chemical Communications **20**: 1423–1425. doi:10.1039/c39900001423.
56. "http://en.wikipedia.org/wiki/Nuclear_magnetic_resonance
57. J.P. Hornak; The Basics of NMR; 1997-2008
58. R.C. Jennison; Fourier Transforms and Convolutions; Pergamon Press, NY;1961
59. E.O.Brigham; The Fast Fourier Transform; Prentice-Hall, Englewood Cliffs, NJ;1974
60. D. Shaw; Fourier Transform NMR Spectroscopy; Elsevier, New York; 1976
61. R. Freeman; A Handbook of Nuclear Magnetic Resonance; Longman Scientific & Technical, Essex, England; 1988
62. 1- J.P. Hornak; The Basics of NMR; 1997-2008
63. C.E. Hayes, W.A. Edelstein, J.F. Schenck; "Radio Frequency Resonators." In Magnetic Resonance Imaging,; ed. by C.L. Partain, R.R. Price, J.A. Patton, M.V. Kulkarni, A.E. James; Saunders, Philadelphia, 1988
64. S.R. Thomas, L.J. Busse, J.F. Schenck "Gradient Coil Technology." In Magnetic Resonance Imaging, ed. by C.L. Partain, R.R. Price, J.A. Patton, M.V. Kulkarni, A.E. JamesSaunders, Philadelphia, 1988.
65. E. Fukushima, S.B.W. Roeder Experimental Pulse NMR Addison-Wesley, Reading, MA 1981.
66. http://www.spectroscopynow.com/userfiles/sepspec/file/specNOW/Tutorials/chem843-2.pdf
67. https://en.wikipedia.org/wiki/Magnetic_resonance_imaging
68. https://en.wikipedia.org/wiki/Liquid_chromatography%E2%80%93mass_spectrometry

ELECTRICAL ENGINEERING

Principles and Special Purpose Applications of Electromagnetic Field and High Voltage

Package 3:Special Purpose Applications-Part Two

Book 8: Electrostatic Precipitator

Dr. Moayad Abdullah AlMayouf

July 2018

Electrostatic Precipitator

CONTENT

1

BASICS AND PRINCIPLES

1.1. Electrostatics

Before the year 1832, when Michael Faraday published the results of his experiment on the identity of electricity, physicists thought "static electricity" was somehow different from other electrical charges. Michael Faraday proved that the electricity induced from the magnet, voltaic electricity produced by a battery, and static electricity is all the same.

Electrostatics deals with the phenomena arising from stationary or slowly moving electric charges.

Since classical antiquity it was known that some materials such as amber attract light particles after rubbing. Electrostatic phenomena arise from the forces that electric charges exert on each other. Even though electrostatically induced forces seem to be rather weak, the electrostatic force between e.g. an electron and a proton, that together make up a hydrogen atom, is about 40 orders of magnitude stronger than the gravitational force acting between them.

Electrostatic phenomena include many examples as simple as the attraction of the plastic wrap to your hand after you remove it from a package. Although charge exchange happens whenever any two surfaces contact and separate, the effects of charge exchange are usually only noticed when at least one of the surfaces has a high resistance to electrical flow. This is because the charges that transfer to or from the highly resistive surface are more or less trapped there for a long enough time for their effects to be observed. These charges then remain on the object until they either bleed off to ground or are quickly neutralized by a discharge.

In principal, the following fundamental laws and parameters characterize and govern the electrostatic principles:

A. Coulomb's law

The fundamental equation of electrostatics is Coulomb's law, which describes the force between two point charges. The magnitude of the electrostatic force between two point electric charges is directly proportional to the product of the magnitudes of each charge and inversely proportional to the square of the distance between the charges Q_1 and Q_2:

$$F = \frac{Q_1 Q_2}{4\pi \varepsilon_0 r^2} \quad (1.1)$$

where ε_0 is the permittivity of free space, a defined value:

$\varepsilon_0 = 1/\mu_0 C^2 = 8.854\ 187\ 817 \times 10^{-12}$ F/ m or (in Amper^2 second^4 kgram^{-1} meter^{-3} or Coulomb^2 Newton^{-1}meter^{-2} or Farad meter^{-1}).

B. The electric field

The electric field (in unit of volt per meter) at a point is defined as the force (newton) per unit charge (coulombs) on a charge at that point:

$$\vec{F} = q\vec{E} \quad (1.2)$$

From this definition and Coulomb's law, it follows that the magnitude of the electric field E created by a single point charge Q is:

$$E = \frac{Q}{4\pi\varepsilon_0 r^2} \quad (1.3)$$

C. Gauss's law

Gauss' law states that "the total electric flux through a closed surface is proportional to the total electric charge enclosed within the surface". The constant of proportionality is the permittivity of free space. Mathematically, Gauss's law takes the form of an integral equation:

$$\int_s \varepsilon_0 \vec{E} \,.\, d\vec{A} = \int_v \rho . dV \quad (1.4)$$

Alternatively, in differential form, the equation becomes

$$\vec{\nabla} . \varepsilon_0 \vec{E} = \rho \quad (1.5)$$

D. Poisson's equation

The definition of electrostatic potential, combined with the differential form of Gauss's law (above), provides a relationship between the potential $\emptyset$ and the charge density ρ:

$$\nabla^2 \emptyset = -\frac{\rho}{\varepsilon_0} \qquad (1.6)$$

This relationship is a form of Poisson's equation. Where ε_0 is Vacuum permittivity.

E. Laplace's equation

In the absence of unpaired electric charge, the equation becomes

$$\nabla^2 \emptyset = 0 \qquad (1.7)$$

which is Laplace's equation.

F. The electrostatic approximation

The validity of the electrostatic approximation rests on the assumption that the electric field is irrotational:

$$\overrightarrow{\nabla} X \vec{E} = 0 \qquad (1.8)$$

From Faraday's law, this assumption implies the absence or near-absence of time-varying magnetic fields:

$$\frac{\partial B}{\partial t} = 0 \qquad (1.9)$$

In other words, electrostatics does not require the absence of magnetic fields or electric currents. Rather, if magnetic fields or electric currents do exist, they must not change with time, or in the worst-case, they must change with time only very slowly. In some problems, both electrostatics and magnetostatics may be required for accurate predictions, but the coupling between the two can still be ignored.

G. Electrostatic potential

Because the electric field is irrotational, it is possible to express the electric field as the gradient of a scalar function, called the electrostatic potential (also known as the voltage). An electric field, *E*, points from regions of high potential, φ, to regions of low potential, expressed mathematically as

$$\vec{E} = -\vec{\nabla}\ \boldsymbol{\Phi} \tag{1.10}$$

The electrostatic potential at a point can be defined as the amount of work per unit charge required to move a charge from infinity to the given point.

H. Electrostatic generators

The presence of surface charge imbalance means that the objects will exhibit attractive or repulsive forces. This surface charge imbalance, which yields static electricity, can be generated by touching two differing surfaces together and then separating them due to the phenomena of contact electrification and the triboelectric effect. Rubbing two nonconductive objects generates a great amount of static electricity. This is not just the result of friction; two nonconductive surfaces can become charged by just being placed one on top of the other. Since most surfaces have a rough texture, it takes longer to achieve charging through contact than through rubbing. Rubbing objects together increases amount of adhesive contact between the two surfaces. Usually insulators, e.g., substances that do not conduct electricity, are good at both generating, and holding, a surface charge. Some examples of these substances are rubber, plastic, glass, and pith. Conductive objects only rarely generate charge imbalance except, for example, when a metal surface is impacted by solid or liquid nonconductors. The charge that is transferred during contact electrification is stored on the surface of each object. Static electric generators, devices which produce very high voltage at very low current and used for classroom physics demonstrations, rely on this effect. Note that the presence of electric current does not detract from the electrostatic forces nor from the sparking, from the corona discharge, or other phenomena. Both phenomena can exist simultaneously in the same system.

1.2. Precipitator

A precipitator is a large, industrial emission-control unit. It is designed to trap and remove dust particles from the exhaust gas stream of an industrial process. Generally, precipitators are used in the following industries:

- Power/Electric
- Cement
- Chemicals
- Metals
- Paper

In many industrial plants, particulate matter created in the industrial process is carried as dust in the hot exhaust gases. These dust-laden gases pass through a precipitator that collects most of the dust. Cleaned gas then passes out of the precipitator and through a stack to the atmosphere.

A natural result from the burning of fossil fuels is the emission of fly ash. Ash is mineral matter present in the fuel. For a pulverized coal unit, 60-80% of ash leaves with the flue gas. Historically, fly ash emissions have received the greatest attention since they are easily seen leaving smokestacks.

Two emission control devices for fly ash are the traditional fabric filters and the more recent electrostatic precipitators. The fabric filters are large bag house filters having a high maintenance cost (the cloth bags have a life of 18 to 36 months, but can be temporarily cleaned by shaking or back flushing with air).

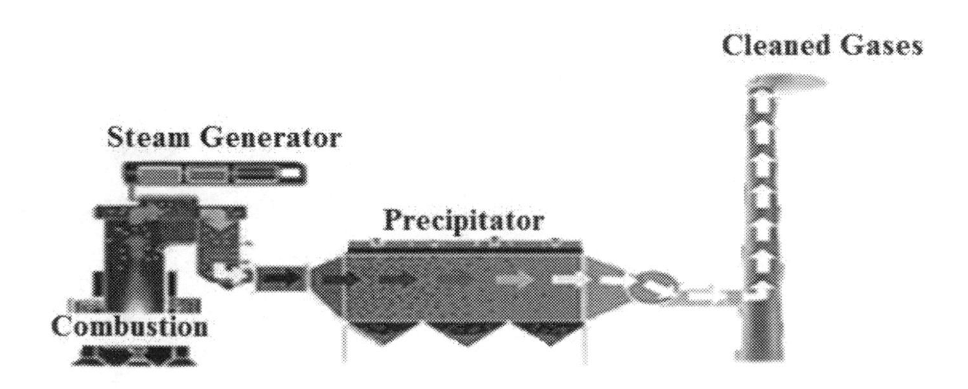

Fig.1.1 Traditional fabric Precipitator

These fabric filters are inherently large structures resulting in a large pressure drop, which reduces the plant efficiency.

Precipitators function by electrostatically charging the dust particles in the gas stream. The charged particles are then attracted to and deposited on plates or other collection devices. When enough dust has accumulated, the collectors are shaken to dislodge the dust, causing it to fall with the force of gravity to hoppers below. The dust is then removed by a conveyor system for disposal or recycling.

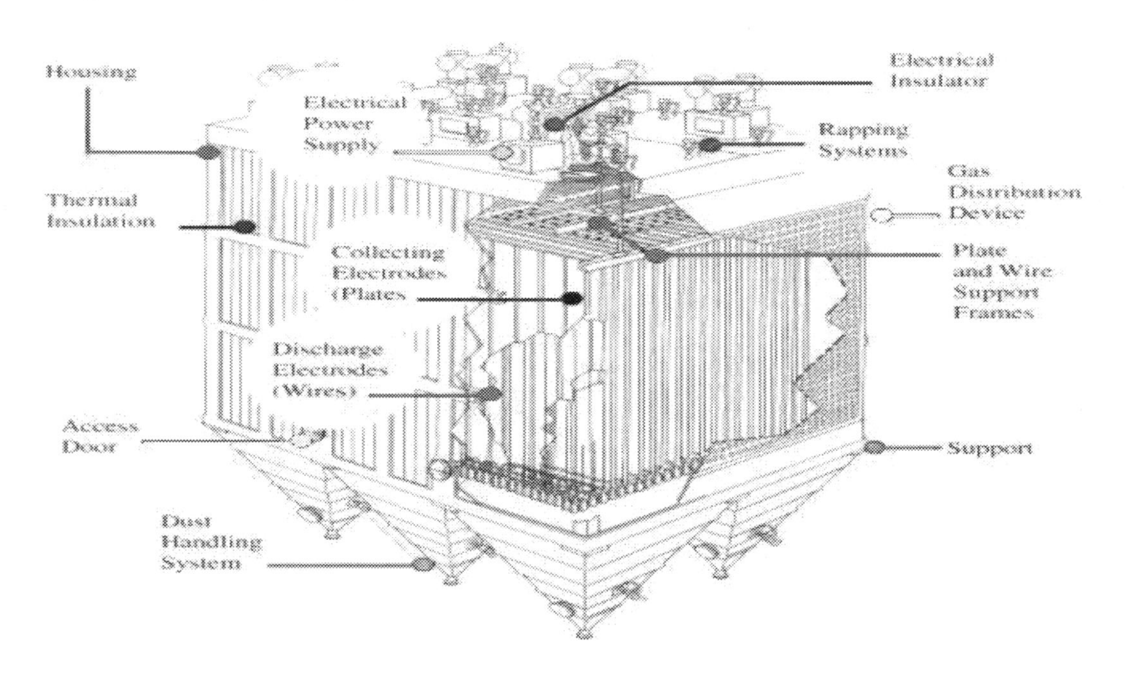

Fig.1.2.Electrostatic precipitator

In principal the categories and types of industrial dust separators /collectors may be classified as following:

A. <u>Inertial separators</u>

Inertial separators separate dust from gas streams using a combination of forces, such as centrifugal, gravitational, and inertial. These forces move the dust to an area where the forces exerted by the gas stream are minimal. The separated dust is moved by gravity into a hopper, where it is temporarily stored.

B. Fabric filters

Commonly known as baghouses, fabric collectors use filtration to separate dust particulates from dusty gases. They are one of the most efficient and cost effective types of dust collectors available and can achieve a collection efficiency of more than 99% for very fine particulates.

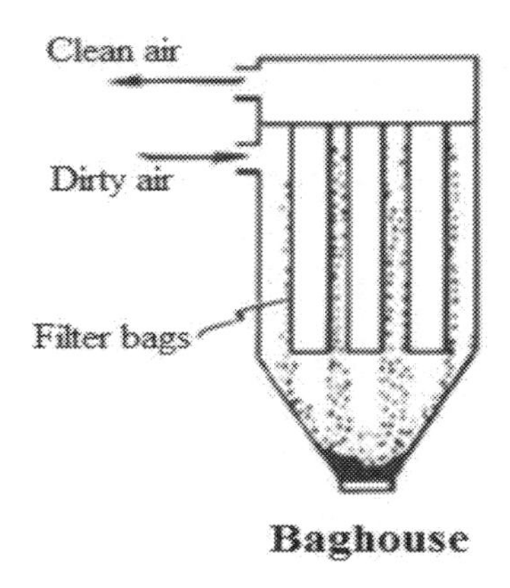

Fig.1.3 Dust collector-Filter

C. Wet scrubbers

Dust collectors that use liquid are commonly known as wet scrubbers. In these systems, the scrubbing liquid (usually water) comes into contact with a gas stream containing dust particles. The greater the contact of the gas and liquid streams, the higher the dust removal efficiency.

A wet electrostatic precipitator (WESP or wet ESP) operates with saturated air streams (100% relative humidity). One type of WESP uses a vertical cylindrical tube with centrally-located wire electrode (gas flowing upward) with water sprays to clean the collected particulate from the collection surface (plates, tubes). The collected water and particulate forms wet film slurry that eliminates the resistivity issues associated with dry ESP's. Another type of WESP (used for coke-oven gas detarring) uses a falling oil film to remove collected material.

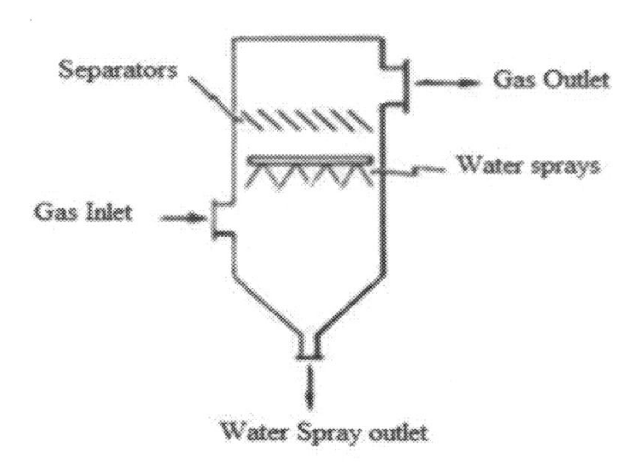

Fig.1.4 Wet Scrubber

D. Electrostatic precipitators

Electrostatic precipitation removes particles from the exhaust gas stream of an industrial process. Often the process involves combustion, but it can be any industrial process that would otherwise emit particles to the atmosphere.

Precipitators function by electrostatically charging the dust particles in the gas stream. The charged particles are then attracted to and deposited on plates or other collection devices. When enough dust has accumulated, the collectors are shaken to dislodge the dust, causing it to fall with the force of gravity to hoppers below. The dust is then removed by a conveyor system for disposal or recycling.

The collected material on the electrodes is removed by rapping or vibrating the collecting electrodes either continuously or at a predetermined interval. Cleaning a precipitator can usually be done without interrupting the airflow. The four main components of all electrostatic precipitators are:

- Power supply unit, to provide high-voltage DC power
- Ionizing section, to impart a charge to particulates in the gas stream
- A means of removing the collected particulates
- A housing to enclose the precipitator zone

Depending upon dust characteristics and the gas volume to be treated, there are many different sizes, types and designs of electrostatic precipitators. Very large power plants may actually have multiple precipitators for each unit.

The following factors affect the efficiency of electrostatic precipitators:

- Larger collection-surface areas and lower gas-flow rates increase efficiency because of the increased time available for electrical activity to treat the dust particles.
- An increase in the dust-particle migration velocity to the collecting electrodes increases efficiency. The migration velocity can be increased by:
    - Decreasing the gas viscosity
    - Increasing the gas temperature
    - Increasing the voltage field

Generally, the electrostatic precipitators have collection efficiency of 99%, but do not work well for fly ash with a high electrical resistivity (as commonly results from combustion of low-sulfur coal). In addition, the designer must avoid allowing unburned gas to enter the electrostatic precipitator since the gas could be ignited.

In contrast to wet scrubbers which apply energy directly to the flowing fluid medium, an ESP applies energy only to the particulate matter being collected and therefore is very efficient in its consumption of energy (in the form of electricity).

E. Unit collectors

Unlike central collectors, unit collectors control contamination at its source. They are small and self-contained, consisting of a fan and some form of dust collector. They are suitable for isolated, portable, or frequently moved dust-producing operations, such as bins and silos or remote belt-conveyor transfer points. Advantages of unit collectors include small space requirements, the return of collected dust to main material flow, and low initial cost. However, their dust-holding and storage capacities, servicing facilities, and maintenance periods have been sacrificed.

2

ELECTROSTATIC PRECIPITATOR PRINCIPLES AND DESIGN

2.1. Introduction

The first use of corona to remove particles from an aerosol was by Hohlfeld in 1824. However, it was not commercialized until almost a century later. In 1907 Dr. Frederick G. Cottrell applied for a patent on a device for charging particles and then collecting them through electrostatic attraction - the first electrostatic precipitator. He was then a professor of chemistry at the University of California, Berkeley. Cottrell first applied the device to the collection of sulfuric acid mist and lead oxide fume emitted from various acid-making and smelting activities. Vineyards in northern California were being adversely affected by the lead emissions.

Prof. Cottrell used proceeds from his invention to fund scientific research through the creation of a foundation called Research Corporation in 1912 to which he assigned the patents. At the time of Cottrell's invention, the theoretical basis for operation was not understood. The operational theory was developed later in the 1920's, in Germany.

Electrophoresis is the term used for migration of gas-suspended charged particles in a direct-current electrostatic field. If your television set accumulates dust on the face it is because of this phenomenon (a CRT is a direct-current machine operating at about 35kV).

An electrostatic precipitator (ESP) - or electrostatic air cleaner is a particulate collection device that removes particles from a flowing gas (such as air) using the electrostatic force of an induced charge. Electrostatic precipitators are highly efficient filtration devices that minimally impede the flow of gases through the device, and can easily remove fine particulate matter such as dust and smoke from the air stream.

Electrostatic precipitators use electrostatic forces to separate dust particles from exhaust gases. A number of high-voltage, direct-current discharge electrodes are placed between grounded collecting electrodes. The contaminated gases flow through the passage formed by the discharge and collecting electrodes. Electrostatic precipitators operate on the same principle as home "Ionic" air purifiers.

Electrostatic precipitator is a device which separates particles from a gas stream by passing the carrier gas between pairs of electrodes across which a unidirectional, high-voltage potential is placed. The particles are charged before passing through the field and migrate to an oppositely charged electrode. These devices are very efficient collectors of small particles, and their use in removing particles from power plant plumes and in other industrial applications are widespread.

Particulate matter (particles) is one of the industrial air pollution problems that must be controlled. It's not a problem isolated to a few industries, but pervasive across a wide variety of industries. That's why the U.S. Environmental Protection Agency (EPA) has regulated particulate emissions and why industry has responded with various control devices. Of the major particulate collection devices used today, electrostatic precipitators (ESPs) are one of the more frequently used. They can handle large gas volumes with a wide range of inlet temperatures, pressures, dust volumes, and acid gas conditions. They can collect a wide range of particle sizes, and they can collect particles in dry and wet states. For many industries, the collection efficiency can go as high as 99%. ESPs aren't always the appropriate collection device, but they work because of electrostatic attraction (like charges repel; unlike charges attract).

A typical schematic diagram of Electrostaic Preceitatore is shown in figures 2.1 and 2.2. Electrostatic precipitators are not only used in utility applications but also other industries (for other exhaust gas particles) such as cement (dust), pulp & paper (salt cake & lime dust), petrochemicals (sulfuric acid mist), and steel (dust & fumes).

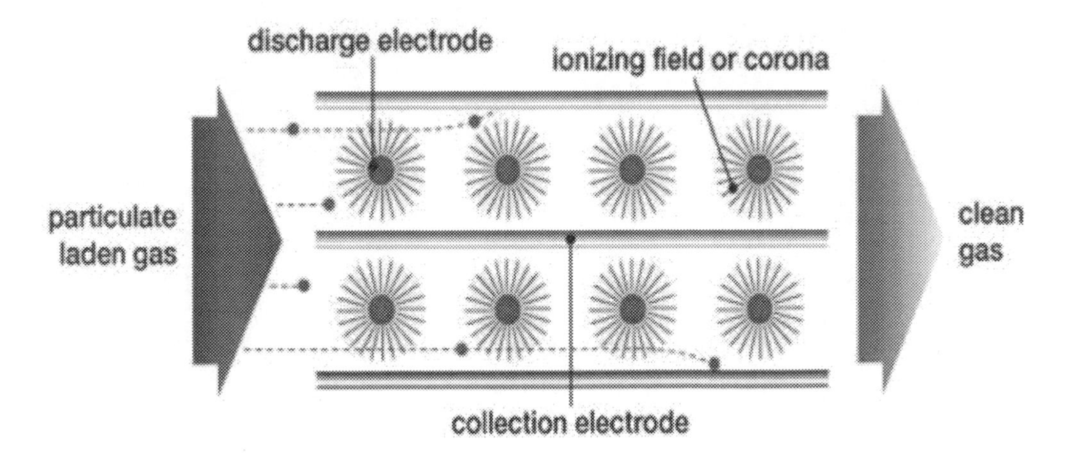

Fig.2.1 Top View of ESP Schematic Diagram [Source: Powerspan Corp.]

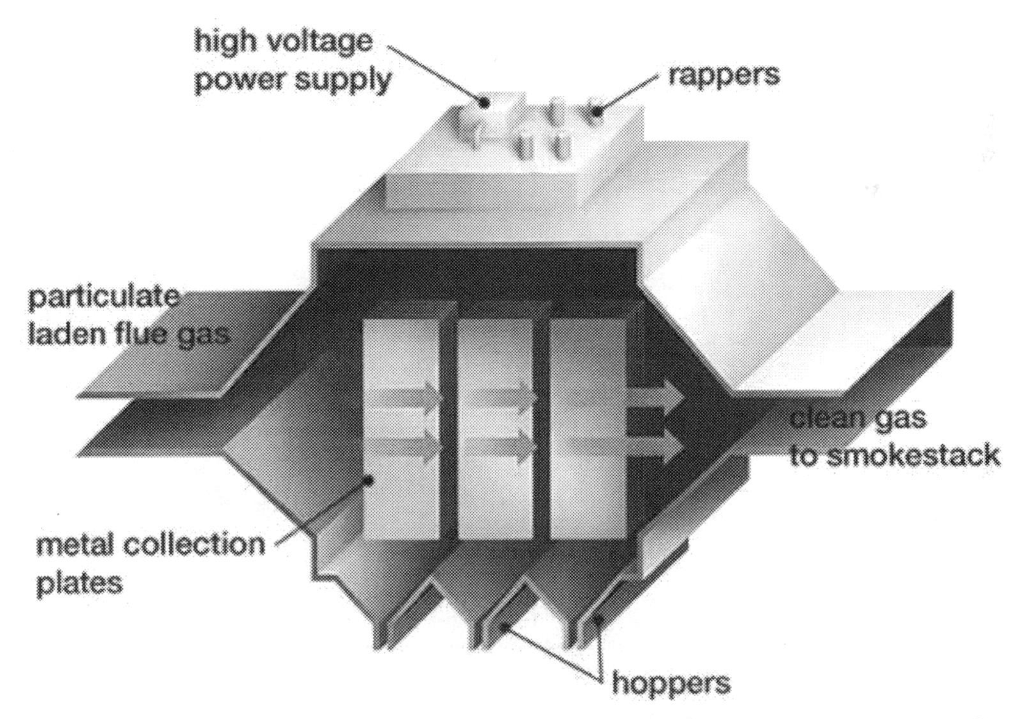

Fig.2.2 Side view of ESP Schematic Diagram [Source: Powerspan Corp.]

2.2. Philosophy and Principles

Every particle either has or can be given a charge - positive or negative. Let's suppose we impart a negative charge to all the particles in a gas stream. Then suppose we set up a grounded plate having a positive charge. What would happen? The negatively charged particle would migrate to the grounded collection plate and be captured. The particles would quickly collect on the plate, creating a dust layer. The dust layer would accumulate until we removed it, which we could do by rapping the plate or by spraying it with a liquid. Charging, collecting, and removing - that's the basic idea of an ESP, but it gets more complicated. The main conceptual theories and principle of Operation precipitation involve the following fields:

A. Particle Charging

Our typical ESP as shown in Figure 2.3 has thin wires called discharge electrodes, which are evenly spaced between large plates called collection electrodes, which are grounded.

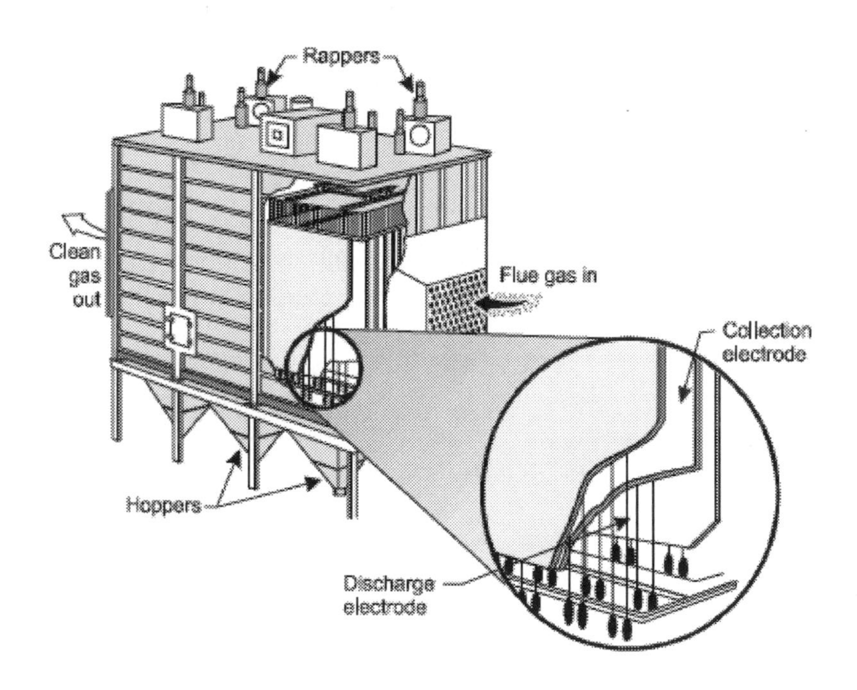

Fig.2.3 Typical dry electrostatic precipitator

A negative, high-voltage, pulsating, direct current is applied to the discharge electrode creating a negative electric field. The electric field strength between the electrodes can be divided into three regions (Figure 2.4). The field is strongest right next to the discharge electrode, weaker in the areas between the discharge and collection electrodes called the inter-electrode region, and weakest near the collection electrode. The region around the discharge electrode is where the particle charging process begins.

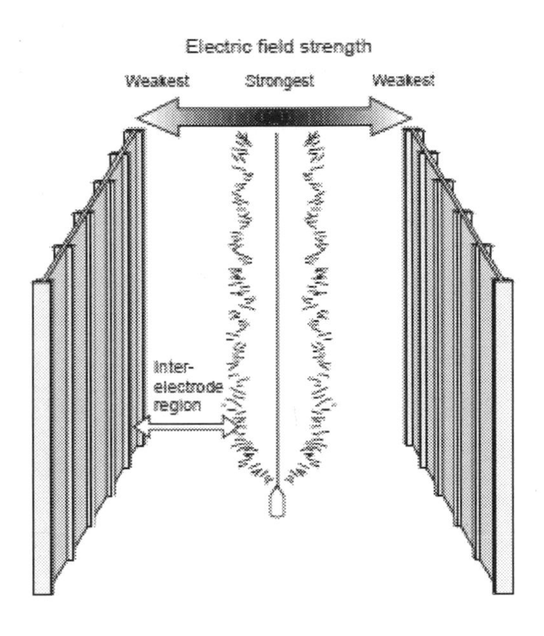

Fig.2.4 ESP electric field

B. Corona Discharge: Free Electron Generation

Several things happen very rapidly (in a matter of a millisecond) in the small area around the discharge electrode. The applied voltage is increased until it produces a corona discharge, which can be seen as a luminous blue glow around the discharge electrode. The free electrons created by the corona are rapidly fleeing the negative electric field, which repulses them. They move faster and faster away from the discharge electrode. This acceleration causes them to literally crash into gas molecules, bumping off electrons in the molecules. As a result of losing an electron (which is negative), the gas molecules become positively charged, that is, they become positive ions (Figure 2.5). So, this is the first thing that happens-gas molecules are ionized, and electrons are liberated. All this activity occurs very close to the discharge electrode.

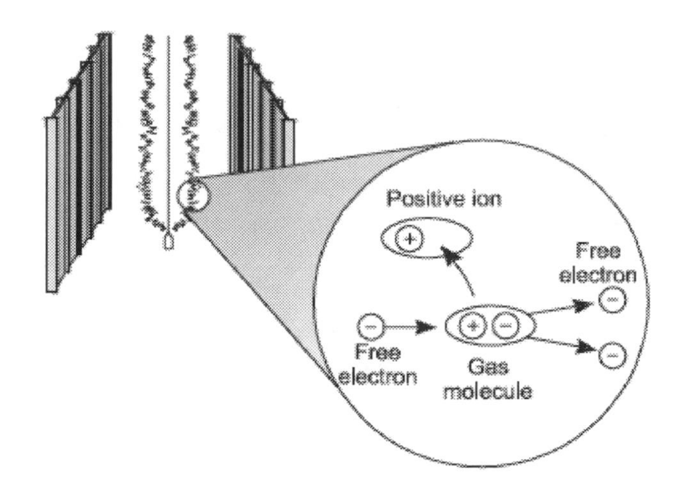

Fig.2.5 Corona generation

This process continues, creating more and more free electrons and more positive ions. The name for all this electron generation activity is avalanche multiplication (Figure 2.6).

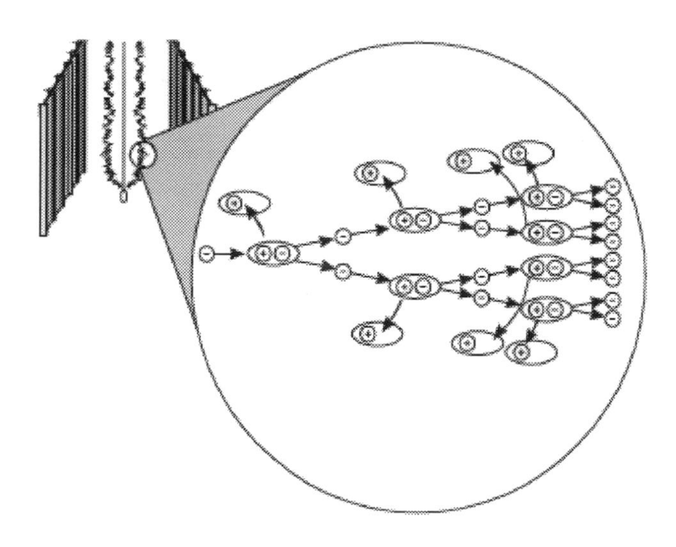

Fig.2.6 Avalanche multiplication of gas molecules

The electrons bump into gas molecules and create additional ionized molecules. The positive ions, on the other hand, are drawn back toward the negative discharge electrode. The molecules are hundreds of times bigger than the tiny electrons and move slowly, but they do pick up speed. In fact, many of them collide right into the metal discharge electrode or the gas space around the wire causing additional electrons to be knocked off. This is called secondary emission. So, this is the second thing that happens. We still have positive ions and a large amount of free electrons.

C. Ionization of Gas Molecules

As the electrons leave the strong electrical field area around the discharge electrode, they start slowing down. Now they're in the inter-electrode area where they are still repulsed by the discharge electrode but to a lesser extent. There are also gas molecules in the inter-electrode region, but instead of violently colliding with them, the electrons kind of bump up to them and are captured (Figure 2.7). This imparts a negative charge to the gas molecules, creating negative gas ions. This time, because the ions are negative, they too want to move in the direction opposite the strong negative field. Now we have ionization of gas molecules happening near the discharge electrode and in the inter-electrode area, but with a big difference. The ions near the discharge electrode are positive and remain in that area. The ions in the middle area are negative and move away, along the path of invisible electric field lines, toward the collection electrode.

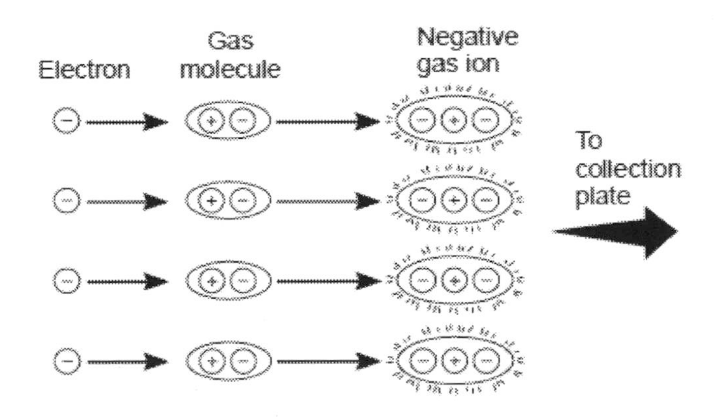

Fig.2.7 Negative gas ions formed in the inter-electrode Region

D. Charging of Particles

These negative gas ions play a key role in capturing dust particles. Before the dust particles can be captured, they must first acquire a negative charge. This is when and where it happens. The particles are traveling along in the gas stream and encounter negative ions moving across their path. Actually, what really happens is that the particles get in the way of the negatively charged gas ions. The gas ions stick to the particles, imparting a negative charge to them. At first the charge is fairly insignificant as most particles are huge compared to a gas molecule. But many gas ions can fit on a particle, and they do. Small particles (less than 1 μm diameter) can absorb “tens” of ions. Large particles (greater than 10 μm) can absorb "tens of thousands" of ions (Turner et al. 1992). Eventually, there are so many ions stuck to the particles, the particles emit their own negative electrical field. When this happens, the negative field around the particle repulses the negative gas ions and no additional ions are acquired. This is called the saturation charge.

Now the negatively-charged particles are feeling the inescapable pull of electrostatic attraction. Bigger particles have a higher saturation charge (more molecules fit) and consequently are pulled more strongly to the collection plate. In other words, they move faster than smaller particles.

Regardless of size, the particles encounter the plate and stick, because of adhesive and cohesive forces. Let's stop here and survey the picture. Gas molecules around the discharge electrode are positively ionized. Free electrons are racing as fast as they can away from the strong negative field area around the discharge electrode. The electrons are captured by gas molecules in the inter-electrode area and impart a negative charge to them. Negative gas ions meet particles and are captured (Figure 2.8). And all this happens in the blink of an eye. The net result is negatively charged particles that are repulsed by the negative electric field around the discharge electrode and are strongly attracted to the collection plate. They travel toward the grounded collection plate, bump into it, and stay there.

More and more particles accumulate, creating a dust layer. This dust layer builds until it is somehow removed. As been said earlier Charging, collecting, and removing is the basic idea of an ESP.

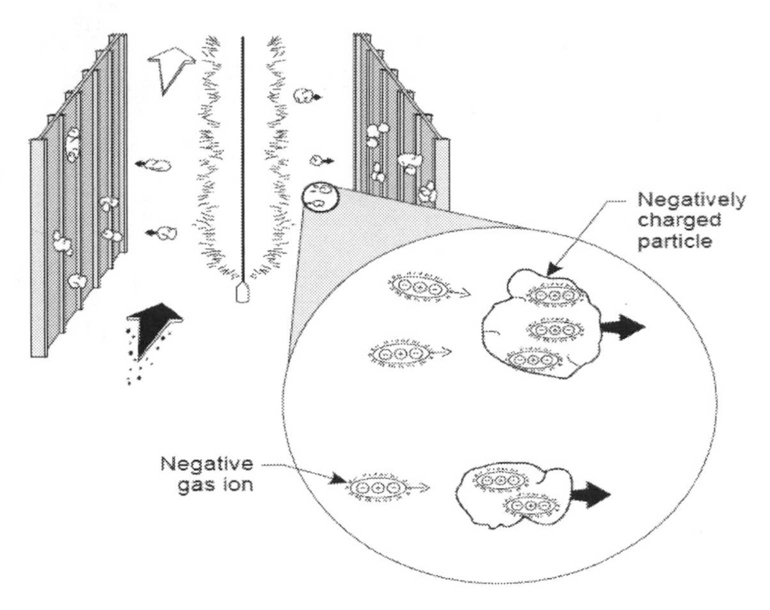

Fig.2.8 Particle charging

E. Particle Charging Mechanisms

Particles are charged by negative gas ions moving toward the collection plate by one of these two mechanisms: field charging or diffusion charging. In field charging (the mechanism described above), particles capture negatively charged gas ions as the ions move toward the grounded collection plate. Diffusion charging, as its name implies, depends on the random motion of the gas ions to charge particles.

In field charging (Figure 2.9), as particles enter the electric field, they cause a local dislocation of the field.

Negative gas ions traveling along the electric field lines collide with the suspended particles and impart a charge to them. The ions will continue to bombard a particle until the charge on that particle is sufficient to divert the electric lines away from it. This prevents new ions from colliding with the charged dust particle. When a particle no longer receives an ion charge, it is said to be saturated. Saturated charged particles then migrate to the collection electrode and are collected.

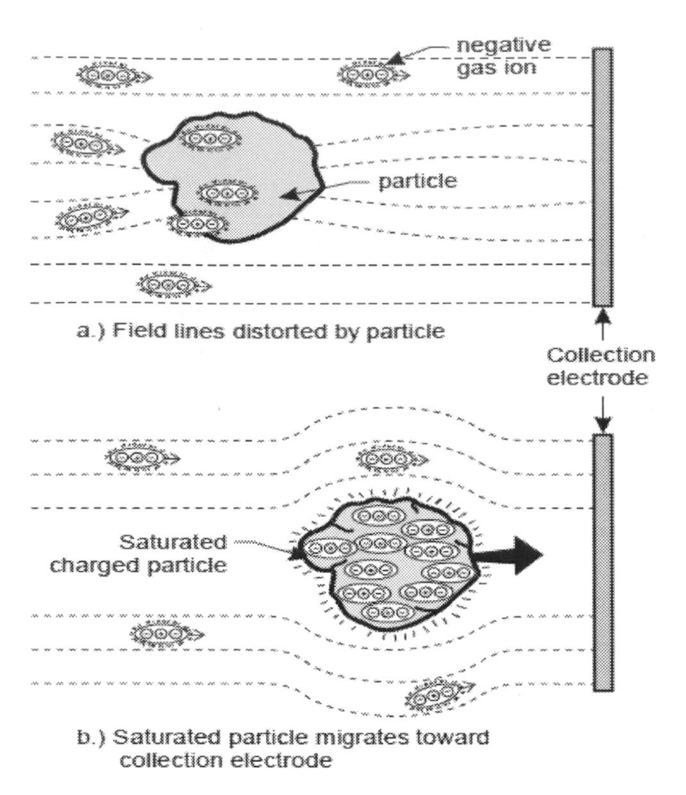

Fig.2.9 Field charging

Diffusion charging is associated with the random Brownian motion of the negative gas ions. The random motion is related to the velocity of the gas ions due to thermal effects: the higher the temperature, the more movement. Negative gas ions collide with the particles because of their random thermal motion and impart a charge on the particles. Because the particles are very small (sub micrometer), they do not cause the electric field to be dislocated, as in field charging. Thus, diffusion charging is the only mechanism by which these very small particles become charged. The charged particles then migrate to the collection electrode.

Each of these two charging mechanisms occurs to some extent, with one dominating depending on particle size. Field charging dominates for particles with a diameter >1.0 micrometer because particles must be large enough to capture gas ions. Diffusion charging dominates for particles with a diameter less than 0.1 micrometer. A combination of these two charging mechanisms occurs for particles ranging between 0.2 and 1.0 micrometer in diameter.

A third type of charging mechanism, which is responsible for very little particle charging is electron charging. With this type of charging, fast-moving free electrons that have not combined with gas ions hit the particle and impart a charge.

F. Electric Field Strength

In the inter-electrode region, negative gas ions migrate toward the grounded collection electrode. A space charge, which is a stable concentration of negative gas ions, forms in the inter-electrode region because of the high electric field applied to the ESP. Increasing the applied voltage to the discharge electrode will increase the field strength and ion formation until sparkover occurs. Sparkover refers to internal sparking between the discharge and collection electrodes. It is a sudden rush of localized electric current through the gas layer between the two electrodes. Sparking causes an immediate short-term collapse of the electric field (Figure 2.10).

For optimum efficiency, the electric field strength should be as high as possible. More specifically, ESPs should be operated at voltages high enough to cause some sparking, but not so high that sparking and the collapse of the electric field occur too frequently. The average sparkover rate for optimum precipitator operation is between 50 and 100 sparks per minute. At this spark rate, the gain in efficiency associated with increased voltage compensates for decreased gas ionization due to collapse of the electric field.

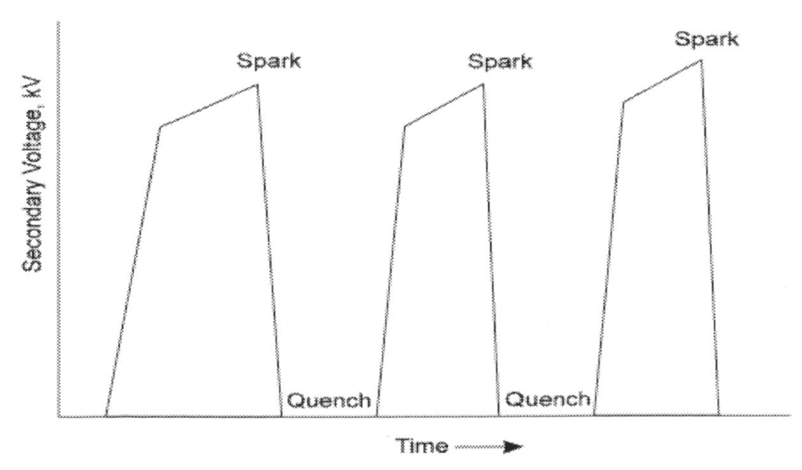

Fig.2.10 Spark generation profile

G. Particle Collection

When a charged particle reaches the grounded collection electrode, the charge on the particle is only partially discharged. The charge is slowly leaked to the grounded collection plate. A portion of the charge is retained and contributes to the inter-molecular adhesive and cohesive forces that hold the particles onto the plates (Figure 2.11). Adhesive forces cause the particles to physically hold on to each other because of their dissimilar surfaces.

Newly arrived particles are held to the collected particles by cohesive forces; particles are attracted and held to each other molecularly. The dust layer is allowed to build up on the plate to a desired thickness and then the particle removal cycle is initiated.

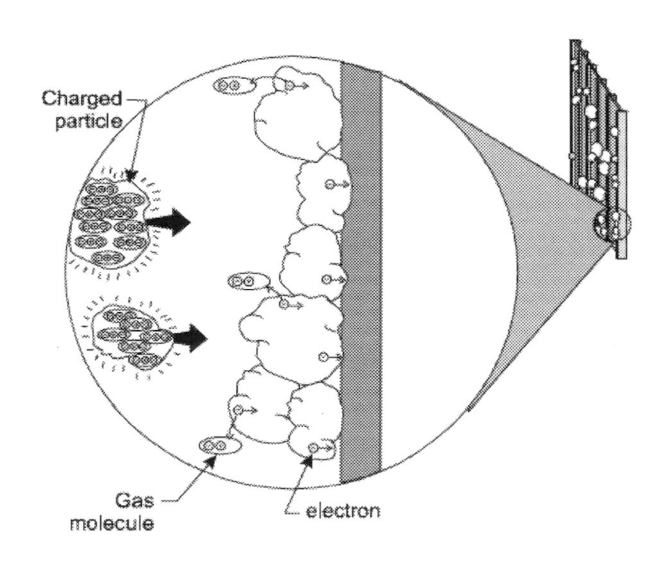

Fig. 2.11 Particle collection at collection electrode

H. Particle Removal

Dust that has accumulated to a certain thickness on the collection electrode is removed by one of two processes, depending on the type of collection electrode. As described in greater detail in the next section, collection electrodes in precipitators can be either plates or tubes, with plates being more common. Tubes are usually cleaned by water sprays, while plates can be cleaned either by water sprays or a process called rapping.

Rapping is a process whereby deposited, dry particles are dislodged from the collection plates by sending mechanical impulses, or vibrations, to the plates. Precipitator plates are rapped periodically while maintaining the continuous flue-gas cleaning process. In other words, the plates are rapped while the ESP is on-line; the gas flow continues through the precipitator and the applied voltage remains constant. Plates are rapped when the accumulated dust layer is relatively thick (0.08 to 1.27 cm or 0.03 to 0.5 in.). Most precipitators have adjustable rappers so that rapper intensity and frequency can be changed according to the dust concentration in the flue gas. Installations where the dust concentration is heavy require more frequent rapping.

Dislodged dust falls from the plates into the hopper. The hopper is a single collection bin with sides sloping approximately 50 to 70^0 to allow dust to flow freely from the top of the hopper to the discharge opening. Dust should be removed as soon as possible to avoid (dust) packing. Packed dust is very difficult to remove. Most hoppers are emptied by some type of discharge device and then transported by a conveyor.

In a precipitator using liquid sprays to remove accumulated liquid or dust, the sludge collects in a holding basin at the bottom of the vessel. The sludge is then sent to settling ponds or lined landfills for proper ultimate disposal. Spraying occurs while the ESP is on-line and is done intermittently to remove the collected particles. Water is generally used as the spraying liquid although other liquids could be used if absorption of gaseous pollutants is also being accomplished.

2.3. Design Methods and Parameters

The precipitator should be designed to provide easy access to strategic points of the collector for internal inspection of electrode alignment, for maintenance, and for cleaning electrodes, hoppers, and connecting flues during outages. The overall design, including the specific components, is based on engineering specifications and/or previous experience with the industrial application using various individual parameters that are appropriate for each specific situation.

To determine and insure the design parameters of Electro Static Precipitators (ESP) various methods of design are usually implemented by the design offices of the Precipitators Manufacturers, these may cover the following procedures:

- Using mathematical equations and models to estimate the collection efficiency or collection area for various configurations.
- Building a pilot-plant to find out the parameters of the full-scale ESP.
- Applying a computer program to test the design features and operating parameters in a simulation of the final design.

These design methods are briefly described as follow:

1. <u>First Method</u>: Using collection efficiency or collection area

The **collection efficiency** of an electrostatic precipitator is strongly dependent on the electrical properties of the particles. A widely taught concept to calculate the collection efficiency is the Deutsch model, which assumes infinite remixing of the particles perpendicular to the gas stream.

Collection efficiency is the primary consideration of ESP design. The collection efficiency and/or the collection area of an ESP can be estimated using several equations. A number of operating parameters can affect the collection efficiency of the precipitator as following:

A. <u>Particle-Migration Velocity</u>

The particle-migration velocity is the speed at which a particle, once charged, migrates toward the grounded collection electrode .Variables affecting particle velocity are particle size, the strength of the electric field, and the viscosity of the gas. Based on the available data of the particle size and electric field strength, the particle-migration velocity can be calculated using either Eqs.2.1 or 2.2 below:

$$w = \frac{d_p E_0 E_p}{4\pi\gamma} \tag{2.1}$$

Where: dp = diameter of the particle, μm
Eo = strength of field in which particles are charged (represented by peak voltage), V/m
Ep = strength of field in which particles are collected (normally the field close to the collecting plates), V/m
γ = gas viscosity, Pa • s (cp)

or $$w = \frac{qE_p}{6\pi\gamma r} \tag{2.2}$$

Where: q = particle charge(s)
r = radius of the particle, μm

Typical particle migration velocity rates, such as those listed in Table 2.1 have been published by various ESP vendors.

Table 2.1 Typical particle-migration velocity rates for various applications

Application	Migration velocity-(cm/sec)
Utility fly ash	4.0-20.4
Pulverized coal fly ash	10.1-13.4
Pulp and paper mills	6.4-9.5
Sulfuric acid mist	5.8-7.62
Cement (wet process)	10.1-11.3
Cement (dry process)	6.4-7.0
Gypsum	15.8-19.5
Smelter	1.8
Open-hearth furnace	4.9-5.8
Blast furnace	6.1-14.0
Hot phosphorous	2.7
Flash roaster	7.6
Multiple-hearth roaster	7.9
Catalyst dust	7.6
Cupola	3.0-3.7

Sources: Theodore and Buonicore 1976; U.S. EPA 1979.

B. *Collection Efficiency*

The simplest form of collection efficiency of the precipitator known as the Deutsch-Anderson equation is given below.

$$\eta = 1 - e^{-w\left(\frac{A}{Q}\right)} \tag{2.3}$$

Where: η = collection efficiency of the precipitator
w = migration velocity, cm/s
A = the effective collecting plate area of the precipitator, m^2
Q = gas flow through the precipitator, m^3/s

This equation is considered under ideal conditions to calculate theoretical collection efficiencies where three significant process parameters are neglected. This approximation can cause the results to be in error by a factor of 2 or more. These three significant process parameters are:

1-The dust re-entrainment; it may occur during the rapping process.
2-The particle size and, consequently, the migration velocity; which are assumed uniform for all particles in the gas stream. Larger particles generally have higher migration velocity rates than smaller particles do.
3-The gas flow rate; it is assumed uniform everywhere across the precipitator and that particle sneakage (particles escape capture) through the hopper section does not occur.

Therefore, this equation should be used only for making preliminary estimates of precipitator collection efficiency.

More accurate estimates of collection efficiency can be obtained by modifying the Deutsch-Anderson equation.

This is accomplished by substituting the effective precipitation rate w_e (calculated from field experience) in place of the migration velocity, w,

$$\eta = 1 - e^{-w_e\left(\frac{A}{Q}\right)} \tag{2.4}$$

The effective precipitation rate represents a semi-empirical parameter that can be used to determine the total collection area necessary for an ESP to achieve a specified collection efficiency required to meet an emission limit.

Using the modified Deutsch-Anderson equation in this manner could be particularly useful when trying to determine the amount of additional collection area needed to upgrade an existing ESP to meet more stringent regulations or to improve the performance of the unit. However, other operating parameters besides collection area play a major role in determining the efficiency of an ESP.

Another modification to the Deutsch-Anderson equation that accounts for non-ideal effects is by decreasing the calculation of collection efficiency by a factor of k. This was devised by Sigvard Matts and Per-Olaf Ohnfeldt of Sweden (Svenska Flaktfabriken) in 1964. The Matts-Ohnfeldt equation is

$$\eta = 1 - e^{-w_k\left(\frac{A}{Q}\right)^k} \qquad (2.5)$$

Where: w_k= average migration velocity, cm/s
k = a constant, usually 0.4 to 0.6

These calculations are used in establishing preliminary design parameters of ESPs.

In an Electric Power Research Institute (EPRI) study, a table was constructed to show the relationship of predicting collection efficiency using the Deutsch-Anderson and Matts- Ohnfeldt equations. This information is given in Table 2.2.

Table 2.2 Collection-efficiency estimations using the Deutsch Anderson and Matts-Ohnfeldt equations

Relative size of ESP (A/Q)	**Deutsch-Anderson (k = 1.0)**	**Matts-Ohnfeldt**		
		k=0.4	k=0.5	k=0.6
1	90	90	90	90
2	99	95.1	96.2	97.2
3	99.9	97.2	98.1	98.8
4	99.99	98.1	99	99.5
5	99.999	98.7	99.6	99.76

When k = 1.0, the Matts-Ohnfeldt equation is the same as the Deutsch Anderson equation. To predict the collection efficiency of an existing ESP when the collection area or gas flow rate is varied, using lower values for k gives more conservative results.

From Table 2.2, you can see that the efficiency estimates calculated using the Matts-Ohnfeldt equation are more conservative than those estimated using the Deutsch-Anderson equation, and may more likely predict how efficiently the ESP will actually operate.

C. ***Specific Collection Area***

The specific collection area (SCA) is defined as the ratio of collection surface area to the gas flow rate into the collector. This ratio represents the A/Q relationship in the Deutsch- Anderson equation and consequently is an important determinant of collection efficiency.

The SCA is given in Equation 2.6.

$$\text{SCA} = \frac{total\, collection\, surface\, area}{Gas\, flow\, rate} \qquad (2.6)$$

Expressed in metric units, $\text{SCA} = \frac{total\, collection\; area\, in\, m^2}{1000\; m^3/h}$

Expressed in English units, $\text{SCA} = \frac{total\, collection\; area\, in\, ft^2}{1000\; ft^3/h}$

For example, if the total collection area of an ESP is 600,000 $ft^2$and the gas flow rate through the ESP is 1,000,000ft^3 /min (acfm), the SCA is 600 ft^2 per 1000 acfm as calculated below.

$$\text{SCA} = \frac{600{,}000\; ft^2}{1000\; (1000\; acfm)} = \frac{600\; ft^2}{(1000\; acfm)}$$

Increases in the SCA of a precipitator design will, in most cases, increase the collection efficiency of the precipitator. Most conservative designs call for an SCA of 20 to 25 m^2 per 1000m^3 /h (350 to 400 ft^2per 1000 acfm) to achieve collection efficiency of more than 99.5%. The general range of SCA is between 11 and 45 m^2per 1000 m^3/hr (200 and 800 ft^2per 1000 acfm), depending on precipitator design conditions and desired collection efficiency.

D. Aspect Ratio

The aspect ratio, which relates the length of an ESP to its height, is an important factor in reducing rapping loss (dust re-entrainment).When particles are rapped from the electrodes, the gas flow carries the collected dust forward through the ESP until the dust reaches the hopper. Although the amount of time it takes for rapped particles to settle in the hoppers is short (a matter of seconds), a large amount of "collected dust" can be re-entrained in the gas flow and carried out of the ESP if the total effective length of the plates in the ESP is small compared to their effective height. For example, the time required for dust to fall from the top of a 9.1-m plate (30-ft plate) is several seconds. Effective plate lengths must be at least 10.7 to 12.2 m (35 to 40 ft) to prevent a large amount of "collected dust" from being carried out of the ESP before reaching the hopper.

The aspect ratio is the ratio of the effective length to the effective height of the collector surface. The aspect ratio can be calculated using Equation 2.7.

$$AR = \frac{effective\ length\ , m(ft)}{effective\ height\ , m(ft)} \qquad (2.7)$$

The effective length of the collection surface is the sum of the plate lengths in each consecutive field and the effective height is the height of the plates. For example, if an ESP has four fields, each containing plates that are 10 feet long, the effective length is 40 feet. If the height of each plate is 30 feet, the aspect ratio is 1.33.

Aspect ratios for ESPs range from 0.5 to 2.0. However, for high-efficiency ESPs (those having collection efficiencies of $> 99\%$), the aspect ratio should be greater than 1.0 (usually 1.0 to 1.5) and in some installations may approach 2.0.

2. <u>Second Method</u>: Using Computer Programs and Models

Engineers can also use mathematical models or computer programs to design precipitators. A mathematical model that relates collection efficiency to precipitator size and various operating parameters can be developed to do the following:

- Design a full-scale ESP from fundamental principles or in conjunction with a pilot plant study.
- Evaluate ESP bids submitted by various manufacturers
- Troubleshoot and diagnose operating problems for existing ESPs

- Evaluate the effectiveness of new ESP developments and technology, such as flue gas conditioning and pulse energizing.

Also, the model can utilize the localized electric field strengths and current densities prevailing throughout the precipitator. These data can be input based on actual readings from operating units, or can be calculated based on electrode spacing and resistivity.

These performance models require detailed information concerning the anticipated configuration of the precipitator and the gas stream characteristics. It is readily apparent that all of these parameters are not needed in each case, since some can be calculated from the others.

The following data is data utilized in the computerized performance model for electrostatic precipitators:

A. <u>ESP Design</u>

- Specific collection area
- Collection plate area
- Collection height and length
- Gas velocity
- Number of fields in series
- Number of discharge electrodes
- Type of discharge electrodes
- Discharge electrode-to-collection plate spacing

B. <u>Particulate Matter and Gas Stream Data</u>

- Resistivity
- Particle size mass median diameter
- Particle size distribution standard deviation
- Gas flow rate distribution standard deviation
- Actual gas flow rate
- Gas stream temperature
- Gas stream pressure
- Gas stream composition

3. Third Method: Pilot Plant

Probably the most reliable method for designing ESPs is to construct and operate a pilot plant. However, time limitations and the expense of construction may make this undesirable.

Construction of a pilot ESP plant is by all means is not an alternative method to the other metods of designing but it is a tool for examining and enhancing the design parameters of the ESP plant. A pilot ESP project can be constructed on an existing industrial process. In this case, a side stream of flue gas is sent to the small pilot ESP. Flue gas sampling gives valuable information such as gas temperature, moisture content, and dust resistivity. Relating these parameters to the measured collection efficiency of the pilot project will help the design engineers plan for scale-up to a full-sized ESP.

A pilot plant on ESP is rather a fairly reliable tool to examin not only the designing of ESP project, but all related issues and components concerning the evaluation parameters for the actual size of the project/plant. These issues and parametes cover the technical parts along the phases of design, construction and operations of the plant/project. Furthermor, issues related to the cost evaluation, time of construction and required resources shall be well judged in more reliable way.

2.4. Design Variables and Factors

The principal design variables are the dust concentration, measured in g/m^3 (lb/ft^3 or gr/ft^3) and the gas flow rate to the ESP, measured in m3/min (ft3/min or acfm). The gas volume and dust concentration (loading) are set by the process exhaust gas flow rate. Once these variables are known, the design of the precipitator for the specific application can be specified. A detailed review of ESP design plans should consider the following factors:

A. **Physical and chemical properties:**
Physical and chemical properties of the dust such as dust type, size of the dust particles, and average and maximum concentrations in the gas stream are important ESP design considerations. The type of dust to be collected in the ESP refers to the chemical characteristics of the dust such as explosiveness. For example, a dry ESP should not be used to collect explosive dust. In this case, it might be a better idea to use a baghouse or scrubber.

Particle size is important; small particles are more difficult to collect and become re-entrained more easily than larger particles. Additional fields may be required to meet regulatory limits.

The dust loading can affect the operating performance. If the dust concentration is too high, the automatic voltage controller may respond by totally suppressing the current in the inlet fields. Suppressed current flow drives the voltage up, which can cause sparking. For this reason, it might be a good idea to install a cyclone or multi-cyclone to remove larger particles and reduce the dust concentration from the flue gas before it enters the ESP. The facility could install a larger ESP (with more plate area), however, this technique would be more costly.

Resistivity is a function of the chemical composition of the dust, the flue gas temperature and moisture concentration. For fly ash generated from coal-fired boilers, the resistivity depends on the temperature and moisture content of the flue gas and on the sulfur content of the coal burned; the lower the sulfur content, the higher the resistivity, and vice versa. If a boiler burns low-sulfur coal, the ESP must be designed to deal with potential resistivity problems. High resistivity can be reduced by spraying water, SO3 or some other conditioning agent into the flue gas before it enters the ESP.

B. **Gas flow rate and gas stream properties:**
Predicting the gas flow rate and gas stream properties is essential for proper ESP design. The average and maximum gas flow rates through the ESP, the temperature, moisture content, chemical properties such as dew point, corrosiveness, and combustibility of the gas should be identified prior to final design. If the ESP is going to be installed on an existing source, a stack test should be performed to determine the process gas stream properties. If the ESP is being installed on a new source, data from a similar plant or operation may be used, but the ESP should be designed conservatively (with a large SCA, a high aspect ratio, and high corona power). Once the actual gas stream properties are known, the designers can determine if the precipitator will require extras such as shell insulation for hot-side ESPs, corrosion-proof coatings, and installation of heaters in hoppers or ductwork leading into and out of the unit.

C. **Discharge electrodes:**
The type of discharge electrodes and electrode support are important. Small-diameter wires should be firmly supported at the top and connected to a weight heavy enough (11.4-kg weights for 9.1-m wires) to keep the wires from swaying. The bottom and top of each wire should be covered with shrouds to help minimize sparking and metal erosion at these points. Newer ESPs are generally using rigid-frame or rigid-electrode discharge electrodes.

D. **Collection electrodes:**
Collection electrodes - type (either tube or plate), shape of plates, size, and mechanical strength - are then chosen. Plates are usually less than 9 m (30 ft) high for high-efficiency ESPs. For ESPs using wires, the spacing between collection plate electrodes usually ranges from 15 to 30 cm (6 to 12 in.). For ESPs using rigid-frame or rigid electrodes, the spacing is typically 30 to 38 cm (12 to 15 inches). Equal spacing must be maintained between plates throughout the entire precipitator. Stiffeners may be used to help prevent the plates from warping, particularly when hot-side precipitators are used.

E. **Electrical Sectionalization:**

Proper electrical sectionalization is important to achieve high collection efficiency in the ESP. Electrical sectionalization refers to the division of a precipitator into a number of different fields and cells, each powered by its own Transformer-Rectifier T-R set. ESPs should have at least three to four fields to attain a high collection efficiency. In addition, the greater the number of fields the better the chance that the ESP will achieve the designed collection efficiency. There should be approximately one T-R set for every 930 to 2970 m^2 (10,000 to 30,000 ft^2) of collection-plate area.

F. **Specific Collection Area (SCA):**

The SCA is the collection area, in m^2 per 1000 m^3/h (ft^2 per 1000 ft^3/min), of flue gas through the precipitator. The typical range for SCA is between 11 and 45 m^2 per 1000 m^3/h (200 and 800 ft^2 per 1000 acfm). The SCA must be large enough to efficiently collect particles (99.5% collection efficiency), but not so large that the cost of the ESP is too high. If the dust has a high resistivity, vendors will generally design the ESP with a higher SCA [usually greater than 22 m^2 per 1000 m^3/h (400 ft^2 per 1000 acfm)] to help reduce resistivity problems.

G. **Aspect ratio:**

Aspect ratio is the ratio of effective length to height of the collector surface. The aspect ratio should be high enough to allow the rapped particles to settle in the hopper before they are carried out of the ESP by the gas flow. The aspect ratio is usually greater than 1.0 for high-efficiency ESPs. Aspect ratios of 1.3 to 1.5 are common, and they are occasionally as high 2.0.

H. **Distribution of gas flow:**

Even distribution of gas flow across the entire precipitator unit is critical to ensure collection of the particles. To assure even distribution, gas should enter the ESP through an expansion inlet plenum containing perforated diffuser plates (see Figure 2.12).

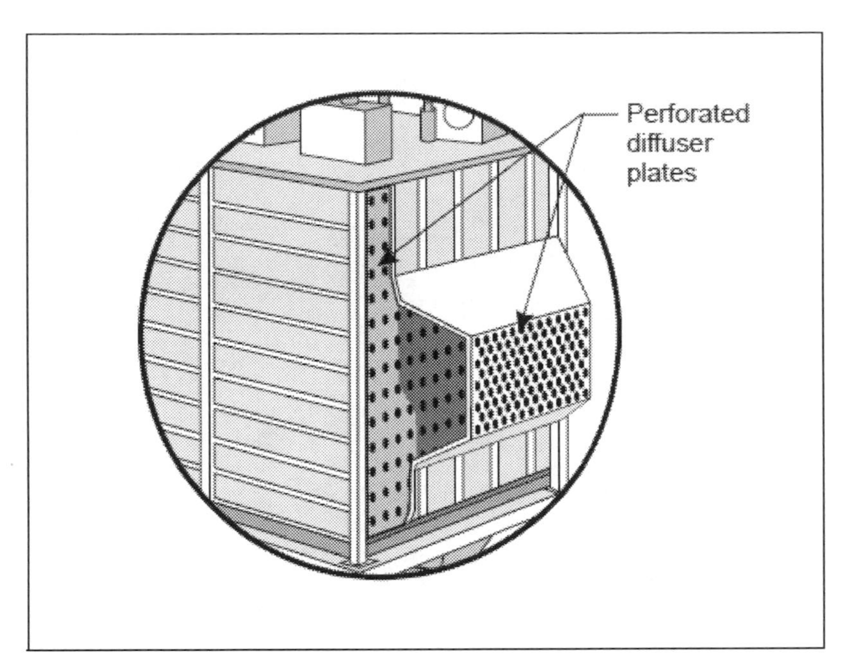

Fig. 2.12 Gas inlet with perforated diffuser plates

In addition, the ducts leading into the ESP unit should be straight as shown in Figure 2.13. For ESPs with straight-line inlets, the distance of **A** should be at least as long as the distance of **B** in the inlet (Katz 1979).

In situations where a straight-line inlet is not possible and a curved inlet must be used (see Figure 2.14.), straightening vanes should be installed to keep the flue gas from becoming stratified.

The gas velocity through the body of the ESP is approximately between 0.6 to 2.4 m/s (2 to 8 ft/ sec). For ESPs having aspect ratios of 1.5, the optimum gas velocity is usually between 1.5 and 1.8m/s (5 and 6 ft/sec). The outlet of the ESP should also be carefully designed to provide even flow of the gas from the ESP to the stack without excessive pressure buildup. This can be done by using an expansion outlet, as shown in Figure 2.15. Figures 2.13 and 2.14 also have expansion outlets.

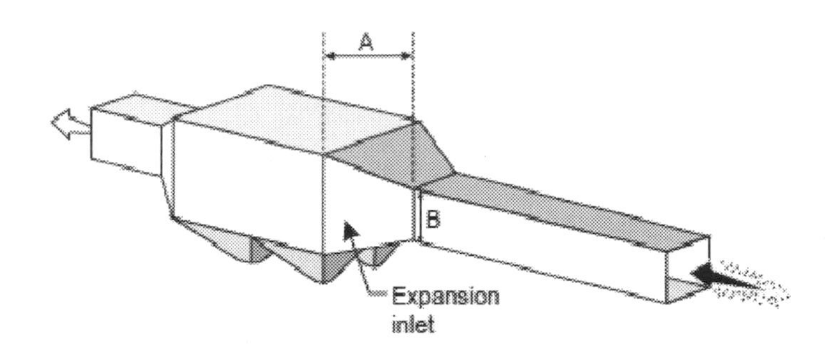

Fig.2.13 Straight-line inlet

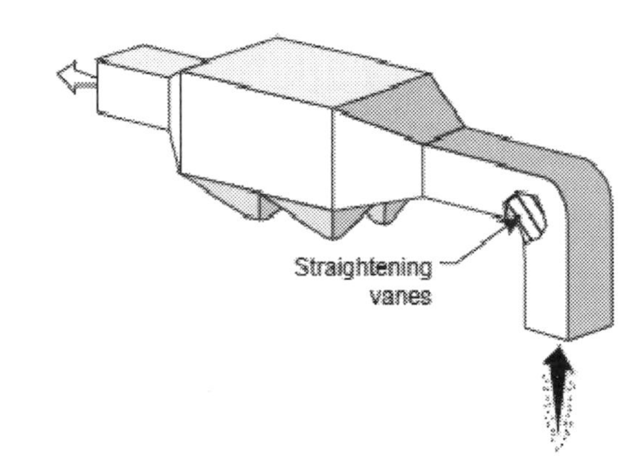

Fig.2.14 Straightening vanes in a curved inlet

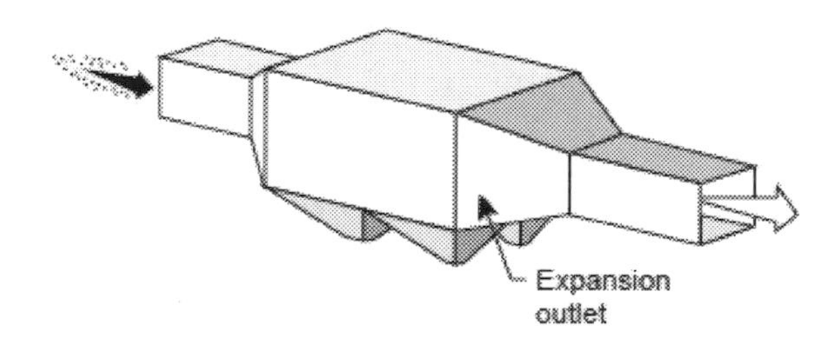

Fig.2.15 ESP with expansion outlet

I. **Hopper and discharge device:**

The hopper and discharge device design including geometry, size, dust storage capacity, number, and location are important so that dust is removed on a routine basis. A well-designed dust hopper is sloped (usually 60°) to allow dust to flow freely to discharge devices. It includes access ports and strike plates to help move dust that becomes stuck. Dust should be only temporarily stored in the hopper and removed periodically by the discharge devices to prevent it from backing up into the ESP where it can touch the plates, possibly causing a cell to short out. In addition to the amount of fly ash present, there are a couple of special considerations to keep in mind when ESPs are used on coal-fired boilers. First, the amount of fly ash in the flue gas can vary depending on what type of coal is burned and the ash content of the coal.

Coal having a higher ash content will produce more fly ash than coal having lower ash values. Consequently, the discharge device must be designed so that the operator can adjust the frequency of fly ash removal. Second, hoppers need to be insulated to prevent ash from "freezing," or sticking, in the hopper.

J. Emission regulations:

Emission regulations in terms of opacity and dust concentration (grain-loading) requirements will ultimately play an important role in the final design decisions. Electrostatic precipitators are very efficient; collection efficiency can usually be greater than 99% if the ESP is properly designed and operated.

2.5. Typical Ranges of Design Parameters

While reviewing a permit for ESP installation, check whether the design specifications are within the range that is typically used by that industry. The ranges of basic design parameters for fly ash precipitators are given in Table 2.3.

Table 2.3. Typical ranges of design parameters for fly ash Precipitators

Parameter	Range (metric units)	Range (English units)
Distance between plates(duct width)	20-30 cm (20-23 cm optimum)	8-12 in. (8-9 in. optimum)
Gas velocity in ESP	1.2-2.4 m/s (1.5-1.8 m/s optimum)	4-8 ft/sec (5-6 ft/sec optimum)
SCA	11-45 m^2/1000 m^3/h (16.5-22.0 m^2/1000 m^3//h optimum)	200-800 ft^2/1000 cfm (300-400 ft^2/1000 cfm optimum)
Aspect ratio (L/H)	1-1.5 (keep plate height less than 9 m for high efficiency)	1-1.5 (keep plate height less than 30 ft for high efficiency)
Particle migration velocity	3.05-15.2 cm/s	0.1-0.5 ft/sec
Number of fields	4-8	4-8
Corona Power / flue gas Volume	59-295 watts/1000 m^3/h	100-500 watts/1000 cfm
Corona Current / ft^2 plate Area	107-860 microamps/m^2	10-80 microamps/ft^2
Plate area per electrical (T-R) set	465-7430 m^2/T-R set (930-2790 m^2/T-R set optimum)	5000-80,000 ft^2/T-R set (10,000- 30,000 ft^2/T-R set optimum)

The designs and sizes the electrostatic precipitator are determined by the designers and manufacturers. However, the operator (or reviewer) needs to check or estimate the collection efficiency and the amount of collection area required for a given process flow rate. Deutsch-Anderson or Matts-Ohnfeldt equations (stated earlier in this chapter) can be used to compute these estimates. These equations are repeated in Table 2.4.

Table 2.4. Equations used to estimate collection efficiency and collection area

Calculation	**Deutsch-Anderson**	**Matts-Ohnfeldt**
Collection efficiency	$\eta = 1 - e^{-w\left(\frac{A}{Q}\right)}$	$\eta = 1 - e^{-w_k\left(\frac{A}{Q}\right)^k}$
Collection area (to meet a required efficiency)	$A = \frac{-Q}{W}[\ln(1-\eta)]$	$A = [-(\frac{Q}{W_k})^k[\ln(1-\eta)]]^{1/k}$
Where:	η = collection efficiency A = collection area w = migration velocity Q = gas flow rate ln = natural logarithm	η = collection efficiency A = collection area w_k = average migration velocity k = constant (usually 0.5) ln = natural logarithm

Example:

As an example of estimation for collection efficiency and collection area, consider the exhaust rate of the gas being processed is given as 1,000,000 ft^3/min. The inlet dust concentration in the gas as it enters the ESP is 8 gr/ft^3. If the emission regulations state that the outlet dust concentration must be less than 0.04 gr/ft^3, how much collection area is required to meet the regulations? Use the Deutsch-Anderson equation for this calculation and assume the migration velocity is 0.3 ft/sec.

1. From Table 2.4, using version of Deutsch-Anderson equation to solve the problem:

$$A = \frac{-Q}{w}[\ln(1-\eta)]$$

Where: A = collection area, ft^2
Q = gas flow rate, ft^3/sec
w = migration velocity, ft/sec
η = collection efficiency
ln = natural logarithm

In this example,

$$Q = 1{,}000{,}000\ ft^3/min = 16{,}667\ ft^3/sec, \text{ and } w = 0.3\ ft/sec$$

2. Calculate the collection efficiency,

$$\eta = \frac{Dust_{in} - Dust_{out}}{Dust_{in}} = \frac{8-0.04}{8} = 0.995 \text{ or } 99.5\%$$

3. Calculate the collection area, A, in ft^2

$$A = \frac{16{,}667}{0.3}[\ln(1-0.995)] = 294{,}358\ ft^2$$

2.6. System Selection

2.6.1 Selection of Precipitators -Dust Collector

The main parameters involved in specifying dust collectors include:

1. The velocity of the air stream created by the vacuum producer;
2. System power, the power of the system motor, usually specified in horsepower;
3. Storage capacity for dust and particles, and
4. Minimum particle size filtered by the unit.

Other considerations when choosing a dust collection system include:

1. Temperature,
2. Moisture content, and
3. Possibility of combustion of the dust being collected.

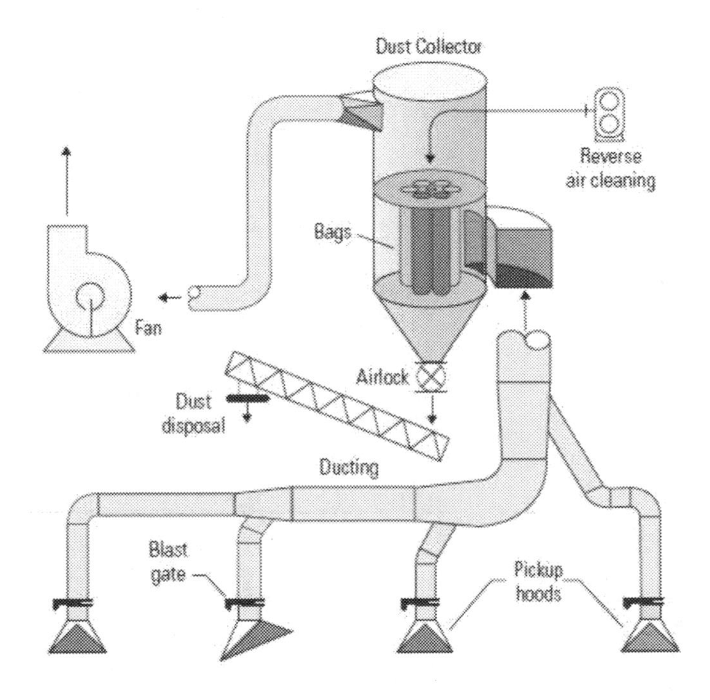

Fig.2.16 Typical Sketch for Dust Collector system

Precipitators vary widely in design, operation, effectiveness, space requirements, construction, and capital, operating, and maintenance costs. Each type has advantages and disadvantages. However, the selection of a dust collector should be based on the following general factors:

- Dust concentration and particle size - For minerals processing operations, the dust concentration can range from 0.1 to 5.0 grains (0.32 g) of dust per cubic feet of air (0.23 to 11.44 grams per standard cubic meter), and the particle size can vary from 0.5 to 100 μm.
- Degree of dust collection required - The degree of dust collection required depends on its potential as a health hazard or public nuisance, the plant location, the allowable emission rate, the nature of the dust, its salvage value, and so forth.
- Characteristics of airstream – E.g. cotton fabric filters cannot be used where air temperatures exceed 180° F (82°C). Also, condensation of steam or water vapor can blind bags. Various chemicals can attach fabric or metal and cause corrosion in wet scrubbers.
- Characteristics of dust – E.g. moderate to heavy concentrations of many dusts (such as dust from silica sand or metal ores) can be abrasive to dry centrifugal collectors. Hygroscopic material can blind bag collectors. Sticky material can adhere to collector elements and plug passages.
- Methods of disposal - Methods of dust removal and disposal vary with the material, plant process, volume, and type of collector used. Collectors can unload continuously or in batches.

2.6.2 Selection of Other related Equipment

The fan and motor system supplies mechanical energy to move contaminated air from the dust-producing source to a dust collector. These equipment should be selected to integrate the performance of the Precipitation - Dust Collection process.

A- **Selection of Fans**

When selecting a fan, the following points should be considered:

- Volume required
- Fan static pressure
- Type of material to be handled through the fan (For example, a radial-blade fan should be used with fibrous material or heavy dust loads, and nonsparking construction must be used with explosive or inflammable materials.)
- Type of drive arrangement, such as direct drive or belt drive

- Space requirements
- Noise levels
- Operating temperature (For example, sleeve bearings are suitable to 250° F/121.1°C; ball bearings to 550° F/287.8°C)
- Sufficient size to handle the required volume and pressure with minimum horsepower
- Need for special coatings or construction when operating in corrosive atmospheres
- Ability of fan to accommodate small changes in total pressure while maintaining the necessary air volume
- Need for an outlet damper to control airflow during cold starts (If necessary, the damper may be interlocked with the fan for a gradual start until steady-state conditions are reached.)

After the above information is collected, the actual selection of fan size and speed is usually made from a rating table published by the fan manufacturer. Such table is known as a multirating table, and it shows the complete range of capacities for a particular size of fan; where the following remarks, also to be considered in reviewing the specification of the fans:

- The multirating table shows the range of pressures and speeds possible within the limits of the fan's construction.
- A particular fan may be available in different construction classes (identified as class I through IV) relating to its capabilities and limits.
- For a given pressure, the highest mechanical efficiency is usually found in the middle third of the volume column.
- A fan operating at a given speed can have an infinite number of ratings (pressure and volume) along the length of its characteristic curve. However, when the fan is installed in a dust collection system, the point of rating can only be at the point at which the system resistance curve intersects the fan characteristic curve.
- In a given system, a fan at a fixed speed or at a fixed blade setting can have a single rating only. This rating can be changed only be changing the fan speed, blade setting, or the system resistance.
- For a given system, an increase in exhaust volume will result in increases in static and total pressures. For example, for a 20% increase in exhaust volume in a system with 5 in. pressure loss, the new pressure loss will be $5 \times (1.20)^2 = 7.2$ in.
- For rapid estimates of probable exhaust volumes available for a given motor size, the equation for brake horsepower, as illustrated, can be useful.

Often, field installation creates airflow problems that reduce the fan's air delivery. The following points should be considered also, when installing the fan:

- Avoid installation of elbows or bends at the fan discharge, which will lower fan performance by increasing the system's resistance.
- Avoid installing fittings that may cause non-uniform flow, such as an elbow, mitred elbow, or square duct.
- Check that the fan impeller is rotating in the proper direction-clockwise or counterclockwise.
- For belt-driven fans;
  - Check that the motor sheave and fan sheave are aligned properly.
  - Check for proper belt tension.
  - Check the passages between inlets, impeller blades, and inside of housing for buildup of dirt, obstructions, or trapped foreign matter.

The operation mechanism of the industrial fans are generally classified in two main categories of design:

A1. Centrifugal fans

Centrifugal fans consist of a wheel or a rotor mounted on a shaft that rotates in a scroll-shaped housing. Air enters at the eye of the rotor, makes a right-angle turn, and is forced through the blades of the rotor by centrifugal force into the scroll-shaped housing. The centrifugal force imparts static pressure to the air. The diverging shape of the scroll also converts a portion of the velocity pressure into static pressure.

There are three main types of centrifugal fans:

- Radial-blade fans - Radial-blade fans are used for heavy dust loads.
- Backward-blade fans - Backward-blade fans operate at higher tip speeds and thus are more efficient. Since material may build up on the blades, these fans should be used after a dust collector.
- Forward-curved-blade fans - These fans have curved blades that are tipped in the direction of rotation. They have low space requirements, low tip speeds, and a low noise factor.

A2. Axial-flow fans

Axial-flow fans are used in systems that have low resistance levels. These fans move the air parallel to the fan's axis of rotation. The screw-like action of the propellers moves the air in a straight-through parallel path, causing a helical flow pattern.

The three main kinds of axial fans are-

- Propeller fans - These fans are used to move large quantities of air against very low static pressures. They are usually used for general ventilation or dilution ventilation and are good in developing up to 0.5 in. wg (124.4 Pa).
- Tube-axial fans - Tube-axial fans are similar to propeller fans except they are mounted in a tube or cylinder. Therefore, they are more efficient than propeller fans and can develop up to 3 to 4 in. wg (743.3 to 995 Pa). They are best suited for moving air containing substances such as condensible fumes or pigments.
- Vane-axial fans - Vane-axial fans are similar to tube-axial fans except air-straightening vanes are installed on the suction or discharge side of the rotor. They are easily adapted to multistaging and can develop static pressures as high as 14 to 16 in. wg (3.483 to 3.98 kPa). They are normally used for clean air only.

B- Selection of Electric Motors

Electric motors are used to supply the necessary energy to drive the fan. Integral -horsepower electric motors are normally three-phase, alternating-current motors. Fractional-horsepower electric motors are normally single-phase, alternating-current motors and are used when less than 1 hp (0.75 kW) is required. Since most dust collection systems require motors with more than 1 hp (0.75 kW), only integral-horsepower motors are discussed here.

The two most common types of integral-horsepower motors used in dust collection systems are:

1. Squirrel-cage motors - These motors have a constant speed and are of a nonsynchronous, induction type.
2. Wound-rotor motors - These motors are also known as slip-ring motors. They are general-purpose or continuous-rated motors and are

chiefly used when an adjustable-speed motor is desired.

Squirrel-cage and wound-rotor motors are further classified according to the type of enclosure they use to protect their interior windings. These enclosures fall into two broad categories:

- Open
- Totally enclosed

Drip-proof and splash-proof motors are open motors. They provide varying degrees of protection; however, they should not be used where the air contains substances that might be harmful to the interior of the motor.

Totally enclosed motors are weather-protected with the windings enclosed. These enclosures prevent free exchange of air between the inside and the outside, but they are not airtight.

Totally enclosed, fan-cooled (TEFC) motors are another kind of totally enclosed motor. These motors are the most commonly used motors in dust collection systems. They have an integral-cooling fan outside the enclosure, but within the protective shield, that directs air over the enclosure.

Both open and totally-enclosed motors are available in explosion-proof and dust-ignition-proof models to protect against explosion and fire in hazardous environments.

Motors are selected to provide sufficient power to operate fans over the full range of process conditions (temperature and flow rate).

3

COMPONENTS OF ESP

All electrostatic precipitators, regardless of their particular designs, contain the following main components:

- Discharge electrodes
- Collection electrodes
- High voltage equipment
- Gas Distribution Systems
- Rappers
- Hoppers
- Ductwork (Shell)

Figure 3.1 shows a typical ESP with wires for discharge electrodes and plates for collection electrodes. This ESP is used to control particulate emissions in many different industries.

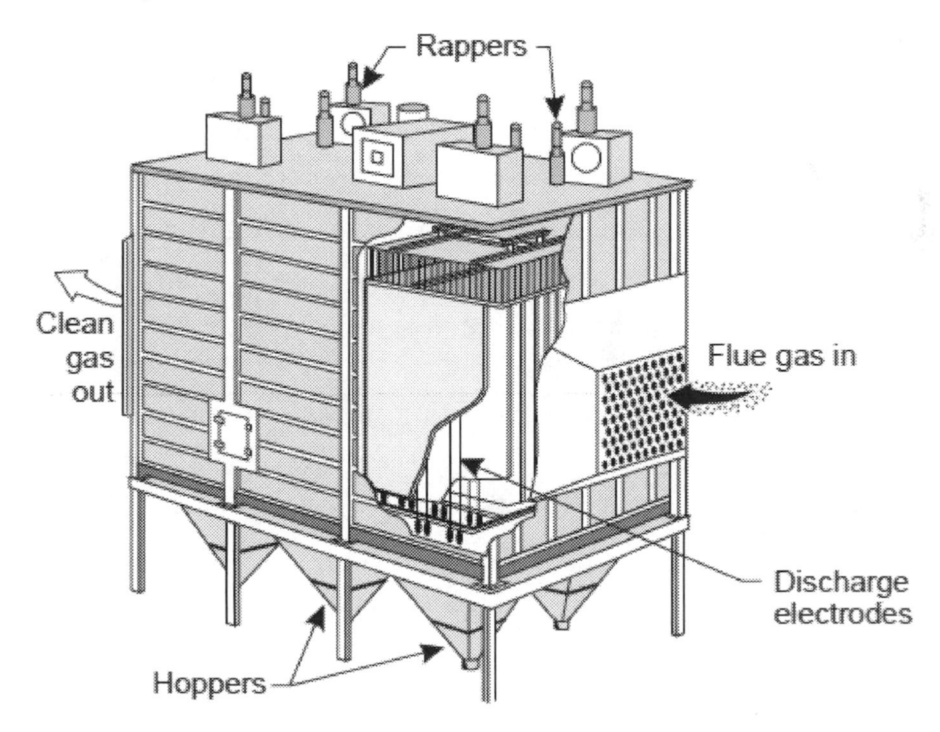

Figure 3.1.Typical dry electrostatic precipitator

These components are briefly described in the following sections.

3.1. Discharge Electrodes

Discharge electrodes are either small-diameter metal wires attached together in rigid frames and hung vertically, or a rigid electrode made from a single piece of fabricated metal

Discharge electrodes emit charging current and provide voltage that generates a strong electrical field that ionizes flue gas between the discharge electrodes and the collecting plates. The electrical field forces dust particles in the gas stream to migrate toward the collecting plates and precipitate onto the collecting plates. Common types of discharge electrodes include:

- Straight round wires
- Twisted wire pairs
- Barbed discharge wires
- Rigid masts
- Rigid frames
- Rigid spiked pipes
- Spiral wires

Discharge electrodes are typically supported from the upper discharge frame and are held in alignment between the upper and lower discharge frames, where High-voltage insulators are incorporated. In weighted wire systems, the discharge electrodes are held taut by weights at the lower end of the wires to help maintain wire alignment and to prevent them from falling into the hopper in the event that the wire breaks (Figure 3.2).

The discharge electrodes in most precipitator designs (prior to the 1980s) are thin, round wires varying from 0.13 to 0.38 cm (0.05 to 0.15 in.) in diameter. The most common size diameter for wires is approximately 0.25 cm (0.1 in.). The wires are usually made from high-carbon steel, but have also been constructed of stainless steel, copper, titanium alloy, and aluminum. The weights are made of cast iron and are generally 11.4 kg (25 lb) or more.

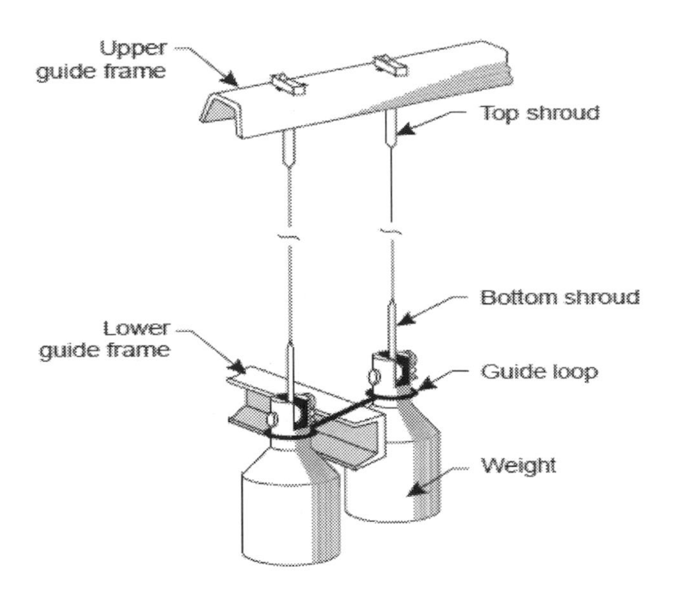

Figure 3.2 Guide frames and shrouds for discharge wires

The size and shape of the electrodes are governed by the mechanical requirements for the system, such as the industrial process on which ESPs are installed and the amount and properties of the flue gas being treated. Most U.S. designs have traditionally used thin, round wires for corona generation. Some designers have also used twisted wire, square wire, barbed wire, or other configurations, as illustrated in Figure 3.3.

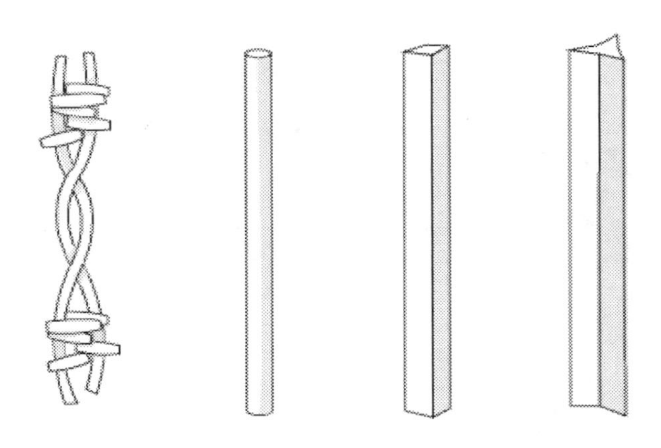

Figure 3.3 Typical wire discharge electrodes

European precipitator manufacturers and most of the newer systems (since the early 1980s) made by U.S. manufacturers use rigid support frames for discharge electrodes. The frames may consist of coiled-spring wires, serrated strips, or needle points mounted on a supporting strip. A typical rigid-frame discharge electrode is shown in Figure 3.4.

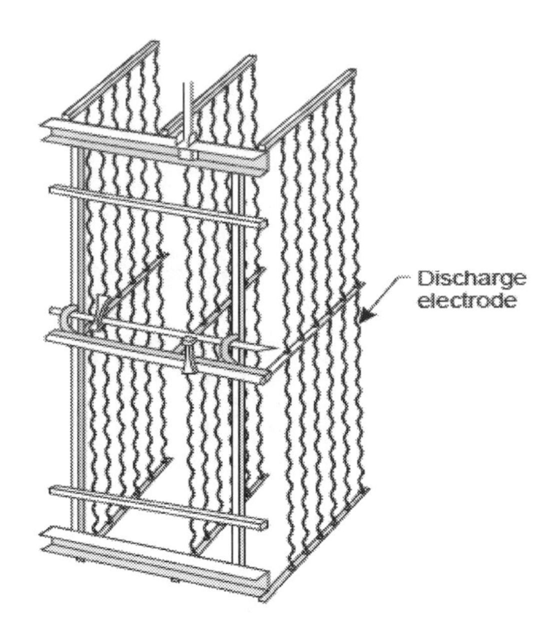

Figure 3.4. Rigid frame discharge electrode design

Another type of discharge electrode is a rigid electrode that is constructed from a single piece of fabricated metal and is shown in Figure 3.5. Both designs are occasionally referred to as rigid-frame electrodes. They have been used as successfully as the older U.S. wire designs. One major disadvantage of the rigid-frame design is that a broken wire cannot be replaced without removing the whole frame.

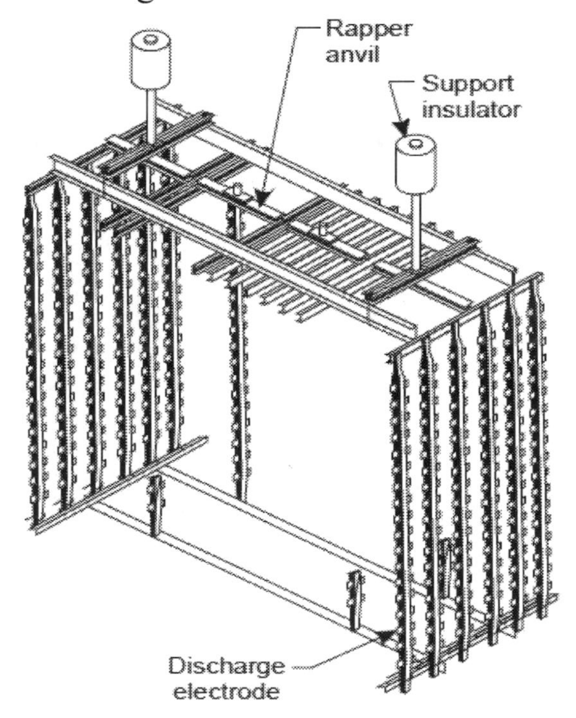

Figure 3.5 Typical rigid discharge electrode

Other manufacturer uses flat plates instead of wires for discharge electrodes. The flat plates, shown in Figure 3.6, increase the average electric field that can be used for collecting particles and provide an increased surface area for collecting particles, both on the discharge and collection plates. The corona is generated by the sharp pointed needles attached to the plates. These units generally use positive polarity for charging the particles. The units are typically operated with low flue gas velocity to prevent particle re-entrainment during the rapping cycle (Turner, et al. 1992).

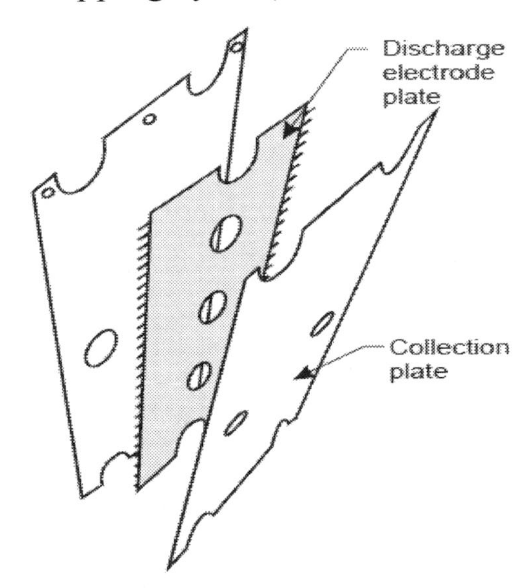

Figure 3.6 Flat-plate discharge electrode

3.2. Collection Electrodes

Collection electrodes collect charged particles. Collection electrodes are either flat plates or tubes with a charge opposite that of the discharge electrodes.

Most precipitators use plate collection electrodes because these units treat large gas volumes and are designed to achieve high collection efficiency.

Collecting plates are designed to receive and retain the precipitated particles until they are intentionally removed into the hopper. Collecting plates are also part of the electrical power circuit of the precipitator. These collecting plate functions are incorporated into the precipitator design. Plate baffles shield the precipitated particles from the gas flow while smooth surfaces provide for high operating voltage.

Collecting plates are suspended from the precipitator casing and form the gas passages within the precipitator. While the design of the collecting plates varies by manufacturer, there are two common designs:

- Plates supported from anvil beams at either end. The anvil beam is also the point of impact for the collecting rapper
- Plates supported with hooks directly from the precipitator casing two or more collecting plates are connected at or near the center by rapper beams, which then serve as impact points for the rapping system

Top, center, or bottom spacer bars may be used to maintain collecting plate alignment and sustain electrical clearances to the discharge system.

The plates are generally made of carbon steel. However, plates are occasionally made of stainless steel or alloy steel for special flue-gas stream conditions where corrosion of carbon steel plates would occur. The plates range from 0.05 to 0.2 cm (0.02 to 0.08 in.) in thickness.

For ESPs with wire discharge electrodes, plates are spaced from 15 to 30 cm apart (6 to 12 in.). Normal spacing for high-efficiency ESPs (using wires) is 20 to 23 cm (8 to 9 in.). For ESPs using rigid-frame or plate discharge electrodes, collection plates are typically spaced 30 to 38 cm (12 to 15 inches) apart. Plates are usually between 6 and 12 m (20 to 40 ft) high.

Collection plates are constructed in various shapes, as shown in Figure 3.7. These plates are solid sheets that are sometimes reinforced with structural stiffeners to increase plate strength. In some cases, the stiffeners act as baffles to help reduce particle re-entrainment losses. This design minimizes the amount of excess rapping energy required to dislodge the dust from the collection plates, because the energy is distributed evenly throughout the plate. The baffles also provide a "quiet zone" for the dislodged dust to fall while minimizing dust re-entrainment.

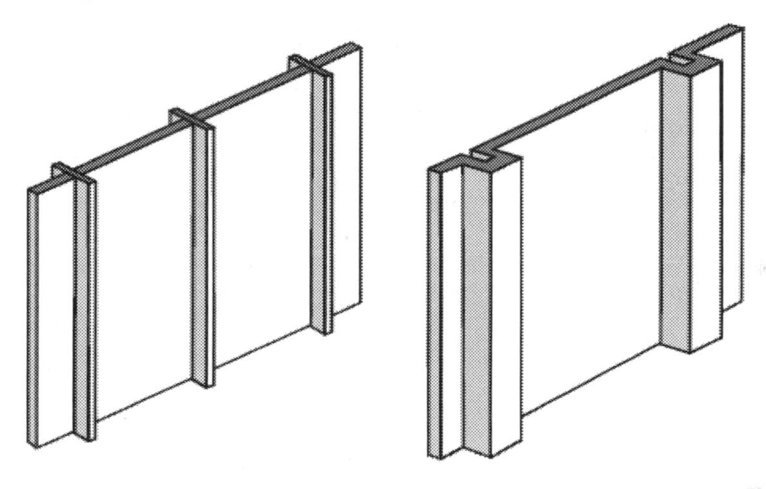

Figure 3.7 Typical collection plates

Tubes are also used as collection electrodes, but not nearly as often. Tubes are typically used to collect sticky particles and when liquid sprays are used to remove the collected particles.

3.3. Power Supplies and Controls

The power supply system is designed to provide voltage to the electrical field (or bus section) at the highest possible level. The voltage must be controlled to avoid causing sustained arcing or sparking between the electrodes and the collecting plates.

High-voltage power supplies and control equipments are used to charge particles in the ESP and to determine and control the strength of the electric field between the discharge and collection electrodes. This is accomplished by using power supply sets consisting of the following basic components and functions:

A. Automatic voltage control
B. Step-up transformer and High-voltage rectifier
C. Sensing device and protection circuitry

A simplified drawing of the circuitry from the primary control cabinet to the precipitator field is shown in Figure 3.8.

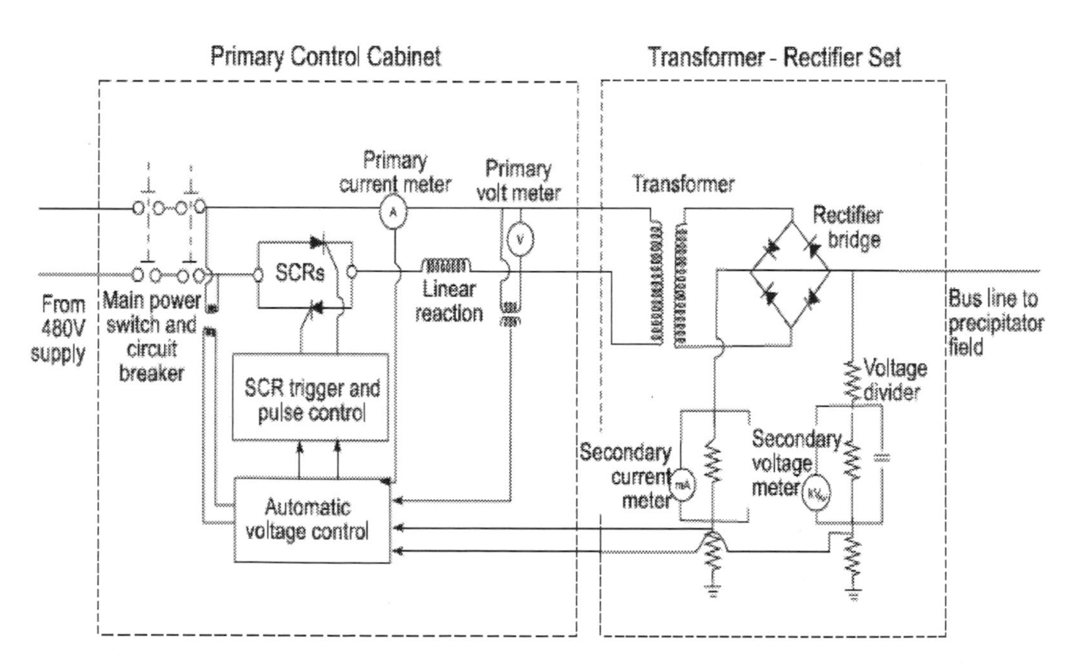

Figure 3.8 Schematic diagram of circuitry associated with precipitators

A. Voltage Controller

To maximize electrostatic precipitator efficiency a voltage controller usually attempts to increase the electrical power delivered to the field. However in some conditions a voltage controller must just maintain power at a constant level. Increased electrical power into the electrostatic precipitator directly correlates with better precipitator performance, but there is a limit. If too much voltages is applied for a given condition (as mentioned in the spark reaction section), a spark over will occur. During a spark over precipitator performance in that field will drop to zero, rendering that field temporarily ineffective.

Automatic voltage control varies the power to the transformer-rectifier in response to signals received from sensors in the precipitator and the transformer-rectifier itself. It monitors the electrical conditions inside the precipitator, avoids causing sustained arcing or sparking between the electrodes and the collecting plates, protects the internal components from arc-over damages, and protects the transformer-rectifier and other components in the primary circuit.

The voltage control device increases the primary voltage applied to the T-R set to the maximum level. As the primary voltage applied to the transformer increases, the secondary voltage applied to the discharge electrodes increases. As the secondary voltage is increased, the intensity and number of corona discharges increase. The voltage is increased until any of the set limits (primary voltage, primary current, secondary voltage, secondary current, or spark rate limits) is reached. Occurrence of a spark counteracts high ESP performance because it causes an immediate, short-term collapse of the precipitator electric field. Consequently, power that is applied to capture particles is used less efficiently. There is, however, an optimum sparking rate where the gains in particle charging are just offset by corona-current losses from sparkover.

The ideal automatic voltage control would produce the maximum collecting efficiency by holding the operating voltage of the precipitator at a level just below the spark-over voltage. Then the control resets to a lower power level, and the power increases again until the next spark occurs.

Automatic Voltage Controller for Electrostatic Precipitators is usually a Silicon-controlled Rectifier (SCR) automatic voltage controller with a linear reactor in the primary side of the transformer. In order to maximize electrostatic precipitator efficiency the following parameters are to be optimized and considered during the operation of the electrostatic precipitator:

- ***Optimize power application***:

  The primary purpose of a voltage controller is to deliver as much useful electrical power to the corresponding electrostatic precipitator field(s) as possible. This is not an easy job; electrical characteristics in the field(s) are constantly changing, which is why a voltage controller is required.

- ***Spark reaction***:

  When the voltage applied to the electrostatic precipitator field is too high for the conditions at the time, a spark over (or corona discharge) will occur. Detrimentally high amounts of current can occur during a spark over if not properly controlled, which could damage the fields. A voltage controller will monitor the primary and secondary voltage and current of the circuit, and detect a spark over condition. Once detected, the power applied to the field will be immediately cut off or reduced, which will stop the spark. After a short amount of time the power will be ramped back up, and the process will start over.

- ***Corona current density***:

  Corona current density should be in the range of 10 - 100 mA/1000 ft2 of plate area. (Calculate this using secondary current divided by collecting area of the electrical field or bus section.) The actual level depends upon:

  - Location of electrical field or bus section to be energized
  - Collecting plate area
  - Gas and dust conditions
  - Collecting electrode and discharge wire geometry

- ***Tripping:***

When a condition occurs that the voltage controller cannot control, often times the voltage controller will trip. A trip means the voltage controller will shut off the individual precipitator power circuit.

B. Transformer-Rectifier sets

The power sets are also commonly called Transformer-Rectifier (T-R) sets. The T-R rating should be matched to the load imposed by the electrical field or bus section. The power supply will perform best when the transformer-rectifiers operate at 70 - 90% of the rated capacity, without excessive sparking. Practical operating voltages for transformer-rectifiers depend on:

- Collecting plate spacing
- Gas and dust conditions
- Collecting plate and discharge electrode geometry

At secondary current levels over 1500 mA, internal impedance of a transformer-rectifier is low, which makes stable automatic voltage control more difficult to achieve. The design of the transformer-rectifier should call for the highest possible impedance that is commensurate with the application and performance requirements. Often, this limits the size of the electrical field or bus section.

It is general practice to add additional impedance in the form of a current-limiting reactor in the primary circuit. This reactor will limit the primary current during arcing and also improve the wave shape of the voltage/current fed into the transformer-rectifier.

In a T-R set, the transformer steps up the voltage from 400 volts to approximately 50,000 volts. This high voltage ionizes gas molecules that charge particles in the flue gas. The rectifier converts alternating current to direct current. Direct (or unidirectional current) is required for electrical precipitation. Most modern precipitators use solid-state silicon rectifiers and oil-filled, high-voltage transformers.

C. Metering and Protection

Measurements on commercial precipitators have determined that the optimum sparking rate is between 50 and 150 sparks per minute per precipitator electrical section. The objective in power control is to maintain corona power input at this optimum sparking rate by momentarily reducing precipitator power whenever excessive sparking occurs.

The most commonly used meters are the following:

- **Primary Voltmeter:** The meter is located across the primary winding of the transformer to measure the a.c. input voltage (220 to 480 volts).
- **Secondary Voltmeter:** This meter is located between the output side of the rectifier and the discharge electrodes to measure the d.c. volts applied to the discharge electrodes.
- **Secondary Ammeter:** This meter is located between the rectifier output and the automatic control module to measure the current supplied to the discharge electrodes in milliamperes.
- **Spark meter:** This meter measures the number of sparks per minute in the precipitator section. Sparks are surges of localized electric current between the discharge electrodes and the collection plate.

The system components are protected by adhering to component limitations. The Transformer Rectifier set (TR set) can be damaged by excessive amounts of current or voltage flowing through it. Each TR set has voltage and current limits established by the manufacturer. These nameplate limit values (typically primary and secondary current and voltage) are programmed into the voltage controller. Through metering circuits, the voltage controller will monitor these values, and ensure these limits are not exceeded.

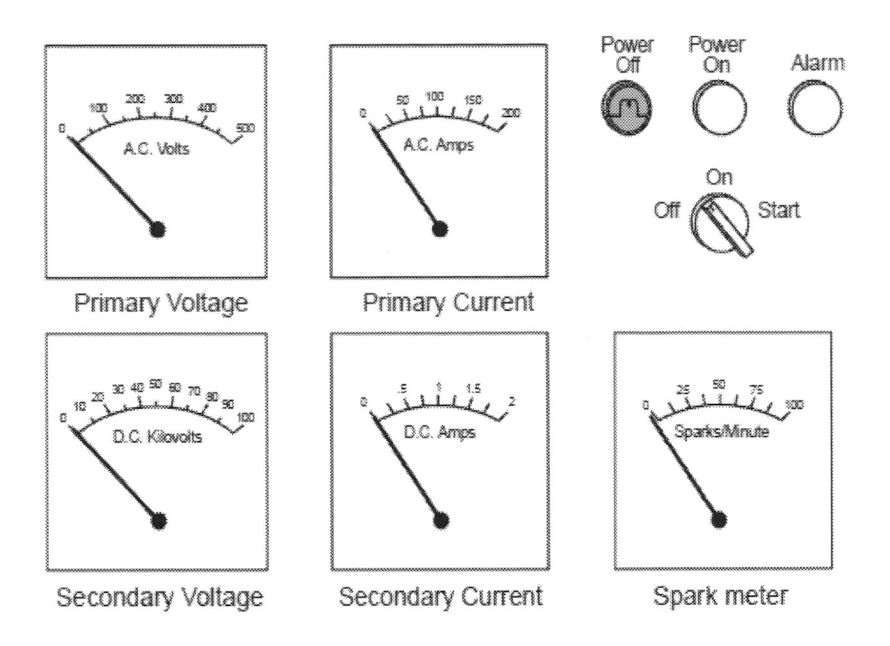

Figure 3.9. Typical gauges (meters) installed on control cabinet for each precipitator field

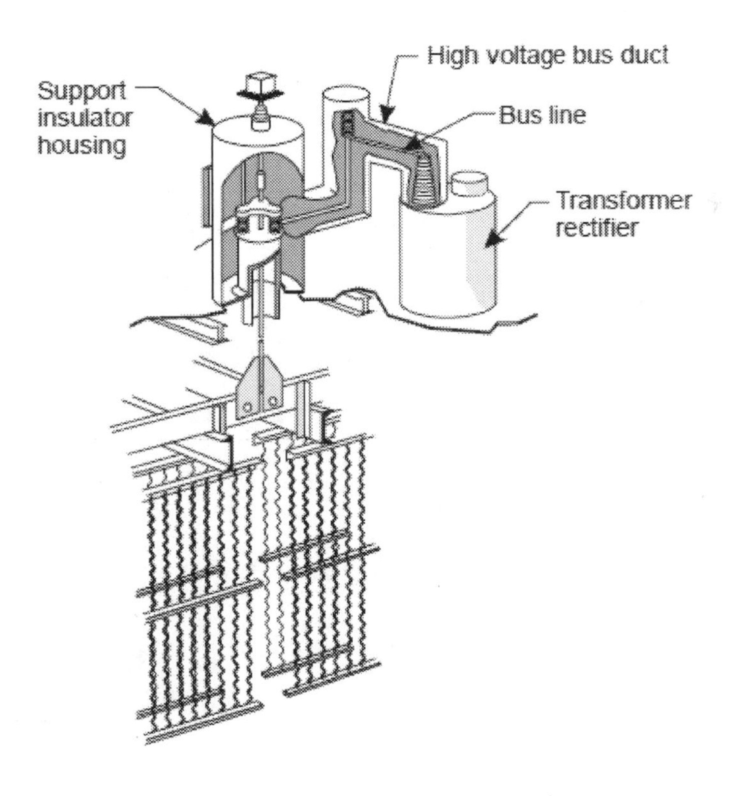

Figure 3.10. High-voltage system

3.4. Gas Distribution Systems

The electrical field or bus section of an electrostatic precipitator is by itself an independent precipitator. Its operation is governed by the inlet gas and dust conditions, as well as the collecting plate and discharge electrode geometries.

Within this electrical field or bus section, one gas passage is also an independent precipitator - governed by the same factors. (Note that the gas passage shares the voltage level with the adjacent gas passages of the same electrical field or bus section, but not the corona current level, which can be different in each gas passage.)

It is significantly importance to have similar gas and dust conditions at the inlet of each electrical field or bus section, and at the inlet of each gas passage of the electrical field or bus section. This can be achieved by having uniformity in the following operational parameters:

- Gas velocity
- Gas temperature
- Dust loading

Gas velocity distribution can be most effectively influenced by the use of gas distribution devices.

The quality of gas velocity distribution can be measured in a scaled-down model of the precipitator and its ductwork, and also in the precipitator itself. Typical criteria are based on ICAC (Institute of Clean Air Companies) recommendations using average gas velocities or on a calculated RMS statistical representation of the gas velocity pattern.

In general, gas distribution devices consist of turning vanes in the inlet ductwork, and perforated gas distribution plates in the inlet and/or outlet fields of the precipitator.

3.5. Rapping Systems

Rappers impart a vibration, or shock, to the electrodes, removing the collected dust. Rappers remove dust that has accumulated on both collection electrodes and discharge electrodes. Occasionally, water sprays are used to remove dust from collection electrodes.

Rappers are time-controlled systems provided for removing dust from the collecting plates and the discharge electrodes as well as for gas distribution devices (optional) and for hopper walls (optional). Rapping systems may be actuated by electrical or pneumatic power, or by mechanical means. Tumbling hammers may also be used to dislodge ash. Rapping methods include:

1- Electric vibrators
2- Electric solenoid piston drop rappers
3- Pneumatic vibrating rappers
4- Tumbling hammers
5- Sonic horns (do not require transmission assemblies)

A rapping system is designed so that rapping intensity and frequency can be adjusted for varying operational conditions. Once the operating conditions are set, the system must be capable of maintaining uniform rapping for a long time. Rapping depends of the configuration of the rapped parts and there are several methods, e.g.:

- Rapping the collection electrodes by **hammer/anvil** system. One rapper system uses hammers mounted on a rotating shaft, as shown in Figure 3.11. As the shaft rotates, the hammers drop (by gravity) and strike anvils that are attached to the collection plates. The frequency of rapping can be changed by adjusting the speed of the rotating shafts. Thus, rapping intensity and frequency can be adjusted for the varying dust concentration of the flue gas.

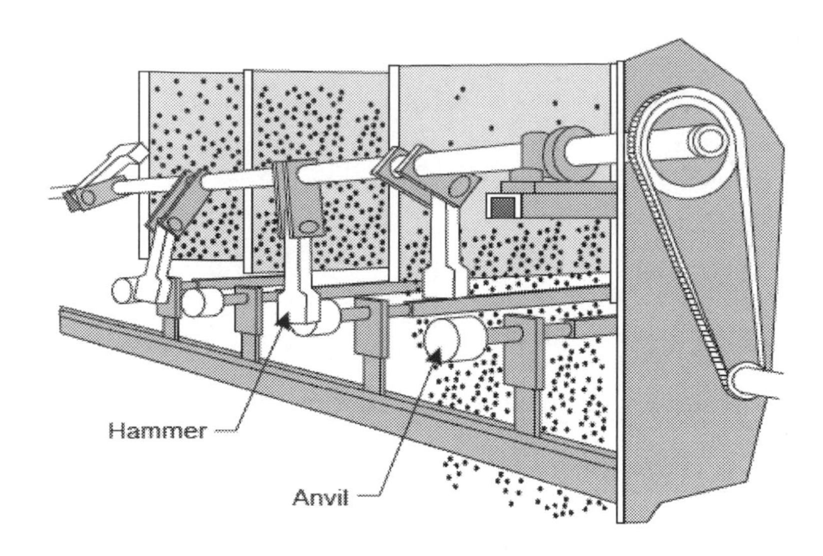

Figure 3.11 Typical hammer/anvil rappers for collection plates

- Rapping the collection electrodes by **magnetic impulse** system. Another rapping system used for many U.S. designs consists of magnetic-impulse rappers to remove accumulated dust layers from collection plates. A magnetic-impulse rapper has a steel plunger that is raised by a current pulse in a coil. The raised plunger then drops back, due to gravity, striking a rod connected to a number of plates within the precipitator as shown in Figure 3.12. Rapper frequency and intensity are easily regulated by an electrical control system. The frequency could be one rap every five minutes or one rap an hour.

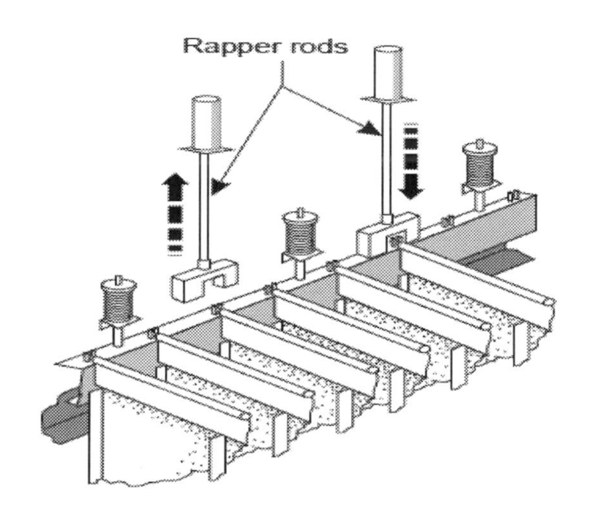

Figure 3.12 Typical magnetic-impulse rappers for collection plates

- Rapping the discharge electrodes of rigid frame by **tumbling hammers** . The tumbling hammers operate similarly to the hammers used to remove dust from collection electrodes. The hammers are arranged on a horizontal shaft. As the shaft rotates, the hammers hit an impact beam which transfers the shock, or vibration, to the center tubes on the discharge system, causing the dust to fall (Figure 3.13).

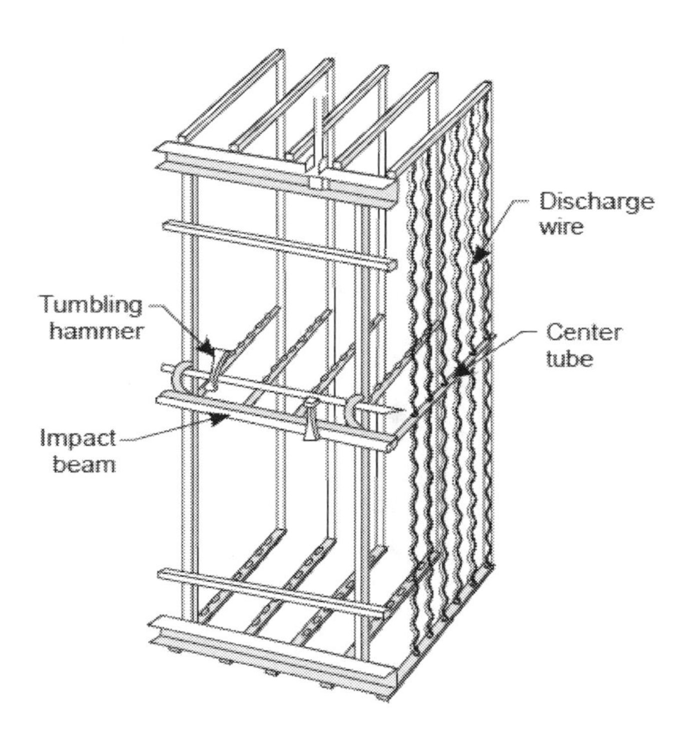

Figure 3.13 Tumbling hammers for rigid-frame discharge electrode

- Rapping wires by **vibrators**. This is usually accomplished by the use of air or electric vibrators that gently vibrate the discharge wires. Vibrators are usually mounted externally on precipitator roofs and are connected by rods to the high-tension frames that support the corona electrodes (Figure 3.14).

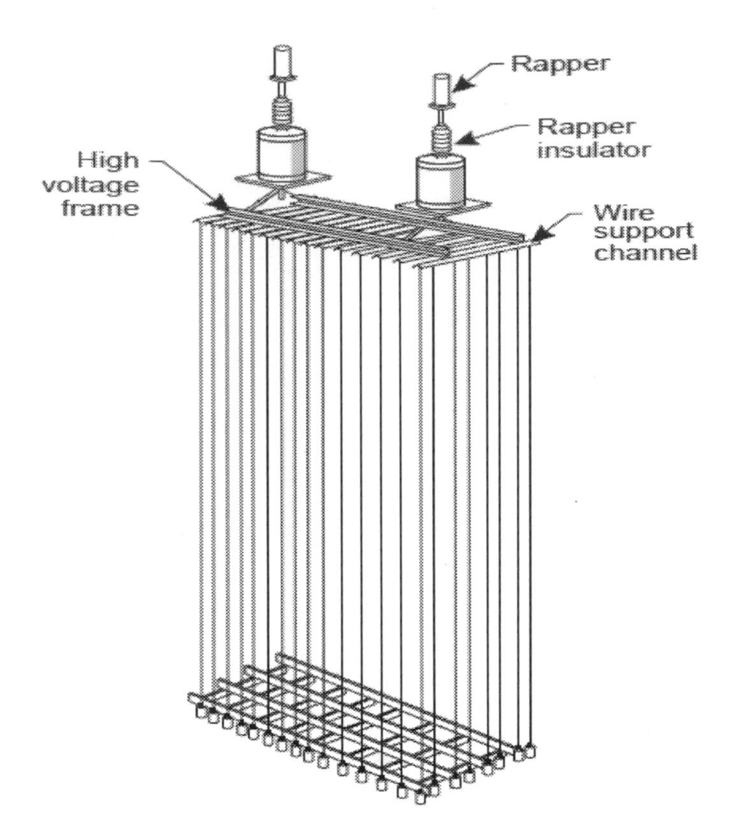

Figure 3.14 Typical electric-vibrator rappers used for wire discharge electrodes

- Liquid sprays are also used (instead of rapping) to remove collected particles from both tubes and plates.

All precipitator rapping systems allow adjustment of rapping frequency, normally starting with the highest frequency (the least time between raps), progressing to the lowest frequency. The times that are actually available may be limited. Rapping systems with pneumatic or electric actuators allow variations of the rapping intensity. Pneumatic or electric vibrators allow adjustments of the rapping time. State-of-the-art rapper controls allow selection of rapping sequences, selection of individual rappers, and provide anti-coincidence schemes which allow only one rapper to operate at a given time.

The adjustment of the rapping system for optimum precipitator performance is a slow process. It requires a substantial amount of time for stabilization after each adjustment.

3.6. Hoppers

Hoppers are located at the bottom of the precipitator. Hoppers are used to collect and temporarily store the dust removed during the rapping process.

When the electrodes are rapped, the dust falls into hoppers and is stored temporarily before it is disposed in a landfill or reused in the process. Dust should be removed as soon as possible to avoid packing, which would make removal very difficult. Hoppers are usually designed with a 50 to 70° (60° is common) slope to allow dust to flow freely from the top of the hopper to the bottom discharge opening.

Some manufacturers add devices to the hopper to promote easy and quick discharge. These devices include strike plates, poke holes, vibrators, and rappers.

Hopper designs also usually include access doors, or ports. Access ports allow easier access for cleaning, inspection, and maintenance of the hopper (Figure 3.15).

Hopper vibrators are occasionally used to help remove dust from the hopper walls. Hopper vibrators are electrically operated devices that cause the side walls of the hopper to vibrate, thereby removing the dust from the hopper walls.

A discharge device is necessary for emptying the hopper and can be manual or automatic. The simplest manual discharge device is the slide gate, a plate held in place by a frame and sealed with gaskets (Figure 3.16).

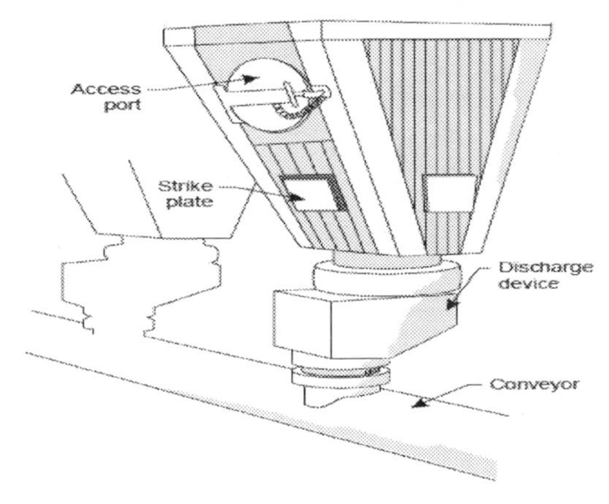

Figure 3.15 Hopper

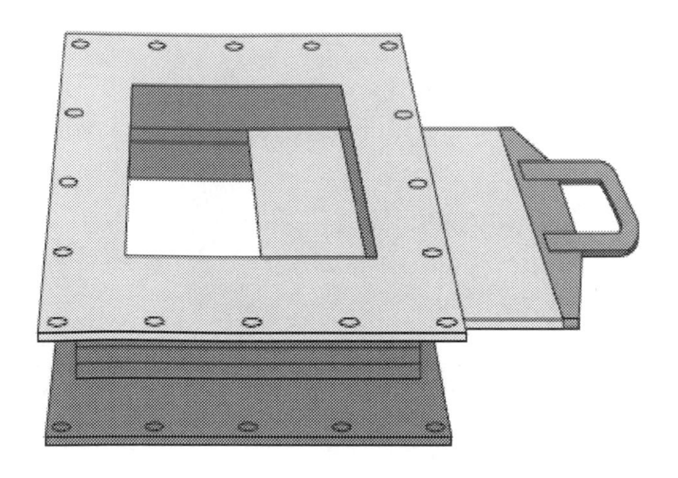

Figure 3.16 Slide-gate

Automatic continuous discharge devices are installed on ESPs that operate continuously. Double-dump valves are shown in Figure 3.17. As dust collects in the hopper, the weight of the dust pushes down the counterweight of the top flap and dust discharges downward.

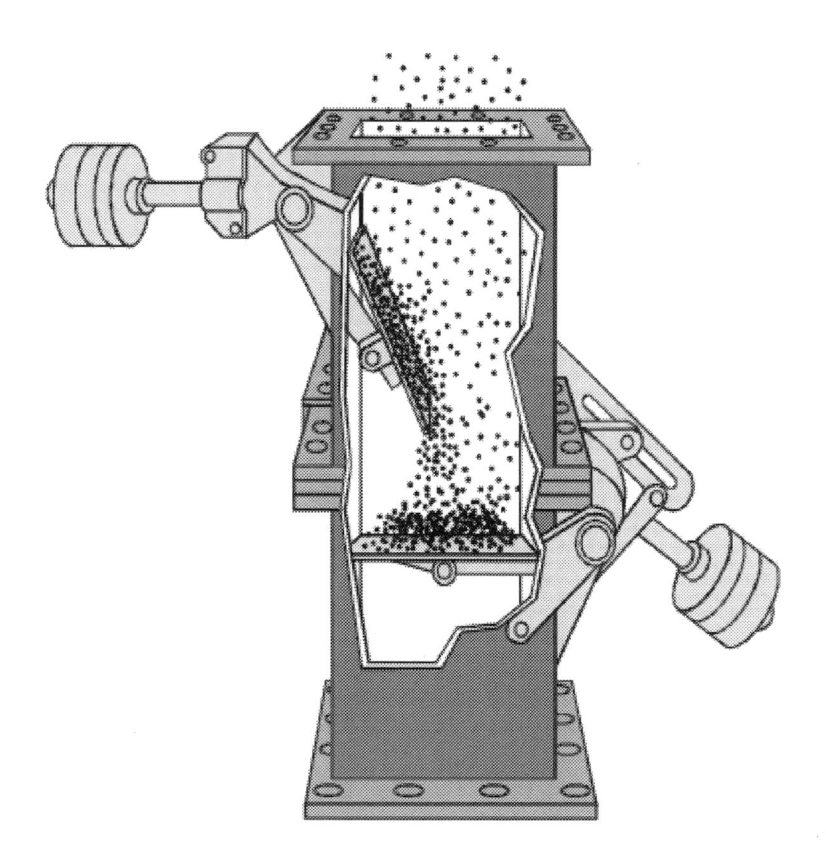

Figure 3.17 Double-dump discharge device

Either the precipitator hopper or the feeder hopper is used for temporarily storing material prior to discharge. Three types of handling systems are in use:

A. Negative pressure or vacuum system; Connects to the hopper by a simple discharge valve.
B. Positive pressure dilute phase system; Uses an airlock-type feeder; the feeder is separated from the hopper by an inlet gate and from the conveying line by a discharge gate.
C. Positive pressure dense phase system; Connects to the hopper with an airlock type feeder.

3.7. Ductwork

The Ductwork (shell) provides the base to support the ESP components and to enclose the unit.

Ductwork connects the precipitator with upstream and downstream equipment. The design of the ductwork takes into consideration the following:

- Low resistance to gas flow; Achieved by selecting a suitable cross-section for the ductwork and by installing gas flow control devices, such as turning valves and flow straighteners .
- Gas velocity distribution; Gas flow control devices are used to maintain good gas velocity distribution.
- Minimal fallout of fly ash; Fallout can be minimized by using a suitable transport velocity.
- Minimal stratification of the fly ash; A suitable transport velocity also reduces fly ash stratification in the gas stream.
- Low heat loss; The goal is to reduce the heat loss of the flue gas to a level that will prevent acid or moisture condensation in the downstream equipment, requiring the use of thermal insulation protected by external siding.

Ductwork structure supports its total load, including wind and snow loads. The design also allows for accumulated fly ash, negative/positive operating pressure, and gas temperature. Expansion joints are used to accommodate thermal growth because precipitator components can expand and contract when the temperature differences between the ESP (400°C) and the ambient atmosphere (20°C) are large (Figure 3.18).

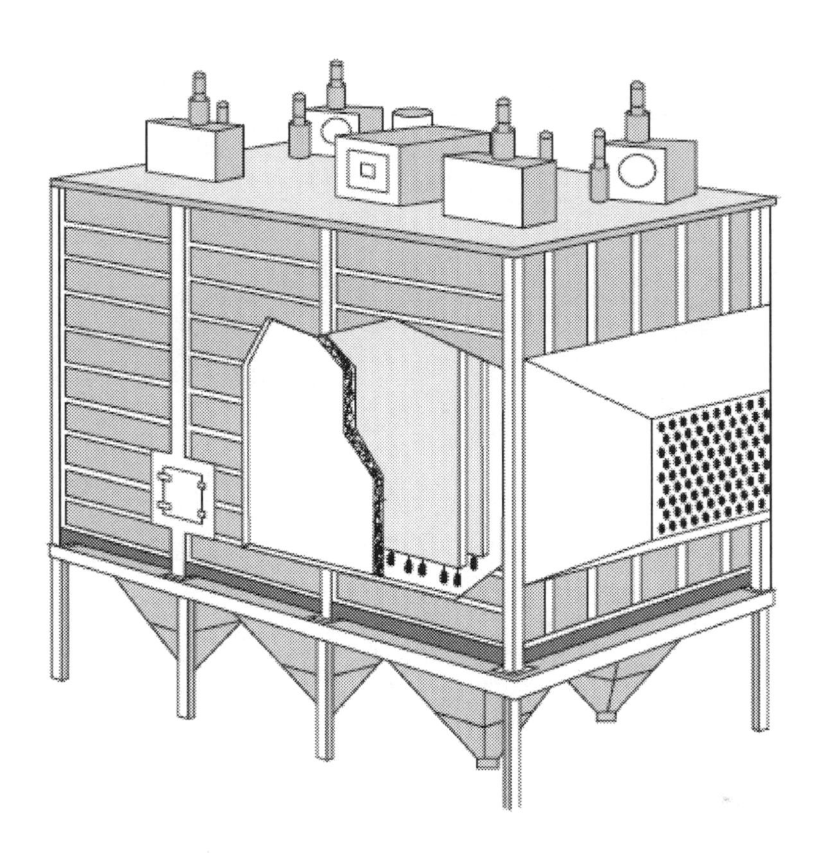

Fig.3.18 EPC Shell

Collection plates and discharge electrodes are normally attached to the frame at the top so that the elements hang vertically due to gravity. This allows the elements to expand or contract with temperature changes without binding or distorting. Shells, hoppers, and connecting flues should be covered with insulation to conserve heat, and to prevent corrosion resulting from water vapor and acid condensation on internal precipitator components. Ash hoppers should be insulated and heated because cold fly ash has a tendency to cake, making it extremely difficult to remove. Insulation material is usually 10 to 15 cm thick.

4

CLASSIFICATION AND TYPES OF EPS

ESPs can be grouped, or classified, according to a number of distinguishing features in their design. These features include the following:

- The structural design and operation of the discharge electrodes (rigid-frame, wires or plate) and collection electrodes (tubular or plate)
- The method of charging (single-stage or two-stage)
- The temperature of operation (cold-side or hot-side)
- The method of particle removal from collection surfaces (wet or dry)

These categories are not mutually exclusive. , an ESP may be referred to as cold-side, tubular, or by some other descriptor. ESP designs usually incorporate a number of ESP features into one unit. For example, a typical ESP used for removing particulate matter from a coal-fired boiler will be a cold-side, single-stage, plate ESP. On the other hand, a hot-side, single-stage, tubular ESP may be used to clean exhaust gas from a blast furnace in a steel mill.

All ESPs, no matter how they are grouped, have similar components and operate by charging particles or liquid aerosols, collecting them, and finally removing them from the ESP before ultimate disposal in a landfill or reuse in the industrial process.

4.1. Tubular and Plate ESPs

A- <u>Tubular Type</u>

Tubular precipitators consist of cylindrical collection electrodes (tubes) with discharge electrodes (wires) located in the center of the cylinder (Figure 4.1). Dirty gas flows into the tubes, where the particles are charged. The charged particles are then collected on the inside walls of the tubes. Collected dust and/or liquid is removed by washing the tubes with water sprays located directly above the tubes. The tubes may be formed as a circular, square, or hexagonal honeycomb with gas flowing upward or downward.

A tubular ESP is tightly sealed to minimize leaks of collected material. Tube diameters typically vary from 0.15 to 0.31 m (0.5 to 1 ft), with lengths usually varying from 1.85 to 4.0m (6 to 15 ft).

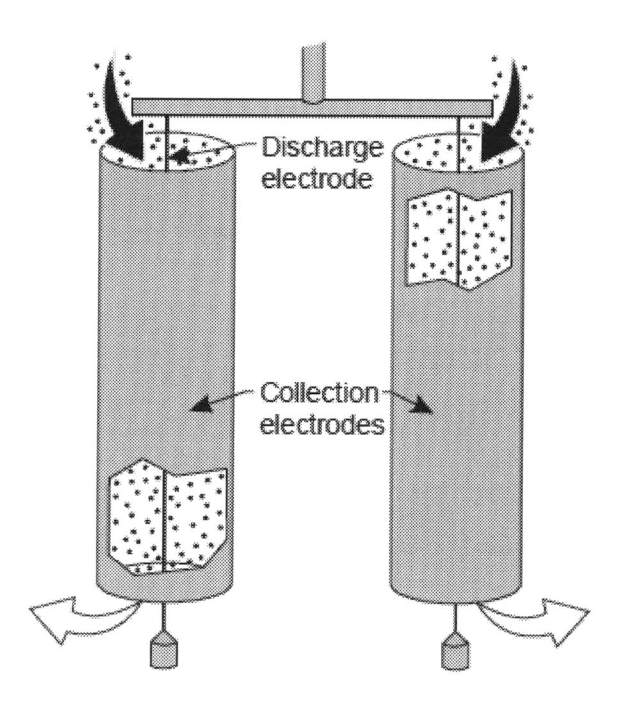

Figure 4.1 Gas flow through a tubular precipitator

Tubular precipitators are generally used for collecting mists or fogs, and are most commonly used when collecting particles that are wet or sticky. Tubular ESPs have been used to control particulate emissions from sulfuric acid plants, coke oven byproduct gas cleaning (tar removal), and iron and steel sinter plants.

B-Plate Type

Plate electrostatic precipitators primarily collect dry particles and are used more often than tubular precipitators. Plate ESPs can have wire, rigid-frame, or occasionally, plate discharge electrodes. Figure 4.2 shows a plate ESP with wire discharge electrodes. Dirty gas flows into a chamber consisting of a series of discharge electrodes that are equally spaced along the center line between adjacent collection plates. Charged particles are collected on the plates as dust, which is periodically removed by rapping or water sprays. Discharge wire electrodes are approximately 0.13 to 0.38 cm (0.05 to 0.15 in.) in diameter. Collection plates are usually between 6 and 12 m (20 and 40 ft) high. For ESPs with wire discharge electrodes, the plates are usually spaced from 15 to 30 cm (6 to 12 in.) apart. For ESPs with rigid-frame or plate discharge electrodes, plates are typically spaced 30 to 38 cm(12 to 15 in.) apart and 8 to 12 m (30 to 40 ft) in height.

Plate ESPs are typically used for collecting fly ash from industrial and utility boilers as well as in many other industries including cement kilns, glass plants and pulp and paper mills.

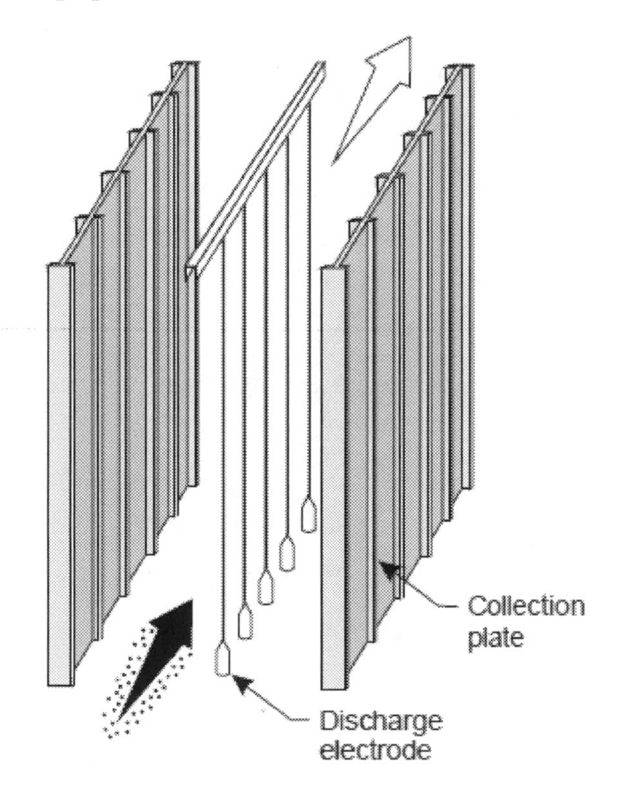

Figure 4.2 Gas flow through a plate precipitator

4.2. Single-stage and Two-stage ESPs

Another method of classifying ESPs is by the number of stages used to charge and remove particles from a gas stream. These are:

- High-voltage, Single-stage precipitators combine an ionization and a collection step. They are commonly referred to as Cottrell precipitators. The high-voltage, single-stage precipitator, is widely used in minerals processing operations.
- Low-voltage, Two-stage precipitators use a similar principle; however, the ionizing section is followed by collection plates. The low-voltage, two-stage precipitator is generally used for filtration in air-conditioning systems

A- Single Stage Type

A **single-stage** precipitator uses high voltage to charge the particles, which are then collected within the same chamber on collection surfaces of opposite charge.

Most ESPs that reduce particulate emissions from boilers and other industrial processes are single-stage ESPs (these units will be emphasized in this course). Single stage ESPs use very high voltage (50 to 70 kV) to charge particles. After being charged, particles move in a direction perpendicular to the gas flow through the ESP, and migrate to an oppositely charged collection surface, usually a plate or tube. Particle charging and collection occurs in the same stage, or field; thus, the precipitators are called single-stage ESPs. The term field is used interchangeably with the term stage and is described in more detail later in this course. Figure 4.1 shows a single stage tubular precipitator. A single-stage plate precipitator is shown in Figure 4.2.

B- Two Stage TYPE

In a **two-stage** precipitator, particles are charged by low voltage in one chamber, and then collected by oppositely charged surfaces in a second chamber.

The two-stage precipitator differs from the single-stage precipitator in both design and amount of voltage applied. The two-stage ESP has separate particle charging and collection stages (Figure 4.3). The ionizing stage consists of a series of small, **positively** charged wires equally spaced 2.5 to

5.1 cm (1 to 2 in.) from parallel grounded tubes or rods. A corona discharge between each wire and a corresponding tube charges the particles suspended in the air flow as they pass through the ionizer. The direct-current potential applied to the wires is approximately 12 to 13 kV.

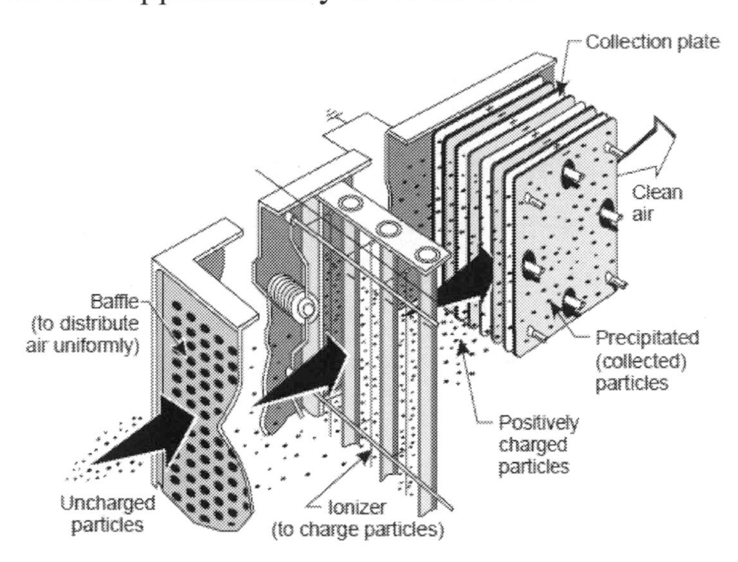

Figure 4.3 Representation of gas flow in a two-stage precipitator

The second stage consists of parallel metal plates less than 2.5 cm (1 in.) apart. The particles receive a positive charge in the ionizer stage and are collected at the negative plates in the second stage. Collected smoke or liquids drain by gravity to a pan located below the plates, or are sprayed with water mists or solvents that remove the particles and cause them to fall into the bottom pan.

Two-stage precipitators were originally designed for air purification in conjunction with air conditioning systems. (They are also referred to as electronic air filters). Two stage ESPs are used primarily for the control of finely divided liquid particles.

Controlling solid or sticky materials is usually difficult, and the collector becomes ineffective for dust loadings greater than 7.35 x 10-3g/m3 (0.4gr/dscf). Therefore, two-stage precipitators have limited use for particulate-emission control. They are used almost exclusively to collect liquid aerosols discharged from sources such as meat smokehouses, pipe-coating machines, asphalt paper saturators, high speed grinding machines, welding machines, and metal-coating operations.

4.3. Cold-side and Hot-side ESPs

Electrostatic precipitators are also grouped according to the temperature of the flue gas that enters the ESP: cold-side ESPs are used for flue gas having temperatures of approximately 204°C (400°F) or less; hot-side ESPs are used for flue gas having temperatures greater than 300°C (572°F).

In describing ESPs installed on industrial and utility boilers, or municipal waste combustors using heat recovery equipment, cold side and hot side also refer to the placement of the ESP in relation to the combustion air preheater. A cold-side ESP is located behind the air preheater, whereas a hot-side ESP is located in front of the air preheater. The air preheater is a tube section that preheats the combustion air used for burning fuel in a boiler. When hot flue gas from an industrial process passes through an air preheater, a heat exchange process occurs whereby heat from the flue gas is transferred to the combustion air stream. The flue gas is therefore "cooled" as it passes through the combustion air preheater. The warmed combustion air is sent to burners, where it is used to burn gas, oil, coal, or other fuel including garbage.

A- **Cold Side Type**

Cold-side ESPs (Figure 4.4) have been used for over 50 years with industrial and utility boilers, where the flue gas temperature is relatively low (less than 204°C or 400°F). Cold-side ESPs generally use plates to collect charged particles. Because these ESPs are operated at lower temperatures than hot-side ESPs, the volume of flue gas that is handled is less. Therefore, the overall size of the unit is smaller, making it less costly. Cold-side ESPs can be used to remove fly ash from boilers that burn high sulfur coal. As explained in later lessons, cold-side ESPs can effectively remove fly ash from boilers burning low-sulfur coal with the addition of conditioning agents.

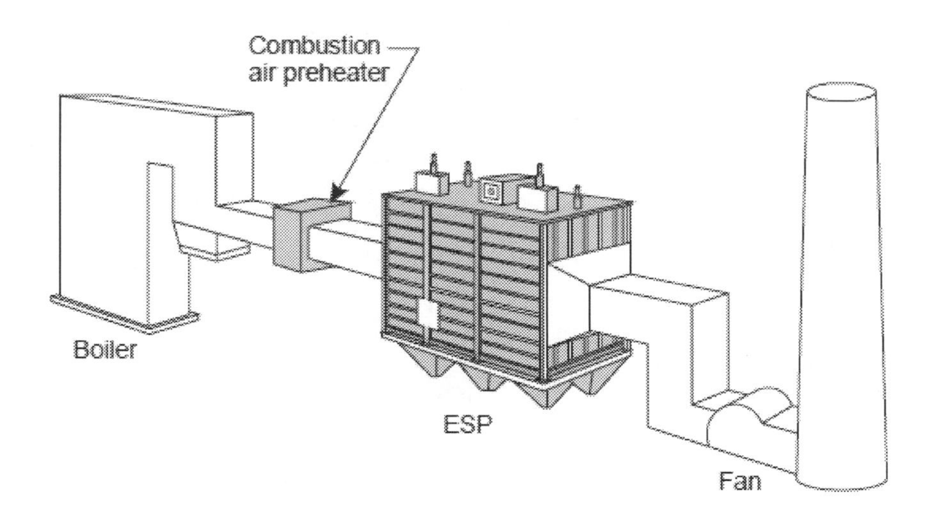

Figure 4.4 Cold-side ESP

B- **Hot Side Type**

Hot-side ESPs (Figure 4.5) are placed in locations where the flue gas temperature is relatively high. Their collection electrodes can be either tubular or plate. Hot-side ESPs are used in high-temperature applications, such as in the collection of cement-kiln dust or utility and industrial boiler fly ash. A hot-side precipitator is located before the combustion air preheater in a boiler. The flue gas temperature for hot-side precipitators is in the range of 320 to 420°C (608 to 790°F).

The use of hot-side precipitators helps reduce corrosion and hopper plugging. However, these units (mainly used on coal-fired boilers) have some disadvantages. Because the temperature of the flue gas is higher, the gas volume treated in the ESP is larger. Consequently, the overall size of the precipitator is larger making it more costly. Other major disadvantages include structural and mechanical problems that occur in the precipitator shell and support structure as a result of differences in thermal expansion.

For years, cold-side ESPs were used successfully on boilers burning high-sulfur coal. However, during the 1970s when utilities switched to burning low-sulfur coal, coldside ESPs were no longer effective at collecting the fly ash. Fly ash produced from low sulfur coal-fired boilers has high resistivity (discussed in more detail later in the course), making it difficult to collect. As you will learn later, high temperatures can lower resistivity. Consequently, hot-side ESPs became very popular during the 1970s for removing ash from coal-fired boilers burning low sulfur coal. However, many of these units did not operate reliably, and therefore, since the 1980s, operators have generally decided to use cold-side ESPs along with conditioning agents when burning low sulfur coal.

Hot-side ESPs are also used in industrial applications such as cement kilns and steel refining furnaces. In these cases, combustion air preheaters are generally not used and hot side just refers to the high flue gas temperature prior to entering the ESP.

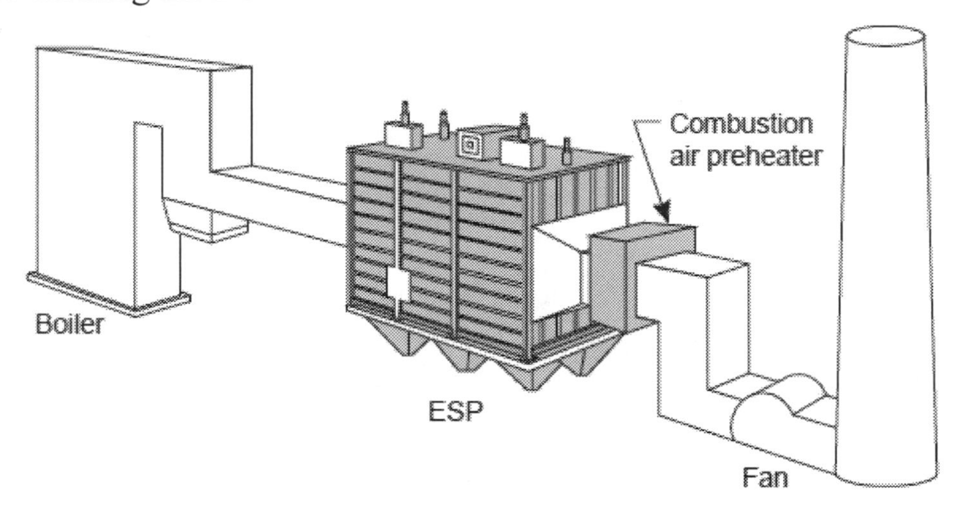

Figure 4.5 Hot-side ESP

4.4. Dry and Wet ESPs

A- Dry ESPs

Most electrostatic precipitators are operated dry and use rappers to remove the collected particulate matter. The term dry is used because particles are charged and collected in a dry state and are removed by rapping as opposed to water washing which is used with wet ESPs. The major portion of this course covers dry ESPs that are used for collecting dust from many industries including steel furnaces, cement kilns and fossil-fuel-fired boilers.

B-Wet ESPs

A wet electrostatic precipitator (WESP or wet ESP) operates with saturated air streams (100% relative humidity). One type of WESP uses a vertical cylindrical tube with centrally-located wire electrode (gas flowing upward) with water sprays to clean the collected particulate from the collection surface (plates, tubes). The collected water and particulate forms a wet film slurry that eliminates the resistivity issues associated with dry ESP's. Another type of WESP (used for coke-oven gas detarring) uses a falling oil film to remove collected material.

Wet ESPs are used for industrial applications where the potential for explosion is high (such as collecting dust from a closed-hood Basic Oxygen Furnace in the steel industry), or when dust is very sticky, corrosive, or has very high resistivity. The liquid flow may be applied continuously or intermittently to wash the collected particles from the collection electrodes into a sump (a basin used to collect liquid). The advantage of using a wet ESP is that it does not have problems with rapping re-entrainment or with back corona.

Figures 4.6 and 4.7 show two different wet ESPs. The casing of wet ESPs is made of steel or fiberglass and the discharge electrodes are made of carbon steel or special alloys, depending on the corrosiveness of the flue gas stream.

In a circular-plate wet ESP, shown in Figure 4.6, the circular collection plates are sprayed with liquid continuously. The liquid provides the electrical ground for attracting the particles and for removing them from the plates. These units can handle gas flow rates of 30,000 to 100,000 cfm. Preconditioning sprays located at the inlet remove some particulate matter prior to the charging stage. The operating pressure drop across these units is typically 1 to 3 inches of water.

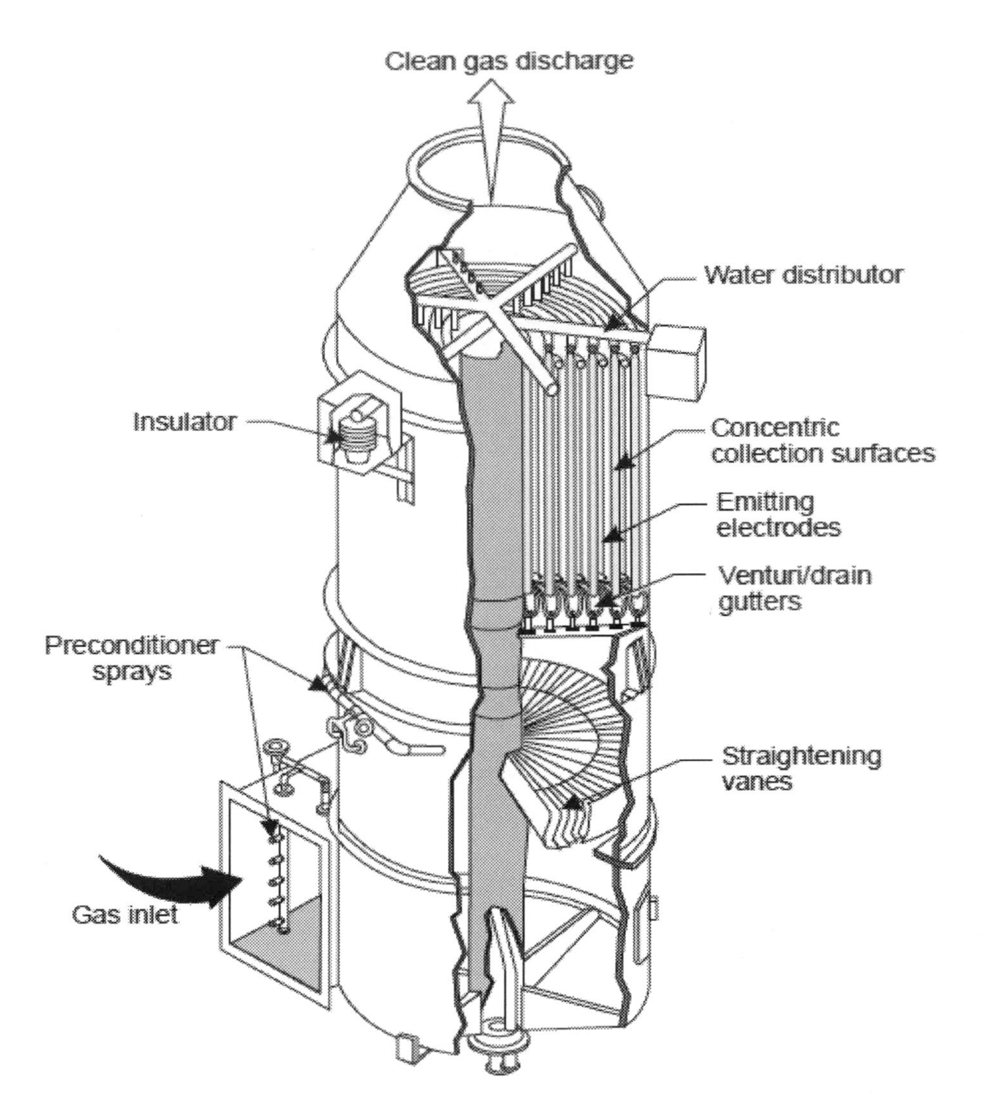

Figure 4.6 Circular-plate wet EPS

Rectangular flat-plate wet ESP, shown in Figure 4.7; operates similarly to circular-plate wet ESP. Water sprays precondition the gas stream and provide some particle removal. Because the water sprays are located over the top of the electrical fields, the collection plates are continuously irrigated. The collected particulate matter flows downward into a trough that is sloped to a drain.

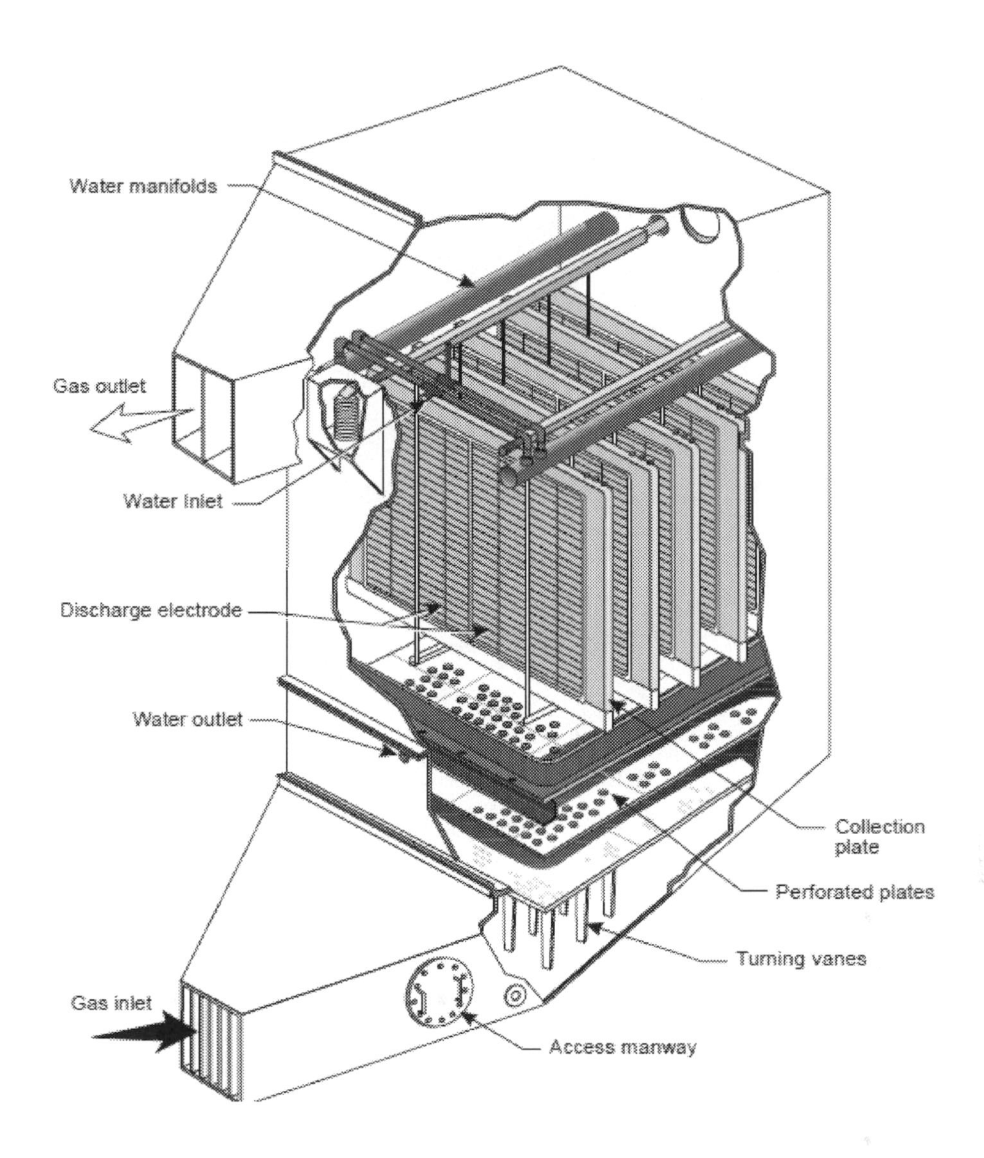

Figure 4.7 Flat-plate modular wet ESP

5

Operation and performance

5.1. Introduction

Depending on the electrostatic precipitator chosen, production, installation and operation startup may take from a few months to one or two years. Proper installation procedures will save time and money, and will also help in future operation and maintenance (O&M) of the ESP.

Good coordination between the ESP designer (vendor) and the installation and maintenance crews will help keep the ESP operating and performing smoothly for years. Occasionally this coordination is overlooked. Because they are so large, ESPs are usually installed by skilled craftsmen who do not work for the ESP vendor, and, therefore, may not be informed of specific installation instructions. Since all design tolerances are critical (especially those affecting discharge and collection electrode alignment), it is imperative that information about the proper installation procedures be transferred from designers to installers.

For optimum operation and performance of EPC Components, the items and parameters presented below should be reviewed and being addressed during installation:

1. ***General Key considerations***

• Easy access to all potential maintenance areas; fans, motors, hoppers, discharge devices, dampers, flue gas flow rate and temperature monitors, insulators, rappers, T-R sets, and discharge and collection electrodes
• Easy access to all inspection and test areas; stack testing ports and continuous emission monitors (opacity monitors)
• Weather conditions; the ESP must be able to withstand inclement weather such as rain or snow

2. ***Uniform flue gas distribution across the entire unit***

Ductwork, turning vanes, baffle plates, and inlets with perforated diffuser plates all affect flue gas distribution. These items are usually installed in the field and should be checked visually. If improperly installed, they induce high airflow regions that decrease collection efficiency and cause re-entrainment of collected dust, especially during rapping cycles.

3. *Complete seal of ESP system from dust pickup to stack outlet*

Air inleakage or outleakage at flanges or collector access points either adds additional airflow to be processed or forces the process gases to bypass the collector. Inleakage to a high-temperature system (hot-side ESP) is extremely damaging, as it creates cold spots which can lead to moisture or acid condensation and possible corrosion. If severe, it can cause the entire process gas temperature to fall below the gas dew point, causing moisture or acid to condense on the hopper walls, the discharge electrode, or collection plates. In addition, air inleakage and moisture condensation can cause caking of fly ash in the hopper, making normal dust removal by the discharge device very difficult. The best way to check for leaks is an inspection of the walls from inside the system during daylight. Light penetration from outside helps to isolate the problem areas.

4. *Proper installation of discharge electrodes and collection plates*

Collection electrodes are usually installed first, and the discharge wires or rigid frames are positioned relative to them. Check each section of electrodes to ensure that the electrodes are plumb, level, and properly aligned.

5. *Proper installation of rappers*

Collection-plate rappers and discharge-electrode rappers should be installed and aligned according to vendor specifications. Check magneticimpulse rappers to see if they strike the support frame on the collection plates. Check hammer and anvil rappers to see if the hammers strike the anvils squarely. Check vibrator rappers installed on discharge wires to make sure they operate when activated. Rapper frequency and intensity can be adjusted later when the unit is brought on-line.

6. *Proper insulation*

Most ESPs use some type of insulation to keep the flue gas temperature high. This prevents any moisture or acids present in the flue gas from condensing on the hoppers, electrodes, or duct surfaces. Because most ESPs are installed in the field, check that all surfaces and areas of potential heat loss are adequately covered.

7. ***Proper installation and operation of discharge devices***

It is important to check the operation of the discharge devices before bringing the ESP on-line to see if they are properly installed. Make sure that the discharge devices are moving in the right direction so they can remove the dust freely from the hopper. A backward-moving screw conveyor can pack dust so tightly that it can bend the screw.

Overfilled hoppers are common operating problems that can be avoided by proper installation and maintenance of discharge devices. Installed as maintenance tools, dust-level detectors in the hoppers can help alert ESP operators that hoppers are nearly full.

8. ***Smoothly running fans***

Check fans for proper rotation, drive component alignments, and vibration. Fans should be securely mounted to a component of sufficient mass to eliminate excessive vibration.

5.2. Operation

As with any air pollution control system, an ESP must be operated and maintained according to the manufacturer's recommendations. Plant personnel must be properly trained to perform these activities with confidence and efficiency. This section reviews the key functions that must be completed to keep the ESP operating as it was intended including installation, startup and shutdown procedures, performance monitoring, routine maintenance and record keeping and problem evaluation.

5.2.1 ESP Pre startup, Startup and Shutdown

In addition to the above items, each ESP should have its own checklist reflecting the unique construction features of that unit. A specific pre startup, startup and shutdown procedure should be supplied by the ESP vendor. Improper pre startup, startup and shutdown can damage the collector. It is imperative for the operator (source) to have a copy of these procedures. Review agency engineers may want to assure that these procedures exist at the sites and that the operators follow the procedures or document reasons for deviations.

1. **Pre startup**

An example of pre startup checklist for the initial startup suggested by Peter Bibbo (1982) is shown in Table 5.1.

2. **Startup**

Startup of an electrostatic precipitator is generally a routine operation. It involves heating a number of components such as support insulators and hoppers. If possible, the ESP should not be turned on until the process reaches steady-state conditions. As described in Lesson 5, this is particularly important for ESPs used on cement kilns burning coal as fuel.

The internal arcing of the ESP could cause a fire or an explosion. When ESPs are used on oil-burning boilers, the boiler should be started with gas or fuel oil. Heavy oil is not a good fuel for startup because tarry particulate emissions can coat collection plates and be difficult to remove. If an ESP is used on a coal-fired boiler, the ESP should not be started until coal firing can be verified. This will help prevent combustible gases from accumulating in the unit and causing explosive conditions. A typical startup procedure for an ESP used on a boiler is given in Table 5.2 (Ref.: Bibbo 1982).

Table 5.1 Pre startup checklist for electrostatic precipitators

Subsystem/Component	Checklist
Collecting plates	1. Free of longitudinal and horizontal bows 2. Free of burrs and sharp edges 3. Support system square and level 4. Spacer bars and corner guides free 5. Free of excessive dust buildup 6. Gas leakage baffles in place and not binding
Discharge electrodes	1. No breaks or slack wires 2. Wires free in guides and suspension weight free on pin 3. Rigid frames square and level 4. Rigid electrodes plumb and straight 5. Free of excessive dust buildup and grounds 6. Alignment within design specifications
Hoppers	1. Scaffolding removed 2. Discharge throat and poke holes clear 3. Level detector unobstructed 4. Baffle door and access door closed 5. Heaters, vibrators, and alarms operational
Top housing or insulator compartments	1. Insulators and bushing clear and dry with no carbon tracks 2. All grounding chains in storage brackets 3. Heaters intact, seal-air system controls, alarms, dampers, and filters in place and operational 4. Seal-air fan motor rotation correct, or vent pipes free 5. All access doors closed
Rappers	1. All swing hammers or drop rods in place and free 2. Guide sleeves and bearings intact 3. Control and field wiring properly terminated 4. Indicating lights and instrumentation operational 5. All debris removed from precipitator 6. All personnel out of unit and off clearances 7. All interlocks operational and locked out a. No broken or missing keys b. Covers on all locks
Transformer-rectifier sets	1. Surge arrestor not cracked or chipped and gap set 2. Liquid level satisfactory 3. High-voltage connections properly made 4. Grounds on: precipitator, output bushings, bus ducts, conduits
Rectifier control units	1. Controls grounded 2. Power supply and alarm wiring properly completed 3. Interlock key in transfer block

Table 5.2 Typical startup procedures for electrostatic precipitators

Normal Operation

Steps	Checklist
1. Preoperational checks (at least 2 hours prior to gas load	1. Complete all maintenance/inspection items. 2. Remove all debris from ESP. 3. Safety interlocks should be operational and all keys accounted for. 4. No personnel should be in ESP. 5. Lock out ESP and insert keys in transfer blocks.
2. Check hoppers (at least one hour prior to gas load)	a. Level-indicating system should be operational. b. Ash-handling system operating and sequence check - leave in operational mode. c. Hopper heaters should be on.
3.Check top housing seal-air system (at least one hour prior to gas load)	a. Check operation of seal-air fan—leave running. b. Bushing heaters should be on.
4.Check rappers (at least one hour prior to gas load)	a. Energize control, run rapid sequence, ensure that all rappers are operational. b. Set cycle time and intensity adjustments, using installed instrumentation—leave rappers operating.
5.***Check T-R sets*** (at least one hour prior to gas load)	a. Check half-wave/full-wave operation (half-wave operation is recommended for filtering fly ash when lignite is burned and a cold-side ESP is used.) b. Keys should be in all breakers. c. Test-energize all T-R sets and check local control alarm functions. d. Set power levels and de-energize all T-R controls. e. Lamp and function-test all local and remote alarms.
6. Gas load *(*After gas at temperature of 200°F has entered ESP for 2 hours)	a. Energize T-R sets. b. Check for normal operation of T-R control. c. Check all alarm functions in local and remote. d. Within 2 hours, check proper operation of ash removal system. e. De-energize bushing heaters after 2 hours (hopper heaters optional).
7.Cold start (when it is not possible to admit flue gas at 200°F for 2 hours prior to energizing controls), proceed as follows:	a. Perform steps 1-6above. Increase rapping intensity 50%. b. Energize T-R sets, starting with inlet field, setting Power trac voltage to a point just below sparking. c. Successively energize successive field as load picks up to maintain opacity, keeping voltage below normal sparking (less than 10 flashes/min on spark indicator). d. Perform step 10d above. e. After flue gas at 200°F has entered ESP for 2 hours, perform steps 6b, c, d, and e above. Set normal rapping.

3. **Shutdown**

When an industrial process is shut down temporarily, the ESP system should be de-energized to save energy. The shutdown of the ESP is usually done by reversing the order of the startup steps. Begin with de-energizing the ESP fields starting with the inlet field to maintain appropriate opacity levels from the stack. The rappers should be run for a short time after the ESP is de-energized so that accumulated dust from the collection plates and discharge wires can be removed. All hoppers should be emptied completely before bringing the unit back on line. A typical shutdown and emergency shutdown procedure for ESPs used on industrial sources is given in Table 5.3 (Ref.: Bibbo 1982).

Table 5.3 Typical shutdown and emergency shutdown procedures for electrostatic precipitators

Steps	Checklist
Typical shutdown (When boiler load drops and total ash quantity diminishes) Note: Normal shutdown is a convenient time to check operation of alarms.	a. De-energize ESP by field, starting with inlet field to maintain opacity limit. b. De-energize outlet field when all fuel flow ceases and combustion air flow falls below 30% of rated flow. c. Leave rappers, ash removal system, seal-air system, and hopper heaters operational. d. Four hours after boiler shutdown, de-energize seal-air system and hopper heaters. Secure ash removal system. e. Eight hours after boiler shutdown, de-energize rappers.
Emergency shutdown	1. De-energize all T-R sets. 2. Follow steps 1c, d, and e above (shutdown).

5.2.2 Operation steps

Typically, six steps take place at the system of electrostatic precipitation, these are:

1- Ionization - Charging of particles
2- Migration - Transporting the charged particles to the collecting surfaces
3- Collection - Precipitation of the charged particles onto the collecting surfaces
4- Charge Dissipation - Neutralizing the charged particles on the

collecting surfaces

5- Particle Dislodging - Removing the particles from the collecting surface to the hopper
6- Particle Removal - Conveying the particles from the hopper to a disposal point

The major precipitator components that accomplish these steps are as follows:

A. Discharge Electrodes
B. Power Components
C. Precipitator Controls
D. Rapping Systems
E. Purge Air Systems
F. Flue Gas Conditioning

5.2.3 Operation Monitoring

As with the operation of any piece of equipment, Operation and performance monitoring and recordkeeping are essential to establishing a good operation and maintenance program. The key to any monitoring program is establishing an adequate baseline of acceptable ranges that is used as a reference point. Then, by monitoring and recording key operating parameters, the operator can identify performance problems, need for maintenance, and operating trends.

Typical Operating parameters that can be monitored include:

• Voltage/current
• Opacity
• Gas temperature
• Gas flow rate and distribution
• Gas composition and moisture

In addition, site-specific data on process operating rates and conditioning system (if used) should also be documented. Operators should not rely on just one parameter as an indicator of performance-trends for a number of parameters gives a clearer picture. The ways these parameters affect performance and the techniques used to measure them are briefly described on *Operation and Maintenance Manual for Electrostatic Precipitators* (*U.S. EPA 1985*), as following:

A. <u>Voltage and Current</u>

Voltage and current values for each T-R set should be recorded; they indicate ESP performance more than any other parameter. Most modern ESPs are equipped with primary voltage and current meters on the low-voltage (a.c.) side of the transformer and secondary voltage and current meters on the high-voltage rectified (D.C.) side of the transformer.

When both voltage and current meters are available on the T-R control cabinet, these values can be multiplied to estimate the power input to the ESP. (Note that the primary current reading is multiplied by the primary voltage reading and the secondary current reading is multiplied by the secondary voltage reading). These values (current times voltage) represent the number of watts being drawn by the ESP and is referred to as the corona power input. In addition, whenever a short term spark occurs in a field it can be detected and counted by a spark rate meter. ESPs generally have spark rate meters to aid in the performance evaluation.

The power input on the primary versus the secondary side of the T-R set will differ because of the circuitry and metering of these values. The secondary power outlet (in watts) is always less than the primary power input to the T-R. The ratio of the secondary power to the primary power will range from 0.5 to 0.9 and average from 0.70 to 0.75 (*U.S. EPA 1985*).

Voltage and current values for each individual T-R set are useful because they inform the operators how effectively each field is operating. However, the trends noted within the entire ESP are more important. T-R set readings for current, voltage, and sparking rates should follow certain patterns from the inlet to the outlet fields. For example, corona power density should increase from inlet to outlet fields as the particulate matter is removed from the gas stream.

The electrical meters on the T-R cabinets are always fluctuating. Normal sparking within the ESP causes these fluctuations in the meter readings. These short term movements of the gauges indicate that the automatic voltage controller is restoring the maximum voltage after shutting down for several milliseconds to quench the spark. When recording values of the electrical data from the T-R meters it is important to note the maximum value that is sustained for at least a fraction of a second.

B. Opacity

In many situations, ESP operation is evaluated in terms of the opacity monitored by a Opacity monitor (transmissometer) on a real-time basis. Under optimum conditions the ESP should be able to operate at some base-level opacity with a minimum of opacity spiking from rapper re-entrainment. A facility can have one or more monitors that indicate opacity from various ESP outlet ducts and from the stack.

An opacity monitor compares the amount of light generated and transmitted by the instrument on one side of the gas stream with the quantity measured by the receiver on the other side of the gas stream. The difference, which is caused by absorption, reflection, refraction, and light scattering by the particles in the gas stream, is the opacity of the gas stream. Opacity is expressed as a percent from 0 to 100% and is a function of particle size, concentration, and path length.

Most of the opacity monitors being installed today are double-pass monitors; that is, the light beam is passed through the gas stream and reflected back across to a transceiver. This arrangement is advantageous for several reasons:

2- Automatic checking of the zero and span of the monitor is possible when the process is operational.
3- The monitor is more sensitive to slight variations in opacity because the path length is longer.
4- The entire electronics package is located on one side of the stack as a transceiver.

Although single-pass opacity monitors are available at a lower cost and sensitivity.

For many sources, dust concentration and opacity correlations can be developed to provide a relative indication of ESP performance. These correlations are very site-specific, but can provide plant and agency personnel with an indication of relative performance for given opacity levels. In addition, site-specific opacity charts can be used to predict deterioration of ESP performance that requires attention by plant personnel. Readings from opacity monitors can also be used to optimize spark rate, voltage/current levels, and rapping cycles, even though the conditions within the ESP are not static. In high-efficiency ESPs, re-entrainment may account for 50 to 70% of the total outlet emissions. Therefore, optimization of the rapping pattern may prove more beneficial than trying to optimize the voltage, current and sparking levels. Dust re-entrainment from rapping must be observed by using the opacity monitor operating in a real-time or non-integrating mode because rapping spikes tend to get smoothed out in integrated averages such as the 6-minute average commonly in use. However, the integrated average does provide a good indication of average opacity and emissions.

When parallel ESPs or chambers are used, an opacity monitor is often placed in each outlet duct, as well as on the stack, to measure the opacity of the combined emissions. Although the stack monitor is commonly used to indicate stack opacity (averaging opacities from different ducts can be difficult), the individual duct monitors can be useful in indicating the performance of each ESP or chamber and in troubleshooting. Although this option is often not required and represents an additional expense, it can be very useful, particularly on relatively large ESPs.

New systems, such as the digital microprocessor design, are available in which the opacity monitor data can be used as input for the T-R controller. In this case, the data are used to control power input throughout the ESP to maintain an opacity level preselected by the source. If the opacity increases, the controller increases power input accordingly until the opacity limit, spark limit, current limit, or voltage limit is reached. This system (often sold as an energy saver because it uses only the power required) can save a substantial quantity of energy:

1- On large, high-efficiency ESPs
2- For processes operating at reduced gas loads.

In many cases, however, reduction of ESP power does not significantly alter ESP performance because dust re-entrainment and gas sneakage constitute the largest sources of emissions; additional power often does not reduce these emissions significantly. In some observed cases, reducing power by one-half did not change the performance. For units typically operated at 1000 to 1500 watts/1000 acfm, operating the ESPs at power levels of 500 to 750 watts/1000 afcm still provide acceptable collection efficiencies.

C. Gas Temperature

Monitoring the temperature of the gas stream can provide useful information concerning ESP performance. Temperature is measured using a thermocouple in conjunction with a digital, analog, or strip chart recorder. Temperature is usually measured using a single point probe or thermocouple. This method has a major limitation in that the probe may be placed at an unrepresentative (stratified) point—one that is not representative of the bulk gas flow. Most ESPs are designed with a minimum of three fields. The gas temperature for each field should be measured at both the inlet and outlet, if possible. Significant temperature changes between the inlet and outlet values may indicate air inleakage problems that should be confirmed by measurement of gas composition.

The gas temperature should be checked once per shift for smaller sources and measured continuously on larger sources and on those sources with temperature-sensitive performance.

Temperature measurement can also be a useful tool in finding excessive inleakage or unequal gas flow through the ESP. Both of these conditions can affect localized gas velocity patterns without noticeably affecting the average velocity within the ESP. Yet, localized changes in gas velocities can reduce ESP performance even though the average gas velocity seems adequate.

D. Gas Flow Rate and Distribution

Gas flow rate determines most of the key design and operating parameters such as specific collection area (ft^2/1000 acfm), gas velocity (ft/sec) and treatment time within the ESP, and specific corona power (watts/1000 acfm). The operator should calculate the flue gas flow rate if the ESP is not operating efficiently. For example, significant variations in oxygen may indicate large swings in the gas flow rate that may decrease ESP performance and indicate the need to routinely determine ESP gas volume. Low SCA (Specific Collection Area) values, high velocities, short gas treatment times (5 seconds or less), and much higher oxygen levels at nearly full load conditions are indicators that excess flue gas flow rate may be causing decreased ESP performance.

Presently, most sources do not continuously measure gas velocities or flow rates. Gas velocities are generally only measured during emission compliance testing or when there is a perceived problem. Because of new technologies and regulations, some of the larger sources are beginning to install continuous flow measurement systems.

Multi-point devices, ultrasonic devices, and temperature-based flow devices can be used to continuously measure gas velocity. These devices must be calibrated to the individual stack where they are installed. Most existing facilities currently use indirect indicators to estimate gas flow rate; these include fan operating parameters, production rates or oxygen/carbon dioxide gas concentration levels. Another important parameter is gas flow distribution through the ESP. Ideally; the gas flow should be uniformly distributed throughout the ESP (top to bottom, side to side).

Actually, however, gas flow through the ESP is not evenly distributed, and ESP manufacturers settle for what they consider an acceptable variation. Standards recommended by the Industrial Gas Cleaning Institute have been set for gas flow distribution. Based on a velocity sampling routine, 85% of the points should be within 15% of the average velocity and 99% should be within 1.4 times the average velocity. Generally, uneven gas flow through the ESP results in reduced performance because the reduction in collection efficiency in areas of high gas flow is not compensated for by the improved performance in areas of lower flow. Also, improper gas distribution can also affect gas sneakage through the ESP. As stated earlier, good gas distribution can be accomplished by using perforated plates in the inlet plenum and turning vanes in the ductwork.

Measurement of gas flow distribution through the ESP is even less common than measuring flue gas flow rate. Because the flow measurements are obtained in the ESP rather than the ductwork (where total gas volumetric flow rates are usually measured), more sensitive instrumentation is needed for measuring the low gas velocities. The instrument typically specified is a calibrated hot-wire anemometer. The anemometer test is usually performed at some mid-point between the inlet and outlet (usually between two fields). Care must be taken to assure that internal ESP structural members do not interfere with the sampling points.

Gas flow distribution tests are conducted when the process is inoperative, and the ESP and ductwork are relatively cool. This often limits the amount of gas volume that can be drawn through the ESP to less than 50% of the normal operating flow; however, the relative velocities at each point are assumed to remain the same throughout the normal operating range of the ESP. A large number of points are sampled by this technique. The actual number depends upon the ESP design, but 200 to 500 individual readings per ESP are not
unusual. By using a good sampling protocol, any severe variations should become readily apparent.

E. Gas Composition and Moisture

The **chemical composition** of both the particulate matter and flue gas can affect ESP performance. In many applications, key indicators of gas composition are often obtained by using continuous emission monitors. However, particle concentration and composition are determined by using intermittent grab sampling.

The operation of an ESP depends on the concentration of electronegative gases O2, H2O, CO and SO2/SO3 to generate an effective corona discharge. Often, sources use continuous monitors to measure these gas concentrations to meet regulatory requirements, or in the case of combustion sources to determine excess air levels (CO2 or O2).

F. Evaluating Air-Load/Gas-Load Voltage-Current (V-I) Curves

In addition to the routine panel meter readings, **other electrical tests** of interest to personnel responsible for evaluating and maintaining ESPs include the air-load and gas-load V-I (voltage- current) tests, which may be conducted on virtually all ESPs. Air-load and gas-load curves are graphs of the voltage (kV) versus the current (mA) values obtained at a set condition (test point). These curves are developed to evaluate ESP performance by comparing the graphs from inlet field to outlet field and over periods in time. Deviation from the normal or previous results can indicate that a problem exists.

Air-load tests are generally conducted on cool, inoperative ESPs through which no gas is flowing. This test should be conducted when the ESP is new, after the first shutdown, and every time off-line maintenance is performed on the ESP. These air-load V-I curves serve as the basis for comparison in the evaluation of ESP maintenance and performance. A typical air-load curve is shown in Fig. 5.1.

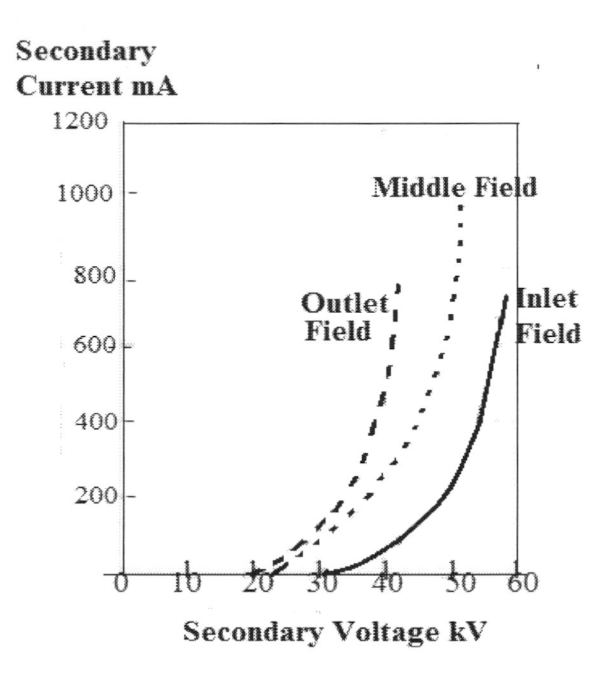

Figure 5.1. Typical air-load voltage and current readings

An air-load V-I curve can be generated with readings from either primary or secondary meters. The following procedures can be used by the ESP operator to develop an air-load curve:

1- Energize a de-energized T-R set on manual control (but with zero voltage and current), and increase the power to the T-R set manually.
2- At corona initiation the meters should suddenly jump and the voltage and near zero current levels should be recorded. It is sometimes difficult to identify this point precisely, so the lowest practical value should be recorded.
3- After corona initiation is achieved, increase the power at predetermined increments [for example, every 50 or 100 milliamps of secondary current or every 10 volts of AC primary voltage (the increment is discretionary)], and record the values for voltage and current.
5- Continue this procedure until one of the following occurs:
    • Sparking

- Current limit is achieved
- Voltage limit is achieved

6- Repeat this procedure for each T-R set.

When the air-load tests have been completed for each field, plot each field's voltage/current curves. When ESPs are equipped with identical fields throughout, the curves for each field should be nearly identical. In most cases, the curves also should be similar to those generated when the unit was new, but shifted slightly to the right due to residual dust on the wires (or rigid frames) and plates of older units. These curves should become part of the permanent record of the ESP.

The use of air-load curves enables plant personnel to identify which field(s) may not be performing as designed. Also, comparison of air-load curves from test runs taken just before and after a unit is serviced will confirm whether the maintenance work corrected the problem(s).

One major advantage of air-load tests is that they are performed under nearly identical conditions each time, which means the curves can be compared. One disadvantage is that the internal ESP conditions are not always the same as during normal operation. For example, misalignment of electrodes may appear or disappear when the ESP is cooled (expansion/contraction), and dust buildup may be removed by rapping during ESP shutdown.

Gas-Load Curves

The gas-load V-I curve, on the other hand, is generated during the normal operation of the process while the ESP is energized. The procedure for generating the gas-load V-I curve is the same as for the air load except that gas-load V-I curves are always generated from the outlet fields first and move toward the inlet. This prevents the upstream flow that is being checked from disturbing the V-I curve of the downstream field readings. Although such disturbances would be short-lived (usually 2 minutes, but sometimes lasting up to 20 minutes), working from outlet to inlet speeds up the process.

The curves generated under gas-load conditions will be similar to air-load curves. Gas load curves will generally be shifted to the left however, because sparking occurs at lower voltage and current when particles are present. The shape of the curve will be different for each field depending on the presence of particulate matter in the gas stream (see Figure 5.2).

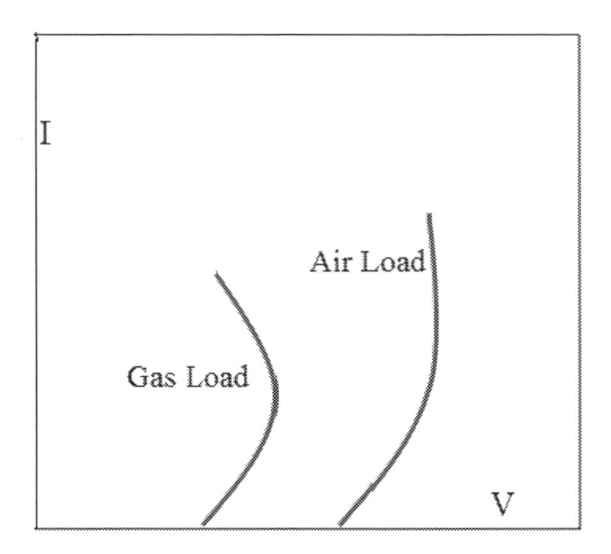

Figure 5.2. Comparison of typical air-load and gas-load

Also, gas-load curves vary from day to day, even minute to minute. Curve positions may change as a result of fluctuations in the following:

- Amount of dust on plates
- Gasflow
- Particulate chemistry loading
- Temperature
- Resistivity

Nonetheless, they still should maintain a characteristic pattern. Gas-load curves are normally used to isolate the cause of a suspected problem rather than being used on a day-today basis; however, they can be used daily if necessary.

5.3. Performance

5.3.1 Performance Requirements

Designing a precipitator for optimum performance requires proper sizing of the precipitator in addition to optimizing precipitator efficiency. While some users rely on the precipitator manufacturer to determine proper sizing and design parameters, others choose to either take a more active role in this process or hire outside engineering firms.

Precipitator performance depends on its size and collecting efficiency. Important parameters include the collecting area and the gas volume to be treated. Other key factors in precipitator performance include the electrical power input and dust chemistry. These parameters briefly described as following:

A. Precipitator sizing

The sizing process is complex as each precipitator manufacturer has a unique method of sizing, often involving the use of computer models and always involving a good dose of judgment. No computer model on its own can assess all the variables that affect precipitator performance.

B. Collecting Efficiency

Based on specific gas volume and dust load, calculations are used to predict the required size of a precipitator to achieve a desired collecting efficiency.

C. Power Input

Power input is comprised of the voltage and current in an electrical field. Increasing the power input improves precipitator collecting efficiency under normal conditions.

5.3.2 Process Variables

Gas characteristics and particle properties define how well a precipitator will perform in a given application. The main process variables to consider are:

A. Gas flow rate;
B. Particle size and size distribution;
C. Particle resistivity; and
D. Gas temperature

Definition and impact of these process variables on the performance of precipitator are considered briefly as following:

A. <u>Gas Flow Rate</u>

The gas flow rate in a power plant is defined by coal quality, boiler load, excess air rate and boiler design. Where there is no combustion, the gas flow rate will have process-specific determinants.

A precipitator operates best with a gas velocity of 3.5 - 5.5 ft/sec. At higher velocity, particle re-entrainment increases rapidly. If velocity is too low, performance may suffer from poor gas flow distribution or from particle dropout in the ductwork.

B. <u>Particle Size</u>

The size distribution in a power plant is defined by coal quality, the coal mill settings and burner design. Particle size for non-combustion processes will have similar determinants.

A precipitator collects particles most easily when the particle size is coarse. The generation of the charging corona in the inlet field may be suppressed if the gas stream has too many small particles (less than 1 μm). Very small particles (0.2 - 0.4μm) are the most difficult to collect because the fundamental field-charging mechanism is overwhelmed by diffusion charging due to random collisions with free ions.

C. <u>Particle Resistivity</u>

Resistivity is resistance to electrical conduction. The higher the resistivity, the harder it is for a particle to transfer its electrical charge. The resistivity of fly ash or other particles is influenced by the chemical composition of the gas stream, particle temperature and gas temperature.

Resistivity should be kept in the range of 108 - 1010 ohm-cm. High resistivity can reduce precipitator performance. For example, in combustion processes, burning reduced-sulfur coal increases resistivity and reduces the collecting efficiency of the precipitator. Sodium and iron oxides in the fly ash can reduce resistivity and improve performance, especially at higher operating temperatures. On the other hand, low resistivity can also be a problem. For example (in combustion processes), unburned carbon reduces precipitator performance because it is so conductive and loses its electrical charge so quickly that it is easily re-entrained from the collecting plate.

Resistivity can be determined as a function of temperature in accordance with IEEE Standard 548. This test is conducted in an air environment containing a specified moisture concentration. The test is run as a function of ascending or descending temperature or both. Data are acquired using an average ash layer electric field of 4 kV/cm. Since relatively low applied voltage is used and no sulfuric acid vapor is present in the environment, the values obtained indicate the maximum ash resistivity.

With particles of high resistivity (cement dust for example) Sulfur trioxide is sometimes injected into a flue gas stream to lower the resistivity of the particles in order to improve the collection efficiency of the electrostatic precipitator.

D. Gas Temperature

The effect of gas temperature on precipitator performance and collecting efficiency can be significant.

Changes in gas temperature can have profound effects on ESP performance and particularly on precipitator collecting efficiency. The temperature variation can be very small (in some cases as little as 15oF) and yet cause a significant change in ESP power levels and opacity. Although gas temperature variations may have some effect on corona discharge characteristics and physical characteristics of the ESP (corrosion, expansion/contraction), their most important effect is on particle resistivity.

For sources with the potential for high resistivity, temperature changes can cause dramatic changes in performance, even when all other parameters seem to be the same.

5.3.3 Performance Improvements

Improvement or optimization of precipitator operation can result in significant savings. Many specific situations encourage a review of precipitator operation to consider the following parameters:

- Deterioration of existing equipment
- Tightening of air pollution emission regulations
- Changes in products and/or production rates
- Frequent forced outages
- De-rating of production

Continuously performance improvement programs are implemented by the concern manufacturers. These may include the following subsystem and issues:

A. Gas Velocity Distribution
B. Corona Power
C. Re-entrainment
D. Equipment Improvements
E. Combustion Process Improvements for Power Plants
F. Process Improvements

Details of these subsystems on the performance improvement of precipitator are considered briefly as following:

A. Gas Velocity Distribution

Efficient precipitator performance depends heavily upon having similar gas conditions at the inlet of each electrical field or bus section and at the inlet of each gas passage of the electrical field or bus section. Uniformity of gas velocity is also desirable - good gas velocity distribution through a precipitator meets the following requirements:

- 85% of all measured gas velocities < 1.15 times the average gas velocity

- 99% of all measured gas velocities < 1.40 times the average gas velocity

The gas velocity distribution in a precipitator can be customized according to the design of the precipitator and the characteristics of the dust particles. Traditionally, precipitators have been designed with uniform gas velocity distribution through the electrical fields, to avoid high-velocity areas that would cause re-entrainment. While this is still a recommended practice, there is an advantage in some cases to developing a velocity profile that brings more particles closer to the hopper.

Gas velocity distribution can be controlled by the following:

- Adding/improving gas flow control devices in the inlet ductwork
- Adding/improving flow control devices in the inlet of the precipitator
- Adding/improving flow control devices in the outlet of the precipitator
- Adding a rapping system to the flow control devices (where applicable)
- Adding/improving anti-sneak baffles at the peripheries of the electrical fields
- Adding/improving hopper baffles
- Eliminating air leakages into the precipitator

B. Corona Power

Precipitator corona power is the useful electrical power applied to the flue gas stream to precipitate particles. Either precipitator collecting efficiency or outlet residual can be expressed as a function of corona power in Watts/1000 acfm of flue gas, or in Watts/1000 ft of collection area.

The separation of particles from the gas flow in an electrostatic precipitator depends on the applied corona power. Corona power is the product of corona current and voltage. Current is needed to charge the particles. Voltage is needed to support an electrical field, which in turn transports the particles to the collecting plates.

In the lower range of collecting efficiencies, relatively small increases in corona power result in substantial increases in collecting efficiency. On the other hand, in the upper ranges, even large increases in corona power will result in only small efficiency increases.

Equally, in the lower range of the corona power levels, a small increase in the corona power results in a substantial reduction in the gas stream particle content. In the upper range of the corona power level, a large increase is required to reduce the particle content.

Optimum conditions depend upon the location of the field (inlet, center and outlet), fly ash characteristics (resistivity) and physical conditions (collecting plates and discharge wires). Corona power levels can be optimized by adjusting or optimizing the following:

Component – Subsystem	Adjusting Parameters
Gas velocity	• Uniformity
Fly Ash	• Particle size • Resistivity
Voltage Controls	• Spark rate setting • Current limits • Voltage limits
Design	• Plate spacing • Collecting plate • discharge electrode design
Rapping System	• Frequency • Intensity
Support Insulator	• Purge air system operation

C. Re-entrainment

Reducing rapping re-entrainment to an acceptable level generally requires a substantial improvement of the gas velocity distribution, the electrical power density and uniformity, as well as an extended optimization program for the collecting-plate rapping system. Re-entrainment of collected particles is the major contributor to particulate emissions of the precipitator. In some cases, re-entrainment accounts for 60 - 80% of the residual.

The major causes of re-entrainment are as follows:

Component-Subsystem	Factors Affecting Re-entrainment
Particles	• Low cohesiveness • Low adhesion to collecting plates • Particle size • Low resistivity
Voltage Controls	• Spark rate setting
Design	• Collecting plate design • Discharge electrode design • Plate spacing
Rapping System	• Frequency • Intensity • Duration (if applicable)
Electrical Field	• Collecting plate and discharge electrode rapping • Sparking • Saltation • Erosion (localized high gas velocity) • Sneakage
Hopper	• Hopper design • Leakage (hopper valve) • Hopper gas flow

D. Equipment Improvements

The objectives of equipment improvements are to optimize corona power, reduce re-entrainment, and optimize gas velocity distribution inside the precipitator. Some important topics to consider when planning equipment improvements include:

1. **Precipitator Size:** When sizing the precipitator, it is important to provide a cross-section that will maintain an acceptable gas velocity. It is also important to provide for enough total discharge wire length and collecting plate area, so that the desired specific corona current and electrical field can be applied.
2. **Gas Velocity Distribution:** Improving gas velocity distribution in the precipitator reduces particle re-entrainment and boosts precipitator efficiency. Typically, a uniform gas velocity is desired, but there are site-specific exceptions. Gas velocity distribution can be modified by using flow control devices and baffles.
3. **Corona Power:** The separation of dust particles from the gas flow in an electrostatic precipitator depends on the applied corona power. Corona power is the product of corona current and voltage. Current is needed to

charge the particles. Voltage is needed to support an electrical field, which in turn transports the particles to the collecting plates.

4. **Sectionalization:** The precipitator is divided into electrical sections that are cross-wise and parallel to the gas flow to accommodate spatial differences in gas and dust conditions. Optimization of corona power involves adjusting the corona power (secondary voltage and current) in each electrical section for optimum conditions.
5. **Particle Re-entrainment**: Minimizing re-entrainment of dust particles is important to improvement of precipitator efficiency. Most precipitator equipment affects the re-entrainment level.
6. **Additional Equipment:** Performance improvement options include the installation of a second precipitator in series with the existing precipitator; using fabric filters downstream of the precipitator; and adding a second particle collector in parallel with the existing collector. Other possibilities include sonic or electrostatic particle upstream of the precipitator; a mechanical upstream collector; or an electrostatically-enhanced or mechanical collector, or a filter downstream of the precipitator.

E. Combustion Process Improvements for Power Plants

Combustion process conditions mainly affect the corona power level. The primary contributors to combustion process conditions and their effects include:

1- Coal

Bituminous coals, sub-bituminous and lignite coals are substantially different from each other in the combustion process. Coal blending is now used for operational and financial benefits. This results in a wide range of boiler and precipitator operating conditions. Precipitating fly ash from difficult coals can be improved with conditioning systems. However, the furnace and its associated equipment can still cause problems in the precipitator, particularly coal mills, burners, and air pre-heaters.

The main parameters and factors effecting combustion process conditions include the following:

— Flue gas flow rate

— Flue gas moisture content
— Fly ash resistivity
— Fly ash inlet loading
— Fly ash particle size

2- Coal mills

The setting of the coal mills and classifiers defines the coal particle size which in turn impacts the fly ash particle size. Larger coal particles are more difficult to combust, but larger fly ash particles are easier to collect in the precipitator.

The main parameters and factors effecting combustion process conditions include the following:

— Fly ash particle size
— Unburned carbon (LOI)

3- Furnace

Base-load operation of the boiler is usually better for precipitator operation than swing-load operation due to more stable operating conditions. Boiler operation at low loads may be as problematic for the precipitator as operating the boiler at its maximum load level, due to fallout of fly ash in the ductwork, low gas temperatures, and deterioration of the quality of the gas velocity distribution. If low load operation cannot be avoided, the installation of additional gas flow control devices in the inlet and outlet of the precipitator may prove beneficial.

The main parameters and factors effecting combustion process conditions include the following:

— Base load/swing load operation
— Flue gas flow rate

4- Burners

The operation of coal burners, together with the setting of the coal mills and their classifiers, affects the percentage of unburned carbon (LOI or UBC) in the fly ash. The use of Lo-NOx burners increases this percentage, and causes re-entrainment and increased sparking in the precipitator. Further, the UBC tends to absorb SO_3, which in turn increases the fly ash resistivity. Over-fire air optimization or coal-reburn systems may reduce UBC in the fly ash.

The main parameters and factors effecting combustion process conditions include the following:

— Flue gas temperature
— Fly ash resistivity
— Unburned carbon (LOI)

5- Air pre-heaters

Regenerative air pre-heaters cause temperature and SO_3 stratification in the downstream gas flow. This problem is more severe in closely coupled systems, where the precipitator is located close to the air pre-heater. Depending upon site-specific conditions, flow mixing devices may be installed in the ductwork to the precipitator, or flue gas conditioning systems may be used to equalize the gas flow characteristics.

The main parameters and factors effecting combustion process conditions include the following:

— Rotation
— Gas flow pattern
— Gas temperature pattern
— SO_3 distribution pattern

F. Process Improvements

Improvement or optimization of precipitator operation can result in significant savings. Many specific situations encourage a review of precipitator operation:

- Deterioration of existing equipment
- Tightening of air pollution emission regulations
- Changes in products and/or production rates
- Frequent forced outages
- De-rating of production

6

Maintenance and Safety Measures

6.1. Preventive Maintenance Checklist

The components of the ESP and their operation are periodically inspected by plant personnel as part of a Preventive Maintenance Program PMP. When the PMP applied, the problems causing EPS performance limitations are detected and corrected before they cause a major shutdown of the ESP. In addition to performing maintenance, keeping records of the actions taken should be an integral part of any maintenance program.

The frequency of inspection of all ESP components should be established by a formal in-house maintenance procedure. Vendors' recommendations for an inspection schedule should be followed. In addition to the daily monitoring of meters and the periodic inspection of ESP components, some operational checks should be performed every shift and the findings should be recorded on a shift data sheet.

A listing of typical periodic maintenance procedures for an ESP used to collect fly ash is given in Table 6.1.

At the end of every shift, these shift data sheets should be evaluated for maintenance needs. These once-per-shift checkpoints include an inspection of rappers, dust discharge systems, and Transformer-Rectifier T-R sets for proper functioning and an indication of which T-R sets are in the "off" position. Rappers that are not functioning should be scheduled for maintenance, particularly if large sections of rappers are out of service. Dust discharge systems should have highest priority for repair; dust should not accumulate in the bottom of the ESP for long periods of time because of the potential for causing severe plate misalignment problems. Hopper heaters can usually be repaired with little difficulty after removing weather protection and insulation. Insulator heaters may be difficult to repair except during short outages. Hopper heaters keep condensation on the insulators to a minimum and help keep the dust warm and free-flowing.

Although an ESP can operate effectively with up to 10% of its wires removed, care must be taken that no more than 5 to 10 wires in any one gas lane are removed. The loss of wires down any one lane can result in a substantial increase in emissions. The only way to adequately track where wires have failed or slipped out of the ESP is with a wire replacement chart.

Good site specific records of both the design and operating history will enable operating personnel to better evaluate ESP performance and problems. Design parameters built into the ESP include the following: the specific collection area (SCA), number of fields, number of T-R sets, sectionalization, T-R set capacity, design velocity and treatment time, aspect ratio and particle characteristics (resistivity). Design records indicate the specific conditions under which the ESP was designed to operate. A comparison between design records and operating records indicate whether operating parameters have changed significantly from the design conditions. Secondly, maintaining proper operating records establishes good baseline information to bracket normal ranges of operation.

Table 6.1 Preventive maintenance checklist for a typical fly ash precipitator

Time Schedule	Actions
Daily	1. Take and record electrical readings and transmissometer data. 2. Check operation of hoppers and ash removal system. 3. Examine control room ventilation system. 4. Investigate cause of abnormal arcing in T-R enclosures and bus duct.
Weekly	1. Check rapper operation. 2. Check and clean air filter. 3. Inspect control set interiors.
Monthly	1. Check operation of standby top-housing pressurizing fan and thermostat. 2. Check operation of hopper heaters. 3. Check hopper level alarm operation.
Quarterly	1. Check and clean rapper and vibrator switch contacts. 2. Check transmissometer calibration.
Semiannual	1. Clean and lubricate access-door dog bolt and hinges. 2. Clean and lubricate interlock covers. 3. Clean and lubricate test connections. 4. Check exterior for visual signs of deterioration, and abnormal vibration, noise, leaks. 5. Check T-R liquid and surge-arrestor spark gap.
Annual	1. Conduct internal inspection. 2. Clean top housing or insulator compartment and all electrical insulating surfaces. 3. Check and correct defective alignment. 4. Examine and clean all contactors and inspect tightness of all electrical connections. 5. Clean and inspect all gasketed connections. 6. Check and adjust operation of switchgear. 7. Check and tighten rapper insulator connections. 8. Observe and record areas of corrosion.
Situational	1. Record air-load and gas-load readings during and after each outage. 2. Clean and check interior of control sets during each outage of more than 72 hours. 3. Clean all internal bushings during outages of more than 5 days. 4. Inspect condition of all grounding devices during each outage over 72 hours. 5. Clean all shorts and hopper buildups during each outage. 6. Inspect and record amount and location of residual dust deposits on electrodes during each outage of 72 hours or longer. 7. Check all alarms, interlocks, and all other safety devices during each outage. Ref.: Bibbo 1982.

6.2. ESP Operating Problems

Evaluating ESP operating problems can be difficult and no single parameter can identify all potential problems; a combination of factors should be considered to accurately pinpoint problems.

For example, although most ESP problems are reflected in the electrical readings, many different problems produce the same characteristics on the meters. In addition, an initial failure or problem can cause a "domino effect" bringing about even more problems and making it difficult to identify the original cause. Table 6.1contains a typical troubleshooting cycle (Szabo and Gerstle 1977) that is useful as a general guide.

The EPA (1985) categorized the major performance problems associated with electrostatic precipitators into the following seven areas:

1- Resistivity,
2- Dust Accumulation,
3- Wire breakage,
4- Hopper pluggage,
5- Misalignment of ESP components,
7- Changes in particle size distribution, and
8- Air inleakage.

These problems are related to design limitations, operational changes, and/or maintenance procedures. The following Identification of these problems and their effect on ESP performance is excerpted from the EPA document titled *Operation and Maintenance Manual for Electrostatic Precipitators* (1985):

A. Problems Related to Resistivity

The resistivity of the collected dust on the collection plate affects the acceptable current density through the dust layer, dust removal from the plates, and indirectly, the corona charging process. High resistivity conditions in utility fly ash applications have received much attention. The optimum resistivity range for ESP operation is relatively narrow; both high and low resistivity cause problems.

At industrial sources where resistivity changes are intermittent, modification of operating procedures may improve performance temporarily. Expensive retrofitting or modifications may be required if the dust resistivity is vastly different than the design range.

A1. <u>High Resistivity</u>

High dust resistivity is a more common problem than low dust resistivity. Particles having high resistivity are unable to release or transfer electrical charge. At the collection plate, the particles neither give up very much of their acquired charge nor easily pass the corona current to the grounded collection plates. High dust resistivity conditions are indicated by low primary and secondary voltages, suppressed secondary currents and high spark rates in all fields. This condition makes it difficult for the T-R controller to function adequately.

Severe sparking can cause excessive charging off-time, spark "blasting" of particulate on the plate, broken wires due to electrical erosion, and reduced average current levels.

The reduced current levels generally lead to deteriorated performance. Because the current level is indicative of the charging process, the low current and voltage levels that occur inside an ESP operating with high resistivity dust generally reflect slower charging rates and lower particle migration velocities to the plate. Particle collection is reduced; consequently, the ESP operates as though it were "undersized." If high resistivity is expected to continue, the operating conditions can be modified or *ESP Operation and Maintenance* conditioning agents can be used to accommodate this problem and thereby improve performance.

High resistivity also tends to promote rapping problems, as the electrical properties of the dust tend to make it very tenacious. High voltage drop through the dust layer and the retention of electrical charge by the particles make the dust difficult to remove because of its strong attraction to the plate. The greater rapping forces usually required to dislodge the dust may also aggravate or cause a rapping re-entrainment problem.

Important items to remember are (1) difficulty in removing the high-resistivity dust is related to the electrical characteristics, not to the sticky or cohesive nature of the dust; and (2) the ESP must be able to withstand the necessary increased rapping forces without sustaining damage to insulators or plate support systems.

A2. Low Resistivity

Low dust resistivity, although not as common, can be just as detrimental to the performance of an ESP as high resistivity. When particles with low resistivity reach the collection plate, they release much of their acquired charge and pass the corona current quite easily to the grounded collection plate. Without the attractive and repulsive electrical forces that are normally present at normal dust resistivities, the binding forces between the dust and the plate are considerably weakened. Therefore, particle reentrainment is a substantial problem at low resistivity, and ESP performance appears to be very sensitive to contributors of reentrainment, such as poor rapping or poor gas distribution.

Since there is lower resistance to current flow for particles with low resistivity (compared to normal or high), lower operating voltages are required to obtain substantial current flow. Operating voltages and currents are typically close to clean plate conditions, even when there is some dust accumulation on the plate. Low-resistivity conditions, are typically characterized by low operating voltages, high current flow, and low spark rates.

Despite the large flow of current under low-resistivity conditions, the corresponding low voltages yield lower particle migration velocities to the plate. Thus, particles of a given size take longer to reach the plate than would be expected. When combined with substantial dust reentrainment, the result is poor ESP performance. In this case, the large flow of power to the ESP represents a waste of power.

Low-resistivity problems typically result from the chemical characteristics of the particulate and not from flue gas temperature. The particulate may be enriched with compounds that are inherently low in resistivity, either due to poor operation of the process or to the inherent nature of the process. Examples of such enrichment include excessive carbon levels in fly ash (due to poor combustion), the presence of naturally occurring alkalis in wood ash, iron oxide in steel-making operations, or the presence of other low-resistivity materials in the dust. Over-conditioning may also occur in some process operations, such as the burning of high-sulfur coals or the presence of high SO3 levels in the gas stream, which lower the inherent resistivity of the dust. In some instances, large ESPs with SCAs greater than 750 ft2/1000 acfm have performed poorly because of the failure to fully account for the difficulty involved in collecting a low-resistivity dust. Although some corrective actions for low resistivity are available, they are sometimes more difficult to implement than those for high resistivity.

A3. Typical High, Normal and Low Resistivity Curves

Evaluating the current and spark rate trends from the inlet to the outlet fields provides a means of evaluating the general resistivity conditions. Moderate dust resistivity conditions, under which ESPs work very well, are indicated by low secondary currents in the inlet field and progressively higher values going toward the outlet. Spark rates under moderate resistivity are moderate in the inlet fields and decrease to essentially zero in the outlet field. High resistivity conditions are indicated by low secondary currents in all of the fields coupled with very high spark rates. Conversely, low resistivity has very high currents and low spark rates in all the fields.

Figure 6.1 shows the typical trend lines for moderate (normal) and high resistivity dusts. As the resistivity goes from moderate to high, the currents decrease dramatically in all of the fields. This is due to the suppressing effect caused by the strong electrostatic field created on the dust layer, and to increased electrical sparking. The decrease in currents is most noticeable in the outlet fields which previously had relatively high currents. Spark rates increase dramatically during high resistivity. Often most of the fields will hit the spark rate limits programmed in by the plant operators.

Once the spark rate limit is sensed by the automatic voltage controllers, it no longer attempts to drive up the voltage. This causes a reduction in the operating voltages of these fields. The overall impact on the opacity is substantially increased emissions. In some cases, puffing again occurs during rapping. This is due to reduced capability of the precipitator fields to collect the slight quantities of particles released during rapping of high resistivity dust.

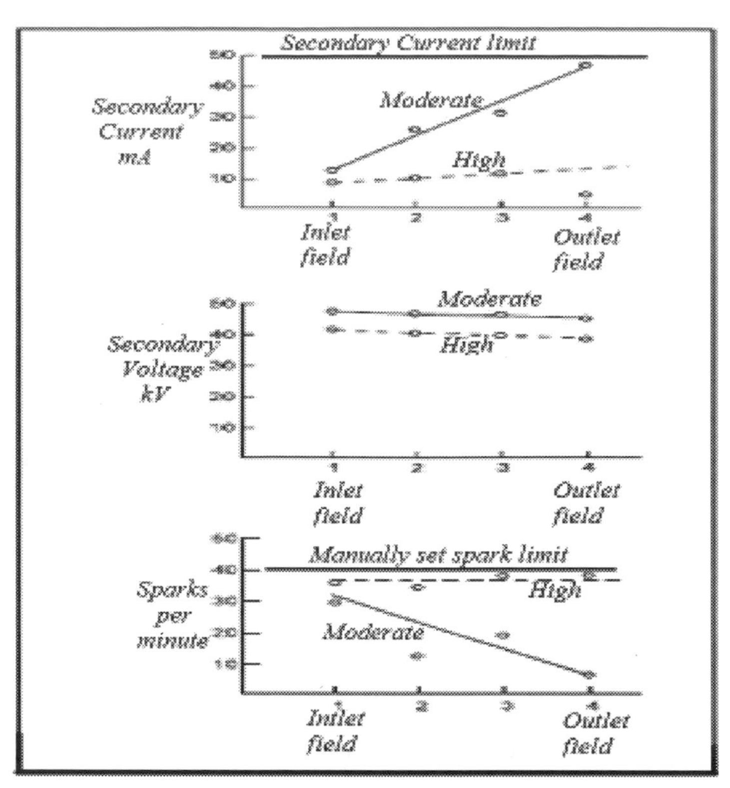

Figure 6.1. Typical T-R set plots - high resistivity versus moderate (normal) Resistivity

Figure 6.2 shows the typical trend lines for moderate (normal) and low resistivity dusts in a four-field ESP. The moderate resistivity dust shows a steady increase of current from the first field to the fourth field, while the secondary current increases rapidly for all fields when the dust exhibits low resistivity. This effect is especially noticeable in the inlet fields which previously had the lowest currents. This increase in current is due simply to the fact that the dust layer's electrostatic field is too weak to significantly impede the charging field created by the discharged electrodes. At low resistivity, the spark rates are generally very low or zero. The voltages in all of the fields are a little lower than normal since the automatic voltage controllers sense that the power supply is at its current limit; therefore, the controller does not attempt to drive the voltage up any further. While the low resistivity conditions persist, there can be frequent and severe puffs (opacity increase) which occur after each collection plate rapper activates.

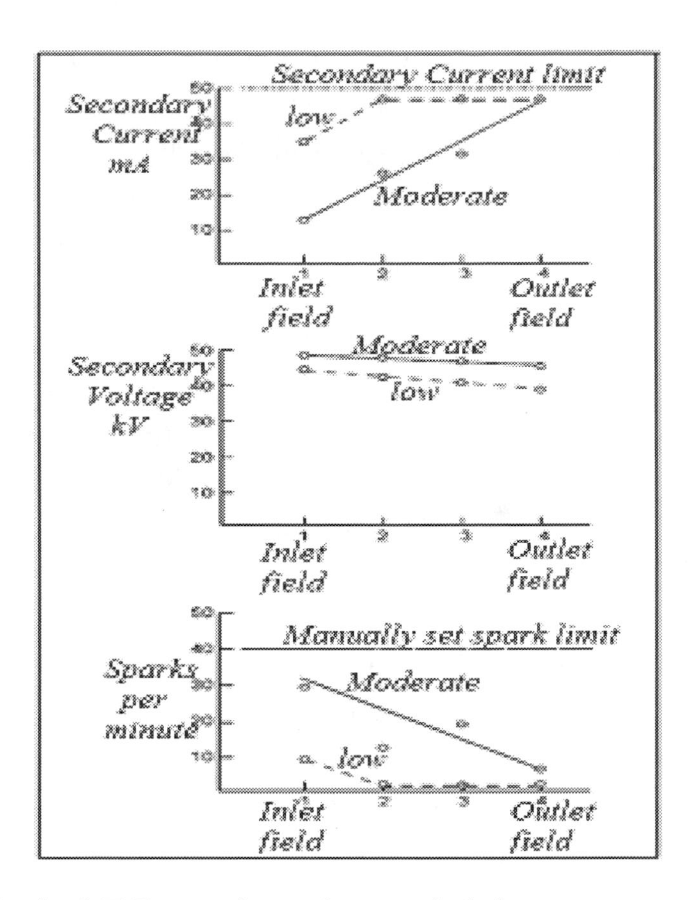

Figure 6.2. Typical T-R set plots - low resistivity versus moderate (normal) resistivity

Using the current, voltage, and spark rate plots is a very good way to use readily available information to evaluate the impossible-to-directly monitor but nevertheless important resistivity conditions. It is possible to differentiate between problems caused by mechanical faults in a single field (such as insulator leakage) and resistivity conditions which inherently affect all of the fields in varying degrees. However, these trend lines are not a perfect analysis tool for evaluating resistivity. A few precipitators never display typical electrical trend lines since they have undersized T-R sets, undersized fields, improperly set automatic voltage controllers, or severe mechanical problems affecting most of the fields.

B. <u>Dust Accumulation</u>

There are three primary causes of dust accumulation on electrodes:

- Inadequate rapping system
- Sticky dust
- Operation at temperatures below the dew point level

The usual cause for buildup of dust on the collection plates or discharge wires is failure of the rapping system or an inadequate rapping system. The rapping system must provide sufficient force to dislodge the dust without damaging the ESP or causing excessive re-entrainment.

The failure of one or two isolated rappers does not usually degrade ESP performance significantly. The failure of an entire rapper control system or all the rappers in one field, however, can cause a noticeable decrease in ESP performance, particularly with high-resistivity dust. Therefore, rapper operation should be checked at least once per day, or perhaps even once per shift. A convenient time to make this check is during routine T-R set readings.

Rapper operation may be difficult to check on some ESPs because the time periods between rapper activation can range from 1 to 8 hours on the outlet field. One method of checking rapper operation involves installing a maintenance-check cycle that allows a check of all rappers in 2 to 5 minutes by following a simple rapping pattern. The cycle is activated by plant personnel, who interrupt the normal rapping cycle and note any rappers that fail to operate. After the check cycle, the rappers resume their normal operation.

Maintenance of rapper operation is important to optimum ESP performance. Excessive dust buildup also may result from sticky dusts or operation at gas dew point conditions. In some cases, the dusts may be removed by increasing the temperature, but in many cases the ESP must be entered and washed out. If sticky particulates are expected (such as tars and asphalts), a wet-wall ESP is usually used because problems can occur when large quantities of sticky particles enter a dry ESP.

Sticky particulates can also become a problem when the flue gas temperature falls below the dew point level. Although acid dew point is usually of greater concern in most applications, moisture dew point is important, too. When moisture dew point conditions are reached, liquid droplets tend to form that can bind the particulate to the plate and wire.

These conditions also accelerate corrosion. Carryover of water droplets or excessive moisture can also cause this problem (e.g., improper atomization of water in spray cooling of the gas or failure of a water wall or economizer tube in a boiler). In some instances the dust layer that has built up can be removed by increasing the intensity and frequency of the rapping while raising the temperature to "dry out" the dust layer. In most cases, however, it is necessary to shut down the unit and wash out or "chisel out" the buildup to clean the plates.

In an operating ESP, differences in the V-I curves can be used to evaluate if a dust buildup problem exists. Buildup of material on the discharge electrodes often means an increase in voltage to maintain a given operating current. The effect of dust buildup on discharge electrodes is usually equivalent to increasing the effective wire diameter. Since the corona starting voltage is strongly a function of wire diameter, the corona starting voltage tends to increase and the whole V-I curve tends to shift to the right (see Figure 6.3). Sparking tends to occur at about the same voltage as it does without dust buildup, unless resistivity is high. This effect on corona starting voltage is usually more pronounced when straight wires are uniformly coated with a heavy dust, and less pronounced on barbed wires and rigid electrodes or when the dust layer is not uniform. Barbed wires and rigid electrodes tend to keep the "points" relatively clean and to maintain a small effective wire diameter and, therefore, a low corona starting voltage. Nevertheless, a higher voltage would still be required to spread the corona discharge over the wire when dust buildup occurs. Thus, buildup on the discharge electrodes would still be characterized by a higher voltage to maintain a given current level.

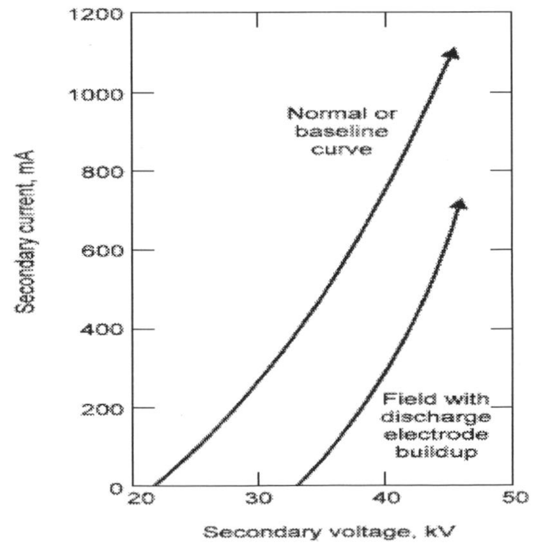

Figure 6.3. V-I curve for a field with excessive wire buildup

C. Wire Breakage

Some ESPs operate for 10 to 15 years without experiencing a single wire breakage. Whereas others experience severe wire breakage problems causing one or more sections to be out of service nearly every day of operation. Much time and effort have been expended to determine the causes of wire breakage. One of the advantages of rigid-frame and rigid electrode ESPs is their use of shorter wires or no wires at all. Although most new ESPs have either rigid frames or rigid electrodes, and some weighted-wire systems have been retrofitted to rigid electrodes, the most common ESP in service today is still the weighted wire.

Wires usually fail in one of three areas: at the top of the wire, at the bottom of the wire, and wherever misalignment or slack wires reduce the clearance between the wire and plate. Wire failure may be due to electrical erosion, mechanical erosion, corrosion, or some combination of these. When wire failures occur, they usually short-out the field where they are located. In some cases, they may short-out an adjacent field as well. Thus, the failure of one wire can cause the loss of particle collection in an entire field or bus section.

In some smaller ESP applications, this can represent one-third to one-half of the charging/collecting area and thus substantially limit ESP performance. One of the advantages of higher sectionalization is that wire failure is confined to smaller areas so overall ESP performance does not suffer as much. Some ESPs are designed to meet emission standards with some percentage of the ESP de-energized, whereas others may not have any margin to cover downtime. Because they receive and remove the greatest percentage of particulate matter, inlet fields are usually more important to ESP operation than outlet fields.

Electrical erosion is caused by excessive sparking. Sparking usually occurs at points where there is close clearance within a field due to a warped plate, misaligned guidance frames, or bowed wires. The maximum operating voltage is usually limited by these close tolerance areas because the spark-over voltage depends on the distance between the wire and the plate. The smaller the distance between the wire and plate, the lower the spark over voltage. Under normal circumstances random sparking does little damage to the ESP.

During sparking, most of the power supplied to energize the field is directed to the location of the spark, and the voltage field around the remaining wires collapses. The considerable quantity of energy available during the spark is usually sufficient to vaporize a small quantity of metal. When sparking continues to occur at the same location, the wire usually "necks down" because of electrical erosion until it is unable to withstand the tension and breaks. Misalignment of the discharge electrodes relative to the plates increases the potential for broken wires, decreases the operating voltage and current because of sparking, and decreases the performance potential of that field in the ESP.

Although the breakage of wires at the top and bottom where the wire passes through the field can be aggravated by misalignment, the distortion of the electrical field at the edges of the plate tends to be the cause of breakage. This distortion of the field, which occurs where the wire passes the end of the plate, tends to promote sparking and gradual electrical erosion of the wires.

Design faults and the failure to maintain alignment generally contribute to **mechanical erosion** (or wear) of the wire. In some designs, the lower guide frame guides the wires or their weight hooks (not the weights themselves) into alignment with the plates. When alignment is good, the guide frame or grid allows the wires or weight hooks to float freely within their respective openings. When the position of the wire guide frame shifts, however, the wire or weight hook rubs the wire frame within the particulate-laden gas stream.

Failures of this type usually result from a combination of mechanical and electrical erosion. Corrosion may also contribute to this failure. Microsparking action between the guide frame and the wire or weight hook apparently causes the electrical erosion. The same type of failure also can occur in some rigid frame designs where the wires ride in the frame.

Another mechanical failure that sometimes occurs involves **crossed wires**. When replacing a wire, maintenance personnel must make sure that the replacement wire does not cross another wire. Eventually, the resulting wearing action breaks one or both wires. If one of the wires does survive, it is usually worn down enough to promote greater sparking at the point of contact until it finally does break. Any wires that are found to be exceptionally long and slack should be replaced; they should not be crossed with another wire to achieve the desired length.

Corrosion of the wires can also lead to wire failures. Corrosion, an electrochemical reaction, can occur for several reasons, the most common being acid dew point. When the rate of corrosion is slow and generally spread throughout the ESP, it may not lead to a single wire failure for 5 to 10 years. When the rate of corrosion is high because of long periods of operating the ESP below the acid dew point, failures are frequent. In these cases the corrosion problem is more likely to be a localized one (e.g., in places where cooling of the gas stream occurs, such as inleakage points and the walls of the ESP). Corrosion-related wire failures can also be aggravated by startup-shutdown procedures that allow the gas streams to pass through the dew point many times. Facilities have mainly experienced wire breakage problems during the initial process shakedown period when the process operation may not be continuous. Once steady operation has been achieved, wire breakage problems tend to diminish at most plants.

Wire crimping is another cause of wire failure. Crimps usually occur at the top and bottom of the wires where they attach to the upper wire frame or bottle weight; however, a crimp may occur at any point along the wire. Because a crimp creates a residual stress point, all three mechanisms (electrical erosion, mechanical erosion, and corrosion) may be at work in this situation. A crimp can:

1. Distort the electric field along the wire and promote sparking;
2. Mechanically weaken the wire and make it thinner;
3. Subject the wire to a stress corrosion failure (materials under stress tend to corrode more rapidly than those not under stress).

Wire failure should not be a severe maintenance problem or operating limitation in a well-designed ESP. Excessive wire failures are usually a symptom of a more fundamental problem.

Plant personnel should maintain records of wire failure locations. Although ESP performance will generally not suffer with up to approximately 10% of the wires removed, these records should be maintained to help avoid a condition in which entire gas lanes may be de-energized. Improved sectionalization helps to minimize the effect of a broken wire on ESP performance, but performance usually begins to suffer when a large percentage of the ESP fields are de-energized.

D. Hopper Pluggage

Perhaps no other problem (except fire or explosion) has the potential for degrading ESP performance as much as **hopper pluggage**. Hopper pluggage can permanently damage an ESP and severely affect both short-term and long-term performance. Hopper pluggage is difficult to diagnose because its effect is not immediately apparent on the T-R set panel meters. Depending on its location, a hopper can usually be filled in 4 to 24 hours. In many cases, the effect of pluggage does not show up on the electrical readings until the hopper is nearly full.

The electrical reaction to most plugged hoppers is the same as that for internal misalignment, a loose wire in the ESP, or excessive dust buildup on the plates. Typical symptoms include heavy or "bursty" sparking in the field(s) over the plugged hopper and reduced voltage and current in response to the reduced clearance and higher spark rate. In weighted-wire designs, high dust levels in the hopper may raise the weight and cause slack wires and increased arcing within the ESP. In many cases, this will trip the T-R set off-line because of over current or under voltage protection circuits. In some situations, the sparking continues even as the dust level exceeds hopper capacity and builds up between the plate and the wire; whereas in others, the voltage continues to decrease as the current increases and little or no sparking occurs. This drain of power away from corona generation renders the field performance virtually useless. The flow of current also can cause the formation of a dust clinker (solidified dust) resulting from the heating of the dust between the wire and plate.

The buildup of dust under and into the collection area can cause the plate or discharge electrode guide frames to shift. The buildup can also place these frames under enough pressure to distort them or to cause permanent warping of the collection plate(s). If this happens, performance of the affected field remains diminished by misalignment, even after the hopper is cleared.

Hopper pluggage can be caused by the following:

- Obstructions due to fallen wires and/or bottle weights
- Inadequately sized solids-removal equipment
- Use of hoppers for dust storage
- Inadequate insulation and hopper heating
- Air inleakage through access doors

Most dusts flow best when they are hot, therefore, cooling the dusts can promote a hopper pluggage problem. Hopper pluggage can begin and perpetuate a cycle of failure in the ESP. For example, there was a case where a severely plugged hopper misaligned both the plates and the wire guide grid in one of the ESP fields. Because the performance of this field had decreased, the ESP was taken off-line and the hopper was cleared. But no one noticed the deteriorated condition of the wire-guide grid. The misalignment had caused the wires and weight hooks to rub the lower guide and erode the metal. When the ESP was brought back online, the guide-grid metal eventually wore through. Hopper pluggage increased as weights (and sometimes wires) fell into the hopper, plugging the discharge opening and causing the hopper to fill again and cause more misalignment. The rate of failure continued to increase until it was almost an everyday occurrence. This problem, which has occurred more than once in different applications, demonstrates how one relatively simple problem can lead to more complicated and costly ones.

In most pyramid-shaped hoppers, the rate of buildup lessens as the hopper is filled due to the geometry of the inverted pyramid. Hopper level indicators or alarms should provide some margin of safety so that plant personnel can respond before the hopper is filled.

When the dust layer rises to a level where it interferes with the electrical characteristics of the field, less dust is collected and the collection efficiency is reduced. Also, re-entrainment of the dust from the hopper can limit how high into the field the dust can go. Although buildups as deep as 4 feet have been observed, they usually are limited to 12 - 18 inches above the bottom of the plates.

E. <u>Misalignment</u>

The electrode misalignment is both a contributor to and a result of component failures. In general, most ESPs are not affected by a misalignment of less than about 3/16 inches. Indeed, some tolerance must be provided for expansion and contraction of the components. Beyond this limit, however, misalignment can become a limiting factor in ESP performance and is visually evident during an internal inspection of the ESP electrodes. Whether caused by warped plates, misaligned or skewed discharge electrode guide frames, insulator failure, or failure to maintain ESP "boxsquareness" misalignment reduces the operating voltage and current required for sparking.

The V-I curve would indicate a somewhat lower voltage to achieve a low current level with the sparking voltage and current greatly reduced. Since the maximum operating voltage/ current levels depend on the path of least resistance in a field, any point of close tolerance will control these operating levels.

F. Changes in Particle Size

Unusually fine particles present a problem under the following circumstances:

1. When the ESP is not designed to handle them
2. When a process change or modification shifts the particle size distribution into the range where ESP performance is poorest.

A shift in particle size distribution tends to alter electrical characteristics and increase the number of particles emitted in the light-scattering size ranges (opacity).

There are two principal charging mechanisms: field charging and diffusion charging. Although field charging tends to dominate in the ESP and acts on particles greater than 1 micrometer in diameter, it cannot charge and capture smaller particles.

Diffusion charging, on the other hand, works well for particles smaller than 0.1 micrometer in diameter. ESP performance diminishes for particulates in the range of 0.2 - 0.9 micrometer because neither charging mechanism is very effective for particles in this range. These particles are more difficult to charge and once charged, they are easily bumped around by the gas stream, making them difficult to collect. Depending upon the type of source being controlled, the collection efficiency of an ESP can drop from as high as

99.9% on particles sized above 1.0 micrometer or below 0.1 micrometer, to only 85 to 90% on particles in the 0.2 - 0.9 micrometer diameter range. If significant quantities of particles fall into this size range, the ESP design must be altered to accommodate the fine particles.

When heavy loadings of fine particles enter the ESP, two significant electrical effects can occur: space charge and corona quenching. At moderate resistivities, the space-charge effects normally occur in the inlet or perhaps the second field of ESPs. Because it takes a longer time to charge fine particles and to force them to migrate to the plate, a cloud of negatively charged particles forms in the gas stream. This cloud of charged particles is called a **space charge**. It interferes with the corona generation process and impedes the flow of negatively charged gas ions from the wire to the collection plate. The interference of the space charge with corona generation is called **corona quenching**. When this occurs, the T-R controller responds by increasing the operating voltage to maintain current flow and corona generation. The increase in voltage usually causes increased spark rates, which may in turn signal the controller to reduce the voltage and current in an attempt to maintain a reasonable spark rate. Under moderate resistivity conditions, the fine dust particles are usually collected by the time they reach the third field of the ESP which explains the disappearance of the space charge in these later fields. The T-R controller responds to the cleaner gas in these later fields by decreasing the voltage level, but the current levels will increase markedly. When quantities of fine particles are being processed by the ESP increasing, the space charging effect may progress further into the ESP.

G. Air In-leakage

In-leakage is often overlooked as an operating problem. In some instances, it can be beneficial to ESP performance, but in most cases its effect is detrimental. In-leakage may occur within the process itself or in the ESP and is caused by leaking access doors, leaking ductwork, and even open sample ports.

In-leakage usually cools the gas stream, and can also introduce additional moisture. Air in-leakage often causes localized corrosion of the ESP shell, plates, and wires. The temperature differential also can cause electrical disturbances (sparking) in the field. Finally, the introduction of ambient air can affect the gas distribution near the point of entry. The primary entrance paths are through the ESP access and hopper doors. In-leakage through hopper doors may re-entrain and excessively cool the dust in the hopper, which can cause both re-entrainment in the gas stream and hopper pluggage. In-leakage through the access doors is normally accompanied by an audible in-rush of air.

In-leakage is also accompanied by an increase in gas volume. In some processes, a certain amount of in-leakage is expected. For example, application of Lungstrom regenerative air heaters on power boilers or recovery boilers is normally accompanied by an increase in flue gas oxygen. For utility boilers the increase may be from 4.5% oxygen at the inlet to 6.5% at the boiler outlet. For other boilers the percentage increase may be smaller when measured by the O2 content, but 20 to 40% increases in gas volumes are typical and the ESP must be sized accordingly. Excessive gas volume due to air in-leakage, however, can cause an increase in emissions due to higher velocities through the ESP and greater re-entrainment of particulate matter. For example, at a kraft recovery boiler, an ESP that was designed for a superficial velocity of just under 6 ft/s was operating at over 12 ft/s to handle an increased firing rate, increased excess air, and in-leakage downstream of the boiler.

Because the velocities were so high through the ESP, the captured material was blown off the plate and the source was unable to meet emission standards.

6.3. Safety Measures

Persons who will be operating and maintaining an ESP must be well trained on all safety aspects to avoid injury. One person at the plant should be assigned the responsibility of constantly checking safety standards and equipment and to train or procure safety training for all those who will work with the ESP. A suggested list of important safety measures and precautions is classified as below:

A. Wiring and controls

1. Prior to startup, double-check that field wiring between controls and devices (T-R sets, rapper prime motors, etc.) is correct, complete, and properly labeled.
2. Never touch exposed internal parts of control system. Operation of the transformer-rectifier controls involves the use of dangerous high voltage.
   Although all practical safety control measures have been incorporated into this equipment, always take responsible precautions when operating it.

- Never use fingers or metal screwdrivers to adjust uninsulated control devices.

B. Access

1. Use a positive method to ensure that personnel are out of the precipitator, flues, or controls prior to energization. Never violate established plant clearance practices.
2. Never bypass the safety key interlock system. Destroy any extra keys. Always keep lock caps in place. Use powdered graphite only to lubricate lock system parts; never use oil or grease. Never tamper with a key interlock.
3. Use grounding chains whenever entering the precipitator, T-R switch enclosure, or bus ducts. The precipitator can hold a high static charge, up to 15 kV, after it is de-energized. The only safe ground is one that can be seen.
4. Never open a hopper door unless the dust level is positively below the door. Do not trust the level alarm. Check from the upper access in the

precipitator. Hot dust can flow like water and severely burn or kill a person standing below the door. Wear protective clothing.

5. Be on firm footing prior to entering the precipitator. Clear all trip hazards. Use the back of the hand to test for high metal temperatures.
6. Avoid ozone inhalation. Ozone is created any time the discharge electrodes are energized. Wear an air-line mask when entering the precipitator, flues, or stack when ozone may be present. Do not use filters, cartridge, or canister respirators.
7. Never poke hoppers with an uninsulatedmetal bar. Keep safety and danger signs in place. Clean, bright signs are obeyed more than deteriorated signs.

C. Fire/explosion

1. In case of boiler malfunction that could permit volatile gases and/or heavy carbon carryover to enter the precipitator, immediately shut down all transformer-rectifier sets. Volatile gases and carbon carryover could be ignited by sparks in the precipitator, causing fire or explosion, damaging precipitator internals.
2. If high levels of carbon are known to exist on the collecting surface or in the hoppers, do not open precipitator access doors until the precipitator has cooled below 52°C (125°F). Spontaneous combustion of the hot dust may be caused by the inrush of air.
3. If a fire is suspected in the hoppers, empty the affected hopper. If unable to empty the hopper immediately, shut down the transformer-rectifier sets above the hopper until it is empty. Use no other method to empty the hopper. Never use water or steam to control this type of fire. These agents can release hydrogen, increasing the possibility of explosion.

7

Examples of ESP Applications

Because ESPs can collect dry particles, sticky or tarry particles, and wet mists, they are used by many different industries to reduce air pollutant emissions, as diverse as chemical production and food processing.

Precipitators are used in many processes to either recover valuable granular solid or powder from process streams, or to remove granular solid pollutants from exhaust gases prior to venting to the atmosphere. Dust collection is an online process for collecting any process-generated dust from the source point on a continuous basis. Dust collectors may be of single unit construction, or a collection of devices used to separate particulate matter from the process air. They are often used as an air pollution control device to maintain or improve air quality.

Mist collectors remove particulate matter in the form of fine liquid droplets from the air. They are often used for the collection of metal working fluids, and coolant or oil mists. Mist collectors are often used to improve or maintain the quality of air in the workplace environment.

Fume and smoke collectors are used to remove sub micrometre size particulate from the air. They effectively reduce or eliminate particulate matter and gas streams from many industrial processes such as welding, rubber and plastic processing, high speed machining with coolants, tempering, and quenching.

Examples of using ESPs include the following:

1- Boilers

ESPs are most widely used for the control of fly ash from industrial and utility boilers and have been used on coal-fired boilers for over 50 years. Particulate matter is generated from boilers when fossil fuels (coal and oil) are burned to generate steam for industrial processes or to produce electric power. Both hot-side and cold-side precipitators are used to control particulate emissions. Other than some construction modifications to account for the temperature difference of the flue gas handled, hot-side and cold-side ESPs are essentially the same. Cold-side ESPs are used most often for collecting fly ash from coal-fired boilers. If the dust has high resistivity, cold-side units are used along with a conditioning agent such as sulfur trioxide.

One technology for reducing sulfur dioxide (SO2) emissions from boilers is dry Flue Gas Desulfurization (FGD). In dry FGD, the flue gas containing SO2 is contacted with an alkaline material to produce a dry waste product for disposal.

This technology consists of three different FGD methods:

- Injection of wet alkaline material (slurry) into a spray dryer with collection of dry particles in an electrostatic precipitator or baghouse,
- Injection of dry alkaline material into the flue gas stream with collection of dry particles in an ESP or baghouse, or
- Addition of alkaline material to the fuel prior to combustion

Spray dryers used in dry FGD are similar to those that have been used for over 40 years in the chemical, food-processing, and mineral preparation industries. Spray dryers are vessels where hot flue gas is contacted with a finely atomized, wet alkaline spray (see Figure 7.1).

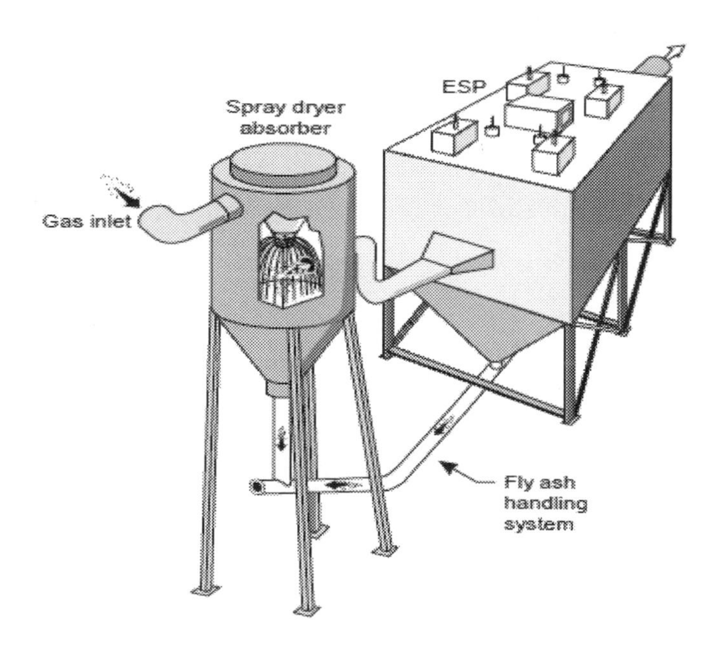

Figure 7.1. Spray dryer with ESP

Flue gas enters the top of the spray dryer and is swirled by a fixed vane ring to cause intimate contact with the slurry spray. Sodium carbonate solutions and lime slurries are the most common alkaline material used. The slurry is atomized into extremely fine droplets by rotary atomizers or two-fluid nozzles. In a rotary atomizer, slurry is broken into droplets by centrifugal force as the atomizer wheel spins at a very high speed. In two-fluid nozzles, slurry is mixed with compressed air, which forms the very small droplets. The high temperature of the flue gas, 120 to 204°C (250 to 400°F), evaporates the moisture from the wet alkaline sprays, leaving a dry, powdered product. The dry product is then collected in an ESP or baghouse.

A number of spray dryer FGD systems have been installed on industrial and utility boilers. They are particularly useful in meeting New Source Performance Standards (NSPS) that require only 70% SO2 removal efficiency for utility boilers burning low-sulfur coal and as retrofit applications for units having to meet higher requirements such as the standards of the 1990 Clean Air Act Amendments (see Table 7.1).

Table 7.1. Commercial spray dryer FGD systems using an ESP or a baghouse

Station or plant	Size (MW)	Installation date	System description	Sorbent	Coal sulfur content (%)	SO_2 emission removal efficiency (%)
Otter Tail Power Company: Coyote Station No. 1, Beulah, ND	410	6/81	Rockwell/Wheelabrator-Frye system: four spray towers in parallel with 3 atomizers in each: reverse air-shaker baghouse with Dacron bags	Soda ash (sodium carbonate)	0.78	70
Basin Electric: Laramie River Station No. 3, Wheatland, WY	500	Spring 1982	Babcock and Wilcox: four spray reactors with 12 "Y-jet" nozzles in each: electrostatic precipitator	Lime	0.54-0.81	85-90
Strathmore Paper Co.: Woronco, MA	14	12/79	Mikropul: spray dryer and pulse jet baghouse	Lime	2-2.5	75
Celanese Corp.: Cumberland, MD	31	2/80	Rockwell/Wheelabrator-Frye system: one spray tower followed by a baghouse	Lime	1-2	85

2- Cement Plants

ESPs are used in cement plants to control particulate emissions from cement kilns and clinker coolers. In a cement plant, raw materials are crushed, ground, blended, and fed into a kiln, where they are heated. The kiln is fired with coal, oil, or gas. The material is heated to a temperature above 1595°C (2900°F), which causes it to fuse. The fused material is called cement clinker. The temperature of the hot, marble-sized, glass-hard clinker is cooled by the clinker cooler. The cooled clinker is then sent to the final grinding mills.

ESPs are frequently used to control kiln emissions because of their ability to handle high-temperature gases. These ESPs are usually hot-side ESPs with collection plates that are rapped or sprayed with water to remove collected dust. The dust generated in the cement kiln frequently has high resistivity. High resistivity can be reduced by conditioning the flue gas with moisture.

Many of the newer cement plants send the high temperature kiln flue gas that contains particulate matter through a cyclone and conditioning tower (uses water to cool the gas temperature) prior to ducting the flue gas to the ESP. The ESP is then operated at a temperature of approximately 150°C (302°F).

A special problem arises during kiln startup due to the fact that the temperature of the kiln must be raised slowly to prevent damage to the heat-resistant (refractory) lining in the kiln. While kilns (especially coal-fired ones) are warming up and temperatures are below those for steady-state operating conditions, complete combustion of the fuels cannot occur, giving rise to combustible gases in the exhaust stream leading into the ESP. Electrostatic precipitators cannot be activated in the presence of combustibles, because the internal arcing of the precipitator could cause a fire or explosion. Use of a cyclone preceding the precipitator helps to minimize the excessive emissions during startup. Periods of excessive emissions during startup, malfunction, or shutdown are specifically exempted from the federal New Source Performance Standards for cement kilns.

ESPs can also be used on clinker coolers. However, the ESP must be carefully designed to prevent moisture in the flue gas from condensing.

Condensed moisture can combine with clinker dust to coat the ESP internals with cement.

3- Steel Mills

ESPs are used in steel mills for reducing particulate emissions from blast furnaces, basic oxygen furnaces, and sinter plants.

In a blast furnace, iron ore is reduced to molten iron, commonly called pig iron. Blast furnaces are large, refractory-lined steel shells. Limestone, iron ore, and coke are charged into the top of the furnace. The gases produced during the melting process contain carbon monoxide and particulate matter. Particulate matter is removed from the blast furnace gas by wet ESPs or scrubbers, so that the gas (CO) can be burned "cleanly" in blast furnace stoves or other processes. Both plate and tube-type ESPs having water sprays to remove dust from collection electrodes are commonly used for cleaning blast furnace gas.

Basic oxygen furnaces (BOFs) refine iron from the blast furnace into steel. A BOF is a pear-shaped steel vessel that is lined with refractory brick. The vessel is charged with molten iron and steel scrap. A water-cooled oxygen lance is lowered into the vessel, where oxygen is blown to agitate the liquid, add intense heat to the process, and oxidize any impurities still contained in the liquid metal. The hot gases generated during the oxygen blow are approximately 1090 to 1650°C (2000 to 3000°F). These are usually cooled by water sprays located in the hood and ducting above the BOF. The cooled gases are then sent to an ESP or scrubber to remove the particulate matter (iron oxide dust). The iron oxide dust can have high resistivity, making the dust difficult to collect in an ESP. This problem can usually be reduced by conditioning the flue gas with additional moisture. Plate ESPs that are rapped or sprayed with water to remove dust from collection plates are commonly installed on BOFs.

In a sinter plant, materials such as flue dusts, iron ore fines (small particles), coke fines, mill scale (waste that occurs from various processing steps), and small scrap are converted into a high-quality blast furnace feed. These materials are first fed onto a traveling grate. The bed of materials is ignited by burning gas in burners located at the inlet of the traveling grate. As the bed moves along the traveling grate, air is pulled down through the bed to burn it, forming a fused, porous, red-hot sinter. The resulting gases are usually sent to an electrostatic precipitator to remove any particulate matter. If oily scrap is used as a feed material, care must be taken to prevent ESP collection plates from being coated with tarry particulate matter. Controlling the amount of oily mill scale and small scrap processed in the sinter plant can help alleviate this problem. Plate ESPs are commonly used in sinter plants.

4- <u>Petroleum Refineries</u>

ESPs are used in petroleum refineries to control particulate emissions from fluid-catalytic cracking units and boilers. In a refinery, heavy crude is broken down into lighter components by various distilling, cracking, and reforming processes. One common process is to "crack" the high-molecular-weight, high-boiling-point compounds (heavy fuel oils) into smaller, lowmolecular- weight, low-boiling-point compounds (gasoline). This is usually done in a fluidcatalytic cracking (FCC) unit.

In an FCC unit, the feed stream (heavy gas oils) is heated and then mixed with a hot catalyst that causes the gas oils to vaporize and crack into smaller hydrocarbon-chain compounds. During this process, the catalyst becomes coated with coke. The coke deposits are eventually removed from the catalyst by a catalyst-regeneration step.

In the regenerator, a controlled amount of air is added to burn the coke deposits off the catalyst without destroying it. The gases in the regenerator pass through cyclones to separate large catalyst particles. The gases can sometimes go to a waste heat boiler to burn any carbon monoxide and organic emissions present in the gas stream. The boiler's exhaust gas still has a high concentration of fine catalyst particles. This flue gas is usually sent to an electrostatic precipitator to remove the very fine catalyst particles.

ESPs can also reduce particulate emissions from boiler exhausts. Oil-fired and, occasionally, coal-fired boilers generate steam that is used in many processes in the refinery. The flue gas from boilers is frequently sent to ESPs to remove particulate matter before the gas is exhausted into the atmosphere. ESPs designed similarly to those used on industrial and utility boilers are used on FCC units and petroleum refinery boilers.

5- Municipal Waste Incinerators

Electrostatic precipitators have been successfully used for many years to reduce particulate emissions from municipal waste incinerators. Municipal incinerators, also commonly called municipal waste combustors (MWCs) are used to reduce the volume of many different solid and liquid wastes. Generally, municipal wastes are composed of combustible materials (e.g. paper, wood, rags, food, yard clippings, and plastic and rubber materials) and noncombustible materials (e.g. rocks, metal, and glass). MWCs burn waste and produce ash residue that is disposed of in landfills.

Both dry and wet plate ESPs are commonly used on municipal incinerators. Collected dust can be removed from collection plates by rapping or by using water sprays. Plate ESPs having rigid frame discharge electrodes are currently being used on MWCs (installed after 1982). The designed collection efficiency is usually in the range of 96 to 99.6%. Dust resistivity can be a problem, particularly if the refuse contains a large quantity of paper products. The dust in the flue gas in this case usually has low resistivity. Resistivity can be adjusted by carefully controlling the temperature and the amount of moisture in the flue gas.

Since the mid-1980s a number of large MWCs (plants having a capacity of 250 to 3000 tons per day) with heat recovery devices have been built. More recent installations have been built with acid gas control systems along with an ESP or baghouse. The ESP (or baghouse) collects acid gas reaction products (mainly calcium chloride and calcium sulfate), unused sorbent material, and fly ash. ESPs are typically designed with 3 to 5 fields and are capable of meeting particulate emission limits of 0.015 gr/dscf and occasionally can achieve limits as low as 0.01 gr/dscf. These units have successfully reduced SO2 by 80% (24 hr avg) and HCl by 90 to 95%.

The acid gas is removed by using dry sorbent injection or spray dryer absorbers. In dry injection systems sorbent is injected (usually hydrated lime) into the furnace or into the ducting prior to the flue gas entering the ESP. Acid gas removal efficiencies of 50% for SO2 and 75% for HCl are routinely achieved (Ref.: Beachler 1992).

A more commonly used acid gas control system is a spray dryer absorber placed ahead of the ESP. These systems have been able to achieve 80%

removal (24 hr avg) for SO2 and 90% removal for HCl. A wet calcium hydroxide slurry is injected into a spray dryer by a rotary atomizer or two-fluid nozzle. The slurry is made by slaking pebble lime (CaO) with water in a paste or detention slaker. The heat of the flue gas evaporates the liquid slurry in the spray dryer and the dry acid gas reaction products along with the particulate matter are collected in the ESP. Background information and data prepared as part of the promulgated NSPS and Emission Guidelines (U.S. EPA 1991) shows very good acid gas removal and particulate emission control for these systems.

6- <u>Hazardous Waste Incinerators</u>

ESPs are used in combination with a number of other Air Pollution Control (APC) devices including wet scrubbers and dry scrubbers (also called spray dryer absorbers) to clean the flue gas generated by burning hazardous wastes.

Some facilities have been designed to use spray dryers to remove the acid gases including HCl, HF, and SO2 followed by the ESP to remove the acid gas reaction salts, any unused sorbent, and particulate matter.

Other facilities have been designed with an APC system consisting of a spray dryer, baghouse, wet scrubber, and a wet ESP (Figure 5.2). The spray dryer cools the flue gas and reduces some of the acid gas components. The baghouse collects the particulate matter (including metals) and the wet scrubber removes HCl (> 99%) and other acid gases.

The wet ESP collects any particulate matter not removed by the baghouse. The wet scrubbing system is a closed loop. The effluent produced in the scrubbers is ultimately sent to the spray dryer to evaporate the liquid, therefore eliminating the need for a waste water treatment system. A number of facilities using this APC system configuration are permitted to burn PCBs and other Toxic Substance Control Act (TSCA) and Resource Conservation and Recovery Act (RCRA) wastes.

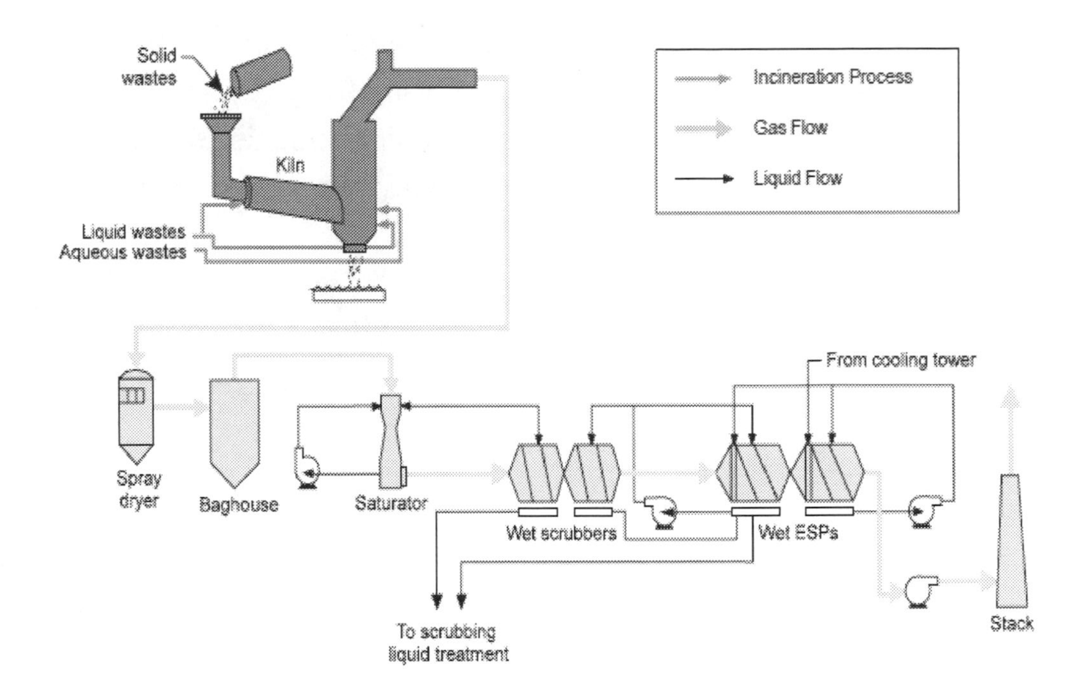

Figure 7.2. APC system for a hazardous waste incinerator consisting of a spray dryer, baghouse, wet scrubbers, and wet ESPs

7- Kraft Pulp and Paper Mills

Plate or tube-type ESPs are used in the kraft pulp and paper industry to reduce particulate emissions from the recovery furnace. In the kraft process of making pulp and paper, chemicals are recovered by using evaporators, recovery furnaces, and reaction tanks. As part of the pulping process, a waste product, black-liquor, is produced. After it is concentrated, the blackliquor concentrate is burned in the recovery furnace to provide heat and steam to various processes in the plant. The recovery furnace is essentially a boiler designed to effectively burn the black-liquor concentrate. The resulting flue gas contains particulate matter that is usually removed by an ESP before it is exhausted into the atmosphere. Dust can be removed from collection electrodes by rapping or by using water sprays.

8- Lead, Zinc, and Copper Smelters

Plate ESPs are used to reduce particulate emissions from a number of processes in the smelting of lead, zinc, and copper metals. Since lead, zinc, and copper are found in sulfide ore deposits, the release of sulfur compounds is a problem during the smelting process. Before being smelted, ore concentrates are often treated, or prepared, by two processes called sintering and roasting. Sintering acts to change the physical form of a material; by taking an ore mixture of large and fine pieces and fusing them into strong and porous products which can be used in the smelting processes. ESPs are commonly used to reduce emissions from lead and zinc sinter plants. ESPs are also effective in reducing emissions from zinc and copper roasters.

Roasters prepare zinc and copper ores by removing unwanted materials such as sulfur. The roasted ore is then sent to other refining processes to produce zinc and copper metals.

9- Other Industries

ESPs are used in many other large and small industrial processes including glass melting, sulfuric acid production, food processing, and chemical manufacturing. Glass melting furnaces usually use hot-side ESPs because the flue gas temperature in this process is approximately 230 to 260°C (450 to 500°F). Sulfuric acid plants usually use plate or tube-type ESPs to collect sulfuric acid mists. Collected mists are removed from collection electrodes by water sprays. Some smaller industries that produce coatings, resins, asphalt, rubber, textiles, plastics, vinyl, and carpet frequently use a small two-stage precipitator to control particles and smoke.

The two-stage ESPs use liquid sprays to remove collected particles, smokes, and oils from the collection plates.

Annex 2/Attachment summarizes the information presented in this chapter for various industries that use ESPs to reduce emissions.

8

Review Exercise and Cases of studies

8.1. Electrostatic Precipitator Operation

*A- **Review Exercise:***

1. In an electrostatic precipitator, the _____________ electrode is normally a small-diameter metal wire or a rigid frame containing wires.
2. The charged particles migrate to the _____________.
3. In a single-stage, high-voltage ESP, the applied voltage is increased until it produces a (an)
   a. Extremely high alternating current for particle charging
   b. Corona discharge, which can be seen as a blue glow around the discharge electrode
   c. Corona spark that occurs at the collection electrode
4. True or False? Particles are usually charged by negative gas ions that are migrating toward the collection electrode.
5. True or False? Large particles move more slowly towards the collection plate than small particles.
6. The average sparkover rate (in sparks per minute) for optimum precipitator operation is between:
   a. 1 - 25
   b. 50 - 100
   c. 100 - 150
   d. 500 - 1,000
7. As dust particles reach the grounded collection electrode, their charge is:
   a. Immediately transferred to the collection plate
   b. Slowly leaked to the grounded collection electrode
   c. Cancelled out by the strong electric field
   8. Particles are held onto the collection plates by:
   a. A strong electric force field
   b. A high-voltage, pulsating, direct current
   c. Intermolecular cohesive and adhesive forces
   d. Electric sponsors
9. Dust that has accumulated on collection electrodes can be removed either by_____________ or a process called _____________.
10. True or False? During the rapping process, the voltage is turned down to about 50% of the normal operating voltage to allow the rapped particles to fall freely into the hopper.

11. _________ electrostatic precipitators are used for removing particulate matter from flue gas that usually has a temperature range of 320 to 420° C (608 to 790° F).
12. In a boiler, hot-side ESPs are located ___________ air preheaters, whereas cold-side ESPs are located ____________ air preheaters.
   a. In front of, behind
   b. Behind, in front of
13. True or False? Wet electrostatic precipitators are used when collecting dust that is sticky or has high resistivity.
14. _______________ESPs are units where particle charging occurs in the first stage, followed by collection in the second stage.

B- <u>Answers:</u>
1. **Discharge**
In an electrostatic precipitator, the discharge electrode is normally a small-diameter metal wire or a rigid frame containing wires.
2. **Collection electrode**
The charged particles migrate to the collection electrode.
3. **b. Corona discharge, which can be seen as a blue glow around the discharge electrode**
In a single-stage, high-voltage ESP, the applied voltage is increased until it produces a corona discharge, which can be seen as a blue glow around the discharge electrode.
4. **True**
Particles are usually charged by negative gas ions that are migrating toward the collection electrode.
5. **False**
Large particles move faster towards the collection plate than small particles. Large particles have a
higher saturation charge than small particles; consequently, large particles are pulled more strongly to the collection plate.
6. **b. 50 - 100**
The average sparkover rate for optimum precipitator operation is between 50 - 100 sparks per minute.
7. **b. Slowly leaked to the grounded collection electrode**
As dust particles reach the grounded collection electrode, their charge is slowly leaked to the grounded collection electrode.
8. **c. Intermolecular cohesive and adhesive forces**
Particles are held onto the collection plates by intermolecular cohesive and adhesive forces.
9. **Water sprays**
Rapping

Dust that has accumulated on collection electrodes can be removed either by water sprays or a process called rapping.

10. **False**

During the rapping process, the voltage is NOT turned down. Rapping occurs while the ESP remains on-line.

11. **Hot-side**

Hot-side electrostatic precipitators are used for removing particulate matter from flue gas that usually has a temperature range of 320 to 420°C (608 to 790°F).

12. **a. In front of, behind**

In a boiler, hot-side ESPs are located in front of air preheaters, whereas cold-side ESPs are located behind air preheaters. Recall that flue gas is cooled as it passes through the combustion air preheater.

13. **True**

Wet electrostatic precipitators are used when collecting dust that is sticky or has high resistivity.

14. **Two-stage**

Two-stage ESPs are units where particle charging occurs in the first stage, followed by collection in the second stage.

8.2. Electrostatic Precipitator Components

*A- **Review Exercise:***

1. List the six major components of an ESP.
2. In many U.S. precipitators, the discharge electrodes are thin wires that are approximately___________ in diameter.
   a. 2.0 in.
   b. 0.1 in.
   c. 0.01 in.
   d. 15.0 in.
3. The discharge wires are hung vertically in the ESP and are held taut and plumb at the bottom by:
   a. A 25-lb weight
   b. Two 25-lb weights
   c. A 50-lb weight
   d. A 5-lb weight
4. True or False? Accumulated dust can be removed from discharge and collection electrodes by rapping.
5. European precipitators and most new U.S.-designed ESPs use ______________ for discharge electrodes.
   a. Wires
   b. Rigid frames
   c. Plates with stiffeners
6. Normal spacing for plates used on high-efficiency wire/plate ESPs is generally:
   a. 0.2 to 0.8 in.
   b. 2 to 4 in.
   c. 8 to 9 in.
   d. 24 to 36 in.
7. Normal spacing for plates used on high-efficiency rigid-frame ESPs is generally:
   a. 2-4 in.
   b. 5-7 in.
   c. 8-9 in.
   d. 12-15 in.
8. In ESPs, plates are usually between _____________ high.
   a. 4 to 12 in.
   b. 20 to 40 ft
   c. 40 to 60 ft
9. Collection electrodes can be constructed in two general shapes: ______________ and_____________.

10. Collected dust is removed from tubular ESPs using:

a. Magnetic impulse rappers
b. Water sprays
c. Hammer and anvil rappers
d. Electric vibrator rappers

11. ESPs control the strength of the electric field generated between the discharge and collection electrodes by using:

a. Transformer-rectifier sets
b. Meters
c. Capacitors
d. Insulators

12. In a T-R set, the transformer ____________ while the rectifier _______________.

a. Steps down the voltage, converts direct current into alternating current
b. Converts alternating current into direct current, steps up the voltage
c. Steps up the voltage, converts alternating current into direct current

13. In the control circuitry on an ESP, the primary voltmeter measures the:

a. Number of sparks
b. Input voltage (in a.c. volts) coming into the transformer
c. Output voltage from the rectifier
d. Operating d.c. voltage delivered to the discharge electrodes

14. The combination of the __________ voltage and current readings gives the power input to the discharge electrodes.

a. Primary
b. Sparking
c. Secondary
d. Tertiary

15. An electric cable that carries high voltage from the T-R set to the discharge electrode is called
a(an):

a. Bus line
b. Pipe
c. Duct
d. Electric vibrator

16. Most precipitators use ______or ___ to remove accumulated dust from collection plates.

a. Air-vibrator rappers (or) water sprays
b. Hammer and anvil (or) magnetic-impulse rappers
c. Electric-vibrator (or) magnetic-impulse rappers

17. Which rappers are commonly used for removing dust from discharge electrodes?

a. Hammer
b. Electric-vibrator and tumbling-hammer
c. Magnetic-impulse
d. Water-spray

18. The dust is temporarily stored in a _____________.
19. A ____________________ discharge device works similarly to a revolving door.
20. A ____________________ uses a screw feeder located at the bottom of the hopper to remove dust from the bin.
21. A ____________________ uses a blower or compressed air to remove dust from the hopper.
22. A ____________________ uses paddles or flaps connected to a drag chain to move dust from the ESP to its final destination.
23. In a precipitator, shells and hoppers should be covered with ____________________ to conserve heat and prevent corrosion.

B- <u>Answers:</u>

1. The six major components of an ESP are discharge electrodes, collection electrodes, high voltage electrical systems, rappers, hoppers, and the shell.

2. **b. 0.1 in.**
In many U.S. precipitators, the discharge electrodes are thin wires that are approximately 0.1 inch in diameter.

3. **a. A 25-lb weight**
The discharge wires are hung vertically in the ESP and are held taut and plumb at the bottom by a 25-lb weight.

4. **True**
Accumulated dust can be removed from discharge and collection electrodes by rapping.

5. **b. Rigid frames**
European precipitators and most new U.S.-designed ESPs use rigid frames for discharge electrodes.

6. **c. 8 to 9 in.**
Normal spacing for plates used on high-efficiency wire/plate ESPs is generally 8 to 9 inches.

7. **d. 12 to 15 in.**
Normal spacing for plates used on high-efficiency rigid-frame ESPs is generally 12 to 15 inches.

8. **b. 20 to 40 ft**
In ESPs, plates are usually between 20 to 40 ft high.

9. **Plates**
Tubes

Collection electrodes can be constructed in two general shapes: plates and tubes.
10. **b. Water sprays**
Collected dust is removed from tubular ESPs using water sprays.
11. **a. Transformer-rectifier sets**
ESPs control the strength of the electric field generated between the discharge and collection electrodes by using transformer-rectifier sets.
12. **c. Steps up the voltage, converts alternating current into direct current**
In a T-R set, the transformer steps up the voltage while the rectifier converts alternating current into direct current.
13. **b. Input voltage (in a.c. volts) coming into the transformer**
In the control circuitry on an ESP, the primary voltmeter measures the input voltage (in a.c. volts) coming into the transformer.
14. **c. Secondary**
The combination of the secondary voltage and current readings gives the power input to the discharge electrodes.
15. **a. Bus line**
An electric cable that carries high voltage from the T-R set to the discharge electrode is called a bus line.
16. **b. Hammer and anvil (or) magnetic-impulse rappers**
Most precipitators use hammer and anvil or magnetic-impulse rappers to remove accumulated dust from collection plates.
17. **b. Electric-vibrator and tumbling-hammer**
For removing dust from discharge electrodes, electric-vibrator rappers (for wires) and tumbling-hammer rappers (for rigid frames) are commonly used.
18. **Hopper**
The dust is temporarily stored in a hopper.
19. **Rotary airlock**
A rotary airlock discharge device works similarly to a revolving door.
20. **Screw conveyor**
A screw conveyor uses a screw feeder located at the bottom of the hopper to remove dust from the bin.
21. **Pneumatic conveyor**
A pneumatic conveyor uses a blower or compressed air to remove dust from the hopper.
22. **Drag conveyor**
A drag conveyor uses paddles or flaps connected to a drag chain to move dust from the ESP to its final destination.
23. **Insulation**
In a precipitator, shells and hoppers should be covered with insulation to conserve heat and prevent corrosion.

8.3. ESP Design Parameters and Their Effects on Collection Efficiency

*A- **Review Exercise:***

1. A charged particle will migrate toward an oppositely charged collection electrode. The velocity at which the charged particle moves toward the collection electrode is called the ____________________ and is denoted by the symbol w.

2. What is the name of the equation given below?

$$\eta = 1 - e^{-w\left(\frac{A}{Q}\right)}$$

a. Johnstone equation
b. Matts-Ohnfeldt equation
c. Deutsch-Anderson equation
d. Beachler-Joseph equation

3. The symbol η in the Deutsch-Anderson equation is the:

a. Collection area
b. Migration velocity
c. Gas flow rate
d. Collection efficiency

4. The Deutsch-Anderson equation does not account for:

a. Dust reentrainment that may occur as a result of rapping
b. Varying migration velocities due to various-sized particles in the flue gas
c. Uneven gas flow through the precipitator
d. All of the above

5. True or False? Using the Matts-Ohnfeldt equation to estimate the collection efficiency of an ESP will give less conservative results than using the Deutsch-Anderson equation.

6. Resistivity is a measure of a particle's resistance to ___________ and ____________________ charge.

7. Dust resistivity is a characteristic of the particle in the flue gas that can alter the____________________ of an ESP.

a. Gas flow rate
b. Collection efficiency
c. Gas velocity

8. Dust particles with ____________________ resistivity are difficult to remove from collection plates, causing rapping problems.

a. Low
b. Normal

c. High

9. High dust resistivity can be reduced by:

a. Adjusting the flue gas temperature

b. Increasing the moisture content of the flue gas

c. Injecting SO3 into the flue gas

d. All of the above

10. True or False? Fly ash that results from burning high-sulfur coal generally has high resistivity.

11. A precipitator is divided into a series of independently energized bus sections called:

a. Hoppers

b. Fields

c. Stages

d. b and c, above

12. A precipitator should be designed with at least ____________________ field(s) to attain a high collection efficiency.

a. One

b. Two

c. Three or four

d. Ten

13. Electrical sectionalization improves collection efficiency by:

a. Improving resistivity conditions

b. Allowing for independent voltage control of different fields

c. Allowing continued ESP operation in the event of electrical failure in one of the fields

d. b and c, above

14. If the design of the precipitator states that 500,000 ft^2 of plate area is used to remove particles from flue gas flowing at 750,000 ft^3/min, what is the SCA of the unit?

a. 0.667 ft2/1000 acfm

b. 667 ft2/1000 acfm

c. 667 acfm/1000 ft2

d. 1.5 acfm/ft2

15. To achieve a collection efficiency greater than 99.5%, most ESPs have a SCA:

a. Less than 250 ft^2/1000 acfm

b. Between 350 and 400 ft^2/1000 acfm

c. Always greater than 800 ft^2/1000 acfm

16. To improve the aspect ratio of an ESP design, the ____________________ of the collection surface should be increased relative to the ____________________ of the plate.

a. Height; length

b. Length; height

17. What should the aspect ratio be for high-efficiency ESPs?
    a. Less than 0.8
    b. Greater than 1.0
    c. Always greater than 1.5

18. In a properly designed ESP, the gas velocity through the ESP chamber will be:
    a. Between 2 and 8 ft/sec
    b. Greater than 20 ft/sec
    c. Approximately between 20 and 80 ft/sec
    d. At least 400 ft^2/1000 acfm

19. In an ESP, the collection efficiency is proportional to the amount of __________ __________ supplied to the unit.

B- Answers:

1. **Migration velocity (or drift velocity)**
The velocity at which the charged particle moves toward the collection electrode is called the migration velocity (or drift velocity) and is denoted by the symbol w.

2. **c. Deutsch-Anderson equation**
It is Deutsch-Anderson equation.

3. **d. Collection efficiency**
The symbol η in the Deutsch-Anderson equation is the collection efficiency.

4. **d. All of the above**
The Deutsch-Anderson equation does not account for the following:
• Dust reentrainment that may occur as a result of rapping
• Varying migration velocities due to various-sized particles in the flue gas
• Uneven gas flow through the precipitator

5. **False**
Using the Matts-Ohnfeldt equation to estimate the collection efficiency of an ESP will give more conservative results than using the Deutsch-Anderson equation because the Matts-Ohnfeldt equation accounts for non-ideal effects.

6. **Accepting**
Releasing
Resistivity is a measure of a particle's resistance to accepting and releasing charge.

7. **b. Collection efficiency**
Dust resistivity is a characteristic of the particle in the flue gas that can alter the collection efficiency of an ESP.

8. **c. High**
Dust particles with high resistivity are difficult to remove from collection plates, causing rapping problems.

9. **d. All of the above**
High dust resistivity can be reduced by the following:

- Adjusting the flue gas temperature
- Increasing the moisture content of the flue gas
- Injecting SO3 into the flue gas

10. **False**
Fly ash that results from burning high-sulfur coal generally has low resistivity. SO3, which lowers the resistivity of fly-ash, normally increases as the sulfur content of the coal increases.
11. **d. b and c, above**
A precipitator is divided into a series of independently energized bus sections called fields or stages.
12. **c. Three or four**
A precipitator should be designed with at least three or four fields to attain a high collection efficiency.
13. **d. b and c, above**
Electrical sectionalization improves collection efficiency by allowing the following:

- Independent voltage control of different fields
- Continued ESP operation in the event of electrical failure in one of the fields

14. **b. 667 ft2/1000 acfm**
If the design of the precipitator states that 500,000 ft^2 of plate area is used to remove particles from flue gas flowing at 750,000 ft^3/min, the SCA of the unit is as follows:
15. **b. Between 350 and 400 ft^2/1000 acfm**
To achieve a collection efficiency greater than 99.5%, most ESPs have a SCA between 350 and 400 ft^2/1000 acfm.
16. **b. Length; height**
To improve the aspect ratio of an ESP design, the length of the collection surface should be increased relative to the height of the plate.
17. **b. Greater than 1.0**
The aspect ratio for high-efficiency ESPs should be greater than 1.0.
18. a. Between 2 and 8 ft/sec
In a properly designed ESP, the gas velocity through the ESP chamber will be between 2 and 8 ft/sec, and most often between 4 and 6 ft/sec.
19. **Corona power**
In an ESP, the collection efficiency is proportional to the amount of corona power supplied to the unit.

8.4. ESP Design Review

*A- **Review Exercise:***

1. Two important process variables to consider when designing an ESP are the gas____________ and the dust ______.
2. In an ESP, the amount of dust coming into the ESP is important. If the dust loading is very high it will:
   a. Suppress the current in the inlet field and cause the controller to drive up the voltage
   b. Increase the current in the inlet field and cause the controller to decrease the voltage
   c. Cause an increase in the dust resistivity
   d. Have no effect on the ESP performance
3. If coal burned in a boiler has a low sulfur content, the resulting dust will usually have___________ resistivity.
   a. High
   b. Low
4. Which of the drawings below shows a good design of an inlet into the ESP?

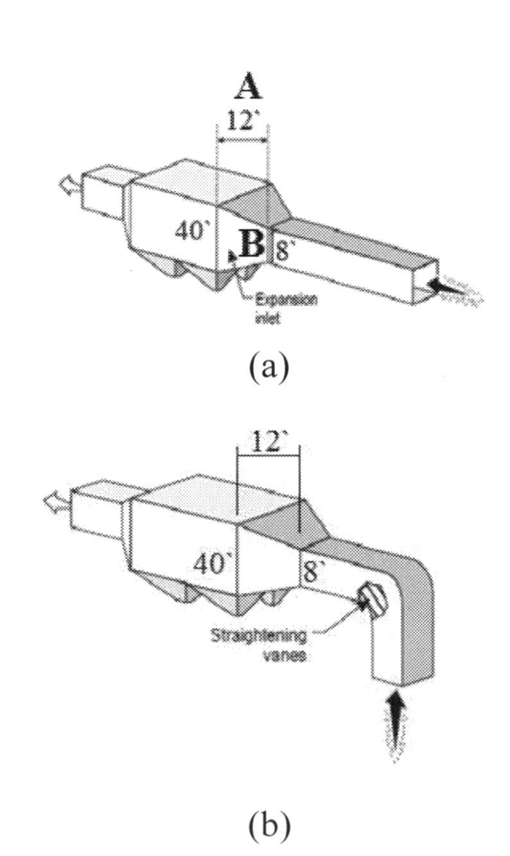

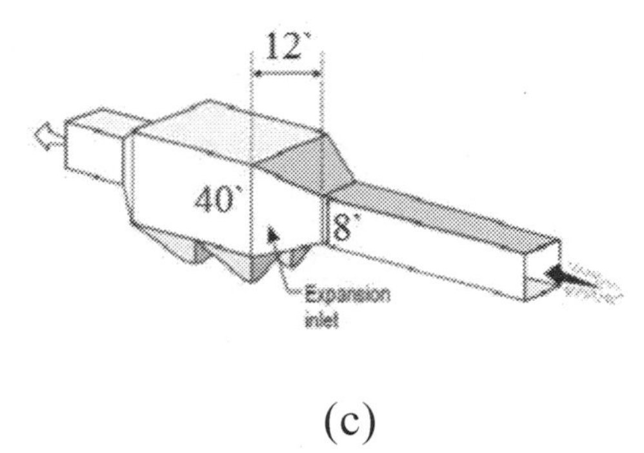

(c)

5. True or False? Dust can be stored in hoppers for any length of time without causing problems.
6. An ESP has a collection area of 750,000 ft2 and filters fly ash from flue gas flowing at 1,500,000 ft3/min. The migration velocity of the dust is 0.25 ft/sec. Estimate the collection efficiency of the ESP using the Deutsch-Anderson equation.

$$\eta = 1 - e^{-w\left(\frac{A}{Q}\right)}$$

7. The design plan states that an ESP will filter fly ash from flue gas that has a dust loading of 2 gr/ft^3 and a flow rate of 2,000,000 acfm (ft^3/min). The dust migration velocity is 0.3 ft/sec. If the regulations state that the emissions must be less than 0.02 gr/ft^3, what is the total collection area needed for the ESP design? Use the Deutsch-Anderson equation.

B- <u>Answers:</u>

1. Flow rate
Concentration
Two important process variables to consider when designing an ESP are gas flow rate and dust concentration.
2. a. Suppress the current in the field and cause the controller to drive up the voltage
In an ESP, the amount of dust coming into the ESP is important. If the dust loading is very high it
will suppress the current in the inlet field and cause the controller to drive up the voltage.
3. a. High
If coal burned in a boiler has a low sulfur content, the resulting dust will usually have high resistivity.
4. c.
The figure in option "c" shows the best inlet design because it has a straight-on inlet and an inlet plenum with a distance of A as long as (or longer than) B. Option "b" is fine if there are straightening vanes in the duct.
5. False

Dust can NOT be stored in hoppers for any length of time without causing problems. Dust should be stored temporarily in the hopper and removed periodically by the discharge device to prevent the dust from backing up into the ESP.

6. 99.94%

Solution:

Calculate the collection efficiency using the Deutsch-Anderson equation:

$$\eta = 1 - e^{-w\left(\frac{A}{Q}\right)}$$

Where: w = 0.25 ft/sec × 60 sec/min = 15 ft/min

A = 750,000 ft^2

Q = 1,500,000 ft^3/min

η = 0.9994 or 99.94%

7. 512,000 ft^2

Solution:

1. Using equation 4-1, calculate the collection efficiency required to meet emissions regulations.

$$\eta = \frac{Dust_{in} - Dust_{out}}{Dust_{in}}$$

η = 0.99 or 99%

2. Calculate the total collection area needed, using the following form of the Deutsch-Anderson equation:

$$A = \frac{-Q}{W}[\ln(1-\eta)]$$

Where: w = 0.3 ft/sec × 60 sec/min = 18 ft/min and Q = 2,000,000 ft^3/min

A = 512,000 ft^2

8.5. Industrial Applications of ESPs

*A- **Review Exercise:***

1. ESPs reduce particulate emissions from which of the following industries?
   a. Utility boilers
   b. Cement kilns
   c. Steel furnaces (basic oxygen furnace) and sinter plants
   d. Municipal waste incinerators
   e. All of the above
2. One technology for reducing both SO_2 gas and particulate emissions involves the injection of a(an)_________slurry in a spray __________ with dry particle collection in an electrostatic precipitator.
3. In a spray dryer, moisture is _______ from the wet alkaline sprays, leaving a______ powdered product.
4. Acid gas and particulate emissions can be controlled by using ______.
   a. Spray dryer absorber and ESP
   b. Dry injection and ESP
   c. a and b, above
5. ESPs should not be activated during the startup of a(an) _______ _____ because of the possibility of a fire or explosion.
6. In a steel mill, which of the following processes would not likely use an ESP to control particulate emissions?
   a. Blast furnace melting
   b. Sinter process
   c. Ingot pouring
   d. Basic oxygen furnace melting and tapping
7. In a municipal incinerator where the burned refuse contains a large quantity of paper products, the resulting dust usually has a ______ resistivity.
   a. High
   b. Low
8. True or False? ESPs are used in petroleum refineries to control particulate emissions from the fluid-catalytic-cracking unit and boiler exhausts.
9. When a spray dryer absorber is used with an ESP to control acid gas and particulate emissions from municipal waste combustors, which of the following is (are) true?
   a. The control system can reduce particulate emissions to a level of less than 0.015 gr/dscf.
   b. The control system can reduce SO_2 by 80%.

c. The dust collected in the ESP hoppers contains calcium chloride which is very hygroscopic
(sticky).
d. All of the above.

10. True or False? Both wet and dry ESPs are used in the pulp and paper industries to remove greater than 99% of the particulate matter from recovery furnace.

B- <u>Answers:</u>

1. **e. All of the above**

ESPs reduce particulate emissions from the following industries:
- Utility boilers
- Cement kilns
- Steel furnaces (basic oxygen furnace) and sinter plants
- Municipal waste incinerators

2. **Alkaline**
Dryer
One technology for reducing both SO_2 gas and particulate emissions involves the injection of an alkaline slurry in a spray dryer with dry particle collection in an electrostatic precipitator.

3. **Evaporated**
Dry
In a spray dryer, moisture is evaporated from the wet alkaline sprays, leaving a dry powdered product.

4. **c. a and b, above**
Acid gas and particulate emissions can be controlled by using either a spray dryer absorber and ESP or dry injection and ESP.

5. **Cement kiln**
ESPs should not be activated during the startup of a cement kiln because of the possibility of a fire or explosion.

6. **c. Ingot pouring**
In a steel mill, ingot pouring would not likely use an ESP to control particulate emissions.

7. **b. Low**
In a municipal incinerator where the burned refuse contains a large quantity of paper products, the resulting dust usually has a low resistivity.

8. **True**
ESPs are used in petroleum refineries to control particulate emissions from the fluid-catalyticcracking unit and boiler exhausts.

9. **d. All of the above**

When a spray dryer absorber is used with an ESP to control acid gas and particulate emissions from municipal waste combustors, the following are true:

- The control system can reduce particulate emissions to a level of less than 0.015 gr/dscf.
- The control system can reduce SO_2 by 80%.
- The dust collected in the ESP hoppers contains calcium chloride which is very hygroscopic (sticky).

10. **True**

Both wet and dry ESPs are used in the pulp and paper industries to remove greater than 99% of the particulate matter from recovery furnaces.

8.6. ESP Operation and Maintenance Goal

*A- **Review Exercise:***

1. Air inleakage at flanges or collector access points in high-temperature systems (hot-side ESPs) may:
   a. Allow dust to settle out quickly into hoppers
   b. Cause acids and moisture to condense on internal components of the ESP
   c. Increase the overall collection efficiency of the unit
2. Gas streams of high temperature should be maintained above the:
   a. Ignition temperature
   b. Gas dew point
   c. Concentration limit
3. Since most ESPs are installed in the field, it is important to check that all surfaces and areas of potential heat loss are adequately covered with:
   a. Paint
   b. Plastic coating
   c. Insulation
   d. Aluminum siding
4. Before the ESP is started, the installation crew should prepare and use a____________________.
5. Which of the following ESP components should be checked before starting the collector?
   a. Hoppers and discharge devices
   b. Rappers
   c. Discharge and collection electrodes
   d. All of the above
6. Two very important parameters monitored by meters on T-R sets and used to evaluate ESP performance are _____and____________________.
7. True or False? Individual T-R set values for voltage and current are important; however, the trends for voltage and current noted within an entire ESP are more valuable in assessing performance.
8. As particulate matter is removed from the gas stream, the _______ should increase from the inlet to the outlet fields.
   a. Opacity
   b. Current density
   c. Rapper intensity
   d. Amperage
9. An opacity monitor (transmissometer) measures:

a. Particle weight
b. Particle size
c. Light differential
d. Primary current

10. True or False? Opacity monitors are useful tools to aid in optimization of spark rate, power levels and rapping cycles in ESPs.
11. True or False? Changes in flue gas temperature generally have little or no effect on particle resistivity.
12. Operating parameters such as specific collection area, superficial velocity, and treatment time are dependent on the ______________.
13. True or False? Because of their open design, gas flow distribution through ESPs are generally very evenly distributed.
14. ____________________ tests are generally conducted on cool, inoperative ESPs through which no gas is flowing.

a. Air Load V-I Curve
b. Gas Load V-I Curve
c. Compliance
d. All of the above

15. True or False? When ESPs are equipped with identical fields, the air-load curves for each field should be very similar.
16. Air Load V-I curves for a given ESP field will generally shift to the _____ if plates are dirty compared to previous tests.

a. Left
b. Right
c. a and b, above

17. Gas-load curves are similar to air-load curves except the gas-load curves are shifted to the______ compared to the air-load curves.

a. Left
b. Right

18. True or False? Gas-load curves generally are identical for a given ESP field on a day-to-day basis.
19. True or False? High dust resistivity is characterized by the tendency toward high spark rates at low current levels.
20. Excessive dust buildup on the collecting plates or discharge wires can be caused by failure of the:

a. Primary and secondary voltage
b. Rapping system
c. Back corona
d. All the above

21. Wire failure can be caused by:

a. Electrical erosion
b. Mechanical erosion

c. Corrosion

d. All of the above

22. True or False? Unlike baghouses, ESPs are not affected by operating temperatures falling below the acid or moisture dew point.

23. True or False? In general, a well-designed ESP can operate effectively with a small percentage (less than 10) of its wires out-of-service.

24. True or False? Dust discharge hopper pluggage is not a major concern for ESPs.

B- <u>Answers:</u>

1. **b. Cause acids and moisture to condense on internal components of the ESP**

Air inleakage at flanges or collector access points in high-temperature systems (hot-side ESPs) may cause acids and moisture to condense on internal components of the ESP.

2. **b. Gas dew point**

Gas streams of high temperature should be maintained above the gas dew point. When the temperature falls below the gas dew point, moisture or acid can condense on ESP components and possibly cause corrosion.

3. **c. Insulation**

Since most ESPs are installed in the field, it is important to check that all surfaces and areas of potential heat loss are adequately covered with insulation.

4. **Checklist**

Before the ESP is started, the installation crew should prepare and use a checklist.

5. **d. All of the above**

The following are some ESP components that should be checked before starting the collector:

- Hoppers and discharge devices
- Rappers
- Discharge and collection electrodes

6. **Voltage**

Current

Two very important parameters monitored by meters on T-R sets and used to evaluate ESP performance are voltage and current.

7. **True**

Individual T-R set values for voltage and current are important; however, the trends for voltage and current noted within an entire ESP are more valuable in assessing performance. T-R set readings for current, voltage, and sparking should follow certain patterns from the inlet to the outlet fields.

8. **b. Current density**

As particulate matter is removed from the gas stream, the current density should increase from the inlet to the outlet fields. The dust concentration in the inlet sections will suppress the current. Increased current density is needed in the outlet sections where there is a greater percentage of very small particles.

9. **c. Light differential**

An opacity monitor (transmissometer) measures light differential. An opacity monitor compares the amount of light generated and transmitted by the instrument on one side of the gas stream with the quantity measured on the other side of the gas stream.

10. **True**

Opacity monitors are useful tools to aid in optimization of spark rate, power levels and rapping cycles in ESPs.

11. **False**

Changes in flue gas temperature have an important effect on particle resistivity. In fact, while gas temperature variations may have some effect on corona discharge characteristics and physical characteristics of the ESP (corrosion, expansion/contraction), their most important effect is on particle resistivity. See Figure 3-1.

12. **Gas flow rate**

Operating parameters such as specific collection area, superficial velocity, and treatment time are dependent on the gas flow rate.

13. **False**

Actually, gas flow through the ESP is not evenly distributed. ESP manufacturers settle for what they consider to be an acceptable variation.

14. **a. Air Load V-I Curve**

Air-Load V-I Curve tests are generally conducted on cool, inoperative ESPs through which no gas is flowing.

15. **True**

When ESPs are equipped with identical fields, the air-load curves for each field should be very similar.

16. **b. Right**

Air Load V-I curves for a given ESP field will generally shift to the right if plates are dirty compared to previous tests. Dirty plates suppress the current. It takes a higher voltage to generate the same amount of current as with a "clean plate" condition.

17. **a. Left**

Gas-load curves are similar to air-load curves except the gas-load curves are shifted to the left compared to the air-load curves. Gas-load curves are generated while the unit is on-line. The curves are generally shifted to the left because sparking occurs at lower voltage and current when particles are present.

18. **False**
Gas-load curves for a given ESP field generally vary on a day-to-day basis. Curve positions can change due to fluctuations in the amount of dust on the plates, gas flow, particulate loadings, temperature, and resistivity.
19. **True**
High dust resistivity is characterized by the tendency toward high spark rates at low current levels.
20. **b. Rapping system**
Excessive dust buildup on the collecting plates or discharge wires can be caused by failure of the rapping system.
21. **d. All of the above**
Wire failure can be caused by the following:

- Electrical erosion
- Mechanical erosion
- Corrosion.

22. **False**
Like baghouses, ESPs are affected by operating temperatures falling below the acid or moisture dew point. At temperatures below the acid or moisture dew point, acid or moisture can condense on ESP components and cause corrosion.
23. **True**
In general, a well-designed ESP can operate effectively with a small percentage (less than 10) of its wires out-of-service.
24. **False**
Dust discharge hopper pluggage is a major concern for ESPs. Hopper pluggage can permanently damage an ESP.

9

Attachments

Attachment-1 summarizes the problems associated with electrostatic precipitators, along with corrective actions and preventive measures; while Attachment-2 summarizes the typical industrial applications of ESPs.

Attachment - 1

Summary of problems associated with electrostatic precipitators

Malfunction	Cause	❖ Effect on electrostatic precipitator efficiency	Corrective action	Preventive measures
1. Poor electrode Alignment	1. Poor design 2. Ash buildup on frame hoppers 3. Poor gas flow	Can drastically affect performance & lower efficiency	Realign electrodes Correct gas flow	Check hoppers frequently for proper operation
2. Broken electrodes	1. Wire not rapped clean, causes an arc which embroglios and burns through the wire 2. Clinkered wire. Causes: a. Poor flow area, distribution through unit is uneven b. Excess free carbon due to excess air above combustion requirements or fan capacity insufficient for demand required c. Wires not properly centered d. Ash buildup, resulting in bent frame, same as (c) e. Clinker bridges	Reduction in efficiency due to reduced power input, bus section	Replace electrode	-Boiler problems; check space between recording steam and air flow pens, pressure gauges, fouled screen tubes -Inspect hoppers; check electrodes frequently for wear; inspect rappers frequently

	the plates and wire shorts out f. Ash buildup, pushes bottle weight up causing sag in the wire g. "J" hooks have improper clearances to the hanging wire h. Bottle weight hangs up during cooling causing a buckled wire i. Ash buildup on bottle weight to the frame forms a clinker and burns off the wire			
3. Distorted or skewed electrode plates	1. Ash buildup in hoppers 2. Gas flow irregularities 3. High temperatures Reduced efficiency Repair or replace plates Correct gas flow Check hoppers frequently for proper operation; check electrode plates during outages	Reduced efficiency Repair or replace plates	Correct gas flow	Check hoppers frequently for proper operation; check electrode plates during outages
4. Vibrating or swinging electrodes	1. Uneven gas flow 2. Broken electrodes	Decrease in efficiency due to reduced power input	Repair electrode	Check electrodes frequently for wear
5. Inadequate level of power input (voltage too low)	1. High dust resistivity 2. Excessive ash on electrodes 3. Unusually fine particle size 4. Inadequate power supply 5. Inadequate sectionalization 6. Improper rectifier and	Reduction in efficiency	Clean electrodes; gas conditioning or alterations in temperature to reduce resistivity; increase sectionalizati	Check range of voltages frequently to make sure they are correct; check insitu resistivity measurement s

	control operation 7. Misalignment of Electrodes		on	
6. Back corona	1. Ash accumulated on electrodes causes excessive sparking requiring reduction in voltage charge	Reduction in efficiency	Same as above	Same as above
7. Broken or cracked insulator or flower pot bushing leakage	1. Ash buildup during operation causes leakage to ground 2. Moisture gathered during shutdown or low-load operation	Reduction in efficiency	Clean or replace insulators and bushings	Check frequently; clean and dry as needed; check for adequate pressurization of top housing
8. Air inleakage through hoppers	1. From dust	conveyor Lower efficiency; dust re-entrained through electrostatic precipitator	Seal leaks	Identify early by increase in ash concentration at bottom of exit to electrostatic precipitator
9. Air inleakage through electrostatic precipitator shell	1. Flange expansion	Same as above; also causes intense sparking	Seal leaks	Check for large flue gas temperature drop across the ESP
10.Gas bypass around electrostatic precipitator • dead passage above plates • around high tension frame	1. Poor design; improper isolation of active portion of electrostatic precipitator	Only few percent drop in efficiency unless severe	Baffling to direct gas into active electrostatic precipitator section	Identify early by measurement of gas flow in suspected areas
11.Corrosion	1. Temperature goes	Negligible until precipitation	Maintain flue gas	Energize precipitator

	below dew point	interior plugs or plates are eaten away; air leaks may develop causing significant drops in performance	temperature above dew point	after boiler system has been on line for ample period to raise flue gas temperature above acid dew point
12.Hopper pluggage	1. Wires, plates, insulators fouled because of low temperature 2. Inadequate hopper insulation 3. Improper maintenance 4. Boiler leaks causing excess moisture 5. Ash conveying system malfunction (gasket leakage, blower malfunction, solenoid valves) 6. Misjudgments of hopper vibrators 7. Material dropped into hopper from bottle weights 8. Solenoid, timer malfunction 9. Suction blower filter not changed	Reduction in efficiency	Provide proper flow of ash	Frequent checks for adequate operation of hoppers. Provide heater thermal insulation to avoid moisture condensation
13. Inadequate rapping, Vibrators fail	1. Ash buildup 2. Poor design 3. Rappers misadjusted	Resulting buildup on electrodes may reduce efficiency	Adjust rappers with optical dust measuring instrument in electrostatic precipitator exit	Frequent checks for adequate operation of rappers

			stream	
14.Too intense rapping	1. Poor design 2. Rappers misadjusted 3. Improper rapping force	Re-entrains ash, reduces efficiency	Same as above	Same as above; reduce vibrating or impact force
15.Control failures	1. Power failure in primary system a. Insulation breakdown in transformer b. Arcing in transformer between high voltage switch contacts c. Leaks or shorts in high-voltage structure d. Insulating field contamination	Reduced efficiency	Find source of failure and repair or replace	Pay close attention to daily readings of control room instrumentati on to spot deviations from normal readings
16.Sparking	1. Inspection door ajar 2. Boiler leaks 3. Plugging of hoppers 4. Dirty insulators	Reduced efficiency	Close inspection doors; repair leaks in boiler; unplug hoppers; clean insulators	Regular preventive maintenance will alleviate these problems

❖ The effects of precipitation problems can be discussed only on a qualitative basis. There are no known emission tests of precipitators to determine performance degradation as a function of operational problems.

Sources: Szabo and Gerstle 1977, and Englebrecht 1980

Attachment-2

Summary of typical ESP applications (by industry)

Industry	Process	Material Collected by ESP	ESP Collection Efficiency	ESP Features	Potential Problems
1. Industrial & utility boilers	Burning fossil Fuels	Fly ash	> 99%	Hot-side and cold-side ESPs	Fly ash from low sulfur coals has high resistivity
	Dry SO2 control systems	Dry, alkaline Product	> 99% (particles); 70-80%(SO2)	Cold-side ESP (usually rigid electrode or rigid frame)	
2. Cement plants	Cement kilns	Particulate emissions	> 99%	Usually hot-side ESPs With collection plates. Rapped or sprayed with water.	Dust often has high resistivity. Combustible gases are present when kiln is warming up.
	Clinker coolers	Particulate emissions	> 99%	Hot-side or cold-side depending on gas temps.	Must prevent moisture in flue gas from condensing
3. Steel mills	Blast furnaces	Carbon monoxide and Particulate matter	> 99% Particulate matter	Wet ESPs Both plate and tube with water sprays.	
	Basic oxygen furnaces	Iron oxide dust	> 99%	Wet or dry plate ESPs	Iron oxide dust can have high resistivity
	Center Plant	Particulate matter	> 99%	Wet or dry plate ESPs	Oily scrap used as feed material can coat plates with tarry substance
4. Petroleum refineries	Fluid-catalytic cracking	Catalyst particles	> 99%	Usually dry ESPs	
	Boiler	Particulate	> 99%	Usually dry	

	Operations	Matte		ESPs	
5. Municipal waste incinerators	Incineration and heat recovery Acid control systems (spray dryer along with ESP)	Particulate matter Acid gas reaction products, unused sorbents	96-99.6% > 99.5 (0.015 gr/dscf) for particulate matter. SO and HCl reduced 80 and 90% respectively	Wet and dry plate ESPs Usually rigid electrode systems (newer facilities)	Low resistivity of dust from paper products
6. Hazardous waste incinerators	Acid control systems (1) Spray dryer and baghouse followed by wet ESPs (2) Spray dryer followed by an ESP	Acid gas reaction products, unused sorbents	> 99% (0.015 gr/dscf) HCl removal efficiency > 95%	Wet ESPs or dry ESPs when used with spray dryer	
7. Kraft pulp and paper mills	Recovery furnace boilers	Particulate matter	> 99%	Wet or dry ESPs	
8. Lead, zinc, copper smelters	Sinter plants Roasting	Particulate matter Particulate Matter	> 99% > 99%	Usually plate ESPs Usually plate ESPs	

10

References

A- Principles, Components, Operation and Applications

1. Faraday, Michael (1839). Experimental Researches in Electricity. London: Royal Inst. e-book at Project Gutenberg
2. Halliday, David; Robert Resnick; Kenneth S. Krane (1992). Physics. New York: John Wiley & Sons. ISBN 0-471-80457-6.
3. Griffiths, David J. (1999). Introduction to Electrodynamics. Upper Saddle River, NJ: Prentice Hall. ISBN 0-13-805326-X.
4. Hermann A. Haus and James R. Melcher (1989). Electromagnetic Fields and Energy. Englewood Cliffs, NJ: Prentice-Hall. ISBN 0-13-249020-X.
5. William J. Beaty, "Humans and sparks; The Cause, Stopping the Pain, and 'Electric People". 1997.
6. "electrostatic precipitator". Compendium of Chemical Terminology Internet edition.
7. Bibbo, P. P. 1982. Electrostatic precipitators. In L. Theodore and A. Buonicore (Eds.), Air Pollution Control Equipment-Selection, Design, Operation and Maintenance (pp.3-44). Englewood Cliffs, NJ: Prentice Hall.
8. Englebrecht, H. L. 1980. Mechanical and electrical aspects of electrostatic precipitator O&M. In R. A.
9. Young and F. L. Cross (Eds.), Operation and Maintenance for Air Particulate Control Equipment (pp. 283-354). Ann Arbor, MI: Ann Arbor Science.
10. Katz, J. 1979. The Art of Electrostatic Precipitators. Munhall, PA: Precipitator Technology.
11. Beachler, D. S., J. A. Jahnke, G. T. Joseph andM.M. Peterson. 1983. Air Pollution Control Systems forSelected Industries-Self-Instructional Guidebook. (APTI Course SI:431). EPA 450/2-82-006. U.S.Environmental Protection Agency.
12. Bethea, R. M. 1978. Air Pollution Control Technology-an Engineering Analysis Point of View. New York: Van Nostrand Reinhold.
13. Katz, J. 1979. The Art of Electrostatic Precipitators. Munhall, PA: Precipitator Technology.
14. Nichols, G. B. 1976, September. Electrostatic Precipitation. Seminar presented to the U.S. Environmental Protection Agency. Research

Triangle Park, NC.
15. Richards, J.R. 1995. Control of Particulate Emissions-Student Manual. (APTI Course 413). U.S. Environmental Protection Agency.
16. Turner, J. H., P. A. Lawless, T. Yamamoto, D. W. Coy, G. P. Greiner, J. D. McKenna, and W.M. Vatavuk. 1992. Electrostatic precipitators. In A. J. Buonicore and W. T. Davis (Eds.), Air Pollution Engineering Manual (pp. 89-113). Air and Waste Management Association. New York: Van Nostrand Reinhold.
17. U.S. Environmental Protection Agency. 1973. Air Pollution Engineering Manual. 2d ed. AP-40.
18. U.S. Environmental Protection Agency. 1985. Operation and Maintenance Manual for Electrostatic Precipitators. EPA 625/1-85/017.
19. White, H. J. 1977. Electrostatic precipitation of fly ash. APCA Reprint Series. Journal of Air Pollution Control Association. Pittsburgh, PA.
20. Beachler, D. S., J. A. Jahnke, G. T. Joseph and M. M. Peterson. 1983. Air Pollution Control Systems for Selected Industries, Self-Instructional Guidebook. (APTI Course SI:431). EPA 450/2-82-006. U.S. Environmental Protection Agency.
21. Bethea, R. M. 1978. Air Pollution Control Technology-an Engineering Analysis Point of View. New York: Van Nostrand Reinhold.
22. Cheremisinoff, P. N., and R. A. Young. (Eds.) 1977. Air Pollution Control and Design Handbook, Part 1. New York: Marcel Dekker.
23. Gallaer, C. A. 1983. Electrostatic Precipitator Reference Manual. Electric Power Research Institute. EPRI CS-2809, Project 1402-4.
24. Hall, H. J. 1975. Design and application of high voltage power supplies in electrostatic precipitation. Journal of Air Pollution Control Association. 25:132.
25. Hesketh, H. E. 1979. Air Pollution Control. Ann Arbor: Ann Arbor Science Publishers.
26. Katz, J. 1979. The Art of Electrostatic Precipitators. Munhall, PA: Precipitator Technology.
27. Richards, J.R. 1995. Control of Particulate Emissions, Student Manual. (APTI Course 413). U.S. Environmental Protection Agency.
28. Szabo, M. F., Y. M. Shah, and S. P. Schliesser. 1981. Inspection Manual for Evaluation of Electrostatic Precipitator Performances. EPA 340/1-79-007.
29. Turner, J. H., P. A. Lawless, T. Yamamoto, D. W. Coy, G. P. Greiner, J. D. McKenna, andW.M. Vatavuk. 1992. Electrostatic precipitators. In A. J. Buonicore and W. T. Davis (Eds.), Air

Pollution Engineering Manual (pp. 89-113).
30. Air and Waste Management Association. New York: Van Nostrand Reinhold.U.S. Environmental Protection Agency. 1985. Operation and Maintenance Manual for Electrostatic Precipitators. EPA 625/1-85/017.
31. White, H. J. 1963. Industrial Electrostatic Precipitation. Reading,MA: Addison-Wesley.White, H. J.
32. Electrostatic precipitation of fly ash. APCA Reprint Series. Journal of Air Pollution Control Association. Pittsburgh, PA. 1977.
33. Beachler, D. S. 1992. Coming clean on waste-to-energy emissions. Chemical Processing Technology International. London.
34. Beachler, D. S., G. T. Joseph, and M. Pompelia. 1995. Fabric Filter Operation Review. (APTI Course SI:412A). U.S. Environmental Protection Agency.
35. Beachler, D. S., J. A. Jahnke, G. T. Joseph andM.M. Peterson. 1983. Air Pollution Control Systems for Selected Industries, Self-Instructional Guidebook. (APTI Course SI:431). EPA 450/2-82-006. U.S. Environmental Protection Agency.
36. Kaplan, S. M., and K. Felsvang. 1979, April. Spray Dryer Absorption of SO2 from Industrial Boiler Flue Gas. Paper presented at 86th National AICHE Meeting. Houston, TX.
37. Pezze, J. 1983. Personal Communication. Pennsylvania Department of Environmental Resources, Pittsburgh, PA.
38. Richards, J. R. 1995. Control of Particulate Emissions, Student Manual. (APTI Course 413). U.S. Environmental Protection Agency.
39. Richards, J. R. 1995. Control of Gaseous Emissions, Student Manual. (APTI Course 415). U.S. Environmental Protection Agency.
40. Szabo,M. F., Y.M. Shah, and S. P. Schliesser. 1981. Inspection Manual for Evaluation of Electrostatic Precipitator Performances. EPA 340/1-79-007.
41. U.S. Environmental Protection Agency. 1976. Capital and Operating Costs of Selected Air Pollution Control Systems. EPA 450/3-76-014.
42. U.S. Environmental Protection Agency. 1980, February. Survey of Dry SO2 Control Systems. EPA 600/ 7-80-030.
43. U.S. Environmental Protection Agency. 1991. Requirements for preparation, adoption, and submittal of implementation plans. In Code of Federal Regulations—Protection of the Environment. 40 CFR 51. Washington, D.C.: U.S. Government Printing Office.
44. U.S. Environmental Protection Agency. 1991. Approval and promulgation of implementation plans. In Code of Federal Regulations—Protection of the Environment. 40 CFR 52.Washington, D.C.: U.S. Government Printing Office.

45.U.S. Environmental Protection Agency. 1991. Standards of performance for new stationary sources— general provisions. In Code of Federal Regulations—Protection of the Environment. 40 CFR 60. Washington, D.C.: U.S. Government Printing Office.
46.Bibbo, P. P. 1982. Electrostatic precipitators. In L. Theodore and A. Buonicore (Eds.), Air Pollution Control Equipment-Selection, Design, Operation and Maintenance (pp.3-44). Englewood Cliffs, NJ: Prentice Hall.
47. Cross, F. L., and H. E. Hesketh. (Eds.) 1975. Handbook for the Operation and Maintenance of Air Pollution Control Equipment. Westport, CT: Technomic Publishing.
48.Englebrecht, H. L. 1980. Mechanical and electrical aspects of electrostatic precipitator O&M. In R. A.
49.Young and F. L. Cross (Eds.), Operation and Maintenance for Air Particulate Control Equipment (pp. 283-354). Ann Arbor, MI: Ann Arbor Science.
50.Katz, J. 1979. The Art of Electrostatic Precipitators. Munhall, PA: Precipitator Technology.
51.Richards, J. R. 1995. Control of Particulate Emissions, Student Manual. (APTI Course 413). U.S. Environmental Protection Agency.
52.Szabo, M. F., and R. W. Gerstle. 1977. Electrostatic Precipitator Malfunctions in the Electric Utility Industry. EPA 600/2-77-006.
53.Szabo,M. F., Y.M. Shah, and S. P. Schliesser. 1981. Inspection Manual for Evaluation of Electrostatic Precipitator Performances. EPA 340/1-79-007.
54.U.S. Environmental Protection Agency. 1985. Operation and Maintenance Manual for Electrostatic Precipitators. EPA 625/1-85/017.
55.U.S. Environmental Protection Agency. 1987, August. Recommended Recordkeeping Systems for Air Pollution Control Equipment. Part I, Particulate Matter Controls. EPA 340/1-86-021.
56.U.S. Environmental Protection Agency. 1993. Monitoring, Recordkeeping, and Reporting Requirements for the Acid Rain Program. In Code of Federal Regulations - Protection of the Environment.

B. Design Review and Parameters

57.Anderson, E. 1924. Report, Western Precipitator Co., Los Angeles, CA. 1919. Transactions of the American Institute of Chemical Engineers. 16:69.
58.Beachler, D. S., J. A. Jahnke, G. T. Joseph and M. M. Peterson. 1983. Air Pollution Control Systems for Selected Industries, Self-

Instructional Guidebook. (APTI Course SI:431). EPA 450/2-82-006. U.S. Environmental Protection Agency.
59. Deutsch, W. 1922. Annals of Physics. (Leipzig) 68:335.
60. Gallaer, C. A. 1983. Electrostatic Precipitator Reference Manual. Electric Power Research Institute. EPRI CS-2809, Project 1402-4.
61. Hall, H. J. 1975. Design and application of high voltage power supplies in electrostatic precipitation. Journal of Air Pollution Control Association. 25:132.
62. Hesketh, H. E. 1979. Air Pollution Control. Ann Arbor: Ann Arbor Science Publishers.
63. Katz, J. 1979. The Art of Electrostatic Precipitators. Munhall, PA: Precipitator Technology.
64. Lawless, P. 1992. ESPVI 4.0, Electrostatic Precipitator V-I and Performance Model: Users' Manual. EPA 600/R-29-104a.
65. Matts, S., and P. O. Ohnfeldt. 1964. Efficient Gas Cleaning with SF Electrostatic Precipitators. Flakten.
66. Richards, J. R. 1995. Control of Particulate Emissions, Student Manual. (APTI Course 413). U.S. Environmental Protection Agency.
67. Rose, H. E., and A. J. Wood. An Introduction to Electrostatic Precipitation in Theory and Practice. London: Constable and Company.
68. Schmidt, W. A. 1949. Industrial and Engineering Chemistry. 41:2428.
69. Theodore, L., and A. J. Buonicore. 1976. Industrial Air Pollution Control Equipment for Particulates. Cleveland: CRC Press.
70. U.S. Environmental Protection Agency. 1978, June. A Mathematical Model of Electrostatic Precipitation (Revision 1). Vol. 1, Modeling and Programming. EPA 600/7-78-llla.
71. U.S. Environmental Protection Agency. 1978, June. A Mathematical Model of Electrostatic Precipitation (Revision 1). Vol. II, User Manual. EPA 600/7-78-lllb.
72. U.S. Environmental Protection Agency. 1979. Particulate Control by Fabric Filtration on Coal-Fired Industrial Boilers. EPA 625/2-79-021.
73. U.S. Environmental Protection Agency. 1980, May. TI-59 Programmable Calculator Programs for Instack Opacity, Venturi Scrubbers, and Electrostatic Precipitators. EPA 600/8-80-024.
74. U.S. Environmental Protection Agency. 1985. Operation and Maintenance Manual for Electrostatic Precipitators. EPA 625/1-85/017.
75. White, H. J. 1963. Industrial Electrostatic Precipitation. Reading,

MA: Addison-Wesley.

76. White, H. J. 1974. Resistivity problems in electrostatic precipitation. Journal of Air Pollution Control Association 24:315-338.
77. White, H. J. 1977. Electrostatic precipitation of fly ash. APCA Reprint Series. Journal of Air Pollution Control Association. Pittsburgh, PA.
78. White, H. J. 1982. Review of the state of the technology. Proceedings of the International Conference on Electrostatic Precipitation. Monterey, CA, October 1981. Air Pollution Control Corporation, Pittsburgh, PA.
79. Beachler, D. S., J. A. Jahnke, G. T. Joseph andM.M. Peterson. 1983. Air Pollution Control Systems for Selected Industries, Self-instructional Guidebook. (APTI Course SI:431). EPA 450/2-82-006. U.S. Environmental Protection Agency.
80. Gallaer, C. A. 1983. Electrostatic Precipitator Reference Manual. Electric Power Research Institute. EPRI CS-2809, Project 1402-4.
81. Katz, J. 1979. The Art of Electrostatic Precipitators. Munhall, PA: Precipitator Technology.
82. Neveril, R. B., J. U. Price, and K. L. Engdahl. 1978. Capital and operating costs of selected air pollution control systems - I. Journal of Air Pollution Control Association. 28:829-836.
83. Richards, J. R. 1995. Control of Particulate Emissions, Student Manual. (APTI Course 413). U.S. Environmental Protection Agency.
84. U.S. Environmental Protection Agency. 1990, January. OAQPS Cost Control Manual. 4th ed. EPA 450/3-90-006.
85. U.S. Environmental Protection Agency. 1991. Control Technology for Hazardous Air Pollutants Handbook. EPA 625/6-91/014.
86. White, H. J. 1977. Electrostatic precipitation of fly ash. APCA Reprint Series. Journal of Air Pollution Control Association. Pittsburgh, PA.

ELECTRICAL ENGINEERING

Principles and Special Purpose Applications of Electromagnetic Field and High Voltage

Package 3:Special Purpose Applications-Part Two

Book 9: Magnetic Bearing

Dr. Moayad Abdullah AlMayouf

July 2018

Magnetic Bearing

Contents

1

Introduction and Development of Bearing

1.1. Introduction

A **bearing** is a device to allow constrained relative motion between two or more parts, typically rotation or linear movement.

Bearings may be classified broadly according to the motions they allow and according to their principle of operation as well as by the directions of applied loads they can handle.

Generally, there are at least six common principles of operation:

- **Sliding bearings,** usually called "bushes", "bushings", "journal bearings", "sleeve bearings", "rifle bearings", or "plain bearings"
- Rolling-element bearings, such as ball bearings and roller bearings
- Jewel bearings, in which the load is carried by rolling the axle slightly off-center
- Fluid bearings, in which the load is carried by a gas or liquid
- Magnetic bearings, in which the load is carried by a magnetic field
- Flexure bearings, in which the motion is supported by a load element which bends.

Some years ago the answer to the question of high-speed machine, would definitely have been that there is a sharp line between motors running at network frequency and motors running faster. However, with the widely spread use of power converters the limit is now moving upwards. During the past ten years turbomolecular pumps operating at 90,000 rpm have been mass-produced showing that the technology is mature. These pumps are often referred to as high-speed turbo pumps.

The term very high speed machines was launched referring to the research developed at MIT on small 0.1 kW turbogenerators operating between 1 and 2 million rpm.

An engineer designing a high-speed machine will have to tackle the questions and problems that are likely to occur connected with high speed technologies such as how to reduce air drag losses and how to eliminate eddy current losses or how to treat problems concerning rotor dynamics and vibration control.

A special type of problem that many engineers regard to be the most difficult one, is the choice of bearings. At high speed the lifetime of a ball bearing is very limited. This may not matter for some applications as is the case for a hand held screwdriver that only operates a few seconds at a time, but for other applications the lifetime is crucial.

In some environments the noise level determines the choice of bearings. If the rotor is not perfectly balanced, or maybe it creeps with time and gets unbalanced, a bearing with high stiffness will transfer the vibrations to the housing and should thus be avoided. In other cases the bearing must be lubrication free, for instance in hydrocarbon free environments like the medical process industry or the vacuum industry. Maintenance free operation is often required, something that lubricated bearings cannot deliver.

Bearings like ceramic ball bearings, air bearings or fluid bearings are often used, all with their special limitations. It is not surprising that Magnetic bearings are increasingly being used for a large variety of applications and are becoming very popular in designing high speed machines.

Their unique features make them attractive for solving classical rotor-bearing problems in a new way and allow novel design approaches for rotating machinery. Classical limitations can be overcome and feature ranges can be extended. Of course, limitations still hold, imposed by principally valid physical constraints and by the actual state of the art.

The advantages of Magnetic Bearings over the conventional types cover the following parameters and performances:

- Contactless support; Magnetic Bearings are free of contact and can be utilized in vacuum techniques, clean and sterile rooms, transportation of aggressive media or pure media
- Speed limit; Highest speeds are possible even till the ultimate strength of the rotor

- Free of lubricant; Absence of lubrication seals allows the larger and stiffer rotor shafts
- Maintenance-free; Absence of mechanical wear results in lower maintenance costs and longer life of the system
- Adaptable stiffness can be used in vibration isolation, passing critical speeds, robust to external disturbances
- Robust against heat, cold, vacuum, steam and chemical substances
- Low bearing losses, the maximum rotation speed is determined by the strength of the rotor materials
- Bearing forces act through an air gap, the rotor can be encapsulated Hermetically

1.2. Development of Conventional Bearing

An early type of linear bearing uses tree trunks laid down under sleds. This technology may date as far back as the construction of the Pyramids of Giza, though there is no definitive evidence. Modern linear bearings use a similar principle, sometimes with balls in place of rollers.

Bearings saw use for holding wheel and axles. The bearings used there were plain bearings that were used to greatly reduce friction over that of dragging an object by making the friction act over a shorter distance as the wheel turned.

The first plain and rolling-element bearings were wood, but ceramic, sapphire, or glass were also used, and steel, bronze, other metals, ceramics, and plastic (e.g., nylon, polyoxymethylene, teflon, and UHMWPE) are all common today. A "jeweled" pocket watch uses stones to reduce friction, and allow more precise time keeping. Even old materials can have good durability. As examples, wood bearings can still be seen today in old water mills where the water provides cooling and lubrication.

Rotary bearings are required for many applications, from heavy-duty use in vehicle axles and machine shafts, to precision clock parts. The simplest rotary bearing is the sleeve bearing, which is just a cylinder inserted between the wheel and its axle. This was followed by the roller bearing, in which the sleeve is replaced by a number of cylindrical rollers. Each roller behaves as an individual wheel. The first practical caged-roller bearing was invented in the mid-1740s by horologist John Harrison for his H3 marine timekeeper. This uses the bearing for a very limited oscillating motion but Harrison also used a similar bearing in a truly rotary application in a contemporaneous regulator clock.

An early example of a wooden ball bearing (see rolling-element bearing), supporting a rotating table, was retrieved from the remains of the Roman Nemi ships in Lake Nemi, Italy. The wrecks were dated to 40 AD. Leonardo da Vinci is said to have described a type of ball bearing around the year 1500. An issue with ball bearings is the balls rub against each other, causing additional friction, but rubbing can be prevented by enclosing the balls in a cage. The captured, or caged, ball bearing was originally described by Galileo in the 1600s. The mounting of bearings into a set was not accomplished for many years after that. The first patent for a ball race was by Philip Vaughan of Carmarthen in 1794.

Friedrich Fischer's idea from the year 1883 for milling and grinding balls of equal size and exact roundness by means of a suitable production machine formed the foundation for creation of an independent bearing industry.

A patent on ball bearings, reportedly the first, was awarded to Jules Suriray, a Parisian bicycle mechanic, on 3 August 1869. The bearings were then fitted to the winning bicycle ridden by James Moore in the world's first bicycle road race, Paris-Rouen, in November 1869. Henry Timken, a 19th century visionary and innovator in carriage manufacturing, patented the tapered roller bearing, in 1898. The following year, he formed a company to produce his innovation. Through a century, the company grew to make bearings of all types, specialty steel and an array of related products and services.

The modern, self-aligning design of ball bearing is attributed to Sven Wingquist of the SKF ball-bearing manufacturer in 1907. Erich Franke invented and patented the wire race bearing in 1934. His focus was on a bearing design with a cross section as small as possible and which could be integrated into the enclosing design. After World War II he founded together with Gerhard Heydrich the company Franke & Heydrich KG (today Franke GmbH) to push the development and production of wire race bearings.
The Timken Company, SKF company, Schaeffler Group , NSK company , NTN Bearing company are now the largest bearing manufacturers in the world.

Today, bearings are used in a variety of applications. Ultra high speed bearings are used in dental hand pieces, aerospace bearings are used in the Mars Rover, and flexure bearings are used in optical alignment systems.

1.3. Development of Magnetic Bearing

Initially, the Magnetic Bearings have been designed to overcome the deficiencies of conventional journal or ball bearings. Mostly in research labs, they showed their ability to work in vacuum with no lubrication and no contamination, or to run at high speed, and to shape novel rotor dynamics. Today, magnetic bearings have been introduced into the industrial world as a very valuable machine element with quite a number of novel features, and with a vast range of diverse applications.

A magnetic bearing is a bearing which supports a load using magnetic levitation. Magnetic bearings support moving machinery without physical contact, for example, they can levitate a rotating shaft and permit relative motion without friction or wear.

It was first proven mathematically in the late 1800s by Earnshaw that using only a magnet to try and support an object represented an unstable equilibrium; however, it was found in the 1930s that by using an electromagnet and measuring the air gap and using it as a feedback parameter, the system could be stabilized.

Earlier patents for magnetic suspensions consists of assemblies of permanent magnets of problematic stability per Earnshaw's Theorem. Early active magnetic bearing patents were assigned to Jesse Beams at the University of Virginia during World War II. These patents are concerned with ultracentrifuges for purification of the isotopes of various elements for the manufacture of the first nuclear bombs, but the technology did not mature until the advances of solid-state electronics and modern computer-based control technology with the work of Habermann and Schweitzer. Extensive modern work in magnetic bearings has continued at the University of Virginia in the Rotating Machinery and Controls Industrial Research Program. The first international symposium for active magnetic bearing technology was held in 1988 with the founding of the International Society of Magnetic Bearings by Prof. Schweitzer (ETHZ), Prof. Allaire (University of Virginia), and Prof. Okada (Ibaraki University).

In 1987 further improved AMB designs were created in Australia by E.Croot but these designs were not manufactured due to expensive costs of production. However, some of those designs have since been used by Japanese electronics companies, they remain a specialty item: where extremely high RPM is required.

Since then there have been many succeeding symposia. Kasarda reviews the history of AMB in depth. The first commercial application of AMB's was with turbomachinery. The AMB allowed the elimination of oil reservoirs on compressors for the NOVA Gas Transmission Ltd. (NGTL) gas pipelines in Alberta, Canada. This reduced the fire hazard allowing a substantial reduction in insurance costs. The success of these magnetic bearing installations led NGTL to pioneer the research and development of a digital magnetic bearing control system as a replacement for the analog control systems supplied by the American company Magnetic Bearings Inc. (MBI). In 1992, NGTL's magnetic bearing research group formed the company Revolve Technologies Inc. to commercialize the digital magnetic bearing technology. This firm was later purchased by SKF of Sweden. The French company S2M, founded in 1976, was the first to commercially market AMB's. Extensive research on magnetic bearings continues at the University of Virginia in the Rotating Machinery and Controls Industrial Research Program.

The evolution of Active Magnetic Bearings AMB may be traced through the patents issued in this field. The table below lists several examples of early U.S. patents for active magnetic bearings.

Table 1.1 Early U.S. Patents in AMB

Inventor(s)	Year	Patent No.	Invention Title
Beams, Holmes	1941	2,256,937	Suspension of Rotatable Bodies
Beams	1954	2,691,306	Magnetically Supported Rotating Bodies
Beams	1962	3,041,482	Apparatus for Rotating Freely Suspended Bodies
Beams	1965	3,196,694	Magnetic Suspension System
Wolf	1967	3,316,032	Poly-Phase Magnetic Suspension Transformer
Lyman	1971	3,565,495	Magnetic Suspension Apparatus
Habermann	1973	3,731,984	Magnetic Bearing Block Device for Supporting a Vertical Shaft Adapted for Rotating at High Speed
Habermann, Loyen, Joli, Aubert	1974	3,787,100	Devices Including Rotating Members Supported by Magnetic Bearings
Habermann, Brunet	1977	4,012,083	Magnetic Bearings
Habermann, Brunet, LeClére	1978	4,114,960	Radial Displacement Detector Device for a Magnetic Bearings

1.4. Classifications and Types of Magnetic Bearing

Magnetic Bearings can be classified according to the following functions and performances:

1- control action
   – Active
   – Passive
   – Hybrid

2- Forcing action
   – Repulsive
   – Attractive

3- Sensing action
   – Sensor sensing
   – Self sensing

4- Load supported
   – Axial or Thrust
   – Radial or Journal
   – Conical

5- Magnetic effect
   – Electro magnetic
   – Electro dynamic

6- Application
   – Precision flotors
   – Linear motors
   – Levitated rotors
   – Bearingless motors
   – Contactless Geartrains

Magnetic bearings exist in several different types, all of them offering noncontact operation. Thus they all have very long lifetime. They are lubrication free and thus maintenance free. They have low stiffness and thus does not transmit vibrations to the housing. They are quiet and they have very low losses, even at very high speed. They are in service in such industrial applications as electric power generation, petroleum refining, machine tool operation and natural gas pipelines. They are also used in the Zippe-type centrifuge used for uranium enrichment. Magnetic bearings are used in turbomolecular pumps where oil-lubricated bearings are a source of contamination. Magnetic bearings support the highest speeds of any kind of bearing; they have no known maximum relative speed.

Active magnetic bearings (AMB) have been successfully used in various applications for several decades. They show great abilities to work under extreme conditions, such as vacuum, high rotation speed or at high temperature [40]. AMB are used today in applications such as turbo-molecular pumps, turbo expanders, textile spindles, machine tool spindles, hard disk drives and magnetically levitated vehicles (MAGLEV).

Now, there are questions coming up about the actual potential of these bearings: what experiences have been made as to the performance, what is the state of the art, what are the physical limits, what can we expect? In particular, there are features such as load, size, stiffness, temperature, precision, speed, losses and dynamics. Even such complex issues as reliability and smartness of the bearing can be seen as features, with increasing importance and growing maturity.

Generally, there are many different types of bearings. Table 1.2 summarized the main classification of bearings and their performances.

Table 1.2. Main classification of bearings and their performances

Type	Description	Performance
Plain bearing	Rubbing surfaces , usually with lubricant	**Friction:** Depends on materials and construction, PTFE has coefficient of friction ~0.05 Stiffness*: Good, provided wear is low, but some slack is normally present **Speed:** Low to very high **Life:** Moderate (depends on lubrication) **Notes:** The simplest type of bearing, widely used, relatively high friction, suffers from stiction in someapplications. Some bearings

		use pumped lubrication and behave similarly to fluid bearings. At high speeds life can be very short.
Type	**Description**	**Performance**
Jewel bearing	Off-center bearing rolls in seating	**Friction:** Low Stiffness*: Low due to flexing **Speed:** Low **Life:** Adequate (requires maintenance) **Notes:** Mainly used in low-load, high precision work such as clocks. Jewel bearings may be very small.
Fluid bearing	Fluid is forced between two faces and held in by edge seal	**Friction:** Zero friction at zero speed, low Stiffness*: Very high **Speed:** Very high (usually limited to a few hundred feet per second at/by seal) **Life:** Virtually infinite in some applications, may wear at startup/shutdown in some cases **Notes:** Mainly used in low-load, high precision work such as clocks. Jewel bearings may be very small.
Magnetic bearings	Faces of bearing are kept separate by magnets (electromagnets or eddy currents)	**Friction:** Zero friction at zero speed, but constant power for levitation, eddy currents are often induced when movement occurs, but may be negligible if magnetic field is quasi-static Stiffness*: Low **Speed:** No practical limit **Life:** Indefinite **Notes:** Often needs considerable power. Maintenance free.
Flexure bearing	Material flexes to give & constrain movement	**Friction:** Very low Stiffness*: Low **Speed:** Very high **Life:** Very high or low depending on materials and strain in application **Notes :** Limited range of movement, no backlash, extremely smooth motion

*Stiffness is the amount that the gap varies when the load on the bearing changes, it is distinct from the friction of the bearing.

2

Principles and Fundamentals of Electromagnetic Bearing

2.1. Features and Milestones of Magnetic bearing

Magnetic bearings support moving machinery without physical contact, for example, they can levitate a rotating shaft and permit relative motion without friction or wear.

Long considered a promising advancement, they are now moving beyond promise into actual service in such industrial applications as electric power generation, petroleum refining, machine tool operation and natural gas pipelines.

Magnetic bearings are most often used to support radial and thrust loads in rotating machinery.

The main features and milestones of Magnetic Bearings are:

1) Magnetic bearing systems incorporate 3 distinct technologies:
    - **Bearings & sensors** ; the electromechanical hardware by which input signals are collected, and supporting forces applied to the machine on which they are installed.
    - **The control system** ; provides the power and control electronics for signal conditioning, calculation of correcting forces, and resultant commands to the power amplifiers for each axis of control.
    - **Control algorithms**; the software programs used in digital magnetic bearing system control including the processing of the input signals after conditioning, and calculation of the command signals to the power amplifiers.

2) Magnetic Bearings may be categorized as following:

- Magnetic Bearings are free of contact and can be utilized in vacuum techniques, clean and sterile rooms, transportation of aggressive media or pure media
- Highest speeds are possible even till the ultimate strength of the rotor
- Absence of lubrication seals allows the larger and stiffer rotor shafts
- Absence of mechanical wear results in lower maintenance costs and longer life of the system
- Adaptable stiffness can be used in vibration isolation, passing critical speeds, and robust to external disturbances

3) Classification of Bearings functions and types are classified according to the following:
   - **Control action** ; Active, Passive, and Hybrid
   - **Forcing action** ; Repulsive and Attractive
   - **Sensing action** ; Sensor sensing and Self sensing
   - **Load supported** ; Axial or Thrust, Radial or Journal , and Conical
   - **Magnetic effect** ; Electro magnetic and Electro dynamic
   - **Application** ; Precision flotors, Linear motors, Bearingless motors, Levitated rotor, and Contactless Geartrains

Although it is beyond the scope of this book to discuss how magnetic bearings are designed, an attempt will be made to introduce some of the characteristics of magnetic bearing-supported systems.

2.2. Operating Principles

Common design configurations of magnetic bearings are shown in Figure 2.1. The coils have virtually infinite life, but the control system can be affected by power outages or component failure; thus auxiliary rolling element bearings must be incorporated into the design . The rolling element bearings operate at half the magnetic bearing air gap.

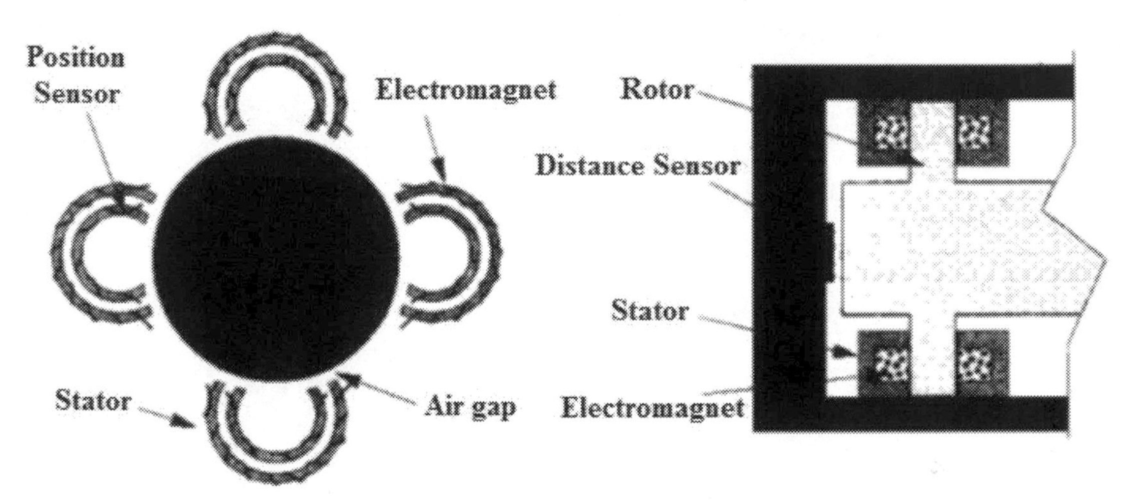

Figure 2.1 Magnetic bearing configurations for supporting radial and thrust loads

Magnetic bearings provide attractive electromagnetic suspension by application of electric current to ferromagnetic materials used in both the stationary and rotating parts (the stator and rotor, respectively) of the magnetic bearing. This creates a flux path that includes both parts, and the air gap separating them, through which non-contact operation is made possible, see figure 2.2.

As the air gap between these two parts decreases, the attractive forces increase, therefore, electromagnets are inherently unstable. A control system is needed to regulate the current and provide stability of the forces, and therefore, position of the rotor.

The control process begins by measurement of the rotor position with a position sensor. The signal from this device is received by the control electronics, which compares it to the desired position, input during machine start-up. Any difference between these two signals results in calculation of the force necessary to pull the rotor back to the desired position. This is translated into a command to the power amplifier connected to the magnetic bearing stator. The current is increased, causing an increase in magnetic flux, an increase in the forces between the rotating and stationary components, and finally, movement of the rotor toward the stator along the axis of control.

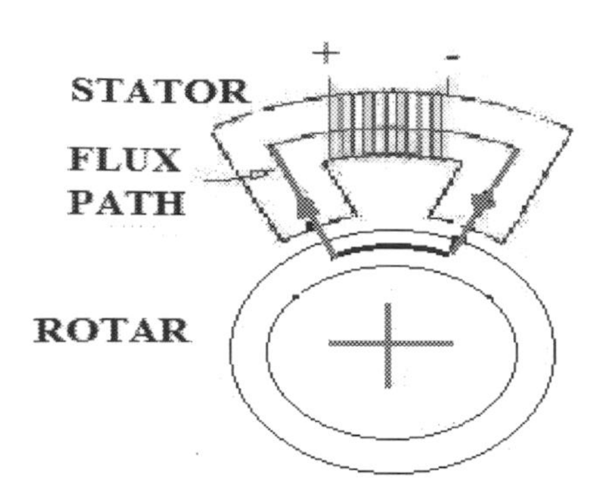

Figure 2.2 Schematic diagram of magnetic circuit

The entire process is repeated thousands of times per second, enabling precise control of machinery rotating at speeds in excess of 100,000 rpm.

Magnetic bearings can also be used to support linear motion devices. A planer version of the horseshoe-shaped magnets often used to support rotating shafts can be used, but this causes vertical motion control to be coupled with angular motion control.

An alternative is to use a number of round bearings, similar to those used to resist shaft thrust loads and shown in Figure 2.3, in a kinematic configuration. This particular bearing has a bias force of about 30 N supplied by the permanent magnet and a control force of about 15 N supplied by the coil.

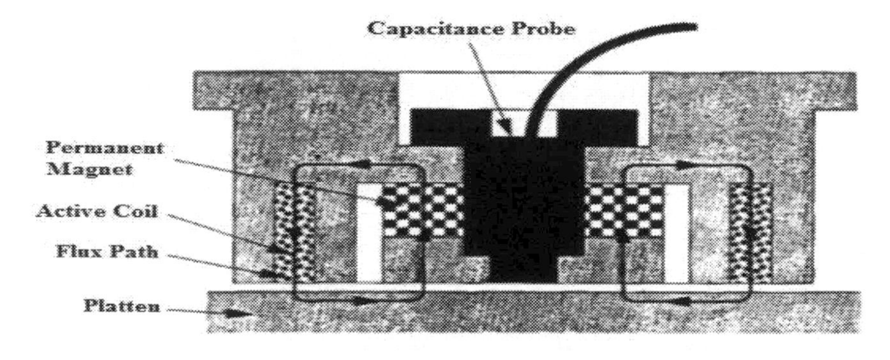

Figure 2.3 Magnetic bearing design used in a kinematic arrangement of magnetic bearing for supporting a precision linear motion

Regardless of the type of load supported, magnetic bearings require a closed-loop control system for stability, as shown schematically in Fig.2.4.

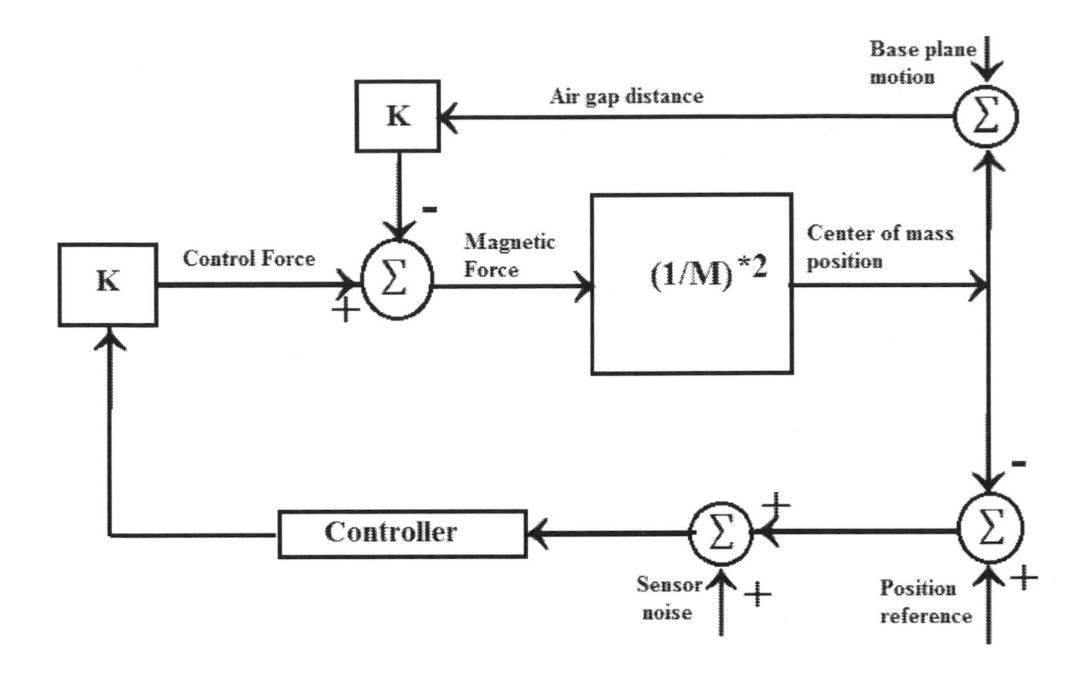

Figure 2.4. schematic closed-loop control system for stability of magnetic bearings

Typically, an analog control loop is used for coarse position control and a digital loop is superimposed on it for fine motion control and compensation for analog component drift.

Some magnetic bearings use a permanent magnet to provide a bias force. The magnetic flux produced by the permanent magnet acts only to levitate a portion of the object's weight in order to minimize the power expended by the active coil. The control force exerted on the object is thus proportional to the current supplied to the coil.

The attraction force between the bearing and the object is produced when energy stored in the magnetic field and in the air gap is transformed into mechanical work.

2.3 Main Performance Parameters

2.3.1 General Properties

In this section the general properties are briefly described while the main properties and performance parameters of the conventional and magnetic bearings are described in more details in the successive sections.

These General Properties of the conventional and magnetic bearings include the following:

1) Speed and Acceleration Limits

Magnetic bearings do not limit the speed or acceleration of components they support. Systems of 100,000 rpm and higher have been built for applications ranging from special pumps to spindles for ultrahigh-peed machining.

2) Range of Motion

Linear motion magnetic bearings can be used to support a carriage that moves linearly. In general, if the coils are stationary, then range of motion will be limited by the variation in force produced as the center of gravity moves with respect to the coils. Magnetically levitated trains have been proposed that use a whole series of coils that are energized as the train passes by them; however, the economics of such a design have yet to be proven. Other designs transfer power to coils on the train via a live rail. Perhaps high temperature superconducting materials would help make magnetically levitated trains an economical reality. Rotary motion magnetic bearing-supported systems are more common in industrial environments and are not motion limited

3) Applied Loads

- Virtually any magnitude load can be supported by a suitable magnetic bearing, depending on how much one wishes to pay and how much room one has.
- Increasing the proportion of the load that is supported by permanent magnets decreases the current that must pass through the coils and the resultant heat generated.

- The magnitude and frequency of applied disturbance forces combined with the controller bandwidth has an effect on achievable resolution as discussed below.

4) Accuracy

Typically, achievable rotational accuracy is 50 μm and 0.1 μm systems have been built. Since magnetic bearings depend on a closed-loop servo-system to achieve stability, the performance of the position sensor and servo-controller will directly affect the accuracy of the system. With new high-speed digital signal processor technology and better sensors for fine-position sensing, there is no reason why nanometer and better accuracy cannot be obtained if one wanted to pay for it. Note that magnetic bearings generate a relatively large amount of heat, so thermal control plays an enhanced role in achieving accuracy with magnetic bearings.

5) Repeatability

- Repeatability of a magnetic bearing system depends on the sensor and control system.
- Impedance probes are commonly used as analog position sensors, so typical repeatability is in the micron range unless precision position sensors are used.

6) Resolution

- Motion control resolution of the bearing gap is also limited by the sensor and control system. There is virtually no mechanical damping in a magnetic bearing-supported system unless the suspended object is also in contact with a viscous fluid.
- Hence disturbance forces acting on the system play an important role in determining motion control resolution of the bearing gap.
- At low frequencies, performance is determined almost completely by the ability of the controller to cancel disturbances.
- The primary control system parameter affecting disturbance cancellation is the controller gain, which determines suspension stiffness.
- The higher the suspension stiffness, the greater the ability to reject force disturbances. Depending on the nature of the source, at high frequencies, the disturbance forces are generally absorbed by the supported object's inertia and internal damping characteristics.

- Since there is no friction with magnetic bearings, motion resolution of an object supported by them is limited only by the actuator, sensor, and control system used.
- Magnetic bearings are efficient only in the attraction mode. In order to obtain high performance in systems with randomly oriented force components, magnetic bearings should be used in an opposed mode design.
- For precision instrument, it is feasible to use gravity to preload the bearing system.

7) Stiffness

- The steady-state stiffness of magnetic bearings can be essentially infinite, depending on how the closed-loop control system is designed.
- Magnetic bearing dynamic stiffness depends on the frequency of the applied load and the bandwidth of the control system.

8) Vibration and Shock Resistance

There are several modes in which a magnetic bearing can be operated in order to actively;

- **Inertial axis control.** The frequency of rotation is measured, amplified, and subtracted from the control signal sent to the coils. This creates a zero stiffness condition for the bearing at the frequency of rotation. As a result, quasi-static and disturbance forces are still resisted, but the rotor is then free to spin about its inertial axis. This virtually eliminates dynamic rotor imbalance.
- **Peak of gain.** Instead of subtracting the rotational frequency component from the control signal, it can be added to achieve very high stiffness at the rotational frequency. This is in effect a feedforward control system which can greatly minimize total radial error motion for low-speed (<1000 rpm) systems.
- **Vibration control.** A magnetic bearing's closed-loop control system can be used in conjunction with feedback from accelerometers to produce forces opposite to those created by vibration. The net effect is to cancel the vibration. Vibration can be reduced by 20 dB using this type of system.
- **Alignment.** Often just achieving proper alignment between rotating components can do wonders to minimize vibration.

- **Dynamic balancing.** Magnetic bearings can be used for in-situ dynamic balancing of components. Rotor speed, angular position, and gap displacement measurement information can be collected and used to determine how to balance the rotor.

9) Damping Capability

- A magnetic bearing's damping capability is attained from the closed-loop control system. Additional magnetic bearing modules can be added at various points along a shaft and used as vibration dampers. In this mode, the gap measurement signal is differentiated and used as a velocity feedback signal.
- There is no friction, static or dynamic, associated with magnetic bearings. However, at high speeds hydrodynamic drag may become a problem if the bearing gap is too small.

10) Thermal Performance

- Magnetic bearings can generate significant amounts of heat and therefore may require external cooling devices, such as recirculating chilled water jackets.
- For systems where the load does not vary greatly, a large percentage of the load can be supported by permanent magnets which minimize coil size and current required to levitate the load.

11) Environmental Sensitivity

- As long as the coils are protected (e.g., hermetically sealed), magnetic bearings can operate in virtually any environment. They have been used successfully in the following environments: air with temperatures ranging from -235 to 450°C, 10-7 torr to 8.5 MPa, water, seawater, steam, helium, hydrogen, methane, and nitrogen.
- One must ensure that in a corrosive environment the system's materials do not fail to perform their structural or sealing functions.

12) Seal-ability

In a normal environment there is really no need to seal magnetic bearings; however, it is a good idea to protect the auxiliary bearings from becoming damaged by contamination.

13) Size and Configuration

Magnetic bearings are typically 2-10 times larger than the rolling element bearings they can replace; however, in many applications, accommodating a magnetic bearing's larger size is not too much of a problem.

14) Weight

Magnetic bearings are very heavy compared to the rolling element bearings they replace. In some applications, such as precision mechanical gyroscopes, the forces encountered by the bearing are so small that the weight of the required bearing is inconsequential anyway. In large stationary industrial applications, such as pipeline compressors or print roll diamond turning machines, bearing weight is not a primary design consideration.

15) Support Equipment

Magnetic bearings require a closed-loop servo system to make them stable. The servo system must have precision displacement sensors to measure the bearing gap and amplifiers for the output signal from the controller. To increase reliability, redundant control and power supply systems are often used on critical magnetic bearing applications such as those used in pipeline compressors.

16) Maintenance Requirements

Magnetic bearings have virtually no maintenance requirements. This makes them especially suitable for equipment that must be kept continually running, such as pipeline compressors.

17) Material Compatibility

Magnetic bearings require wound coils (typically, copper wire) for the stator, and an iron rotor or iron rotor laminations (to minimize eddy current losses). It is possible to hermetically seal the copper windings in a can to protect them from hostile environments. Similarly, it is possible to plate the iron rotor or laminations with a noncorroding material (e.g., chrome or nickel). Since magnetic bearings are noncontact devices and run dry, problems with material compatibility are usually not encountered.

18) Required Life

Since magnetic bearings are noncontact devices, they can have essentially infinite life.

19) Availability, Designability, and Manufacturability

Magnetic bearings are generally custom designed for the application and are thus as yet not available off the shelf except for some pre-engineered complete spindle assemblies. There are many successful commercial applications of magnetic bearings, such as: Pipeline compressor (Speed: 5250rpm, Radial load: 14kN, Axial load: 50kN, and drive power reduced drastically), Print roller diamond turning machine (Speed: 0-1500rpm, Radial load: 18kN, and high rotational accuracy-better than 1μm independent of load), and Turbomolecular vacuum pump (Speed: 0-30000rpm, Radial load: 75kN, and reduced pump size by allowing higher speed)

20) Cost

Magnetic bearings are probably the most expensive type of bearing one can use; however, for the problems they solve, effective system cost can be low compared to design solutions that use other bearings.

2.3.1 Load and Forces

Bearings vary greatly over the size and directions of forces that they can support. Forces can be predominately radial, axial (thrust bearings) or moments perpendicular to the main axis.

The term **load** already, as simple as it seems, touches upon basic properties of magnetic bearings. The load capacity depends on the arrangement and geometry of the electromagnets, as well as the magnetic properties of the material, of the power electronics, and of the control laws . Furthermore, carrying a load is not just a static behavior – usually it has strong dynamic requirements. Magnetic forces are generated in magnetic fields. Magnetic fields themselves can be generated by a current, or a permanent magnet. A basic set-up of an Active Magnetic Bearing AMB with main elements is shown in Figure 2.5.

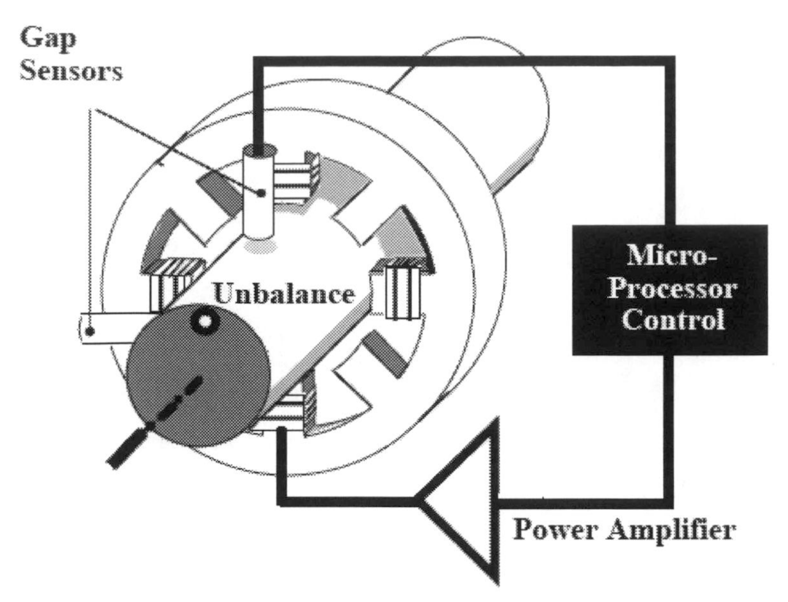

Figure 2.5. Basic set-up of an active magnetic bearing carrying a rotor load

A rotation-symmetrical *magnetic field* H is generated around a straight conductor with a constant current i (Figure 2.6a). The contour integral around the conductor says that

$$\int H\,\mathrm{ds} = I \tag{2.1}$$

This means that the magnitude of the magnetic field in Figure 2.6a is $H = i/2\pi r$. The magnetic field is independent of the material around the conductor. If the integration path encompasses several current loops, as is the case with the air coil

in Figure 2.6b, then the integral yields

$$\int Hds = ni \tag{2.2}$$

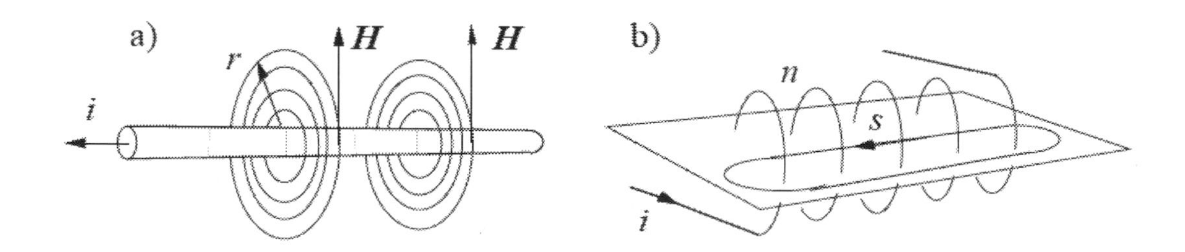

Fig. 2.6 a) Conductor with magnetic field b)Air coil

In magnetic bearing technology electromagnets or permanent magnets cause the magnetic flux to circulate in a magnetic loop. The magnetic flux Φ can be visualized by magnetic field lines. Each field line is always closed. The density of these lines represents the flux density B. The magnetic field H is linked to the flux density B, ie. Magnetic induction, by

$$B = \mu_0 \mu_r(H) H \quad (2.3)$$

Here, $\mu_0 = 4\pi 10^{-7}$ (H/m) stands for the magnetic field constant of the vacuum, and μ_r is the relative permeability depending on the medium the magnetic field acts upon. μ_r equals 1 in a vacuum, and also approximately in air. By using ferromagnetic material, where μ_r is generally >>1, the magnetic loop can be concentrated in that core material. The behaviour of ferromagnetic material, is usually visualized in a B-H diagram (Figure 2.7), showing the well-known phenomena of hysteresis and saturation. Saturation means, as a consequence, that the flux density B does not increase much more beyond B_{sat} even when the magnetic field H and the generating current i is further increased. The current, corresponding to that saturation limit, be i_{sat}.

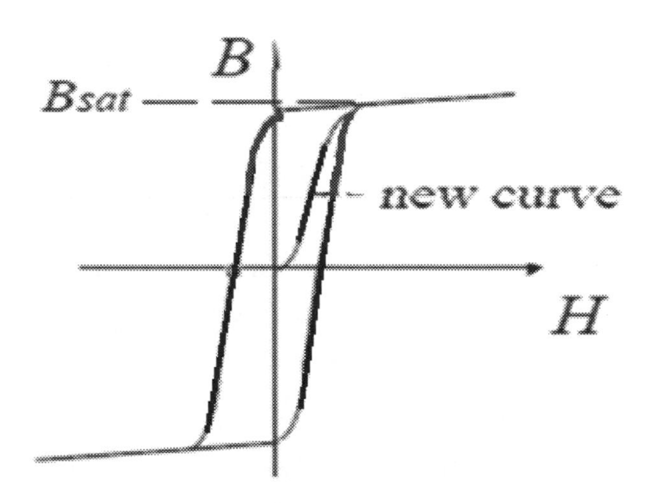

Fig. 2.7 B-H diagram hysteresis loop

For deriving the force in an AMB let us consider Figure 2.8. It shows a single two-pole magnetic bearing element, as part of a complete bearing ring of Figure 2.8, indicating the path of the magnetic flux Φ.

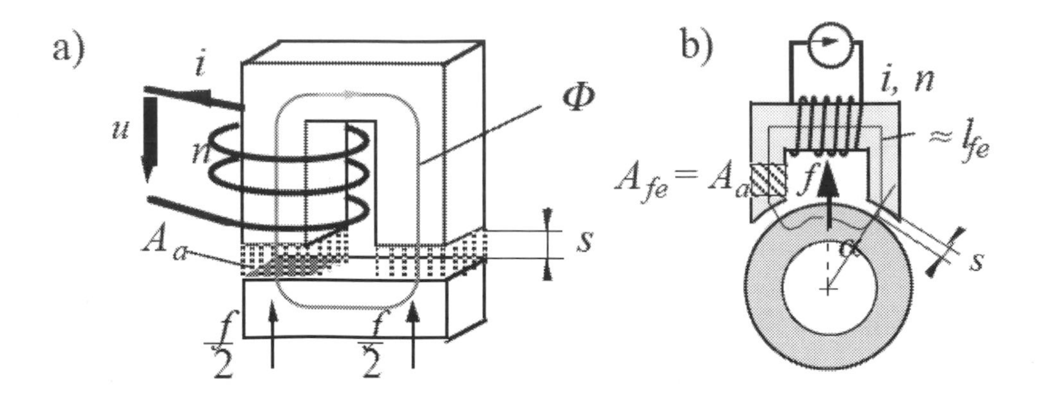

Figure 2.8 a) Force of a magnet b) Geometry of a radial bearing

The usual assumptions hold, i.e.:

- The iron part l_{fe} in the magnetic loop is neglected,
- The relations for static fields hold as the frequencies for the alternating current are not too high,
- The flux Φ is homogeneous in the iron core and the air gap, and
- The cross-sectional areas are the same (*Afe=Aa*).

Then, the induction $B = B_a$ is the same along the magnetic loop. It is proportional to the current *i* until the saturation induction B_{sat} is reached. A further increase of the current beyond i_{sat} does not increase the induction much further beyond B_{sat}. The force *f* exerted can be derived by considering the energy W_a stored in the air gap between rotor and magnet

$$W_a = \frac{1}{2} B_a H_a V_a = \frac{1}{2} B_a H_a A_a 2s \quad (2.4)$$

The force acting on the ferromagnetic body is generated by a charge of the field energy in the air gap, as function of the body position. For small displacements d*s* the magnetic flux ($B_a A_a$) remains constant. When the air gap increases by d*s* the Volume $Va = 2sA_a$ increases, and the energy W_a in the field increases by d*W*. This energy has to be provided mechanically, i.e. an attractive force has to be overcome. Thus

$$f = \frac{dWa}{ds} = B_a H_a A_a = \frac{B_a{}^2 A_a}{\mu_0} \quad (2.5)$$

In the range, where the induction B_a is proportional to the magnetic field H_a and the currrent i, i.e. below saturation, the force as a function of coil current i and air gap s for the arrangement of Figure 2.8a is

$$f = \mu_0 A_a \left(\frac{ni}{2s}\right)^2 = \frac{1}{2} \mu_0 A_a n^2 \frac{i^2}{s^2} = k \frac{i^2}{s^2} \qquad (2.6)$$

Equation (1) shows the quadratic dependence of the force on the current and the inversely quadratic dependence on the air gap. In the case of a real radial bearing magnet, the force of both magnetic poles affect the rotor with an angle α (figure 2.8b), as opposed to the model of the U-shaped magnet of Figure 2.8a. In the case of a radial bearing with four pole pairs α equals for instance 22.5°with cos α =0.92. Considering α we obtain

$$f = \frac{1}{2} \mu_0 A_a n^2 \frac{i^2}{s^2} \cos \alpha = k \frac{i^2}{s^2} \cos \alpha \qquad (2.7)$$

The force increases with the maximum admissible "magnetomotive force" ni_{max} , i.e. the product of the maximum current i_{max} and winding number n. This value is subject to design limitations. As a consequence, the maximum value for the force depends on the winding cross section, the mean winding length and the possible heat dissipation, or the available amount of cooling, respectively. Therefore, one limitation for a high static load is the adequate dissipation of the heat generated by the coil current due to the Ohm resistance of the windings. This "soft" limitation can be overcome by a suitable design.

Assuming that this problem has been adequately considered, then the current i_{max}will eventually reach a value where the flux generated will cause saturation, and then $i_{max} = i_{sat}$, and the carrying force has reached its maximal value f_{max}. Any overload beyond that physically motivated "hard" limitation of the carrying force f_{max}will cause the rotor to break away from its centre position and touch down on its retainer bearings.

In order to compare the carrying performance of different bearing sizes, the carrying force is related to the size of the bearing, or more precisely, to the projection of the bearing area db (Figure 2.9),leading to the specific carrying force. Let us assume that the pole shoe width p equals the leg width c.

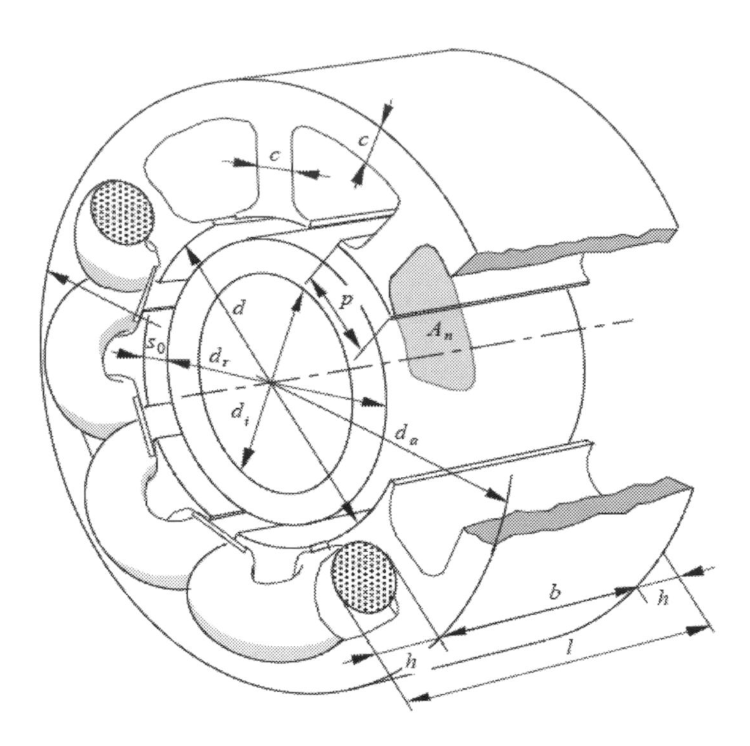

Fig. 2.9 Geometry of a radial bearing magnet

d	*Inner diameter(bearing diameter)*	d_a	*Outer diameter*
dr	*Rotor diameter*	d_j	*Shaft diameter*
c	*Leg width*	l	*Bearing length*
h	*Winding head height*	p	*Pole shoe width*
A_n	*Slot cross section(winding space)*	S_0	*Nominal air gap*
b	*Bearing width(magnetically active part)*		

On the bearing diameter d we have one eighth of the circumference per pole at our disposal. Using half of that for the pole shoe width p, the pole shoe surface is given by

$$A_a = \frac{d\pi}{8} 0.5b \tag{2.8}$$

With actually available Si-alloyed transformer sheets, which are used for bearing magnets, a maximum flux density $B_{max} \approx 1.5$ Tesla $< B_{sat}$ is recommended. Inserting this value for B_a in equation (4), and considering that the forces of both poles do not act perpendicularly, but at an angle of $\pi/8$, we obtain with A_a from equations (8) and (4) the specific carrying force:

$$\frac{f_{\max}}{db} = \frac{B_{\max}{}^{2}\,\pi}{\mu_0\,8}\,0.5\cos\frac{\pi}{8} = \frac{1.5^{2}\,\pi}{\mu_0\,8}\,0.5\cos(22.5°) = 32\,\frac{N}{\mathrm{Cm}^2} \quad (2.9)$$

Based on this result, an estimation of the carrying force $f_{\max}$ can be determined from Figure 2.10. The specific load of 32 N/Cm^2 (or 0.32 MPa) is considerably lower than that for oil lubricated bearings.

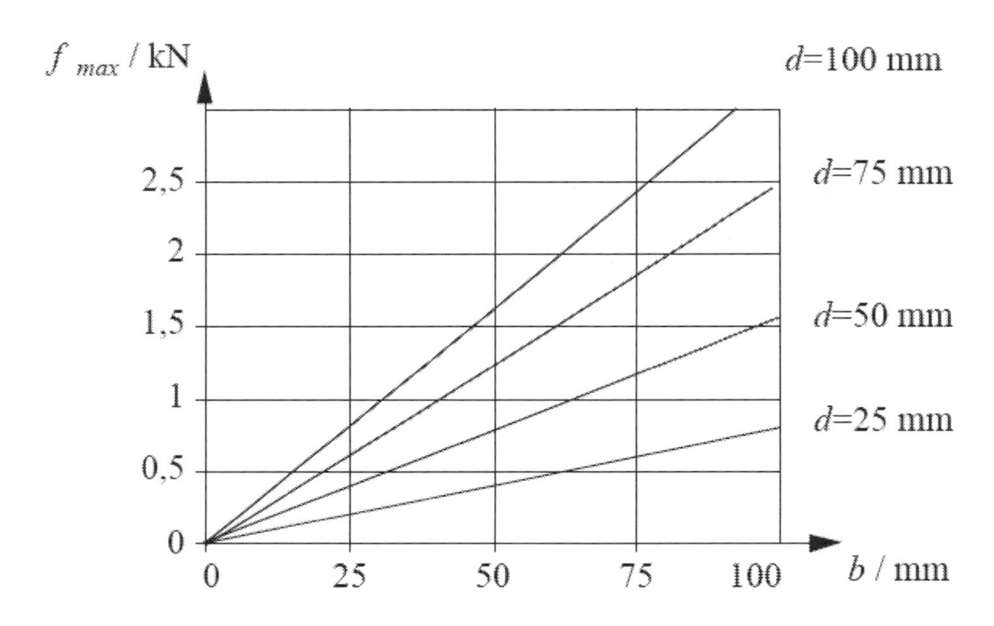

Fig. 2.10 Carrying force of radial bearings having width b and diameter d at specific carrying force of 32 N/ mm^2

2.3.2. Static and Dynamic Stiffness and bandwidth

The stiffness of a bearing characterizes its spring-like behavior, i.e. the stiffness is the ratio of the supported load with respect to the resulting displacement of that load. In classical bearings the stiffness stems for example from the elasticity of the oil film or the deformation of balls and inner ring of a ball bearing. In the Active Magnetic Bearings AMB the force is generated by a control current, which can be adjusted to the needs and opens a novel way of shaping the stiffness and even the overall dynamic behavior, and thus the term “stiffness” may not be the best way to describe the performance of an AMB, but is still used for comparison reasons with classical bearings.

One way of obtaining a high static stiffness is by applying PID-control. In that way, the current ***i***, and implicitly the force as well, is shaped depending on the displacement x of the rotor within the air gap as

$$i = Px + I\int xdt + D\mathrm{d}x/\mathrm{d}t \qquad (2.10)$$

The integral part of the PID control brings the position $\boldsymbol{x}$ to the same value before and after the load step, and thus the rotor shows a behavior that cannot be obtained with classical bearings (Figure 2.11). The time constant for that compensation cannot be made arbitrarily small without increasing the stiffness constant P and the damping constant D. Looking at the bearing only, its static stiffness has become infinite as under stationary load there is no displacement – of course, as long as the maximum carrying force is respected. For comparison, the behavior without compensating integrator is shown, as it would correspond to a classical bearing with spring/damper characteristics as well.

The steady-state stiffness of magnetic bearings can be essentially infinite, depending on how the closed-loop control system is designed. Magnetic bearing dynamic stiffness depends on the frequency of the applied load and the bandwidth of the control system.

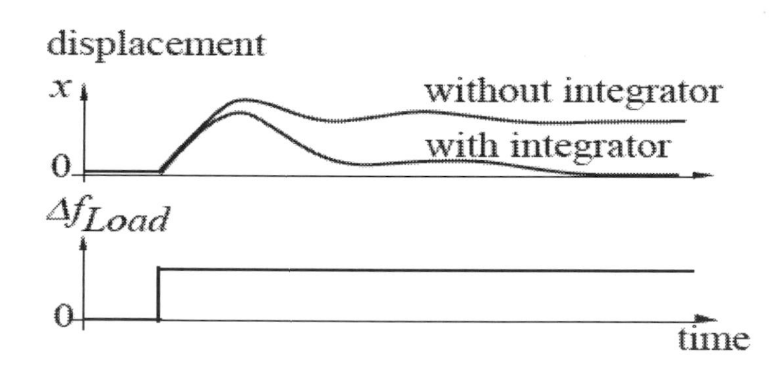

Figure 2.11. Step response of rotor position to a change in load force with PD and PID control.

The term dynamic stiffness characterizes the fact that in an AMB the force depends on the control current, and it is frequency dependent, as well as the displacement. There are limitations on the frequency range which will be explained subsequently. The current is generated in a power amplifier, and it makes sense to look at the electromagnet and power amplifier as a unit, the whole unit being termed magnetic actuator. The output voltage of the amplifier is limited to a value $\pm Up$, which is based on the design of the amplifier. If all the voltage is used to overcome the resistance Rcu of the copper windings in the coils a maximum current I_o can be provided.

Figure 2.12. visualizes the amplitude I_x for the current generated in the bearing windings vs. the frequency ω in the case of an amplifier output voltage $Up \sin \omega t$. The dropdown at higher frequencies is caused by the inductance L of the coils leading to a cut-off frequency $\omega c = Rcu/L$.

If now the amplitude of the current Ix is limited to 0.5 i_{max}, taking into account that usually half of the admissible current i_{max} is being used for pre-magnetization and for carrying constant loads and the other half is being used for generating dynamic forces, we obtain the operating range of the actuator. The highest frequency where the actuator can still operate with its maximal current is called power bandwidth ω_{pbw}. The bandwidth can be enhanced by increasing the power of the amplifier. At higher frequencies the output voltage of the amplifier runs into saturation, and its dynamic behavior becomes nonlinear. The signal bandwidth describes the linear behavior of the actuator and, of course, only holds for the small signal behavior within the operating range The signal bandwidth of the actuator usually goes at least one decade beyond the cut-off frequency ωc.

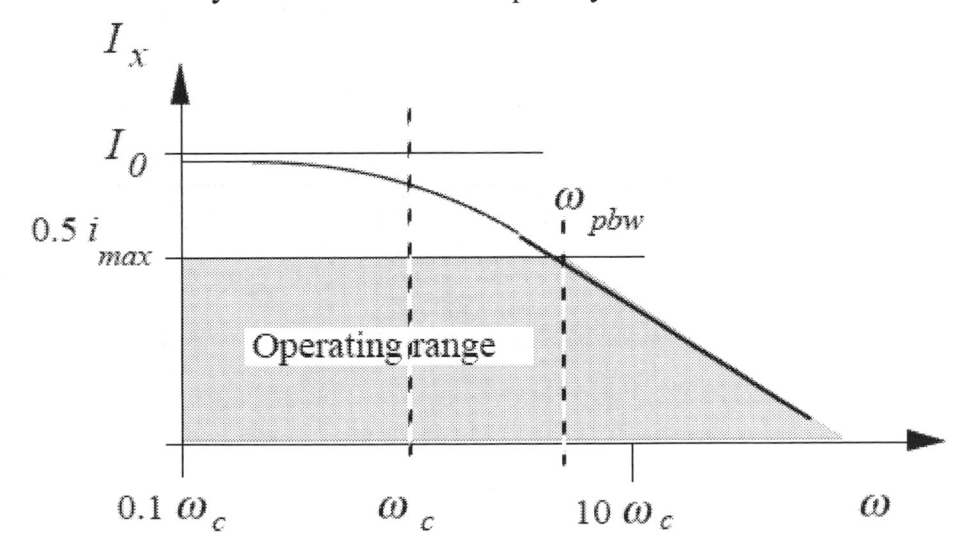

Figure 2.12. Operating range of a magnetic bearing actuator (power amplifier with bearing magnet)

When designing a magnetic bearing system one has to estimate the amplifier power P_{max} necessary to achieve the power bandwidth ω_{pbw} for the maximal force amplitude f_{max} over a nominal air gap s_0. It can be shown that for a bearing with two magnets, such as in Figure 2.8b, differentially arranged as in Figure 2.5, the relation holds

$$\omega_{pbw} = 0.92 \frac{P_{max}}{s_0 f_{max}} \quad (2.11)$$

As an example, a force of 1000N can be generated over an air gap of 0.3 mm with a 1kW amplifier up to a frequency of about 500 Hz. Most of that power is used for the dynamics which can be seen as an inductive load. The actual energy loss, see section on “losses”, is much less.

2.3.3. Speed

Common motions permitted by bearings are:

- Axial rotation e.g. shaft rotation
- Linear motion e.g. drawer
- Spherical rotation e.g. ball and socket joint
- Hinge motion e.g. door

Different bearing types have different operating speed limits. Speed is typically specified as maximum relative surface speeds, often specified ft/s or m/s. Rotational bearings typically describe performance in terms Some applications apply bearing loads from varying directions and accept only limited play or "slop" as the applied load changes of the product ***DN*** where ***D*** is the diameter (often in mm) of the bearing and ***N*** is the rotation rate in revolutions per minute.

Generally there is considerable speed range overlap between bearing types. Plain bearings typically handle only lower speeds, rolling element bearings are faster, followed by fluid bearings and finally magnetic bearings which are limited ultimately by centripetal force overcoming material strength.

The features characterizing a high-speed rotor can be looked at under various aspects. The term “high-speed” can refer to the rotational speed, the circumferential speed of the rotor in a bearing, the circumferential speed of the rotor at its largest diameter, or the fact that a rotor is running well above its first critical bending frequency. The requirements on the AMB and the design limitations can be very different.

Rotational speed

A record from about 50 years ago are the 300 kHz rotation speed that have been realized in physical experiments for testing the material strength of small steel balls (about 1 to 2 mm in diameter) under centrifugal load . In today's industrial applications rotational speeds that have been realized are in the range of about 3kHz for a grinding spindle, or about 5kHz for small turbo-machinery. Problems arise from eddy current and hysteresis losses in the magnetic material, air losses, and the related requirements for power generation and adequate heat dissipation if the rotor runs in vacuum.

Circumferential speed

The circumferential speed of the rotor in a bearing at its largest diameter, or the fact that a rotor is running well above its first **critical** bending frequency. The requirements on the AMB and the design limitations can be very different.

The circumferential speed is a measure for the centrifugal load and leads to specific requirements on design and material .The centrifugal load, Figure 2.13, leads to tangential and radial stresses in the rotor, given by

$$\sigma_t = \frac{1}{8} \rho\, \Omega^2 \left[(3+v)(r_i^2 + r_a^2) + (3+v)\frac{r_i^2 r_a^2}{r^2} - (1+3v) r^2 \right] \quad (2.12A)$$

$$\sigma_r = \frac{1}{8} \rho\, \Omega^2 (3+v)\left[(r_i^2 + r_a^2) - \frac{r_i^2 r_a^2}{r^2} - r^2 \right] \quad (2.\ 12B)$$

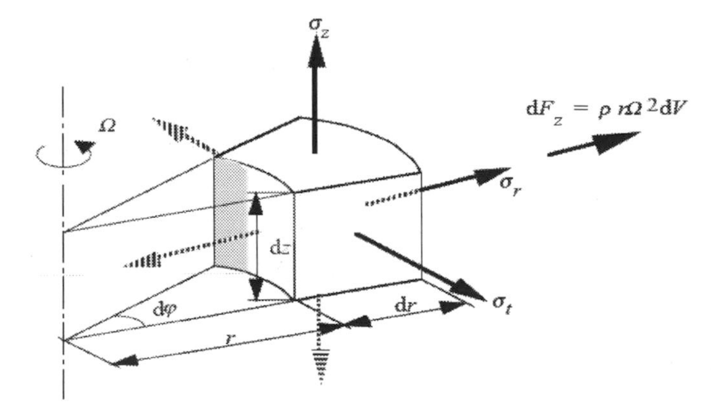

Figure 2.13. Centrifugal loads acting on the volume element of a rotor

Numerous lab experiments have been performed. Rotor speeds of up to 340 m/s in the bearing area can be reached with iron sheets from amorphous metal (metallic glass), having good magnetic and mechanical properties .The theoretical value for the achievable speed v_{max}lies much higher.

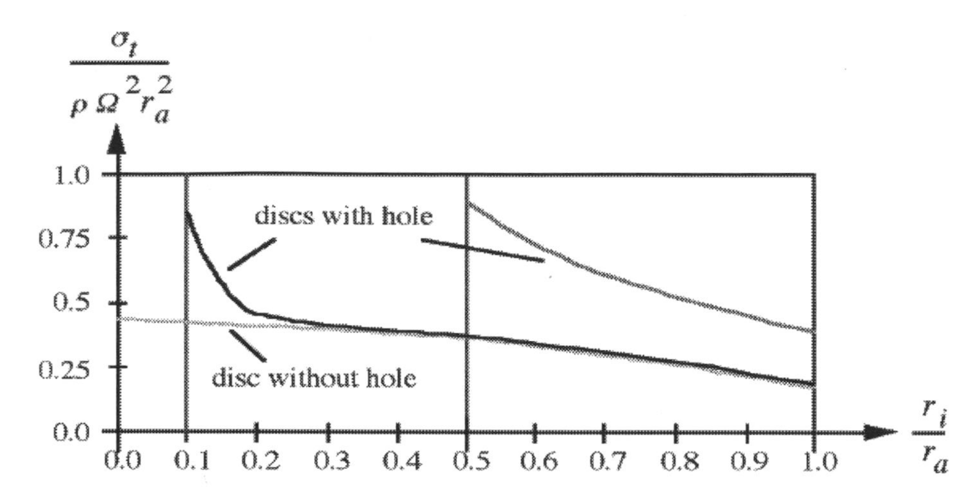

Figure 2.14. Tangential stress distribution for a disc with and without hole in the center

It can be derived from Equation 2.13 (σ_s is the yield strength, ρ is the density of the material), and the according values for some materials are given in table 2.3.

$$f_{max} = (r_a \Omega)_{max} = \sqrt{\frac{8\sigma_s}{(3+v)\,\rho}} \qquad (2.13)$$

Table 2.3. Achievable circumferential speeds for a full disc

Material	vmax / [m/sec]
Steel	576
Bronze	434
Aluminium	593
Titanium	695
Soft ferromagnetic sheets	565
Amorphous metal	826

In industrial applications the speed usually is limited not by the bearings themselves, but by the mechanical design of the motor drive. Figure 2.15 gives a survey on various AMB applications that have been realized conventionally. For high speeds permanent magnet synchronous drives are used where the rotor is wound with carbon fibers, allowing speeds of about 280 m/s.

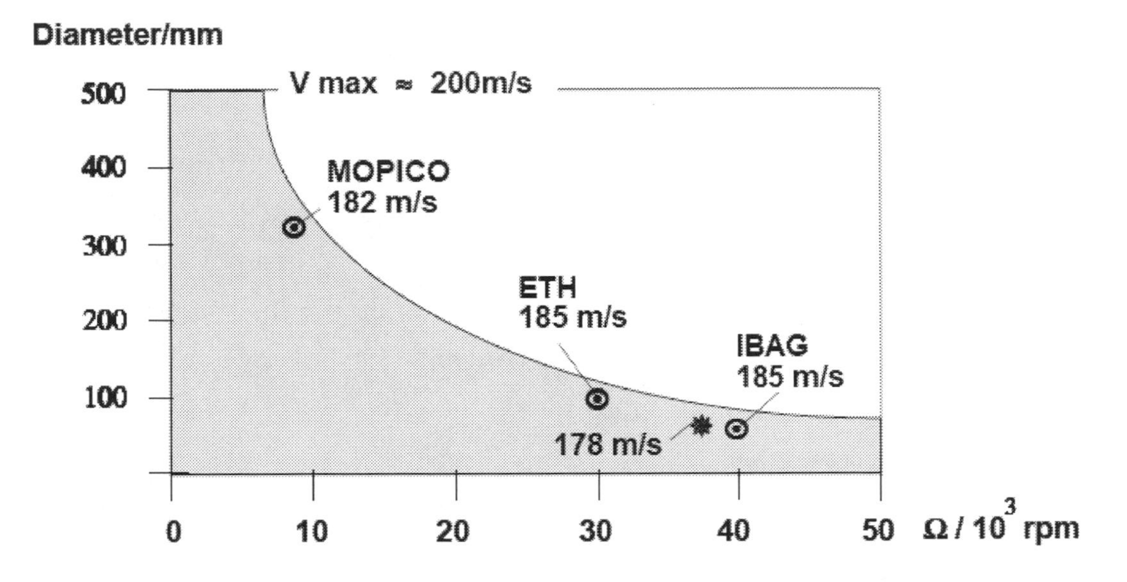

Figure 2.15. Maximal diameter of the (asynchronous) motor drive in function of the rotor speed (* indicates the rotor of Figure 2.15 broken at 178 m/s)

Supercritical speed

A rotor may well have to pass one or more critical bending speeds in order to reach its operational rotation speed. In classical rotor dynamics this task is difficult to achieve. In AMB technology it is the controller that has to be designed carefully to enable a stable and well-damped rotor behavior. Passing just the first critical elastic speed is state of the art and can be very well done with AMB. This has been shown even with an automated controller design, based on self-identification and subsequent self-tuning with the H∞-method. In many lab experiments two critical speeds have been passed, too, using various design methods, for example . Three elastic modes have been dealt with in , using additional notch filters and a zero-pole cancelling filter. It is felt that further research in developing methods for the design of robust control for highly elastic systems is necessary.

2.3.4. Losses

Conventional bearings are simply a hole of the correct shape containing the relatively moving part, and use surfaces in rubbing contact, often with a lubricant such as oil or graphite. They are very widely used. Particularly with lubrication they often give entirely acceptable life and friction.

However, reducing friction in conventional bearings is often important for efficiency, to reduce wear and to facilitate extended use at high speeds and to avoid overheating and premature failure of the bearing. Essentially, a bearing can reduce friction by virtue of its shape, by its material, or by introducing and containing a fluid between surfaces or by separating the surfaces with an electromagnetic field.

- **By shape**, gains advantage usually by using spheres or rollers, or by forming flexure bearings.
- **By material**, exploits the nature of the bearing material used. (An example would be using plastics that have low surface friction.)
- **By fluid**, exploits the low viscosity of a layer of fluid, such as a lubricant or as a pressurized medium to keep the two solid parts from touching, or by reducing the normal force between them.
- **By fields**, exploits electromagnetic fields, such as magnetic fields, to keep solid parts from touching.

Combinations of these can even be employed within the same bearing. An example of this is where the cage is made of plastic, and it separates the rollers/balls, which reduce friction by their shape and finish.

With **contact-free rotors** there is no friction in the magnetic bearings. The operation of active magnetic bearings causes much less losses than operating conventional ball or journal bearings, but, nevertheless, the losses have to be taken into account, and sometimes they lead to limitations . Losses can be grouped into losses arising in the stationary parts, in the rotor itself, and losses related to the design of the control.

The iron losses depend on the rotor speed, the material used for the bearing bushes, and the distribution of flux density B over the circumference of the bushes. The breaking torque caused by the iron losses consists of a constant component of hysteresis loss and a component of eddy-current loss, which grow proportionally to the rotational speed.

The iron losses in the rotor can limit operations, as, in particular in vacuum applications, it can be difficult to dissipate the generated heat.

The hysteresis losses *Ph* arise if at re-magnetization the *B-H*-curve in the diagram of Figure 3 travels along a hysteresis loop. At each loop the energy diminishes by *Wh* = *Vfe ABH*. Here, *ABH* stands for the area of the hysteresis loop, and *Vfe* for the volume of the iron. Consequently, the hysteresis losses are proportional to the frequency of re-magnetization *fr*.

The area of the hysteresis loop depends on the material of the magnet and on the amplitude of the flux density variation. For iron and flux densities between 0.2 and 1.5 Tesla the classical relation

$$Ph = kh\, fr \,.\, Bm^{1.6}\, Vfe \qquad (2.14)$$

holds, where the material constant *kh* has to be derived from loss measurements and from the area of the hysteresis loop, respectively. It is obvious that soft magnetic material with a very small loop area will reduce these losses. Experimentally derived data are presented in.

The eddy-current losses *Pe* arise when the flux density within the iron core changes. A compact core (Figure 2.16a) acts like a short circuit winding and generates large eddy currents. The eddy-current losses can be reduced by dividing the iron core in insulated sheets (figure 2.16b), or in particles (sinter cores). The smaller these divisions, the smaller the eddy-current losses. Losses in laminated iron can be calculated approximately, if the flux in the sheets is sinusoidal and distributed evenly.

$$Pe = \frac{1}{6p}\, \pi^2\, e^2 fr^2\, Bm^2\, Vfe \qquad (2.15)$$

Here, **ρ** is the specific electric resistance of the iron, *e* stands for thickness of the sheets, *fr* for re-magnetization frequency, and *Bm* for the maximum flux density or the amplitude of the flux density.

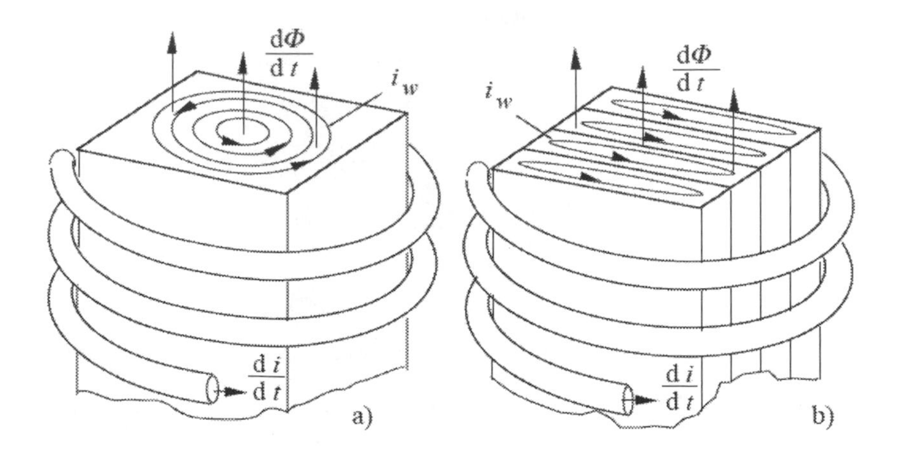

Figure 2.16 Reducing the eddy current losses by dividing the iron core (a) into sheets(b)

Losses in the stationary parts of the bearing come mainly from copper losses in the windings of the stator and from losses in the amplifiers. The copper losses are a heat source, and, if no sufficient cooling is provided, can limit the control current and hence the maximal achievable carrying force.

Losses in the rotor part are more complex and lead to more severe limitations. These losses comprise iron losses caused by hysteresis and eddy currents, and air drag losses. The losses heat up the rotor, cause a breaking torque on the rotor, and have to be compensated by the drive power of the motor. The relations of the various losses with respect to one another are shown in Figure 2.18. In general, the eddy current losses are the largest ones.

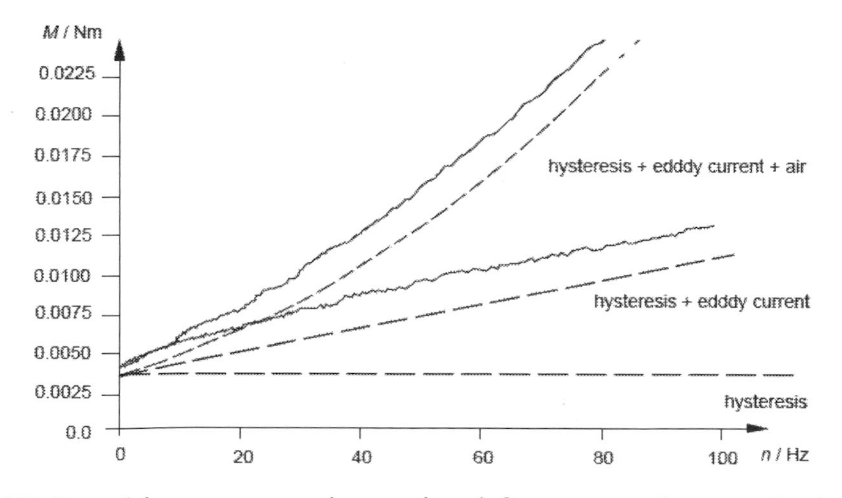

Figure 2.17. Breaking torques determined from experiments (—) compared to calculated values (---)

The flux density on the rotor surface, and the inherent losses, depend on the structural shape of the bearing. When the bearing has a format as shown in Figure 2.18a, the iron is re-magnetized twice upon one revolution. Eddy-current losses can be kept low as far as the rotor can be sheeted easily, i.e. built as a stack of punched circular lamination sheets. If, on the contrary, the bearing has a format as shown in Figure 2.18b, the iron passes below poles with equal polarity, which keeps the hysteresis losses smaller than with format a). However, it is almost impossible to laminate rotor b). Still, the use of format b) is certainly recommended.

The use of iron free magnetic bearings and drives, based on Lorentz forces, has been investigated mainly for precision bearings, where the influence of hysteresis would be detrimental to precision control. This approach will be dealt with in the next section under the heading "precision".

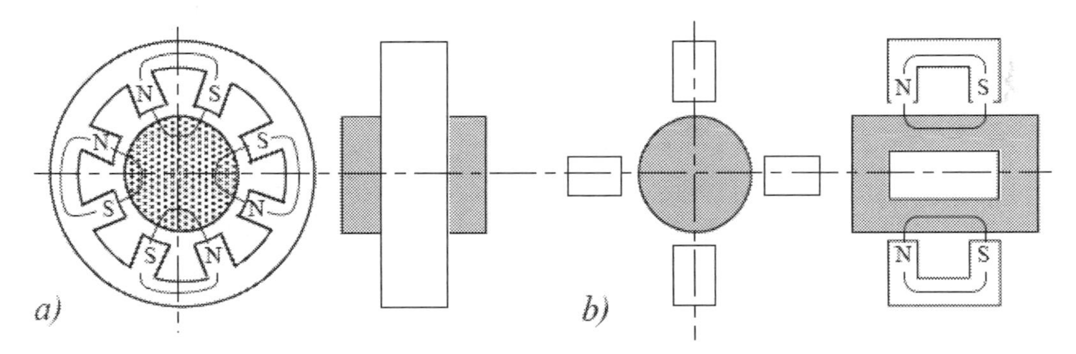

Figure 2.18. Construction models of radial bearings
a) Field lines perpendicular to the rotor axis(heteropolar bearing)
b) Field lines parallel to the rotor axis(homopolar bearing)

The air losses can be predominant at high rotation speeds, and for special applications, such as flywheels for energy storage. They can be calculated or rather estimated by dividing the rotor into sections with similar air-friction conditions, in order to take into account different rotor geometries. Thus, a simple cylindrical rotor is divided, for instance, into cylinders without sheathing, including frontal areas / cylinder front areas within the axial bearing / cylinders within the bearing and the motor / cylinders within the back-up bearing. The breaking torque elements have now to be calculated and then added. Very small air gaps increase the air drag.

The concept of "zero power" control is another way of reducing the losses by reducing the control current itself as much as possible. The static magnetic field, for compensating the static load or for pre-magnetization, is supplied by permanent magnets. The control current is only used for stabilizing the rotor hovering. The rotor is expected to rotate about its main axis of inertia, thus performing a so-called permanent, force-free rotation. The control required for that kind of operation needs information about the periodic parts of the disturbances acting on the rotor, which have to filtered out or compensated for in the sensor signals. The approach is very useful in cases where the energy losses have to be kept minimal, for example for energy flywheels, and where the residual vibratory motions of the geometric rotor axis are not so important.

2.3.5. Other performance parameters and features

A. Size

In principle, there appears to be no upper limit for the bearing size, it can be adapted to any load. Problems arising with assembling the bearing lead to special design variations, where the bearing is separated in two halves, or the single magnets are even treated individually.

Small bearings are of special interest to micro-techniques. Potential applications are video heads, medical instruments, hard disk drives, and optical scanners. The challenge lies in simplifying the design and in the manufacturing process. For more clarifications two cases are presented as in the following examples:

Example One : In the design, fabrication, and testing of a millimeter-level magnetically suspended micro-motor is described. The micro-motor has a stator outside diameter of 5 mm, rotor outside diameter of 1.5 mm, air gap of 0.1mm, and thickness of 0.5 mm. The stator and rotor are manufactured by both the EDM and UD LIGA methods. The micro-motor with UD LIGA-made rotor has been successfully suspended and spun up to 600 rpm. The proposed micro inductive sensor is made from thick photoresist lithography and electroplating processes has advantages of low cost, good manufacturability, and ease of assembly.

Example Two : In a very flat micro-motor is presented with an overall thickness of about 2mm, the thickness of the rotor is 150 to 340 mm .It was produced by chemical etching of non grain oriented silicon steel. The air gap is 30 mm. Regarding a future integration of the complete motor with means of thin film technology, inductive sensors were implemented. Different designs of three micro-rotors are shown in Figure 2.19.

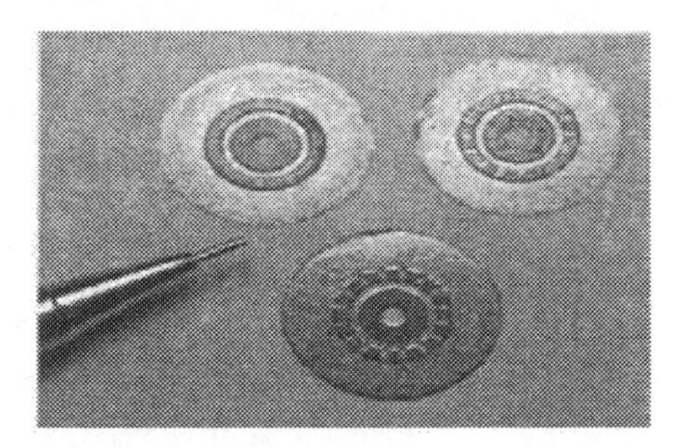

Figure 2.19. Different designs of Bearings: Diameter 15 mm, thickness 150 to 340 mm, total mass about 270 mg

B. Precision

Precision in rotating machinery means most often how precise can the position of the rotor axis be guaranteed. This has consequences for machining tools, and for the surface quality of parts that are being machined by grinding, milling or turning. In addition, the question of how precise can magnetic bearings become in principle, is of interest for applications such as optical devices, optical scanner, wafer stepper, or lithography. These machines and processes are key elements for semiconductor industry. Active magnetic bearings levitate an object, rotating or not, with feedback control of measured displacement sensor signal. The performance of AMB systems is therefore directly affected by the quality of a sensor signal. Precision control is facilitated by the absence of hysteresis and of deformation-prone heat sources, which touches upon material and design aspects.

C. Play and Elasticity

Some applications apply bearing loads from varying directions and accept only limited play or "slop" as the applied load changes. One source of motion is gaps or "play" in the bearing. As example, a 10 mm shaft in a 12 mm hole has 2 mm play. A second source of motion is elasticity in the bearing itself. As example, the balls in a ball bearing are like stiff rubber, and under load deform from round to a slightly flattened shape. The race is also elastic and develops a slight dent where the ball presses on it.

Allowable play varies greatly depending on the use. As example, a wheelbarrow wheel supports radial and axial loads. Axial loads may be hundreds of Newton force left or right, and it is typically acceptable for the wheel to wobble by as much as 10 mm under the varying load. In contrast, a lathe may position a cutting tool to ±0.02 mm using a ball lead screw held by rotating bearings. The bearings support axial loads of thousands of Newton in either direction, and must hold the ball lead screw to ±0.002 mm across that range of loads.

D. Life

Fluid and magnetic bearings can potentially give indefinite life. Rolling element bearing life is statistical, but is determined by load, temperature, maintenance, vibration, lubrication and other factors.

For plain bearings some materials give much longer life than others. Some of the John Harrison clocks still operate after hundreds of years because of the lignum vitae wood employed in their construction, whereas his metal clocks are seldom run due to potential wear.

E. Maintenance

Most bearings in high cycle operations need periodic lubrication and cleaning, and may require adjustment to minimize the effects of wear. Many bearings require periodic maintenance to prevent premature failure, although some such as fluid or magnetic bearings may require little maintenance.

2.4. Categories and Types of Magnetic Bearings

A magnetic bearing is a contact free bearing wherein the load is carried by magnetic forces. Furthermore the magnetic field is generated in such a way that it provides necessary stiffness and damping for the rotor to be levitated safely during operation.

Usually, magnetic bearings are divided into two categories, often referred to as either active or passive depending on whether an electronic control system is being used or not.

This nomenclature has been around for quite a while even though it is rather unfortunate and sometimes gives rise to misunderstandings. It would certainly be more clarifying to classify magnetic bearings according to the physical cause of the force, as H. Bleuler has done. He divides the bearings in two groups, reluctance force bearings and electrodynamic bearings. Then each group is divided into five respectively four categories depending on how the magnetic force is generated.

Some bearing categories are of only limited academic interest, while others have found many useful applications. Based on Bleulers classification some of the most technically and commercially important bearing types will be presented below. But first we need to take a closer look at the general stability criterion for magnetic bearings.

2.4.1 The Earnshaw Stability Criterion

As early as 1842 Earnshaw was able to show that it is impossible for an object to be suspended in stable equilibrium purely by means of magnetic or electrostatic forces.

Let V be the scalar magnetic potential at a point, proportional to the scalar product of $\bar{B}$ at that point and the constant magnetization $\overline{m}$ for a single magnetic pole on the suspended object. Then according to the Laplace Eq.(2.16) the sum of the second derivatives is zero,

$$\frac{d^2V}{dx^2} + \frac{d^2V}{dy^2} + \frac{d^2V}{dz^2} = 0 \qquad (2.16)$$

In order for a magnetic pole to experience no magnetic force and have no tendency to move, it is necessary that

$$\frac{dV}{dx} = \frac{dV}{dy} = \frac{dV}{dz} = 0 \qquad (2.17)$$

This is the condition for a maximum or a minimum in the potential V. Only at a potential minimum is the equilibrium stable. The condition for a potential minimum is that its second derivative is positive. At best, two of the second derivatives can be made to be positive, then, according to the Laplace equation, the third has to be negative. In that direction the equilibrium will be unstable. The theorem has been derived for a single pole. However, recently Lang was able to show that the theorem can be applied to a magnet as a whole.

In order to achieve stability in all directions it is necessary to add a dynamic term, which can be either an electromagnetic or a mechanical one. Adding an electromagnetic term will exchange the Laplace equation into the Poisson equation, which is used in active magnetic bearings and electrodynamic bearings. When a mechanical force derivative is added it can be done directly into Eq. (2.16). For the bearing to be of the non-contact type the preferred mechanical forces are either gyroscopic or fluid dynamic.

Gravitation is for instance not possible to utilize as a means of stabilizing the bearing, since it has only negligable derivatives. Say that we want to compensate the vertical magnetic force F_y with a gravitational force -mg. Then the total potential energy V_{tot} should be used instead of the magnetic potential V, the latter being denoted V_{mag} below to avoid confusion. Unfortunately this does not change anything in the Laplace equation (2.16) since the second term remains unchanged as shown below:

$$\frac{d^2V_{tot}}{dy^2} = \frac{d^2}{dy^2}\int F_{y,tot}dy = \frac{d^2}{dy^2}\int (F_{y,mag} - mg)dy$$

$$= \frac{dF_{y,mag}}{dy} - \frac{dmg}{dy} = \frac{dF_{y,mag}}{dy} = \frac{d^2V_{mag}}{dy^2} \qquad (2.18)$$

Thus, gravity does not improve the stability of a levitating magnet.

What is said above can be applied to other constant forces as well, like air drag in a windmill. They cannot be fully levitated by permanent magnets alone. However, other air forces like the wedge effect in an air bearing can be used for stabilization since that force has a space derivative defined by the airgap.

2.4.2 Classical Active Magnetic Bearings, AMB

An active magnetic bearing is a typical example of an advanced mechatronics product. Most AMBs on the market today utilize attracting reluctance forces generated by strong stator electromagnets acting on a ferromagnetic rotor. The magnetic equilibrium reached in this way is inherently unstable. Thus an analog or digital controller is used to stabilize the bearing. Contactless sensors are used to measure the rotor position. Any deviation from the desired position results in a control signal that via the controller and the amplifiers changes the current in the electromagnet so that forces are generated to pull the rotor back so that stability is maintained.

Currently only attractive forces are used, but theoretically also repulsive forces could be used for actively controlled levitation. This is the reason why active bearings are sometimes referred to as attracting bearings, while passive bearings are often referred to as repulsive.

The advantages with active bearings over passive bearings are obvious as they offer the possibility to change and adapt the controller algorithms according to machine dynamics. The disadvantages are equally clear and are related to the cost of the bearing. Also reliability is a problem as the controller is dependent on high power quality and thus need energy back up.

Today active magnetic bearings are found in a variety of applications like compressors, gas turbines, motors, flywheels, gyroscopes, fans and machine tool spindles.

2.4.3 Permanent Magnet Bearings, PMB

Permanent magnet bearings are based on the principle of the repulsion or attraction of two permanent magnets. Radial and axial bearings are realizable by a suitable arrangement. In Fig. 2.20 different structures of permanent magnet bearings are shown. Their magnets all have the form of a ring. Axially magnetized rings are preferred, since they cost less to produce than rings with radial magnetization.

Usually rare earth magnets are used as material for the rings since they pro vide especially high energy densities. In the absence of extra precaution the maximum speed is limited by the permanent magnets, since they are brittle and lack mechanical strength.

One of the first successful commercial products where magnetic bearings offered true added value to the customer was the turbomolecular vacuum pump. With lubrication free magnetic bearings it was possible to manufacture hydrocarbon free pumps – a great achievement at that time.

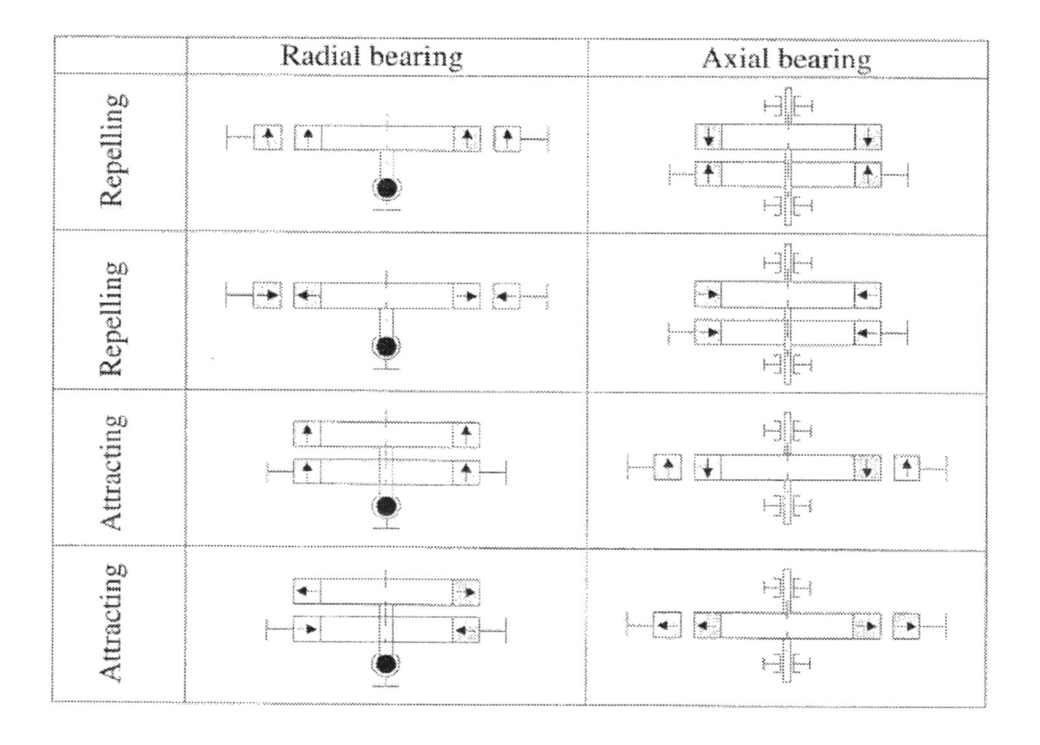

Figure 2.20 Structural forms of magnet bearings
(Jean-Paul Yonnet, Permanent Magnet Bearings and Couplings, IEEE Transactions on Magnetics, Vol. Mag-17, No 1, January 1981)

The first mass produced pump of this kind had one passive magnetic bearing on the high vacuum side and one ball bearing on the exhaust side. The magnetic bearing consisted of concentrically arranged repulsive ring magnets, Fig. 2.21. To the left a bearing with a single magnet pair, and to the right two oppositely directed pairs are used. The latter configuration gives almost 3 times the stiffness compared to the former one. An optimization of PMB geometry was recently done by Lang & Fremerey, who showed that many thin magnets should be used instead of a few thick ones.

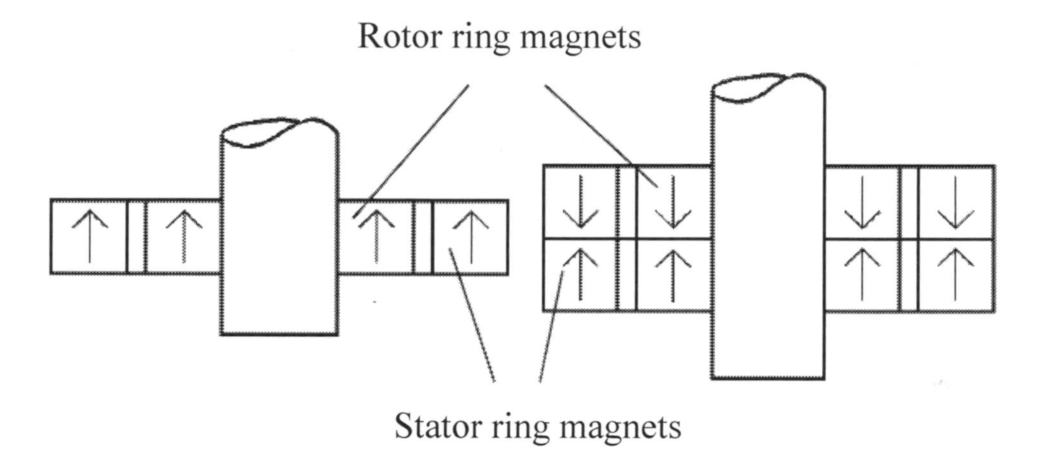

Fig. 2.21. Passive magnetic radial bearings

For low speed applications it is convenient to let the inner rings be mounted on the rotor, while for high speed applications it is better to let the outer rings rotate due to the centrifugal forces, in which case these forces are used more advantageously in keeping the rotating magnets in place.

When the term passive magnetic bearing is mentioned in this report as well as in most literature this is the magnet arrangement usually referred to. Attractive magnetic forces from permanent magnets could have been used as well, but as the geometry for such bearings is slightly more complicated they are seldom used.

All passive bearings of these kinds have one thing in common. According to the Earnshaw theorem they are unstable in at least one direction. For the pump mentioned above the repulsive bearing is strongly unstable in the axial direction. This is compensated for by the use of one ball bearing that prevents the bearing from moving axially.

The next progress was made when the axial stability from the ball bearing could be exchanged for an active magnetic axial bearing. In this case two repulsive radial bearings were used. This concept was originally developed by Gilbert and was later used by S2M in Vernon for flywheels for gyroscopic stabilization of satellites. Later, Boden & Fremerey in Jülich modified this concept for use in the turbomolecular vacuum pumps, X-ray tubes and other industrial applications. In 1999 Fremerey and Kolk presented a 500 Wh flywheel for energy storage based on this concept.

2.4.4 Superconducting bearings, SCB

A superconductor has by definition two properties that makes it ideal for magnetic bearings. The first one is the Meissner effect which makes the superconductor act like a perfect diamagnet. A permanent magnet placed above a superconducting surface will "see" a mirror image under the surface polarized so that the force is repulsive. This levitation is stable in the vertical direction, and indifferent in the horizontal direction. If the real magnet moves the image moves too, thus maintaining the repulsive force vertical. Fig.2.22 illustrates the Meissner effect on the Repulsive force.

The modern high temperature superconductors have a built in defect that at a first glance makes them even more interesting for magnetic bearings. If the magnet is placed close to the surface before the superconductor has been cooled down to its superconducting state, not all the flux is expelled when the temperature is decreased again. This method is called field quenching. Some of the field is "frozen" into the superconductor due to an effect called flux pinning. This frozen field stabilizes the magnet in both radial and axial direction. If the surface is turned upside down the magnet does not fall down.

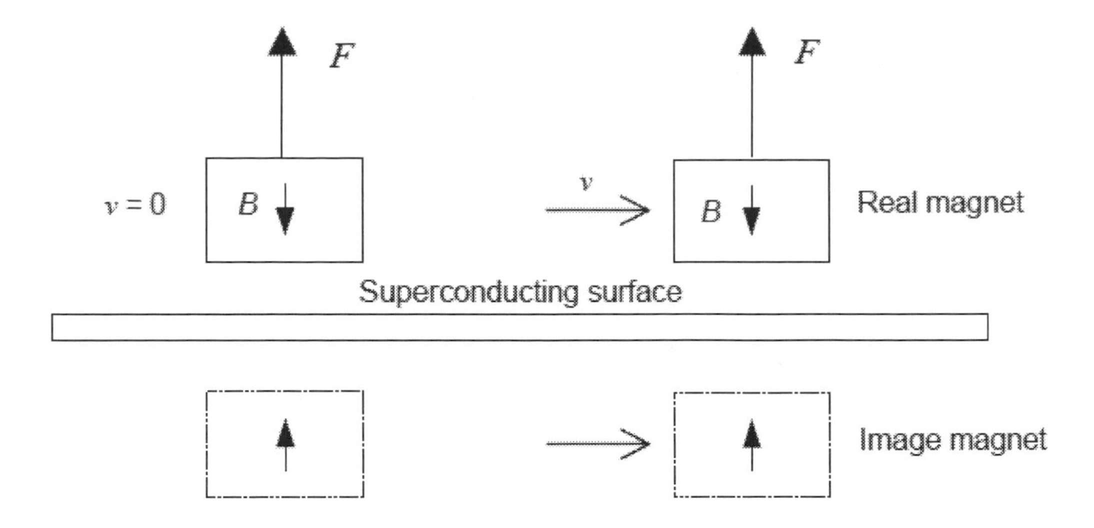

Fig. 2.22. Repulsive force due to the Meissner effect

In 1990 , this effect was evaluated for use in magnetic bearings, but it was found that the long time stability of such bearings was limited. It was then predicted that when used in a rotating machine exposed to vibrations, then according to the hysteresis loop caused by flux pinning the bearing would slowly move in the direction of the static load and eventually touchdown. Later research on superconducting bearings at Argonne national lab and many other places have proved this to be true. This is regarded to be a major drawback for this bearing technology.

The second effect defining a superconductor is the zero resistance. This effect has been used to make strong electromagnets for magnetically levitated test trains. In this case the bearing principle is not the superconducting one, but the electrodynamic one. The superconductors are just used to replace the permanent magnets that would otherwise have been used. In this case the track is made of aluminium, not superconductors.

It is theoretically possible, but hardly practical, to use the superconductor as the conducting member in an electrodynamic bearing. Due to bad mechanical properties and due to large hysteresis losses this may be found not to work properly. As is nowadays well known, the AC losses in a superconductor are quite large.

3

Magnetic Bearing Analysis

3.1. Introduction and Steps of Analysis

In the following sections analytical expressions for bearing forces and losses, as well as stiffness- and damping coefficients are presented .The equations of motion will be derived, in order to study system dynamics and stability. The analysis is applicable to bearings with at least two rows of magnets, and to bearings with/without pole shoes.

The analysis of the bearing is more like the analysis of an electrical machine. Thus an important goal of this analysis is to convert the electromagnetic properties of the bearing into useful mechanical and rotor dynamic data as stated in table 3.1.

Table 3.1. List of electrical bearing properties and their related mechanical ones.

Electrical properties	Mechanical properties
Induced voltage	Speed, rotational and lateral
Current	Force
Current space gradient	Stiffness, damping
Resistance	Losses
Phase lag	Force angle, cross coupling
Demagnetization	Lifetime

The analysis will be performed according to the following steps:

1. First the magnetic circuit is investigated, without any rotating electric conductor, to find an expression for the space gradient of the flux density, $\partial B_r/\partial \rho$.
2. Then the conducting cylinder, to which one rotating and one translating coordinate system are fixed, is introduced into the airgap, and the flux density time derivative in the rotating frame, $\partial \tilde{B}_r/ \partial t$, is calculated around the rotor surface.

3. Now the shapes of the eddy current paths have to be found. From the flux density change, the induced voltage $\tilde{E}(t)$ around an arbitrary eddy current path of area A can be calculated by integration as

$$\tilde{E}(t) = \frac{\partial \tilde{\phi}_r}{\partial t} = \iint_A \frac{\partial \tilde{B}_r(t)}{\partial t} dA \quad \text{(A)}$$

   Finding the areas A that maximizes the amplitude of the induced voltage, to a great extent facilitates the search for the main eddy current circuits. However, since an infinite number of paths are possible, some initial help from 3D-FEM simulations will help to find the main pattern.
4. Once the shape of the eddy current paths are known, the resistance R can be calculated. This also requires an estimation of the skin depth, which has to be analyzed separately.
5. When the shapes of the eddy current paths are known, the magnetic circuits and their corresponding reluctance models have to be found in order to calculate the inductance L.
6. Using the resistance and the inductance, the currents in the rotating frame can be found from the induced voltage $\tilde{E}(t)$ as

$$\tilde{\imath}(t) = \hat{\imath}\sin(\omega t - \varphi) = \frac{\tilde{E}(t)}{R + j\omega L} \quad \text{(B)}$$

7. The **currents** can now be transformed to the **non-rotating frame**, where they appear to be fixed in space and constant, for each given set of eccentricity and lateral velocity. Using the phasor notation one can write

$$\tilde{\imath}(t) = \vec{I}(\varphi) \quad \text{(C)}$$

8. The bearing force is achieved by calculating the **Lorenz force**, which can be found by integration in the non-rotating frame over the conducting cylinder volume V_{cyl},

$$\bar{F} = \iiint_{V_{cyl}} (\bar{J} \times \bar{B}) \, dV \quad \text{(D)}$$

9. The bearing losses equal the Ohmic losses, which can be calculated from the currents and the resistance.

10. From the bearing force derivatives, the stiffness and damping coefficients can be derived.
11. A stability analysis can now be performed using the coefficients above.
12. Finally, the equations of motion have been formulated in a general way so that engineers can simulate the dynamic behavior of their own different applications.

The above steps of analysis are covered in further details in the following sections.

3.2. Magnetic Circuit

The magnetic circuit from the permanent magnet rings is illustrated in Fig. 3.1, where the stator magnets and pole shoes are shown for an inner rotor bearing. The rotor, which is non-magnetic, is not shown. Since the bearing is homopolar, it is convenient to introduce cylindrical coordinates. In Fig. 3.2 the radial flux component is plotted schematically versus cylinder coordinate ρ_0 at $z = 0$, where the subscript 0 indicates that it is a stator fixed coordinate system.

The axial magnetic flux density B from the magnets is concentrated in the pole shoes and leaves the pole shoe inner and outer surface in radial direction. To avoid too much leakage on the outside, the diameter of the iron pole shoe can be reduced as in Fig. 3.1. For the analysis the radial flux B_r will be defined positive in positive ρ_0 - direction. The analysis will first be performed for an inner rotor bearing, and then for the other two types of bearings.

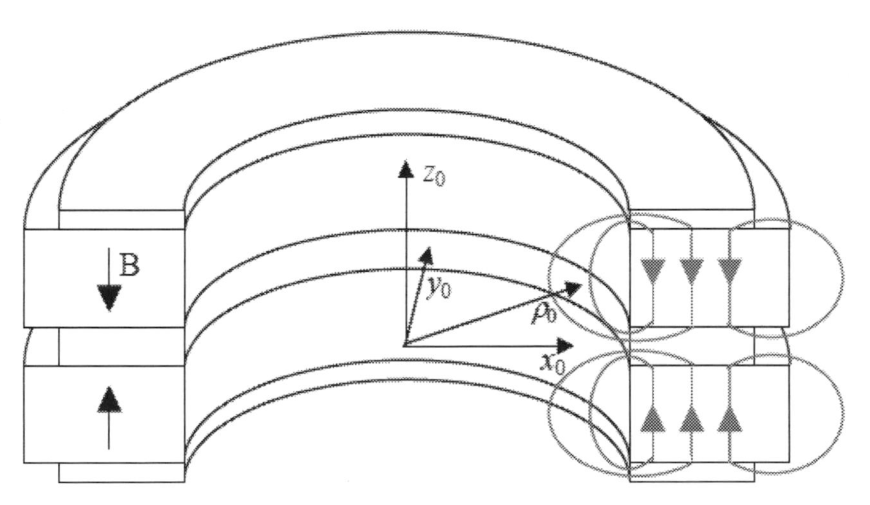

Figure 3.1 Magnetic Circuit

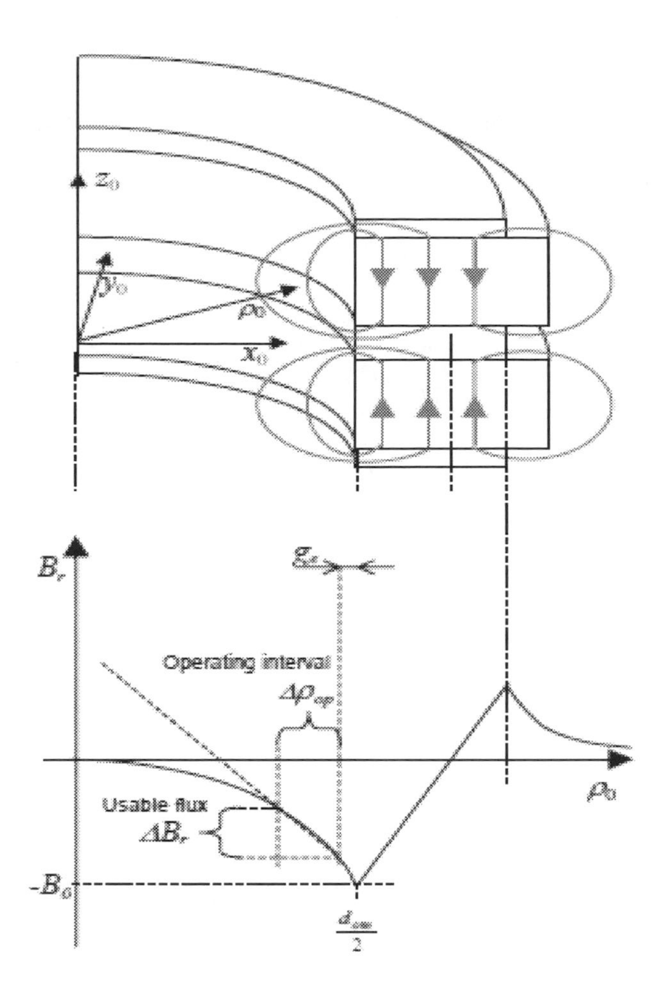

Figure 3.2. Radial flux density distribution. Inner rotor bearing

3.2.1 Magnetic circuit of an inner rotor bearing

Consider Fig. 3.2, the radial air gap flux density B_r has its maximum absolute value $B_0 = |B_r|$ at the inner surface of the pole shoe at $\rho_0 = d_{om}/2$, and it approaches zero towards the center of the stator, where it finally vanishes.

Since the cylinder has a limited wall thickness t_r we can only use a limited volume of the magnetic flux. The operating interval $\Delta\rho_{0p}$ is defined by

$$\Delta\rho_{0p} = t_r + 2\, g_a \qquad (3.1)$$

where g_a is the available airgap defined by

$$g_a = g - g_e \qquad (3.2)$$

Here g_e is the emergency airgap required by the emergency bearing and g *is the* total airgap. Normally the emergency bearing clearance is set to half the airgap, thus Eq. (3.1) becomes

$$\Delta\rho_{0p} = t_r + 2\,g$$

One of the most important quantities when calculating eddy currents is the radial flux gradient $\delta B_r/\delta\rho_0$.

Since $\Delta\rho_{0p}$ is small compared to the inner diameter of the magnet, d_{0m}, we will assume that the flux gradient is linear in this region. Thus the expression becomes

$$\frac{\delta B_r(\rho_0)}{\delta\rho_0} \approx \frac{\Delta B_r(\rho_0)}{\Delta\rho_{0p}} = \frac{B_r\left(\frac{d_{om}}{2}-\Delta\rho_{0p}-g_e\right)-B_r\left(\frac{d_{om}}{2}-g_e\right)}{t_r + 2\,g_a} \qquad (3.3)$$

3.2.2 Magnetic circuit of an outer rotor bearing

The analysis of an outer rotor bearing is basically identical to an inner rotor bearing, with the difference that the rotor is on the outside of the magnet, and that the flux in that region is positive, given the magnets having the same orientation. There is another difference as well. The leakage flux this time occurs at the inner diameter of the magnet. Thus the utilization of the total flux, and thus also of the magnets, is better in the outer rotor case.

However, since the value of the flux density, $B_r\ (\rho_0)$, is either measured or taken from the FEM software rather than from a reluctance model, there is no real difference in the analytical part between the bearing types. Thus the flux density change becomes

$$\frac{\delta B_r(\rho_0)}{\delta\rho_0} = \frac{\Delta B_r(\rho_0)}{\Delta\rho_{0p}} = \frac{B_r\left(\frac{D_{im}}{2}+g_e\right) + B_r\left(\frac{D_{im}}{2}+\Delta\rho_{0p}+g_e\right)}{t_r + 2\,g_a} \qquad (3.4)$$

Concerning leakage flux it is worth noting that it is possible to eliminate leakage flux almost completely, using radial magnets. But such magnets are still considerably more expensive than axially magnetized ones, and their use is not yet to be regarded as a viable solution for mass production.

3.2.3 Magnetic circuit of an intermediate rotor bearing

For the intermediate bearing described in Fig. 3.3, any of Eq. (3.3) and (3.4) can be used, since an intermediate rotor bearing can be said to consist of both an outer and an inner rotor bearing. However, due to symmetry it is possible to simplify the analysis for the case of the intermediate rotor bearing.

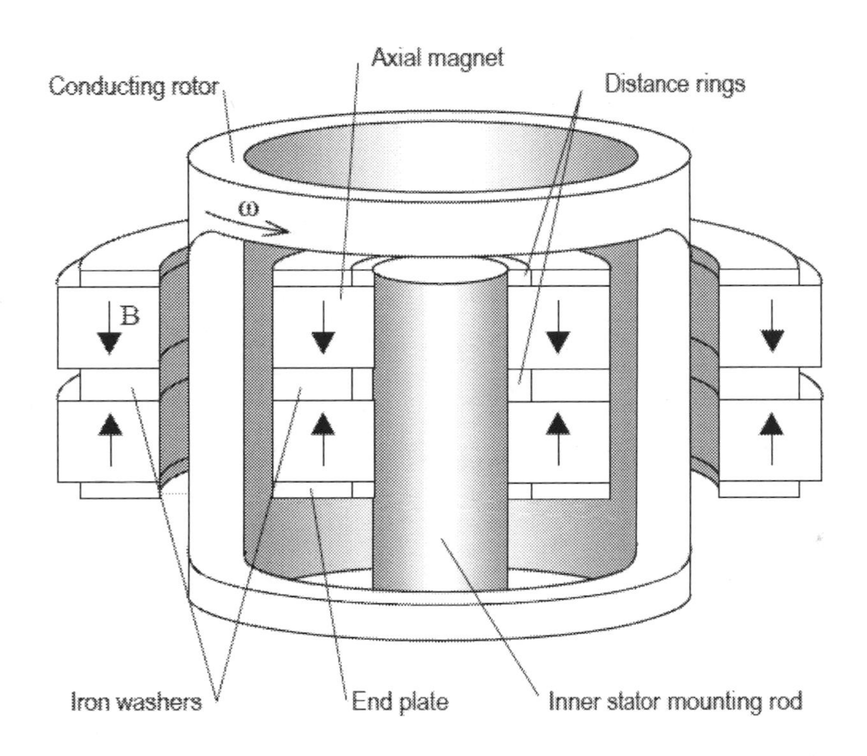

Figure 3.3 Description of the intermediate rotor bearing

Consider Fig. 3.4 where the radial flux density of the bearing is shown. The inner and outer magnets help each other to linearize the flux in the whole airgap. The flux gradient can be derived directly from Fig. 3.4, and is found to be

$$\frac{\delta B_r(\rho_0)}{\delta \rho_0} = \frac{\Delta B_r(\rho_0)}{\Delta \rho_0} = \frac{-2B_0}{t_r + 2g} \tag{3.5}$$

where the total airgap g has been used instead of the available airgap g_a used in Eq. (3.3) and (3.4). (Whenever in doubt though, on whether the flux really is linear or not, the latter equations should be used).

The gradient is higher for this bearing than for the other bearings, since the null radial flux limit lies approximately in the middle between the magnets.

However, it should be noted that for the intermediate bearing, B_0 is lower than for the other two bearing types, because the repulsive poles weaken the flux to some extent, so the flux gradient is not double as high as one might think.

Until now the radial flux and its gradients have been derived. They are of importance in determining the induced voltages. The axial flux plays a different role, and is of importance to the bearing forces. To generalize, the flux is axial in the magnets, and in the airgap besides the magnets. In the intermediate bearing the flux lines are compressed between the magnets so that the axial flux is basically parallel to the magnets, but for the inner and outer rotor bearings, the flux more resembles the shape of half circles, as shown in Fig. 3.2.

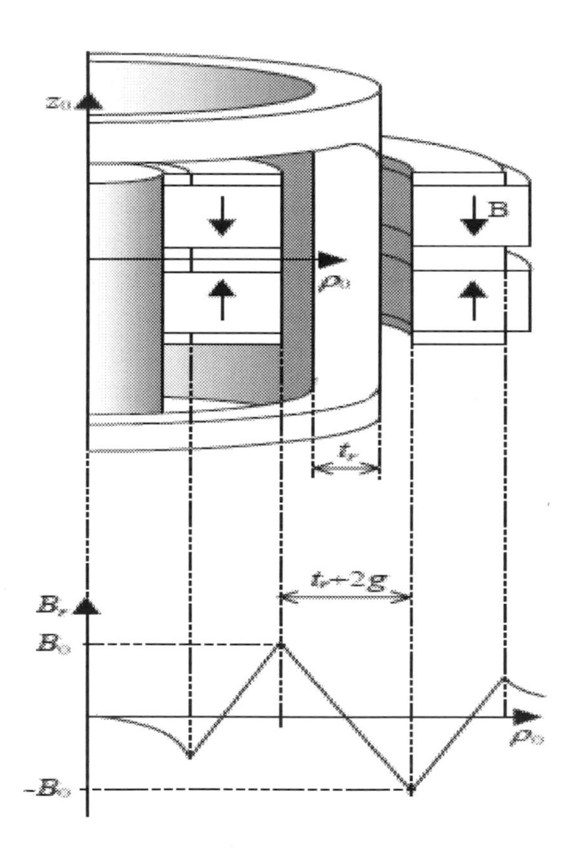

Figure 3.4. Radial flux density. Intermediate rotor bearing

3.3. Eddy Current Circuits

Previously the magnetic circuit has been analyzed with special emphasis on the radial flux gradients in the airgap. The conducting rotor will now be introduced into the airgap, and the eddy current paths will be identified and analyzed. As already stated, eddy currents are induced in a conductor only when an arbitrary conductor area experiences a change of flux. Thus for a conductor rotating or moving in a homopolar stationary field, a necessary but not sufficient condition for such a change of flux to exist, is that at least one of the following sub conditions are met:

1. The conductor spins about an axis, which is not concentric with the center of the homopolar flux.
2. The conductor moves about any of its remaining five degrees of freedom.
3. The magnets experience various kinds of mechanical or magnetic in-homogeneities. However , In reality, perfectly homopolar magnets cannot be manufactured.

Sub condition 1 and 2 may interact in such a way as to produce no net change of flux. In the following paragraphs, eddy currents resulting from these flux changes will be analyzed in detail.

3.3.1 Induced eddy currents

In this section eddy currents induced by eccentric operation will be derived. These currents are related to the bearing load, and are responsible for bearing properties like stiffness and cross coupling stiffness.

For the analysis it will be assumed that the center of the homopolar flux coincides with the bearing center, and that the lateral motion of the rotor is zero. Further, the magnets will be assumed to be homogenous and perfectly circular.

A perspective view is given in Fig. 3.5. In order to improve the viewing angle the coordinate system has been turned so that the y-axis points to the left. Accordingly, the rotor is displaced to the right.

Since the shape of the eddy currents by no means is obvious, we shall analyze them a bit further. Before doing so we need to introduce a coordinate frame $x_1y_1z_1$ translated in the xy plane by the vector $\Delta\bar{r}_0$, Fig. 3.5. We also need a rotating coordinate frame $x_ry_rz_1$ fixed to the rotor in which we will calculate the eddy currents. The angle between x_1 and x_r is ωt. For convenience we introduce a second cylindrical coordinate frame $\rho\phi z_1$ where ϕ is defined as the positive angle between x_1 and $\bar{\rho}$.

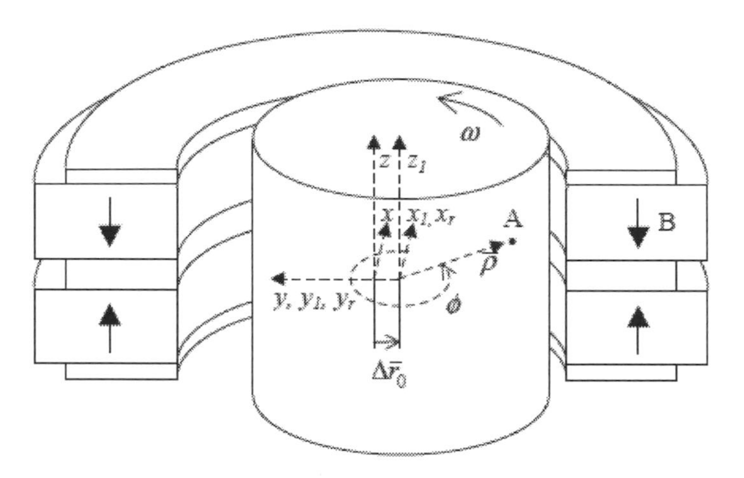

Fig. 3.5. Definition of coordinate systems at time ωt=0

Let's first study the case when the rotor is centered, as in Fig. 3.6a. On the conducting cylinder surface there are no points that can see any change in flux density as it rotates in the homogenous magnetic field. Thus according to Lenz's law no voltage is induced in the conductor anywhere. Consequently no eddy current path exists over which there is any induced EMF, and so no currents will be induced. This is true regardless of rotor shape. It does not need to be circular, but in the analysis it will be assumed to be so in order to find simple expression for the resistance.

In Fig. 3.6b the rotor is displaced downwards in the negative y-direction and the corresponding eddy currents are shown in Fig. 3.6d.

Now let us focus on point A in Fig. 3.6c, which is spinning around axis z_1 at speed ω. When the rotor is not centered, point **A** on the rotor surface will experience not only the homopolar flux, but also a superposed diametrical two-pole flux. Point A will thus see an almost sinusoidal flux with a constant bias.

To find the eddy current amplitude and phase we need to know the time derivatives of the normal flux penetrating the surface. Let us first make a few simplifications concerning the eccentric motion of point **A**.

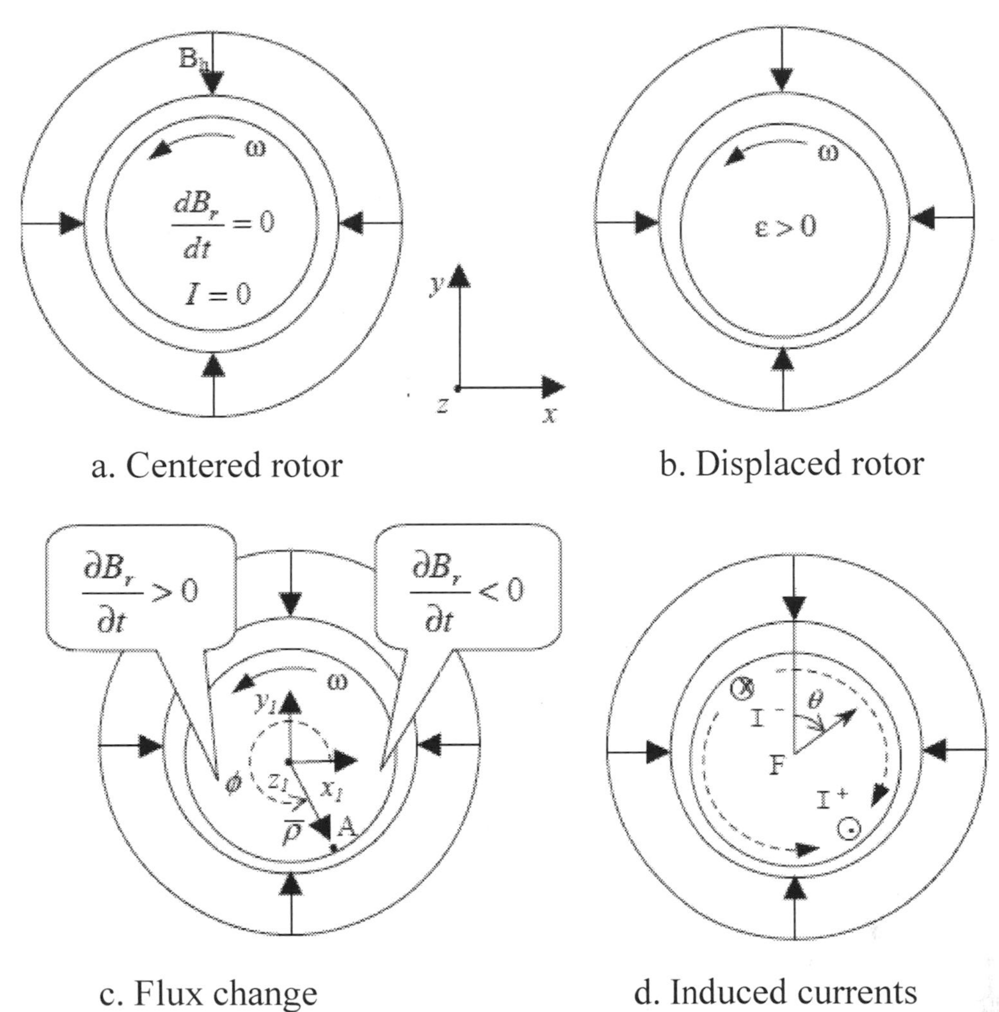

Fig.3.6 Flux and eddy current distributions with rotor eccentricity

Assume a position for point **A** so that $\phi = 0$ at time $t = 0$. Also assume a certain displacement Δr_0 in the negative y- direction. At this instance of time the position of A expressed in the stator coordinate system xyz is $(\rho, -\Delta r_0, 0)$.

Since $|\Delta \bar{r_0}| \ll |\bar{\rho}|$ we can assume that $|\bar{r}| \approx |\bar{\rho}|$ for $\omega t = 0$ and $\omega t = \pi$.

With the simplification above we can write the radial position for point A as it spins around axis z_1 as

$$|\bar{r}| \approx |\bar{\rho}| - |\Delta \bar{r_0}| \sin(\omega t) \qquad (3.6)$$

Or

$$r \approx \rho - \Delta r_0 \sin(\omega t) \tag{3.7}$$

Now the time derivative in point A of the flux from an intermediate bearing, using Eq. (3.5) and (3.7) is

$$\frac{\partial B_r}{\partial t} = \frac{\partial B_r}{\partial r} \cdot \frac{\partial r}{\partial t} \approx \frac{-2B_0}{t_r + 2g} \cdot \frac{\partial(\rho - \Delta r_0 \sin(\omega t))}{\partial t}$$

$$= \omega\, \Delta r_0 \frac{-2B_0}{t_r + 2g} \cos(\omega t) \tag{3.8}$$

For a given time t_0 the radial speed $\partial r/\partial t$ varies sinusoidally around the rotor perimeter of the cylinder, so an expression valid for all points around the rotor has the form

$$\frac{\partial B_r}{\partial t} \approx \omega\, \Delta r_0 \frac{-2B_0}{t_r + 2g} \cos(\omega t_0 + \psi) \tag{3.9}$$

where ψ is the angle in the rotating coordinate frame between axis x_r and the point in question.

Fig. 3.7 shows the radial (normal) flux versus x_r around the rotor surface for a given time $t_0 = 0$.

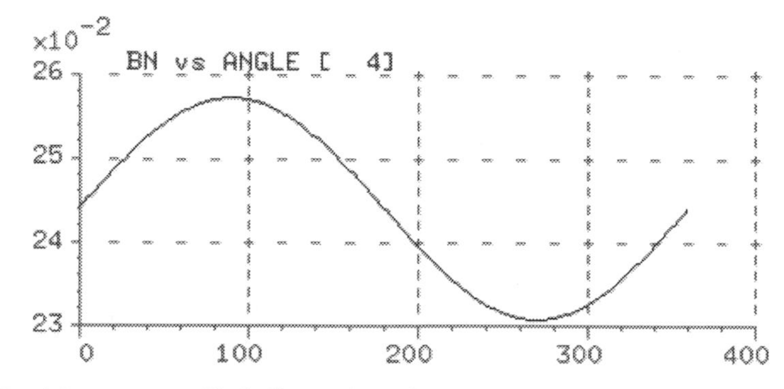

Fig. 3.7. Air gap radial flux density penetrating the rotor surface
(Calculation is based on using the 3D-FEM software Mega)

The normal flux B_n in Fig. 3.7 penetrating the inner rotor surface can be found by integrating Eq. 3.9. and setting the integration constant equal to the homopolar flux component, B_h, which in this case is 244 mT. Thus

$$B_n = B_r = \int \frac{\partial B_r(t)}{\partial t}\, dt = \Delta r_0 \frac{-2B_0}{t_r + 2\,g} \cos(\omega t_0 + \psi) + B_h$$

Where the homopolar flux component on the surface of the rotor is given by

$$|B_h| = B_0\,(1 - \frac{-2g}{t_r + 2\,g})$$

where the sign is positive for the inner surface of the rotor, and negative for the outer. For the induced voltage only the time derivative in (3.9) is important. The flux itself is significant only for the generation of forces, and then particularly the axial flux, not so much the radial one.

The induction bearing may be compared to the induction generator due to their obvious geometrical similarities. Thus the induction bearing rotor will behave as a 2-pole induction rotor made up of an infinite number of bars (See figures 3.2 and 3.3) with a dc-current applied to the stator winding. For simplicity, the calculation of the axial eddy current i_z (rotor bar currents) can be performed assuming that the current paths is concentrated to four rotor bars, (two for each eddy current circuit,) connected at each end to short circuit rings. Since the model is reduced to 4 bars, this simplification means that the axial bar currents $i_z$equals the tangential short circuit current i_{sc}, which will also be denoted i_t. Thus

$$i = i_z = i_{sc}$$

Generally, the current paths, specially the bar currents, are more rounded than is possible in a squirrel cage. The rounded “real” current paths are shown in Fig. 3.8a,b, and in the analysis a compromise will be used, which is shown in detail in Fig.3.12a,b, in which the short circuit rings, region 1 and 2_s , and the bars, region 3, are straight, while the corners are rounded, region 2_c .

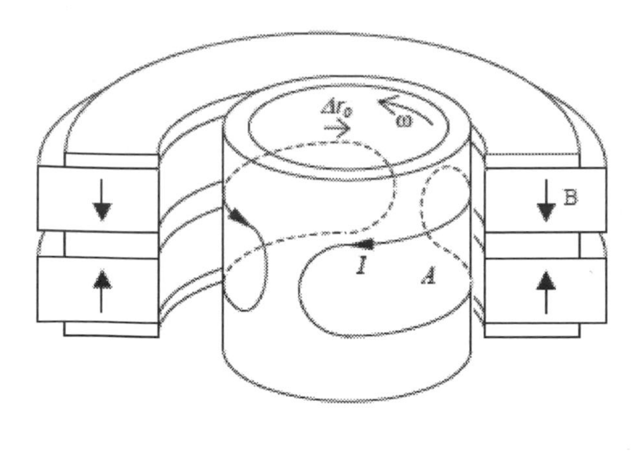

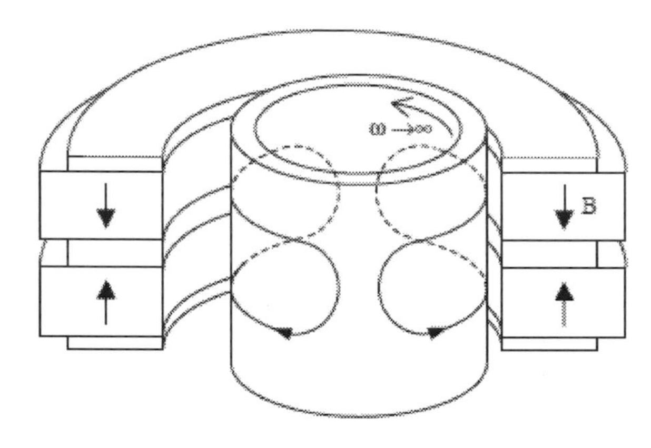

Fig. 3.8a,b. Restoring eddy current circuits

The eddy current paths encircle the area A, Fig. 3.8a and 3.9, and the current amplitude i depends on the flux change across this area as

$$i = \frac{E}{R+j\omega l} = \frac{\partial \phi_r/\partial t}{(R+j\omega l)} = \frac{1}{(R+j\omega l)} \iint_n \frac{\partial B_r}{\partial t}\, \mathrm{d}A \tag{3.10}$$

where E is the induced voltage, and R and L are the resistance and the inductance of the whole eddy current circuit. The influence on these variables will be analyzed one by one in the following subchapters.

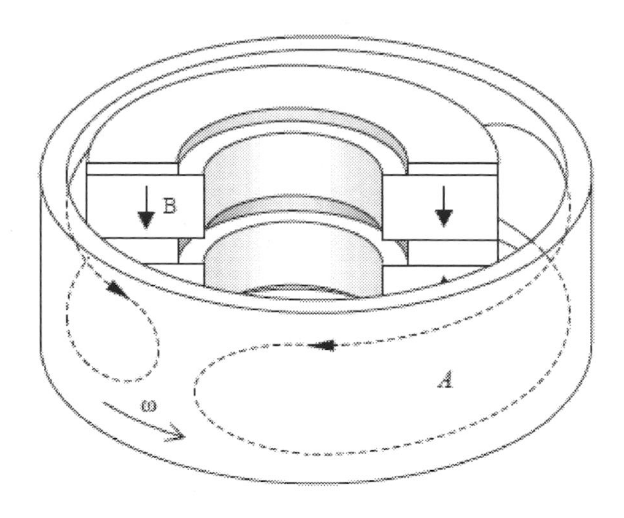

Fig. 3.9. Eddy currents in a flywheel bearing

3.3.2 Induced voltage

To calculate the induced voltage, the eddy current path has to be defined, in terms of variables. The variables are defined in Fig. 3.10, where a 4-row intermediate rotor bearing with three identical eddy current paths is shown. The rotor is rotating and is displaced in the negative y-direction, giving rise to induced voltages and currents. To the right the short-circuit currents and bar currents are displayed. To the left the four bars in the model are displayed for two different positions. The black bars show the position for maximum induced voltage, and the bars in gray show the same bars in the position for maximum current. Since the voltage is to be studied, the focus will be on the black bars and the area *A*.

The area *A* is defined inside each circuit, and it is limited in axial direction by the upper and lower integration variables z_u and z_l, shown to the right, and in tangential direction by the upper and lower angular integration variables ψ_u and ψ_l. The latter are shown to the left.

Using the expression from Eq. (3.9) for the flux density change, the integral for the time t_0 is found to be

$$E = \iint_A \frac{\partial B_r}{\partial t}\, dA = \int_{z_l}^{z_u}\int_{\psi_l}^{\psi_u} \frac{\partial B_r}{\partial t}\, rd\,\psi dz = 2l_w \int_{\psi_l}^{\psi_u} \frac{\partial B_r}{\partial t}\, rd\,\psi$$
$$= 2\omega\Delta r_0 \frac{2l_w}{t_r + 2g} \int_{\psi_l}^{\psi_u} B_0 \cos(\omega t_0 + \psi)\, rd\,\psi \qquad (3.11)$$

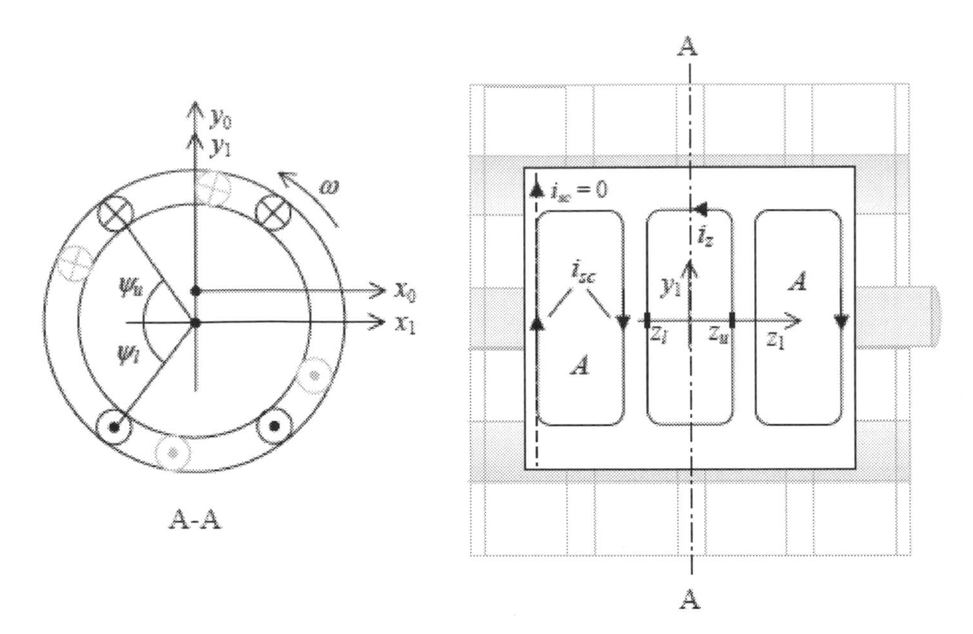

Fig. 3.10. Definition of integration limits for the calculation of flux change within the area A. The black currents to the left shows the bar position for maximum voltage, and the grey currents show the position for maximum current

where the integration boundaries comes from the eddy current path shown in Fig. 3.10. Observe that the result from the integration of the z-coordinate equals the pole width $2l_w$, which is less than the integration boundaries $z_u - z_l$ since there is no radial flux outside the pole.

The upper and lower angular integration limits ψ_u and $\psi_l$depend on the width of the axial eddy current path, the rotor bars. The latter is investigated in the next section, and is illustrated in Fig. 3.12a,b, in which the bars are referred to as "region 3" and the width of the bars is denoted w_3. Using *w3* the integration limits ψ_u and ψ_l can be derived from Fig. 3.10 left. The width w_3 [m] expressed in angular coordinates is w_3/r [rad]. Thus

$$\psi_u = \frac{\pi}{2} - \frac{w_3/2}{r} = \frac{\pi}{2} - \frac{r/\pi}{r} = \frac{\pi}{2} - \frac{1}{\pi}, \text{ and}$$

$$\psi_l = -\frac{\pi}{2} + \frac{w_3/2}{r} = -\frac{\pi}{2} + \frac{r/\pi}{r} = -\frac{\pi}{2} + \frac{1}{\pi}$$

the integral in Eq. (3.11) has its maximum value for $t_0 = 0$, so evaluating the integral for $t_0 = 0$, using the integration limits above results in the amplitude of the induced voltage

$$\hat{E} = \omega \Delta r_0 \frac{8 l_w r B_0 \cos(\frac{1}{\pi})}{t_r + 2g} \quad (3.12)$$

Inserting it into Eq. (3.10) results in the current amplitude

$$\hat{\imath}_z = \hat{\imath}_{z,e} = \omega \Delta r_0 \frac{8 l_w r B_0 \cos(\frac{1}{\pi})}{(t_r + 2g)\sqrt{R^2 + (\omega l)^2}} \quad (3.13)$$

In literature on rotor dynamics the radial displacement Δr_0 is often denoted *e*. It is often expressed in terms of eccentricity $\mathcal{E}$, which is a dimensionless unit related to the airgap defined by

$$\mathcal{E} = \frac{e}{g}$$

Expressing Eq. (3.13) in terms of eccentricity results in the final expression for the current amplitude:

$$\hat{\imath}_z = \hat{\imath}_t = \omega \mathcal{E} g \frac{8 l_w r B_0 \cos(\frac{1}{\pi})}{(t_r + 2g)\sqrt{R^2 + (\omega L)^2}} \quad (3.14)$$

The axial current i_z turns direction into a tangential current i_t at the cylinder ends, or, if many magnets are used, at each symmetry line B-B, see Fig. 3.12 right. The tangential current in the bearing cylinder corresponds to the current in the short circuit ring in an induction generator.
The phase shift φ , and the force angle θ are then

$$\varphi = \arctan \frac{\omega L}{R} \quad \text{, and} \quad (3.15)$$

$$\theta = \arctan \frac{R}{\omega L} \quad (3.16)$$

Equation (3.14) represents a current that initially increases linearly with frequency ω and then asymptotically reaches its maximum value.

The force angle in Eq. (3.16) starts from π/2 at low frequency, which then asymptotically reaches zero at high speed. This explains the angle θ in Fig. 3.6d.

An attempt to illustrate the eddy currents is made in Fig. 3.8-3.9. Fig. 3.8a and 3.9 shows the two main eddy current circuits at low speed, and Fig. 3.8b shows the theoretical current distribution at infinitely high speed.

The pole area A, which is used in Eq. 3.10 to 3.11, refers to the whole area which is enclosed by the eddy currents. But the integral in Eq. (3.11) only refers to the part of the area, effective area A_{eff}, where the flux is radial and contributing to the induced voltage.

When pole shoes are used, the simple model assuming that all radial flux goes through the pole shoe, gives accurate results. The currents will basically follow the geometry of the magnets, which makes it easy to estimate the shape of the current path, however, one should consider that in reality the current distribution is not discrete as illustrated in Fig. 3.8 and 3.9, but sinusoidally distributed, and the shape is rounded or elliptic. In order to avoid elliptical integrals, and focus more on the understanding of the bearing, a discrete and somewhat simplified current path will be used, which is shown in Fig. 3.12 and 3.13. An expression for the effective pole area A_{eff} based on this current path is

$$A_{eff} \approx 2l_w\left(\frac{\pi\, d_{rc}}{2} - w_3\right) \qquad (3.17)$$

where $2l_w$ is the pole width and $w3$ is defined in Eq. (3.18). Finally d_{rc} is the diameter of the rotor cylinder centerline, defined as

$$d_{rc} = \frac{(D_r + d_r)}{2}$$

Equation (3.17) is valid for inner, outer as well as intermediate bearings, all with iron poles.

This expression for the area is the same for all pole shoes, regardless if they belong to a 2-row bearing or to the additional pole shoe inserted when more magnets are added. For the end plate, or end pole shoe, the pole width is normally the half, that is l_w instead of $2l_w$, but this is of less interest since the end poles are not enclosed by any currents, unless the rotor is very long. Now consider a bearing without pole shoes. The magnets are mounted with the poles directly opposing one another, as in Fig.3.11a,b. Since there are no pole shoes, the area *A* in which the flux is normal is not that well defined, and the direction of the flux lines have to be carefully studied. There exists no region without radial flux, so there will be no difference between the total area *A* and the effective area A_{eff}. There is also a more pronounced difference between the area in the 2-row bearing, Fig. 3.11a, and the area for the additional row, Fig. 3.11b, than for the case when pole shoes are used. An estimation would be to exchange *lw* in Eq. (3.17) for $l_m/2$ in the 2-row case and $l_m//4$ in the latter case.

For the additional row in Fig. 3.11b this approximation is well motivated by the fact that by symmetry the flux lines change direction from upward to downward over the length of one magnet. The flux is parallel with the magnets when crossing symmetry line B-B. Thus there is a change of angle with 180 degrees over the length l_m. If the "pole" is now defined as the area where the flux angle is less than 45 degrees, then the width of this area is $2.\,l_m/4$ which should be compared to the pole width $2l_w$ in Eq. (3.17).

In the 2-row case, Fig. 3.11a, there is only 1 symmetry line A-A, thus allowing the flux lines to expand to the sides, forming a magnetic 4-pole. This allows a wider pole area *A*. A rough estimation, at least for quadratic magnet cross sections, is to set the pole width to $2.\,l_m/2$, since this will result in four equal pole areas.

Now having determined the shape and the area of the eddy current circuits, expressions for the resistance *R* and the inductance *L* can finally be found. In order to find a formula for the resistance it will also be necessary to study the influence of skin effect (See section 3.3.3 Circuit Elements).

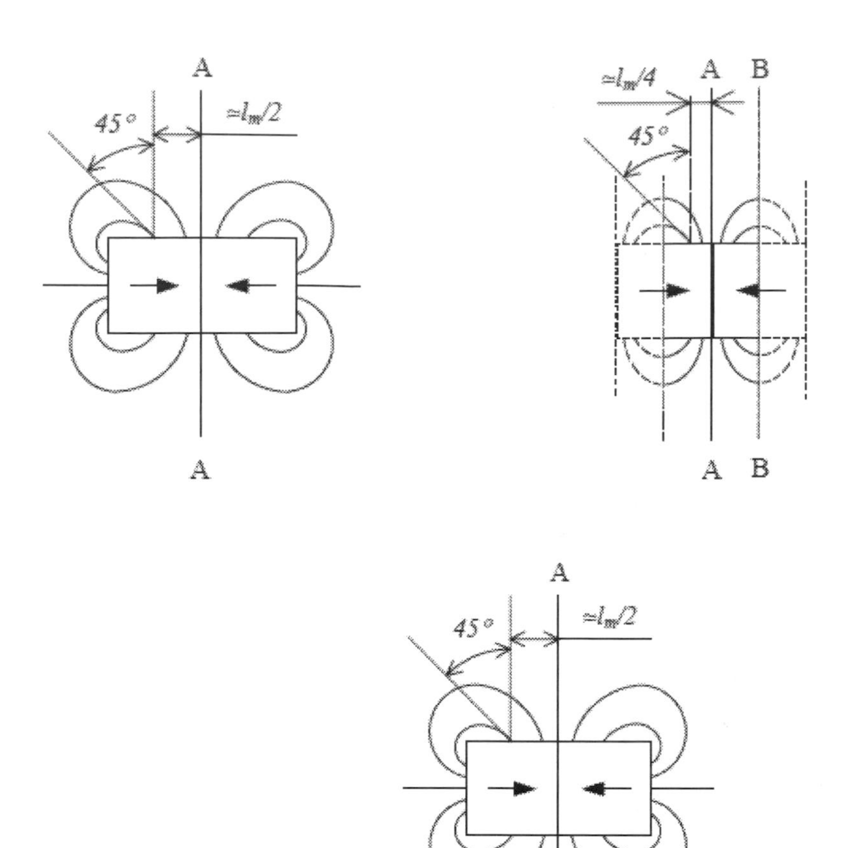

Fig. 3.11a. Expanding flux lines in a 2-row bearing without pole shoe

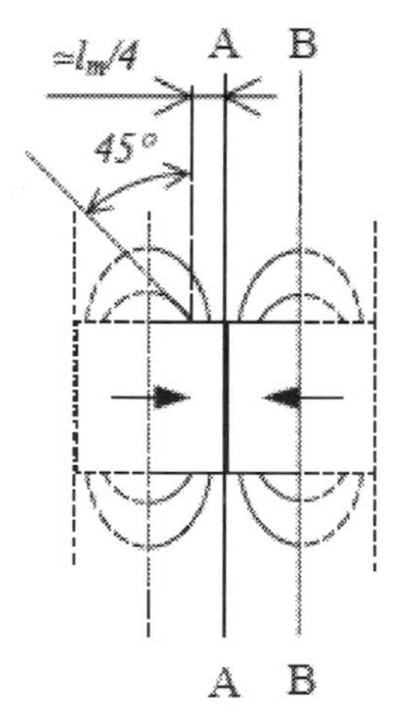

Fig. 3.11b. Compressed flux lines for an additional magnet row without pole shoes

3.3.3 Circuit elements

A. Resistance

The resistance R of the eddy current circuits can be derived using the simplified model shown in Fig. 3.12. One has to bear in mind that in reality the currents are not bound to these discrete paths. But at this stage of the development phase it is more appropriate to use this simple model which gives a better understanding of the physics than would the typical series expansions resulting from a solution to the diffusion equation with proper boundaries. The model is derived for two cases:

- the 2-row bearing with pole shoes, shown in Fig. 3.12a, and
- the case for additional rows including pole shoes, Fig. 3.12b.

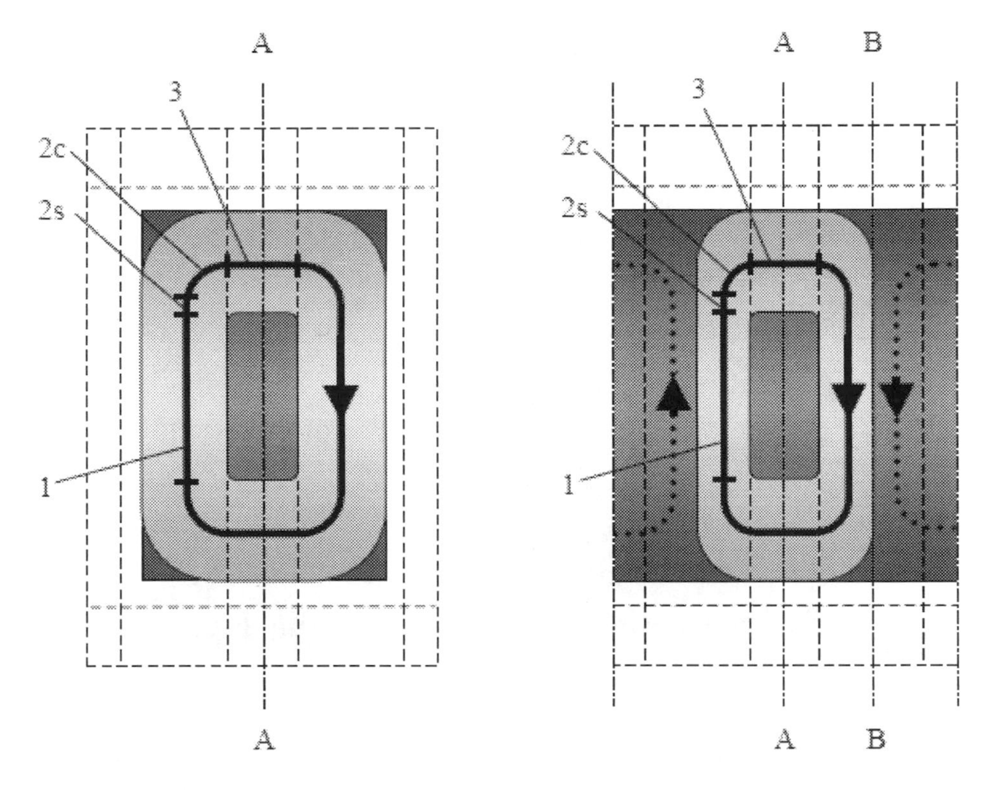

Fig. 3.12a,b. Eddy current path in a 2-row bearing (left) and due to an additional row (right). Both bearings have pole shoes. The short circuit paths (1) are connected both to each other forming a full short circuit ring (the other half not shown), and to the bars (3) via the rounded corner (2c).

Each of these cases can be recalculated for corresponding bearings without pole shoes by modifying the pole width. Consider again Fig. 3.12a. The conducting cylinder, is somewhat shorter than the stator. The latter has dashed contours and consists of magnets and pole shoes. On the cylinder one of the eddy currents is shown in gray. This current blocks the view of the other current, which is located on the reverse side of the cylinder.

In Fig. 3.12b a part of an infinitely long bearing is shown, in order to study the current induced due to one additional magnet and pole shoe. In this case there is not only the symmetry line A-A, but also the symmetry line B-B, which narrows the available space for the eddy currents. Thus the resistance in this case will be higher than for the circuit in Fig. 3.12a.

The symmetry line B-B exists only in infinitely long bearings. However, in order to reduce the number of possible cases in the model, a bearing with any number of rows will be regarded as consisting of only these two types of currents. This approximation will be best for either 2-row bearings or bearings with many rows. The largest deviation will be found in 3-row bearings.

Now consider Fig. 3.13, which shows a linearized model of the eddy currents in Fig. 3.12, omitting some corner effects. The same linear model can be used for both types of currents, just taking into consideration that the width of region 1, w_1, is different for the two cases, which also affects the length of region 2, l_2. The length of each current segment is denoted l_1 to l_3, and the widths are correspondingly denoted w_1 and w_3. w_2 is not used since it is not constant. Instead the average $(w_1 + w_3)/2$ is used. Region 2 will be divided into a straight part of length l_{2s} and a circular part of length l_{2c}. The lengths and widths are defined geometrically from Fig. 3.13 and are summed up in Eq. (3.18).

$$\begin{cases} l_1 = \pi r - 2w_3 \\ l_2 = l_{2s} + l_{2c} = \frac{w_3}{2} - r_2 + \frac{2\pi r_2}{4} \quad where\ r_2 = \frac{w_1}{2} \\ l_3 = 2 \\ w_1 \begin{cases} \frac{l_r}{2} - l_w\ for\ 2-row\ bearing \\ \frac{l_m}{2}\ for\ additional\ bearing \end{cases} \\ w_3 = \frac{2r}{\pi} \end{cases} \tag{3.18}$$

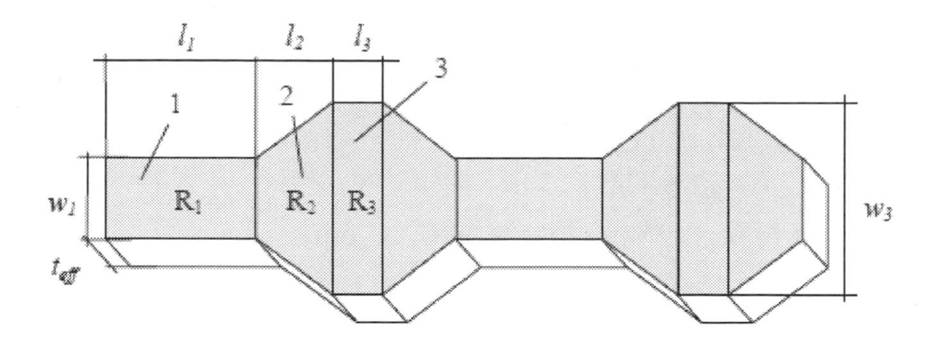

Fig. 3.13. Linear model of the eddy current path shown in Fig. 3.12a,b. Numbers refer to the different regions

In these formulas r refers to the radius of the center of the rotor cylinder,

$$r = r_{rc} = \frac{R_r + r_r}{2} \quad , \tag{3.19}$$

and l_2 is a quarter of the circumference of an ellipse. However, since an ellipse does not have any simple analytical expression for this length, an approximation is used consisting of a straight part l_{2s} and a circular part l_{2c} with the radius r_2. Of the same reason, the width w_2 is taken as an average as shown in the linearized model in Fig. 3.13.

The width w_3 is by no means obvious and requires an explanation. It could have been set to one quarter of the circumference of the rotor, which is true in that aspect that the whole area actually carries eddy currents. There exists no area inside the current loop as in Fig. 3.12a,b where the current is completely absent. However, in reality the current density is sinusoidally distributed, which means that the main part of the currents in region 3 passes through the upper and the lower part of Fig. 3.12, which will result in higher losses compared to the constant current distribution in the model. To require equal losses for both approaches is a way to define w_3. Setting the currents equal in both cases results in

$$I_z = \int_A j_z \, . dA$$

$$I_z = t_{eff} \int_0^{\pi/2} \hat{j}_z \sin \propto r d \propto$$

$$I_z = \hat{j}_z \, r \, t_{eff} \tag{3.20}$$

where ∝ is an integration angle which is zero on top of Fig. 3.12 and follows the circumference of the rotor. This angle is π/2 turned counterclockwise from the load angle θ. Also the losses shall be equal, thus

$$R_3 \, I_z{}^2 = \int_A \rho j_z{}^2 \, dV$$

$$\rho \frac{l_3}{t_{eff} \; w_3} I_z{}^2 = \rho \int_0^{\pi/2} (\hat{j}_z \sin \propto)^2 \, l_3 \, t_{eff} \, r d \propto \tag{3.21}$$

Solving, using Eq. (3.20), gives

$$w_3 = \frac{2r}{\pi} \tag{3.22}$$

The total resistance R of the linearized circuit can now be written as

$$R = 2\,(R_1 + 2R_2 + R_3) = 2\frac{\rho}{t_{eff}}\left(\frac{l_1}{w_1} + 4\frac{l_2}{w_1 + w_3} + \frac{l_3}{w_3}\right) \tag{3.23}$$

where t_{eff} is the effective thickness of the circuit due to the skin depth. Using the expression from Eq. (3.18), the resistance in each eddy current circuit of a 2-row bearing with pole shoes is found to be

$$R_{2-row} = \frac{\rho}{t_{eff}}\left(4\,\frac{r\left(\pi - \frac{4}{\pi}\right)}{l_r - 2l_w} + \frac{r\frac{8}{\pi} + (\,l_r - 2l_w)(\pi - 2)}{\frac{2r}{\pi} + \frac{l_r}{2} - l_w} + 2\pi\frac{l_w}{r}\right) \tag{3.24}$$

and the resistance in each additional circuit, which is created when more magnet rows are added to the bearing, to be

$$R_{add} = \frac{\rho}{t_{eff}}\left(4\,\frac{r\left(\pi - \frac{4}{\pi}\right)}{l_m} + \frac{r\frac{8}{\pi} + \,l_m\,(\pi - 2)}{\frac{2r}{\pi} + \frac{l_m}{2}} + 2\pi\frac{l_w}{r}\right) \tag{3.25}$$

B. Skin depth

In the previous subsection the effective thickness t_{eff} of the conducting cylinder has been used in the expressions for resistance. At low speed, the effective thickness equals the material thickness t_r of the rotor cylinder, but as speed increases the skin effect decreases the effective thickness of the circuit. An often-used definition of t_{eff} relates to the skin depth δ_{skin} as

$$t_{eff} = \begin{cases} t_r \,, & t_r \leq \delta_{skin} \\ \delta_{skin}, & t_r > \delta_{skin} \end{cases} \tag{3.26}$$

Where

$$\delta_{skin} = \sqrt{\frac{2}{\omega\sigma\mu_0}} \tag{3.27}$$

This relation is applicable to outer and inner rotor bearings. For the intermediate rotor bearing, having two airgaps with magnets on both sides of the cylinder, the effective thickness is accordingly

$$t_{eff} = \begin{cases} t_r \,, & t_r \leq 2\delta_{skin} \\ 2\delta_{skin} \,, & t_r > 2\delta_{skin} \end{cases} \tag{3.28}$$

The model above, which is often used in literature, has the clear disadvantage that the function derivative is discontinuous, leading to curves for forces and losses with discontinuous slope. A better estimation for the inner and outer rotor bearing is the gradually increasing t_{eff} according to the exponential function

$$t_{eff} = \delta_{skin}\,(1 - e^{-t_r/\delta_{skin}}) \tag{3.29}$$

and for the intermediate bearing the function is, of the same reason as above,

$$t_{eff} = 2\delta_{skin}\,(1 - e^{-t_r/2\delta_{skin}}) \tag{3.30}$$

Results of using these two different approximations (Eqs. (3.28) and (3.30) on bearing stiffness are discussed in section 3.6.1 Stiffness.

C. Inductance

The inductance L of the eddy current circuit is strongly dependent on whether iron pole shoes are used or not. This factor is sometimes neglected in literature on eddy current devises, probably due to the fact that the pole shoes do not to a great extent affect the air gap flux density, since the magnets are operating in repulsive mode. However, in generator theory it is well known that the magnetizing inductance and the armature inductance can vary a lot.

In the case under consideration, the eddy current circuits should be treated as the armature winding. Normally the phase angle between the armature current and the magnetizing current is of importance, but since the bearing is homopolar and the effective airgap is constant, this does not affect the value of the inductance.

Consider first the case when pole shoes are not used. If the relative permeability of the magnets is close to unity, then the influence of the magnets is negligible. Thus the inductance can be calculated as that from an air wound coil, and it will be the same for as well outer, inner and intermediate bearings.

By using the inductance formula for two infinitely long conductors , the inductance of the tangential parts of the current circuits is given as below:

$$L = l_t \frac{\mu_0}{\pi} \left(\frac{1}{2} + \ln\frac{l_0}{r_0}\right) \qquad (3.31)$$

where l_t is the length of the tangential sections of the coil, which in Fig. 3.12b are the vertical parts of the current paths. If the length $l_1 + 2l_2$ is projected on to the vertical symmetry line A-A the vertical length is found to be $l_1 + w_3$. The radius r_0 of the conductor is approximated to half the width w_1 , thus for the bearing without pole shoes $r_0 = l_m/8$. Finally the distance between the centers of the conductors l_0 is $3l_m/4$, which is found from the model in Fig.3.12b applied to the magnets without pole shoes in Fig. 3.11b.

Thus the inductance of one eddy current circuit from an additional magnet is

$$L_{add} = (l_1 + w_3)\frac{\mu_0}{\pi}\left(\frac{1}{2} + \ln\frac{\frac{3l_m}{4}}{\frac{l_m}{8}}\right)$$
$$= \frac{\mu_0}{\pi}(l_1 + w_3)\left(\frac{1}{2} + \ln 6\right) \quad (3.32)$$

Similarly, by applying the model in Fig. 3.12a to the magnets in Fig. 3.11a the corresponding value for the 2-row bearing with a rotor length l_r the inductance of one eddy current circuit in a 2-row bearing without pole shoes is

$$L_{2-row} = (l_1 + l_3)\frac{\mu_0}{\pi}\left(\frac{1}{2} + \ln\frac{l_r + l_m}{l_r - l_m}\right) \quad (3.33)$$

where the length $w_1 = (l_r - l_m)/2$, $l_0 = l_r - w_1$ and finally $r_0 = w_1/2$.

These relations can be found explicitly from Fig. 3.14a,b. Eq. 3.33 is valid for rotors approximately equally long as the two magnets, that is

$$l_r \approx (2 \pm 0.5)l_m \quad (3.34)$$

The important condition is that the flux has to be axial $\pm 45°$ within the circuit region 1 of width w_1.

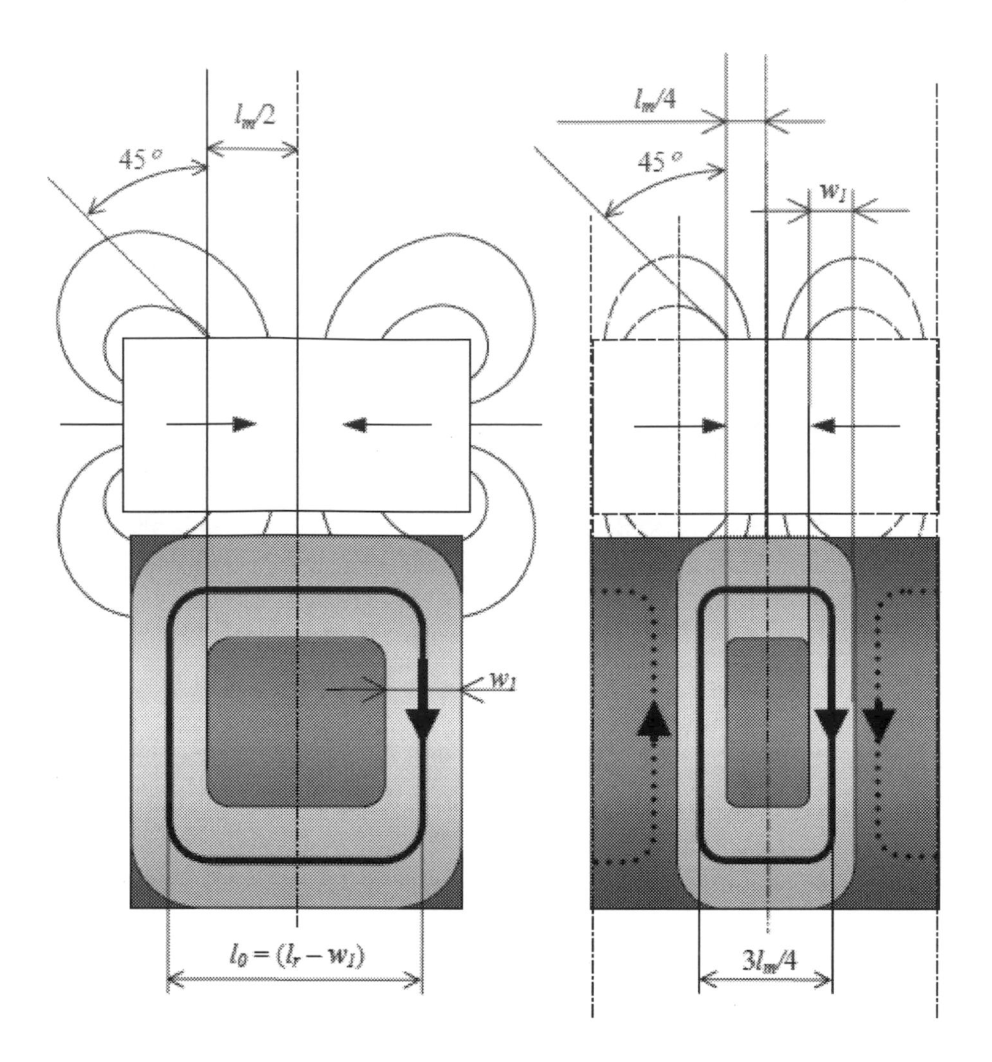

Fig. 3.14a,b. Eddy current paths in bearings without pole shoes. Compare with Fig. 3.12a,b showing bearings with pole shoes

If **pole shoes are used**, the inductance is dramatically increased. A 2D-model of the inductance can be established as follows:

Consider Fig. 3.15 above in which an intermediate bearing is shown. The magnetic flux (arrows) from the two opposite eddy current loops (currents perpendicular to the picture) is shown. The flux follows the iron outer disc, penetrates the effective airgap including the copper cylinder twice (cylinder not shown for clarity), and uses the inner iron pole as a return path.

Since it is a 2D model, it does not take leakage between adjacent poles in the axial direction into account, and there is also some stray inductance between adjacent iron poles that is neglected.

The inductance depends on the number of turns, *n*, and the reluctance R_m as

$$L = \frac{n^2}{R_m} = \frac{n^2}{R_{Fe,o}+R_{Fe,i}+2R_{g,eff}} \approx \frac{n^2}{2R_{g,eff}} \tag{3.35}$$

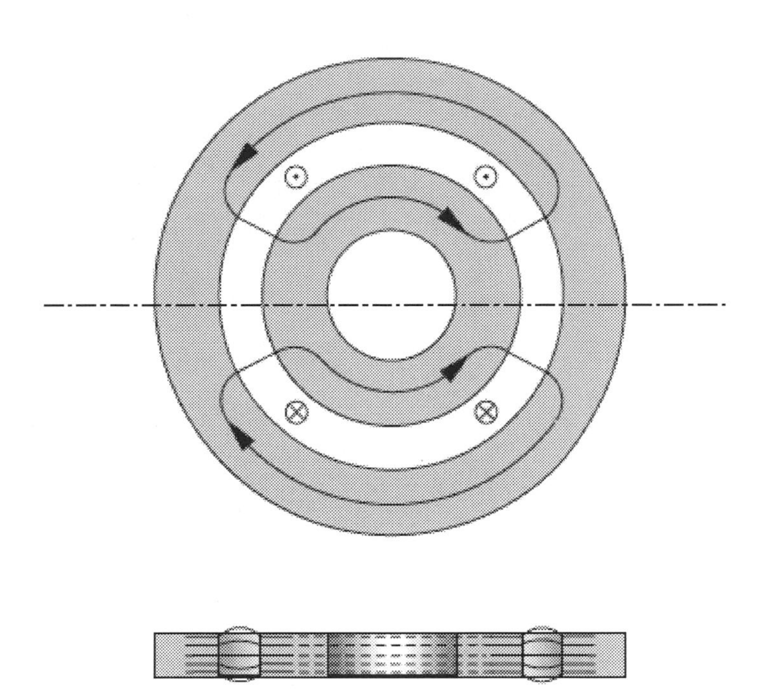

Fig. 3.15a,b. Enhanced 2D-model of the inductance. For clarity the copper cylinder is not shown. a) Main flux in one pair of pole shoes in an intermediate bearing. b) Fringing between poles

where the reluctance in the outer and inner pole shoes, $R_{Fe,o}$ and $R_{Fe,i}$, have been considered to be small in comparison with the effective airgap reluctance $R_{g,eff}$, which is

$$R_{g,eff} = \frac{l_{g,eff}}{\mu_0 A} = \frac{2g+t_r}{2\mu_0 l_w(l_1 + w_3)} \tag{3.36}$$

where $l_{g,eff}$ is the effective airgap including two mechanical airgaps each of length g and the copper rotor thickness t_r. Using the fact that the winding has two turns in parallel, one in each airgap, the inductance will be:

$$L = 4\,\mu_0\, l_w \frac{l_1 + w_3}{2g+t_r} \tag{3.37}$$

This far the model is a pure 2D model, but by modifying the airgap width in the axial direction it is possible to compensate for fringing, which in this case is a 3D effect. The washers have a thickness l_w or $2l_w$, and in Eq. (3.37) the same values are also used for the axial length of the airgap. In reality the flux lines have shapes more like arcs as illustrated in Fig. 3.15b. By exchanging l_w for the average effective airgap width $l_{w,eff}$ fringing can be accounted for in the model. Typically fringing increases the effective airgap width with approximately 10%, so

$$l_{w,eff} = 1.1\, l_w \tag{3.38}$$

Naturally, fringing can be calculated more accurately, which might be necessary in future work. Then leakage inductance between the poles should also be taken into account. Using the simplification in Eq. (3.38), the final expression for the inductance is found to be

$$L = 4\, \mu_0\, l_{w,eff} \frac{l_1 + w_3}{2g + t_r} \tag{3.39}$$

A drawback of this 2D model is that the inductance lack a speed dependence. At high speed the skin effect will prevent some of the flux penetration. Thus the inductance will decrease at high speed, which this model does not take into consideration. In section 3.6.1 the analytical model is compared to the FEM model as given in Fig. 3.24, it can be seen that the slope of the stiffness curve is slightly higher for the FEM model, which indicates that the inductance is decreasing somewhat at very high speed. On the other hand, when applying a model of a variable inductance, it is also important to reduce the flux change, since this is also related to the skin effect.

The total impedance and its components are shown in Fig. 3.16 for a 2- row bearing with pole shoes. It is evident that the reactance ωL is the dominant part at high speed. For this particular bearing the resistance and the reactance equals at about 30.000 rpm. The increase in resistance is due to the reduced effective thickness caused by skin effect at high speed.

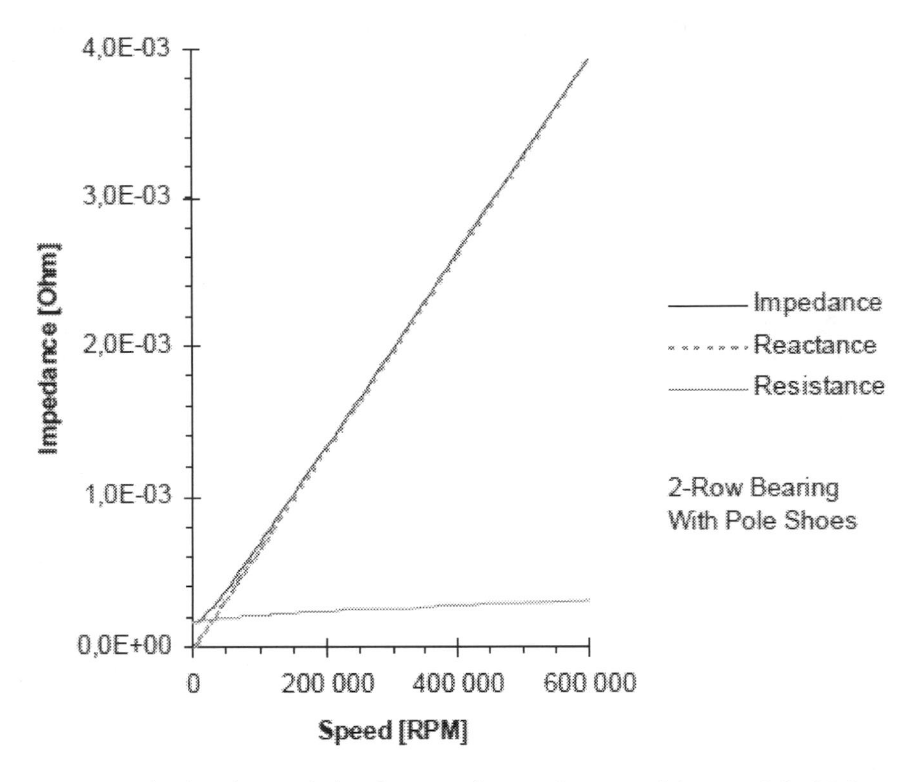

Fig. 3.16. Analytical model of rotor impedance. Above 30.000 rpm the reactance ωL is the dominant part

3.3.4 Eddy currents induced by lateral motion

Previously eddy currents induced due to eccentric operation have been studied in section 3.3.1. The second condition during which eddy currents can be induced is when the conductor moves about any of its remaining five degrees of freedom in such a way that the conductor surface experiences a varying magnetic flux density. It can be axial movements in the z-direction, tilting movements or lateral movements in the x-y plane. The latter category is of particular interest in rotor dynamics, and will be studied below in more detail.

The types of lateral motion that will be studied are

- Constant speed
- Vibrations
- Low frequency whirl
- High frequency whirl
- Synchronous whirl

A. Constant speed

To understand the rotors different motions, and their effect on the eddy currents, consider first the simplest case when the rotor moves with constant velocity. Clearly, the practical use of this example is very limited, since the rotor will quickly reach to the bearing boundaries, normally the emergency bearing, and bounce back.

Fig. 3.17a shows a rotor, which is not rotating, but moving in the bearing with constant radial velocity v. A point on the surface of the rotor, heading in the direction of motion, will experience a flux density change as given earlier in Eq. (3.8)

$$\frac{\partial B_r}{\partial t} = \frac{\partial B_r}{\partial r} \cdot \frac{\partial r}{\partial t}$$

where the radial flux derivative is already determined in Eq. (3.5)

$$\frac{\partial B_r}{\partial r} = \frac{-2B_0}{t_r + 2g}$$

and the radial velocity has a maximum value equal to *v* in the direction of motion and then varies around the perimeter according to

$$\frac{\partial r}{\partial t} = v \cos(\phi - \phi_0) \qquad (3.40)$$

Where ϕ_0 is the angle along the direction of motion. For coordinate references see Fig. 3.5.

Inserting Eq. (3.5) and (3.40) into Eq. (3.8) gives the final expression for the flux density change:

$$\frac{\partial B_r}{\partial t} = \frac{\partial B_r}{\partial r} \cdot \frac{\partial r}{\partial t} = \frac{-2B_0}{t_r + 2g} \cdot v \cos(\phi - \phi_0) \qquad (3.41)$$

Integrating the flux density change over the area A results in the flux change

$$\frac{\partial \phi_r}{\partial t} = \int_A \frac{\partial B_r}{\partial t}\, dA = \frac{2B_0 . v}{t_r + 2g} \int_{\psi_l}^{\psi_u} 2\, l_w \cos(\phi)\, r d\phi$$

$$= v \frac{8 l_w r B_0}{t_r + 2g} \cos(\frac{1}{\pi}) \quad (3.42)$$

where the integration limits are the same as in Eq. (3.12). Inserting into Eq. (3.10) gives the expression for the velocity-induced current.

$$\hat{\imath}_{z,v} = v . \frac{8 l_w r B_0 \cos(\frac{1}{\pi})}{(t_r + 2g)\sqrt{R^2 + (\omega L)^2}} \quad (3.43)$$

This current is a pure damping current, and will be used later when calculating the damping coefficient for this particular case when the rotor is not rotating, and having constant velocity *v*.

The resistance *R* and the inductance *L* are identical to the ones calculated previously concerning the eccentric rotation. Comparing Eq. (3.43) and (3.13) shows that

$$\hat{\imath}_{z,v} = \frac{v}{\omega \Delta r_0} \hat{\imath}_{z,e} \quad (3.44)$$

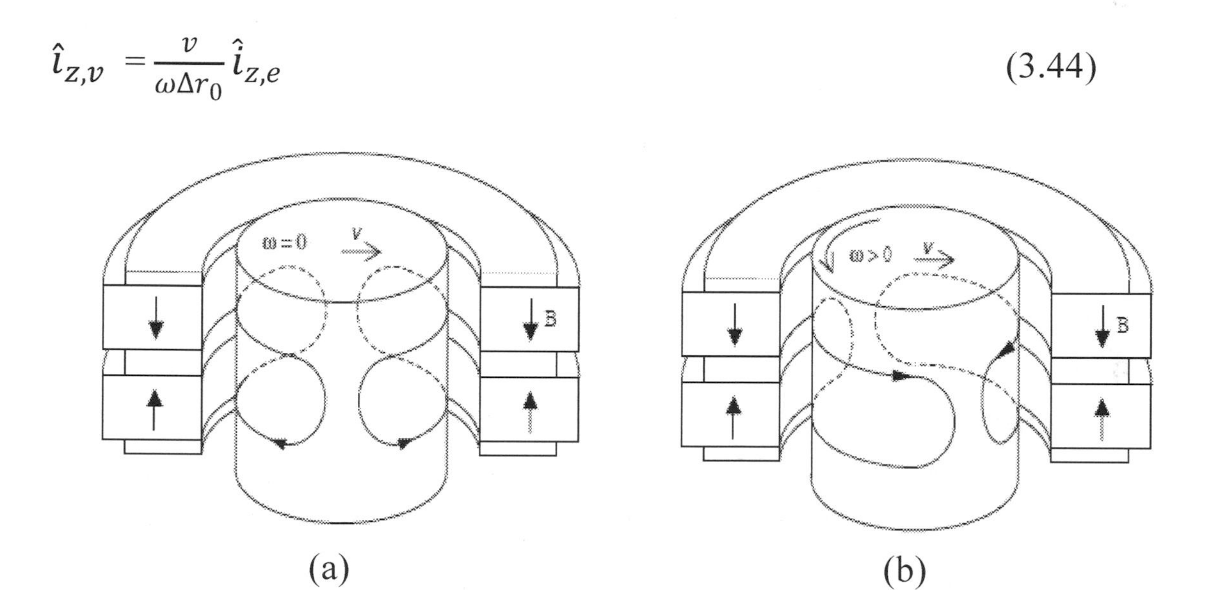

Fig. 3.17. Internal damping circuit when rotor moves with speed v. a) Zero rotational speed. b) Non-zero rotational speed ω.

Since the frequency $\omega$is zero, Eq. (3.43) can be simplified to

$$\hat{\imath}_{z,v} = v . \frac{8 l_w r B_0 \cos(\frac{1}{\pi})}{(t_r + 2g)R} \quad (3.45)$$

This is the only expression in which the inductance is of no importance for the induced currents and forces.

B. Vibrations

Rotor vibrations is understood as a reciprocating motion along any vector in the x-y plane. For instance a vibration with frequency ω_t parallel to the y-axis passing through the origin will cause the rotor axis z_1 to move according to

$$\Delta\overline{r_0} = a_v \, sin(\omega_v t)\, \hat{y} \qquad (3.46)$$

where a_v is the amplitude of the vibration. The velocity v_0 is found by derivation to be

$$\overline{v_0} = a_v\omega_v \, sin(\omega_v t)\, \hat{y} \qquad (3.47)$$

B1. Zero rotational speed

At zero rotational speed the damping currents are directed so as to produce a force directed oppositely towards the direction of motion. If the inductance is negligible, which is the case for low enough frequencies, the damping force is proportional to the lateral speed v_0, and Eq. (3.45) from the previous subsection can be used as a good low frequency approximation. A mechanical interpretation of this damping effect resembles that of a perfect viscous one.

At higher frequencies the inductance is not negligible. This delays the current and the damping force. From control theory it is well known that a delayed damping action produces stiffness. Thus the damping effect of these eddy currents will gradually turn into stiffness when the frequency increases. This can be shown by inserting the speed v_0 in Eq. (3.43). It results in the delayed damping current $i_{z,f}$ due to vibration of frequency ω_v, which is

$$\hat{\imath}_{z,f} = a_v\omega_v \cdot \frac{8 l_w r B_0 \cos(\frac{1}{\pi})}{(t_r + 2\,g)\sqrt{R^2 + (\omega L)^2}} \qquad (3.48)$$

Where the phase lag with respect to velocity is as given in Eq.(3.15)

$$\varphi = \arctan \frac{\omega L}{R}$$

and since velocity v_0 comes π/2 ahead of the displacement, it is clear that the phase lag with respect to displacement is 0 at very high frequencies. Forces produced by this current will be in phase with, and proportional to, the displacement, properties which define bearing stiffness.

Important conclusions from this section are that a non-rotating bearing
- has certain high-frequency stiffness, and
- has certain damping at all frequencies.

B2. Non-zero rotational speed

At non-zero speed currents induced by vibrations will be dragged along with the rotor surface due to the phase delay which occurs in the now rotating coordinate system. Thus the damping force will have a tangential component like the restoring force. We can call it a "phase delayed internal eddy current damping". (It should be noted that this damping is not identical to the term "rotating damping" used in rotordynamics.

The directions of the velocity dependent force F_v is shown in Fig. 3.18b, and is compared to the eccentricity induced force F_r shown in Fig. 3.18a. It is to be noted that the damping eddy current circuit has a forward shift compared to the restoring circuit. This comes from the fact that speed is a displacement derivative and lies π/2 ahead of the displacement itself. The flux change in the rotating frame due to lateral velocity $\overline{v_0} = (v_{0x} + v_{0y})$ is

$$\frac{\partial B_r}{\partial t} = \frac{\partial B_r}{\partial r} \cdot \left|\frac{\partial \bar{r}}{\partial t}\right| = \frac{\partial B_r}{\partial r} \cdot \left|v_{0x}\cos(\omega t)\hat{x} + v_{0y}sin(\omega t)\hat{y}\right| \quad (3.49)$$

where Eq. (3.47) has been used and applied to both the *x*- and *y*- directions, and where

$$\frac{\partial B_r}{\partial r} = \frac{-2B_0}{t_r + 2\,g}$$

, as given in Eq. (3. 5)

Assuming movements only in the y-direction, the velocity function given in Eq. (3.47), becomes

$$\bar{v}_{0x} = 0 \text{ and } \bar{v}_{0y} = a_v\omega_v\cos(\omega_v t)\hat{y}$$

Inserting this into Eq. (3.49) results in the expression for the flux density change due to velocity,

$$\frac{\partial B_r}{\partial r} = \frac{-2B_0}{t_r + 2g} \cdot a_v \omega_v \cos(\omega_v t) \sin \omega t \qquad (3.50)$$

Since the vibration is always combined with displacement $\Delta_{y0}=a_v \sin(\omega_v t)$, another term needs to be added to the final expression. This term is the speed
$\Delta_{r0} \omega \cos(\omega t)$ from Eq. (3.8) where Δ_{r0}in this case is Δ_{y0}. Inserting into Eq. (3.50) gives$(\omega_v t)$

$$\frac{\partial B_r}{\partial t} = \frac{-2B_0}{t_r + 2g} \cdot (\omega a_v \sin(\omega_v t) \cos \omega t + a_v \omega_v \cos(\omega_v t) \sin \omega t) \qquad (3.51)$$

In Eq. (3.51) above the vibrations and displacements are limited to the y-axis. Note that there are two frequencies involved, vibrational frequency ω_v and rotational frequency ω and simple expressions can only be achieved when they are either equal, or one of them can be neglected.

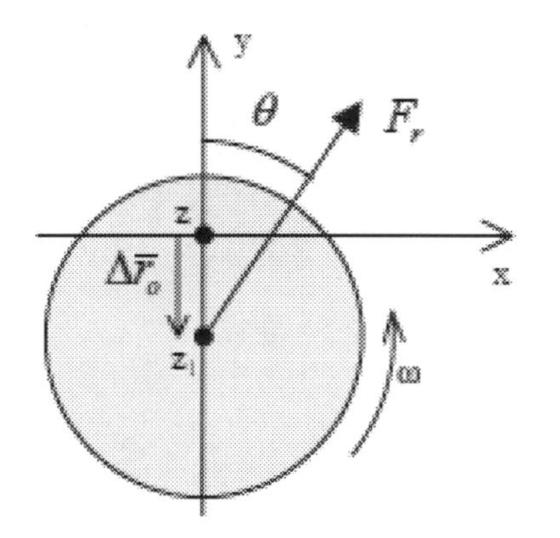

Fig. 3.18a. Force due to eccentricity

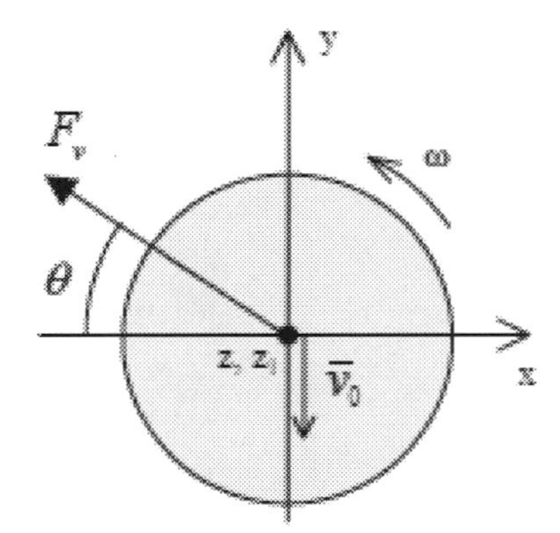

Fig. 3.18b. Force due to lateral speed

In the following subsections these special cases will be studied more in detail.

C. <u>Low frequency whirl</u>

Rotor vibrations normally have the shape of elliptical or circular orbits, a superposition of two perpendicular vibrations with equal frequency ω_w but not necessarily equal amplitude. The rotor is said to whirl, and this motion can be either forward or backward with respect to the direction of rotation.

Thus the whirl frequency ω_W will induce damping eddy currents in the rotor similar to the time delayed damping currents described in the previous section. The magnetic circuits of these damping eddy currents are the same as the ones shown in Fig. 3.17, where two cases are shown. The first case above shows the rotor having no rotational speed, and the second case shows an example where the rotor is spinning at velocity ω. One just needs to consider that the velocity is no longer parallel to the displacement as in the previous case, also illustrated in Fig. 3.19a. If the orbits are circular, the speed and displacement are perpendicular, as shown in Fig. 3.19b.

Calculating the currents and forces from this damping is not trivial. However, Filatov has done a study on the dynamics of slow lateral motion of a conductor superimposed on high-speed rotation. Applying his results to the phase shift and the force angle in the special case when the whirl frequency ω_W is much lower than the rotational frequency ω, results in damping forces F_v that are perpendicular to the force F_r induced by eccentricity when displacement and velocity have the same direction, which is illustrated in Fig. 3.19a. Applied to the case of low frequency whirl, they turn out to be oppositely directed, Fig. 3.19b.

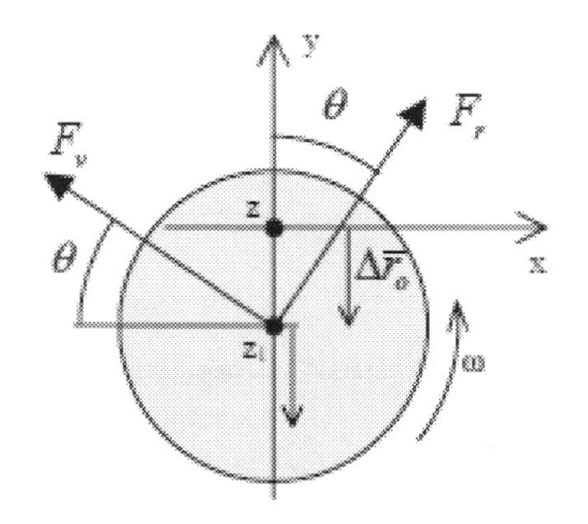

Fig. 3.19a. Forces due to eccentricity and velocity

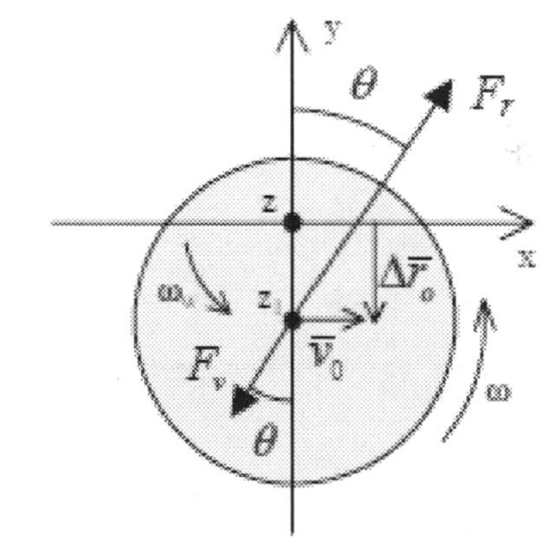

Fig. 3.19b. Forces due to low frequency whirl

D. High frequency whirl

If the vibrational frequency ω_f is much higher than the rotational speed ω, then there is not much time for the currents to be dragged around by the rotation. The situation is more similar to the first case with a non-rotating but vibrating rotor. Causes for high frequency vibration can be interaction with turbine blades, disturbances from cutting or grinding tools in machine spindles, motor cogging if applicable and inverter switching. The motion patterns from these disturbances have no simple shapes, but are more stochastic to their nature. They usually excite harmonics related to the bending critical frequencies of the shaft or turbine blades, and at least the ones belonging to the shaft will have orbital shapes, which might be referred to as high frequency whirl.

When active magnetic bearings are used, it has proven to be quite difficult to damp these vibrations, since a high frequency cut off filter is normally used before the damping algorithm to prevent the controller from saturating the amplifiers. The sensor position is also critical, since it is axially displaced with respect to the bearing, and thus may not see the vibration modes that are present in the bearing.

The induction bearing provides both damping and high frequency stiffness, and has potential to complement active magnetic bearings in these kinds of applications.

E. Synchronous whirl

The term synchronous whirl will be used when the whirl frequency ω_w equals the rotational frequency ω. Reviewing Eq. (3.51)

$$\frac{\partial B_r}{\partial t} = \frac{-2B_0}{t_r + 2\,g} \cdot (\omega a_v \sin(\omega_v t) \cos \omega t + a_v \omega_v \cos(\omega_v t) \sin \omega t)$$

which is valid for a one-dimensional vibration along the y-axis, and remembering that whirl is a superposition of two perpendicular vibrations, observe that

$$\omega_w = \omega \Rightarrow \frac{\partial B_r}{\partial t} = \frac{-2B_0}{t_r + 2\,g} .0 = 0 \quad , \qquad (3.52)$$

which means that during synchronous whirl no flux change occurs on the rotor surface in the rotating coordinate frame. Thus no forces, no losses and no damping occur, and these rotor vibrations cannot be transmitted to the stator via the induction bearing. Instead they are transmitted via the external damper, which is not described in this dissertation since there are several available concepts to choose among, some of them having nothing to do with eddy currents.

Important types of synchronous whirl occur due to

- Unbalance
- rigid body critical speeds
- bending critical speeds.

The inability to transfer forces from synchronous whirl has advantages and disadvantages. A great advantage is of course reduced costs of rotor balancing, and quiet operation. When for instance active magnetic bearings are used on unbalanced rotors it is often important to filter away the synchronous frequency from the controller. This can be done using either a variable notch filter, or by using a feed forward algorithm. With induction bearings this filtering is done automatically.

A disadvantage is that passing the bending critical speed is impossible without additional damping, so the requirements regarding the external damper might be quite high. However, passing the rigid body modes is not a problem, as will be explained in section 3.6.

3.3.5 Stray eddy currents

Any inhomogeneities in the permanent magnets, it may come from sintering or from magnetization, will most likely cause a non-rotationally symmetric magnetic field, which in turn will induce stray eddy currents. These currents will have a shape depending on the inhomogeneity. The most common one is misalignment between the geometric and the magnetic axis. Local flux dips, or random change of magnetic orientation around the perimeter of the magnet rings are also common.

Stray eddy currents may also result from in-homogeneities or remnant magnetic flux in the iron pole shoes. Iron screws and pins close to the airgap can also alter the flux so as to induce unwanted eddy currents.

Many FEM simulations on different kinds of in-homogeneities have been performed. They all show that the influence on restoring forces is very limited, whereas the influence on losses is large.

When ordering magnets it is very important to specify the maximum allowed misalignments and local flux dips depending on application loss requirements. Today not many magnet manufacturers undertake such considerations.

Though, even in the case of not perfectly manufactured magnets, the induction bearing still produce very low losses compared to many other magnetic bearing types.

4

Performance Parameters

4.1. Bearing Forces

This far we have analyzed the magnetic flux and the eddy current circuits. By forming the product ($J \times B$) along the eddy current circuit in Fig. 3.12a,b, the Lorenz force distribution is achieved. In Fig. 3.12a,b it is important to note that the magnetic flux is radial in cut A-A while it is axial in cut B-B. This comes automatically from the flux lines in Fig. 3.1. In Fig. 4.1a,b this is illustrated from another viewing angle.

Thus, in the linear model of the current path in Fig. 3.13, the current is tangential and the flux is axial in region 1. In region 3 the flux is radial and the current is axial. Since the length l_1 normally is much longer than the length l_3, the main force is generated in region 1. In Fig. 4.1c below the force distribution from region 1 is illustrated.

The resultant force F_{res} in Fig. 4.1d is obtained by integrating the force contributions over the whole cylinder. The force angle θ is the same as the one defined in Eq. (3.16), and depends on speed, resistance and inductance. If we know the point of attack, or any other point on the line defined by the vector F_{res} , the brake torque M_z can be calculated. Due to symmetry this line has to pass the center of the homopolar flux, which is the origin of the stator.

If the rotor is displaced in the negative *y*-direction the torque can easily be calculated as the product of the x-component of the force, F_x with the displacement Δ_{r0}.

If the bearing is not strictly homopolar, like for instance if the magnets are not perfectly axially magnetized, the point of attack is somewhere else and the brake torque will be larger.

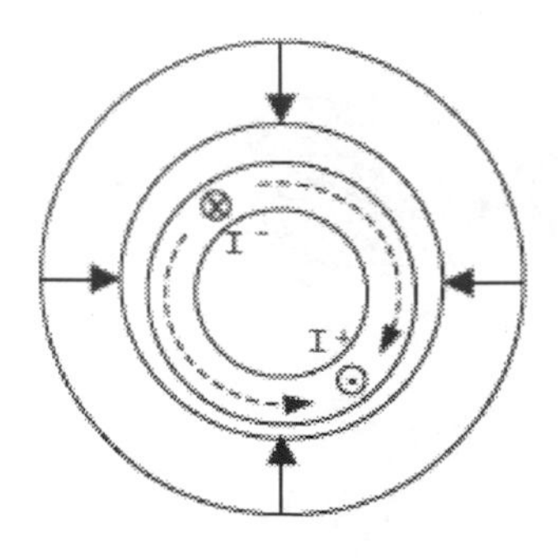

a. Same as Fig. 3.64a.
Cut A-A, radial flux.

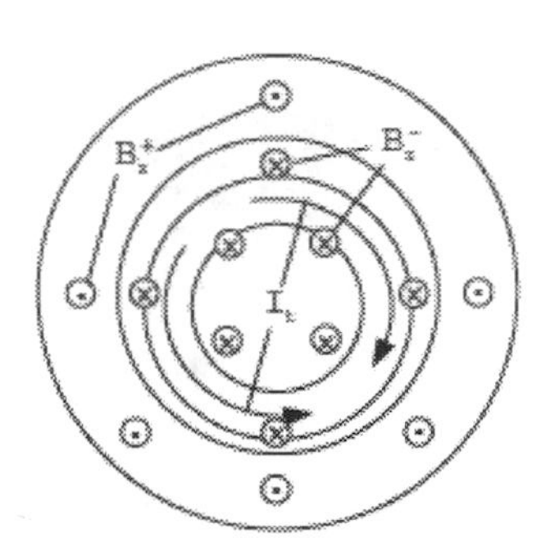

b. Cut B-B, axial flux.

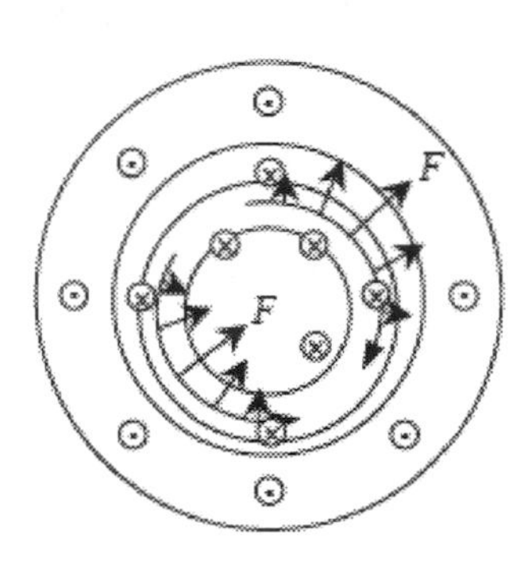

c. J × B force distribution

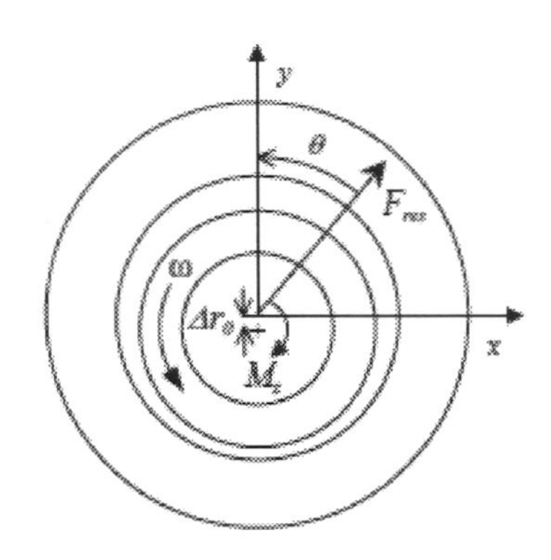

d. Resultant force and brake torque

Fig. 4.1 Flux and Force distribution with rotor eccentricity
Region 1: the short circuit ring in an induction generator
Region 3: the active part or the rotor bars,
see Fig. 3.13 for different number of regions

In Fig. 3.8a, b the currents are plotted for different speeds and thus different force angle. In Fig. 3.8a the speed is low and the load angle is large. See also Fig. 4.2. Thus the force angle is also large and the bearing force has a large tangential force component, which has a strongly destabilizing effect on the rotor. In Fig. 3.8b the speed is infinitely high, and the force angle is zero so that no tangential force component is present. Thus the force is totally restoring.

Unfortunately infinite speed is out of reach, so there will always be a certain tangential force component. However, by using iron pole shoes, or washers, between the magnets, the inductance is increased and thus the relationship between restoring force and tangential force is increased. This reduces the requirements for the external damper that has to be added due to stability considerations.

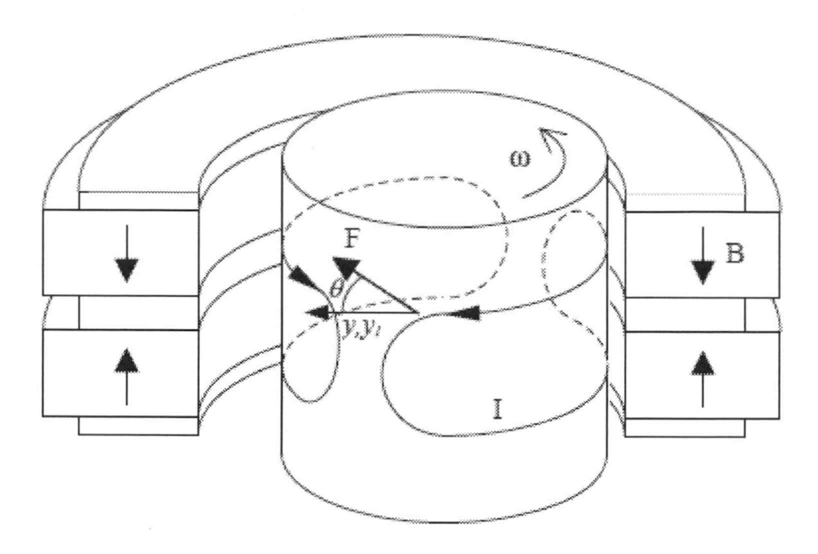

Fig. 4.2. Resulting force between eddy currents and magnets

In the following the force from one magnet will be calculated using the simplification that the force contribution comes solely from the tangential currents and that the flux is axial in this region.

The force acting on the rotor can be calculated as the volume integral of the Lorenz forces. Starting with the intermediate magnet rows, each such row induces 2 eddy current circuits, each consisting of two identical tangential parts. Thus the tangential parts can be integrated separately. Using the current *I*, the resistance *R* and the inductance *L* from Eq. (3.25) and (3.39) the force contribution from one tangential current part can now be calculated as

$$\bar{F}_t = \int_V \bar{J} \text{x} \bar{B} \; dV \tag{4.1}$$

The integral can be simplified using the fact that we know the force angle. Thus an angle α can be introduced measured from the force angle and integrated from -π/2 to π /2. Projecting the force to this line by multiplying with cos**α** gives the integral

$$F_t = \int_{-\pi/2}^{\pi/2} \hat{j}_t \cos\alpha \, . B_z \cos\alpha \, . \frac{l_m}{2} t_{eff} \, r d\alpha \tag{4.2}$$

where j_t is the tangential current density which has its maximum at $\alpha = 0$. At this point the total current I equals

$$I_t = \hat{j}_t \, t_{eff} \frac{l_m}{2} \tag{4.3}$$

which should be the same current as in Eq. (3.20). Thus Eq. (4.2) can be written as

$$F_t = I_t B_z \int_{-\pi/2}^{\pi/2} \cos^2 \alpha \, . \, r d\alpha = \frac{\pi}{2} I_t B_z \, r \tag{4.4}$$

Remembering that the total force F_{add} from one added intermediate magnet row comprises the sum of four tangential current parts, results in

$$F_{add} = 4 \, F_t = 2 \, \pi \, I_t B_z \, r \tag{4.5}$$

The force direction is along the force angle, which is given by Eq. (3.16).

$$\theta_{add} = \arctan \frac{R}{\omega L}$$

For 2-row bearings, and for the two end magnets when the bearing consists of more than two rows, the total force from these two magnets F_{2-row} and the force angle θ_{2-row} can be calculated in the same way. One only has to remember that the resistance R and the inductance L has to be taken from Eq. (3.24) and (3.33) instead.

Thus the total bearing force F_B for an intermediate rotor bearing with n rows, is

$$\bar{F}_B = \bar{F}_{2-row} + (n-2)\bar{F}_{add} \, , \quad n \geq 2 \, , \tag{4.6}$$

where vector notation has been used since the forces F_{2-row} and F_{add} are not acting in the same direction. It has a restoring force component $F_{R,B}$ directed oppositely towards the eccentricity

$$F_{R,B} = F_{2-row} \cos \theta_{2-row} + (n-2)\, F_{add} \cos \theta_{add} \quad (4.7)$$

and a tangential force component $F_{T,B}$which is

$$F_{T,B} = F_{2-row} \sin \theta_{2-row} + (n-2)\, F_{add} \sin \theta_{add} \quad (4.8)$$

Finally the force angle θ_B of the total bearing can be calculated as

$$\theta_B = \frac{F_{T,B}}{F_{R,B}} \quad (4.9)$$

which is illustrated in Fig. 4.3 below.

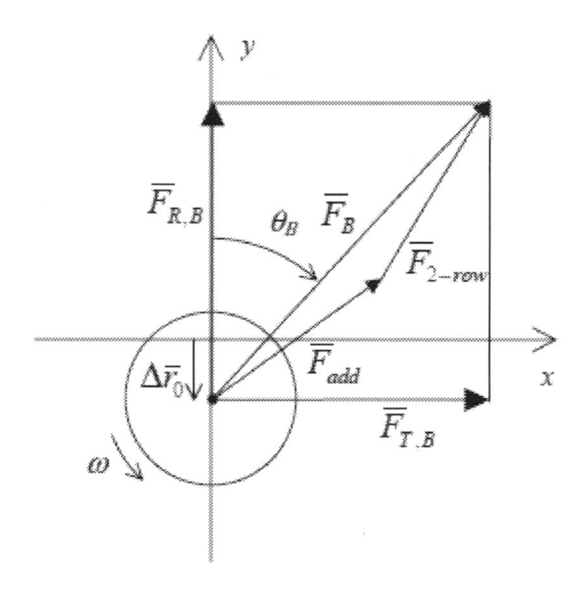

Fig. 4.3. Force components from a 3-row bearing

4.2. Bearing Losses

Bearing losses P_B are composed of eddy current losses P^e_B and fluid film losses P^f_B. Often magnetic bearings are used in high vacuum, where losses from the gas are very low, while they may be high when used in a high viscosity process fluid. Thus

$$P_B = P^e_B + P^f_B \tag{4.10}$$

These losses are to some degree connected, since the cooling of the rotor is related to the fluid film losses.

The eddy current losses consist of contributions from all different currents described in the preceding sections, and before adding their contributions, it is important to find out if the currents overlap each other, and what phase angle they have, otherwise it is not possible to know if they increase linearly or quadratically when added, or if they maybe cancel each other out. The synchronous whirl described earlier is an interesting example where losses due to imbalance are completely cancelled out.

4.2.1 Losses due to eccentricity

Losses due to eccentricity is the most important loss component, since it is related to the bearing load and cannot be reduced by manufacturing magnets with higher precision. Contributions from the different magnets do not interact in this case since they are geometrically separated. The currents might interact due to mutual inductance though. However, once they are determined, the losses can be added.

For a 2-row bearing there are two eddy currents, and the losses due to eccentricity is

$$P^l_{2-row} = 2\, R_{2-row} + I^2_{2-row} \tag{4.11}$$

where the index l stands for load, and where the current I_{2-row} and the resistance R_{2-row} have to be calculated according to Eq. (3.20) and (3.24).

Similarly, the losses from an additional intermediate row $P^{l}{}_{add}$ can be calculated using the resistance and current for this particular case.

$$P^{l}{}_{add} = 2\, R_{add}\, I^{2}_{add} \tag{4.12}$$

Altogether, the losses due to eccentricity for a bearing with n magnet rows is Then

$$P^{l}{}_{B} = P^{l}{}_{2-row} + (n-2)\, P^{l}{}_{add} \;,\; n \geq 2 \tag{4.13}$$

Losses due to eccentricity can be alternatively calculated using the mechanical power from the torque produced by the tangential force $F_{T,B}$.

$$P^{l}{}_{B} = M\omega + \Delta\, r_{0}\, F_{T,B} \cdot \omega \tag{4.14}$$

Losses defined in this way are sometimes referred to as "rotating damping losses". However, it is important to remember that the tangential force is not a constant, but a function of the impedance ωL, so the traditional usage of this term is not applicable in the stability analysis of eddy current devices, without taking special notice.

4.2.2 Losses due to lateral velocity

Eddy currents induced due to lateral velocity can be calculated in the same way as above, using the currents from Eq. (3.43) or (3.48) instead. Like previously, the current do not overlap each other from different magnets, since their shape are identical to the ones above, so they can be calculated as

$$P^{v}{}_{B} = P^{v}{}_{2-row} + (n-2)\, P^{v}{}_{add} \;,\; n \geq 2 \tag{4.15}$$

These currents always overlap the currents from eccentricity, so they can not be simply added, they have to be integrated. They can help to completely eliminate the losses from eccentricity, as in the case mentioned earlier regarding synchronous whirl, but they can also add quadratically. The worst case would be that of synchronous backward whirl, where the losses would be

$$P_{B} = 2(\, P^{l}{}_{B} + P^{v}{}_{B}\,) \tag{4.16}$$

which is a very unlikely situation. Backward whirl may happen during a severe emergency bearing touch down, but even in this case it is not synchronous. Thus Eq. (4.16) may be used as an upper limit for bearing losses caused by load and lateral velocity.

4.2.3 Losses due to magnet inhomogeneity

Eddy currents induced due magnet inhomogeneity may have any shape depending on the type of magnet imperfection that is studied. Below some of them will be listed.

- Angular misalignment
- Mechanical eccentricity
- Surface roughness
- Varying density.

Angular misalignment is a severe type of imperfection. It causes a radial two pole flux superimposed on the expected flux, which should be a homopolar axial flux diverging into a radial flux. These eddy currents are similar to the ones caused by eccentricity, and cause bearing forces and losses as if there actually were some radial displacement of the rotor. Some 3D-FEM calculations were performed, and show that a misalignment of 3 degrees equals approximately a bearing operating at maximum eccentricity. The bearings used in the experiments have a misalignment of less than 0.5 degree.

Mechanical eccentricity occurs when the magnets are mounted on one surface, and the other surface is facing the airgap, that is the inner and outer diameters are not concentric. The best mechanical tolerances available today are ± 0.05 mm for ring magnets in mass production, which is surprisingly bad.

Neither losses due to angular misalignments nor due to mechanical eccentricity can be simply added to the other losses, since the currents are very similar and might be overlapping.

Surface roughness and locally varying density causes high frequency small eddy currents, very unlike the ones caused by a two-pole flux. These losses add very well to the other losses, since most of the currents are not overlapping. The high frequency currents are bound to the surface of the conducting cylinder, due to smaller skin depth. The losses might also be quite large due to the frequency. Surface roughness is primarily caused by diamond thread cutting or by the modern water cutting method. For use in magnetic bearings this method should be followed by grinding, which unfortunately is quite expensive.

Since it is very difficult to know from one measurement which of the above mentioned magnet imperfections that cause the losses, the sum of these losses will be denoted P^i, where i stands for inhomogeneity. The total bearing losses P_B have to be found by integration using Eq. (4.17) below:

$$P_B = \iiint_V \rho \, (I^l + I^v + I^i)^2 dV \qquad (4.17)$$

Likely, the current distribution I^i is not known, but instead P^i might have been measured. Then the value of the integral has to be within the boundaries

$$0 \leq P_B \leq 3(P_B^l + P_B^v + P_B^i) \qquad (4.18)$$

where the upper limit comes from the worst case when all currents have the same frequency, phase and geometric shape. As before, the subscripts *B, 2 row* or *add* tells what part of the bearing the losses belong to.

P^i may unfortunately be a function of as well eccentricity as lateral velocity. Thus this function has to be found by several measurements, which is a tedious job. However, during the experimental phase of the project, a more practical definition of the losses due to inhomogeneity was found.

4.2.4 Magnetic offset

The first two magnet flaws, angular misalignment and mechanical non concentricity, acts as an eccentricity and causes a magnetic offset. That is, the mechanical center of the bearing may be offset a certain distance from the magnetic center of the bearing, where the magnetic center is defined as the point where the radial forces are zero, regardless of rotational speed. This point is easy to find experimentally, and at this point, or at least close to it, the losses also have a minimum. These losses will be denoted the "no load losses", or "residual losses", P^r. If the bearing eccentricity is redefined to be measured from the magnetic center, the residual losses do not include eccentricity terms, and may be added to the load losses. The losses, thus consist of load losses and residual losses

$$P_B \cong (P_B^l + P_B^r) \tag{4.19}$$

A few words on eccentricity. When many magnets are used, the magnetic center may differ between all magnets. The magnetic center for the whole bearing is the point where the total radial forces are zero. At this point all individual eccentricities and their loss contributions are treated as a part of the total bearing residual loss, P_B^r .

The velocity losses P_B^v are kept zero during the experiments. They may still not be added to the other losses, but have to be integrated. Thus the upper limit for the total losses in Eq. (3.17) can be rewritten

$$0 \leq P_B \leq 3(P_B^l + P_B^v + P_B^r) \tag{4.20}$$

4.3. Stiffness and Damping

4.3.1 Stiffness

Stiffness is defined as the negative derivative of the force and its components with regard to displacement. For a simple one degree of freedom mass and spring system, as the one shown in Fig. 4.4, a displacement x results in a spring force F_x, and the spring constant or spring stiffness k is then defined as

$$k = \frac{F_x}{x} \quad (4.21)$$

If the mass is released, it will oscillate with the natural frequency ω_n,

$$\omega_n = \sqrt{\frac{k}{m}} \quad (4.22)$$

For at two degree of freedom mass and spring system, like for a bearing rotor moving in the x and y directions, where the force is not parallel to the displacement, three different stiffness definitions will be used.

A. In-plane stiffness, K

The derivative of the total bearing force F_B with respect to displacement Δr_0 is by Filatov defined as the "in-plane" stiffness, K. Though this definition is not used in literature, it has proven to be very useful in the analysis of electro dynamic bearings. It is responsible for the load capability of the bearing.

$$K = \frac{dF_B}{d(\Delta r_0)} \quad (4.23)$$

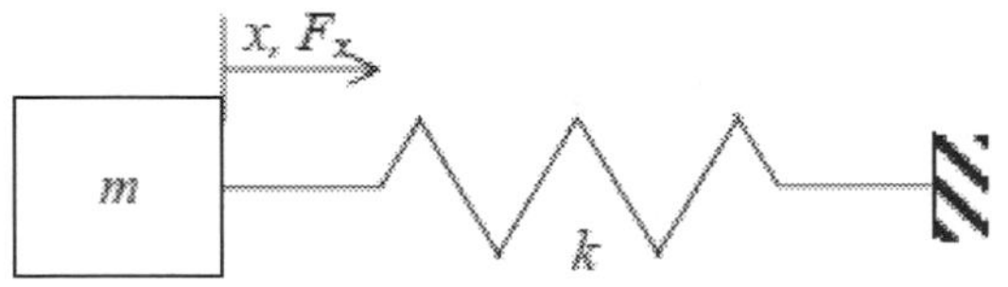

Fig. 4.4. Simple mass-spring system

B. Rotor dynamic stiffness, k

In rotor dynamics usually the restoring force component is regarded in the definition of stiffness, k. It can be expressed in several ways,

$$k = \frac{-dF_{R,B}}{d(\Delta r_0)} = \frac{-dF_B}{d(\Delta r_0)} \cos\phi = K \cos\phi \tag{4.24}$$

The rotor dynamic stiffness determines the natural frequencies of the system.
The first critical speed, provided a rigid shaft and symmetric bearing arrangement, occurs at

$$\omega_n = \sqrt{\frac{2k}{m}} \tag{4.25}$$

where 2 identical bearings each having stiffness k are assumed.

C. Cross coupling stiffness, k_c

In the same way as above, stiffness can also be defined using the tangential force component. This is referred to as the cross coupling stiffness, $\boldsymbol{k_c}$, and it plays an important role in the stability analysis of the bearing. Thus

$$k_c = \frac{-dF_{T,B}}{d(\Delta r_0)} = \frac{-dF_B}{d(\Delta r_0)} \sin\phi = K \sin\phi \tag{4.26}$$

Fig. 4.5 shows the in-plane stiffness calculated using the two different skin depth approximations from Eq. (3.28) and (3.30) respectively.

Fig. 4.6 shows the calculated in-plane stiffness for bearings with different number of rows, and for bearings with and without pole shoes (Fig. 4.7). In Fig. 4. 8 the corresponding force angle is calculated. All calculated data are compared with FEM results.

Fig. 4.9 shows a comparison of the three different stiffness concepts defined above. Observe that both the in-plane stiffness *K* and the rotor dynamic stiffness k goes asymptotically towards an upper limit when speed increases, while the destabilizing cross-coupling stiffness k_c has a local maximum, and then decreases with speed. This is one of the reasons why it is so easy to establish stable levitation using electro dynamic systems.

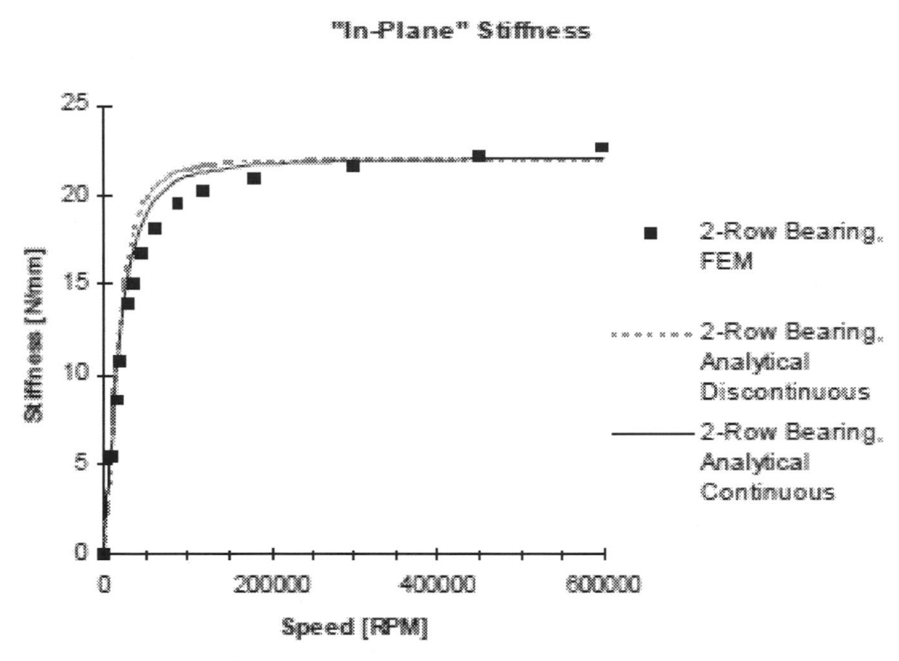

Fig. 4.5. Stiffness calculated using Eq. (3.28) and (3.30). FEM results shown as reference

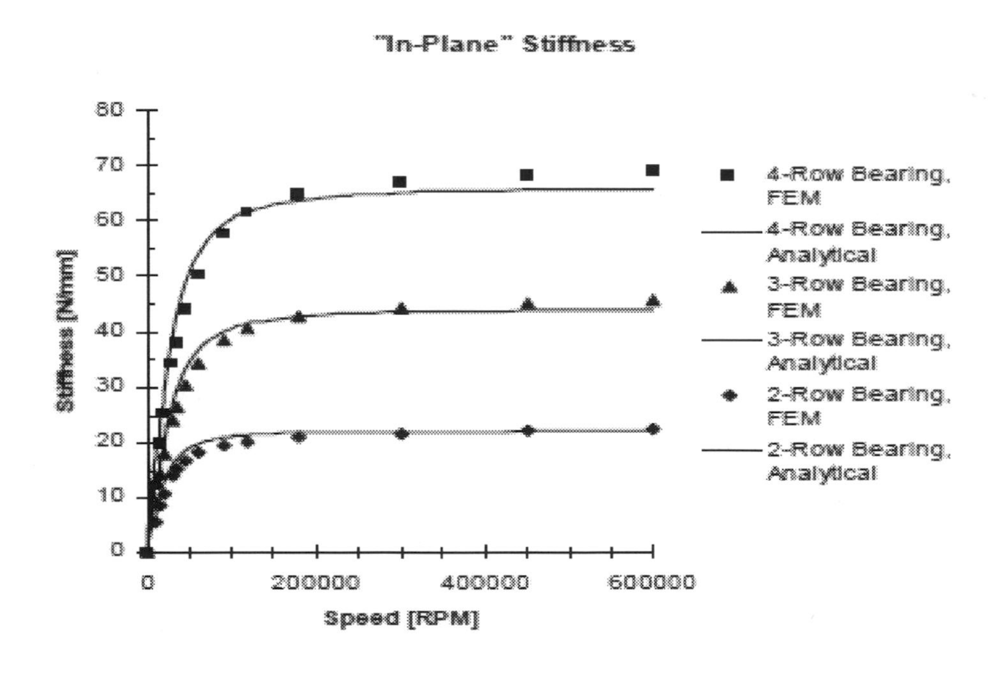

Fig. 4.6. In-plane stiffness versus speed for different intermediate rotor bearings

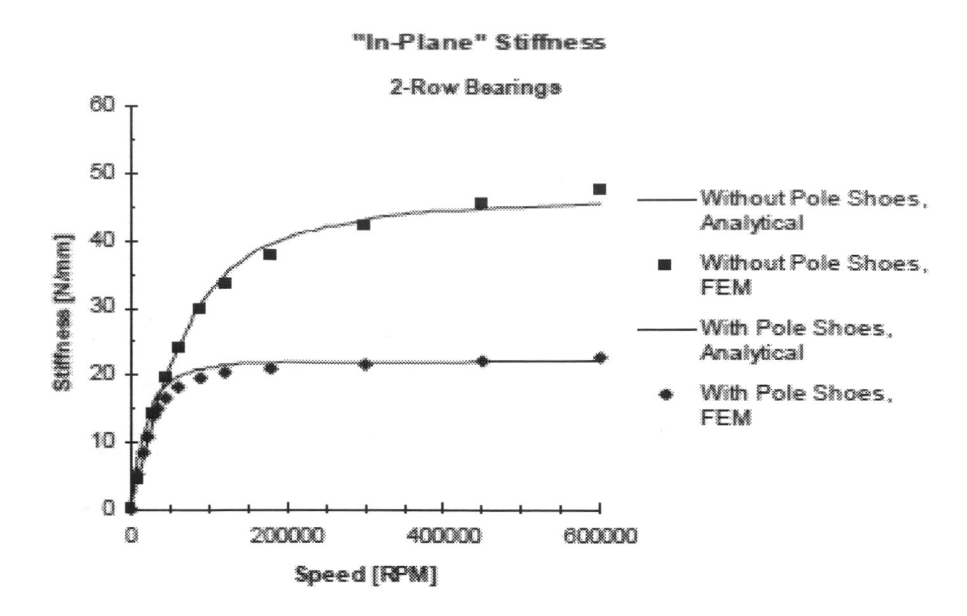

Fig. 4.7. In-plane stiffness versus speed for intermediate rotor bearings with and without pole shoes

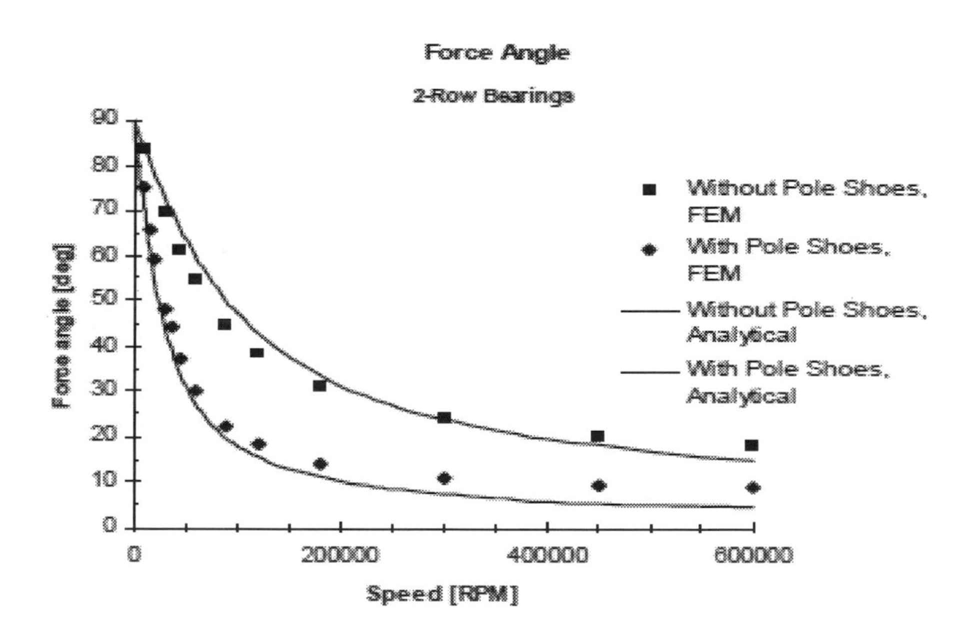

Fig. 4.8. Corresponding force angle for the bearings in Fig. 4.7.

4.3.2 Damping

In the same way as for the definition of stiffness, the damping coefficient can be defined in different ways, since it does not act parallel and opposite to the velocity.

A. In-plane damping, C

The derivative of the total bearing force F_B with respect to lateral velocity v is a damping coefficient C taking the total damping force into consideration. It acts in the same plane, but not parallel to, the velocity, and will be denoted "In-plane damping coefficient".

$$C = -\frac{dF_B}{dv} \tag{4.27}$$

Recalling Eq.(3.44), that

$$\hat{\imath}_{z,v} = \frac{v}{\omega \Delta r_0} \hat{\imath}_{z,e}$$

means that the damping coefficient can be written

$$C = -\frac{K}{\omega} \tag{4.28}$$

B. Rotor dynamic damping coefficient, c

In rotor dynamics usually the oppositely directed force component is regarded in the definition of damping, c. It can be expressed in several ways,

$$C = -\frac{dF_B}{dv} sin\phi = \frac{K}{\omega} sin\phi \tag{4.29}$$

C. Cross coupling damping coefficient, C_c

In the same way as above, stiffness can also be defined using the tangential force component. This is referred to as the cross coupling damping coefficient, C_c, where

$$C_c = -\frac{dF_B}{dv} cos\phi = \frac{K}{\omega} cos\phi \quad (4.30)$$

The cross coupling damping force is directed outward in forward whirl and inward, that is centering, in backward whirl.

D. Internal and external damping

Damping is often referred to as either internal or external, with the coefficients d_i and d_a. Some writers prefer the notation rotating and non-rotating damping and use the coefficients C_r and C_n. A complete theory is built up around these terms. The induction bearing has a rotating conductor, and thus the damping according to this nomenclature is rotating, or internal.

Assuming that the induction bearing stator is non-conducting and does not have any hysteresis, then the external or non-rotating damping is zero. It is easy to build in some non-rotating damping in the bearing, but in the analysis it will be assumed that this has not been done.

Below is an attempt to clarify the different damping notations being used.

$$\begin{cases} C_n = d_a \approx 0 \\ C = \frac{K}{\omega} sin\theta = (C_n + C_r) \approx C_r = d_i \\ C_c = \frac{K}{\omega} cos\theta \end{cases} \quad (4.31)$$

According to the rather extensive and dominant theory on rotating damping, it is also possible to explain the cross coupling stiffness using the rotating damping. Thus

$$k_c = \omega\, d_i = \omega\, d_r = \omega\, c \quad (4.32)$$

and the tangential force can be expressed as

$$F_T = k_c r = r\omega\, c \qquad (4.33)$$

However, the theory completely fails to explain the bearing stiffnesses k and K and the restoring force $F_{R,B}$ in Eq. 4.7. The theory also has difficulties to handle the fact that the damping coefficient c decreases with speed. Thus this theory is seldom referred to among researchers in the field of these bearings.

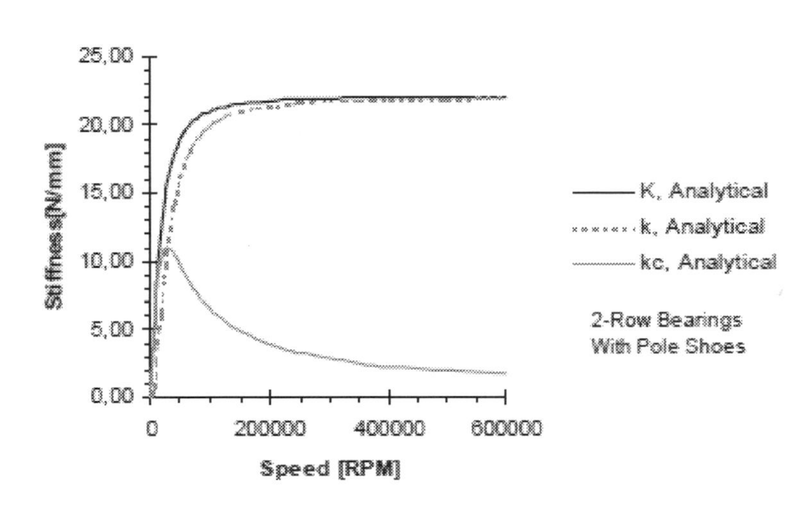

Fig.4.9.Calculated stiffness coefficients for 2-row intermediate bearings with pole shoes

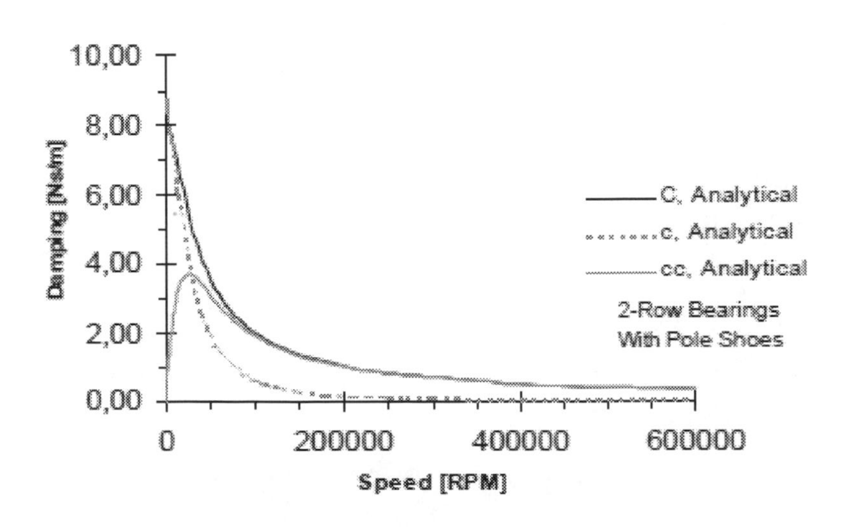

Fig.4.10. Corresponding damping coefficients for the bearings in Fig.4.9

4.4. Rotor Dynamics and Stability

Let us study a few aspects on rotor dynamics, to make it easier to compare induction bearings with other bearing types.

A bearing is said to be isotropic if the stiffness is the same in all directions. Tilt pad bearings are for instance not always isotropic. Neither are all active magnetic bearings. Homopolar bearings are.

High stiffness is required in a tool spindle in order to obtain accuracy, while low stiffness is necessary in a centrifuge to avoid vibrations due to imbalance forces. Basically a shaft rotating at a certain speed rotates about its geometric center if the stiffness is high, while at low stiffness it rotates about its center of gravity. A rotor accelerating from standstill to high speed will pass through the first rigid body critical speed, which is the speed where the rotor changes from rotating about the geometrical axis to rotating about its center of gravity. Below the critical speed the rotor is said to run sub-critically, and above that speed it runs super-critically.

The first critical speed ω for a rigid rotor with two flexible radial bearings equally distributed from the center of gravity, the bearings not sitting to close to each other, can be calculated from

$$\omega_n = \sqrt{\frac{2k}{m}}$$, as given in Eq. (4.25)

where ***m*** is the mass of the rotor. At this frequency the rotor vibrates so that the axis of rotation is always parallel to the original direction. This mode is called the cylindrical mode.

If the speed is increased another mode will be introduced, the so-called conical mode. The names of the nodes refer to the shape of the volumes that are built up by the vibrating shaft. The vibrational frequency for the conical mode depends not on the mass of the rotor as for the parallel mode, but on the moments of inertia. If this moment is very high the conical mode will appear before the cylindrical mode.

Normally a rotor is not as symmetric as described above. This will most likely result in two conical modes, one for each bearing.

At high speed things get more complicated. The two conical modes are split up into four modes due to gyroscopic effects. Depending on factors like rotor diameter and stiffness several elastic body modes will also be added. A convenient way to represent the different possible frequencies is to map them in a Campbell diagram like in Fig. 4.11. In this diagram the operating line can be drawn, and each crossing with a line referring to a specific vibration mode will represent a critical speed. Some of these modes are more likely to be introduced than others.

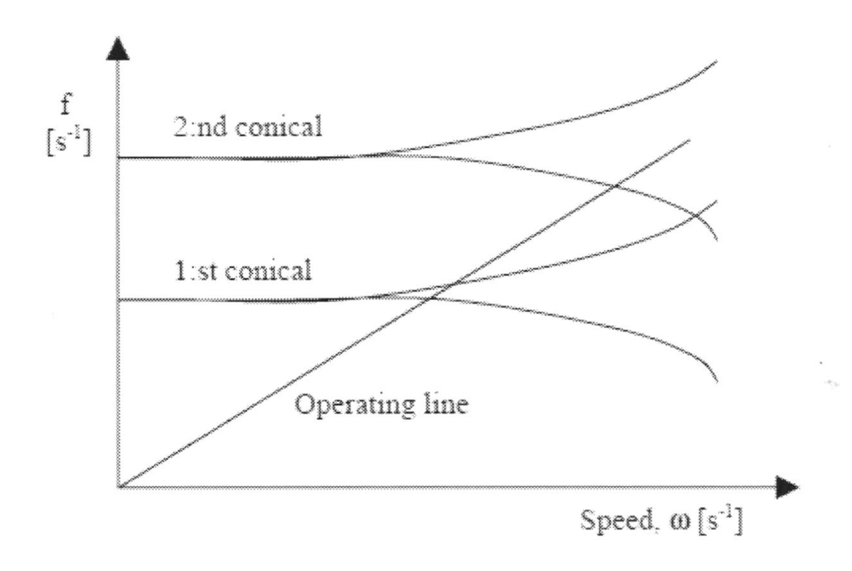

Fig. 4.11. Modal analysis of a rigid unsymmetrical beam

If a mode is introduced, the vibration level is normally very high, unless the damping is very good or the acceleration or retardation is high.

Active Magnetic Bearings AMB offer a very interesting way to eliminate this problem, as the stiffness can be instantaneously altered, as **i** Fig. 4.12. It is thus possible to temporarily lower the stiffness before passing through a critical speed so that the critical speed is reduced under the actual speed. When the speed is higher the stiffness is raised again. In addition, with AMB the damping can be individually programmed for each bending mode for maximum system performance.

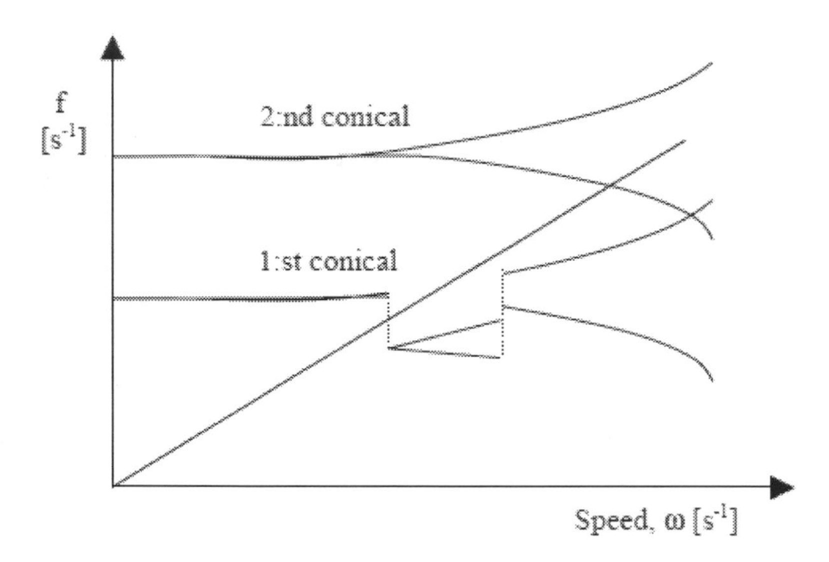

Fig. 4.12. Active control of first conical modes

Induction bearings offer another possibility to avoid the problem. As the stiffness gradually increases from zero to an asymptotic value when the speed is increased, it is possible to design a bearing so that the rotor always runs supercritically, even at considerably low speed. We call this “resonant free operation”. Fig. 4.13. shows how this is possible.

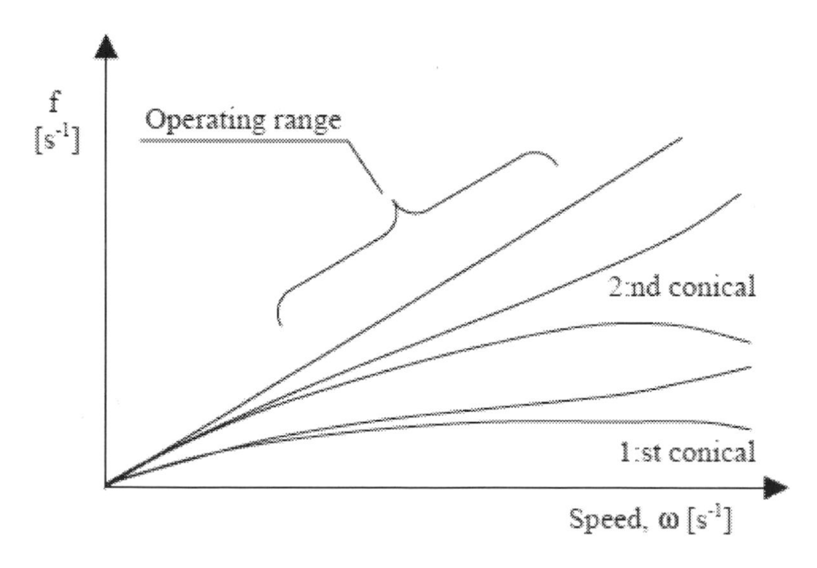

Fig. 4.13. Resonant free operation with induction bearings

4.4.1 Equations of motion

To study the static equilibrium position of the rotor and its stability, the equations of motion are required. For a two degree of freedom system, thus neglecting gyroscopic effects, and also neglecting effects of imbalance, the system becomes

$$\begin{cases} m\,\ddot{x} = -kx - d\dot{x} - k_c y + d_c\dot{y} + f_{e,x}(t) \\ m\,\ddot{y} = k_c x - d_c\dot{x} - ky - d\dot{y} + f_{e,y}(t) \end{cases} \tag{4.34}$$

where $f_e(t)$ represent all external forces acting on the system.

The stiffness and damping coefficients k and d and their related cross coupling terms k_c and dc can be found from approximation-mathematical simulation methods for the bearing model geometry such as finite methods or from experimental validation of the prototypes.

These four coefficients can be reduced to two, using the similarities from Eq. (4.24) and (4.26). Thus

$$\begin{cases} m\,\ddot{x} = (-K\cos\theta)\,x + \left(-\frac{K}{\omega}\sin\theta\right)\dot{x} + (-K\sin\theta)\,y + \left(\frac{K}{\omega}\cos\theta\right)\dot{y} \\ m\,\ddot{y} = (K\sin\theta)\,x + \left(-\frac{K}{\omega}\cos\theta\right)\dot{x} + (-K\cos\theta)\,y + \left(-\frac{K}{\omega}\sin\theta\right)\dot{y} \end{cases}$$
(4.35)

where the influence from external forces has been neglected. For a certain speed ω, the variables K and θ are fixed, and can be calculated analytically from Eq. (4.24) and (4.26). Also can be found from approximation-mathematical simulation methods for the bearing model geometry such as finite methods or from experimental validation of the prototypes.

Peripheral damping

In all rotating machines, there is always a certain damping from sources that are not always known, like the foundation. Often there are also additional vibration dampers installed. Adding the damping coefficient c_p to the equations of motion will take care of the dynamic effects of this peripheral damping.

Peripheral stiffness

Usually designers of electro dynamic bearings use peripheral bearings as well, providing either positive or negative stiffness k_p . Filatov uses a radial electro dynamic bearing in combination with a passive magnetic axial bearing, which provides negative radial stiffness, and thus has a destabilizing effect on the system. The author has used a passive radial bearing providing positive stiffness in combination with an electro dynamic radial bearing in the pump prototype 1, and in prototype 2, the scales, additional stiffness comes from the springs. Thus in order to provide a general system applicable for most types of bearing and damper combinations, the peripheral stiffness as well as the peripheral damping has to be added to the equations of motion, which is done in Eq. (4.36).

$$\begin{cases} m\,\ddot{x} = (-k_p - K\cos\theta)\,x + (-c_p - \frac{K}{\omega}\sin\theta\,)\dot{x} + (-K\sin\theta\,)\,y + (\frac{K}{\omega}\cos\theta\,)\dot{y} \\ m\,\ddot{y} = (K\sin\theta\,)\,x + (-\frac{K}{\omega}\cos\theta\,)\,\dot{x} + (-k_p - K\cos\theta\,)\,y + (-c_p - \frac{K}{\omega}\sin\theta\,)\dot{y} \end{cases}$$

(4.36)

The equation system above can be rewritten in state space equations and be processed in for instance a software to find the bearing dynamics in the time domain for most applications. The external forces $f_e(t)$ may be added to perform harmonic excitations or to find step response and so on.

4.4.2 Static Equilibrium

When the rotor is exposed to a static load W, like gravity, the equilibrium position will move to a new equilibrium position. This position is found by setting all time derivatives in Eq. (4.34) or (4.35) to zero and introducing the force

$$f_{e,y} = W \tag{4.37}$$

Solving the equations of motion shows that the new equilibrium will move along a line defined by the force angle θ. This line is in fluid film bearing dynamics known as the static load line, and is usually derived for either constant speed or constant load. Fig. 4.14 below shows the static equilibrium and the load line derived for constant speed. The displacement along the static load line is

$$\Delta r_0 = \frac{W}{K} \tag{4.38}$$

The new equilibrium position (x, y) is, when the load is directed downward as in Fig. 4.14,

$$(x,y) = (\Delta r_0 \sin\theta, -\Delta r_0 \cos\theta) = \left(\frac{W}{K}\sin\theta\ ,\ -\frac{W}{K}\cos\theta\right) \tag{4.39}$$

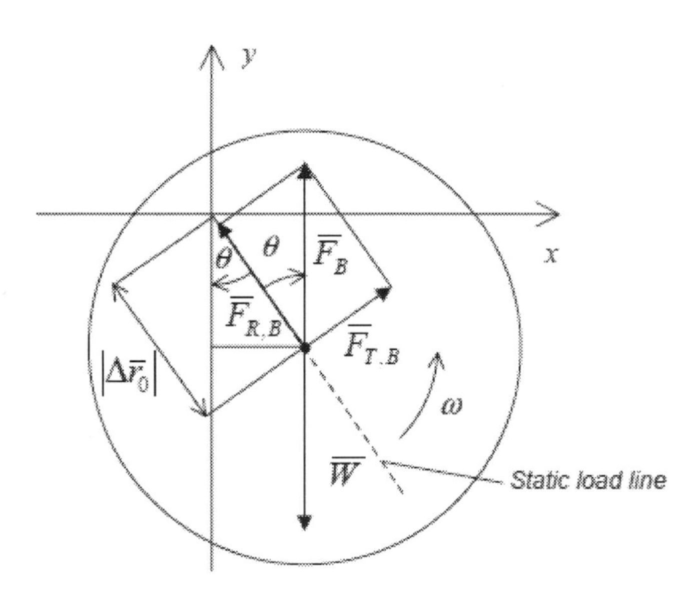

Fig. 4.14. Static load equilibrium

4.4.3 Load Range

When load is increased, the eccentricity will eventually grow so large that emergency bearing contact will take place. To calculate this force it is convenient to use the "in plane stiffness" K. The eccentricity takes place along the static load line. The load and the bearing force are identical but oppositely directed in the equilibrium position. Using Eq. (4.38) and replacing the displacement $\Delta r_0$with the available airgap g_a gives the bearing force at emergency bearing contact

$$F_{B,max} = Kg_a \tag{4.40}$$

Emergency bearings are normally mounted so as to limit rotor movements to a region within 50% of the total airgap g. Thus

$$g_a = 0.50 \text{ g} \tag{4.41}$$

Note that according to the static equilibrium shown in Fig. 4.14, both the restoring force and the tangential force components are used to carry the load.

Since bearings force is speed dependent, the maximum load should not be applied until the machine has accelerated above the speed for which the maximum load was calculated. If the maximum load is due to the weight of the machine, the load shall be calculated for the highest speed at which safe emergency bearing operation can be guaranteed.

If **higher load** is desired, it is possible to add permanent magnets to form a "static unloader". Such devises can carry very large weights without causing much negative stiffness in any direction. However, no such study has ever been performed to find the maximum levels. Likely, there exists no upper limit for a given maximum allowed negative stiffness.

This implicates that very large machines like stationary flywheels can be operated according to this principle.

4.4.4 Stability

The fact that a static equilibrium position exists, as described previously, does not guarantee that this equilibrium is stable.

The standard procedure to analyze the stability of the system defined by the set of equations of motion given in Eq. (4.36), is the Routh-Hurwitz procedure. Filatov has performed a stability analysis according to this procedure, and though his bearing and all coefficients are very unlike the induction bearing presented here, the system of equations is identical. Thus his results can be used.

To solve the Routh-Hurwitz requirements for stability, Filatov made a simplification based on

$$c \gg \frac{K}{\omega m} \tag{4.42}$$

so that the rotational speed dependent terms can be neglected. He also assumed that the lateral velocity v_0 is small compared to the rotational velocity $\omega\Delta r_0$ and that the lateral acceleration $\dot{v}_0$ is small compared to ωv_0 .

If the peripheral stiffness k_p is set to zero, which means that electrodynamic bearing is responsible for all system stiffness, then he was able to show that the system is stable, if the damping coefficient from the peripheral dampers

$$c_p > \sqrt{\frac{K}{m}} \frac{sin\theta}{\sqrt{cos\theta}} \tag{4.43}$$

where it should be pointed out that in his system the mass m is suspended in only one radial bearing of stiffness K.

Since θ decreases with speed, and K increases to an asymptotic value, it means that the system is always stable at high speed, and always unstable at low speed. To improve low speed properties, either more efficient dampers, or a bearing with higher stiffness or lower force angle is required.

Sometimes, like in prototype 1 and 2, it is appropriate to add peripheral stiffness k_p to improve low speed stability. Using this method the author has achieved stability at all speeds, neglecting bending critical speeds.

5

Active Magnetic Bearing AMB

5.1. AMB Components and Working Principle

This chapter does not provide a complete theory about Active Magnetic Bearing AMB. Nevertheless, an overview of the AMB theory is given in order to give the bases for the Operation and Performance of AMB.

In this section a general overview of active magnetic bearings is provided, the basic working principle, the main components as well as AMB configurations.

Figure 5.1 shows the basic components of a magnetic bearing and its working principle. A magnetic rotor is suspended by an electromagnet. In order to get an active control of the rotor, its position is measured by a position sensor. The position signal is then treated by a controller, which gives a current set point. This signal is then amplified by the power amplifier, in order to get the necessary actuator current. A closed loop control is thus realized and the system can be stabilized.

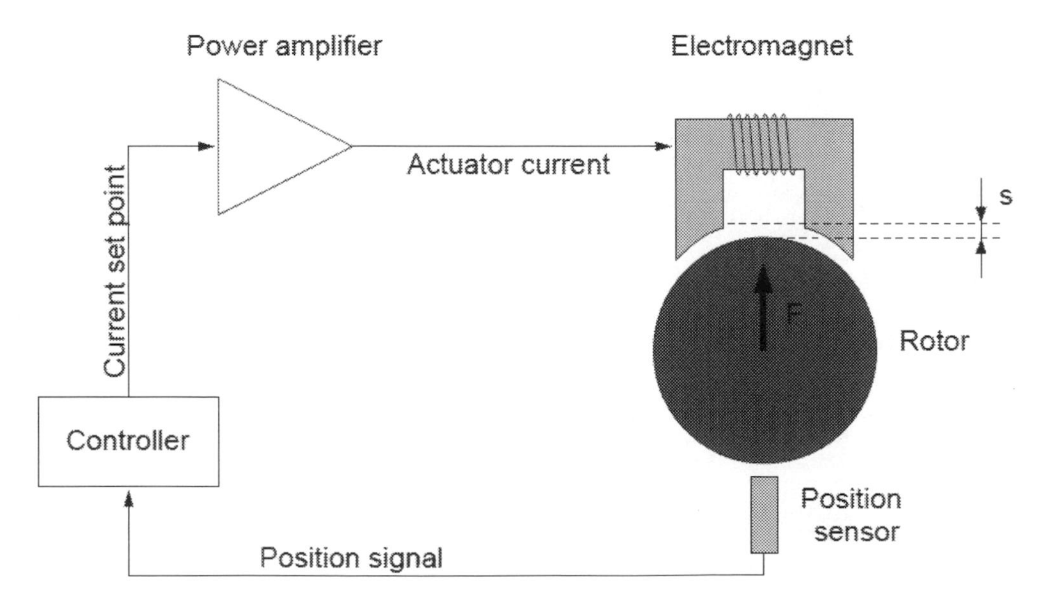

Figure 5.1. Function principle of the active magnetic bearing.

This single actuator enables the levitation along only one axis and only in one direction. In AMB systems, several actuators are used in order to control the rotor levitation along several degrees-of-freedom (DOF). Actuators are typically arranged as pairs facing each-other.

Electromagnets are composed of a soft magnetic core and electrical coils. They look somewhat like the stator of an electrical motor. The components and parts of the AMB are briefly described as below:

1- Iron Core

The iron core is a material conducting the magnetic field to the air gap. Its magnetic permeability has to be high, as well as its magnetic saturation. In order to minimize eddy current losses, the core usually consists of insulated lamination sheets.

A lot of AMB radial bearings have four magnetic poles. That means that the bearing is able to attract the rotor along two orthogonal axis, and in both directions. Figure 5.2. shows the configuration of an heteropolar NS-SN-NS-SN radial bearing. Doing so brings the advantages that the axis are decoupled.

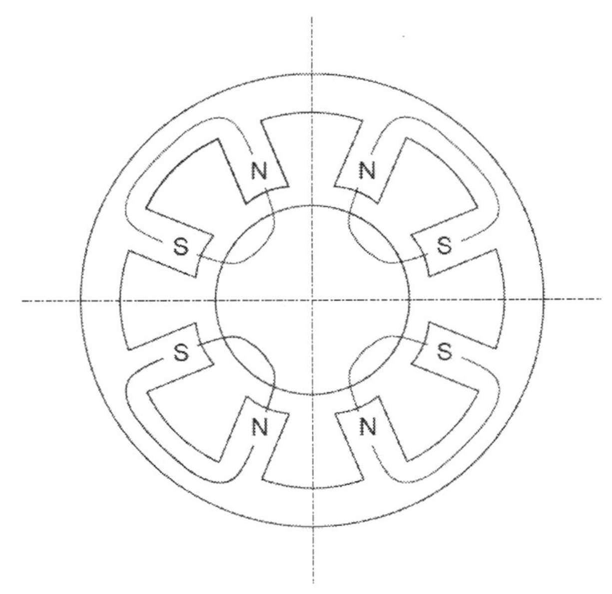

Figure 5.2. Magnetic poles configuration for a NS-SN-NS-SN heteropolar radial bearing.

Actually only three poles are sufficient for a radial bearing. This configuration is not commonly used because of the coupling between the axis. The tendency in jet engine industry is to build radial actuators with 6 poles. These bearings are able to guarantee the levitation even in case of total failure of an actuator axis.

2- Windings

The current through the winding is the source of magnetic field. The winding is made of an insulated conductor wound on the soft magnetic core. In order to improve the efficiency of the AMB, the conductor has to have a low electrical resistance and must be wound with a high fill-factor.

3- Rotor

The rotor, in standard constructions, is realized with a lamination packet shrinked on a nonmagnetic shaft. Tight manufacturing tolerances are needed in order to avoid unbalances. The mechanical properties of the rotor lamination have to be good, in order to overcome the centrifugal stress due to high speed rotation.

4- Position Sensors

In most applications, there are position sensors in AMBs. Since AMBs are actively controlled regarding to the sensor signal, the control performance strongly depends on the sensor performance. The use of the different sensor families depends strongly on the media through which or on which they are to measure. The different measuring methods are classified and defined briefly as following :

A. Inductive sensors

Inductive sensors consist of a coil with an alternating current passing through it. The coil is surrounded by a ferrite core. The rotor must be magnetically conducting at the measuring position. The impedance of the coil depends on the distance to the rotor. This can be evaluated electronically.

B. Eddy-current sensors

Eddy-current sensors are essentially built up in the same way as the inductive sensors. The two sensor types differ in the level of excitation frequency and the choice of the rotor material. Typically the excitation frequency lies between 100kHz and 2MHz. The rotor material should be nonmagnetic with high electrical conductivity. Usually the measurement is done on aluminum, which ideally satisfies these conditions. The eddy currents induced in the aluminum extract energy from the electric excitation circuit. This energy represents a measure for the gap width between rotor and sensor. Because of the higher excitation frequency the bandwidth of eddy-current sensors is larger than that of inductive sensors. Both sensor types have about the same immunity to disturbance.

C. Capacitive sensors

The design of a capacitive sensor is based on having two small excited electrodes facing the rotor and separated by small airgaps forming two series capacitors with the rotor. The equivalent circuit of the capacitive sensor configuration is a arrangement as a series circuit of two capacitors. The sensor is driven by an alternating current. The amplitude of the resulting voltage between the two electrodes is proportional to the distance between them. The resolution of this type of sensor is very high (0.2μm with a measuring range of 0.5mm). Capacitive sensors are sensitive to pollution, which influences the permittivity in the air gap.

D. Magnetic sensors

In a ferromagnetic circuit with an air gap, a constant current generates a magnetic field. The magnetic flux through the front faces of the air gap depends on the length of the air gap. It could be measured by a Hall probe. Magnetic sensors are by definition sensitive to external magnetic fields.

E. Optical sensors

The principle of a simple optical sensor is based on transmitting a light beam towards a detector. Transmitter, detector and rotor are placed in such a way that the rotor partially interrupts the light beam. Therefore the received light quantity is a measure for the position of the rotor. All optical methods of position sensing are sensitive to contamination. Therefore they are unsuited for many applications.

5- Controller

The heart of an active magnetic bearing system is the control electronics. Today controllers are mainly based on digital technology. Advantages of digital technology are:

- Flexibility in the development, belated changes can be made simply
- Complex controller structures, even time variant systems are realizable
- All quantities are accessible for monitoring or other processes

- Calibration of the control parameter
- Fast adaptation of the control software to new systems
- Cheap reproduction of the software

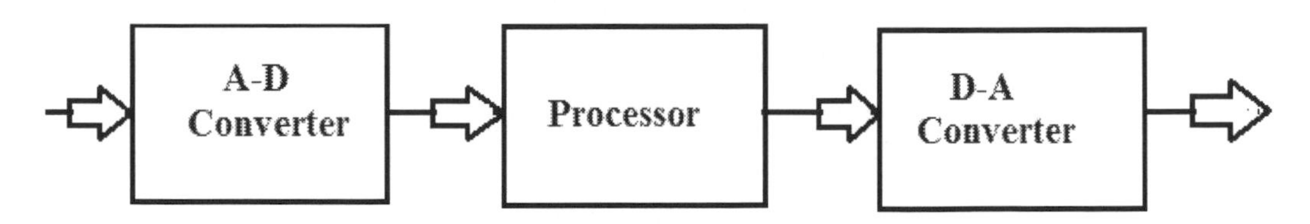

Figure 5.3 Rough structure of a digital control components/system

For real time processing, Digital Signal Processors (DSPs) are used. Analog to digital and digital to analog converters are used before and after the processor. Different methods such as Proportional Development PD, Proportional Integral Derivative PID, optimal output feedback or observer based state feedback are in use, where:

i. **Proportional - Derivative PD:** The aim of using PD controller is to increase the stability of the system by improving control since it has an ability to predict the future error of the system response. In order to avoid effects of the sudden change in the value of the error signal, the derivative is taken from the output response of the system variable instead of the error signal. Therefore, D mode is designed to be proportional to the change of the output variable to prevent the sudden changes occurring in the control output resulting from sudden changes in the error signal. In addition D directly amplifies process noise therefore D-only control is not used.

ii. **Proportional-Integral-Derivative (PID) control**: PID is the most common control algorithm used in industry and has been universally accepted in industrial control. The popularity of PID controllers can be attributed partly to their robust performance in a wide range of operating conditions and partly to their functional simplicity, which allows engineers to operate them in a simple, straightforward manner. As the name suggests, PID algorithm consists of three basic coefficients; proportional, integral and derivative which are varied to get optimal response. A PID controller attempts to correct the error between a measured process variable and a desired set point by calculating and then outputting a corrective action that can adjust the process accordingly.

6- Power Amplifiers

The power amplifiers convert the control signals into control currents. Switching amplifiers are usually used because of their low losses. The amplifier is often the limitating component in an AMB system. Amplifiers with voltage or flux density control may improve AMB performance in certain cases, but are not yet widely used. Power amplifiers in a magnetic bearing system serve to generate the control current ic which corresponds to the desired signal id (usually voltages). Like the signal processing electronics the power electronics can be divided into two families:

- Linear or Analog power amplifiers; the bearing current can be considered as ideally smoothed. Unfortunately analog power amplifiers have large power losses.
- Switched power amplifiers; Instead of a transistor, which is operated linearly, this type of switched amplifier consists of a structure with 4 switches which are simultaneously open or closed.

In order to minimize rotor losses configurations with homopolar radial bearings are possible. Other AMB configurations using permanent magnets have been developed.

The major application of AMBs nowadays are turbo-molecular pumps. The advantages of AMBs compared to conventional bearings can be very significant. They work without emission of lubricant, without emission of particles due to friction and without wear and are moreover very reliable. A lot of systems are running for more than ten years without maintenance. All these advantages make AMBs being a highly attractive alternative to the existing jet engine ball bearings.

5.2. Operation and Performance of AMB

5.2.1 Basic Operation

As been described earlier, an active magnetic bearing (AMB) consists of an electromagnet assembly, a set of power amplifiers which supply current to the electromagnets, a controller, and gap sensors with associated electronics to provide the feedback required to control the position of the rotor within the gap. The power amplifiers supply equal bias current to two pairs of electromagnets on opposite sides of a rotor. This constant tug-of-war is mediated by the controller which offsets the bias current by equal but opposite perturbations of current as the rotor deviates by a small amount from its center position.

The gap sensors are usually inductive in nature and sense in a differential mode. The power amplifiers in a modern commercial application are solid state devices which operate in a Pulse Width Modulation (PWM) configuration. The controller is usually a microprocessor or DSP.

It is difficult to build a magnetic bearing using permanent magnets due to the limitations imposed by Earnshaw's theorem, and techniques using diamagnetic materials are relatively undeveloped. As a result, most magnetic bearings require continuous power input and an active control system to hold the load stable. Because of this complexity, the magnetic bearings also typically require some kind of back-up bearing in case of power or control system failure.

Two sorts of instabilities are very typically present with magnetic bearings. Firstly attractive magnets give an unstable static force, decreasing with greater distance, and increasing at close distances. Secondly since magnetism is a conservative force, in and of itself it gives little if any damping, and oscillations may cause loss of successful suspension if any driving forces are present, which they very typically are.

An active magnetic bearing (AMB) system supports a rotating shaft, without any physical contact by suspending the rotor in the air, with an electrically controlled (or/and permanent magnet) magnetic force. It is a mechatronic product which involves different fields of engineering such as Mechanical,

Electrical, Control Systems, and Computer Science etc. AMBs are typical mechatronic devices, and one of the most attractive features of such devices is their ability of internal information processing.

The machine is termed smart if it uses its internally measured signals to optimize its state. Such a smart machine makes use of the built-in active control to incorporate additional or higher performance functions. Thus, the machine may acquire higher precision and the ability for self-diagnosis, it can calibrate itself, it can give a prognosis about its future ability to function in a satisfactory way, or about its remaining lifetime, and possibly, it can suggest a correction measure, a therapy, or even induce it itself. It is the mechatronic structure of the machine, the built-in control, its sensors, processors, actuators, and above all, its software which enable these novel features. This is a way to design machines and products with higher performance, less maintenance costs, longer lifetime, and an enhanced customer attraction. In this respect, AMBs already show promising features, but they have by no means reached their full potential. Subsequently an outlook on the concept of such a smart machine is given.

The smart machine in Figure 5.4. consists of three main parts. One is the "Actual Mechatronic System", the actual machine with its process, sensors, actuators and the controller. In our case this would be the rotor of a machine tool or a turbo-machine in magnetic bearings. The second part is the "Mechatronic System Model", a software representation of the actual machine. Of course, setting up such a model may not be simple, and that is why identification techniques are an important tool in this technology. The model, or a part of it, will be used, too, for designing the control of the actual machine. The third part describes the "Smart Machine Management". It indicates the additional functions that can be conceived by making "smart" use of the available information .

At first, data have to be collected from the actual machine and its sensors and from the model as well which runs in parallel to the actual machine.

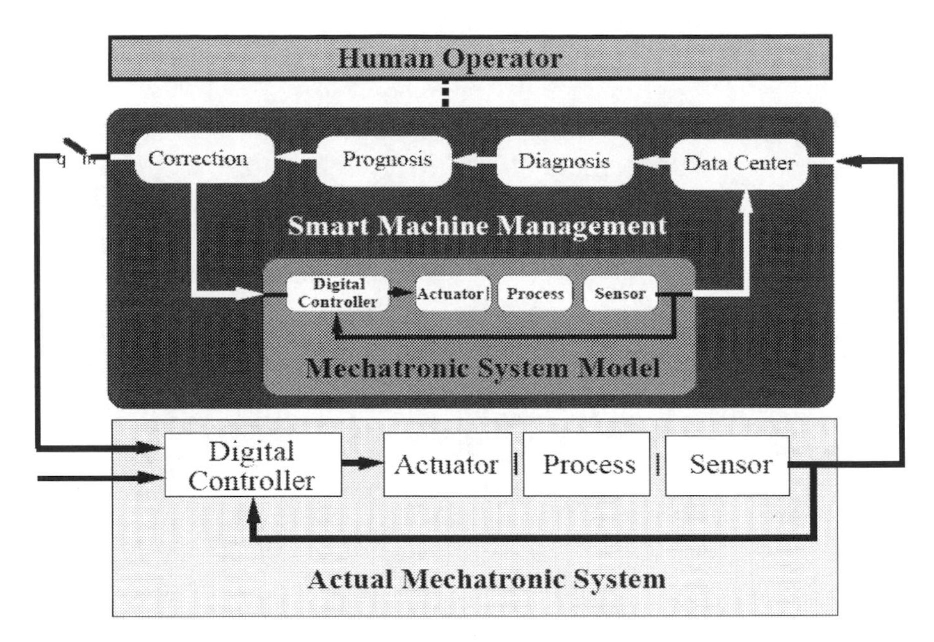

Figure 5.4. A concept diagram for the structure of a smart machine

Based on this information a diagnosis of the present state will be possible. It can be a model based diagnosis, and furthermore, due to the built-in control loop, self-diagnosis or active diagnosis will be possible, i.e., it will be possible to check assumptions about parameters by creating suitable test signals in the model, and in the actual system. This approach could further improve identification procedures, and it will be of interest for reliability management, for finding out about failures in mechanical components such as cracks in the rotor, or about failures in electrical components, for example in sensors. Based on the results of diagnosis a prognosis about the future behaviour of the machine or about the need for and the consequences of corrective measures can be derived. Such corrections, for example, may be the compensation of an unbalance, special procedures for passing critical speeds, changing the feed of a machine tool during the manufacturing of delicate parts by taking into account the cutting forces or tool wear , or it may even lead to a self-tuning of the parameters of the control loop .

In particular, the reliability of magnetic bearings, which is one of the most important aspects, could be further improved by an appropriate management of the available information and by creating an analytical redundancy in addition to any hardware redundancy.

5.2.2 Performance and specification criteria

As in the conventional bearings, active magnetic bearings' characteristics can be sufficiently described to allow assessments of rotor-bearing performance via rotor-dynamic analysis. Approximate equivalent stiffness and damping coefficients can be developed for magnetic bearings that allow conventional rotor-dynamic analyses including undamped and damped critical speed analyses and forced response analyses to be conducted. Although this approach allows some insight into rotor-bearing system performance, it partially masks the true nature of magnetic bearings as feedback control systems.

Since the objective of magnetic suspension applications in rotating machinery is to stably locate the rotor during machine operation, a magnetic bearing system is, in essence, a position control system that suffers some complications from rotor rotation. The accuracy of the rotor position control is determined by position sensor measurement accuracy, controller gain, controller bandwidth, and disturbance forces acting on the rotor. However, the fundamental challenge to successful magnetic bearing application is achievement of system stability. Stability of the rotor-bearing system is related to the vibration performance of the machine, an important consideration in specification of any type of bearing system.

Accordingly, it would seem that the most meaningful specification of magnetic bearing system performance would be in classical control parameters related to frequency response, relative stability, and stability robustness to parameter variations. However, this requires complete knowledge of the characteristics of the turbo-machinery manufacturer's rotor which comprises the plant of the feedback control system.

Ideally, performance criteria for a magnetic bearing controller could be specified by the classical control system specifications of gain and phase margin (relative stability), synchronous frequency response, and stability robustness to plant parameter variations. However, as can be concluded from the above discussion, there are a multitude of options for stabilizing controllers and actuator designs for a given rotor system.

Identification of suitable criteria is difficult in rotor- bearing systems where rotor open-loop behavior will vary widely across the class of turbomachinery and instability can arise from various sources. This condition makes specification of controller performance requirements difficult.

The problem of practical system performance specifications can be resolved by resorting to the traditional frequency response parameter used for performance evaluation of turbomachinery rotor bearing systems: vibration amplitude. Although vibration standards are often invoked to minimize damaging dynamic loading of conventional bearing systems, satisfaction of appropriate vibration standards for magnetic bearing systems ensures that the previously mentioned attributes of closed loop stability at synchronous and nonsynchronous frequencies have been met, and that the system is relatively insensitive to disturbances. However, vibration standards do not explicitly indicate relative stability margins and they do not ensure stability robustness. They also unfairly burden the magnetic bearing manufacturer with the task of demonstrating acceptable performance over a rotor system for which he is not completely responsible.

The performance aspects of magnetic bearing and recommended specification criteria can be summarized as follows:

1- The bearing system shall permit the driven equipment to run continuously at all speeds within the operating speed range at the maximum unbalance permitted by ISO 1940 for the appropriate class of rotor without exceeding a double amplitude of vibration measured adjacent to the individual bearings of $2780/\sqrt{f}$ μm or 50 μm, whichever is less, and where f is the frequency of vibration in cpm.
2- The controller may use analog or digital technology of any of a variety of algorithms provided it meets the above requirements for vibration and is stable (repeatable) with time.
3- Output or full state feedback using position, velocity, current, flux, or force is permissible provided it meets the above requirements for vibration and is stable with time.
4- Power amplifiers shall be low distortion, linear or switching type with a bandwidth consistent with the above vibration requirements. Bearing bandwidth shall be consistent with the above vibration requirements.

5- The load capacity of the bearing system shall be a minimum of the total of static plus dynamic loads for the application where static and dynamic loads are as defined below. Static loads consist of weight and steady state aerodynamic, hydrodynamic, or electromagnetic loads of the driven equipment and dynamic loads consist of varying components of the above, plus unbalance loads of the driven equipment.

6- The required dynamic capacity of radial bearings is specified at synchronous frequency and shall be a minimum of 2.5 times the appropriate ISO grade to which the rotor is balanced. The required capacity of axial bearings shall be a minimum of the total of specified static and dynamic loads as indicated above.

7- The radial position sensors shall feature a linearity and sensitivity that is independent of environmental temperature and that is consistent with stable operation of the magnetic bearing-rotor system. This generally requires linearity for radial sensors to within 2% over the magnetic air gaps and a minimum sensitivity of 10 V over the air gap involved. These requirements may be relaxed for axial sensors. To reduce plant measurement disturbance, the mounting of the radial position sensors shall ensure a concentricity with respect to the radial bearing bore of 50 μm per meter of bearing diameter with a lower limit of 15 μm.

8- All magnetic bearing components shall be compatible with the specified application operating environment of temperature, pressure, humidity, and process fluid.

9- Temperature increase due to joule, hysteresis, eddy current, and windage losses within the magnetic bearing components shall not unduly compromise the performance or the life of the bearing system. Also, operating stress levels due to centrifugal, thermal, electromagnetic, and mechanical loads shall not unduly compromise the performance or the life of the bearing system.

10- Before design approval, the magnetic bearing manufacturer should be required to conduct and present sufficient analyses to demonstrate that the above requirements are met. However, the equipment manufacturer is ultimately responsible for satisfactory performance of the complete machine with the magnetic bearing system.

5.3. Characteristics of Magnetic Bearing Systems

A minimum standardization of performance characteristics is applied so often to minimize the difficulty encountered by turbomachinery manufacturers confronted with a new bearing technology. The difficulty of characterizing performance of the complete magnetic bearing system without the rotor is circumvented for specification of bearing static and dynamic load capacity and vibration response of the completed installation.

The following subsections are directed at an examination of the characteristics of magnetic bearing systems important to overall performance of the rotor-bearing system. This will permit development of performance criteria for inclusion in a standard specification for magnetic bearing application in turbomachinery.

5.3.1. Open-Loop behavior of rotor system

For a general rotor system, a minimum of five axes of motion control are required: four axes of radial motion control, two at each of two radial bearings, and an axial axis of control provided by a thrust bearing. The basic system that is required for active control of one axis of the rotor will consist of at least one state variable sensor, a magnetic control coil, a coil driver, and the electrical network to connect it all together, see sketch given in fig.5.5 which shows an example of basic operation and control for a single axis . These elements introduce their own individual characteristics to overall system performance including frequency response limits, eddy current losses, and saturation which must be accounted for in the overall specification of the system. In addition, mechanical resonances of the stator and rotor will interact with the performance of the control system.

Specification of magnetic bearing system performance parameters starts with a discussion of the frequency behavior of the plant of the feedback control system. The components in the upper half of Fig.5.6. define the plant in one axis of control as consisting of the suspended rotor, the magnetic negative stiffness, the magnetic bearing, and the power amplifier.

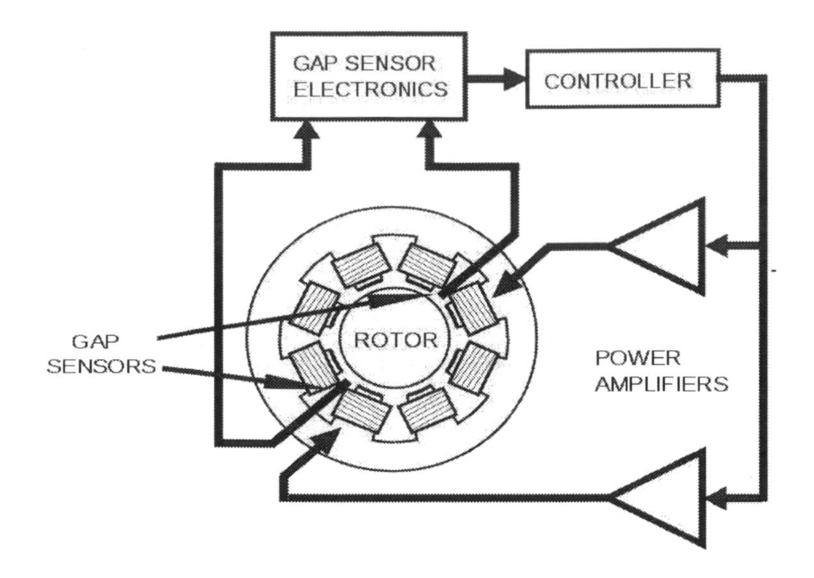

Fig. 5.5 Basic operation and control system for a single axis

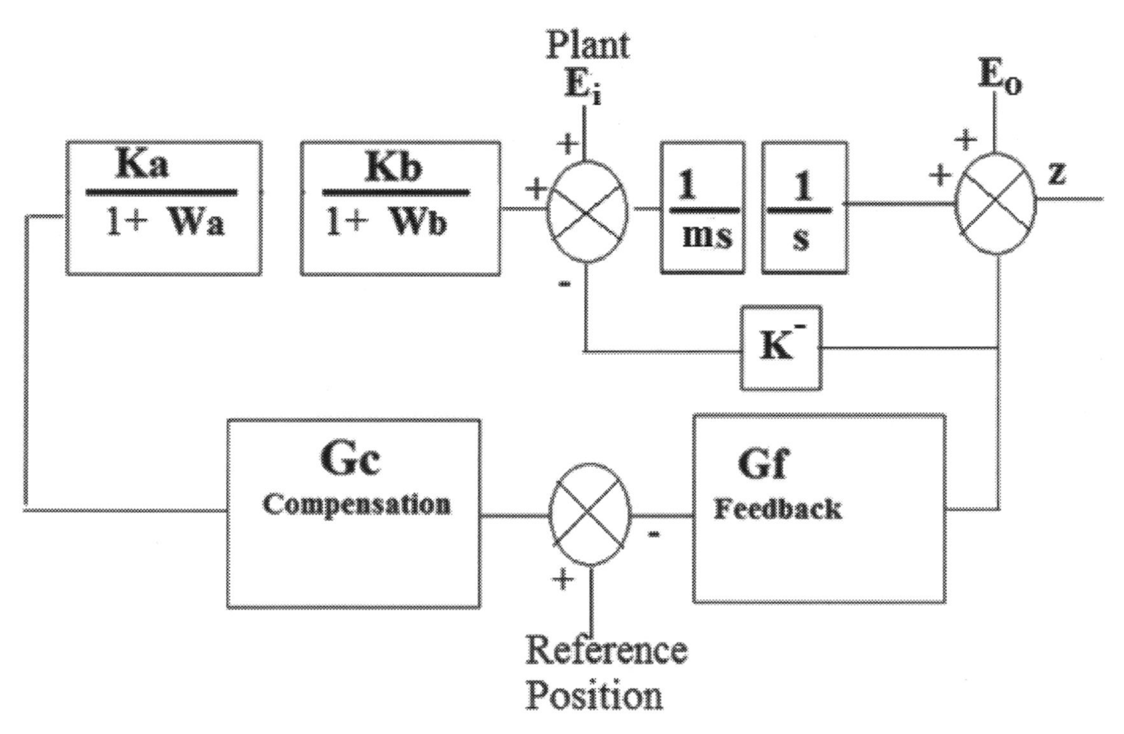

Fig. 5.6 Magnetic Bearing System for one axis control

There are two types of disturbances acting on the plant: Ei and Eo. The first, Ei, is the external forces acting on the rotor including rotating unbalance caused by the misalignment of elastic and inertial axes and aerodynamic or hydrodynamic forces. These forces are summed with the magnetic bearing control forces before passing through the rotor to affect the rotor output state variable of position.

The second type, Eo, usually ignored in conventional treatments, is the measurement disturbance created by an imperfect state variable measurement. For a position sensor looking at an imperfect shaft surface, the result is a corrupted position measurement that consists of the true center of mass position and components at harmonics of the rotational speed. This corrupted measurement is used as a feedback signal for the magnetic bearing hence producing a disturbance that excites rigid and flexible modes of the rotor system.

The power amplifier and the magnetic bearing have been included in the definition of the plant to emphasize their influence on the control system performance. Mechanical resonances in the rotor are unavoidable and they impose constraints on the frequency dependent characteristics of the power amplifier and the magnetic bearing.

The power amplifiers and magnetic bearing contribute phase lag to the open-loop transfer function for the plant requiring stabilization at high frequencies where structural interaction can occur. Core losses in the magnetic bearing contribute to phase lag by opposing flux changes in the bearing at a frequency below that normally associated with an L-R circuit. Accordingly, the bandwidth of the bearing should be as high as practical as will be shown subsequently. The magnetic bearing represented in Fig. 5.6. as a single lag with a break frequency of Wb and a roll-off thereafter which is first order (slope of – l or less).

The power amplifier is also represented as a first order lag with a break frequency that should be as high as practical for best performance. Furthermore, the coil driver should have a characteristic which is independent of load inductance and resistance. Unfortunately, this constraint is coupled with power limitations and the complications of design offered by pulse width modulation of the current to the bearing. Preferred designs should operate as a true trans-conductance amplifier yielding a current output for a voltage input. The amplifier total harmonic distortion at rated output should be as small as possible to avoid nonlinear resonance effects.

Considering a simple rotor system supported by an uncompensated magnetic bearing, the equation of motion for one axis of control is :

$$f = m.\ddot{x} - k^{-} . x \tag{5. 1}$$

Where
m = rotor mass
k^{-} = magnetic bearing negative stiffness
x = rotor displacement
$\ddot{x}$= rotor acceleration
f = external force

Equation (5.1) above applies directly to the axial control axis. For a radial control direction, Equation (5.1) applies if k^{-} is taken as the effective or combined negative stiffness of all magnetic bearings in the control direction. Equation (5.1) assumes no cross-coupled motion between axes.

In Laplacian notation, Equation (5.1) can be expressed as:

$$F(s) = m\,.\,s^2\,X(s) + k^{-}.X(s) \tag{5. 2}$$

or as $$\frac{X(s)}{F(s)} = \frac{1}{m\,.s^2 + k^{-}} \tag{5.3}$$

where **s** = jω and the capital letters in functions of **s** denote Laplace transforms. Accordingly, the rotor transfer function, X(s)/F(s), is shown for simplicity in Fig. 5.5. as $1/m\,.s^2$ with negative feedback of the magnetic bearing uncompensated negative stiffness. Thus, this representation is strictly correct for rigid body motion only but flexible mode effects are added subsequently. The $1/m\,.s^2$ transfer function shows that the rotor acts as a low pass filter rejecting high frequency disturbances.

Flexible mode effects of cross-coupled motion of real rotor systems can be accounted for by incorporating multi degrees of freedom and including gyroscopic and internal damping parameters in the model. Using complex notation to allow a more compact form, the equation of radial motion can be written:

$$[M]\{\ddot{z}\} + [D - j\omega G]\{\dot{z}\} + [K - j\omega D]\{z\} = \{f\} \qquad (5.4)$$

Where
[M] = rotor mass matrix
[D] = rotor damping matrix
[G] = rotor gyroscopic matrix
[K] = rotor stiffness matrix
{f} = rotor external force vector
{z} = rotor displacement vector $[\ldots.\ x_i + jy_i \ldots.][\ldots\ \beta_i + j\alpha_i\ \ldots.]T$
x_i = translational displacement of the ith mass along the x- axis
β_i = angular displacement of the ith mass about the x- axis
y_i = translational displacement of the ith mass along the y- axis
α_i = angular displacement of the ith mass about the y- axis
ω = frequency

The force developed by a magnetic bearing operating on a rotor through an air gap is:

$$F = \frac{\partial F}{\partial I}\partial I + \frac{\partial F}{\partial x}\partial x \qquad (5.5)$$

where x = rotor displacement.
I = bearing current

or expressed as a stiffness,

$$K = K^{+} + K^{-} \qquad (5.6)$$

Where the first term is the bearing positive "current" stiffness and the second term is the bearing negative "position" stiffness caused by the increasing attractive force as the gap is reduced. This must be overcome for stable suspension since the homogeneous solution to Equation (5.1) is in the form of hyperbolic functions indicating that x grows with time. This can be accomplished for a given rotor position in the air gap by ensuring that the first term in Equation (5.6) is larger, i.e., $K^{+} > K^{-}$.

5.3.2 Bearing system load capacity

The magnetic bearing actuator design must be consistent with frequency and voltage characteristics of the controller in order to develop peak dynamic forces as required over the control bandwidth to maintain contact free suspension. Furthermore, this must be done within current limitations without excessive bearing inductance.

For a four quadrant bearing system, two different operating modes are possible: Class A and Class B. Class A takes advantage of opposing quadrants to react the dynamic load. By neglecting the reluctance of the magnetic core, this configuration can develop a peak dynamic force given by:

$$F = \frac{I_{max} V_{max} S_i N_{pp}}{2 S \varepsilon_o \omega} \tag{5.7}$$

Where all variables are in consistent MKS units:

I_{max} = maximum current available from amplifier
V_{max} = maximum voltage available from amplifier
S_i = Inductive area of a single magnetic bearing pole
N_{pp} = number of pole pairs in the bearing quadrant
S = projected area of magnetic bearing quadrant
ε_o = magnetic bearing air gap
ω = frequency of disturbance

Implicit in the development of the above is that the saturation current of the bearing coincides with the amplifier current rating; a usual condition for cost effective designs.

Class B control utilizes current modulation to only one quadrant to react to the dynamically varying load while the opposing quadrant is energized with a small bearing DC current. For horizontal installation, only the upper quadrants modulate the load. The maximum dynamic force developed in a Class B configuration neglecting magnetic core reluctance is :

$$F = \frac{I_{max} V_{max} S_i N_{pp}}{2.83 S \varepsilon_o \omega} \frac{1}{\sqrt{X_s}} \tag{5.8}$$

where the factor of safety over the static load rating X_S has been purposely introduced to illustrate the reduction in dynamic load capacity incurred by over specification of X_S.

5.3.3. Limitations

Limitations in AMB arise from two reasons: the state of the actual technology in design and material, and the basic physical relations. There are various issues characterized the design and performance of the AMB , which they may help to make substantiated design decisions. These issues are summarized as below:

1- The maximal load depends on design.
2- The specific load depends on the available ferromagnetic material and its saturation properties, and is therefore limited to 32 to 60 N/cm^2.
3- The frequency and the amplitude of disturbances acting on the rotor, such as unbalance forces, that can be adequately controlled, depend on the design of the power amplifier (power and band width).
4- The maximally achieved rotation speed is about 300 kHz in physical experiments. For industrial applications values of about 6 kHz have been realized.
5- Circumferential speeds, causing centrifugal loads, are limited by the strength of material. Values of about 250 to 300 m/s have been realized with actual design.
6- Supercritical speed means that one or more critical speeds can be passed by the elastic rotor. It appears to be difficult to pass more than two or three.
7- The size of the bearing depends on design and manufacturability. There are large bearings with dimensions and loads in meters and tons. The smallest bearings actually built have dimensions in the range of mm, with a thickness being as small as 150 μm.
8- High temperature bearings have been realized, running in experiments at an operating temperature of 600°C °F (1100°F). For ferromagnetic material the Curie temperature would be a physical limit.
9- The losses of magnetic bearings at operating speed are much smaller than that of classical bearings. Eddy current losses will limit the

rotation frequency of massive rotors (heating up, driving power), the air drag will be crucial at high circumferential speeds (driving power).

10- A high precision of the position of the rotor axis (in the range of mm) requires high resolution sensors and adequate signal processing to separate disturbance signals from the desired ones.

11- A very high precision, aimed at for non rotating suspension and position serving (in the range of nm), requires iron free magnetic paths to avoid hysteresis effects, and adequate sensing.

12- The information processing within the AMB system can be used to make the rotating machinery smart. Actual limits have not yet been determined. The advanced information processing within the bearing system, extending the smartness of the rotating machinery, will be a promising research area.

13-The potential and limitations of high temperature super-conductors, as an extension or an alternative to AMBs, is not yet sufficiently known.

5.4. Controllability and Compensation of the AMB

5.4.1 Background

For any active magnetic bearing system, the basic operation and control philosophy is in dealing with functioning and controlling the operation of the main components of the system which consist of electromagnet, rotor, sensor, controller, and amplifier, as shown in fig.5.5 the sketch diagram.

A first classification of magnetic bearings according to the physical operation principle allows to pick out two main families:

a) Active Magnetic Bearings, making use of an electronic control unit to regulate the current flowing in the coils of the actuators. They need external source of energy.

b) Passive Magnetic Bearings: they do not need any electronic equipment.

The control of the mechanical structure is achieved without the introduction of any external energy source. They exploit the reluctance force or the Lorentz force due to the generation of eddy currents developed in a conductor in a relative motion in a magnetic field.

Active Magnetic Bearings require sensors an electronic equipment but, although more expensive respect to classical ball bearings, they offer several technological advantages:

- The absence of all fatigue and tribology issues due to contact: it allows the use of such bearings in vacuum systems, in clean and sterile rooms, or for the transport of aggressive or very pure media, and at high temperatures;
- No lubrication needed;
- No contamination by the dust created by friction between the rotor and the stator;
- Low bearing losses: at high operating speeds are 5 to 20 times less than in conventional ball or journal bearings, result in lower operating costs;
- Viscous friction can be avoided if the rotor is confined in high vacuum;
- Low vibration level;

- Dynamics adaptable to the desired application by tuning of the control loop;
- Precise positioning of the rotor due to the control loop: this is mainly determined by the quality of the measurement signal within the control loop. Conventional inductive sensors, for example, have a measurement resolution of about 1 ÷ 1000μm of a millimeter;
- Achievable fast positioning and/or high rotational speed of the rotor;
- The small sensitivity to the operating conditions;
- The predictability of the behavior.

5.4.2 Controller Requirements

In order to specify the Controller Requirements , there are Generally a primary and secondary tasks which cover the following functions:

1- Primary tasks:

- coordinate transformation of sensor signals
- collection of any other parameters needed by control algorithm
- generation of control current (or flux) *requests*: execute control algorithm
- coordinate transformation, biasing of amplifier signals

2- Secondary tasks:

- permit modification of control algorithm
- implement diagnostic measurements

Two main families of control architecture can be listed for active magnetic bearings:

- Decentralized control: the action of each actuator is independent from the others and exploits a dedicated control law and sensors information;
- Centralized control: actuators are coupled as well as sensors information. A single control action is devoted to feed power drivers.

Ideally, performance criteria for a magnetic bearing controller could be specified by the classical control system specifications of gain and phase margin (relative stability), synchronous frequency response, and stability robustness to plant parameter variations. However, there are a multitude of options for stabilizing controllers and actuator designs for a given rotor system. Identification of suitable criteria is difficult in rotor- bearing systems where rotor open-loop behavior will vary widely across the class of turbomachinery and instability can arise from various sources.

For a magnetic bearing, the stiffness and dynamic range are important features. High stiffness may make AMB system unstable, while low stiffness may make it unable to suspend the rotor.

Another requirement of the controller is that it must accommodate disturbances, Ei in Fig. 5.6, at both synchronous and nonsynchronous frequencies. Nonsynchronous disturbances can come from a variety of sources including coupling misalignment, seal and impeller hydrodynamic effects, rubs, etc. The controller must possess good synchronous response characteristics since the primary exciting mechanism in most turbo-machinery applications will be mass unbalance. The corresponding synchronous bearing forces must be consistent with the dynamic load capacity of the magnetic bearings. Accordingly, a relatively flat synchronous response is desirable.

There are two approaches to implement the AMB controller. One is analog approach, the other is digital approach. The design of the analog controllers is relatively familiar to most of designers and have been implemented by various applications. Generally the implementation of analog controllers is featured by:

- Restricted to performing linear math, implementation of linear filters and some nonlinearity using diodes, analog multipliers.
- Using op-amps, resistors and capacitors. Inductors are avoided because of size, weight, non-ideality.
- Problems: component drift, reliability of potentiometers, general characteristics of filters are fixed by circuits.
- Advantages: inexpensive, compact, fundamental simplicity

The digital controllers are suitable for optimal control of magnetic bearing system because its properties can be changed easily by modifying its software. Therefore, it is an approach which makes the design of controller easier. However, the rotor state variables of velocity and acceleration are not naturally accessible in a magnetic bearing suspension and must be generated with more hardware than a displacement feedback system. Generally the implementation of digital controllers is featured by:

- Fastest when performing linear math ,can implement nonlinear equations too, but at greater computational expense.

- Using digital signal processors, A/D converters, D/A converters
- Problems: electrical complexity, higher incidence of component failure, cost.
- Advantages: easier to modify algorithms, implement complex algorithms, perform data acquisition

By digital controller the proportional path is used to adjust the stiffness, the differentiator is used to adjust the damping of the rotor and the low pass digital filter with certain gain is used to adjust the total stiffness and damping. The transfer function of this controller is given by signal processing system.

In order to filter the high frequency noise, a low pass filter with small time constant is used before the signal is sent to the power amplifier. The effect of the low pass filter to the performance of the system can be ignored, because the small time constant causes a large pole point. The power amplifier is used as control coil driver, which can control the position of the rotor, as well as position adjustment.

The active magnetic bearings system is a closed loop control system. For this system, the stability is an essential feature. By adjusting the magnitude and phase characteristics of the controllers, a stable closed loop control system can be obtained. But it may be difficult to obtain a system with a wide stable range.

The closed loop control system consists of an eddy current displacement sensors, magnitude and phase controller, power amplifiers, bias coils and control coils. For the digital approach, the controllers are implemented with real-time filters and other software packages in the digital signal processing (DSP).

The negative stiffness of the bearing is a challenging control problem. Here again, state variable feedback and digital technology offers a solution. A velocity observer implemented into a model based compensator can recover rotor velocity without differentiation of the rotor position signal and ensure system stability despite the magnetic bearing negative stiffness.

Alternatively, flux feedback can minimize the negative stiffness with less control complexity. Since the magnetic force of attraction is given by:

$$F = \frac{B^2 S}{2\,\mu_0} \qquad (5.9)$$

Where B = magnetic induction
S = bearing projected surface area
μ_0= permeability of free space

it is evident that a linear force vs. rotor displacement is possible if $B \sim \sqrt{x}$. This latter relationship is possible by measuring B. The result is a bearing force characteristic which is independent of the bearing air gap; the unstable effect largely disappears. In fact, reductions in the negative stiffness on the order of five to ten times are achievable.

Some bearing designs utilize linearization of the bearing stiffness by the use of large DC current biases alone to minimize the effect of negative stiffness. This scheme does not eliminate the unstable effect of the negative stiffness and the remaining nonlinearity introduces resonant frequencies into the dynamic system which are subharmonic and ultraharmonic to the main resonance (pole) predicted on the basis of linear theory. These nonlinear resonances can cause further restrictions on stabilizing controllers particularly when the bearing is operating near saturation.

A review of the optimal control technologies seems to indicate that the Linear Quadratic Regulator (LQR) offers everything desirable in a magnetic suspension: good stability, robustness, and flat frequency response properties.
The optimal performance of the full state feedback controller using variable gain to account for variation in plant with rotational speed appears ideally suited to magnetic bearing applications.

5.4.3. Compensation

For recent mechanical systems, smaller size, lighter weight, higher speed, and higher accuracy have been required. Magnetic bearings are suitable to meet those requirements. A magnetic bearing suspends a rotor without mechanical contact by employing magnetic force. However, magnetic force has a significant nonlinear relationship with the air gap between the rotor and the electromagnet, and versus the current which flows in the electromagnet.

Compensation has been required in order to compensate for non-linearity over a wide range of air gaps, without using a sensor except in the case where a displacement sensor was used for magnetic bearings.

It is possible to design and obtain (by using linear control system theory) a stable control system. However, when the rotor deviates a great deal from the center, linearity is lost and stable control is not available. Therefore, while a rotor is moving normally around the center, the stable control status could be retained. However, when large vibration is generated by external force or at startup, normal operation is very difficult to achieve.

To solve this problem, three compensation methods to compensate for the electromagnet's non-linear characteristics within the linear system are considered and these are briefly described below:

A. Bias Method

The most remarkable feature of this method is that a command signal added to the bias f_s is multiplied by another, separate signal which corresponds to the air gap between the electromagnet and the rotor. This air gap signal I_1 or I_2 is obtained by adding/subtracting a displacement sensor signal x (which detects rotor deviation X) to/from signal x_0 corresponding to air gap X_0 in the center position.

The two power amplifier commands are

$I_{s1} = x_1(f_0 + f_s)$ and $I_{s2} = x_2(f_0 - f_s)$

When the displacement sensor's sensing factor is K_p , two air gap signals are expressed:

$x_1 = K_p X_1$ and $x_2 = K_p X_1$

According to these relations, the relation between controller command f_s and force F is obtained, and expressed as:

$$F=F_1 - F_2 = K[\left(\frac{K_1 x_1\sqrt{f_s}}{X_1}\right)^2 - \left(\frac{K_2 x_2\sqrt{f_s}}{X_2}\right)^2] \quad (5.10)$$

$$F= 4 K K_1^2 K_p^2[(f_0 + f_s)^2) - (f_0 - f_s)^2)] \quad (5.11)$$

$$F=k f_s \quad \text{Where } (k = 4K K_1^2 K_p^2 f_0) \quad (5.12)$$

This equation does not have a variable rotor deviation $\overline{X}$, that is to say, this equation is effective with or without considering a rotor levitating position. Thus, stable controllability is assured wherever a rotor is located only if the control system can be stabilized in the center position. This is because loop gain in the control system is not changed even if the rotor levitating position is changed.

B. Square Root Method

This method improves upon the ordinary method. A multiplier is provided in the output stage of the distributor. Then a command signal from the square root circuit is multiplied by a signal which corresponds to the air gap. The square root circuit just mentioned is provided in place of a bias addition circuit as in the Bias method. The two power amplifier commands are:

$$I_{s1}= x_1\sqrt{f_s} \quad (f_s \geq 0)$$
$$I_{s2}= x_2 \sqrt{-f_s} \quad (f_s< 0)$$

From these equations and displacement sensor's sensing factor K_p, the relation between controller command f_s and force F can be obtained as follows:

$$F=K \left(\frac{K_1 x_1\sqrt{f_s}}{X_1}\right)^2 \quad (f_s \geq 0 \text{ only electromagnet 1 works}) \quad (5.13)$$

$$F=K \left(\frac{K_1 x_2\sqrt{-f_s}}{X_2}\right)^2 \quad (f_s < 0 \text{ only electromagnet 1 works.}) \quad (5.14)$$

$F = k\, f_s$ Where $(k = K\, K_1^2\, K_p^2)$ (5.15)

Since this equation does not have a variable rotor deviation $\bar{x}$, it can be established regardless of the rotor levitating position. Therefore, once an initial setting is stabilized

Because of the fact that two electromagnets are applied with current, Bias method cannot be employed for a bearing with only one electromagnet, while Square Root method can be applied to bearings with either one or two electromagnets. When an external disturbance load is zero, the controller command and power amplifier command become zero and no current flows. Therefore, Joule heat becomes zero and heat loss will not be generated by eddy current in the rotor when the rotor rotates at high speed.

C. Combination Method

In Bias method, since current is always applied to two electromagnets, Joule heat cannot be avoided. Also heat loss is generated by eddy current in a rotor when the rotor rotates at high speed. In B method, due to a square root circuit, input-output ratio of the circuit becomes larger when the command is close to zero. A slight variation of the controller command value will change greatly the electromagnet coil's current. Thus, response of the force affecting a rotor becomes less when the controller command is close to zero because of electric circuit and electromagnet inductance.

In other words, one of the Bias method features is that there is no problem with reduced response. The combination method (C method), combining A and B methods and with intermediate features, is slightly more effective. In C method, two current commands I_{s1} and I_{s2} are in accordance with the following equations, corresponding to controller command f_s:

1) $-1 \le f_s < -a$

$$I_{s1} = 0$$
$$I_{s2} = x_2\sqrt{-f_s}$$

2) $-a \le f_s < a$

$$I_{s1} = \frac{x_1\sqrt{a}}{2}\left(1 + \frac{f_s}{a}\right)$$
$$I_{s2} = \frac{x_2\sqrt{a}}{2}\left(1 - \frac{f_s}{a}\right)$$

3) $a \leq f_s \leq 1$ $\quad I_{s1} = x_1\sqrt{f_s}$

$I_{s2} = 0$

Where $0 < a < 1$

From these equations and displacement sensor's sensing factor K_p, the relation between controller command f_s and force F can be obtained as follows:

$$F = F_1 - F_2 = k\, f_s \qquad (5.16)$$

Where ($k = K\, K_1^{\,2}\, K_p^{\,2}$)

Since this equation does not have variable $\overline{X}$ as do the other two methods, it can be established regardless of the rotor levitating position.

6

Design of Magnetic Bearings

6.1. Background

Areas involved in the design of magnetic bearing systems can be classified as following:

1- Mechanical design (Modal frequencies)
2- Magnetic actuator design (Bearing magnet, Coil design and Power amplifier)
3- Control system design (Controller design and Sensor design)

According to the known technology, magnetic bearings can be classified for their design according to the purpose of the levitated object as following:

1- Precision flotors (precision stages, isolation bases, isolation springs) using Levitation force or/and Propulsion force.
2- Linear motors (Contactless sliders, maglev trains and conveyors) which is performed on having Levitation and Propulsion forces
3- Levitated rotors (gas turbines, energy storage flywheels, high speed spindles, balancing and vibration control of rotors). Rotor levitated by Radial and Axial Active Magnetic Bearings.
4- Bearingless motors (canned pumps, compact pumps, blood pumps, spindle drives, semiconductor process). To run as Radial load, Thrust load or Torque.
5- Contactless Gears and Couplers(Regulated torque or transmission)

As for any electromagnetic design process the optimal design is carried out in two steps:

- Modeling the magnetic circuit to determines the accuracy of achieving the objective.
- Optimization of the parameters to determine the efficiency of the achieving the objective.

This chapter gives an introduction to AMB design parameters and features. The magnetic model & Current-Force relationship are presented in mathematical form, well-known linear AMB equation. Finally an overview of the losses in AMB and high temperature aspects in AMB Design are given.

6.2. Design Parameters and features

Millstones and steps of designing induction bearings are briefly given in following simple hands on rules:

1- Choose configuration:
   a- Inner rotor, with ring magnets on the outside.
   b- Outer rotor, with ring or disc magnets on the inner stator.
   c- Intermediate rotor type, comprising a tubular rotor with magnets both on the inside and on the outside of the rotor.

2- Choose rotor material with as high conductivity as possible.

3- Decide the lowest rotational speed during continuous operation.
   Let the wall thickness of the cylinder equal one skin depth or more for this frequency. For intermediate rotors do not use too thick walls. A thickness of between one and one and a half skin depth is recommended for best use of the magnets.
   Choose the pole width of the magnets, including pole shoes, to two times the sum of the airgap and the skin depth at this frequency.

4. Decide the maximum rotational speed.
   Choose the maximum rotor diameter with respect to constraints like centrifugal acceleration or other design criteria. Use bandage only on outer rotor type bearing in order to keep the airgap small.

5. Choose magnet arrangement.
   Flux concentration with pole shoes gives high stiffness at low speed. Other advantages are simple magnet arrangement, flat stiffness versus speed curve and small side forces.

6. Choose magnet cross section depending on optimization criteria.
   Maximum stiffness: Let the width of the magnet equal at least two times the pole width plus two times the axial length of a pole shoe, if the latter are used. Economic stiffness: Let the width be about 60% of the value above. This will result in maximum stiffness per kilo magnet.

7. Choose number of magnets for desired stiffness.
   Adjust by changing diameter or bearing width.

8. Choose type of bearing end.
   Reduce the axial length of the end pole shoe to 50%.
   If there is place enough, let this pole shoe cover the airgap and some of the conducting cylinder.

Do not make the cylinder slightly longer than the stator package. Make it either somewhat shorter or much longer.

6.3. Magnetic Model

A. Magnetomotive Force

The magnetic field is generated by an electrical current. Ampère's circuital law (equation 6.1) gives the relation between the magnetic field H [A/m] and the current sum enclosed by the closed integration path L [m].

$$\oint H \cdot dl = Ni \tag{6.1}$$

where N is the turn number of the coil and i the current going through, as drawn on figure 6.1.

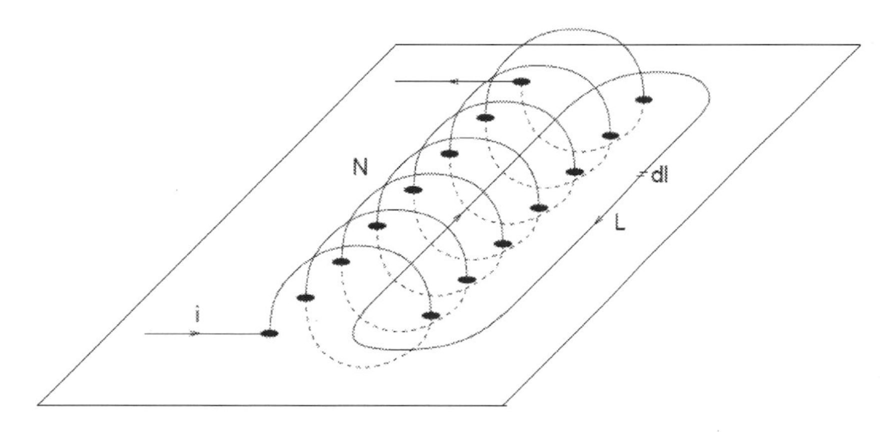

Figure 6.1. Air solenoid

B. Soft Magnetic Material Magnetization

The magnetic flux density B [T] is given by the following relation:

$$B = \mu_r \mu_o H \tag{6.2}$$

Here μ_o represents the permability of air ($\mu_o = 4\pi.10{-}7$ [H/m]) and μ_r reepresents the iron core relative permeability.

When a magnetic field is applied on a material, the generated magnetic flux densitiy B depends on the material properties. **Diamagnetic** materials do have $\mu_r < 1$,while for **Paramagnetic** $\mu_r > 1$, which means that the field B generated is higher than the flux density in the vacuum.

Ferromagnetic material, which are used for the realization of magnets, have a μ_r ≫1, and are available for high temperature applications (Here high temperature means T ≫373°K. A confusion should not be done with high temperature superconductor magnets, which have today a maximum operating temperature of about 130°K). However the temperature has to be kept lower than the Curie temperature, beyond which the materials loose ferromagnetic properties. Ferromagnetic materials show non-linear characteristics. Their magnetization depends on their history, and can be represented on a B–H diagram (figure 6.2). When an unmagnetized material sample is put whithin a homogeneous field H, the magnetization flux B increases rapidly, according to a new magnetization curve. If H keeps increasing, the material reaches its magnetic saturation, and B then only increases with a slope μ_o.

When the outer field is reduced to $H = 0$, the flux density does not run reversibly along the new curve. The plotted relationship will follow a similar curve as for the magnetization, but back toward the zero flux density. This behavior is called magnetic hysteresis. At $H = 0$, the flux density on the demagnetization curve is not zero. The material has been magnetized and the remaining flux density B_R is called remanent flux density or remanence. In order to get $B = 0$, a field H_C, called the coercitive field, is necessary.

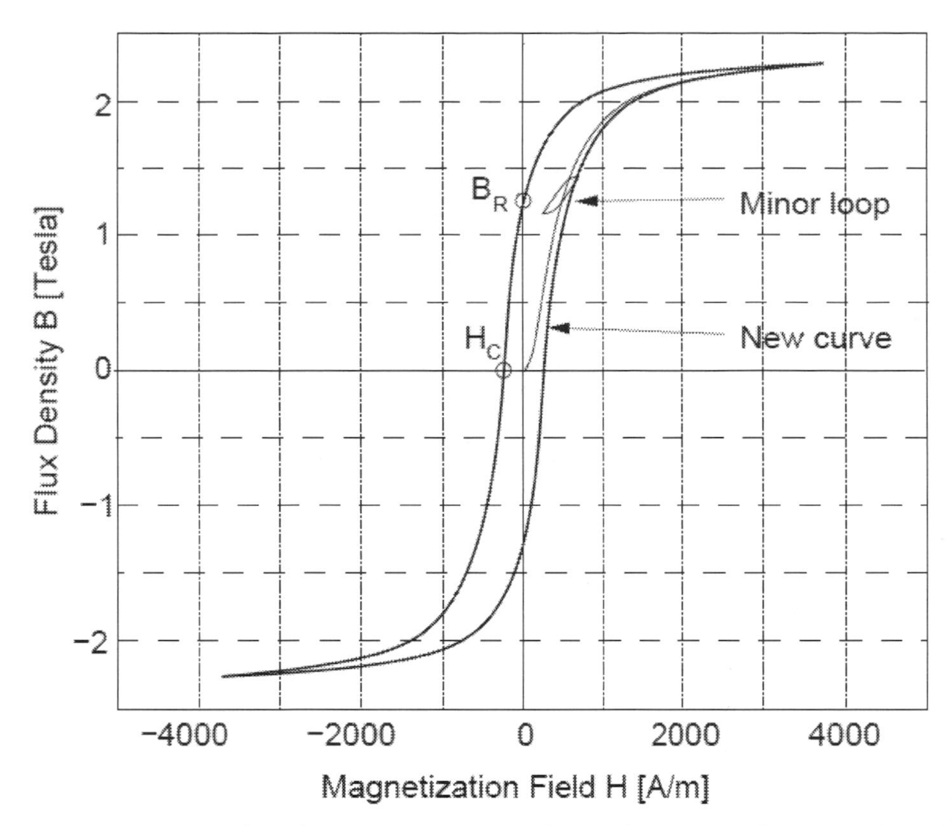

Figure 6.2. Magnetization curve B-H of a soft magnetic material (here AFK 502).

C. Force of a Single Actuator

Figure 6.3 shows a magnet and the main geometrical characteristics required to calculate the AMB force.

Typically it can be assumed that the magnetic flux Φ [T$\cdot m^2$], in the stator cross-section A_{fe} [m^2] is constant along the whole loop. Also it can be assumed that

$A_{fe} = A_a$, where A_a is the cross-section of the air-gap. Thus, the flux and flux densitybecome:

$$\Phi = B_{fe}A_{fe} = B_a A_a \tag{6.3}$$

$$B_{fe} = B_a = B \tag{6.4}$$

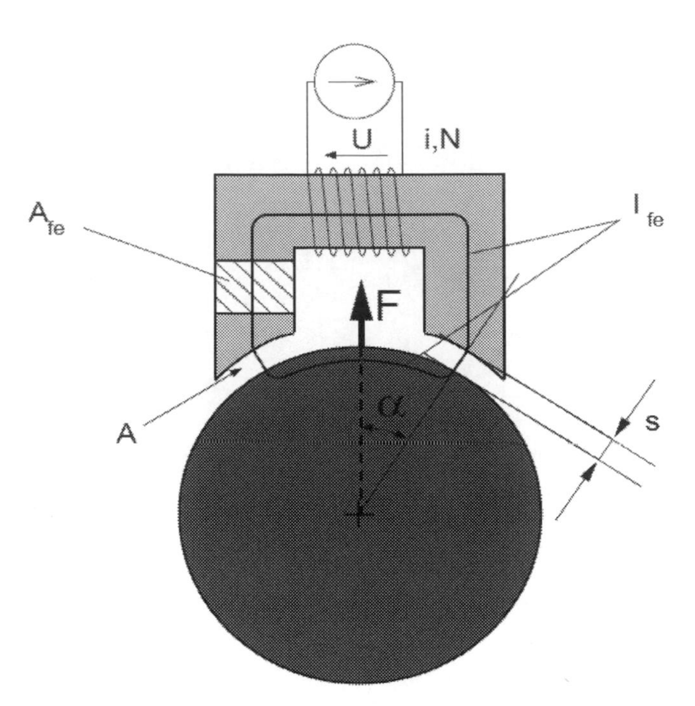

Figure 6.3. Magnetic force of an AMB

The magnetic path can be decomposed into two parts, the field in the air, and in the soft magnetic material.

$$Ni = H_{fe} l_{fe} + 2H_a s \quad (6.5)$$

where s [m] is the air gap between stator and rotor. The average magnetic path length in the lamination is l_{fe} [m].

From equations 6.5 and 6.2, and after solving for B the relation 6.6 gives the flux density.

$$B = \mu_o \frac{Ni}{(\frac{l_{fe}}{\mu_r} + 2s)} \quad (6.6)$$

Since the iron relative permeability $\mu_r \gg 1$, the magnetization of the iron can often be neglected, and the relation can be simplified:

$$B \approx \mu_o \frac{Ni}{2s} \quad (6.7)$$

Now the AMB force is obtained by considering the energy stored in the air gap and is given by equation 6.8:

$$F = \frac{{B_a}^2 A_a}{\mu_o} \tag{6.8}$$

The angle between the force direction and the center of the cross section A is determined by α. For common four poles radial AMBs, which means eight actuator teeth, α is 22.5°. Equations 6.7 and 6.8 yield the force for one actuator:

$$F = \frac{1}{4}\mu_o\, N^2 A_a \frac{i^2}{s^2}\cos\alpha = k\frac{i^2}{s^2}\cos\alpha \tag{6.9}$$

where $k = \frac{1}{4}\mu_o\, N^2 A_a$ (6.10)

D. Inductance L in the Magnetic Circuit

The static inductance L of the magnetic circuit is given by:

$$L = \frac{N\phi}{i} \tag{6.11}$$

which, with equations 6.3, 6.6 and 6.11, yields

$$L = \frac{\mu_0\, N^2 A_a}{\left(\frac{l_{fe}}{\mu_r} + 2s\right)} \tag{6.12}$$

6.4. Current - Force Relationship

In order to be able to produce magnetic force along two opposite directions, the actuators are usually arranged in pairs (figure 6.4). This enables a full control of the rotor along one axis.

For a pair of magnets, the force F_x [N] represents the force difference between the positive and the negative directions. For this case, the actuator currents are defined as the sum of a bias current i_0 [A] and a control current i_x [A] for thc positive actuator, and the difference $i_0 - i_x$ for the negative actuator. The air gaps are defined by the deviation x [m] and the nominal air gap s_0 [m], thus the terms (s_x- x) and ($s_0 + x$) are inserted.

$$F_x = F_+ - F_- = k\left(\frac{(i_0 + i_x)^2}{(s_0 - x)^2} + \frac{(i_0 - i_x)^2}{(s_0 + x)^2}\right)\cos\alpha \quad (6.13)$$

with $k = \frac{1}{4}\mu_o N^2 A_a$

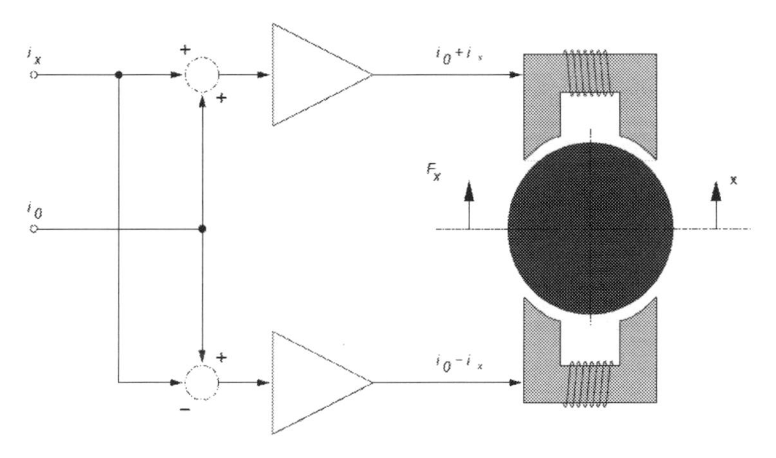

Figure 6.4. Pair of electromagnets for positive and negative force.

By considering that $x \ll s_0$ and $i_x \ll i_0$, equation 6.13 can be linearized using Taylor's series. It yields the typical AMB relation 6.14:

$$k_s = \frac{4ki_0}{s_0^{\,2}}(\cos\alpha)\, i_x + \frac{4ki_0^{\,2}}{s_0^{\,3}}(\cos\alpha)\, x = k_i\, i_x + k_s\, x \quad (6.14)$$

$$k_s = k_i\,\frac{i_0}{s_0} \quad (6.15)$$

The coefficient k_s is the open-loop stiffness. It is less than zero, so the AMB has an inherent negative stiffness. This negative stiffness leads to an open-loop instability. The force-current coefficient k_i is the actuator gain.

6.5. Types of Losses

The equations presented previously do not take the losses into account. Generally, in an electromagnetic system the following types of losses are considered :

- Hysteresis losses
- Eddy current losses
- Copper losses

Magnetic losses P_{fe} [W] are due to hysteresis and eddy currents losses and are often united in the iron losses. They depend on the rotor speed, the magnetic material, and the direction of the flux vector B [T] over the rotor circumference. Iron losses occur mainly in the rotor, whereas the copper losses dominate in the stator.

Magnetic bearings are contact-free, consequently there is no friction. The breaking torque in AMBs is however much smaller than the torque due to friction losses in conventional bearings.

A. Hysteresis Losses

Hysteresis losses are due to the re-magnetization of the magnetic material in the B – H curve. Hysteresis losses can be described as "frictional heat of the elemental mag nets" during a magnetization cycle. The area surrounded by the hysteresis curve corresponds to the energy expenditure during such a cycle.

The power loss due to hysteresis is approximately proportional to the square of the maximum induction and proportional to the remagnetization frequency:

$$P_h \; \alpha \; B_{max}{}^2 \cdot f_r$$

In more analytical form, at each loop the energy diminishes by $W_h = V_{fe}$ A_{BH}, with A_{BH} [T·H] area of the hysteresis loop, V_{fe} [m3] volume of the iron. Losses are proportional to the frequency of re-magnetization f_r [Hz], if the flux in sheets is sinusoidal and distributed evenly.

$$P_h = k_h \, f_r \, B_m{}^{1.6} \, V_{fe} \qquad (6.16)$$

B. Eddy Current Losses

When the flux density in a conductive material changes, eddy currents are generated. They are minimized by using an iron core made of insulated sheets or particles.

Eddy currents themselves generate a magnetic field. It acts against the initial field according to Lenz's law. Therefore, eddy currents delay the set-up of the magnetic field and reduce the dynamics of the electromagnetic system.

Following the law of induction an alternating magnetic field induces voltages. If the material is an electrical conductor these voltages cause currents(eddy) in the exposed material given by

$$P_e = \frac{1}{\rho} f_r^{\,2} \phi^2$$

Consequently, in more analytical form the eddy currents losses P_e [W] in laminated iron is given by,

$$P_e = \frac{1}{6\rho} \pi^2 d^2 f_r^{\,2} B_m^{\,2} V_{fe} \tag{6.17}$$

where ρ [Ω/m] is the specific electrical resistance of the iron, *d* [m] stands for the thickness of the sheets, f_r [Hz] for re-magnetization frequency, V_{fe} [m3] volume of the iron, B_m [T] for the maximum flux density or the amplitude of the flux density respectively.

Therefore, the losses due to eddy currents are proportional to the thickness *d* of the sheets.

$$P_e \ \alpha \ d^2 \tag{6.18}$$

Thin sheets (down to 150μm) are commonly used for stator and rotor.

C. Copper/Resistance Losses

The main losses in the stator are resistance losses in the actuator windings. They are dependent on the square of the coil current and the wire resistance:

$$P_{Cu} = R\, i^2 \tag{6.19}$$

The electrical resistance R [Ω] of a wire is given by:

$$R = \rho\, \frac{A}{L} \tag{6.20}$$

ρ[Ω/m] is the resistivity, A [m^2] the cross sectional area of the wire and l [m] the length of the wire. For3 T>0°C:

$$\rho = \rho(T) = \rho_0(1 + \alpha_1 T + \alpha_2 T^2) \tag{6.21}$$

Where T [°C] is the temperature, α_1, α_2 [K^{-1}] are temperature coefficients. The electrical resistance is considered as linear over a small temperature range, typically from 0°C to 100°C. On a large temperature range, the electrical resistance temperature dependence has to be considered as quadratic.

D. Other Losses in AMB

- **Air Losses**

Air losses are dependent on the square of the rotation speed. Experimentally they have been determined by comparing losses in air and vacuum environments.

- **Amplification Losses**

Losses in AMB systems are generated by the digital circuitry and the power amplifiers. Switching amplifiers are usually used. The efficiency of the power amplifier is situated around 95-98%. For example, typical amplifiers, are able to deliver 50V and 6A, the power losses per actuator is about 12W.

6.6. High Temperature Aspects in AMB Design

The design of all magnetic bearing components should be compatible with the specified application operating environment of temperature, pressure, humidity, and process fluid. Temperature increase due to joule, hysteresis, eddy current, and windage losses within the magnetic bearing components shall not unduly compromise the performance or the life of the bearing system. Also, operating stress levels due to centrifugal, thermal, electromagnetic, and mechanical loads shall not unduly compromise the performance or the life of the bearing system.

Before design approval, the magnetic bearing manufacturer should be required to conduct and present sufficient analyses to demonstrate that the above operating requirements are met.

This section gives an overview of construction difficulties related to high temperature electromechanical system. This section starts with physical aspects, then chemical aspects, to conclude with more practical points for high temperature electromechanical designs. The physical, mechanical, electrical and chemical aspects and properties which are required for AMB design under high temperature operation environment are briefly described as following:

1) Thermal expansion

Thermal expansion or more precisely difference of thermal expansion between different materials is a very relevant topic. Several ceramics and metal are necessary to build mechatronic systems. In order to ensure a good alignment at both room and high temperature and to avoid high mechanical stress or clearance with fastening parts, precautions are needed.

Since the thermal expansion of the FeCo stator laminations (λ_{FeCo} =9.8x10^{-6} K^{-1}) is smaller than the one of the housing e.g. Hastelloy X (λ_{Hast} =15.2x10^{-6} K^{-1}) , a diametrical bearing clearance between housing and stator (approximately 0.37 mm) appears at 550°C.The most common way to guarantee centering, or at least mechanical stability of the system, is to clamp the laminations between two rings having the same thermal expansion coefficient as the housing. The laminations have to be pressed axially sufficiently strong, in order to produce friction force. By this way no lamination can move in the radial direction due to the AMB force.

2) Mechanical properties

Elastic modulus of metals changes with temperature and decreases until liquefaction down to zero. Designing and controlling a system, which works from room to high temperature must take into account these stiffness changes. All the computations have to be verified for all temperature ranges. Materials should not remain under high stress. Creeping becomes important at high temperature. Creep aspects must be taken into account during the design process. All metallic parts have to fulfill their function for a minimal known time at a given temperature.

3) Electrical Properties

In electromechanical systems good electrical conductors are needed. The temperature coefficients of electrical resistance are relevant as well , see equations 6.20 and 6.21.

In order to get information about the behavior of the materials at high temperature, wounded coils can be put into a furnace for long term measurements. During this time, electrical resistance and inductance are measured. Tests to be done at three different temperatures: 500, 650 and 800°C. The tests should be performed with the chosen material for at least 100 hours at different temperatures to prove their reliability. For comparison purpose, resistance values are always given relatively to the resistance after 2 hours at the given temperature.

4) Magnetic Properties

Magnetic properties depend on the temperature in a non-linear way. Iron is no longer ferromagnetic when heated up above the so-called *Curie temperature*. This temperature is actually not set so precisely. Properties decrease slowly in function of the temperature and above a specific point they decrease dramatically; this point is the *Curie temperature*. Special FeCo alloys have been developed for high temperature applications. They feature high saturation and good high temperature capability. They need a heat treatment to reach the best magnetic properties and they show some ageing.

5) Electrical Insulations

Over temperatures of 260°C only few plastics can survive, thus electrical insulations have to be realized with other materials. Good electrical insulators are also ceramics and glass. Most of the ceramics are stable at temperature over 1000°C and some glasses can achieve work in such conditions. Nevertheless formability of these non-ductile materials makes them complex to use.

6) Oxidation

The working atmosphere plays an important role on the material behaviour. Oxidation occurs of course at room temperature, but at high temperature, this reaction is much accelerated. Using appropriate materials or working in non-oxidant atmosphere slow down this phenomena. Each part of the system has to be well chosen to keep the life span long enough.

7) Diffusion

When two components are in contact, chemical reactions take place between them. Many materials are inert at room temperature, but no longer at high temperature. For example, diffusion in nickel-clad copper wire shows changing conductive properties at high temperature with the time.

7

Applications of Magnetic Bearings

7.1. Fields of applications

Active magnetic bearings will most likely do the job for most applications, AMB can make a difference in both new designs and redesigns of existing ones.

The prospect of oil-free operation, active vibration control, on-line balancing and fitness for high rotational speed to reduce size of the equipment surely will raise heads among designers of turbo-machinery, while increased reliability and remote monitoring will thrill plant managers.

Active magnetic bearing (AMB) technology has been developed in for numerous different applications. The first commercial product with a rotor suspended on AMB was offered to the market in 1973.

Magnetic bearing advantages include very low and predictable friction, ability to run without lubrication and in a vacuum. Magnetic bearings are increasingly used in industrial machines such as compressors, turbines, pumps, motors and generators. Magnetic bearings are commonly used in watt-hour meters by electric utilities to measure home power consumption. Magnetic bearings are also used in high-precision instruments and to support equipment in a vacuum, for example in flywheel energy storage systems. A flywheel in a vacuum has very low windage losses, but conventional bearings usually fail quickly in a vacuum due to poor lubrication. Magnetic bearings are also used to support maglev trains in order to get low noise and smooth ride by eliminating physical contact surfaces. Disadvantages include high cost, and relatively large size. Turbines is another group of applications where magnetic bearings offer interesting design possibilities.

A new application of magnetic bearings is their use in artificial hearts. The use of magnetic suspension in ventricular assist devices was pioneered by Prof. Paul Allaire and Prof. Houston Wood at the University of Virginia culminating in the first magnetically suspended ventricular assist centrifugal pump (VAD) in 1999.

In an ever changing world, limits of current designs are constantly pushed further, requiring genuinely new ideas and technologies to meet new targets. The fields of applications of Magnetic Bearings can be categorized into the following Industries:

- •Semiconductor Industry
- •Bio-medical Engineering
- •Vacuum Technology
- •Structural Isolation
- •Rotor Dynamics
- •Maglev Transportation
- •Precision Engineering
- •Energy Storage
- •Aero Space
- •Turbo Machines

The Main applications cover the following systems and components:

- •Turbo molecular pumps
- •Blood pumps
- •Molecular beam choppers
- •Epitaxy centrifuges
- •Contact free linear guides
- •Variable speed spindles
- •Pipeline compressor
- •Elastic rotor control
- •Test rig for high speed tires
- •Magnarails and maglev systems
- •Gears, Chains, Conveyors, etc
- •Energy Storage Flywheels
- •High precision position stages
- •Active magnetic dampers
- •Smart Aero Engines
- •Turbo machines

In general, the applications of AMB may split in four different domains ,

- Light industry applications
- Heavy industry applications
- Aerospace applications,
- Machine tool applications,

A brief description is given in the following sections where we shall go through the main applications of each domain.

A. **Light industry applications**

The light industry applications include many different types of machines which their common characteristic is to have a light rotor (less than 50 kg) and not to be a machine tool spindle.

Some of the applications are under development such as the X-ray tube applications, others are commercial products but manufactured in limited quantities and on a customized design basis such as the pumps for liquid helium, others are standard products, manufactured on a large scale basis such as the turbo molecular vacuum pumps.

The major reasons to use AMB for the suspension of the rotating anode of X-ray tubes are:

- higher speed of rotation, so capability to accept higher flux of X- rays enabling better image,
- unlimited life,
- capability to move axially the rotor and to adjust the focusing point of the electron beam on the anode, and
- no vibration, no noise.

For liquid helium pumps, it is obvious that the major advantage of the AMB is its capability to operate in a 4 K environment. Also many other reasons such as:

- very limited heat input (less than 1 W) in the liquid helium,
- high speed, so better efficiency, and
- no thermal barrier, so shorter shaft.

For turbo-molecular vacuum pumps the main reasons for using AMB are:

- no pollution of the vacuum by any lube oil or grease (the auxiliary bearings are dry lubricated) ,
- high speed and so better level of vacuum, and
- no vibrations.

There are many other light industry applications of the AMB such as: neutron cropper, liquid metal centrifuges, medical centrifuges, rotating mirrors, textile electro spindle but today they are either at the development stage or represent a very limited market. But this may change in the future.

This domain is the largest one for the application of the active magnetic bearings and is also the one which has grown most drastically. The major reasons for utilizing AMB for these machines are:

- completely oil-free machines (elimination of the lube oil system),
- capability to operate at speeds higher than the third critical speed (first shaft bending mode),
- no process gas pollution by lube oil and elimination of fire hazard,
- vibration-free equipment

B. Heavy industry applications

The heavy industry applications are those related to turbo machines and electrical machines where the rotor mass exceeds 50 kg and for ground (or sea) based applications. These machines are mostly:

1) Compressors and drivers: The major world OEMs have compressors equipped with active magnetic bearings.

2) Turbo expanders: In addition to the reasons already mentioned one major advantage of the AMB is its capability to withstand a large range of temperature (-150·C on the turbine side, +150°C on the compressor side) and temperature gradients.

3) Other applications: The application concerns on the new machines. Development is going on for high speed electrical motors (with low speed electrical motors, a multiplying gear is necessary and so reintroduces the lube oil). The main users are in the field of power station, pipelines, refineries and chemical plants. The most spectacular one is related to high capacity turbo generators (800 -1,500 MW) for power plants.

The characteristics of the active magnetic bearings are such that this new technology represents a revolution in the suspension of rotating shafts. The applications already made, clearly indicate that this technology is no more at the infancy stage.

Of course, development will continue and the AMB technology will incorporate the new materials and components such as high permeability laminations and microprocessors .

But one can say that the AMB technology has reached the level of an industrial technology applicable to industrial machines operating 24 hours a day all year long.

C. Aerospace applications

This domain is mostly under development. The only commercial applications in this area today are:

- Momentum wheel for satellites, and
- Turbo-molecular vacuum pumps.

The main reasons for using active magnetic bearings on momentum wheels for satellites are:

- capability to operate in vacuum,
- no friction and so very limited power consumption, and
- unlimited life.

The turbo-molecular vacuum pumps such as the one installed on the European space laboratory are very similar to the regular heavy industrial ones. The major development is related to the application of the AMB to:

- rocket engines,
- turbo-pumps, and
- aircraft engines.

The main reasons for applying AMB to rocket engine turbo-pumps are:

- better capability to withstand high temperature (steam) and low temperature (liquid hydrogen) and temperature gradients,
- higher speed of rotation, so better performances of the machine,
- better control of the rotor dynamics, and
- no wear, no lubrication, better reliability.

D. Machine tool applications

The main reasons for using AMB on machine tool spindles are:

- high speed, so higher metal removal rate, no wear, better reliability,
- no vibrations, better accuracy, and
- permanent monitoring of the operating conditions (adaptive control).

These products are now standard products manufactured on an industrial basis.

7.2. Examples of AMB applications

A. High Temperature AMB

The application of active magnetic bearings (AMBs) for gas turbines and aircraft engines would open large potentials for novel design. In order to utilize the full advantages of active magnetic bearings, operation in gas turbine and aircraft engines requires that the magnetic bearing should work properly at high temperatures. Challenges in designing such bearings consist in material evaluation, manufacturing process and high temperature displacement sensor development. High temperature active magnetic bearings (HT AMBs) are under development in various places. See an example of test rig for a high temperature active magnetic bearing in Figure 7.1, Operating in a containment heated up to 550°C, and at rotor speeds at 30’000 rpm.

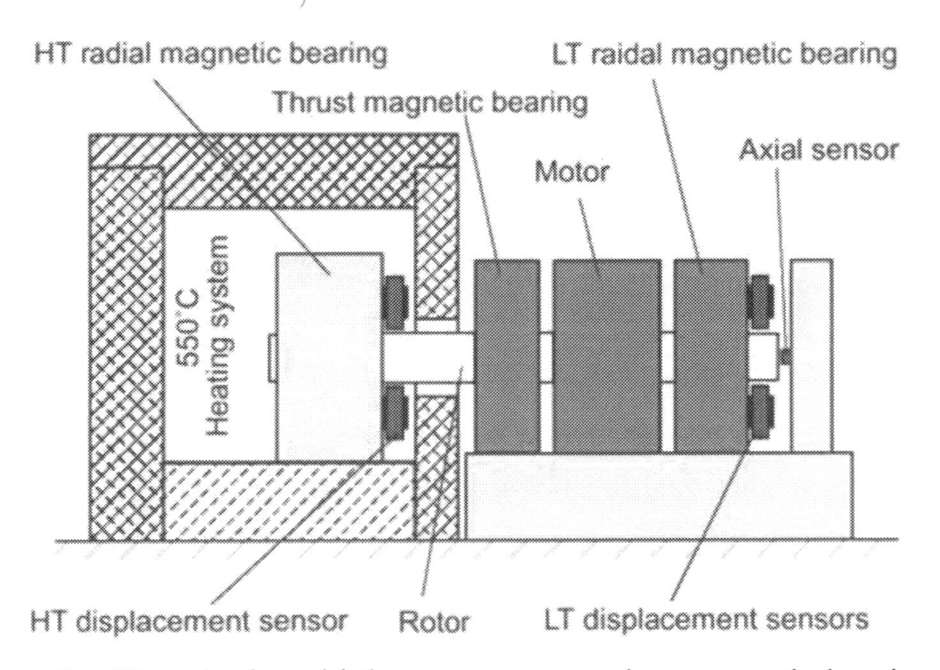

Figure 7.1. Test rig for a high temperature active magnetic bearing

Such a performance cannot be reached by any other kind of bearing. The soft magnetic materials for stator and rotor are cobalt based alloys such as Hiperco 50 and Hiperco 50-HS, the windings are made of ceramic coated copper wire with high temperature potting materials. Functional tests were quite successful, but the long-term exposure to high temperature needs further research, as the actual results are not yet convincing. In addition, heat dissipation of the internally generated losses under heavy bearing loads have shown to need special attention.

B. Magnetic Bearing Wheel for Space Application

Almost all spacecraft in orbit are equipped with ball bearing momentum or reaction wheels serving as actuators in the attitude control system. Amongst the various on board system components the wheels have been identified as one of the main sources of vibration noise due to static and dynamics unbalance movements, and the mechanical imperfections of the bearing rotating parts.

The magnetic bearing wheels having all the well-known special advantages as higher speed, no contact and lubrication. By using the magnetic bearings ,it is possible to suppress vibrations in the actively controlled degrees of freedom. Vibration suppression in this context means an isolation of the stator from the rotor in the frequency range of the disturbances rather than the suppression of unbalance movement of the rotor.

Magnetic bearing wheels should be on the first glance ideally suited to exhibit very low noise figures, because they have no mechanical contact between stator and rotor parts. But unfortunately there are several possible sources of disturbing forces and torques in magnetic bearing wheels as well. Some of these sources are due to unfavorable design details and should be foreseen and avoided in time. Some others are of principle nature but can be overcome by control means.

Magnetic bearing momentum wheels can be used as gyroscopic actuators in the attitude control system of spacecraft. If the wheel provides a vernier gimballing capability, three axis attitude control for roll and yaw angles is possible with one wheel only.

In magnetic bearings - as far as no permanent magnets are involved for direct attraction/repulsion - the bearing forces between the stator and the rotor part are fully controllable. They can be designed frequency dependent and with a time varying characteristic for example. Unbalance forces are transmitted to the stator only at an amount proportional to the suspension current. Applying a special control law, the disturbance component in the suspension current can be attenuated while keeping an overall stable levitation of the rotor. This in fact isolates the unbalance forces from the stator and the satellite structure.

C. Multi Stage Boiler Feed Pump

Magnetic bearing usage in pumps is a natural development when one considers the successes of these bearings in large centrifugal compressors. For pumps of significant power level, the elimination of lubrication systems that would otherwise be needed for conventional bearings and the attendant reduction of maintenance are benefits that should be claimed.

Application to multistage pumps with orders of magnitude greater power is the next step and is a further test of the feasibility of magnetic bearings, because such pumps often operate with close internal clearances and flexible shafts. Typical of such pumps are those used for boiler feed water in the electric utility industry.

The design of a multistage boiler feed pump is a compromise between allowing sufficient clearance for shaft flexure and thermal distortion and minimizing clearances to increase the efficiency of the impeller ring and the shaft sealing systems. Ideally the bearings would be mounted closer together than is current practice, with a closer tolerance on shaft movement. If submerged magnetic bearings were used, then this could be accomplished by eliminating the outboard (non-drive end) seal and putting the inboard bearing between the first-stage impeller and the inboard seals.

D. Titanium Powder Production

Plasma rotary electrode process (PREP) is being used for production of titanium alloy powder. In this technique, titanium billet is rotated and one end of the billet is melted by plasma arc, and droplets are sprayed by centrifugal force and cooled in an inert atmosphere. The rotating mechanism is like a cantilever arm with a heavy weight top and the critical speed is lowered by the weight of the top. Also there are many vibration sources in this system. To avoid this irrational problem, adopted the magnetic bearing was adopted as the drive mechanism in place of the motor and gear train.

The use of the magnetic bearing resulted in an increase of allowable unbalance and a large decrease of the vibration of the machine. These improvements made it possible to produce fine and rapidly solidified particles with a higher productivity.

Titanium alloys are chemically active and react with refractory materials , and ordinary powder production technique such as gas atomizing cannot be used. Therefore, Rotating Electrode Process (REP) and plasma rotating electrode process (PREP) are developed by Nuclear Metals Inc. and being used for production of titanium.

E. Flywheels

Flywheels are an ancient means of storing kinetic energy, the most widely used application being the potter's disk. Increasingly efficient frequency converters in conjunction with modern high speed motor-generator technology and advanced flywheel materials have opened new possibilities for flywheel in applications such as

- Energy Storage/Recovery in Regenerative Systems
- Load Leveling
- Power Conditioning
- Uninterruptible Power supplies

Magnetic bearings levitate the flywheel without friction and are, therefore, the bearings of choice to optimize overall flywheel efficiency. From the depicted steam engine to highly dynamic energy storage solutions, flywheels have come a long way. Today, flywheels operate in near vacuum conditions to reduce windage losses and are designed for dynamic and cyclical energy storage and retrieval. Further advantages are: absence of aggressive or toxic chemicals; long life cycle; and maintenance-free operation.

Figure 7.2. Flywheel running with magnetic bearing

F. Compressors & Expanders

RMG Co. integrates generator and turbo expander without the need for a gearbox between expander wheel and generator. Electricity is generated by expanding natural gas coming at high pressure from the transportation network prior to distribution to consumers. Typically, two units are connected in series or parallel depending on inlet pressure and gas flow-rate. The design challenge consisted in building a sealed and gearless machine running at high speed in order to achieve high power density. The Active Magnetic Bearings operate in pure natural gas atmosphere. The shaft was designed to be sub-critical at 32000 rpm, allowing for a very compact machine.

Technical Data

Speed:	32'000 rpm
Rotor weight:	110 kg
Radial bearing:	ø110mm/2200N
Motor power:	550 kW

Figure 7.3. Integrated generator and turbo expander by RMG

G. Turbo Blowers

This radial turbo blower for CO2 gas is at the heart of very powerful lasers for cutting metal sheets up to 25 mm thickness. The design challenge was to minimize hydrocarbon contamination leading to a construction where Active Magnetic Bearing operate on the process side requiring leak-tight construction. Contamination throughout gassing was also a major concern that had to be addressed by focusing on choice of materials and assembly processes. To date this blower has an installed base of over 8000 units, highlighting MECOS experience with handling volume production.

Technical Data

Speed:	54'000rpm
Rotor weight:	3.6 kg
Radial bearing:	ø48mm/230N
Motor power:	12 kW

Cleanness and reliability are of very high importance for this application. In close team cooperate ion with our OEM customer we developed a system, which has been manufactured since 1999 in volume production for a worldwide market. Magnetic Bearings are oil-free, maintenance-free, allow for high speed and, by using a digital control system, adaptive algorithms as well as switching between control schemes can be accomplished easily.

Figure 7.4. Turbo Blower for Power Laser

H. Turbo Booster Pumps

This turbo booster pump designed by JTEKT Co. Ltd. of Osaka, Japan (former Koyo Seiko Co. Ltd.) produces an ultra-high vacuum and enhances flow rates in semiconductor plants. The Active Magnetic Bearing system features MECOS advanced control hardware and software allowing seamless integration of the equipment into the customer's plant.

Technical Data

Speed: 24'000rpm
Motor power: 0.8 kW

Figure 7.5. Vacuum Turbo booster pump, JTEKT Inc. Ltd, Osaka Japan

7.3. Special Design Applications

1) Magnetic Bearings for new machine designs

If magnetic bearings are considered early when developing a new machine, completely new designs are possible. Often there is no need for a shaft at all. The bearings and the motor can then be integrated into the load, thus reducing the size and the mechanical complexity of the whole machine. This concept is especially useful for gasturbines, centrifuges, pumps, flywheels, gyroscopes and other direct driven applications.

A flywheel without shaft is shown in Fig. 7.6, and has obvious advantages over conventional flywheel designs. The drive and the bearings can be integrated in the design without the need for a long thin shaft. In this particular design two radial homopolar induction bearings and one passive axial bearing of permanent magnet type are used. For efficiency reasons, an outer rotor permanent magnet machine is used for motor and generator mode.

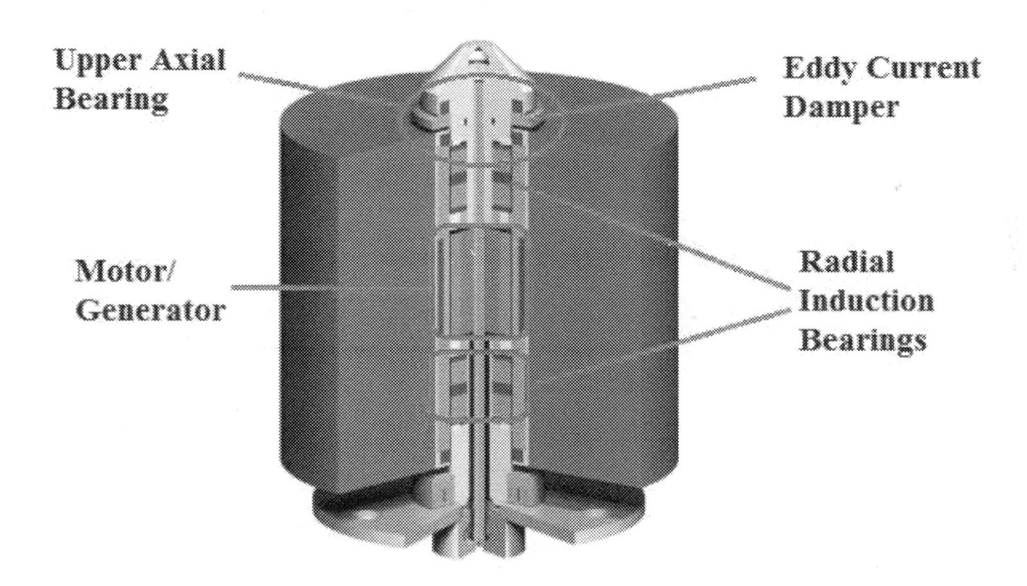

Fig. 7.6. High speed flywheel for energy storage.
Similar flywheels are manufactured by AFSTrinity Flywheel Corporation.

The shaft is often the weak point in conventional flywheels, due to bending problems at high speed and due to the very large torque when the machine operates as a generator. Especially flywheels used for power quality applications, which have to deliver very high power during short periods, suffer from these torque peaks.

2) Turbomolecular Pump

A **turbomolecular pump** is a type of vacuum pump, superficially similar to a turbopump, used to obtain and maintain high vacuum. These pumps work on the principle that gas molecules can be given momentum in a desired direction by repeated collision with a moving solid surface. In a turbopump, a rapidly spinning turbine rotor 'hits' gas molecules from the inlet of the pump towards the exhaust in order to create or maintain a vacuum.

The turbomolecular pump was invented in 1958 by Becker, based on the older molecular drag pumps developed by Gaede in 1913, Holweck in 1923 and Siegbahn in 1944.

Fig. 7.7. Interior view of a turbomolecular pump

Turbomolecular pumps must operate at very high speeds, and the friction heat buildup imposes design limitations. Some turbomolecular pumps use magnetic bearings to reduce friction and oil contamination. Because the magnetic bearings and the temperature cycles allow for only a limited clearance between rotor and stator, the blades at the high pressure stages are somewhat degenerated into a single helical foil each. Maximum pressure

3) Zippe-type centrifuge

The Zippe-type centrifuge is a device designed to collect **Uranium-235**. It was developed in the **Soviet Union** by a team of 60 German scientists captured after **World War II**, working in detention. The centrifuge is named after the team's lead experimenter, **Gernot Zippe**.

Natural uranium consists of three isotopes; the majority (99.274 percent) is U-238, while approximately 0.72 percent is U-235 and the remaining 0.0055 percent is U-234. If natural uranium is enriched to contain 3 percent U-235, it can be used as fuel for light water nuclear reactors. If it is enriched to contain 90 percent Uranium-235 , it can be used for nuclear weapons.

Enriching uranium is difficult because the isotopes are very similar in weight: U-235 is only 1.26% lighter than u-238. It requires a centrifuge that can spin at 1,500 revolutions per second (90,000 RPM). For comparison, automatic washing machines operate at only about 12 to 25 revolutions per second during the spin cycle.

The device has a hollow, cylindrical rotor filled with gaseous uranium in the form of its hexafluoride. A pulsating magnetic field at the bottom of the rotor, similar to that used in an electric motor, is able to spin it quickly enough that the U-238 is thrown towards the edge. The lighter U-235 collects in the center. The bottom of the gaseous mix is heated, producing convection currents that move the U-238 down. The U-235 moves up, where scoops collect it.

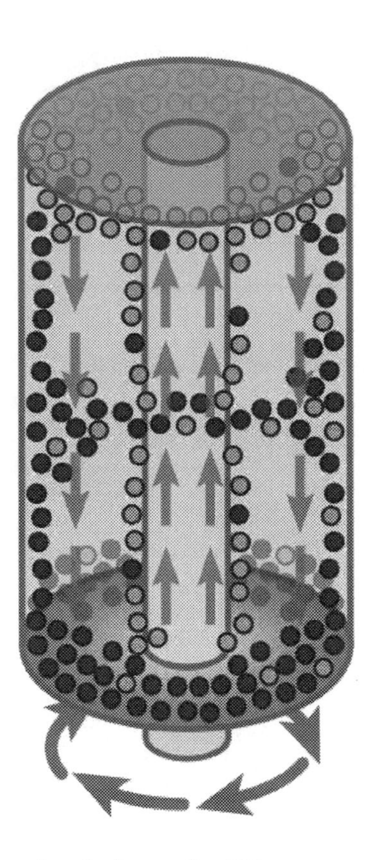

Fig. 7.8. Diagram of the principles of a Zippe-type gas centrifuge with U-238 represented in dark blue and U-235 represented in light blue.

To reduce friction, the rotor spins in a vacuum. A magnetic bearing holds the top of the rotor steady, and the only physical contact is the needle-like bearing that the rotor sits on.

8

Attachments

Annex 1 Earnshaw's Theorem

A1.1. Background and Definition

Earnshaw's theorem states that a collection of point charges cannot be maintained in a stable stationary equilibrium configuration solely by the electrostatic interaction of the charges. This was first proven by Samuel Earnshaw in 1842. It is usually referenced to magnetic fields, but originally applied to electrostatic fields. It applies to the classical inverse-square law forces (electric and gravitational) and also to the magnetic forces of permanent magnets and paramagnetic materials (but not diamagnetic materials).

Informally, the case of a point charge in an arbitrary static electric field is a simple consequence of Gauss's law. This means that the force field lines around the particle's equilibrium position should all point inwards, towards that position. If all of the surrounding field lines point towards the equilibrium point, then the divergence of the field at that point must be negative (i.e. that point acts as a sink). However, Gauss's Law says that the divergence of any possible electric force field is zero in free space. In mathematical notation, an electrical force $\mathbf{F}(\mathbf{r})$ deriving from a potential $U(\mathbf{r})$ will always be divergenceless (satisfy Laplace's equation):

$$\nabla \cdot \mathbf{F} = \nabla \cdot (-\nabla U) = -\nabla^2 U = 0.$$

Therefore, there are no local minima or maxima of the field potential in free space, only saddle points. A stable equilibrium of the particle cannot exist and there must be an instability in at least one direction.

This theorem also states that there is no possible static configuration of ferromagnets which can stably levitate an object against gravity, even when the magnetic forces are stronger than the gravitational forces.

Earnshaw's theorem has even been proven for the general case of extended bodies, and this is so even if they are flexible and conducting, provided they are not diamagnetic, as diamagnetism constitutes a (small) repulsive force, but no attraction.

There are, however, several exceptions to the rule's assumptions which allow magnetic levitation.

Earnshaw's theorem, in addition to the fact that configurations of classical charged particles orbiting one another are also unstable due to electromagnetic radiation, mean that even dynamic systems of charges are unstable, long term. This, for quite some time led to the puzzling question of why matter stays together as much evidence was found that matter was held together electromagnetically.

A1.2. Proofs for Magnetic Dipoles

While a more general proof may be possible, three specific cases are considered here. The first case is a magnetic dipole of constant magnitude that has a fast (fixed) orientation. The second and third cases are magnetic dipoles where the orientation changes to remain aligned either parallel or anti-parallel to the field lines of the external magnetic field. In paramagnetic and diamagnetic materials the dipoles are aligned parallel and anti-parallel to the field lines, respectively.

The proofs considered here are based on the following principles.

The energy U of a magnetic dipole with a magnetic dipole moment **M** in an external magnetic field **B** is given by

$$U = -\mathbf{M} \cdot \mathbf{B} = -(M_x B_x + M_y B_y + M_z B_z).$$

The dipole will only be stably levitated at points where the energy has a minimum. The energy can only have a minimum at points where the Laplacian of the energy is greater than zero. That is, where

$$\nabla^2 U = \frac{\partial^2 U}{\partial x^2} + \frac{\partial^2 U}{\partial y^2} + \frac{\partial^2 U}{\partial z^2} > 0.$$

Finally, because both the divergence and the curl of a magnetic field are zero (in the absence of current or a changing electric field), the Laplacians of the individual components of a magnetic field are zero. That is,

$$\nabla^2 B_x = 0, \nabla^2 B_y = 0, \nabla^2 B_z = 0.$$

For a magnetic dipole of fixed orientation (and constant magnitude) the energy will be given by

$$U = -\mathbf{M} \cdot \mathbf{B} = -(M_x B_x + M_y B_y + M_z B_z),$$

where M_x, M_y and M_z are constant. In this case the Laplacian of the energy is always zero,

$$\nabla^2 U = 0,$$

so the dipole can have neither an energy minimum or an energy maximum. That is, there is no point in free space where the dipole is either stable in all directions or unstable in all directions.

Magnetic dipoles aligned parallel or anti-parallel to an external field with the magnitude of the dipole proportional to the external field will correspond to paramagnetic and diamagnetic materials respectively. In these cases the energy will be given by

$$U = -\mathbf{M} \cdot \mathbf{B} = -k\mathbf{B} \cdot \mathbf{B} = -k(B_x^2 + B_y^2 + B_z^2),$$

where k is a constant greater than zero for paramagnetic materials and less than zero for diamagnetic materials.

In this case, it will be shown that

$$\nabla^2 (B_x^2 + B_y^2 + B_z^2) \geq 0,$$

which, combined with the constant k, shows that paramagnetic materials can have energy maxima but not energy minima and diamagnetic materials can have energy minima but not energy maxima. That is, paramagnetic materials can be unstable in all directions but not stable in all directions and diamagnetic materials can be stable in all directions but not unstable in all directions. Of course, both materials can have saddle points.

Finally, the magnetic dipole of a ferromagnetic material (a permanent magnet) that is aligned parallel or anti-parallel to a magnetic field will be given by

$$\mathbf{M} = k\frac{\mathbf{B}}{|\mathbf{B}|},$$

so the energy will be given by

$$U = -\mathbf{M}\cdot\mathbf{B} = -k\frac{\mathbf{B}}{|\mathbf{B}|}\cdot\mathbf{B} = -k\frac{(B_x^2+B_y^2+B_z^2)}{(B_x^2+B_y^2+B_z^2)^{1/2}} = -k(B_x^2+B_y^2+B_z^2)^{1/2};$$

but this is just the square root of the energy for the paramagnetic and diamagnetic case discussed above and, since the square root function is monotonically increasing, any minimum or maximum in the paramagnetic and diamagnetic case will be a minimum or maximum here as well. There are, however, no known configurations of permanent magnets that stably levitate so there may be other reasons not discussed here why it is not possible to maintain permanent magnets in orientations anti-parallel to magnetic fields.

Earnshaw's theorem was originally formulated for electrostatics (point charges) to show that there is no stable configuration of a collection of point charges. The proofs presented here for individual dipoles should be generalizable to collections of magnetics dipoles because they are formulated in terms of energy which is additive. A rigorous treatment of this topic, however, is currently beyond the scope of this article.

- **Fixed-orientation magnetic dipole**

It will be proven that at all points in free space

$$\nabla\cdot(\nabla U) = \nabla^2 U = \frac{\partial^2 U}{\partial x^2} + \frac{\partial^2 U}{\partial y^2} + \frac{\partial^2 U}{\partial z^2} = 0.$$

The energy U of the magnetic dipole $\mathbf{M}$ in the external magnetic field $\mathbf{B}$ is given by

$$U = -\mathbf{M} \cdot \mathbf{B} = -(M_x B_x + M_y B_y + M_z B_z).$$

The Laplacian will be

$$\nabla^2 U = -\left(\frac{\partial^2 (M_x B_x + M_y B_y + M_z B_z)}{\partial x^2} + \frac{\partial^2 (M_x B_x + M_y B_y + M_z B_z)}{\partial y^2} + \frac{\partial^2 (M_x B_x + M_y B_y + M_z B_z)}{\partial z^2}\right).$$

Expanding and rearranging the terms (and noting that the dipole $\mathbf{M}$ is constant) we have

$$\nabla^2 U = -\left(M_x \left(\frac{\partial^2 B_x}{\partial x^2} + \frac{\partial^2 B_x}{\partial y^2} + \frac{\partial^2 B_x}{\partial z^2}\right) + M_y \left(\frac{\partial^2 B_y}{\partial x^2} + \frac{\partial^2 B_y}{\partial y^2} + \frac{\partial^2 B_y}{\partial z^2}\right) + M_z \left(\frac{\partial^2 B_z}{\partial x^2} + \frac{\partial^2 B_z}{\partial y^2} + \frac{\partial^2 B_z}{\partial z^2}\right)\right)$$

or

$$\nabla^2 U = -(M_x \nabla^2 B_x + M_y \nabla^2 B_y + M_z \nabla^2 B_z),$$

but the Laplacians of the individual components of a magnetic field are zero in free space (not counting electromagnetic radiation) so

$$\nabla^2 U = -(M_x 0 + M_y 0 + M_z 0) = 0,$$

which completes the proof.

❖ Magnetic dipole aligned with external field lines

The case of a paramagnetic or diamagnetic dipole is considered first. The energy is given by

$$U = -k(B_x^2 + B_y^2 + B_z^2).$$

Expanding and rearranging terms,

$$\nabla^2 (B_x^2 + B_y^2 + B_z^2) = 2[|\nabla B_x|^2 + |\nabla B_y|^2 + |\nabla B_z|^2 + B_x \nabla^2 B_x + B_y \nabla^2 B_y + B_z \nabla^2 B_z];$$

but since the Laplacian of each individual component of the magnetic field is zero,

$$\nabla^2(B_x^2 + B_y^2 + B_z^2) = 2[|\nabla B_x|^2 + |\nabla B_y|^2 + |\nabla B_z|^2];$$

and since the square of a magnitude is always positive,

$$\nabla^2(B_x^2 + B_y^2 + B_z^2) \geq 0.$$

As discussed above, this means that the Laplacian of the energy of a paramagnetic material can never be positive (no stable levitation) and the Laplacian of the energy of a diamagnetic material can never be negative (no instability in all directions).

Further, because the energy for a dipole of fixed magnitude aligned with the external field will be the square root of the energy above, the same analysis applies.

❖ Laplacian of individual components of a magnetic field

It is proved here that the Laplacian of each individual component of a magnetic field is zero. This shows the need to invoke the properties of magnetic fields that the divergence of a magnetic field is always zero and the curl of a magnetic field is zero in free space. (That is, in the absence of current or a changing electric field.) See Maxwell's equations for a more detailed discussion of these properties of magnetic fields.

Consider the Laplacian of the x component of the magnetic field

$$\nabla^2 B_x = \frac{\partial^2 B_x}{\partial x^2} + \frac{\partial^2 B_x}{\partial y^2} + \frac{\partial^2 B_x}{\partial z^2} = \frac{\partial}{\partial x}\frac{\partial}{\partial x}B_x + \frac{\partial}{\partial y}\frac{\partial}{\partial y}B_x + \frac{\partial}{\partial z}\frac{\partial}{\partial z}B_x.$$

Because the curl of **B** is zero,

$$\frac{\partial B_x}{\partial y} = \frac{\partial B_y}{\partial x},$$

and

$$\frac{\partial B_x}{\partial z} = \frac{\partial B_z}{\partial x},$$

so we have

$$\nabla^2 B_x = \frac{\partial}{\partial x}\frac{\partial}{\partial x} B_x + \frac{\partial}{\partial y}\frac{\partial}{\partial x} B_y + \frac{\partial}{\partial z}\frac{\partial}{\partial x} B_z.$$

But since B_x is continuous, the order of differentiation doesn't matter giving

$$\nabla^2 B_x = \frac{\partial}{\partial x}\left(\frac{\partial B_x}{\partial x} + \frac{\partial B_y}{\partial y} + \frac{\partial B_z}{\partial z}\right) = \frac{\partial}{\partial x}(\nabla \cdot \mathbf{B}).$$

The divergence of **B** is constant (zero, in fact) so

$$\nabla^2 B_x = \frac{\partial}{\partial x}(\nabla \cdot \mathbf{B} = 0) = 0.$$

The Laplacian of the y component of the magnetic field B_y field and the Laplacian of the z component of the magnetic field B_z can be calculated analogously. Alternatively, one can use the identity

$$\nabla^2 \mathbf{B} = \nabla(\nabla \cdot \mathbf{B}) - \nabla \times (\nabla \times \mathbf{B}),$$

where both terms in the parentheses vanish.

9

References

1. Schweitzer, G., Bleuler, H. and Traxler, A., 2003, "ActiveMagnetic Bearings: Basics, Properties and Applications of Active Magnetic Bearings", Authors Working Group, www.mcgs.ch reprint.
2. Chiba, A., Fukao, T., Ichikawa, O., Oshima, M., Takemoto, M. and Dorrell, D.G., 2005, "Magnetic Bearings & Bearingless Drives", Newnes, Elsevier.
3. Maslen, E., 2000, "Magnetic Bearings", University of Virginia.
4. Groom N.J. and Bloodgood, V.D. Jr., 2000, "A Comparison of Analytical and Experimental Data for a Magnetic Actuator", NASA-2000-tm210328.
5. Bloodgood, V.D. Jr., Groom, N.J. and Britcher, C.P., 2000, "Further development of an optimal design approach applied to axial magnetic bearings", NASA-2000-7ismb-vdb.
6. Schweitzer G, Bleuler H, Traxler A. (1994) Active Magnetic Bearings, Zurich, vdf Hochschulverlag AG.
7. Traxler A. (1985) Eigenschaften und Auslegung von beruehrungsfreien elektromagnetischen Lagern. Diss. ETH Zurich Nr. 7851.
8. Beams JW, Young JL, Moore JW. (1946) The production of high centrifugal fields. J. Appl. Phys., 886-890.
9. Larsonneur R. (1990) Design and control of active magnetic bearing system for high speed rotation. Diss. ETH Zurich Nr. 9140.
10. Viggiano F. (1992) Aktive magnetische Lagerung und Rotorkonstruktion elektrischer Hochgeschwindigkeitsantriebe. Diss ETH Zurich Nr. 9746.
11. Loesch F. (2001) Identification and automated controller design for active magnetic bearing systems. Diss. ETH Zurich Nr. 14474.
12. Fujiwara H, Matsushita O, Okubo H. (2000) Stability Evaluation of High Frequency Eigen Modes for Active Magnetic Bearing Rotors. *Proc. 7th Internat. Symp. on Magnetic Bearings*, ETH Zurich, August 23-25, 83-88.
13. Pang DC, Lin JL, Shew BY, Zmood RB. (2000) Design, fabrication, and testing of a millimeter-level magnetically suspended micro-motor. *Proc. 7th Internat. Symp. on Magnetic Bearings*, ETH Zurich, August 23-25, 105-110.
14. Boletis A, Vuillemin R, Aymon C, Kuemmerle M, Aeschlimann B, Moser R, Bleuler H, Fulin E, Bergqvist J. (2000) A six magnetic

actuators integrated micro-motor. *Proc. 7th Internat. Symp. on Magnetic Bearings, ETH Zurich*, August 23-25, 101-103.

15. Mekhiche M, Nichols S, Oleksy J, Young J, Kiley J, Havenhill D. (2000) 50 krpm, 1,100°F magnetic bearings for jet turbine engines. *Proc. 7th Internat. Symp. on Magnetic Bearings*, ETH Zurich, August 23-25, 123-128.
16. Kondoleon AS, Kelleher WP. (2000) Soft magnetic alloys for high temperature radial magnetic bearings. *Proc.7th Internat. Symp. on Magnetic Bearings*, ETH Zurich, August 23-25, 111-116.
17. Ohsawa M, Yoshida K, Ninomiya H, Furuya T, Marui E. (1998) High Temperature Blower for Molten Carbonate Fuel Cell Supported by Magnetic Bearings. *Proc. 6th Internat. Symposium on Magnetic Bearings*, MIT Cambridge , August 5-7, 32-41.
18. Xu L, Zhang J. (2002) A Study on High Temperature Displacement Sensor. To appear in *IEEE Trans. On Instrumentation and Measurement.*
19. Xu L, Wang L, Schweitzer G. (2000) Development for magnetic bearings for high temperature suspension. *Proc. 7th Internat. Symp. on Magnetic Bearings*, ETH Zurich, August 23-25, 117-123.
20. Mizuno T, Higuch T. (1994) Experimental measurement of rotational losses in magnetic bearings. *Proc. 4th Internat. Symp. on Magnetic Bearings*, ETH Zurich, August 23-26, 591-595.
21. Allaire PE, Kasarda MEF, Fujita LK: (1998) Rotor power losses in planar radial magnetic bearings – effects of number of stator poles, air gap thickness, and magnetic flux density. *Proc. 6th Internat. Symp. on Magnetic Bearings*, MIT Cambridge, August 5-7, 383-391.
22. Meeker D, Maslen E, Kasarda M. (1998) Influence of actuator geometry on rotating losses in heteropolar magnetic bearings. *Proc. 6th Internat. Symp. on Magnetic Bearings*, MIT Cambridge, August 5-7, 392-401.
23. Ahrens M, Kucera L. (1996) Analytical calculation of fields, forces and losses of a radial magnetic bearing with rotating rotor considering eddy currents. *Proc. 5th Internat. Symp. on Magnetic Bearings*, Kanazawa, August 28- 30, 253-258.
24. Mack M. (1967) Luftreibungsverluste bei elektrischen Maschinen kleiner Baugrössen. Diss. TH Stuttgart.
25. Fremerey J. K. (1988) Radial Shear Force Permanent Magnet Bearing System with Zero-power axial Control and Passive Radial Damping. In G Schweitzer (Ed) *Magnetic Bearings*, Springer-Verlag, 25-32.
26. Yakushi K, Koseki T, Sone S. (2000) Three Degree-of-Freedom Zero Power Magnetic Levitation Control by a 4-Pole Type Electromagnet. *Proc. Internat. Power Electronics Conference* IPEC, Tokyo 2000.

27. Mizuno T. (2000) A unified transfer function approach to control design for virtually zero power magnetic suspension. *Proc. 7th Internat. Symp. on Magnetic Bearings*, ETH Zurich, August 23-25, 117- 123.
28. Jeon S, Ahn HJ, Han DC. (2000) New Design of Cylindrical Capacitive Sensor for On-line Precision Control of AMB Spindle. *Proc. 7th Internat. Symp. on Magnetic Bearings*, ETH Zurich, August 23-25, 495-500.
29. Kim WJ, Trumper DL. (1998) Six-degree-of-freedom planar positioner with linear magnetic bearings/motor. *Proc. of the 6th International Symp. on Magnetic Bearings*, MIT Cambridge, 641-649.
30. Holmes M, Trumper DL, Hocken R. (1998) Magnetically-suspended stage for accurate positioning of large samples in scanned probe microscopy. *Proc. 6th Internat. Symp. on Magnetic Bearings*, MIT Cambridge, August 5-7, 123-132.
31. Molenaar A, Zaaijer EH, van Beek HF. (1998) A novel low dissipation long stroke planar magnetic suspension and propulsion stage. *Proc. of the 6th International Symp. on Magnetic Bearings*, MIT Cambridge, 650-659.
32. Schweitzer G. (2000) Magnetic bearings as a component of smart rotating machinery. *Proc. 5th Internat. Conf. on Rotor Dynamics* IFToMM, Darmstadt, Sept. 7-10, 3-15.
33. Nordmann R, Ewins D, et al. (2001) IMPACT - Improved Machinery Performance using Active Control Technology. BRITE/EURAM Project No. BRPR-CT97-0544, Jan. 1998-March 2001.
34. Mueller M. (2002) On-line-Process Monitoring in High Speed Milling with an Active Magnetic Bearing Spindle. Diss ETH Zurich, to appear.
35. R. Field, V. Iannello, A Reliable Magnetic Bearing System for Turbomachinery. *Proc. of the 6th International Symp. on Magnetic Bearings*, MIT Cambridge, 42-51.
36. Charles, D., *Spinning a Nuclear Comeback*, Science, Vol. 315, (30 March 2007)
37. Beams, J. , *Production and Use of High Centrifugal Fields*, Science, Vol. 120, (1954)
38. Beams, J. , *Magnetic Bearings*, Paper 810A, Automotive Engineering Conference, Detroit, Michigan, USA, SAE (Jan. 1964)
39. Habermann,H. , Liard, G. *Practical Magnetic Bearings* , IEEE Spectrum, Vol. 16, No. 9, (September 1979)
40. Schweitzer, G. , *Characteristics of a Magnetic Rotor Bearing for Active Vibration Control*, Paper C239/76, First International Conference on Vibrations in Rotating Machinery, (1976)
41. E. Croot, *Australian Inventors Weekly*, NSW Inventors Association, Vol. 3, (April 1987)

42. Kasarda, M. *An Overview of Active Magnetic Bearing Technology and Applications*, The Shock and Vibration Digest, Vol.32, No. 2: A Publication of the Shock and Vibration Information Center, Naval Research Laboratory, (March 2000)
43. "Design and Analysis of a Novel Low Loss Homopolar Electrodynamic Bearing." Lembke, Torbjörn. PhD Thesis. Stockholm: Universitetsservice US AB, 2005. ISBN 91-7178-032-7
44. "3D-FEM Analysis of a Low Loss Homopolar Induction Bearing" Lembke, Torbjörn. 9th International Symposium on Magnetic Bearings (ISMB9). Aug. 2004.
45. Seminar at KTH – the Royal Institute of Technology Stockholm. Feb 24. 2010
46. Robert M. Besançon, ed (1990). "Vacuum Techniques" (3rd edition ed.). Van Nostrand Reinhold, New York. pp. pp. 1278-1284.
47. Charles, D., *Spinning a Nuclear Comeback,* Science, Vol.315, (30March 2007)
48. Beams, J. , *Production and Use of High Centrifugal Fields*, Science, Vol. 120, (1954)
49. Beams, J. , *Magnetic Bearings*, Paper 810A, Automotive Engineering Conference, Detroit, Michigan, USA, SAE (Jan. 1964)
50. Habermann,H. , Liard, G. *Practical Magnetic Bearings* , IEEE Spectrum, Vol. 16, No. 9, (September 1979)
51. Schweitzer, G. , *Characteristics of a Magnetic Rotor Bearing for Active Vibration Control*, Paper C239/76, First International Conference on Vibrations in Rotating Machinery, (1976)
52. E. Croot, *Australian Inventors Weekly*, NSW Inventors Association, Vol. 3, (April 1987)
53. Kasarda, M. *An Overview of Active Magnetic Bearing Technology and Applications*, The Shock and Vibration Digest, Vol.32, No. 2: A Publication of the Shock and Vibration Information Center, Naval Research Laboratory, (March 2000)
54. The Zippe Type - The Poor Man's Bomb, BBC Radio 4, 19 May 2004
55. Tracking the technology, Nuclear Engineering International, 31 August 2004
56. Slender and Elegant, It Fuels the Bomb, New York Times, March 23, 2004
57. The Gas Centrifuge and Nuclear Proliferation, Marvin Miller, October 22, 2004
58. The Zippe Type - The Poor Man's Bomb, BBC Radio 4, 19 May 2004
59. Tracking the technology, Nuclear Engineering International, 31 August 2004

60. The Gas Centrifuge and Nuclear Proliferation, Marvin Miller, October 22, 2004
61. Samuel Earnshaw, "On the Nature of the Molecular Forces which Regulate the Constitution of the Luminiferous Ether," *Trans. Camb. Phil. Soc.*, V7, pp. 97-112 (1842).
62. W. T. Scott, "Who Was Earnshaw?", *American Journal of Physics*, V27, p. 418 (1959).
63. "Design and Analysis of a Novel Low Loss Homopolar Electrodynamic Bearing." Lembke, Torbjörn. PhD Thesis. Stockholm: Universitetsservice US AB, 2005. ISBN 91-7178-032-7
64. "3D-FEM Analysis of a Low Loss Homopolar Induction Bearing" Lembke, Torbjörn. 9th International Symposium on Magnetic Bearings (ISMB9). Aug. 2004.
65. Seminar at KTH – the Royal Institute of Technology Stockholm. Feb 24. 2010
66. Robert M. Besançon, ed (1990). "Vacuum Techniques" (3rd edition ed.). Van Nostrand Reinhold, New York. pp. pp. 1278-1284.
67. Charles, D., *Spinning a Nuclear Comeback,* Science, Vol.315, (30March 2007)
68. Beams, J. , *Production and Use of High Centrifugal Fields*, Science, Vol. 120, (1954)
69. Beams, J. , *Magnetic Bearings*, Paper 810A, Automotive Engineering Conference, Detroit, Michigan, USA, SAE (Jan. 1964)
70. Habermann,H. , Liard, G. *Practical Magnetic Bearings* , IEEE Spectrum, Vol. 16, No. 9, (September 1979)
71. R. Field, V. Iannello, A Reliable Magnetic Bearing System for Turbomachinery. *Proc. of the 6th International Symp. on Magnetic Bearings*, MIT Cambridge, 42-51.
72. Schweitzer G, Active magnetic bearings - chances and limitations International Centre for Magnetic Bearings, ETH Zurich, CH-8092 Zurich, schweitzer@ifr.mavt.ethz.ch
73. Eric Maslen, Magnetic Bearings ,University of Virginia Department of Mechanical, Aerospace and Nuclear Engineering Charlottesville, Virginia,Revised June 5, 2000
74. Reistad, K., et. al.: Magnetic Suspension for Rotating Equipment, Phase I Project Final Report, National Science Foundation (March 1980).
75. Schweitzer, *G.:* Stabilization of Self Excited Rotor Vibrations by an Active Damper, SpringerVerlag, New York (1975).
76. Yocum, J. F., and Slafer, L. I., Control System Design in the Presence of Severe Structural Dynamics Interactions, Journal of Guidance and Control, Vol. I, No.2, (March-April. 1978).

77. Maslen, E. R., et. al.: Practical Limits to the Performance of Magnetic Bearings: Peak Force, Slew Rate, and Displacement Sensitivity.
78. Johnson, B. *G.:* Active Control of a Flexible, Two-Mass Rotor: The Use of Complex Notation, Sc. D. Thesis, Massachusetts Institute of Technology (September 1986).
79. Chen, R. M.: Specification and Performance of Radial Active Bearings with Velocity and Acceleration Observers, Ph.D. Thesis, Rensselaer Polytechnic Institute (1986).
80. Maslen, E. H., and Bielk, J. R.: Implementing Magnetic Bearings in Discrete Flexible Structure Models (June 8, 1989).
81. Humphris,R.R., Allaire,P.E., et al, "Effect of Control Algorithms on Magnetic Journal Bearing Properties," J. of Eng. for Gas Turbines and Power, Vol.108, No.4, pp.624-632 (1986)
82. Kanemitsu,Y., Ohsawa and Watanabe,K., "Active Control of a Flexible Rotor by an Active Bearing," Prot. of the 1st. Intern. Symp., ETH Zuri<h, Switzerland, pp.367-380 (1988)
83. Bleuler,H., "Controlling Magnetic Bearing Systems with a Digital Signal Processor." Proc. of the 1st. Intern. Symp., ETH Zurich, Switzerland, pp.381,390 (1988)
84. Okada,Y., Nagai,B. and Shimane,T., "Digital Control of Magnetic Bearing with Rotationally Synchronized Interruption,"Proc. of the 1st. Intern. Symp., ETH Zurich, Switzerland, pp.357-364 (1988)
85. Larsonneur,R., Herzog,R., "Optimal Design of Structure Predefined Discrete Control for Rotor in Magnetic Bearings", Proc. of the 1st. Intern. Symp., ETH Zurich, Switzerland, pp.347-356 (1988)
86. Matsushita,O., Bleuler,H., Sugaya,T. and Kaneko,R., "Modeling for Flexible Mechanical Systems", Proc. of the 1st. Intern. Symp., ETH Zurich, Switzerland, pp.337-346 (1988)
87. Gibbs, Philip & Geim, Andre. "Is Magnetic Levitation Possible?". High Field Magnet Laboratory. http://www.hfml.ru.nl/levitation-possible.html. Retrieved 2009-09-08.
88. Samuel Earnshaw, "On the Nature of the Molecular Forces which Regulate the Constitution of the Luminiferous Ether," *Trans. Camb. Phil. Soc.*, V7, pp. 97-112 (1842).
89. W. T. Scott, "Who Was Earnshaw?", *American Journal of Physics*, V27, p. 418 (1959).
90. Earnshaw, S., “On the Nature of the Molecular Forces which Regulate the Constitution of the Luminferous Ether”, Trans. Camb. Phil. Soc., 7, pp 97-112 (1842)
91. J. Powell and G. Danby, “Electromagnetic Inductive Suspension and Stabilization System for a Ground Vehicle,” U.S. Patent 3 470 828, Oct.7, 1969.

92. Richards, P.L., Tinkham, M., “Magnetic Suspension and Propulsion Systems for High-Speed Transportation”, J. Appl. Phys. Vol. 43, No. 6, Jun. 1972, pp. 2680-2690.
93. Sacerdoti, G., Catitti, A., Soglia, L. “ Self Centering Rotary Magnetic Suspension Device”, US Patent No: 3811740, Publication date 1974- 05-21.
94. Lembke, T., “ Design of Magnetic Induction Bearings”, M.Sc. thesis, Chalmers Univesity of Technology, Göteborg 1990.
95. Lembke, T., “Eddy Induced Magnetic Bearing”, European patent No. 0594033, Publication date 1994-04-27.
96. Post, R., “Dynamically stable Magnetic Suspension/Bearing System”.US patent No. 5495221, Publication date 1996-02-27.
97. Lefevre, L., “Helical Air-gap Winding Permanent Magnet Machine”, Electrical Machines and Drives, Royal Institute of Technology,Stockholm 1998.
98. Lang, M. and Fremerey, J.K., “Optimization of Permanent-Magnet Bearings”, 6th International Symposium on Magnetic Suspension Technology, Turin, October 2001.
99. Filatov, A., “”Null-E” Magnetic Bearings”, Ph.D. dissertation, University of Virginia, August 2002.
100. Post, R.F.; Fowler, T.K.; Post, S.F., "A high-efficiency electromechanical battery", Proceedings of the IEEE , Vol. 8(1993), pp.462 – 474
101. Schweitzer G., Bleuler H., Traxler A., “Active Magnetic Bearings“, vdf Hochschulverlag AG an der ETH Zürich, 1994.
102. Filatov A. V., Maslen EH, Gillies G. T., “A method of noncontact suspension of rotating bodies using electromagnetic forces”, Journal of Applied Physics, Vol. 91 (2002) pp. 2355-2371
103. Filatov A. V., Maslen, E. H., and Gillies, G. T., “Stability of an Electrodynamic Suspension” Journal of Applied Physics, Vol. 92 (2002), pp. 3345-3353.
104. Lembke T.,”Induction Bearings –A Homopolar Concept for High Speed Machines”, Dep. of Electrical Machines and Power Electronics, Royal Institute of Technology, Stockholm, 2003
105. Lembke T., “3D-FEM Analysis of a Low Loss Homopolar Induction Bearing”, International Symposium on Magnetic Bearings, ISMB9, Lexington, KY, USA 2004.
106. Krämer E., “Dynamics of Rotors and Foundations”, Springer Verlag, Berlin, 1993
107. Lembke T., “Assessment and Experimental Verification of the Homopolar Induction Bearing”, Internal report for Magnetal AB, Uppsala, 2005.

108. J. K. Fremerey, M. Kolk: A 500-Wh power flywheel on permanent magnet bearings, Proceedings of the 5th International Symposium on Magnetic Suspension Technology, 1-3. 12. 1999, Santa Barbara, CA
109. Boden K., Fremerey J.K., "Industrial Realization of the "System KFA-Jülich" Permanent Magnet Bearing Lines" Proceedings of MAG ′92, Alexandria, Virginia, July 29-31, 1992
110. Engström J., "On Design of Slotless Permanent Magnet Machines", Licentiate thesis, Department of Electrical Machines and Drives, Royal Institute of Technology, 1999
111. Slemon G. R., "Electric Machines and Drives", ISBN 0-201-57885-9, Addison-Wesley Publishing Company, 1992.
112. "M3A 100kW Flywheel Power System" Data Sheet, AFS Trinity Power Corporation 2005, www.afstrinity.com
113. "Annual Report 2001", Magnetal AB, Uppsala, 2001
114. Lang, M. "Berechnung und Optimierung von passive permanent magnetischen Lagern für rotierende Mashinen, VDI-Verlag, Düsseldorf, 2004, ISBN 3-18-335721-6
115. Mega Manual, Department of Electronic and Electrical Engineering, University of Bath, Claverton Down, Bath BA2 7AY, United Kingdom
116. Lembke. T. "Compliant foil fluid film bearing with eddy current damper", US Patent 6,469,411
117. R. G. Gilbert: "Magnetic suspension", US Patent 2,946,930, 1955
118. Anton, V.L. , 2000, "Analysis and initial synthesis of a novel linear actuator with active magnetic suspension", 0-7803-8486-5/04/$20.00 © 2004 IEEE
119. Chee, K.L., 1999, "A Piezo-on-Slider Type Linear Ultrasonic Motor for the Application of Positioning Stages", Proceedings of the 1999IEEE/ASME.
120. Shyh-Leh, C., 2002, "Optimal Design of a Three-Pole Active Magnetic Bearing".
121. Torbjörn A. Lembke, 2005, "Design and Analysis of a Novel Low Loss Homopolar Electrodynamic Bearing", Doctoral Thesis in Electrical Machines and Drives, Stockholm, Sweden 2005.
122. Luc BURDET, 2006, "active magnetic bearing design and characterization for high temperature applications", THÈSE NO 3616 (2006), Lausanne, EPFL 2006.
123. Eric Maslen ,2000, " Magnetic Bearings", University of Virginia, Department of Mechanical, Aerospace, and Nuclear Engineering Charlottesville, Virginia, Revised June 5, 2000.
124. T. Higuchi,*200,* "Proceedings of the Second International Symposium on Magnetic Bearings", SEIKEN SYMPOSIUM JULY 12-14, 1990,TOKYO, JAPAN.

125. Jeffrey Hillyard, 2006, "Magnetic Bearings", Joint Advanced Student School, St. Petersburg, Russia, Department of Mechanical Engineering, Technical University of Munich.
126. Felix Betschon, 2009, " Design Principles of Integrated Magnetic Bearings ", Diss. ETH Nr. 13643, dissertation for the degree of Doctor of Technical Sciences, submitted to the Swiss Federal Institute of Technology, Zurich.
127. Lawrence A. Hawkins, Brian T. Murphy, and John Kajs , 1999 , " APPLICATION OF PERMANENT MAGNET BIAS MAGNETIC BEARINGS TO AN ENERGY STORAGE FLYWHEEL", Center for Electromechanics, University of Texas, Austin, TX 78712, Fifth Symposium on Magnetic Suspension Technology, December 1-3, 1999.

Printed in the United States
By Bookmasters